ISBN 978-1-5278-0211-7
PIBN 10902507

INSTITUTION

OF

MECHANICAL ENGINEERS.

ESTABLISHED 1847.

PROCEEDINGS.

1892.

PUBLISHED BY THE INSTITUTION,

19 Victoria Street, Westminster, S.W.

CONTENTS.

1892.

PAST-PRESIDENTS.

GEORGE STEPHENSON, 1847-48. (*Deceased* 1848.)

ROBERT STEPHENSON, F.R.S., 1849-53 (*Deceased* 1859.)

SIR WILLIAM FAIRBAIRN, BART., LL.D., F.R.S., 1854-55. (*Deceased* 1874.)

SIR JOSEPH WHITWORTH, BART., D.C.L., LL.D., F.R.S., 1856-57, 1866. (*Deceased* 1887.)

JOHN PENN, F.R.S., 1858-59, 1867-68. (*Deceased* 1878.)

JAMES KENNEDY, 1860. (*Deceased* 1886.)

THE RIGHT HON. LORD ARMSTRONG, C.B., D.C.L., LL.D., F.R.S., 1861-62, 1869.

ROBERT NAPIER, 1863-65. (*Deceased* 1876.)

JOHN RAMSBOTTOM, 1870-71.

SIR WILLIAM SIEMENS, D.C.L., LL.D., F.R.S., 1872-73. (*Deceased* 1883.)

SIR FREDERICK J. BRAMWELL, BART., D.C.L., F.R.S., 1874-75.

THOMAS HAWKSLEY, F.R.S., 1876-77.

JOHN ROBINSON, 1878-79.

EDWARD A. COWPER, 1880-81.

PERCY G. B. WESTMACOTT, 1882-83.

SIR LOWTHIAN BELL, BART., F.R.S., 1884.

JEREMIAH HEAD, 1885-86.

SIR EDWARD H. CARBUTT, BART., 1887-88.

CHARLES COCHRANE, 1889.

JOSEPH TOMLINSON, 1890-91.

Institution of Mechanical Engineers.

OFFICERS.

1892.

PRESIDENT.

William Anderson, D.C.L., F.R.S., ... Woolwich.

PAST-PRESIDENTS.

The Rt. Hon. Lord Armstrong, C.B., D.C.L., LL.D., F.R.S., Newcastle-on-Tyne.
Sir Lowthian Bell, Bart., F.R.S., ... Northallerton.
Sir Frederick J. Bramwell, Bart., D.C.L., F.R.S., ... London.
Sir Edward H. Carbutt, Bart., ... London.
Charles Cochrane, ... Stourbridge.
Thomas Hawksley, F.R.S., ... London.
Jeremiah Head, ... Middlesbrough.
John Ramsbottom, ... Alderley Edge.
John Robinson, ... Leek.
Joseph Tomlinson, ... London.
Percy G. B. Westmacott, ... Newcastle-on-Tyne.

VICE-PRESIDENTS.

Sir James N. Douglass, F.R.S., ... London.
Sir Douglas Galton, K.C.B., D.C.L., F.R.S., ... London.
Alexander B. W. Kennedy, F.R.S., ... London.
Edward B. Marten, ... Stourbridge.
Sir James Ramsden, ... Barrow-in-Furness.
E. Windsor Richards, ... Low Moor.

MEMBERS OF COUNCIL.

John A. F. Aspinall, ... Horwich.
William Dean, ... Swindon.
John Hopkinson, Jun., D.Sc., F.R.S., ... London.
Samuel W. Johnson, ... Derby.
Arthur Keen, ... Birmingham.
William Laird, ... Birkenhead.
John G. Mair-Rumley, ... London.
Francis C. Marshall, ... Newcastle-on-Tyne.
Henry D. Marshall, ... Gainsborough.
Edward P. Martin, ... Dowlais.
William H. Maw, ... London.
T. Hurry Riches, ... Cardiff.
William H. White, C.B., F.R.S., ... London.
J. Hartley Wicksteed, ... Leeds.
Thomas W. Worsdell, ... Arnside.

TREASURER.

Harry Lee Millar.

SECRETARY.

Alfred Bache,
Institution of Mechanical Engineers, 19 *Victoria Street*, *Westminster*, *S.W.*
[Telegraphic address :—*Mech, London.*]

Institution of Mechanical Engineers.

ESTABLISHED 1847.

LIST OF MEMBERS,

WITH YEAR OF ELECTION.

[Telegraph Address and Telephone No. appended within brackets.]

1892.

HONORARY LIFE MEMBERS.

1890. H. R. H. Albert Edward, Prince of Wales, K.G., K.T., K.P., G.C.B., G.C.S.I., &c., Marlborough House, Pall Mall, London, S.W.

1892. Field Marshal H.R.H. the Duke of Cambridge, K.G., K.T., K.P., G.C.B., G.C.S.I., &c., Gloucester House, Park Lane, London, W.

1883. Abel, Sir Frederick Augustus, K.C.B., D.C.L., D.Sc., F.R.S., The Imperial Institute, Imperial Institute Road, London, S.W.; and 40 Cadogan Place, London, S.W. [*Imperial Institute, London.* 8743.]

1878. Crawford and Balcarres, The Right Hon. the Earl of, K.T., F.R.S., 2 Cavendish Square, London, W.; Haigh Hall, Wigan; and Observatory, Dunecht, Aberdeen.

1889. Eiffel, Gustave, 37 Rue Pasquier, Paris.

1888. Haughton, Rev. Samuel, M.D., D.C.L., LL.D., F.R.S., Trinity College, Dublin.

1883. Kennedy, Professor Alexander Blackie William, F.R.S., 19 Little Queen Street, Westminster, S.W. [*Kinematic, London.*]

1878. Rayleigh, The Right Hon. Lord, F.R.S., 4 Carlton Gardens, London, S.W.; and Terling Place, Witham, Essex.

1888. Rosse, The Right Hon. the Earl of, K.P., D.C.L., LL.D., F.R.S., Birr Castle, Parsonstown, Ireland.

MEMBERS.

1890. Abbott, Arthur Harold, care of Messrs. Octavius Steel and Co., Calcutta, India: (or care of F. C. Abbott, 101 Lambeth Palace Road, London, S.E.)

1878. Abbott, Thomas, Newark Boiler Works, Newark. [*Abbott, Newark.*]

1883. Abbott, William Sutherland, Locomotive Superintendent and Assistant Engineer, Alagoas Railway, Maceio, Brazil: (or care of George S. Abbott, 9 Disraeli Road, Upton, London, E.)

1861. Abel, Charles Denton, Messrs. Abel and Imray, 20 Southampton Buildings, London, W.C. [*Patentable, London.* 2729.]

1874. Abernethy, James, F.R.S.E., 4 Delahay Street, Westminster, S.W.

1892. Acland, Captain Francis Edward Dyke, Maxim-Nordenfelt Guns and Ammunition Co., 32 Victoria Street, Westminster, S.W.

1876. Adams, Henry, 60 Queen Victoria Street, London, E.C. [*Viburnum, London.*]

1879. Adams, William, Locomotive Superintendent, London and South Western Railway, Nine Elms, London, S.W.

1848. Adams, William Alexander, Gaines, Worcester.

1881. Adams, William John, 35 Queen Victoria Street, London, E.C. [*Packing, London.* 1854.]

1871. Adamson, Joseph, Messrs. Joseph Adamson and Co., Hyde, near Manchester. [*Adamson, Hyde.*]

1886. Adamson, Thomas Alfred, 27 Leadenhall Street, London, E.C.

1851. Addison, John, Colehill Cottage, Fulham, London, S.W.

1889. Addy, George, Waverley Works, Effingham Road, Sheffield. [*Milling, Sheffield.*]

1887. Ahmed Bey, Colonel, Imperial Naval Arsenal, Constantinople.

1891. Ahrbecker, Henry Conrad Vandepoel, Morts Dock and Engineering Co., Balmain, Sydney, New South Wales.

1886. Aisbitt, Matthew Wheldon, 53 Mount Stuart Square, Cardiff. [*Aisbitt, Cardiff.*]

1886. Albright, John Francis, Messrs. R. E. Crompton and Co., 4 Mansion House Buildings, Queen Victoria Street, London, E.C.; and Savernake Lodge, Chelmsford.

1885. Alderson, George Beeton, Messrs. Allen Alderson and Co., Alexandria, Egypt; Norland House, Ramleh, Alexandria, Egypt: (or care of Messrs. Stafford Allen and Sons, 7 Cowper Street, Finsbury, London, E.C.)

1881. Alexander, Edward Disney, Engineer's Department, London County Council; and 8 Sheen Park, Richmond, Surrey.

1875. Allan, George, New British Iron Works, Corngreaves, near Birmingham; and Corngreaves Hall, near Birmingham.

1885. Allcard, Harry, Messrs. Easterbrook Allcard and Co., Albert Works, Penistone Road, Sheffield.

1884. Allen, Alfred Evans, 37 Wellington Street, Hull.

1891. Allen, Marcus, Union Brass and Iron Works, Great Ancoats Street, and Phœnix Iron Works, Jersey Street, Manchester. [*Valves, Manchester. Nat.* 60.]

1881. Allen, Percy Ruskin, Woodberrie Hill, Loughton, Essex.

1884. Allen, Samuel Wesley, Exchange Buildings, Mount Stuart Square, Cardiff.

1885. Allen, William Henry, Messrs. W. H. Allen and Co., York Street Works, Lambeth, London, S.E. [*Pump, London.*]

1882. Allen, William Milward, Principal Assistant Engineer, Engine Boiler and Employers' Liability Insurance Co., 12 King Street, Manchester.

1877. Alley, Stephen, Messrs. Alley and MacLellan, Sentinel Works, Polmadie Road, Glasgow. [*Alley, Glasgow.* 673.]

1865. Alleyne, Sir John Gay Newton, Bart., Chevin, Belper.

1884. Alleyne, Reynold Henry Newton, Messrs. Scriven and Co., Leeds Old Foundry, Marsh Lane, Leeds.

1872. Alliott, James Bingham, Messrs. Manlove Alliott and Co., Bloomsgrove Works, Ilkeston Road, Nottingham. [*Manloves, Nottingham.*]

1891. Allott, Charles Sneath, 46 Brown Street, Manchester.

1876. Allport, Charles James, Whitehall Club, Parliament Street, Westminster, S.W.

1871. Allport, Howard Aston, Dodworth Grove, Barnsley.

1884. Almond, Harry John, Cartagena and Herrerias Steam Tramways, 43 Muralla del Mar, Cartagena, Spain: (or care of Messrs. G. and W. Almond, 67 Willow Walk, London, S.E.)

1885. Amos, Ewart Charles, Mansion House Chambers, 11 Queen Victoria Street, London, E.C.; and Eastdene, St. James' Road, Sutton, Surrey. [*Drilling, London.*]

1867. Amos, James Chapman, West Barnet Lodge, Lyonsdown, Barnet.

1891. Anderson, Alexander Southerland, Chief Engineer, Ordnance Department, Ordnance Factory, Cawnpore, India.

1880. Anderson, Edward William, Messrs. Easton and Anderson, Erith Iron Works, Erith, S.O., Kent; and Roydon Lodge, Erith, S.O., Kent.

1890. Anderson, Herbert William, Messrs. Hilton Anderson and Co., Manor Works, Halling, near Rochester.

1892. Anderson, John Wemyss, Kilbourn Refrigerator Co., 59 Commercial Road, Liverpool. [*Ammonia, Liverpool.* 1509.]

1856. Anderson, William, D.C.L., F.R.S., Director-General of Ordnance Factories, Royal Arsenal, Woolwich; and Lesney House, Erith, S. O., Kent.

1891. Anderson, William, Messrs. Head Wrightson and Co., Teesdale Iron Works, Stockton-on-Tees.

1892. Andrew, Thomas, Rand Club, Johannesburg, Transvaal, South Africa.

1885. Anson, Frederick Henry, 15 Dean's Yard, Westminster, S.W.

1867. Appleby, Charles James, Messrs. Appleby Brothers, 22 Walbrook, London, E.C. [*Millwright, London.*]; and East Greenwich Works, London, S.E.

1883. Appleby, Percy Vavasseur, Messrs. Appleby Brothers, 22 Walbrook, London, E.C.

1874. Aramburu y Silva, Fernando, Messrs. Aramburu and Sons, Cartridge Manufacturers, Calle de la Virgen de las Azucenas, Madrid: (or care of Manuel Cardenosa, 86 Great Tower Street, London, E.C.)

1891. Archbold, John, Eastwood Collieries, Eastwood, R.S.O., Notts.

1881. Archbold, Joseph Gibson, Manager, Blyth Dry Dock, Blyth, Northumberland.

1889. Archer, Charles Frederick, Messrs. Joseph Richmond and Co., 30 Kirby Street, Hatton Garden, London, E.C.

1874. Archer, David, Oldbury Railway-Carriage and Wagon Co., Oldbury, near Birmingham; and 275 Pershore Road, Birmingham.

1883. Arens, Henrique, Messrs. Arens and Irmaos, Engineering Works, Rio de Janeiro, Brazil: (or care of Messrs. Marshall Sons and Co., Britannia Iron Works, Gainsborough.)

1882. Armer, James, Messrs. John Birch and Co., 11 Queen Street Place, London, E.C.; and 13 Clifton Road, Brockley, London, S.E.

1859. Armitage, William James, Farnley Iron Works, Leeds.

1858. Armstrong, The Right Hon. Lord, C.B., D.C.L., LL.D., F.R.S., Elswick, Newcastle-on-Tyne; and Cragside, Morpeth.

1866. Armstrong, George, Great Western Railway, Locomotive Department, Stafford Road Works, Wolverhampton.

1882. Armstrong, George Frederick, F.R.S.E., Professor of Engineering, The University, Edinburgh.

1876. Armstrong, William, Jun., Mining Engineer, Wingate Colliery, County Durham.

1870. Armstrong, William Irving, Timber Works and Saw Mills, 17 North Bridge Street, Sunderland.

1873. Arnold, David Nelson, Messrs. Willans Arnold and Co., Spanish Steel Works, Sheffield; and Friars Close, Barnwell, Northamptonshire.

1887. Arrol, Sir William, Dalmarnock Iron Works, Glasgow.

1887. Arteaga, Alberto de, Libertad 1357, Buenos Aires, Argentine Republic: (or care of M. Raggio-Carneiro, 129A Winchester House, Old Broad Street, London, E.C.)

1873. Ashbury, Thomas (*Life Member*), 5 Market Street, Manchester; and Asb Grove, Victoria Park, Longsight, Manchester. [*Thomas Ashbury, Manchester*.]

1888. Ashby, George, Tardeo, Bombay, India.

1890. Ashley, Thomas James, Messrs. Fawcett Preston and Co., Phœnix Foundry, Liverpool.

1884. Ashwell, Frank, Victoria Foundry, Sycamore Lane, Leicester. [*Iron, Leicester*. 100.]

1891. Ashworth, Henry, Ollerton, Bolton.

1890. Askham, John Unwin, Messrs. Askham Brothers and Wilson, Yorkshire Steel Works, Napier Street, Sheffield.

1890. Askham, Philip Unwin, Messrs. Askham Brothers and Wilson, Yorkshire Steel Works, Napier Street, Sheffield.

1881. Aspinall, John Audley Frederick, Chief Mechanical Engineer, Lancashire and Yorkshire Railway, Horwich, near Bolton; and Fern Bank, Heaton, Bolton.

1891. Asplen, Bernard, Southall: (or care of W. W. Asplen, Foxton Hall, Royston, Cambridgeshire.)

1877. Astbury, James, Smethwick Foundry, near Birmingham.

1890. Aston, John W., Chief Teacher of Mechanical Science, Municipal School of Art, Birmingham; and Messrs. G. E. Belliss and Co., Ledsam Street, Birmingham.

1886. Atkey, Albert Reuben, Robin Hood Cycle Co., Upper Parliament Street, Nottingham.

1889. Atkinson, Alexander, N. W. Hotel, Lahore, Punjâub, India: (or care of Messrs. Grindlay and Co., 55 Parliament Street, Westminster, S.W.)

1875. Atkinson, Edward (*Life Member*), The Projectile Company, New Road, Wandsworth Road, London, S.W.

1890. Atkinson, Edward Turner, London County Council, Spring Gardens, London, S.W.

1892. Atkinson, James, British Gas Engine and Engineering Co., Albion Works, Mansfield Road, Gospel Oak, London, N.W.; and 3 Nassington Road, Hampstead, London, N.W. [*Differential, London.*]

1892. Ault, Edwin, Great George Street Chambers, Westminster, S.W.

1892. Austin, James Meredith, care of Hermann Kühne, 102 Old Broad Street, London, E.C.; and 64 Wharton Road, Addison Park, London, W.

1882. Aveling, Thomas Lake, Messrs. Aveling and Porter, Rochester. [*Aveling, Rochester*.]

1891. Bagshaw, Walter, Victoria Foundry, Batley.

1890. Bailey, Charles Stuart, Russell Mine, Montgomery County, North Carolina, United States.

1886. Bailey, William, 14 Delahay Street, Westminster, S.W.

1885. Bailey, William Henry, Albion Works, Salford, Manchester. [*Beacon, Salford.*]

1880. Baillie, Robert, Messrs. Westwood Baillie and Co., London Yard Iron Works, Poplar, London, E.

1887. Baillie, Robert Alexander, Messrs. Westwood Baillie and Co., London Yard Iron Works, Poplar, London, E.

1872. Bailly, Philimond, 282 Rue Royale, Bruxelles, Belgium.

1890. Bain, George, Locomotive Department, Egyptian Government Railways, Cairo, Egypt.

1880. Bain, William Neish, 40 St. Enoch Square, Glasgow; and Collingwood, 7 Aytoun Road, Pollokshields, Glasgow. [*Glacis, Glasgow.*]

1869. Bainbridge, Emerson, Nunnery Colliery Offices, New Haymarket, Sheffield.

1873. Baird, George, St. Petersburg; and Fulmer, Slough.

1890. Baker, Sir Benjamin, K.C.M.G., F.R.S. (*Life Member*), 2 Queen Square Place, Westminster, S.W.

1875. Bakewell, Herbert James, Engineer, Department of the Controller of the Navy, Admiralty, Whitehall, London, S.W.

1877. Bale, Manfred Powis, Appold Street, Finsbury, London, E.C.

1884. Balmokand, Lala, Executive Engineer, Public Works Department, Punjaub, India; care of Lala Shamba Das, Said Mitha, Lahore, Punjaub, India.

1887. Bamlett, Adam Carlisle, Agricultural Engineering Works, Thirsk.

1892. Banister, George Henry, Carriage Department, Royal Arsenal, Woolwich.

1888. Baraclough, William Henry, Triangle Chambers, Martineau Street, Birmingham.

1885. Barker, Tom Birkett, Scholefield Street, Birmingham.

1882. Barlow, Henry Bernoulli, 4 Mansfield Chambers, 17 St. Ann's Square, Manchester. [*Monopoly, Manchester.*]

1880. Barlow-Massicks, Thomas, Millom Iron Works, Millom, Cumberland.

1891. Barnes, John Edward Lloyd, Messrs. Sloan and Lloyd Barnes, Castle Chambers, 26 Castle Street, Liverpool. [*Technical, Liverpool.*]

1881. Barnett, John Davis, Assistant Mechanical Superintendent, Grand Trunk Railway, Stratford, Ontario, Canada.

1887. Barningham, James, 41 Victoria Buildings, Victoria Street, Manchester.

1884. Barr, Archibald, D.Sc., Professor of Engineering, The University, Glasgow.

1878. Barr, James, care of William McConnell, Underwood House, Paisley.

1882. Barrett, John James, 5 Chinchpoogly Road, Bombay, India.

1885. Barrie, William, Superintendent Engineer, Nippon Yusen Kaisha Steam Ship Co., Tokyo, Japan.

1887. Barringer, Herbert, 88 Bishopsgate Street Within, London, E.C.

1862. Barrow, Joseph, Messrs. Thomas Shanks and Co., Johnstone, near Glasgow. [*Shanks, Johnstone.*]

1867. Barrows, Thomas Welch, Messrs. Barrows and Stewart, Portable Engine Works, Banbury. [*Barrows, Banbury.*]

1871. Barry, John Wolfe, 21 Delahay Street, Westminster, S.W. [*Wolfebarry, London.* 3024.]

1883. Bartlett, James Herbert, Middlesbrough, Kentucky, United States.

1887. Bate, Capt. Charles McGuire, R.E., War Office, Whitehall, London, S.W.

1885. Bateman, Henry, Superintending Engineer, Rangoon Tramways, Rangoon, India.

1891. Bates, Henry, Messrs. Hulse and Co., Ordsal Tool Works, Regent Bridge, Salford, Manchester; and 30 Halliwell Terrace, Trafford Road, Salford, Manchester.

1891. Battle, Arthur Edwin, 359–361 Collins Street, Melbourne, Victoria: (or care of F. G. Battle, Whitehall, Potterhanworth, Lincoln.) [Reuter's Agency: *Battle, Melbourne.*]

1892. Baxter, Peter Macleod, Departmento de Talleres, Minas de Rio Tinto, Huelva, Spain.

1889. Bayford, William James, Engineer and Manager, Messrs. Meakin and Co., Brewers, Delhi, India.

1872. Bayliss, Thomas Richard, Belmont, Northfield, Birmingham.

1891. Baynes, John, Midland Railway-Carriage and Wagon Co., Suffolk House, Laurence Pountney Hill, London, E.C.

1877. Beale, William Phipson, Q C., 12 Old Square, London, W.C.; and 19 Upper Phillimore Gardens, Kensington, London, W.

1887. Beardmore, William, Parkhead Forge and Steel Works, Glasgow.

1891. Beatty, Hazlitt Michael, Locomotive Superintendent, Western Railway, Cape Government Railways, Salt River, near Cape Town, Cape Colony; and Rosclare Camp Ground, Rondebosch, near Cape Town, Cape Colony.

1880. Beaumont, William Worby, 163 Strand, London, W.C.; and Melford, Palace Road, Tulse Hill, London, S.W.

1859. Beck, Edward (*Life Member*), Dallam Forge, Warrington; and Springfield, Warrington.

1873. Beck, William Henry, 115 Cannon Street, London, E.C.

1887. Beckwith, George, Enfield House, Fairlop Road, Leytonstone, London, N.E.

1891. Beckwith, George Charles, 17 Wind Street, Swansea. [*Beckwith, Swansea.*]

1875. Beckwith, John Henry, Managing Director, Messrs. Galloways, Knott Mill Iron Works, Manchester.

1882. Bedson, Joseph Phillips, Parkhurst, Middlesbrough.

1875. Beeley, Thomas, Engineer and Boiler Maker, Hyde Junction Iron Works, Hyde, near Manchester. [*Beeley, Hyde.*]

1884. Beetlestone, George John, Sudbrook Works, near Chepstow.

1888. Beldam, Asplan, 77 Gracechurch Street, London, E.C.

1885. Bell, Charles Lowthian, Clarence Iron Works, Middlesbrough; and Linthorpe, Middlesbrough. [*Bells, Middlesbrough.* 5510.]

1858. Bell, Sir Lowthian, Bart., F.R.S., Clarence Iron Works, Middlesbrough; Rounton Grange, Northallerton; and Reform Club, Pall Mall, London, S.W. [*Sir Lowthian Bell, Middlesbrough.*]

1880. Bell, William Henry, Vale Rectory, Guernsey.

1879. Bellamy, Charles James, 195 Earl's Court Road, South Kensington, London, S.W.

1868. Belliss, George Edward, Steam Engine and Boiler Works, Ledsam Street, Birmingham. [*Belliss, Birmingham.*]

1878. Belsham, Maurice, Messrs. Price and Belsham, 52 Queen Victoria Street, London, E.C.

1880. Benham, Percy, Messrs. Benham, 66 Wigmore Street, London, W. [*Benham, London.* 7065.]

1887. Bennetts, Edward John, Battery Reef Gold Mine, P.O. Box 96, Krugersdorp, Transvaal, South Africa.

1890. Berkley, James Eustace, Locomotive and Carriage Superintendent, H.H. the Nizam's Guaranteed State Railways, Secunderabad, India.

1878. Berrier-Fontaine, Marc, Directeur des Constructions navales, Toulon, France: (or care of Messrs. P. S. King and Son, Canada Buildings, King Street, Westminster, S.W.)

1890. Bertram, Alexander, Wigan Coal and Iron Works, Wigan.

1891. Bertram, David Noble, Messrs. Bertrams, St. Katherine's Works, Sciennes, Edinburgh.

1887. Bertram, William, 29 Bachelor's Walk, Dublin.

1861. Bessemer, Sir Henry, F.R.S., Denmark Hill, London, S.E.

1891. Best, Francis Edward, 1 Swan Walk, Chelsea Embankment, London, S.W.

1891. Bevis, Alfred William, The Nest, Harborne, Birmingham.

1866. Bevis, Restel Ratsey, Messrs. Laird Brothers, Birkenhead Iron Works, Birkenhead; and Manor Hill, Birkenhead.

1892. Bickle, Thomas Edwin, Messrs. Bickle and Co., Great Western Docks, Plymouth. [*Engineers, Plymouth.* 176.]

1885. Bicknell, Arthur Channing, 42 Pelham Street, South Kensington, London, S.W.

1883. Bicknell, Edward, care of Bank of Bengal, Calcutta, India: (or 8 Canynge Square, Clifton, Bristol.)

1884. Bika, Léon Joseph, Locomotive Engineer-in-Chief, Belgian State Railway, 29 Rue des Palais, Bruxelles, Belgium.

1888. Billinton, Robert John, Locomotive Superintendent, London Brighton and South Coast Railway, Brighton.

1890. Bingham, Charles Henry, Messrs. Walker and Hall, Electro Works, Sheffield.

1887. Binnie, Alexander Richardson, Engineer, London County Council, Spring Gardens, London, S.W.; and 14 Campden Hill Gardens, Notting Hill, London, W.

1877. Birch, Robert William Peregrine, 5 Queen Anne's Gate, Westminster, S.W.

1891. Bird, George, Messrs. James Bartle and Co., Western Iron Works, Notting Hill, London, W.

1847. Birley, Henry, 6 Brentwood, Pendleton, R.O., Manchester.

1888. Birtwistle, Richard, Messrs. S. S. Stott and Co., Laneside Foundry, Haslingden, Manchester.

1879. Black, William, Messrs. Black Hawthorn and Co., Gateshead. [*Blackthorn, Newcastletyne.*]

1891. Black, William, 72 Bute Road, Cardiff.

1891. Blackburn, Arthur Henry, Fuel Economizer Co., Matteawan, New York, United States.

1891. Blackburn, George William, Messrs. T. Green and Son, Smithfield Iron Works, Leeds.

1890. Blackburn, John, Resident Engineer, Colne Valley Water Works, Bushey, Watford.

1862. Blake, Henry Wollaston, F.R.S., Messrs. James Watt and Co., 90 Leadenhall Street, London, E.C.

1886. Blandford, Thomas, Corbridge, R.S.O., Northumberland.

1881. Blechynden, Alfred, Naval Construction and Armaments Works, Barrow-in-Furness.

1892. Blechynden, John, Kobe Paper Mill, Kobe, Japan.

1867. Bleckly, John James, Bewsey Iron Works, Warrington; and Daresbury Lodge, Altrincham.

1882. Blundstone, Samuel Richardson, Catherine Chambers, 8 Catherine Street, Strand, London, W.C.

1863. Boeddinghaus, Julius, Electrotechniker, Düsseldorf, Germany.

1884. Bone, William Lockhart, Works of the Ant and Bee, West Gorton, Manchester.

1892. Booth, John William, Union Foundry, Rodley, near Leeds.

1890. Booth, Robert, 110 Cannon Street, London, E.C.

1883. Booth, William Stanway, Messrs. John Jameson and Son, Bow Street Distillery, Dublin.

1892. Borns, George Maximilian, Citizen Street, Goulburn, New South Wales: (or care of Messrs. Maw Dredge and Hollingsworth, 35 and 36 Bedford Street, Strand, London, W.C.)

1880. Borodin, Alexander, General Manager, Russian South Western Railways, Kieff, Russia.

1888. Borrows, William, Messrs. Edward Borrows and Sons, Providence Foundry, Sutton, St. Helen's, Lancashire.

1891. Boswell, Samuel, Messrs. Galloways, Knott Mill Iron Works, Manchester.

1885. Boughton, Henry Francis, Parade Chambers, Sheffield; and The Poplars, Stairfoot, Barnsley. [*Boughton, Barnsley. Nat.* 625, *Sheffield.*]

1888. Boulding, Sidney, Messrs. Green and Boulding, 21 Featherstone Street, London, E.C.

1886. Boult, Alfred Julius, Messrs. W. P. Thompson and Boult, 323 High Holborn, London, W.C.

1888. Boultbee, Frederic Richard, 46 Queen Victoria Street, London, E.C.; and 4 Loris Road, Shepherd's Bush, London, W.

1878. Bourdon, François Edouard, 74 Faubourg du Temple, Paris: (or care of Messrs. Negretti and Zambra, Holborn Viaduct, London, E.C.)

1886. Bourne, Thomas Johnstone, Colonial College, Hollesley Bay, Woodbridge; and Clyde Villa, Southborough, Tunbridge Wells.

1879. Bourne, William Temple, Messrs. Bourne and Grove, Bridge Steam Saw Mills, Worcester.

1891. Bousfield, John Ebenezer, 4 South Street, Finsbury, London, E.C. [*Invention, London.* 169.]

1879. Bovey, Henry Taylor, Professor of Engineering, McGill University, Montreal, Canada.

1880. Bow, William, Messrs. Bow McLachlan and Co., Thistle Engine Works, Paisley. [*Bow, Paisley.*]

1888. Bowen, Edward (*Life Member*), Locomotive and Carriage Superintendent, Porto Alegre and New Hamburg Railway, Rio Grande do Sol, Brazil: (or care of Benjamin Packham, 122 Upper Lewes Road, Brighton.)

1858. Bower, John Wilkes (*Life Member*), Meredale, Rugby Road, Leamington Spa.

1892. Bowker, Arthur F., Engineer and Manager, Mid-Kent Water Works, High Street, Snodland, S.O., Kent.

1890. Boyd, John White, Superintendent Engineer, Hong Kong and Whampoa Dock Co., Hong Kong, China.

1884. Boyer, Robert Skeffington, 46 Mount Stuart Square, Cardiff.

1882. Bradley, Frederic, Sandhills, Liverpool; Clensmore Foundry, Kidderminster; and Wolverley House, Southport.

1889. Bradley, Isaac, Manager, Gatling Gun Works, Perry Barr, Birmingham.

1878. Braithwaite, Charles C., 35 King William Street, London Bridge, London, E.C.

1875. Braithwaite, Richard Charles, Messrs. Braithwaite and Kirk, Crown Bridge Works, Westbromwich. [*Braithwaite, Westbromwich.*]

1854. Bramwell, Sir Frederick Joseph, Bart., D.C.L., LL.D., F.R.S., 5 Great George Street, Westminster, S.W. [*Wellbram, London.* 3060.]

1892. Brand, David Jollie, Messrs. Brand and Dryburgh, Cleveland Foundry and Engine Works, Townsville, North Queensland.

1888. Bratt, Augustus Hicks Henery, Le Kueh Coal and Iron Mines, Kiangsu, North China; Astor House, Shanghai, China: (or care of Messrs. David Owen and Co., 50 Exchange Chambers, Bixteth Street, Liverpool.)

1885. Brearley, Benjamin J., Union Plate Glass Works, St. Helen's.

1889. Brebner, Samuel Gordon, Chief Mechanical Engineer, Small Arms Ammunition Factory, Kirkee, Poona, India.

1868. Breeden, Joseph, New Mill Works, Fazeley Street, Birmingham. [*Breeden, Birmingham.*]

1891. Brewster, Edwin Henry George, 12 Delahay Street, Westminster, S.W.

1890. Brewster, Walter Seckford, Messrs. Beyer Peacock and Co.; Halstead Lodge, Ocean Street, Woollahra, Sydney, New South Wales.

1883. Bricknell, Augustus Lea, The Limes, 2 Horsford Road, Brixton Hill, London, S.W.

1891. Bridie, Ronald Hope, 90 Cannon Street, London, E.C. [*Enwheeling, London.*]

1887. Brier, Henry, Scotch and Irish Oxygen Co., Rosehill Works, Polmadie, Glasgow.

1889. Briggs, Charles, Jun., care of Robert Briggs, Howden.

1881. Briggs, John Henry, Messrs. Simpson and Co., Engine Works, 101 Grosvenor Road, Pimlico, London, S.W.; and Howden.

1889. Bright, Philip, Messrs. J. Tylor and Sons, 2 Newgate Street, London, E.C.

1886. Bright, William, Manager, Fairwood Tin-Plate Works, Gowerton, R.S.O., Glamorganshire.

1891. Broadbent, William, Messrs. Thomas Broadbent and Sons, Central Iron Works, Huddersfield. [*Broadbent, Huddersfield.* 102.]

1891. Brock, Cameron William Harrison, 3 Prince's Mansions, 68 Victoria Street, Westminster, S.W.

1865. Brock, Walter, Messrs. Denny and Brothers, Engine Works, Dumbarton.

1890. Brodie, John Alexander, 3 Cook Street, Liverpool.

1852. Brogden, Henry (*Life Member*), Hale Lodge, Altrincham, near Manchester.

1890. Brogden, Thomas, Messrs. Appleby and Brogden, Sandside, Scarborough.

1892. Bromiley, William J., Messrs. Dobson and Barlow, Kay Street Machine Works, Bolton.

1892. Brooke, John Walter, Adrian Iron Works, Lowestoft.

1892. Brooke, Robert Grundy, Messrs. Holden and Brooke, St. Simon's Works, Salford, Manchester. [*Influx, Manchester.*]

1884. Brook-Fox, Frederick George, Executive Engineer, East Coast Railway, Chicacole, Ganjam District, Madras Presidency, India: (or care of Messrs. H. S. King and Co., 65 Cornhill, London, E.C.)

1880. Brophy, Michael Mary, Messrs. James Slater and Co., 251 High Holborn, London, W.C.

1891. Brotherhood, Arthur Maudslay, 15 and 17 Belvedere Road, Lambeth, London, S.E.

1874. Brotherhood, Peter, 15 and 17 Belvedere Road, Lambeth, London, S.E.; and 94 Cromwell Road, South Kensington, London, S.W. [*Brotherhood, London.*]

1886. Brown, Andrew, Messrs. T. Cosser and Co., McLeod Road Iron Works, Kurrachee, India: (or care of P. B. Brown, 187 Sutherland Road, Sheffield.)

1866. Brown, Andrew Betts, F.R.S.E., Messrs. Brown Brothers and Co., Rosebank Iron Works, Edinburgh.

1891. Brown, Arthur Mogg, P.O. Box 1260, Johannesburg, Transvaal, South Africa

1885. Brown, Benjamin, Widnes Foundry, Widnes.

1879. Brown, Charles, Engineer, Bâle, Switzerland: (or care of Dr. Gardiner Brown, 9 St. Thomas' Street, London Bridge, London, S.E.)

1880. Brown, Francis Robert Fountaine, Mechanical Superintendent, Intercolonial Railway of Canada, Moncton, New Brunswick, Canada.

1889. Brown, Frederick Alexander William, Lieutenant R.A., Inspector of Ordnance Machinery, Mauritius: (or 42 Fermoy Road, St. Peter's Park, London, W.)

1888. Brown, Frederick Gills, care of Commissioner, Australian Irrigation Colonies, 35 Queen Victoria Street, London, E.C.; and 9 Baskerville Road, Wandsworth Common, London, S.W.

1881. Brown, George William, Messrs. Huntley Boorne and Stevens, Reading Tin Works, Reading.

1887. Brown, James, Sir W. G. Armstrong Mitchell and Co., Elswick Engine Works, Newcastle-on-Tyne.

1892. Brown, James Fiddes, 147 Woodbridge Road, Ipswich.

1884. Brown, Oswald, 2 Victoria Mansions, 28 Victoria Street, Westminster, S.W. [*Acqua, London.*]

1890. Brown, Robert, Manor House Engine Works, Far Cotton, Northampton.

1888. Brown, William, Messrs. W. Simons and Co., London Works, Renfrew.

1892. Brown, William, Messrs. Siemens Brothers and Co., Woolwich.

1887. Browne, Frederick John, Messrs. Austin Wood Browne and Co., Austin Foundry, Parkfield Street, Islington, London, N.

1874. Browne, Tomyns Reginald, Deputy Locomotive Superintendent, East Indian Railway, Jamalpur, Bengal, India: (or care of Messrs. W. Watson and Co., 27 Leadenhall Street, London, E.C.)

1874. Bruce, Sir George Barclay, 2 Westminster Chambers, 3 Victoria Street, Westminster, S.W.

1889. Bruce, Robert, 30 Great St. Helen's, London, E.C. [*Tangential, London.*]

1867. Bruce, William Duff, Vice-Chairman, Port Commission, Calcutta; and 23 Roland Gardens, South Kensington, London, S.W.

1888. Bruff, Charles Clarke, Coalport China Co., Coalport, near Ironbridge, Salop.

1873. Brunel, Henry Marc, 21 Delahay Street, Westminster, S.W. [3024.]

1870. Brunlees, Sir James, F.R.S.E., 12 Victoria Street, Westminster, S.W.

1892. Brunlees, John, 12 Victoria Street, Westminster, S.W. [3245.]

1891. Brunner, Adolphus, 59 Königin Strasse, Munich, Bavaria: (or care of L. F. Brunner, 257 Romford Road, Forest Gate, London, E.)

1887. Brunton, Philip George, Resident Engineer, Department of Roads and Bridges, Public Works Office, Sydney, New South Wales: (or care of J. D. Brunton, 19 Great George Street, Westminster, S.W.)

1884. Bryan, William B., Engineer, East London Water Works, Lea Bridge, Clapton, London, N.E.

1892. Buckley, John T., Messrs. Stevenson and Co., Canal Foundry, Preston; and Poulton Street, Kirkham, near Preston.

1873. Buckley, Robert Burton, Superintending Engineer, East Indian Railway, Arrah, Bengal, India; 27 Queensberry Place, South Kensington, London, S.W.: (or care of H. Burton Buckley, 1 St. Mary's Terrace, Paddington, London, W.)

1877. Buckley, Samuel, Messrs. Buckley and Taylor, Castle Iron Works, Oldham.

1886. Buckney, Thomas, Messrs. E. Dent and Co., 61 Strand, London, W.C.

1887. Buckton, Walter, 27 Ladbroke Square, London, W.

1878. Buddicom, Harry William, Plas-Derwen, Abergavenny.

1882. Budge, Enrique, Engineer-in-Chief, Harbour Works, Valparaiso, Chile: (or care of Messrs. Rose-Innes Cox and Co., 4 Fenchurch Avenue, London, E.C.)

1881. Bulkley, Henry Wheeler, 149 Broadway, New York, United States.

1884. Bullock, Joseph Howell, General Manager, Pelsall Coal and Iron Works, near Walsall; and Glenhurst, Lichfield Road, Walsall.

1882. Bulmer, John, Spring Garden Engineering Works, Pitt Street, Newcastle-on-Tyne.

1891. Bumsted, Francis Dixon, Cannock Chase Foundry and Engine Works, Hednesford, near Stafford.

1884. Bunning, Charles Ziethen, The Borax Co., 9 Mehmet Ali Pacha Khan, Constantinople.

1884. Bunt, Thomas, Superintendent Engineer, Kiangnan Arsenal, Shanghai, China: (or care of R. Pearce, Lanarth House, Holders Hill, Hendon London, N.W.)

1884. Bunting, George Albert, Locomotive Superintendent, Costa Rica Railways, San José, Costa Rica: (or Gaisford House, Gaisford Street, Kentish Town, London, N.W.)

1885. Burder, Walter Chapman, Messrs. Messenger and Co., Loughborough.

1891. Burgess, Francis Chassereau Boughey, Office of Director General of Railways, Technical Section, Simla, India.

1881. Burn, Robert Scott, Oak Lea, Edgeley Road, near Stockport.

1878. Burnett, Robert Harvey, Messrs. Beyer Peacock and Co., Gorton Foundry, Manchester.

1878. Burrell, Charles, Jun., Messrs. Charles Burrell and Sons, St. Nicholas Works, Thetford. [*Burrell, Thetford.*]

1885. Burrell, Frederick John, Messrs. Charles Burrell and Sons, St. Nicholas Works, Thetford. [*Burrell, Thetford.*]

1887. Burstal, Edward Kynaston, Messrs. Stevenson and Burstal, 38 Parliament Street, Westminster, S.W.

1890. Burstall, Henry Robert John, 19 Little Queen Street, Westminster, S.W.; and 76 King's Road, London, N.W.

1877. Burton, Clerke, Burghill, Charlton Kings, Cheltenham.

1884. Butcher, Joseph John, Abendroth and Root Manufacturing Co., 28 Cliff Street, New York, United States.

1882. Butler, Edmund, Kirkstall Forge, near Leeds. [*Forge, Kirkstall.*]

1892. Butler, Henry William, Messrs. Ransomes and Rapier, Waterside Iron Works, Ipswich.

1884. Butler, Hugh Myddleton, Kirkstall Forge, near Leeds.

1891. Butler, James, Victoria Iron Works, Halifax; and Longfield, Halifax.

1888. Butter, Frederick Henry, Carriage Department, Royal Arsenal, Woolwich; and 4 Hanover Road, Brookhill Park, Plumstead.

1891. Butter, Henry Joseph, Messrs. Tannett Walker and Co., Leeds; and Claremont, Burrage Road, Plumstead.

1892. Byrne, Francis Furlong, 37 College Green, Dublin.

1887. Caiger, Emery John, Messrs. E. J. Caiger and Co., 77 Billiter Buildings, Billiter Street, London, E.C. [*Caiger, London.*]

1886. Cairnes, Frederick Evelyn, care of Messrs. Pearson and Son, Yabsley Street, Poplar, London, E.

1889. Callan, William, River Plate Fresh Meat Co., 2 Coleman Street, London, E.C.

1886. Cambridge, Henry, Stuart Chambers, Mount Stuart Square, Cardiff.

1877. Campbell, Angus, Logie, Mussoorie, N. W. Provinces, India.

1880. Campbell, Daniel, Messrs. Campbell and Schultz, Botolph House, 10 Eastcheap, London, E.C. [*Duke, London.* 1893.]

1869. Campbell, James, Hunslet Engine Works, Leeds. [*Engineco, Leeds.*]

1882. Campbell, John, Messrs. R. W. Deacon and Co., Kalimaas Works. Soerabaya, Java: (or care of R. Campbell, Slamat Cottage, Mount Vernon, Glasgow.)

1892. Campbell, William Walker, Messrs. Campbell and Calderwood, Soho Engine Works, Paisley.

1885. Capito, Charles Alfred Adolph, 12 Prince's Street, Hanover Square, London, W.; and 9 Belgrave Terrace, Lee, London, S.E.

1892. Capper, David Sing, Professor of Mechanical Engineering, King's College, Strand, London, W.C.

1860. Carbutt, Sir Edward Hamer, Bart., 19 Hyde Park Gardens, London, W.; and Nanhurst, Cranley, Guildford.

1878. Cardew, Cornelius Edward, Locomotive and Carriage Superintendent, Burma State Railways, Insein, Burma; care of Messrs. King King and Co., Bombay, India: (or care of Rev. J. H. Cardew, Wingfield Rectory, Trowbridge.)

1875. Cardozo, Francisco Corrêa de Mesquita (*Life Member*), Messrs. Cardozo and Irmâo, Pernambuco Engine Works, Pernambuco, Brazil: (or care of Messrs. Fry Miers and Co., 8 Great Winchester Street, London, E.C.)

1878. Carlton, Thomas William, 1 Canfield Gardens, Priory Road, West Hampstead, London, N.W.

1887. Carlyle, Thomas, Lieutenant R.A., Inspector of Ordnance Machinery, Singapore: (or 72 Glyndon Road, Plumstead.)

1892. Carnegie, David, Royal Laboratory, Royal Arsenal, Woolwich.

1869. Carpmael, Frederick, 106 Croxted Road, West Dulwich, London, S.E.

1866. Carpmael, William, 24 Southampton Buildings, London, W.C. [*Carpmael, London.* 2608.]

1877. Carr, Robert, 1 West Pier, London Docks, London, E.

1892. Carrack, Charles, Messrs. Crossley Brothers, 5 Hounds Gate, Nottingham.

1884. Carrick, Henry, Messrs. Carrick and Wardale, Redheugh Engine Works, Gateshead; and Newbrough Lodge, Fourstones, R.S.O., Northumberland. [*Wardale, Gateshead.*]

1888. Carrick, Samuel Stewart, Superintendent Engineer, Shaw Savill and Albion Steamship Co., 34 Leadenhall Street, London, E.C.

1874. Carrington, William T. H., 72 Mark Lane, London, E.C.

1885. Carter, Herbert Fuller, Consulting Mechanical Engineer, Guanajuato, Mexico: (or care of H. Maynard Carter, 126 Wool Exchange, Basinghall Street, London, E.C.)

1877. Carter, William, Manager, The Hydraulic Engineering Company Chester.

1891. Carter, William Charles, Mansion House Chambers, Queen Victoria Street, London, E.C. [*Tympanum, London.*]

1890. Carver, Henry Clifton, Messrs. Coates and Carver, 3 Cross Street, Manchester.

1888. Castle, Frank, Royal College of Science, Exhibition Road, South Kensington, London, S.W.

1891. Caswell, Samuel John, 31 Sakai Machi, Kobe, Japan.

1892. Causer, William George, Brighton Villa, Handsworth, R.O., Birmingham.

1883. Cawley, George, 358 Strand, London, W.C.

1892. Chadwick, Osbert, C. M. G., Crown Agents' Department, Colonial Office, Downing Street, London, S.W.; and 11 Airlie Gardens, Kensington, London, W.

1876. Challen, Stephen William, Messrs. Taylor and Challen, Derwent Foundry, 60 and 62 Constitution Hill, Birmingham. [*Derwent, Birmingham.*]

1892. Chalmers, George, St. John del Rey Mining Co., 28 Tower Chambers, Finsbury Pavement, London, E.C.

1886. Chalmers, John Reid, 18 Hemingford Road, Barnsbury, London, N.

1884. Chamberlain, John, care of J. Chamberlain, 188 West Ferry Road, Millwall, Poplar, London, E.

1890. Chandler, Noel, Cannock Chase Foundry and Engine Works, Hednesford, near Stafford.

1887. Chapman, Alfred Crawhall, 2 St. Nicholas' Buildings, Newcastle-on-Tyne.

1888. Chapman, Arthur, Messrs. Marillier and Edwards, 1 Hastings Street, Calcutta, India; Great Eastern Hotel, Calcutta, India: (or St. Andrew's Cottage, Bury St. Edmund's.)

1882. Chapman, Hedley, Messrs. Chapman Carverhill and Co., Scotswood Road, Newcastle-on-Tyne.

1866. Chapman, Henry, 69 Victoria Street, Westminster, S.W. [*Tubalcain, London.*]; and 10 Rue Laffitte, Paris.

1878. Chapman, James Gregson, Messrs. Fawcett Preston and Co., Phœnix Foundry, Liverpool; and 25 Austinfriars, London, E.C. [*Fawcett, London.*]

1887. Chapman, Joseph Crawhall, 70 Chancery Lane, London, W.C.

1885. Charnock, George Frederick, Engineering Department, Technical College, Bradford.

1877. Chater, John, Messrs. Henry Pooley and Son, 89 Fleet Street, London, E.C.

1890. Chater, John Richard, Madras Railway, Madras, India; and care of John Chater, 223 Peckham Rye, London, S.E.

1891. Chatterton, Alfred, Professor of Engineering, College of Engineering, Madras, India.

1887. Chatwin, James, Victoria Works, Great Tindal Street, Ladywood, Birmingham.

1867. Chatwood, Samuel, Lancashire Safe and Lock Works, Bolton; and High Lawn, Broad Oak Park, Worsley, near Manchester.

1873. Cheesman, William Talbot, Hartlepool Rope Works, Hartlepool.

1881. Chilcott, William Winsland, The Terrace, H.M. Dockyard, Sheerness.

1877. Chisholm, John, General Manager, Messrs. A. and J. Stewart and Clydesdale, 41 Oswald Street, Glasgow; and 9 Corunna Street, Dumbarton Road, Glasgow.

1888. Chubb, Thomas Lyon, Locomotive Carriage and Wagon Superintendent, Ferro Carril del Oeste, Talleres Tolosa, Buenos Aires, Argentine Republic.

1880. Churchward, George Dundas, Locomotive Superintendent, China Railway Company, care of H.B.M.'s Consulate, Tientsin, North China: (or care of A. W. Churchward, London Chatham and Dover Railway, Queenborough Pier, Queenborough.)

1891. Clark, Augustus, Bowman's Heirs, Pernambuco, Brazil.

1871. Clark, Christopher Fisher, Mining Engineer, Garswood Coal and Iron Co., Park Lane Collieries, Wigan; and Cranbury Lodge, Park Lane, Wigan. [*Park Lane, Wigan.*]

1878. Clark, Daniel Kinnear, 8 Buckingham Street, Adelphi, London, W.C.

1867. Clark, George, Southwick Engine Works, near Sunderland.

1889. Clark, Thomas Alexander, Superintendent of Workshops, George Heriot's Hospital School, Edinburgh.

1889. Clarke, Francis, Dane John Iron Works, Canterbury.

1885. Clarke, Leslie, 132 Westbourne Terrace, Hyde Park, London, W.

1891. Clarkson, Thomas, Fairholme, Egmont Road, Sutton, Surrey.

1892. Clay, Charles Butler, Engineer and District Manager, National Telephone Co., Arcade Chambers, Sunderland.

1886. Clayton, Samuel, St. Thomas' Engine Works, Sunbridge Road, Bradford.

1882. Clayton, William Wikeley, Messrs. Hudswell Clarke and Co., Railway Foundry, Jack Lane, Leeds. [*Loco, Leeds.* 504.]

1890. Cleathero, Edward Thomas, Messrs. Cleathero and Nichols, Phœnix Works, Boleyn Road, Kingsland, London, N.; and 16 Tollington Place, London, N.

1890. Cleaver, Arthur, Engineer, Nottingham Laundry Co., Sherwood, near Nottingham; and Hornby House, Sherwood, near Nottingham.

1890. Cleland, William, Sheffield Testing Works, Blonk Street, Sheffield.

1871. Cleminson, James, Dashwood House, 9 New Broad Street, London, E.C. [*Catamarca, London.*]

1873. Clench, Frederick, Messrs. Robey and Co., Globe Iron Works, Lincoln. [*Robey, Lincoln.*]

1885. Clifton, George Bellamy, Great Western Railway Electric Light Works, 150 Westbourne Terrace, Paddington, London, W.

1885. Close, John, Jun., York Engineering Works, Leeman Road, York.

1885. Clutterbuck, Herbert, Engineers' Department, London County Council, Spring Gardens, London, S.W.

1882. Coates, Joseph, Messrs. Robey and Co., Globe Iron Works, Lincoln.

1881. Cochrane, Brodie, Mining Engineer, Aldin Grange, Durham.

1858. Cochrane, Charles, Woodside Iron Works, near Dudley; and Green Royde. Pedmore, near Stourbridge.

1887. Cochrane, George, Resident Engineer, London Hydraulic Power Works, 46 Holland Street, Blackfriars Road, London, S.E.

1885. Cochrane, John, Grahamston Foundry and Engine Works, Barrhead, near Glasgow. [*Cochrane, Barrhead.*]

1869. Cochrane, Joseph Bramah, Woodside Iron Works, near Dudley.

1868. Cochrane, William, Mining Engineer, Elswick Colliery, Elswick, Newcastle-on-Tyne; and Oakfield House, Gosforth, Newcastle-on-Tyne.

1864. Coddington, William, M.P., Ordnance Cotton Mill, Blackburn; and Wycollar, Blackburn.

1889. Coey, Robert, Assistant Locomotive Engineer, Great Southern and Western Railway, Inchicore Works, near Dublin.

1889. Colam, William Newby, Billiter Buildings, Billiter Street, London, E.C. [*Colam, London.*]

1884. Cole, Charles, Messrs. Cole Booth and Co., Vulcan Works, Dudley Hill, Bradford.

1892. Cole, Henry Aylwin Bevan, 79½ Gracechurch Street, London. E.C. [*Carbuncle, London.*]

1878. Coles, Henry James, Sumner Street, Southwark, London, S.E.

1877. Coley, Henry, Mansion House Chambers, Queen Victoria Street, London, E.C.

1892. Collen, Robert Henry, 10 Dover Road, Northfleet, S.O., Kent.

1884. Collenette, Ralph, Bowling Iron Works, Bradford.

1884. Colquhoun, James, General Manager, Tredegar Iron Coal and Steel Works, Tredegar.

1884. Coltman, John Charles, Messrs. Huram Coltman and Son, Engineering Works, Meadow Lane, Loughborough.

1878. Colyer, Frederick, 18 Great George Street, Westminster, S.W.

1888. Combe, Abram, Messrs. Combe Barbour and Combe, Falls Foundry, Belfast.

1889. Common, John Freeland Fergus, 4 Bute Crescent, Cardiff.

1881. Compton-Bracebridge, John Edward, Messrs. Easton and Anderson, 3 Whitehall Place, London, S.W.

1888. Constantine, Ezekiel Grayson, 32 Victoria Street, Manchester. [*Constant, Manchester.*]

1886. Conyers, Sidney Ward, Railway Construction Branch, Public Works Department, Sydney, New South Wales.

1874. Conyers, William, National Mortgage and Agency Co. of New Zealand, Melbourne, Victoria.

1888. Cook, John Joseph, Messrs. Robinson Cooks and Co., Atlas Foundry, St. Helen's, Lancashire.

1892. Cooke, Rupert Thomas, Stockton Malleable Iron Works, Stockton-on-Tees.

1877. Cooper, Arthur, North Eastern Steel Co., Royal Exchange, Middlesbrough.

1883. Cooper, Charles Friend, Messrs. Paterson and Cooper, Telegraph Works, Pownall Road, Dalston, London, E. [*Patella, London.* 1140.]

1877. Cooper, George, Pencliffe, Alleyne Road, West Dulwich, London, S.E.

1891. Cooper, Myles, 36 Victoria Street, Manchester.

1874. Cooper, William, Neptune Engine Works, Hull. [*Neptune, Hull.*]

1881. Coote, Arthur, Messrs. R. and W. Hawthorn Leslie and Co., Hebburn, Newcastle-on-Tyne.

1881. Copeland, Charles John, 11 Redcross Street, Liverpool.

1885. Coppée, Evence, 223 Avenue Louise, Bruxelles, Belgium.

1892. Corin, Philip Burne, Messrs. J. M. B. Corin and Son, Anchor Foundry, Penzance.

1878. Cornes, Cornelius, 6 Norfolk Crescent, Bath [*Stothert, Cornes, Bath.*]; and 76 Cannon Street, London, E.C. [*Stothert, Cornes, London.*]

1848. Corry, Edward, 9 New Broad Street, London, E.C.

1881. Cosser, Thomas, McLeod Road Iron Works, Kurrachee, India : (or care of Messrs. Ironside Gyles and Co., 1 Gresham Buildings, Guildhall, London, E.C.)

1883. Cotton, Henry Streatfeild, London Hydraulic Power Co., Palace Chambers, Bridge Street, Westminster, S.W.

1887. Coulman, John, Hull and Barnsley Railway, Spring Head Works, Hull.

1868. Coulson, William, Carlton Grove, Carlton Miniott, Thirsk.

1878. Courtney, Frank Stuart, 3 Whitehall Place, London, S.W.; and 76 Redcliffe Square, South Kensington, London, S.W.

1875. Coward, Edward, Messrs. Melland and Coward, Cotton Mills and Bleach Works, Heaton Mersey, near Manchester.

1875. Cowen, Edward Samuel, Messrs. G. R. Cowen and Co., Beck Works, Brook Street, Nottingham; and 9 The Ropewalk, Nottingham. [*Cowen, Nottingham.* 87.]

1880. Cowper, Charles Edward, 6 Great George Street, Westminster, S.W.

1847. Cowper, Edward Alfred, 6 Great George Street, Westminster, S.W.

1892. Cowper-Coles, Sherard Osborn, London Metallurgical Co., 80 Turnmill Street, London, E.C.

1878. Coxhead, Frederick Carley, 27 Leadenhall Street, London, E.C.

1887. Crabbe, Alexander, 4 Hawkhill Place, Dundee.

1882. Craven, John, Messrs. Smith Beacock and Tannett, Victoria Foundry, Leeds.

1890. Craven, Thomas Edwin, 20 Victoria Chambers, South Parade, Leeds.

1866. Craven, William, Vauxhall Iron Works, Osborne Street, Manchester.

1889. Cribb, Frederick James, Messrs. Marshall Sons and Co., Britannia Iron Works, Gainsborough.

1884. Crighton, John, Union Engineering Co., Pollard Street East, Manchester.

1873. Crippin, Edward Frederic, Mining Engineer, Brynn Hall Colliery, Ashton, near Wigan.

1883. Croft, Henry, Chemauns, Vancouver Island.

1878. Crohn, Frederick William, 14 Burney Street, Greenwich, London, S.E.

1877. Crompton, Rookes Evelyn Bell, Arc Works, Chelmsford; and Mansion House Buildings, Queen Victoria Street, London, E.C. [*Crompton, Chelmsford.*]

1884. Crook, Charles Alexander, Telegraph Construction and Maintenance Works, Enderby's Wharf, East Greenwich, London, S.E.

1881. Crosland, James Foyell Lovelock, Chief Engineer, Boiler Insurance and Steam Power Co., 67 King Street, Manchester.

1891. Crosland, Joseph, Messrs. Seebohm and Dieckstahl, Dannemora Steel Works, Sheffield; and Stanley Avenue, Birkdale, Southport.

1865. Cross, James, Messrs. John Hutchinson and Co., Alkali Works, Widnes; and Eirianfa, Llangollen.

1890. Cross, Robert James, Consulting Engineer, Great Western Steamship Co.; care of Messrs. Mark Whitwill and Son, Bristol.

1871. Crossley, William, 153 Queen Street, Glasgow. [*Crossley, Glasgow.* 584.]

1875. Crossley, William John, Messrs. Crossley Brothers, Great Marlborough Street, Manchester. [*Crossleys, Openshaw.*]

1882. Cruickshank, William Douglass, Chief Government Engineer Surveyor Marine Board, Sydney, New South Wales.

1889. Cullen, William Hart, Resident Engineer, The Aluminium Co., Oldbury, near Birmingham.

1875. Curtis, Richard, Messrs. Curtis Sons and Co., Phœnix Works, Chapel Street, Manchester. [*Curtius, Manchester.*]

1887. Cutler, George Benjamin, Messrs. Samuel Cutler and Sons, Providence Iron Works, Millwall, London, E. ; and 4 Westcombe Park, Blackheath, London, S.E.

1876. Cutler, Samuel, Messrs. Samuel Cutler and Sons, Providence Iron Works, Millwall, London, E. [*Cutler, Millwall.* 5059.]; and 16 Great George Street, Westminster, S.W.

1888. Dadabhoy, Cursetjee, Messrs. Shapurji Sorabji and Co., Bombay Foundry and Engine Works, Khetwady, Bombay, India; and Cumbala Hill, Bombay, India.

1864. Daglish, George Heaton, Rock Mount, St. Anne's Road, Aigburth, near Liverpool. [*Daglish, Aigburth.* 2717.]

1891. Daglish, Harry Bolton, Messrs. Robert Daglish and Co., St. Helen's Engine and Boiler Works, St. Helen's, Lancashire.

1883. D'Albert, Charles, Messrs. Hotchkiss and Co., 6 Route de Gonesse, St. Denis, near Paris; and 16 Rue des Chesneaux, Montmorency, Seine-et-Oise, France.

1890. Dalby, William Ernest, Engineering Department, The University, Cambridge.

1889. Dalgarno, James Robert, Danesford, Countess Wells Road, Mannofield, Aberdeen.

1881. D'Alton, Patrick Walter, London Electric Supply Corporation, Stowage Wharf, Deptford, London, S.E.

1866. Daniel, Edward Freer, Messrs. Worthington and Co., The Brewery, Burton-on-Trent; and 89 Derby Street, Burton-on-Trent.

1866. Daniel, William, Messrs. John Fowler and Co., Steam Plough and Locomotive Works, Leeds; and Fern Bank, Horsforth, Leeds.

1891. Daniels, Thomas, Messrs. Nasmyth Wilson and Co., Patricroft, Manchester.

1888. Darbishire, James Edward, 110 Cannon Street, London, E.C.

1878. Darwin, Horace (*Life Member*), The Orchard, Huntingdon Road, Cambridge.

1873. Davey, Henry, Messrs. Hathorn Davey and Co., Sun Foundry, Dewsbury Road, Leeds [*Sun Foundry, Leeds*]; and 3 Prince's Street, Westminster, S.W.

1884. Davidson, James Young, 13 Fairlawn Avenue, Acton Green, Chiswick, London, W.

1888. Davidson, Samuel Cleland, Sirocco Works, Bridge End, Belfast.

1880. Davies, Charles Merson, Locomotive Carriage and Wagon Superintendent, Bengal-Nagpur Railway, Nagpur, Central Provinces, India; and Laurieville, Queen's Drive, Crosshill, Glasgow.

1885. Davies, Edward John Mines, 16 Camden Street, Oakley Square, London, N.W.

1891. Davies, John Hubert, P.O. Box 455, Johannesburg, Transvaal, South Africa.

1874. Davis, Alfred, 28 St. Ermin's Mansions, Westminster, S.W. [*Sivad, London.*]

1868. Davis, Henry Wheeler, 53 New Broad Street, London, E.C.

1876. Davis, Joseph, Lancashire and Yorkshire Railway, Engineer's Office, Manchester.

1877. Davison, John Walter, Likova Damba, Nijni Novgorod, Russia: (or Park Cottage, South Hill Road, Milton, near Gravesend.)

1884. Davison, Robert, Caledonian Railway, Locomotive Department, St. Rollox, Glasgow.

1873. Davy, David, Messrs. Davy Brothers, Park Iron Works, Sheffield. [*Motor, Sheffield.*]

1892. Davy, William James, 41 Wandsworth Bridge Road, Fulham, London, S.W.

1883. Daw, James Gilbert, Messrs. Nevill Druce and Co., Llanelly Copper Works, Llanelly.

1874. Daw, Samuel, Staffa Lodge, South Park Hill Road, Croydon.

1879. Dawson, Bernard, 110 Cannon Street, London, E.C. [*Crocus, London.*]; and The Laurels, Malvern Link, Malvern. [*Heather, Malvern Link.*]

1875. Dawson, Edward, Messrs. Forster Brown and Rees, Guild Hall Chambers, Cardiff.

1890. Day, George Cameron, Messrs. Day Summers and Co., Northam Iron Works, Southampton.

1886. Dayson, William Ogden, Ebbw Vale Steel Iron and Coal Works, Ebbw Vale, R.S.O., Monmouthshire.

1874. Deacon, George Frederick, Victoria Mansions, 32 Victoria Street, Westminster, S.W.

1880. Deacon, Richard William, Messrs. Samuel Fisher and Co., Nile Foundry, Birmingham; and 19 Clarendon Road, Edgbaston, Birmingham.

1868. Dean, William, Locomotive Superintendent, Great Western Railway, Swindon.

1887. Deas, James, Clyde Navigation, Glasgow.

1866. Death, Ephraim, Messrs. Death and Ellwood, Albert Works, Leicester.

1884. Decauville, Paul, Portable Railway Works, Petit Bourg, Seine-et-Oise, France. [*Decauville, Corbeil.*]

1890. Deeley, Richard Mountford, Locomotive Department, Midland Railway, Derby; and 10 Charnwood Street, Derby.

1877. Dees, James Gibson, 36 King Street, Whitehaven.

1889. Defries, Wolf, Messrs. Defries and Sons, 147 Houndsditch, London, E. [*Defries, London.*]

1882. Denison, Samuel, Messrs. Samuel Denison and Son, Old Grammar School Foundry, North Street, Leeds. [*Weigh, Leeds.* 221.]

1892. Dennis, George D., Superintendent Engineer to William Whiteley, 147 Queen's Road, London, W.

1888. Dent, Charles Hastings, London and North Western Railway, Lime Street Station, Liverpool.

1883. Dick, Frank Wesley, Palmers Shipbuilding and Iron Works, Jarrow.

1891. Dick, John Norman, Government Marine Surveyor, Penang, Straits Settlements.

1890. Dickinson, Alfred, Engineer, South Staffordshire Tramways, Darlaston, Wednesbury.

1891. Dickinson, Douglas Holt, The Wood, Maybury, near Woking.

1891. Dickinson, James Clark, Palmer's Hill Engine Works, Sunderland.

1880. Dickinson, John, Palmer's Hill Engine Works, Sunderland. [*Bede, Sunderland.*]

1892. Dickinson, Richard Elihu, Bowling Iron Works, Bradford.

1892. Dickinson, Richard Henry, Locomotive Superintendent, Birmingham Central Tramways, Kyotts Lake Depôt, Birmingham.

1875. Dickinson, William, Messrs. Easton and Anderson, 3 Whitehall Place, London, S.W.

1888. Dickson, George Manners, Assistant Engineer, Calcutta Water Works, Municipal Office, Calcutta, India.

1886. Dixon, Robert, Messrs. Dixon and Corbitt, Teams Hemp and Wire Rope Works, Gateshead. [*Dixon, Gateshead.*]

1883. Dixon, Samuel, Messrs. Kendall and Gent, Victoria Works, Springfield, Salford, Manchester. [*Tools, Manchester.*]

1887. Dixon, William Basil, Earle's Shipbuilding and Engineering Works, Hull.

1872. Dobson, Benjamin Alfred, Messrs. Dobson and Barlow, Kay Street Machine Works, Bolton. [*Dobsons, Bolton.*]

1873. Dobson, Richard Joseph Caistor, Suiker Fabrick, Kalibayor, Banjoemas, Java: (or care of Charles E. S. Dobson, 4 Chesterfield Buildings, Victoria Park, Clifton, Bristol.)

1880. Dodd, John, Messrs. Platt Brothers and Co., Hartford Iron Works, Oldham.

1868. Dodman, Alfred, Highgate Foundry, Lynn. [*Dodman, Lynn.*]

1889. Dolby, Ernest Richard, 8 Prince's Street, Westminster, S.W.

1880. Donald, James, Superintendent Engineer, Messrs. James Fisher and Sons, Fisher's Buildings, Barrow-in-Furness.

1876. Donaldson, John, Messrs. John I. Thornycroft and Co., Steam Yacht and Launch Builders, Church Wharf, Chiswick, London, W.; and Tower House, Turnham Green.

1873. Donkin, Bryan, Jun., Messrs. B. Donkin and Co., Southwark Park Road, Bermondsey, London, S.E.

1891. Donovan, Edward Wynne, Messrs. J. S. Leach and Co., Mount Street Works, Harpurhey, Manchester.

1865. Douglas, Charles Prattman, Consett Iron Works, near Blackhill, County Durham; and Parliament Street, Consett, County Durham.

1879. Douglass, Sir James Nicholas, F.R.S., Trinity House, London, E.C. [2242.]; and Stella House, Dulwich, London, S.E.

1879. Douglass, William, Chief Engineer to the Commissioners of Irish Lights, Westmoreland Street, Dublin.

1891. Douglass, William James, Messrs. Douglass Brothers, Globe Iron Works, Blaydon-on-Tyne, R.S.O., County Durham.

1887. Douglass, William Tregarthen, 17 Victoria Street, Westminster, S.W.

1857. Dove, George, Messrs. Cowans Sheldon and Co., St. Nicholas Engine and Iron Works, Carlisle; and Viewfield, Stanwix, near Carlisle.

1873. Dove, George, Redbourn Hill Iron and Coal Co., Frodingham, near Doncaster [*Redbourn, Frodingham.*]; and Hatfield House, Hatfield, near Doncaster.

1866. Downey, Alfred C., Messrs. Downey and Co., Coatham Iron Works, Middlesbrough; and Belle Vue, Marton Road, Middlesbrough.

1881. Dowson, Joseph Emerson, 3 Great Queen Street, Westminster, S.W. [*Gaseous, London.*]

1880. Doxford, Robert Pile, Messrs. William Doxford and Sons, Pallion Shipbuilding and Engine Works, Sunderland.

1874. Dredge, James, 35 Bedford Street, Strand, London, W.C. [3663.]

1890. Drewet, Tom, Government Senior Inspector of Steam Boilers, Town Custom House, Bombay, India.

1886. Drummond, Dugald, Locomotive Superintendent, Caledonian Railway, St. Rollox Works, Glasgow.

1889. Drummond, Richard Oliver Gardner, De Beer's Diamond Mining Co., Kimberley, South Africa.

1877. Dübs, Charles Ralph, Messrs. Dübs and Co., Glasgow Locomotive Works, Glasgow.

1877. Dübs, Henry John Sillars, Messrs. Dübs and Co., Glasgow Locomotive Works, Glasgow.

1885. Duckering, Charles, Water Side Works, Rosemary Lane, Lincoln.

1880. Duckham, Frederic Eliot, Engineer, Millwall Docks, London, E.

1881. Duckham, Heber, 182 Lewisham Road, London, S.E. [*Duckham, London.*]

1879. Duncan, David John Russell, Messrs. Duncan Brothers, 2 Victoria Mansions, 28 Victoria Street, Westminster, S.W. [*Doucine, London.*]; and Kilmux, Leven.

1886. Duncan, Norman, Mechanical Engineer to the Municipality, Rangoon, British Burmah, India.

1892. Dunlop, James, Victoria Jubilee Technical Institute, Byculla, Bombay, India.

1870. Dunlop, James Wilkie, 39 Delancey Street, Regent's Park, London, N.W.

1881. Dunn, Henry Woodham, Charlcombe Grove, Lansdown, Bath.

1890. Dunn, Hugh Shaw, Engineer, Caprington Collieries, Kilmarnock.

1885. Durham, Frederick William, 27 Leadenhall Street, London, E.C. [*Oilring, London.*]; and Glemham Lodge, New Barnet.

1886. Duvall, Charles Anthony, Messrs. E. Bennis and Co., Lancashire Stoker Works, Deansgate Foundry, Bolton.

1887. Dymond, George Cecil, Messrs. W. P. Thompson and Co., 6 Lord Street, Liverpool.

1865. Dyson, Robert, Messrs. Owen and Dyson, Rother Iron Works, Rotherham.

1880. Eager, John Edward, Messrs. William Crichton and Co., Engineering and Shipbuilding Works, Abo, Finland.

1869. Earnshaw, William Lawrence, Superintending Marine Engineer, South Eastern Railway, Folkestone.

1858. Easton, Edward, 11 Delahay Street, Westminster, S.W.

1884. Eastwood, Charles, Manager, Linacre Gas Works, Liverpool.

1892. Eastwood, Thomas Carline, Messrs. Eastwood Swingler and Co., Victoria and Railway Iron Works, Derby. [*Swingler, Derby.*]

1888. Eaton-Shore, George, Borough Engineer, Temple Chambers, Crewe.

1875. Eaves, William, Engineer, Messrs. John Brown and Co., Atlas Steel and Iron Works, Sheffield.

1878. Eckart, William Roberts, Messrs. Salkeld and Eckart, 632 Market Street, P. O. Box 1844, San Francisco, California, United States.

1868. Eddison, Robert William, Messrs. John Fowler and Co., Steam Plough and Locomotive Works, Leeds.

1886. Ede, Francis Joseph, Messrs. Ede Brothers, Silchar, Cachar, India.

1887. Edlin, Herbert William, P.O. Box 199, Cape Town, Cape Colony: (or The Limes, Ellerton Road, Surbiton, R. O., Kingston-on-Thames.)

1883. Edmiston, James Brown, Marine Superintending Engineer, Messrs. Hamilton Fraser and Co., K Exchange Buildings, Liverpool; and Ivy Cottage, Highfield Road, Walton, Liverpool.

1871. Edwards, Edgar James, 12 Dartmouth Street, Westminster, S.W.; and 42 Rye Hill Park, Peckham, London, S.E.

1877. Edwards, Frederick, 62 Bishopsgate Street Within, London, E.C.

1888. Ellery, Henry George, 7 Fernbank Road, Redland, Bristol.

1875. Ellington, Edward Bayzand, Hydraulic Engineering Works, Chester; and Hydraulic Engineering Co., Palace Chambers, 9 Bridge Street, Westminster, S.W.

1859. Elliot, Sir George, Bart., Houghton-le-Spring, near Fence Houses. [*Elliot Company, London.*]

1892. Elliott, Archibald Campbell, D.Sc., Professor of Engineering, University College of South Wales and Monmouthshire, Cardiff.

1883. Elliott, Henry John, Assistant Manager, Elliott's Metal Works, Selly Oak, near Birmingham. [*Elmeco, Birmingham.*]

1869. Elliott, Henry Worton, Selly Oak Works, near Birmingham. [*Elmeco, Birmingham.*]

1882. Elliott, Thomas Graham, Messrs. Fairbairn Naylor Macpherson and Co., Wellington Foundry, Leeds.

1892. Ellis, Joseph S., Messrs. Edward Finch and Co., Bridge Works, Chepstow; and Myrtle Cottage, Chepstow.

1880. Ellis, Oswald William, 6 Grosvenor Place, Jesmond, Newcastle-on-Tyne. [*Robey, Newcastle-on-Tyne.*]

1885. Elsworthy, Edward Houtson, Messrs. Richardson and Cruddas, Byculla Iron Works, Bombay, India; and 91 King Henry's Road, London, N.W.

1875. Elwell, Thomas, 223 Avenue de Paris, Plaine St. Denis, Seine, France.

1878. Elwin, Charles, London County Council, Spring Gardens, London, S.W.

1889. Emett, George Henry Hawkins, Hope Foundry, Dewsbury.

1890. English, Lt.-Colonel Thomas, Palmer's Ordnance Works, Jarrow: (or care of W. Stamm, 30 Victoria Street, Westminster, S.W.)

1885. Errington, William, Salisbury Buildings, Bourke Street. Melbourne, Victoria.

1891. Esson, David Duncan, 47 Kennington Oval, London, S.E.

1890. Esson, John, Messrs. Blaikie Brothers, Footdee Iron Works, Aberdeen.

1889. Etches, Harry, Waterous Engine Works Co., Brantford, Ontario, Canada.

1884. Etherington, John, 39A King William Street, London Bridge, London, E.C.

1887. Evans, Arthur George, Palace Chambers, 9 Bridge Street, Westminster, S.W.

1884. Evans, David, Messrs. Bolckow Vaughan and Co., Cleveland Iron and Steel Works, South Bank, R.S.O., Yorkshire.

1888. Evans, Joseph, Culwell Foundry, Wolverhampton.

1892. Evanson, Frederic Macdonnell, Alexandra Hotel, Dale Street, Liverpool.

1887. Everard, John Breedon, 6 Millstone Lane, Leicester.

1887. Everitt, Nevill Henry, Messrs. Thomas Piggott and Co., Atlas Works, Birmingham; and Knowle Hall, Warwickshire.

1881. Ewen, Thomas Buttwell, Messrs. Ewen and Mitton, Smithfield Works, Sherlock Street, Birmingham.

1891. Ewing, James Alfred, F.R.S., Professor of Mechanism and Applied Mechanics, Engineering Department, The University, Cambridge; and Langdale Lodge, Cambridge.

1890. Exton, George Gaskell, Messrs. Chubb and Son, 128 Queen Victoria Street, London, E.C.

1868. Fairbairn, Sir Andrew, Messrs. Fairbairn Naylor Macpherson and Co., Wellington Foundry, Leeds; and Askham Richard, York.

1875. Farcot, Jean Joseph Léon, Messrs. Farcot and Sons, Engine Works, 13 Avenue de la Gare, St. Ouen, France.

1880. Farcot, Paul, Messrs. Farcot and Sons, Engine Works, 13 Avenue de la Gare, St. Ouen, France.

1881. Farrar, Sidney Howard, Messrs. Howard Farrar and Co., Port Elizabeth, South Africa; and care of Messrs. F. A. Robinson and Co., 69 Cornhill, London, E.C.

1882. Fawcett, Thomas Constantine, White House Engineering Works, Leeds. [*Fawcett, Leeds.*]

1884. Fearfield, John Piggin, Lace Machine Works, Stapleford, near Nottingham; and The Ferns, Stapleford, near Nottingham. [*Fearfield, Nottingham.*]

1882. Feeny, Victor Isidore, 60 Queen Victoria Street, London, E.C. [*Victor Feeny, London.*]

1876. Fell, John Corry, 1 Queen Victoria Street, London, E.C.; and Excelsior Works, Old Street, London, E.C.

1877. Fenton, James, Passlands, 278 Upper Richmond Road, Putney, London, S.W.

1869. Fenwick, Clennell, Victoria Docks Engine Works, Victoria Docks, London, E. [*Clennell, London.*]

1892. Fenwick, James, 19 Bridge Street, Sydney, New South Wales. [1038.]

1870. Ferguson, Henry Tanner, Wolleigh, Bovey Tracey, near Newton Abbot.

1881. Ferguson, William, Harbour Board, Wellington, New Zealand: (or care of Montgomery Ferguson, 81 James Street, Dublin.)

1866. Fiddes, Walter, Clapton Villa, Belgrave Road, Tyndall's Park, Bristol.

1867. Field, Edward, Chandos Chambers, 22 Buckingham Street, Adelphi, London, W.C.

1888. Field, Howard, Messrs. John Bell and Son, 118A Southwark Street, London, S.E.; and Hillcote, Buckhurst Hill, S.O., Essex.

1884. Fielden, Joseph Petrie, Messrs. Thomas Robinson and Son, Railway Works, Rochdale.

1874. Fielding, John, Messrs. Fielding and Platt, Atlas Iron Works, Gloucester. [*Atlas, Gloucester.*]

1891. Finlayson, Finlay, Messrs. Miller and Co., Vulcan Foundry, Coatbridge.

1891. Firth, George Henry, Messrs. Richard Johnson and Nephew, Bradford Iron Works, Manchester.

1887. Firth, William, Water Lane, Leeds.

1888. Fischer, Gustave Joseph, Tramway Construction Branch, Public Works Department, Beresford Chambers, Castlereagh Street, Sydney, New South Wales.

1889. Fisher, Henry Bedwell, Locomotive Works, London Brighton and South Coast Railway, Brighton.

1884. Fisher, Henry Oakden, Ty Mynydd, Radyr, near Cardiff.

1888. FitzGerald, Maurice Frederick, Professor of Engineering, Queen's College, Belfast.

1877. Flannery, James Fortescue, 9 Fenchurch Street, London, E.C. [2283.]

1883. Fletcher, George, Masson and Atlas Works, Litchurch, Derby.

1872. Fletcher, Herbert, Ladyshore Colliery, Little Lever, Bolton; and The Hollins, Bolton.

1867. Fletcher, Lavington Evans, Chief Engineer, Manchester Steam Users' Association, 9 Mount Street, Albert Square, Manchester. [*Steam Users', Manchester.*]

1892. Focken, Charles Frederick, Messrs. Apcar and Co., Raddah Bazar, Calcutta, India; and care of Institute of Engineers and Shipbuilders, Hong Kong, China.

1859. Fogg, Robert, 11 Queen Anne's Gate, Westminster, S.W.

1887. Foley, Nelson, Engineering Manager, Società Industriale Napoletana Hawthorn-Guppy, Naples, Italy.

1886. Folger, William Mayhew, Commander, United States Navy, Bureau of Ordnance, Naval Department, Washington, D.C., United States.

1877. Forbes, Daniel Walker, Smithfield Works, New Road, Blackwall London, E.

1882. Forbes, David Moncur, Engineer, H. M. Mint, Bombay.

1892. Forbes, Percy Alexander, Messrs. Lambert Brothers, Tube Mills, Iron and Brass Works, Walsall.

1882. Forbes, William George Loudon Stuart, Superintendent of General Workshops, H. M. Mint, Calcutta.

1892. Forrest, Hilary Sheldon, Messrs. Dobson and Barlow, Kay Street Machine Works, Bolton.

1888. Forster, Alfred Llewellyn, Assistant Engineer, Newcastle and Gateshead Water Works, Newcastle-on-Tyne.

1888. Forster, Edward John, Malta Villa, West Smethwick, Birmingham.

1882. Forsyth, Robert Alexander, Courtway, Gold Tops, Newport, Monmouthshire.

1889. Foster, Ernest Howard, Messrs. Henry R. Worthington, 86 Liberty Street, New York, United States.

1886. Foster, Frederick, Messrs. Barnett and Foster, Niagara Works, Eagle Wharf Road, New North Road, London, N. [*Drinks, London.* 306.]

1889. Foster, Herbert Anderton (*Life Member*), Messrs. John Foster and Son, Black Dike Spinning Mills, Queensbury, near Bradford.

1888. Foster, James, Lily Bank, St. Andrew's Drive, Pollokshields, Glasgow.

1884. Foster, John Slater, Messrs. Jones and Foster, 39 Bloomsbury Street, Birmingham.

1882. Fothergill, John Reed, Superintendent Marine Engineer, 1 Bathgate Terrace, West Hartlepool.

1877. Foulis, William, Engineer, Glasgow Corporation Gas Works, 42 Virginia Street, Glasgow.

1885. Fourny, Hector Foster, French Chambers, Queen's Dock-Side, Hull. [*Veritas, Hull.*]

1866. Fowler, George, Basford Hall, near Nottingham.

1847. Fowler, Sir John, Bart., K.C.M.G., 2 Queen Square Place, Westminster, S.W.

1885. Fowler, William Henry, 6 Victoria Station Approach, Manchester; and The Poplars, New Moston, near Manchester.

1866. Fox, Sir Douglas, 2 Victoria Mansions, 28 Victoria Street, Westminster, S.W.

1875. Fox, Samson, Leeds Forge, Leeds.

1884. Frampton, Edwin, General Engine and Boiler Co., Hatcham Iron Works, Pomeroy Street, New Cross Road, London, S.E. [*Oxygen, London*. 8007.]

1888. Francken, William Augustus, care of Messrs. Grindlay and Co., 55 Parliament Street, London, S.W.

1885. Franki, James Peter, Morts Dock and Engineering Co., Morts Bay, Sydney, New South Wales: (or care of Messrs. Goldsbrough Mort and Co., 149 Leadenhall Street, London, E.C.)

1877. Fraser, John Hazell, Messrs. John Fraser and Son, Millwall Boiler Works, London, E.; and 110 Cannon Street, London, E.C.

1888. Frenzel, Arthur Benjamin, 93 Liberty Street, New York, United States.

1891. Frier, John Drummond, 18 Almack Road, Clapton, London, N.E.

1876. Frost, William, Manager, Carlisle Steel and Engine Works, Sheffield; and Barnsley Road, Sheffield.

1866. Fry, Albert, Bristol Wagon Works, Lawrence Hill, Bristol.

1891. Fuller, Charles Frederick, 171 Queen Victoria Street, London, E.C.

1884. Furness, Edward, Knollcroft, Knoll Road, Bexley, S.O., Kent.

1890. Gadd, William, Assistant Locomotive Engineer, Waterford and Limerick Railway, Limerick.

1866. Galloway, Charles John, Managing Director, Messrs. Galloways, Knott Mill Iron Works, Manchester. [*Galloway, Manchester.*]

1862. Galton, Sir Douglas, K.C.B., D.C.L., F.R.S., 12 Chester Street, Grosvenor Place, London, S.W.

1884. Ganga Ram, Rai Bahadur, Executive Engineer, Public Works Department, Amritsar, Punjaub, India: (or care of Messrs. Thomas Wilson and Co., 24 Rood Lane, London, E.C.)

1891. Garrard, Charles Riley, Abingdon Works, Bath Street, Birmingham.

1882. Garrett, Frank, Messrs. Richard Garrett and Sons, Leiston Works, Leiston, R.S.O., Suffolk. [*Garrett, Leiston.*]

1867. Gauntlett, William Henry, 33 Albert Terrace, Middlesbrough. [*Pyrometer, Middlesbrough.*]

1888. Gaze, Edward Henry James, 4 Victoria Drive, Mount Florida, Glasgow.

1888. Geddes, Christopher, 23 Brunswick Street, Liverpool.

1880. Geoghegan, Samuel, Messrs. A. Guinness Son and Co., St. James' Gate Brewery, Dublin. [*Guinness, Dublin.*]

1887. Gibb, Andrew, Managing Engineer, Messrs. Rait and Gardiner, Millwall Docks, London, E.; and 30 South Street, Greenwich, London, S.E.

1871. Gibbins, Richard Cadbury, Berkley Street, Birmingham. [*Gibbins, Birmingham.*]

1883. Gilchrist, Percy Carlyle, F.R.S. (*Life Member*), Palace Chambers, 9 Bridge Street, Westminster, S.W. [*Gilchrist, London*]; and Frognal Bank, Finchley New Road, Hampstead, London, N.W.

1856. Gilkes, Edgar, Westholme, Grange-over-Sands, viâ Carnforth, Lancashire.

1880. Gill, Charles, Messrs. Young and Gill, Engineering Works, Java; and Java Lodge, Beckenham.

1889. Gill, Frederick Henry, Messrs. Alexander Penney and Co., 107 Fenchurch Street, London, E.C.

1884. Gimson, Arthur James, Messrs. Gimson and Co., Engine Works, Vulcan Street, Leicester. [*Gimson, Leicester.* 6.]

1881. Girdwood, William Wallace, Indestructible Packing Works, 9 Lea Place, East India Dock Road, Poplar, London, E.

1874. Gjers, John, Messrs. Gjers Mills and Co., Ayresome Iron Works, Middlesbrough.

1887. Gledhill, Manassah, Sir Joseph Whitworth and Co., Openshaw, Manchester.

1880. Godfrey, William Bernard, 23 St. Swithin's Lane, London, E.C.

1888. Goff, John, Messrs. Salt and Co., The Brewery, Burton-on-Trent.

1882. Goldsmith, Alfred Joseph, Lillington, Moray Street, New Farm, Brisbane, Queensland.

1879. Goldsworthy, Robert Bruce, Messrs. Thomas Goldsworthy and Sons, Britannia Emery Mills, Hulme, Manchester. [*Goldsworthy, Manchester.*]

1867. Gooch, William Frederick, Vulcan Foundry, Newton-le-Willows, Lancashire.

1877. Goodbody, Robert, Messrs. Goodbody, Clashawaun Jute Factory, Clara, near Moate, Ireland.

1869. Goodeve, Thomas Minchin, 5 Crown Office Row, Temple, London, E.C.

1875. Goodfellow, George Ben, Messrs. Goodfellow and Matthews, Hyde Iron Works, Hyde, near Manchester. [*Goodfellow, Hyde.*]

1884. Goodger, Walter William, 34 Dairy House Road, Derby.

1890. Goodman, John, Professor of Engineering, Yorkshire College, Leeds.

1885. Goodwin, Arnold, Jun., 56 Sumner Street, Southwark, London, S.E.

1889. Goold, William Tom, 39 Queen Victoria Street, London, E.C.; and 18 The Hawthorns, Finchley, London, N.

1865. Göransson, Göran Fredrick, Sandvik Iron Works, near Gefle, Sweden: (or care of James Bird, 118 Cannon Street, London, E.C.)

1887. Gordon, Alexander, Niles Tool Works, and Messrs. Gordon and Maxwell, Hamilton, Ohio, United States.

1875. Gordon, Robert, 6 Pilmour Place, St. Andrews.

1888. Gore, Arthur Saunders, Sherborne Metal Works, Sherborne Street, Birmingham.

1879. Gorman, William Augustus, Messrs. Siebe and Gorman, 187 Westminster Bridge Road, London, S.E. [*Siebe, London.*]

1880. Gottschalk, Alexandre, 13 Rue Auber, Paris.

1877. Goulty, Wallis Rivers, Messrs. Wheatley Kirk, Price, and Goulty, Albert Chambers, Albert Square, Manchester. [*Indicator, Manchester.*]

1887. Gourlay, Charles Gershom, Messrs. Gourlay Brothers and Co., Dundee Foundry, Dundee.

1890. Grace, Robert William, Latrobe Steel Works, Latrobe, Westmoreland Co., Pennsylvania, United States.

1878. Grafton, Alexander, Vulcan Works, Bedford. [*Grafton, Bedford.*]

1886. Grant, Percy, Sola Works, Ferro Carril del Sud, Buenos Aires, Argentine Republic: (or care of John M. Grant, 136 Sutherland Avenue, Maida Vale, London, W.)

1891. Gray, George Macfarlane (*Life Member*), Imperial Chinese Customs, Hong Kong, China.

1865. Gray, John Macfarlane, Chief Examiner of Engineers, Marine Department, Board of Trade, St. Katharine Dock House, Tower Hill, London, E.; and 1 Claremont Road, Forest Gate, London, E. [*Yarg, London.*]

1876. Gray, John William, Engineer, Corporation Water Works, Broad Street, Birmingham.

1879. Gray, Thomas Lowe, Lloyd's Register, 2 White Lion Court, Cornhill, London, E.C.; and Rokesley House, St. Michael's Road, Stockwell, London, S.W.

1879. Greathead, James Henry, 8 Victoria Chambers, 15 Victoria Street, Westminster, S.W.

1861. Green, Sir Edward, Bart., Messrs. E. Green and Son, Phœnix Works, Wakefield.

1888. Green, Henry Joseph Kersting, Messrs. Barry and Co., 5 Lyons Range, Calcutta, India; 13 Garden Reach, Calcutta, India: (or care of Messrs. J. B. Barry and Son, 110 Cannon Street, London, E.C.)

1871. Greener, John Henry, 15 Walbrook, London, E.C.

1890. Greening, William Alfred, 88 Bishopsgate Street Within, London, E.C.

1878. Greenwood, Arthur, Messrs. Greenwood and Batley, Albion Works, Leeds.

1874. Greenwood, William Henry, Birmingham Small Arms and Metal Co., Adderley Park Works, Birmingham.

1879. Grenville, Robert Neville, Butleigh Court, Glastonbury.

1892. Gresham, Harry Edward, Messrs. Gresham and Craven, Craven Iron Works, Salford, Manchester. [*Brake, Manchester.* 613.]

1880. Gresham, James, Messrs. Gresham and Craven, Craven Iron Works, Salford, Manchester. [*Brake, Manchester.* 613.]

1883. Grew, Frederick, 12 Stockleigh Road, St. Leonard's-on-Sea.

1874. Grew, Nathaniel, Dashwood House, 9 New Broad Street, London, E.C.

1884. Griffiths, James E., Messrs. Griffiths and James, 2 Bute Crescent, Cardiff.

1873. Griffiths, John Alfred, Matlock, Charters Towers, Queensland: (or care of Thomas Griffiths, Alderley Edge, Manchester.)

1889. Grimshaw, James Walter, Resident Engineer, Harbours and Rivers Department, Sydney, New South Wales; and Australian Club, Sydney, New South Wales.

1891. Groom, Richard Alfred, Shropshire Works, Wellington, Salop.

1879. Grose, Arthur, Messrs. Grose Norman and Co., Reliance Works, Northampton.

1886. Grove, David, 24 Friedrich Strasse, Berlin.

1870. Guilford, Francis Leaver, Messrs. G. R. Cowen and Co., Beck Works, Brook Street, Nottingham. [*Cowen, Nottingham.* 87.]

1884. Gulland, James Ker, Diamond Drill Co., 8 Victoria Street, Westminster, S.W. [*Gulland, London.*]

1886. Guy, Charles Williams, 123 Oakfield Road, Penge, London, S.E.

1870. Gwynne, James Eglinton Anderson (*Life Member*), Essex Street Works, Strand, London, W.C. [*Gwynnegram, London.*]

1870. Gwynne, John, Hammersmith Iron Works, Hammersmith, London, W.; and 89 Cannon Street, London, E.C.

1888. Hadfield, Robert Abbott, Hecla Foundry Steel Works, Sheffield. [*Hadfield, Sheffield.*]

1884. Hall, Albert Francis, George F. Blake Manufacturing Co., 111 Federal Street, Boston; and 3 Cordis Street, Charlestown, Boston, Massachusetts, United States.

1892. Hall, George Edward, Mechanical Superintendent, Lighting Department, Salford Corporation, Wilburn Street, Salford, Manchester.

1879. Hall, John Francis, Norbury, Pittsmoor, Sheffield.

1881. Hall, John Percy, Managing Director, Messrs. John Penn and Sons, Greenwich, London, S.E.

1882. Hall, John Willim, Ivy House, Bilston.

1890. Hall, Oscar Standring, Messrs. Robert Hall and Sons, Hope Foundry, Bury.

1874. Hall, Thomas Bernard, 119 Colmore Row, Birmingham; and Ingleside, Sandon Road, Edgbaston, Birmingham.

1871. Hall, William Silver, Messrs. Takata and Co., Ginza San Chome 18, Banchi, Tokio, Japan; and 88 Bishopsgate Street Within, London, E.C.

1889. Hall-Brown, Ebenezer, Messrs. Hall-Brown Buttery and Co., Helen Street Engine Works, Govan, Glasgow. [*Triple, Glasgow.* 1843.]

1880. Hallett, John Harry, 115 Bute Docks, Cardiff. [*Consulting, Cardiff.*]

1871. Halpin, Druitt, 9 Victoria Chambers, 17 Victoria Street, Westminster, S.W. [*Halpin, London.* 3075, care of Victoria Chambers Co.]

1875. Hammond, Walter John, The Grange, Knockholt, near Sevenoaks.

1886. Hanbury, John James, Resident Engineer and Locomotive Superintendent, Metropolitan Railway, Neasden, London, N.W.

1870. Hannah, Joseph Edward, Water Works, Winnipeg, Manitoba, Canada.

1892. Hansell, Robert Blackwell, General Manager, Mount Vernon Siemens Steel Works, near Glasgow; and 3 Maule Terrace, Partick, Glasgow.

1888. Harada, Torazo, Superintending Engineer, Osaka Shipping Co., Osaka, Japan.

1891. Harcourt, Otto Simon Henry, Clarence Iron Works, Leeds.

1888. Harding, Thomas Walter, Tower Works, Leeds.

1874. Harding, William Bishop, IX Ker Rakos utcza 5 ik. sz., lsö. Emelet, Budapest, Hungary.

1881. Hardingham, George Gatton Melhuish, 191 Fleet Street, London, E.C. [*Hardingham, London.*]

1883. Hardy, John George, 13 Riemergasse, Stadt, Vienna.

1869. Harfield, William Horatio, Mansion House Buildings, Queen Victoria Street, London, E.C.

1887. Hargraves, Richard, 3 London Road, Blackburn.

1887. Hargreaves, John Henry, Messrs. Hick Hargreaves and Co., Soho Iron Works, Crook Street, Bolton.

1884. Harker, Harold Hayes, Locomotive Superintendent, Minas and Rio Railway, Cruzeiro, Rio de Janeiro, Brazil: (or care of Jesse T. Curtis, Hill Street, Poole.)

1888. Harker, William, Messrs. Richard Schram and Co., 17A Great George Street, Westminster, S.W.

1888. Harland, Sir Edward James, Bart., M.P., Messrs. Harland and Wolff, Belfast; and Baroda House, Kensington Palace Gardens, London, W.

1891. Harris, Gordon, Messrs. Merryweather and Sons, Fire-Engine Works, Greenwich Road, London, S.E.

1879. Harris, Henry Graham, Messrs. Bramwell and Harris, 5 Great George Street, Westminster, S.W.

1885. Harris, John Henry, Worthington Pumping Engine Co., 153 Queen Victoria Street, London, E.C. [*Tuneharp, London.*]

1873. Harris, Richard Henry, 63 Queen Victoria Street, London, E.C.; and Oak Hill, Surbiton, R.O., near Kingston-on-Thames.

1877. Harris, William Wallington, Messrs. A. M. Perkins and Son, 6 Seaford Street, Regent Square, London, W.C.; and 24 Alexandra Villas, Hornsey Park, London, N.

1892. Harrison, Abraham Wyke, Lion Street, Abergavenny.

1885. Harrison, Frederick Henry, Lincoln Malleable Iron Works, Lincoln. [*Malleable, Lincoln.*]

1888. Harrison, George, 21 Hillsboro Road, East Dulwich, London, S.E.

1889. Harrison, Gilbert Harwood, Lieutenant R.E., Assistant Inspector of Gun Carriages, Royal Arsenal, Woolwich.

1885. Harrison, Joseph, Royal College of Science, Exhibition Road, South Kensington, London, S.W.

1891. Harrison, Joseph Hutchinson, 2 Exchange Place, Middlesbrough; and Clifford Villa, Coatham, Redcar.

1887. Harrison, Thomas Henry, Messrs. Davey Paxman and Co., 139 Queen Victoria Street, London, E.C.; and 22 Granville Villas, Earlsfield Road, Wandsworth, London, S.W.

1890. Harrison, William Robert, Manager, Hull Cart and Wagon Co., Newington Iron Works, Hull.

1883. Hart, Frederick, 36 Prospect Street, Poughkeepsie, New York, United States: (or care of A. Pye-Smith, Messrs. Samuel Osborn and Co., 2 Victoria Mansions, 28 Victoria Street, Westminster, S.W.)

1882. Hart, Norman, 56 Crouch Hall Road, Crouch End, London, N.

1872. Hartnell, Wilson, Benson's Buildings, Park Row, Leeds.

1882. Harvey, Charles Randolph, Messrs. G. and A. Harvey, Albion Machine Works, Govan, near Glasgow.

1892. Harvey, Edward Cartwright, Engineer, Geldenhuis Estate and Gold Mining Co., P. O. Box 1022, Johannesburg, Transvaal, South Africa.

1892. Harvey, Francis Haniel, Messrs. Harvey and Co., Hayle Foundry, Hayle, Cornwall.

1886. Harvey, John Boyd, Ashfield, Totnes.

1883. Harvey, Robert, 1 Palace Gate, London, W.

1878. Harwood, Robert, Soho Iron Works, Bolton.

1881. Haslam, Sir Alfred Seale, Union Foundry, Derby. [*Zero, Derby.*]

1885. Hatton, Robert James, Henley's Telegraph Works, North Woolwich, London, E.

1857. Haughton, S. Wilfred (*Life Member*), Greenbank, Carlow, Ireland.

1878. Haughton, Thomas, 110 Cannon Street, London, E.C. [*Haughnot, London.*]

1885. Haughton, Thomas James, Waterside, Ferry Road, Teddington, S.O., Middlesex.

1892. Hawkins, Rupert Skelton, Locomotive and Carriage Department, Indian Midland Railway, Jhansi, India.

1861. Hawkins, William Bailey, 39 Lombard Street, London, E.C.

1870. Hawksley, Charles, 30 Great George Street, Westminster, S.W.

1891. Hawksley, George William, Brightside Boiler and Engine Works, Savile Street East, Sheffield. [*Hawksley, Sheffield.* 337.]

1856. Hawksley, Thomas, F.R.S., 30 Great George Street, Westminster, S.W.

1873. Hay, James A. C., Superintending Engineer and Constructor of Shipping to the War Department, Royal Arsenal, Woolwich.

1882. Hayes, Edward, Watling Works, Stony Stratford. [*Hayes, Stony Stratford.*]

1879. Hayes, John, 30 and 31 St. Swithin's Lane, London, E.C.

1880. Hayter, Harrison, 33 Great George Street, Westminster, S.W.

1885. Head, Archibald Potter, Queen's Square, Middlesbrough.

1888. Head, Harold Ellershaw, 24 Auriol Road, West Kensington, London, W.

1869. Head, Jeremiah, Queen's Square, Middlesbrough [*Head, Middlesbrough.*]; and 26 Lombard Street, London, E.C.

1873. Headly, Lawrance, Exchange Iron Foundry and Implement Works, Corn Exchange Street, Cambridge; and Yorke House, Newmarket Road, Cambridge. [*Vanes, Cambridge.*]

1857. Healey, Edward Charles, 163 Strand, London, W.C.

1890. Heap, Ray Douglas Theodore, Messrs. Crompton and Co., 4 Mansion House Buildings, Queen Victoria Street, London, E.C.; and 37 Sinclair Road, Kensington, London, W.

1872. Heap, William, 9 Rumford Place, Liverpool. [*Metal, Liverpool.* 809.]

1889. Heath, George Wilson, Messrs. Heath and Co., Observatory Works, Crayford, Kent.

1888. Heatly, Harry, Messrs. Heatly and Gresham, 7 Hastings Street, Calcutta, India.

1875. Heenan, Richard Hammersley, Messrs. Heenan and Froude, Newton Heath Iron Works, near Manchester; and The Manor House, Wilmslow, near Manchester. [*Spherical, Newton Heath.*]

1879. Hele-Shaw, Henry Selby, Professor of Engineering, University College, Liverpool.

1869. Henderson, David Marr, Engineer-in-Chief, Imperial Maritime Customs Service of China, Shanghai, China.

1883. Henderson, John Baillie, Engineer to the Queensland Government, Water Supply Department, Brisbane, Queensland.

1891. Henderson, Thomas, 6 and 8 Trueman Street, Liverpool.

1883. Henderson, William, Perth Road, Stanley, R.S.O., Perthshire.

1878. Henesey, Richard, Messrs. Donald Henesey and Couper, Ripon Iron Works, Frere Road, Bombay, India.

1888. Henning, Gustavus Charles, 726 Temple Court, 5 Beekman Street, New York, United States.

1879. Henriques, Cecil Quixam, Messrs. John H. Wilson and Co., Sandhills, Liverpool. [*Engineers, Liverpool.*]

1875. Hepburn, George, Redcross Chambers, Redcross Street, Liverpool. [*Hepburn, Liverpool.*]

1891. Hepburn, Thomas, Chief Engineer, Gunpowder Factory, Kirkee, Poona, India.

1876. Heppell, Thomas, Mining Engineer, Ouston Collieries, Chester-le-Street.

1892. Herbert, Alfred, Machine-Tool Works, Coventry. [*Lathe, Coventry.* 52.]

1884. Hernu, Arthur Henry, 69 Victoria Street, Westminster, S.W.

1884. Hervey, Matthew Wilson, Assistant Engineer, West Middlesex Water Works, Hammersmith, London, W.

1879. Hesketh, Everard, Messrs. J. and E. Hall, Iron Works, Dartford. [*Hesketh, Dartford.*]

1872. Hewlett, Alfred, Haseley Manor, Warwick.

1887. Hibbert, George, Hibbert's Works, Bank Road, Gateshead.

1871. Hick, John, Mytton Hall, Whalley, near Blackburn.

1885. Hicken, Thomas, 1519 Calle Brandzen, Barraccas al Norte, Buenos Aires, Argentine Republic: (or care of Miss Hicken, Bourton, near Rugby.)

1879. Higson, Jacob, Mining Engineer, Crown Buildings, 18 Booth Street, Manchester.

1889. Hill, Arthur Ripley, Messrs. Hill Brothers, Nevins Foundry, Hunslet, Leeds.

1885. Hill, Robert Anderson, Royal Mint, Little Tower Hill, London, E.

1890. Hiller, Edward George, Chief Engineer, National Boiler Insurance Co., 22 St. Ann's Square, Manchester.

1882. Hiller, Henry, Consulting Engineer, National Boiler Insurance Co., 22 St. Ann's Square, Manchester; and Athelney, Stanley Road, Alexandra Park, Manchester.

1873. Hilton, Franklin, General Manager, Ebbw Vale Steel Iron and Coal Works, Ebbw Vale, R.S.O., Monmouthshire.

1887. Hindson, William, Messrs. J. Abbot and Co., Park Works, Gateshead.

1891. Hodge, Arthur, Trewirgie, Redruth.

1891. Hodges, Frank Grattidge, Carriage and Wagon Works, Midland Railway, Derby.

1870. Hodges, Petronius, 142 Burngreave Road, Sheffield.

1880. Hodgson, Charles, Messrs. Saxby and Farmer, Railway Signal Works, Canterbury Road, Kilburn, London, N.W.

1889. Hodgson, George Herbert, Thornton Road, Bradford.

1892. Hodgson, Henry Edwin, Brookhouse Iron Works, Cleckheaton, S.O., Yorkshire.

1891. Hogarth, Thomas Oswald, Great Western Railway Works, Swindon.

1889. Hoggins, Alfred Farquharson, 25 Kestrel Avenue, Herne Hill, London, S.E.

1866. Holcroft, Thomas, Bilston Foundry, Bilston.

1886. Holden, James, Locomotive Superintendent, Great Eastern Railway, Stratford Works, London, E.

1884. Holland, Calvert Bernard, General Manager, Ebbw Vale Steel Iron and Coal Works, Ebbw Vale, R.S.O., Monmouthshire.

1886. Hollis, Charles William, Messrs. Ketton and Hollis, Meadow Tool Works, Mayfield Grove, Nottingham.

1885. Hollis, Henry William, Thornville House, Darlington.

1891. Holman, Hugh Wilson, Messrs. E. J. Caiger and Co., 77 Billiter Buildings, Billiter Street, London, E.C. [*Caiger, London.*]

1892. Holmström, Carl Albert, Maxim-Nordenfelt Guns and Ammunition Co., 32 Victoria Street, Westminster, S.W.

1883. Holroyd, John, Tomlinson Street, Hulme, Manchester. [*Knit, Manchester.*]

1885. Holroyd, John Herbert, West's Patent Press Company, Gadag, Dharwar District, India.

1863. Holt, Francis, Locomotive Department, Midland Railway, Derby.

1873. Holt, Henry Percy, The Cedars, Didsbury, Manchester.

1890. Holt, Robert, Lecturer in Engineering, Walker Engineering Laboratories, Brownlow Hill, Liverpool.

1890. Holt, William Procter, Messrs. Beyer Peacock and Co., Gorton Foundry, Manchester.

1888. Homan, Harold, Messrs. Homan and Rodgers, 10 Marsden Street, Manchester. [*Namoh, Manchester.* 637.]

1867. Homer, Charles James, Mining Engineer, Ivy House, Stoke-upon-Trent.

1890. Hooker, Benjamin, Pear Tree Court, Farringdon Road, London, E.C.

1892. Hope, John Basil, Locomotive Department, North Eastern Railway, Leeds.

1866. Hopkins, John Satchell, Jesmond Grove, Highfield Road, Edgbaston, Birmingham.

1885. Hopkinson, Charles, Werneth Chambers, 29 Princess Street, Manchester.

1856. Hopkinson, John, Inglewood, St. Margaret's Road, Bowdon, near Altrincham.

1874. Hopkinson, John, Jun., D.Sc., F.R.S., Messrs. Chance Brothers and Co., Lighthouse Works, near Birmingham; and 3 Westminster Chambers, 5 Victoria Street, Westminster, S.W. [3092.]

1877. Hopkinson, Joseph, Messrs. Joseph Hopkinson and Co., Britannia Works, Huddersfield.

1890. Hopper, Allan, Messrs. William Hopper and Co., Moscow, Russia.

1890. Hopper, James Russell, Messrs. William Hopper and Co., Moscow, Russia.

1889. Hopwood, John, Locomotive Superintendent, Argentine Great Western Railway, Mendoza, Argentine Republic.

1891. Hornbrook, Raymond Hillman, Imperial Revenue Cutter "Fei Hoo," care of Imperial Maritime Customs, Shanghai, China.

1880. Hornsby, James, Messrs. Richard Hornsby and Sons, Spittlegate Iron Works, Grantham. [*Hornsbys, Grantham.*]

1889. Horsfield, Cooper, Messrs. Holroyd Horsfield and Wilson, Larchfield Foundry, Hunslet Road, Leeds.

1891. Horsfield, Ralph, Messrs. Kirk and Horsfield, Chapel-en-le-Frith, near Stockport.

1873. Horsley, Charles, 22 Wharf Road, City Road, London, N.

1892. Horsnell, Daniel, 79 Farringdon Road, London, E.C.

1868. Horton, Enoch, Alma Works, Darlaston, near Wednesbury.

1871. Horton, George, 4 Cedars Road, Clapham Common, London, S.W.

1886. Hosgood, John Howell, Locomotive and Hydraulic Superintendent, Barry Dock and Railways, Barry, near Cardiff.

1889. Hosken, Richard, Severn Tunnel Works, Sudbrook, near Chepstow.

1873. Hoskin, Richard, 8 Norfolk Street, Sheffield.

1888. Hosking, Thomas, Messrs. T. and J. Hosking, Dockhead Iron Works, 53 Parker's Row, Bermondsey, London, S.E.

1892. Houghton, Francis Gassiot, 17 Victoria Street, Westminster, S.W.

1866. Houghton, John Campbell Arthur, Woodside Iron Works, near Dudley.

1889. Houghton, Thomas Harry, 58 Pitt Street, Sydney, New South Wales: (or care of Messrs. James Simpson and Co., 101 Grosvenor Road, Pimlico, London, S.W.) [*Expansion, Sidney.*]

1887. Houghton-Brown, Ernest, Messrs. Houghton-Brown Brothers, Kingsbury Iron Works, Ballspond, London, N.

1891. How, William Field, Mutual Life Buildings, George Street, Sydney, New South Wales. [*Alaska, Sydney.*]

1864. Howard, Eliot, Messrs. Hayward Tyler and Co., 84 Upper Whitecross Street, London, E.C.

1879. Howard, James Harold, Britannia Iron Works, Bedford; and Kempston Grange, Bedford.

1882. Howard, John William, 78 Queen Victoria Street, London, E.C.

1885. Howarth, William, Manager, Oldham Boiler Works, Oldham. [*Boilers, Oldham.*]

1861. Howell, Joseph Bennett, Messrs. Howell and Co., Brook Steel Works, Brookhill, Sheffield. [*Howell, Sheffield.*]

1877. Howell, Samuel Earnshaw, Messrs. Howell and Co., Brook Steel Works, Brookhill, Sheffield. [*Howell, Sheffield.*]

1892. Howitt, James John, Messrs. Bowman Thompson and Co., Lostock Gralam, Northwich.

1882. Howl, Edmund, Messrs. Lee Howl and Co, Tipton. [*Howl, Tipton.*]

1877. Howlett, Francis, Messrs. Henry Clayton Son and Howlett, Atlas Works, Woodfield Road, Harrow Road, London, W. [*Brickpress, London.*]

1891. Hoy, Henry Albert, Locomotive Works, Lancashire and Yorkshire Railway, Horwich, near Bolton.

1887. Hoyle, James Rossiter, Messrs. Thomas Firth and Sons, Norfolk Works, Sheffield.

1891. Hubback, Charles Arbuthnot, 9 Church Crescent, St. Albans.

1882. Hudson, John George, Messrs. Hick Hargreaves and Co., Soho Iron Works, Crook Street, Bolton; and Glenholme, Bromley Cross, Bolton.

1884. Hudson, Robert, Gildersome Foundry, near Leeds [*Gildersome, Leeds.* 14.]; and Weetwood Mount, Headingley, near Leeds. [454.]

1881. Hughes, Edward William Mackenzie, Locomotive and Carriage Superintendent, East Coast State Railway, Waltair, Vizagapatam, Madras Presidency, India: (or Ericstane, Helensburgh.)

1867. Hughes, George Douglas, Messrs. G. D. Hughes and Son, Queen's Foundry, London Road, Nottingham.

1889. Hughes, John, Messrs. Hughes and Lancaster, 16 Great George Street, Westminster, S.W.

1871. Hughes, Joseph, Kingston, Wareham.

1891. Hughes, Robert M., Folkestone Technical School, Folkestone.

1892. Hullah, Arthur, Victoria Jubilee Technical Institute, Byculla, Bombay, India: (or care of Walter Hunter, 12 Chetwynd Terrace, Meadow Road, Leeds.)

1883. Hulse, Joseph Whitworth, Messrs. Hulse and Co., Ordsal Tool Works, Regent Bridge, Salford, Manchester. [*Esluh, Manchester.*]

1864. Hulse, William Wilson, Ordsal Tool Works, Regent Bridge, Salford, Manchester. [*Esluh, Manchester.*]

1890. Humphries, Edward Thomas, Messrs. Edward Humphries and Co., Atlas Iron Works, Pershore.

1866. Humphrys, Robert Harry, Messrs. Humphrys Tennant and Co., Deptford Pier, London, S.E.

1882. Hunt, Reuben, Aire and Calder Chemical Works, Castleford, near Normanton.

1885. Hunt, Richard, Messrs. Thomas Hunt and Sons, Albion Iron Works, 132 Bridge Road West, Battersea, London, S.W.

1856. Hunt, Thomas, Egerton Mount, Heaton Chapel, R.O., Stockport.

1874. Hunt, William, Alkali Works, Lea Brook, Wednesbury; Hampton House, Wednesbury; and Aire and Calder Chemical Works, Castleford, near Normanton.

1889. Hunter, Charles Lafayette, Engineer, Bute Docks, Cardiff.

1886. Hunter, John, Messrs. Campbells and Hunter, Dolphin Foundry, Saynor Road, Hunslet, Leeds.

1877. Hunter, Walter, Messrs. Hunter and English, High Street, Bow, London, E. [*Venator, London.*]

1888. Huxley, George, 20 Mount Street, Manchester.

1885. Hyland, John Frank, Railway Contractor, São Carlos do Pinhal, Estado de São Paulo, Brazil: (or care of Messrs. Lewis and Hyland, New Rents, Ashford, Kent.)

1877. Imray, John, Messrs. Abel and Imray, 20 Southampton Buildings, London, W.C.

1882. Ingham, William, Assistant Engineer, National Boiler Insurance Co., 22 St. Ann's Square, Manchester.

1888. Ingleby, Joseph, 20 Mount Street, Manchester.

1872. Inman, Charles Arthur, Messrs. Clay Inman and Co., Birkenhead Forge, Beaufort Road, Birkenhead.

1883. Instone, Thomas, 22 Leadenhall Buildings, Leadenhall Street, London, E.C.

1892. Irons, Thomas, Manager, Messrs. Hudson Brothers, Clyde Engineering Works, Granville, New South Wales.

1887. Ivatt, Henry Alfred, Locomotive Engineer, Great Southern and Western Railway, Inchicore Works, near Dublin.

1887. Ivatts, Lionel Edward, Paseo Salamanca, F 2° Derecha, San Sebastian, Spain.

1884. Jacks, Thomas William Moseley, Patent Shaft Works, Wednesbury; and Woodgreen, Wednesbury.

1859. Jackson, Matthew Murray, 47 Norton Road, West Brighton, Brighton; and care of Messrs. Howard and Pitcairn, 155 Fenchurch Street, London, E.C.

1847. Jackson, Peter Rothwell, Salford Rolling Mills, Manchester; and Blackbrooke, Pontrilas, R.S.O., Herefordshire. [*Jacksons, Manchester.*]

1873. Jackson, Samuel, C.I.E., care of E. Jackson, Beltwood, Ranmoor, Sheffield.

1886. Jackson, Thomas, 131 Hyde Park Road, Headingley, Leeds.

1889. Jackson, William, Thorn Grove, Mannofield, Aberdeen.

1876. Jacobs, Charles Mattathias, 88 Bishopsgate Street Within, London, E.C. [*Vexillum, London.*]

1878. Jakeman, Christopher John Wallace, Manager, Messrs. Merryweather and Sons, Tram Locomotive Works, Greenwich Road, London, S.E.

1889. James, Charles William, New York Oxygen Co., 366–372 First Avenue, New York, United States.

1877. James, Christopher, 4 Alexandra Road, Clifton, Bristol.

1877. James, John William Henry, 2 Victoria Mansions, 28 Victoria Street, Westminster, S.W.

1889. James, Reginald William, 1 Queen Victoria Street, London, E.C.

1879. Jameson, George, Glencormac, Bray, Ireland.

1881. Jameson, John, Messrs. Jameson and Schaeffer, Akenside Hill, Newcastle-on-Tyne. [*Jameson, Newcastle-on-Tyne.* 226.]

1888. Jaques, Lieut. William Henry, Secretary to Ordnance Committee, United States; and South Bethlehem, Pennsylvania, United States.

1888. Jeejeebhoy, Piroshaw Bomanjee, 17 Church Street, Bombay, India.

1880. Jefferies, John Robert, Messrs. Ransomes Sims and Jefferies, Orwell Works, Ipswich.

1881. Jefferiss, Thomas, Messrs. Tangyes, Cornwall Works, Soho, near Birmingham. [*Tangyes, Birmingham.*]

1877. Jeffreys, Edward Homer, Hawkhills, Chapel Allerton, Leeds.

1884. Jenkins, Alfred, Wharncliffe, Victoria Road, Penarth.

1880. Jenkins, Rhys, Patent Office, 25 Southampton Buildings, London, W.C.

1892. Jenkins, William John, Albion Iron Works, Miles Platting, Manchester.

1878. Jensen, Peter, 77 Chancery Lane, London, W.C. [*Venture, London.*]

1889. Jessop, George, London and Leicester Steam-Crane and Engine Works, Leicester.

1886. Jewell, Henry William, Messrs. Jewell and Son, City Foundry, Winchester.

1863. Johnson, Bryan, Hydraulic Engineering Works, Chester; and 9 Upper Northgate Street, Chester.

1885. Johnson, John Clarke, Messrs. James Russell and Sons, Crown Tube Works, Wednesbury.

1890. Johnson, John William, care of Baron Knoop, Grande Loubianka, Moscow, Russia.

1891. Johnson, Lacey Robert, Master Mechanic, Pacific Division, Canadian Pacific Railway, Vancouver, British Columbia.

1888. Johnson, Lawrence Potter, Assistant Locomotive Superintendent, Burma State Railway, Insein, British Burma.

1882. Johnson, Samuel, Manager, Globe Cotton and Woollen Machine Works, Rochdale.

1887. Johnson, Samuel Henry, Engineering Works, Carpenter's Road, Stratford, London, E.; and The Warren Hill, Loughton, Essex.

1861. Johnson, Samuel Waite, Locomotive Superintendent, Midland Railway, Derby.

1888. Johnson, William, Castleton Foundry and Engineering Works, Armley Road, Leeds.

1891. Johnston, Andrew, Bank Buildings, Hong Kong, China. [*Marine, Hong Kong.*]

1872. Joicey, Jacob Gowland, Messrs. J. and G. Joicey and Co., Forth Banks West Factory, Newcastle-on-Tyne. [*Engines, Newcastle-on-Tyne.*]

1882. Jolin, Philip, 35 Narrow Wine Street, Bristol; and 2 Elmdale Road, Redland, Bristol.

1891. Jones, Charles Frederick, Messrs. Greaves Cotton and Co , Empress Mill, Delisle Road, Parel, Bombay, India.

1871. Jones, Charles Henry, Assistant Locomotive Superintendent, Midland Railway, Derby.

1873. Jones, Edward, Kirkstall, Aldridge Road, Perry Barr, Birmingham.

1884. Jones, Felix, Messrs. Jones and Foster, 39 Bloomsbury Street, Birmingham.

1878. Jones, Frederick Robert, Superintending Engineer, Sirmoor State, Nahan, near Umballa, Punjaub, India: (or care of Messrs. Richard W. Jones and Co., Newport, Monmouthshire.)

1867. Jones, George Edward, District Locomotive Superintendent, North Western Railway, Multan, Punjaub, India: (or care of Mrs. Edward Jones, 9 Sydenham Villas, Cheltenham.)

1878. Jones, Harry Edward, Engineer, Commercial Gas Works, Stepney, London, E.

1881. Jones, Herbert Edward, Locomotive Department, Midland Railway, Manchester.

1890. Jones, Morlais Glasfryn, 6 Delahay Street, Westminster, S.W.

1882. Jones, Samuel Gilbert, Hatherley Court, Gloucester.

1887. Jones, Thomas, Central Board School, Deansgate, Manchester.

1872. Jones, William Richard Sumption, Rajputana State Railway, Ajmeer, India: 32 Madeley Road, Ealing, London, W.: (or care of Messrs. Henry S. King and Co., 45 Pall Mall, London, S.W.)

1883. Jordan, Edward, Manager, Cardiff Junction Dry Dock and Engineering Works, Cardiff.

1891. Jordan, Henry George, Jun., Municipal Technical School, Princess Street, Manchester.

1880. Joy, David, 8 Victoria Chambers, 15 Victoria Street, Westminster, S.W.; and Manor Road House, Beckenham.

1891. Judd, Joseph Henry, Head Master, Technical and Manual Instruction School, York Place, Brighton.

1878. Jüngermann, Carl, Maschinenbau Actien Gesellschaft Vulcan, Bredow bei Stettin, Germany.

1884. Justice, Howard Rudulph, 55 and 56 Chancery Lane, London, W.C. [*Syng, London.* 2504.]

1889. Kanthack, Ralph, 21 Golden Square, Regent Street, London, W. [*Kanthack, London.*]

1888. Kapteyn, Albert, Westinghouse Brake Co., Canal Road, York Road, King's Cross, London, N.

1882. Keeling, Herbert Howard, Merlewood, Eltham.

1869. Keen, Arthur, London Works, near Birmingham. [*Globe, Birmingham.*]

1883. Keen, Francis Watkins, Patent Nut and Bolt Works, Westbromwich.

1867. Kellett, John, Clayton Street, Wigan.

1873. Kelson, Frederick Colthurst, Angra Bank, Waterloo Park, Waterloo, near Liverpool.

1881. Kendal, Ramsey, Locomotive Department, North Eastern Railway, Gateshead.

1879. Kennedy, Professor Alexander Blackie William, F.R.S., 19 Little Queen Street, Westminster, S.W. [*Kinematic, London.*]

1863. Kennedy, John Pitt, Bombay Baroda and Central Indian Railway, 45 Finsbury Circus, London, E.C.; and 29 Lupus Street, St. George's Square, London, S.W.

1892. Kennedy, Thomas, The Glenfield Engineering Works, Kilmarnock.

1868. Kennedy, Thomas Stuart, Parkhill, Wetherby.

1875. Kenrick, George Hamilton, Messrs. A. Kenrick and Sons, Spon Lane, Westbromwich; and Whetstone, Somerset Road, Edgbaston, Birmingham.

1892. Kensington, Frederick, 2 Copthall Buildings, London, E.C.

1866. Kershaw, John, Marazion, St. Leonard's-on-Sea.

1884. Kershaw, Thomas Edward, Chilvers Coton Foundry, Nuneaton.

1890. Key, George Andrew, General Manager, Wallsend Pontoon Works, Bute Docks, Cardiff.

1885. Keydell, Amandus Edmund, Lloyd's Register of Shipping, Dundee.

1885. Keyworth, Thomas Egerton, Ferro Carril Buenos Aires y Rosario, Campana, Buenos Aires, Argentine Republic: (or care of J. R. H. Keyworth, 25 Park Road South, Birkenhead.)

1885. Kidd, Hector, Colonial Sugar Refining Co., Sydney, New South Wales.

1888. Kikuchi, Kyozo, Superintendent Engineer, Hirano Spinning Mill, Osaka, Japan.

1872. King, William, Engineer, Liverpool United Gas Works, Duke Street, Liverpool.

1889. Kirby, Frank Eugene, Constructing Engineer, Detroit Dry Dock Co., Detroit, Michigan, United States.

1872. Kirk, Alexander Carnegie, LL.D., Messrs. Robert Napier and Sons, Lancefield House, Glasgow; and Govan Park, Govan, Glasgow.

1877. Kirk, Henry, Messrs. Kirk Brothers and Co., New Yard Iron Works, Workington. [*Kirks, Workington.*]

1884. Kirkaldy, John, 40 West India Dock Road, London, E. [*Compactum, London.*]

1875. Kirkwood, James, Chief Inspector of Machinery for Pei Yang Squadron; care of Commissioner of Customs, Kowloon, Hong Kong, China: (or Melita Cottage, Denny.)

1864. Kirtley, William, Locomotive Superintendent, London Chatham and Dover Railway, Longhedge Works, Wandsworth Road, London, S.W. [3005.]

1859. Kitson, Sir James, Bart., Monk Bridge Iron Works, Leeds.

1868. Kitson, John Hawthorn, Airedale Foundry, Leeds. [*Airedale, Leeds.*]

1874. Klein, Thorvald, Suffolk House, 5 Laurence Pountney Hill, London, E.C.

1889. Knap, Conrad, 11 Queen Victoria Street, London, E.C.

1891. Knight, Bertrand Thornton, Engineer, Royal Siamese State Railways, Bangkok, Siam: (or care of Major Knight, Swansea.)

1886. Knight, Charles Albert, Babcock and Wilcox Boiler Co., 107 Hope Street, Glasgow.

1890. Knight, James Percy, Messrs. A. W. Robertson and Co., Custom House Engine Works, Victoria Docks, London, E.

1889. Knox, James, Civil and Mechanical Engineer, Auckland, New Zealand: (or care of E. D. Knox, 53 Belsize Park Gardens, South Hampstead, London, N.W.)

1881. Laing, Arthur, Deptford Shipbuilding Yard, Sunderland.

1872. Laird, Henry Hyndman, Messrs. Laird Brothers, Birkenhead Iron Works, Birkenhead. [*Laird, Birkenhead.* 4003.]

1872. Laird, William, Messrs. Laird Brothers, Birkenhead Iron Works, Birkenhead. [*Laird, Birkenhead.* 4003.]

1883. Lake, William Robert, 45 Southampton Buildings, London, W.C. [*Scopo, London.*]

1878. Lambourn, Thomas William, Naughton Hall, near Bildeston, S.O., Suffolk.

1881. Langdon, William, Locomotive Superintendent and Chief Mechanical Engineer, Rio Tinto Railway and Mines, Huelva, Spain: (or care of T. C. Langdon, Tamar Terrace, Launceston.)

1881. Lange, Frederick Montague Townshend, 21 Rue Wissocq, Boulogne-sur-mer, France.

1877. Lange, Hermann Ludwig, Manager, Messrs. Beyer Peacock and Co., Gorton Foundry, Manchester.

1879. Langley, Alfred Andrew, 33 Chester Terrace, Regent's Park, London, N.W.

1879. Lapage, Richard Herbert, Elmwood, Surbiton.

1890. Last, Arthur John, Oulton, Abbeville Road, Balham, London, S.W.

1888. Latham, Baldwin, 7 Westminster Chambers, 13 Victoria Street, Westminster, S.W.; and Duppas House, Old Town, Croydon.

1890. Laurie, Leonard George, Mill Parade, Newport, Monmouthshire.

1881. Lavalley, Alexander, 48 Rue de Provence, Paris.

1867. Lawrence, Henry, The Grange Iron Works, Durham.

1874. Laws, William George, Borough Engineer and Town Surveyor, Town Hall, Newcastle-on-Tyne; and 5 Winchester Terrace, Newcastle-on-Tyne. [*Engineer, Newcastle-on-Tyne.*]

1882. Lawson, Frederick William, Messrs. Samuel Lawson and Sons, Hope Foundry, Leeds.

1870. Layborn, Daniel, Messrs. Daniel Layborn and Co., Dutton Street, Liverpool.

1883. Laycock, William S., Messrs. Samuel Laycock and Sons, Horse-hair Cloth Works, Sheffield; and Ranmoor, Sheffield.

1860. Lea, Henry, Messrs. Henry Lea and Thornbery, 38 Bennett's Hill, Birmingham. [*Engineer, Birmingham.* 113.]

1892. Lea, Richard Henry, Manager, Messrs. Singer and Co.'s Cycle Works, Coventry.

1889. Leaf, Henry Meredith, Messrs. Crompton and Co., Mansion House Buildings, Queen Victoria Street, London, E.C.

1883. Leavitt, Erasmus Darwin, Jun., 604 Main Street, Cambridgeport, Massachusetts, United States.

1890. Ledingham, John Machray, Royal Laboratory, Royal Arsenal, Woolwich.

1886. Lee, Charles Eyre, 18 Newhall Street, Birmingham.

1887. Lee, Cuthbert Ridley, Messrs. J. Coates and Co., Suffolk House, Laurence Pountney Hill, London, E.C.

1862. Lee, J. C. Frank, 9 Park Crescent, Portland Place, London, W.

1892. Lee, Richard John, Messrs. Harrison Lee and Sons, City Foundry, Limerick.

1890. Lee, Samuel Edward, Messrs. Harrison Lee and Sons, City Foundry, Limerick.

1863. Lees, Samuel, Messrs. H. Lees and Sons, Park Bridge Iron Works, Ashton-under-Lyne.

1889. Legros, Lucien Alphonse, 57 Brook Green, Hammersmith, London, W.

1883. Lennox, John, 2 Victoria Mansions, 28 Victoria Street, Westminster, S.W.

1858. Leslie, Andrew, Coxlodge Hall, Newcastle-on-Tyne.

1888. Leslie, Sir Bradford, K.C.I.E., Tarrangower, Willesden Lane, Brondesbury, London, N.W.

1883. Leslie, Joseph, Marine Engineer, Messrs. Apcar and Co., Raddah Bazar, Calcutta.

1888. Letchford, Joseph, Manager, Messrs. David Munro and Co., Stuart Street, Melbourne, Victoria; care of Richard Speight, Glenroy Park, Hampton Street, Middle Brighton, Melbourne, Victoria: (or care of James Letchford, 370 Wandsworth Road, London, S.W.)

1878. Lewis, Gilbert, 37 Monton Street, Moss Side, Manchester.

1884. Lewis, Henry Watkin, Llwyn-yr-eos, Abercanaid, near Merthyr Tydfil.

1872. Lewis, Richard Amelius, Messrs. John Spencer and Sons, Tyne Hæmatite Iron Works, Scotswood-on-Tyne.

1887. Lewis, Rowland Watkin, Messrs. Edwin Lewis and Sons, Britannia Boiler Tube Works, Wolverhampton.

1884. Lewis, Sir William Thomas, Bute Mineral Estate Office, Aberdare; and Mardy, Aberdare.

1880. Lightfoot, Thomas Bell, Cornwall Buildings, 35 Queen Victoria Street, London, E.C. [*Separator, London.*]; and 7 Eastcombe Villas, Charlton Road, Blackheath, London, S.E.

1887. Lindsay, Joseph, Messrs. Urquhart Lindsay and Co., Blackness Foundry, Dundee.

1891. Lindsay, William Robertson, Messrs. W. B. Thompson and Co., Lilybank Engine Works, Dundee.

1890. Lineham, Wilfrid James, Professor of Engineering and Mechanical Science, The Goldsmiths' Institute, New Cross, London, S.E.; and Jesmond, Leyland Road, Lee, London, S.E.

1856. Linn, Alexander Grainger, 121 Upper Parliament Street, Liverpool.

1876. Lishman, Thomas, Mining Engineer, Hetton Colliery, near Fence Houses.

1881. List, John, Superintendent Engineer, Messrs. Donald Currie and Co., Orchard Works, Blackwall, London, E.; and 3 St. John's Park, Blackheath, London, S.E.

1885. Lister, Frank, Messrs. Lister and Co., Beechcliffe, Keighley; and Oaklands, Keighley.

1890. Lister, Robert Ramsbottom, Messrs. Beyer Peacock and Co., Gorton Foundry, Manchester.

1866. Little, George, Messrs. Platt Brothers and Co., Hartford Iron Works, Oldham.

1890. Livens, Frederick Howard, Messrs. Ruston Proctor and Co., Sheaf Iron Works, Lincoln.

1886. Livsey, John Edward, Demonstrator in Mechanics and Mathematics, Royal College of Science, Exhibition Road, South Kensington, London, S.W.

1867. Lloyd, Charles, National Conservative Club, 9 Pall Mall, London, S.W.

1871. Lloyd, Francis Henry, James Bridge Steel Works, near Wednesbury [*Steel, Wednesbury*]; and Stowe Hill, Lichfield.

1854. Lloyd, George Braithwaite (*Life Member*), Edgbaston Grove, Birmingham.

1882. Lloyd, Robert Samuel, Messrs. Hayward Tyler and Co., 84 Upper Whitecross Street, London, E.C.

1852. Lloyd, Samuel, The Farm, Sparkbrook, Birmingham.

1890. Locke, Arthur Guy Neville, 12 Dartmouth Street, Westminster, S.W.

1879. Lockhart, William Stronach, 67 Granville Park, Blackheath, London, S.E.

1881. Lockyer, Norman Joseph, care of Sir A. M. Rendel, 8 Great George Street, Westminster, S.W.

1884. Logan, Andrew Linton, Railway Signal Works, Worcester.

1890. Logan, John Walker, Messrs. Richard Hornsby and Sons, P. O. Box 216, Johannesburg, Transvaal, South Africa; and care of William Weatherley, 51 Newlands, Lincoln.

1883. Logan, Robert Patrick Tredennick, Engineer's Office, Great Northern Railway of Ireland, Dundalk.

1874. Logan, William, Mining Engineer, Langley Park Colliery, Durham.

1884. Longbottom, Luke, Locomotive Carriage and Wagon Superintendent, North Staffordshire Railway, Stoke-on-Trent.

1880. Longridge, Michael, Chief Engineer, Engine and Boiler Insurance Co., 12 King Street, Manchester.

1856. Longridge, Robert Bewick, Managing Director, Engine and Boiler Insurance Co., 12 King Street, Manchester; and Yew Tree House, Tabley, near Knutsford.

1875. Longridge, Robert Charles, Kilrie, Knutsford.

1880. Longworth, Daniel, 2 Charleville Road, Park Road, Dublin.

1887. Lorrain, James Grieve, Norfolk House, Norfolk Street, London, W.C. [*Lorrain, London.*]

1888. Low, David Allan, Lecturer on Engineering, The People's Palace Technical Schools, Mile End Road, London, E.

1861. Low, George, Bishop's Hill Cottage, Ipswich.

1885. Low, Robert, 11 Queen Victoria Street, London, E.C. [*Armabantur, London.*]; Powis Lodge, Vicarage Park, Plumstead.

1884. Lowcock, Arthur, Coleham Foundry, Shrewsbury.

1884. Lowdon, John, General Manager, Barry Graving Dock and Engineering Co., Exchange Buildings, Cardiff. [*Bardock, Cardiff.*]

1891. Lowdon, Thomas, Kingsland Crescent, Barry Docks, B.O., near Cardiff.

1873. Lowe, John Edgar, Messrs. Bolling and Lowe, 2 Laurence Pountney Hill, London, E.C. [*Bird, London.* 1530.]

1873. Lucas, Arthur, 27 Bruton Street, New Bond Street, London, W.

1889. Lucy, Arthur John, Messrs. Turner Morrison and Co., Sugar Works, Cossipore, Calcutta, India.

1886. Lucy, William Theodore, care of Frank Hudson, Central Uruguay Railway, Monte Video, Uruguay: English Club, Monte Video, Uruguay: (or Thornleigh, Woodstock Road, Oxford.)

1877. Lupton, Arnold, Professor of Mining Engineering, Yorkshire College, Leeds; and 6 De Grey Road, Leeds. [*Arnold Lupton, Leeds.* 330.]

1887. Lupton, Kenneth, Messrs. K. and H. Lupton, Vulcan Works, Burgess, Coventry.

1878. Lynde, James Henry, Buckland, Ashton-on-Mersey, near Manchester.

1889. Macallan, George, Works Manager, Great Eastern Railway, Stratford Works, London, E.

1890. Macau, Richard Thompson, Dawlish House, Willesden, London, N.W.

1892. Macbean, John James, Messrs. Howarth Erskine and Co., Singapore, Straits Settlements.

1888. Macbeth, John Bruce King, 44 Tamarind Lane, Bombay, India: (or care of Norman Macbeth, Heaton, Bolton.)

1883. Macbeth, Norman, Messrs. John and Edward Wood, Victoria Foundry, Bolton.

1890. MacBrair, William Maxwell, 34 Crookes Road, Sheffield.

1884. MacCarthy, Samuel, Messrs. Lloyd and Lloyd, 90 Cannon Street, London, E.C.; and 18 Adelaide Road, Brockley, London, S.E.

1877. MacColl, Hector, Messrs. MacIlwaine and MacColl, Ulster Iron Works, Belfast.

1879. Macdonald, Augustus Van Zundt, Locomotive Engineer, Addington, Christchurch, New Zealand.

1889. Macdonald, James Alexander, Broad Oaks Iron Works, Chesterfield.

1892. Machado, Dr. Antonio Augusto, Manager, Companhia Metropolitana, Engineering and Boiler Works, Bahia, Brazil: (or care of Messrs. Heuser Humble and Co., 1 Fowkes Buildings, Great Tower Street, London, E.C.)

1892. Mackay, Charles O'Keefe, Locomotive Department, Lancashire and Yorkshire Railway, Horwich, near Bolton.

1890. Mackay, Joseph, Bangkok Dock Co., Bangkok, Siam: (or care of Messrs. John Birch and Co., 10 Queen Street Place, London, E.C.) [*Mackay, Bangkok.*]

1885. Mackenzie, John William, Messrs. Wheatley and Mackenzie, 40 Chancery Lane, London, W.C.; and Northfield, Oxford Road, Upper Teddington, S.O., Middlesex.

1875. Maclagan, Robert, care of Dr. Maclagan, 9 Cadogan Place, Belgrave Square, London, S.W.

1889. MacLay, Alexander, Professor of Mechanical Engineering, Glasgow and West of Scotland Technical College, 38 Bath Street, Glasgow.

1886. MacLean, Alexander Scott, Messrs. Alexander Scott and Sons, Sugar Refinery, Berry-yards, Greenock; and 31 Bank Street, Greenock.

1877. MacLellan, John A., Messrs. Alley and MacLellan, Sentinel Works, Polmadie Road, Glasgow. [*Alley, Glasgow.* 673.]

1888. Macleod, Arthur William, Schwebo Mining Syndicate, Kyouk Myoung Post Office, Upper Burmah.

1864. Macnab, Archibald Francis, Inspecting and Examining Engineer, Government Marine Office, Tokyo, Japan.

1865. Macnee, Daniel, 2 Westminster Chambers, 3 Victoria Street, Westminster, S.W. [*Macnee, London.*]; and Rotherham.

1884. Macpherson, Alexander Sinclair, Messrs. Fairbairn Naylor Macpherson and Co., Wellington Foundry, Leeds.

1892. Mactear, James, F.R.S.E., 28 Victoria Street, Westminster, S.W. [*Celestine, London.* 3066.]

1879. Maginnis, James Porter, 9 Carteret Street, Queen Anne's Gate, Westminster, S.W. [*James Maginnis, London;* and *Offsett, London.*]

1891. Mahon, Reginald Henry, Captain R.A., Superintendent H.M. Shell Factory, Cossipore, Calcutta, India.

1873. Mair-Rumley, John George, Messrs. Simpson and Co., Engine Works, 101 Grosvenor Road, Pimlico, London, S.W. [*Aquosity, London.*]

1884. Mais, Henry Coathupe, 61 Queen Street, Melbourne, Victoria.

1879. Malcolm, Bowman, Locomotive Superintendent, Belfast and Northern Counties Railway, Belfast.

1891. Manisty, Edward, Dundalk Iron Works, Dundalk, Ireland; and 24A Bryanston Square, London, W.

1888. Mano, Bunji, Professor of Mechanical Engineering, Imperial University, Tokyo, Japan.

1875. Mansergh, James, 3 Westminster Chambers, 5 Victoria Street, Westminster, S.W.

1891. Manson, James, Locomotive Superintendent, Glasgow and South Western Railway, Kilmarnock.

1862. Mappin, Sir Frederick Thorpe, Bart., M.P., Messrs. Thomas Turton and Sons, Sheaf Works, Sheffield; and Thornbury, Sheffield.

1878. Marié, Georges, Engineer, Chemins de fer de Paris à Lyon et à la Méditerranée, Bureaux du Matériel, Boulevart Mazas, Paris.

1891. Marks, Edward Charles Robert, 13 Temple Street, Birmingham.

1888. Marks, George Croydon, 13 Temple Street, Birmingham. [*Pumps, Birmingham.*]

1884. Marquand, Augustus John, Pierhead Chambers, Cardiff. [*Martial, Cardiff.*]

1887. Marriott, William, Engineer and Locomotive Superintendent, Eastern and Midlands Railway, Melton Constable, Norfolk.

1887. Marsden, Benjamin, Messrs. S. Marsden and Son, Screw-Bolt and Nut Works, London Road, Manchester.

1871. Marsh, Henry William, Winterbourne, near Bristol.

1875. Marshall, Rev. Alfred (*Life Member*), The Vicarage, Feckenham, Redditch.

1865. Marshall, Francis Carr, Messrs. R. and W. Hawthorn Leslie and Co., St. Peter's Works, Newcastle-on-Tyne.

1890. Marshall, Frank Herbert, Ormesby Iron Works, Middlesbrough.

1885. Marshall, Henry Dickenson, Messrs. Marshall Sons and Co., Britannia Iron Works, Gainsborough. [*Marshalls, Gainsborough.* 6648.]

1871. Marshall, James, Messrs. Marshall Sons and Co., Britannia Iron Works, Gainsborough. [*Marshalls, Gainsborough.* 6648.]

1885. Marshall, Jenner Guest, Messrs. Chance Brothers and Co., Glass Works, near Birmingham; and Westcott Barton Manor, Oxfordshire.

1877. Marshall, William Bayley, Richmond Hill, Edgbaston, Birmingham. [*Augustus, Birmingham.*]

1847. Marshall, William Prime, Richmond Hill, Edgbaston, Birmingham. [*Augustus, Birmingham.*]

1859. Marten, Edward Bindon, Pedmore, Stourbridge. [*Marten, Stourbridge.* 8504.]

1853. Marten, Henry John, The Birches, Codsall, near Wolverhampton; and 4 Storey's Gate, Westminster, S.W.

1881. Martin, Edward Pritchard, Dowlais Iron Works, Dowlais.

1878. Martin, Henry, 8 East Park Terrace, Southampton.

1888. Martin, Henry James, Tresleigh House, Walters Road, Swansea.

1889. Martin, The Hon. James, Messrs. James Martin and Co., Phœnix Foundry, Gawler, South Australia: (or care of J. C. Lanyon, 27 Gresham House, Old Broad Street, London, E.C.)

1892. Martin, Thomas George, Messrs. James McGowan and Co., Wapping Wall, London, E.

1886. Martin, William Hamilton, Engineering Manager, The Scheldt Royal Shipbuilding and Engineering Works, Flushing, Holland.

1882. Martindale, Warine Ben Hay, 38 Parliament Street, Westminster, S.W.; and Overfield, Bickley, R.S.O., Kent.

1882. Masefield, Robert, 14 Markham Square, Chelsea, London, S.W.

1884. Massey, George, Post Office Chambers, Pitt Street, Sydney, New South Wales.

1890. Massey, Stephen, Messrs. B. and S. Massey, Openshaw, Manchester.

1892. Masterton, John Fraser, Messrs. Neilson and Co., Hyde Park Locomotive Works, Glasgow.

1867. Mather, William, M.P., Messrs. Mather and Platt, Salford Iron Works, Manchester. [*Mather, Manchester.*]

1883. Mather, William Penn, Queen Dyeing Co., Providence, Rhode Island, United States.

1882. Matheson, Henry Cripps, care of Messrs. Russell and Co., Hong Kong, China: (or care of Messrs. Matheson and Grant, 32 Walbrook, London, E.C.)

1891. Mathewson, Jeremiah Eugene, Tilghman's Sand-Blast Co., Bellefield Works, Bellefield Lane, Sheffield.

1875. Matthews, James, 22 Ashfield Terrace East, Newcastle-on-Tyne.

1886. Matthews, Robert, Parrs House, Heaton Mersey, near Manchester.

1875. Mattos, Antonio Gomes de, Messrs. Maylor and Co., Engineering Works, 136 Rua da Sande, Rio de Janeiro, Brazil: (or care of Messrs. Fry Miers and Co., Suffolk House, 5 Laurence Pountney Hill, London, E.C.)

1853. Maudslay, Henry (*Life Member*), Westminster Palace Hotel, 4 Victoria Street, Westminster, S.W.: (or care of John Barnard, 47 Lincoln's Inn Fields, London, W.C.)

1873. Maw, William Henry, 35 Bedford Street, Strand, London, W.C. [3663.]

1884. Maxim, Hiram Stevens, Maxim Nordenfelt Guns and Ammunition Co., Victoria Mansions, 32 Victoria Street, Westminster, S.W.

1859. Maylor, William, Chesterleigh, Albemarle Road, Beckenham.

1874. McClean, Frank, Norfolk House, Norfolk Street, Strand, London, W.C.

1891. McCredie, Arthur Latimer, 250 Pitt Street, Sydney, New South Wales. [*Ebony, Sydney*. 63.]

1892. McDonald, John, Locomotive Works, Imperial Government Railways, Tokyo, Japan.

1878. McDonald, John Alexander, Assistant Engineer for Roads and Bridges, Public Works Office, Sydney, New South Wales: (or care of James E. McDonald, 4 Chapel Street, Cripplegate, London, E.C.)

1865. McDonnell, Alexander, 2 Victoria Mansions, 28 Victoria Street, Westminster, S.W.; and The Cedars, Norwood Green, Southall.

1891. McFarlane, George, Bank of Scotland Buildings, 24 George Square, Glasgow. [*Asphodel, Glasgow.*]

1881. McGregor, Josiah, Crown Buildings, 78 Queen Victoria Street, London, E.C. [*Sahib, London.*]

1892. McGregor, Peter (*Life Member*), Imperial Maritime Customs, Kowloon, Hong Kong, China.

1892. McIntosh, William Forbes, Messrs. Douglas Lapraik and Co., Hong Kong, China.

1889. McIntyre, John Henry A., Lecturer on Mechanical Engineering, Allan Glen's School, Glasgow.

1881. McKay, John, 13 Grey Street, Newcastle-on-Tyne.

1880. McLachlan, John, Messrs. Bow McLachlan and Co., Thistle Engine Works, Paisley. [*Bow, Paisley.*]

1888. McLaren, Henry, Messrs. J. and H. McLaren, Midland Engine Works, Leeds.

1882. McLaren, Raynes Lauder, 96 Addison Road, Kensington, London, W.

1888. McLarty, Farquhar Matheson, Penang Foundry, Penang: (or care of William Bow, Thistle Engine Woıks, Paisley.)

1879. McLean, William Leckie Ewing, Renfrew Forge and Steel Co., Porterfield, Renfrew.

1885. McNeil, John, Messrs. Aitken McNeil and Co., Helen Street, Govan, Glasgow. [*Colonial, Glasgow.*]

1891. Meade, Thomas de Courcy, Engineer, Hornsey District Board, 99 Southwood Lane, Highgate, London, N [*Umpireship, London.* 7581.]

1882. Meats, John Tempest, Mason Machine Works, Taunton, Massachusetts, United States.

1881. Meik, Charles Scott, care of P. Walter Meik, 16 Victoria Street, Westminster, S.W.

1858. Meik, Thomas, 21 York Place, Edinburgh.

1887. Melhuish, Frederick, Assistant Engineer, Southwark and Vauxhall Water Works, Southwark Bridge Road, London, S.E.

1891. Melville, William Charles, Superintendent Engineer, Liverpool Steam Tug Co., 44 Chapel Street, Liverpool.

1888. Melville, William Wilkie, 284 Ivydale Road, Nunhead, London, S.E.

1878. Menier, Henri, 56 Rue de Châteaudun, Paris.

1876. Menzies, William, Messrs. Menzies and Co., 50 Side, Newcastle-on-Tyne. [*William Menzies, Newcastle-on-Tyne. G.P.O.* 200. *Nor. Dis.* 1144.]

1875. Merryweather, James Compton, Messrs. Merryweather and Sons, Fire-Engine Works, Greenwich Road, London, S.E.; and 63 Long Acre, London, W.C. [*Merryweather, London.*]

1891. Metcalfe, Frederick Spencer, Pumping Station, Sewage Works, Burton-on-Trent.

1881. Meysey-Thompson, Arthur Herbert, Messrs. Hathorn Davey and Co., Sun Foundry, Dewsbury Road, Leeds.

1877. Michele, Vitale Domenico de, 14 Delahay Street, Westminster, S.W.; and Higham Hall, Rochester.

1884. Middleton, Reginald Empson, 17 Victoria Street, Westminster, S.W.

1891. Middleton, Robert, Sheepscar Foundry, Leeds.

1891. Middleton, Robert Thomas, Superintendent of Bridge Works, Bombay Baroda and Central India Railway, Bombay, India.

1886. Midelton, Thomas, Aylesbury, Albemarle Street, North Kingston, Sydney, New South Wales.

1862. Miers, Francis C., Messrs. Fry Miers and Co., Suffolk House, 5 Laurence Pountney Hill, London, E.C.; and Eden Cottage, West Wickham Road, Beckenham. [*Foundation, London.* 1920.]

1864. Miers, John William, 74 Addison Road, Kensington, London, W.

1874. Milburn, John, Hawkshead Foundry, Quay Side, Workington.

1887. Miles, Frederick Blumenthal, Messrs. Bement Miles and Co., Callowhill and Twenty-first Streets, Philadelphia, United States.

1889. Miller, Adam, 205 Mansion House Chambers, 11 Queen Victoria Street, London, E.C.

1885. Miller, Harry William, New Chimes Gold Mining Co., P.O. Box 1083, Johannesburg, Transvaal, South Africa.

1886. Miller, John Smith, Messrs. Smith Brothers and Co., Hyson Green Works, Nottingham.

1887. Miller, Thomas Lodwick, 7 Tower Buildings N., Water Street, Liverpool.

1885. Millis, Charles Thomas, Principal, Educational Department, Borough Road Polytechnic, London, S.E.

1887. Milne, William, Locomotive Superintendent, Natal Government Railways, Durban, Natal.

1856. Mitchell, Charles, Sir W. G. Armstrong Mitchell and Co., Low Walker, Newcastle-on-Tyne; and Jesmond Towers, Newcastle-on-Tyne.

1892. Mitcheson, George Arthur, Longton, Staffordshire. [*Mitcheson, Longton.* 445.]

1870. Moberly, Charles Henry, 13 Belmont Park, Lee, London, S.E.

1885. Moir, James, Superintendent Engineer, Bombay Steam Navigation Co., Frere Road, Bombay.

1879. Molesworth, Sir Guilford Lindsay, K.C.I.E., The Manor House, Bexley, S.O., Kent.

1882. Molesworth, James Murray, Aberdeen House, Upper Holly Walk, Leamington.

1881. Molinos, Léon, 48 Rue de Provence, Paris.

1885. Monk, Edwin, care of Josiah McGregor, Crown Buildings, 78 Queen Victoria Street, London, E.C.

1884. Monroe, Robert, Manager, Penarth Slipway and Engineering Works, Penarth Dock, Penarth.

1872. Moon, Richard, Jun., Penyvoel, Llanymynech, Montgomeryshire.

1884. Moore, Benjamin Theophilus, Longwood, Bexley, S.O., Kent.

1876. Moore, Joseph, 1099 Adeline Street, Oakland, San Francisco, California; (or care of Ralph Moore, Government Inspector of Mines, 13 Clairmont Gardens, Glasgow.)

1880. Moreland, Richard, Messrs. Richard Moreland and Son, 3 Old Street, St. Luke's, London, E.C. [*Expansion, London.*]

1889. Morgan, David John, Merchants' Exchange, Cardiff.

1885. Morgan, Thomas Rees, Morgan Engineering Works, Alliance, Ohio, United States.

1887. Morison, Donald Barns, Messrs. T. Richardson and Sons, Hartlepool Engine Works, Hartlepool.

1888. Morris, Charles, 5 Mangoe Lane, Calcutta, India.

1874. Morris, Edmund Legh, New River Water Works, Finsbury Park, London, N.

1890. Morris, Francis Sanders, Chandos Chambers, 22 Buckingham Street, Adelphi, London, W.C.

1890. Morris, John Alfred (*Life Member*), Empire Works, 78 Great Bridgewater Street, Manchester.

1892. Morton, David Home, 95 Bath Street, Glasgow.

1858. Mountain, Charles George, 204 St. Vincent Street, Glasgow.

1886. Mountain, William Charles, Messrs. Ernest Scott and Co., Close Works, Newcastle-on-Tyne; and 9 St. George's Terrace, Jesmond, Newcastle-on-Tyne.

1884. Mower, George A., Crosby Steam Gage and Valve Co., 75 Queen Victoria Street, London, E.C. [*Crosby, London.*]

1885. Mudd, Thomas, Manager, Messrs. William Gray and Co., Central Marine Engineering Works, West Hartlepool.

1873. Muir, Alfred, Messrs. William Muir and Co., Britannia Works, Sherborne Street, Strangeways, Manchester.

1873. Muir, Edwin, 37 Brown Street, Manchester.

1876. Muirhead, Richard, Kentish Engineering Works, Maidstone. [*Muirhead, Maidstone.*]

1890. Müller, Henry Adolphus, Locomotive Superintendent, Municipal Railway, 3 North Road, Entally, Calcutta, India.

1890. Mumford, Charles Edward, St. Andrew's Works, Bury St. Edmunds.

1890. Munro, John, Merchant Venturers' Technical School, Unity Street, Bristol.

1890. Munro, Robert Douglas, Chief Engineer, Scottish Boiler Insurance and Engine Inspection Co., 13 Dundas Street, Glasgow.

1889. Münster, Bernard Adolph, Engineer, Yokohama, Japan.

1891. Murdoch, Robert Macmillan, Phœnix Metal Die and Engineering Co., 40 Princes Street, Stamford Street, Blackfriars, London, S.E.

1890. Murray, Alexander John, Chief Mechanical Engineer, Government Gun-Powder Factory, Kirkee, Bombay, India.

1890. Murray, Kenneth Sutherland, Brin's Oxygen Works, 69 Horseferry Road, Westminster, S.W.

1881. Musgrave, James, Messrs. John Musgrave and Sons, Globe Iron Works, Bolton. [*Musgrave, Bolton.*]

1882. Musgrave, Walter Martin, Messrs. John Musgrave and Sons, Globe Iron Works, Bolton. [*Musgrave, Bolton.*]

1888. Myers, William Beswick (*Life Member*), 14 Victoria Street, Westminster, S.W.

1870. Napier, James Murdoch, Messrs. David Napier and Son, Vine Street, York Road, Lambeth, London, S.E.

1889. Nash, Thomas, Sheffield Testing Works, Blonk Street, Sheffield; and Guzerat House, Nether Edge, Sheffield.

1888. Nathan, Adolphus, Messrs. Larini Nathan and Co., Milan; and 15 Via Bigli, Milan, Italy.

1861. Naylor, John William, Messrs. Fairbairn Naylor Macpherson and Co., Wellington Foundry, Leeds.

1883. Neate, Percy John, 16 The Banks, High Street, Rochester.

1889. Needham, Joseph Edward, Patent Office, 25 Southampton Buildings, London, W.C.

1892. Nelson, Arthur David, Hay and Lackey Streets, Sydney, New South Wales. [*Nelson, Sydney.* 160.]

1884. Nelson, John, Contractors' Office, Dringhouses, York.

1887. Nelson, Sidney Herbert, Messrs. Samuel Worssam and Co., Oakley Works, King's Road, Chelsea, London, S.W.

1881. Nesfield, Arthur, Messrs. Simpson Mackirdy and Co., 29 South Castle Street, Liverpool.

1882. Nettlefold, Hugh, Screw Works, 16 Broad Street, Birmingham. [*Nettlefolds, Birmingham.*]

1890. Newton, Percy, Vassall Lodge, Addison Road, Kensington, London, W.

1884. Nicholls, James Mayne, Locomotive Superintendent, Nitrate Railways, Iquique, Chili.

1884. Nicholson, Henry, care of G. H. Hill, Albert Chambers, Albert Square, Manchester.

1891. Nicholson, Thomas, Crownpoint Boiler Works, St. Marnock Street, Crownpoint Road, Glasgow.

1877. Nicolson, Donald, 16 St. Helen's Place, London, E.C.

1886. Noakes, Thomas Joseph, Messrs. Thomas Noakes and Sons, 35 and 37 Brick Lane, Whitechapel, London, E.

1884. Noakes, Walter Maplesden, 73 Clarence Street, Wynyard Square, Sydney, New South Wales.

1882. Nordenfelt, Thorsten, 9 Victoria Street, Westminster, S.W.

1892. Norris, William, Messrs. Robey and Co., Globe Iron Works, Lincoln.

1868. Norris, William Gregory, Coalbrookdale Iron Works, Coalbrookdale, Shropshire.

1883. North, Gamble, Pisagua, Chile: (or care of B. Depledge, Woolpack Buildings, 3 Gracechurch Street, London, E.C.)

1891. North, George, 61 and 62 Gracechurch Street, London, E.C. [*Spero, London.*]

1882. North, John Thomas, Messrs. North Humphrey and Dickenson, Engineering Works, Iquique, Chile; Woolpack Buildings, 3 Gracechurch Street, London, E.C.; and Avery House, Avery Hill, Eltham.

1878. Northcott, William Henry, General Engine and Boiler Co., Hatcham Iron Works, Pomeroy Street, New Cross Road, London, S.E.; and 7 St. Mary's Road, Peckham, London, S.E. [*Oxygen, London.* 8007.]

1888. Norton, William Eardley, 8 Great George Street, Westminster, S.W.

1882. Nunneley, Thomas, Wanstead House, Burley, Leeds.

1885. Oakes, Sir Reginald Louis, Bart., York Engineering Works, Leeman Road, York.

1887. O'Brien, Benjamin Thompson, 45 Fern Grove, Liverpool.

1887. O'Brien, John Owden, Messrs. W. P. Thompson and Co., Ducie Buildings, 6 Bank Street, Manchester.

1890. Ockendon, William, Messrs. John Brown and Co., Atlas Steel and Iron Works, Sheffield.

1868. O'Connor, Charles, 15 Wesley Street, Waterloo, near Liverpool.

1888. O'Donnell, John Patrick, 70 and 71 Palace Chambers, 9 Bridge Street, Westminster, S.W.; and Avondale, College Road, Bromley, Kent. [*ODonnell, London.* 3059.]

1887. O'Flyn, John Lucius, Messrs. L. and H. Gucret and Co., Exchange, Cardiff.

1889. Ogden, Fred, Patent Office, 25 Southampton Buildings, London, W.C.

1886. Ogle, Percy John, 4 Bishopsgate Street Within, London, E.C. [*Oglio, London.* 2463.]

1875. Okes, John Charles Raymond, 39 Queen Victoria Street, London, E.C. [*Oaktree, London.*]

1882. Orange, James, Messrs. Danby Leigh and Orange, Hong Kong, China: (or care of Mrs. Mary Orange, 2 West End Terrace, Jersey.)

1885. Ormerod, Richard Oliver, 35 Philbeach Gardens, South Kensington, London, S.W.

1867. Oughterson, George Blake, care of Peter Brotherhood, Belvedere Road, Lambeth, London, S.E.

1889. Owen, Thomas, Midland Railway, Derby.

1868. Paget, Arthur, Loughborough. [*Paget Company, Loughborough.*]

1877. Panton, William Henry, Messrs. Dorman Long and Co., Middlesbrough.

1877. Park, John Carter, Locomotive Engineer, North London Railway, Bow, London, E.

1872. Parker, Thomas, Locomotive Carriage and Wagon Superintendent, Manchester Sheffield and Lincolnshire Railway, Gorton, near Manchester.

1888. Parker, Thomas, Jun., Locomotive Department, Manchester Sheffield and Lincolnshire Railway, Gorton, near Manchester.

1891. Parker, Thomas, Chief Engineer, Electric Construction Corporation, Wolverhampton.

1879. Parker, William, 108 Fenchurch Street, London, E.C.; and Ash Dune, Catford Hill, London, S.E. [11053.]

1871. Parkes, Persehouse, Messrs. Persehouse Parkes and Co., 21 Drury Buildings, 21 Water Street, Liverpool. [*Fibrous, Liverpool.*]

1884. Parlane, William, Manager, Hong Kong Ice Company, Hong Kong, China.

1892. Parratt, William Heather, Enmore Plantation, East Coast, Demerara, British Guiana.

1892. Parrott, Thomas Henry, Messrs. G. E. Belliss and Co., Ledsam Street, Birmingham.

1886. Parry, Alfred, Messrs. Parry and Co., Vulcan Iron Works, Calcutta, India: (or care of Messrs. J. D. Brown and Co., 5 and 7 Fenchurch Street, London, E.C.)

1889. Parry, Evan Henry, Eagle Chambers, Adelaide Street, Swansea.

1878. Parsons, The Hon. Richard Clere, Messrs. Bateman Parsons and Bateman, 2 Albany Buildings, 39 Victoria Street, Westminster, S.W. [*Outfall, London.* 3233]; and Oak Lea, Wimbledon Park, Surrey.

1886. Passmore, Frank Bailey, Mansion House Chambers, 11 Queen Victoria Street, London, E.C. [*Knarf, London.*]

1880. Paterson, Walter Saunders, Bombay Burmah Trading Corporation, Rangoon, British Burmah, India: (or care of Messrs. Wallace Brothers, 8 Austin Friars, London, E.C.)

1877. Paton, John McClure Caldwell, Messrs. Manlove Alliott and Co., Bloomsgrove Works, Ilkeston Road, Nottingham. [*Manloves, Nottingham.*]

1891. Paton, Robert J., Companhia McHardy, Campinas, São Paulo, Brazil.

1881. Patterson, Anthony, Dowlais Iron Works, Dowlais.

1883. Pattison, Giovanni, Messrs. C. and T. T. Pattison, Engineering Works, Naples. [*Pattison, Naples.*]

1891. Pattison, Joseph, 115 Bute Docks, Cardiff.

1891. Paul, Matthew, Jun., Messrs. Matthew Paul and Co., Levenford Works, Dumbarton.

1891. Paulson, Scott, Box 455, Johannesburg, Transvaal, South Africa: (or care of Dr. Paulson, Mountsorrel, near Loughborough.)

1872. Paxman, James Noah, Messrs. Davey Paxman and Co., Standard Iron Works, Colchester. [*Paxman, Colchester.*]

1880. Peache, James Courthope, Messrs. Willans and Robinson, Ferry Works, Thames Ditton.

1890. Peacock, Francis, Locomotive Superintendent, Smyrna and Cassaba Railway, Smyrna, Asia Minor.

1890. Peacock, James Albert Wells, Assistant Locomotive Superintendent, Smyrna and Cassaba Railway, Smyrna, Asia Minor.

1869. Peacock, Ralph, Messrs. Beyer Peacock and Co., Gorton Foundry, Manchester.

1879. Pearce, George Cope, Ryefields, Ross.

1873. Pearce, Richard, Carriage and Wagon Superintendent, East Indian Railway, Howrah, Bengal, India.

1884. Pearson, Frank Henry, Earle's Shipbuilding and Engineering Works, Hull.

1885. Pearson, Henry William, Engineer, Bristol Water Works, Small Street, Bristol.

1870. Pearson, Thomas Henry, Moss Side Iron Works, Ince, near Wigan.

1888. Peel, Charles Edmund, Quay Parade, Swansea.

1884. Penn, George Williams, Lloyd's Bute Proving House, Cardiff.

1873. Penn, John, M.P., Messrs. John Penn and Sons, Marine Engineers, Greenwich, London, S.E.

1873. Penn, William, Messrs. John Penn and Sons, Marine Engineers, Greenwich, London, S.E.

1874. Pepper, Joseph Ellershaw, Clarence Iron Works, Leeds.

1874. Percy, Cornelius McLeod, King Street, Wigan.

1879. Perkins, Stanhope, Healey Terrace, Fairfield, near Manchester.

1882. Perry, Alfred, Messrs. Chance Brothers and Co., Lighthouse Works, near Birmingham.

1890. Perry, Weston Alcock, Phosphor-Bronze Co., Birmingham; and Kenwood, St. Peter's Road, Birmingham.

1881. Philipson, John, Messrs. Atkinson and Philipson, Carriage Manufactory, 27 Pilgrim Street, Newcastle-on-Tyne. [*Carriage, Newcastle-on-Tyne.* 415.]

1885. Phillips, Charles David, Emlyn Engineering Works, Newport Monmouthshire. [*Machinery, Newport, Mon.*]

1885. Phillips, Henry Parnham, District Locomotive Superintendent, Eastern Bengal State Railway, Calcutta, India; and 71 Grosvenor Street, London, W.

1878. Phillips, John, 4 Corona Road, Burnt Ash Hill, Lee, London, S.E.

1885. Phillips, Lionel, Mining Engineer, Bultfontein Diamond Mine, Kimberley, South Africa; and care of H. Eckstein, Box 149, Johannesburg, Transvaal, South Africa.

1879. Phillips, Robert Edward, Royal Courts Chambers, 70 and 72 Chancery Lane, London, W.C.; and Rochelle, Selhurst Road, South Norwood, London, S.E. [*Phicycle, London.*]

1890. Phillips, Walter, 108 Fenchurch Street, London, E.C. [*Philology, London.* 11053.]

1882. Phipps, Christopher Edward, Locomotive Superintendent, Madras Railway, Perambore Works, Madras; and care of Gerald E. Phipps, Clewer House, St. Albans.

1876. Piercy, Henry James Taylor, Messrs. Piercy and Co., Broad Street Engine Works, Birmingham. [*Piercy, Birmingham.* 20.]

1877. Pigot, Thomas Francis, 41 Upper Mount Street, Dublin.

1888. Pilkington, Herbert, 15 Sedgeley Road, Tipton.

1883. Pillow, Edward, 2 Carlton Terrace, Mill Hill Road, Norwich.

1892. Pinder, Charles Ralph, Messrs. E. W. Tarry and Co., P. O. Box 254, Johannesburg, Transvaal, South Africa.

1876. Pinel, Charles Louis, Messrs. Lethuillier and Pinel, 26 Rue Meridienne, Rouen, France. [*Lethuillier Pinel, Rouen.*]

1892. Pirie, George, Bengal Ordnance Department, Dum Dum, near Calcutta, India.

1882. Pirrie, John Sinclair, Austral Otis Elevator and Engineering Works, South Melbourne, Victoria: (or care of Messrs. John Birch and Co., 11 Queen Street Place, London, E.C.)

1888. Pirrie, Norman, 17 Granville Square, London, W.C.

1888. Pirrie, William James, Messrs. Harland and Wolff, Belfast.

1883. Pitt, Walter, Messrs. Stothert and Pitt, Newark Foundry, Bath. [*Stothert, Bath.*]

1887. Place, John, Linotype Co., 6 Serjeants' Inn, Fleet Street, London, E.C.

1871. Platt, James, Messrs. Fielding and Platt, Atlas Iron Works, Gloucester. [*Atlas, Gloucester.*]

1883. Platt, James Edward, Messrs. Platt Brothers and Co., Hartford Iron Works, Oldham.

1867. Platt, Samuel Radcliffe (*Life Member*), Messrs. Platt Brothers and Co., Hartford Iron Works, Oldham.

1878. Platts, John Joseph, Resident Engineer, Odessa Water Works, Odessa, Russia.

1869. Player, John, Clydach Foundry, near Swansea.

1892. Pogson, Alfred Lee, Engineer-in-Chief, Harbour Trust Board and Works, Madras, India.

1888. Pogson, Joseph, Manager and Engineer, Huddersfield Corporation Tramways, Huddersfield.

1890. Poke, George Henry, Chief Engineer, Government Gun-Carriage Factory, Colaba, Bombay, India.

1886. Pollock, James, 22 Billiter Street, London, E.C. [*Specific, London.*]

1876. Pooley, Henry, Messrs. Henry Pooley and Son, Albion Foundry, Liverpool. [*Pooley, Liverpool.*]

1890. Potter, William Henry, Parliament Chambers, Upper Parliament Street, Nottingham.

1864. Potts, Benjamin Langford Foster, 55 Chancery Lane, London, W.C.; and 117 Camberwell Grove, London, S.E.

1878. Powell, Henry Coke, Tintern House, 64 Burnt Ash Hill, Lee, London, S.E.

1890. Powell, James Richard, Pierhead Chambers, Cardiff.

1874. Powell, Thomas, Brynteg, Neath.

1891. Powles, Henry Handley Pridham, Faraday House, Charing Cross Road, London, W.C.

1867. Pratchitt, John, Messrs. Pratchitt Brothers, Denton Iron Works, Carlisle.

1865. Pratchitt, William, Messrs. Pratchitt Brothers, Denton Iron Works, Carlisle.

1892. Pratt, Middleton, 6 Richmond Terrace, New Brighton, near Birkenhead.

1885. Pratten, William John, Messrs. Harland and Wolff, Belfast.

1890. Preece, William Henry, F.R.S., General Post Office, St. Martin's-le-Grand, London, E.C.

1882. Presser, Ernest Charles Antoine, 4 Salesas, Madrid.

1877. Price, Henry Sherley, Messrs. Wheatley Kirk, Price, and Goulty, 49 Queen Victoria Street, London, E.C. [*Indices, London.*]

1866. Price, John, General Manager, Messrs. Palmer's Shipbuilding and Iron Works, Jarrow; and 6 Osborne Villas, Jesmond, Newcastle-on-Tyne.

1890. Price, John, Inspecting Engineer, Workington.

1889. Price, John Bennett, Messrs. Charles Macintosh and Co., Cambridge Street, Manchester; and 13 Fishergate Hill, Preston.

1859. Price-Williams, Richard, 4 Victoria Mansions, 32 Victoria Street, Westminster, S.W. [*Spandrel, London.*]

1886. Price-Williams, Seymour William, 3 Westminster Chambers, 5 Victoria Street, Westminster, S.W.

1874. Prosser, William Henry, Messrs. Harfield and Co., Blaydon-on-Tyne.

1885. Pudan, Oliver, Chief Engineer's Office, Cambria Iron Works, Johnstown, Pennsylvania, U.S.: (or 15 Princes Street, Yeovil.)

1890. Pugh, Charles Henry, Whitworth Works, Rea Street South, Birmingham.

1887. Pullen, William Wade Fitzherbert, 100 Llandaff Road, Cardiff.

1884. Puplett, Samuel, Albany Buildings, 47 Victoria Street, Westminster, S.W.

1866. Putnam, William, Darlington Forge, Darlington.

1887. Pyne, Thomas Salter, care of H.H. the Ameer of Afghanistan, Kabul: (or care of E. C. Clarke, Foreign Office, Government of India, Simla or Calcutta, India: or care of Edmund Neel, C.I.E., India Office, Whitehall, London, S.W.)

1892. Quentrall, Thomas, H.M. Inspector of Mines, Kimberley, South Africa.

1870. Radcliffe, William, Camden House, 25 Collegiate Crescent, Sheffield.

1878. Radford, Richard Heber, 15 St. James' Row, Sheffield. [*Radford, Sheffield.*]

1868. Rafarel, Frederic William, Cwmbran Nut and Bolt Works, near Newport, Monmouthshire.

1884. Rafarel, William Claude, Barnstaple Foundry and Engineering Works, Victoria Road, Barnstaple. [*Rafarel, Barnstaple.*]

1885. Rainforth, William, Jun., Britannia Iron Works, Lincoln. [*Rainforths, Lincoln.*]

1878. Rait, Henry Milnes, Messrs. Rait and Gardiner, 155 Fenchurch Street, London, E.C. [*Repairs, London.*]

1892. Ramsay, William, Superintendent Engineer, Scottish Oriental Steamship Co., Hong Kong, China.

1847. Ramsbottom, John, Fernhill, Alderley Edge, Cheshire.

1866. Ramsden, Sir James, Abbot's Wood, Barrow-in-Furness; and Reform Chambers, 105 Pall Mall, London, S.W.

1860. Ransome, Allen, 304 King's Road, Chelsea, London, S.W. [*Ransome, London.*]

1886. Ransome, James Edward, Messrs. Ransomes Sims and Jefferies, Orwell Works, Ipswich. [*Ransomes, Ipswich.*]

1873. Rapier, Richard Christopher, Messrs. Ransomes and Rapier, Waterside Iron Works, Ipswich; and 32 Victoria Street, Westminster, S.W. [*Ransomes Rapier, Westminster.*]

1888. Rapley, Frederick Harvey, Messrs. Taite and Carlton, 63 Queen Victoria Street, London, E.C.

1889. Ratcliffe, James Thomas, Baumwoll-Manufactur von Izr. K. Poznanski, Lodz, Russian Poland.

1883. Rathbone, Edgar Philip, Standard Bank Buildings, and P. O. Box 963, Johannesburg, Transvaal, South Africa. [*Viking, Johannesburg.*]

1867. Ratliffe, George, 7A Laurence Pountney Hill, London, E.C.

1862. Ravenhill, John R., Delaford, Iver, near Uxbridge.

1872. Rawlins, John, Manager, Metropolitan Railway-Carriage and Wagon Works, Saltley, Birmingham. [*Metro, Birmingham.*]

1883. Reader, Reuben, Phœnix Works, Cremorne Street, Nottingham.

1887. Readhead, Robert, Messrs. John Readhead and Co., West Docks, South Shields.

1882. Reay, Thomas Purvis, Messrs. Kitson and Co., Airedale Foundry, Leeds.

1881. Redpath, Francis Robert, Canada Sugar Refinery, Montreal, Canada. [*Redpath, Montreal.*]

1883. Reed, Alexander Henry, 64 Mark Lane, London, E.C. [*Wagon, London.*]

1870. Reed, Sir Edward James, K.C.B., M.P., F.R.S., Broadway Chambers, Westminster, S.W. [*Carnage, London.*]

1891. Reed, Thomas Alfred, 2 Bute Crescent, Cardiff.

1884. Rees, William Thomas, Mining Engineer, Gadlys Cottage, Aberdare.

1890. Reeves, Frank, Assistant Engineer, Ferro Carril de Buenos Aires al Pacifico, 228 Calle Piedad, Buenos Aires, Argentine Republic.

1891. Reid, Hugh (*Life Member*), Messrs. Neilson and Co., Hyde Park Locomotive Works, Glasgow.

1883. Reid, James, Messrs. Neilson and Co., Hyde Park Locomotive Works, Glasgow.

1889. Rendell, Alan Wood, Locomotive Superintendent, East Indian Railway, Jumalpore, Bengal, India: (or 21A Goldhurst Terrace, South Hampstead, London, N.W.

1890. Rendell, Samuel, Messrs. Beyer Peacock and Co., Gorton Foundry, Manchester; and 306 Fairfield Road, Fairfield, near Manchester.

1859. Rennie, George Banks, 20 Lowndes Street, Lowndes Square, London, S.W.
1879. Rennie, John Keith, 49 Queen's Gate, London, S.W.
1881. Rennoldson, Joseph Middleton, Marine Engine Works, South Shields. [*Rennoldson, South Shields.* 11.]
1876. Restler, James William, Engineer, Southwark and Vauxhall Water Works, Southwark Bridge Road, London, S.E.
1883. Reunert, Theodore, Box 209, Kimberley, South Africa; Box 92, Johannesburg, Transvaal, South Africa: (or care of Messrs. Findlay, Durham and Brodie, 43–46 Threadneedle Street, London, E.C.)
1862. Reynolds, Edward, Messrs. Vickers Sons and Co., River Don Works, Sheffield.
1879. Reynolds, George Bernard, Manager, Warora Colliery, Warora, Central Provinces, India.
1890. Rice, Thomas Sydney, Aldermary House, 60 Watling Street, London, E.C. [*Ricto, London.*]
1866. Richards, Edward Windsor, Low Moor Iron Works, near Bradford.
1882. Richards, George, Suffolk House, Laurence Pountney Hill, London, E.C.
1884. Richards, Lewis, West Cumberland Iron and Steel Works, Workington; and Derwent View, Workington.
1863. Richardson, The Hon. Edward, C.M.G., Wellington, New Zealand.
1892. Richardson, Harry Alfred, Messrs. Hick Hargreaves and Co., Soho Iron Works, Crook Street, Bolton.
1865. Richardson, John, Methley Park, near Leeds.
1873. Richardson, John, Messrs. Robey and Co., Globe Iron Works, Lincoln.
1891. Richardson, John Scott, Southern Railway of Peru, Arequipa, Peru: (or care of J. W. Champness Richardson, Lindum, 23 Coleridge Road, Crouch End, London, N.)
1887. Richardson, Thomas, Messrs. T. Richardson and Sons, Hartlepool Engine Works, Hartlepool.
1859. Richardson, William, Messrs. Platt Brothers and Co., Hartford Iron Works, Oldham.
1884. Riches, Charles Hurry, Assistant Locomotive Superintendent, Taff Vale Railway, Cardiff.
1890. Riches, Glenford Mitchell, 40 Sydney Road, South Norwood, London, S.E.
1874. Riches, Tom Hurry, Locomotive Superintendent, Taff Vale Railway, Cardiff. [*Locomotive, Cardiff.*]
1889. Richmond, Joseph, New Sun Iron Works, Burdett Road, Bow, London, E.; and 30 Kirby Street, Hatton Garden, London, E.C.
1873. Rickaby, Alfred Austin, Bloomfield Engine Works, Sunderland. [*Rickaby, Sunderland.*]
1879. Ridley, James Cartmell, Swalwell Steel Works, Newcastle-on-Tyne.

1887. Riekie, John, District Locomotive Superintendent, North Western Railway, Quetta, Beluchistan, India.

1874. Riley, James, General Manager, Steel Company of Scotland, 150 Hope Street, Glasgow.

1884. Ripper, William, Professor of Mechanical Engineering, The Technical School, St. George's Square, Sheffield.

1889. Riva, Enrico, Locomotive and Carriage Superintendent, Rete Adriatica, Ferrovie Meridionale, Florence, Italy.

1879. Rixom, Alfred John, 108 Park Road, Loughborough.

1891. Roberts, Hugh Jorwerth, Messrs. Burn and Co., Howrah Iron Works, Howrah, Calcutta, India: (or care of R. P. Roberts, 3 Osborne Road, Liscard, near Liverpool.)

1887. Roberts, Thomas, Locomotive Engineer, Government Railways, Adelaide, South Australia.

1879. Roberts, Thomas Herbert, Mechanical Superintendent, Chicago and Grand Trunk Railway, Detroit, Michigan, United States.

1887. Roberts, William, Mount Erica, Sylvan Road, Upper Norwood, London, S.E.

1879. Robertson, William, Newlyn, Eton Avenue, Hampstead, London, N.W.

1883. Robins, Edward, 22 Conduit Street, London, W.

1890. Robinson, Frederick Arthur, Messrs. F. A. Robinson and Co., 69 Cornhill, London, E.C.

1874. Robinson, Henry, Professor of Surveying and Civil Engineering, King's College, Strand, London, W.C.; and 7 Westminster Chambers, 13 Victoria Street, Westminster, S.W.

1876. Robinson, James Salkeld, Messrs. Thomas Robinson and Son, Railway Works, Rochdale. [*Robinson, Rochdale.*]

1859. Robinson, John, Messrs. Sharp Stewart and Co., Atlas Works, Glasgow; and Westwood Hall, Leek, near Stoke-upon-Trent.

1886. Robinson, John, Barry Dock and Railways, Barry, near Cardiff.

1878. Robinson, John Frederick, Messrs. Sharp Stewart and Co., Atlas Works, Glasgow.

1891. Robinson, John George, Locomotive and Carriage Engineer, Waterford and Limerick Railway, Limerick.

1892. Robinson, Leslie Stephen, 28 Victoria Street, Westminster, S.W.

1890. Robinson, Sydney Jessop, Messrs. W. Jessop and Sons, Brightside Steel Works, Sheffield.

1878. Robinson, Thomas Neild, Messrs. Thomas Robinson and Son, Railway Works, Rochdale. [*Robinson, Rochdale.*]

1891. Roche, Francis James, Shanghai Water Works, Shanghai, China.

1890. Rochford, John, Commissioners of Irish Lights, Westmoreland Street, Dublin.

1888. Rock, John William, Exchange Corner, Pitt Street, Sydney, New South Wales: (or care of E. G. Rock, The Red House, Ingatestone.)

1879. Rodger, William, care of Messrs. C. H. B. Forbes and Co., 3 Elphinstone Circle, Bombay: (or care of Messrs. Duncan Stewart and Co., London Road Iron Works, Glasgow.)

1892. Rodgers, John, Messrs. J. S. Rodgers and Sons, Newcastle, New South Wales.

1884. Rodrigues, José Maria de Chermont, Rua de S. Pedro 54 sobrado, Rio de Janeiro, Brazil.

1872. Rofe, Henry, 8 Victoria Street, Westminster, S.W.

1885. Rogers, Henry John, Watford Iron Works, Watford. [*Engineer, Watford.*]

1887. Rogers, Horace Wyon, 43 Upper Thames Street, London, E.C.

1892. Ronald, Henry, Small Arms Ammunition Factory, Dum Dum, near Calcutta, India: (or care of James Ronald, 10 Campbell Terrace, Plumstead.)

1889. Rosenthal, James Hermann, Babcock and Wilcox Boiler Co., 114 Newgate Street, London, E.C.

1881. Ross, William, Messrs. Ross and Walpole, North Wall Iron Works, Dublin. [*Iron, Dublin.* 311.]

1856. Rouse, Frederick, Locomotive Department, Great Northern Railway, Peterborough.

1878. Routh, William Pole, Sutton Court, Sutton, Surrey.

1888. Rowan, James, Messrs. David Rowan and Son, Elliot Street, Glasgow.

1892. Rowe, Almond, Senior Government Marine Surveyor, Singapore, Straits Settlements.

1891. Rowland, Bartholomew Richmond, Messrs. Luke and Spencer, Ardwick, Manchester.

1867. Ruston, Joseph, Messrs. Ruston Proctor and Co., Sheaf Iron Works, Lincoln; and 6 Onslow Gardens, South Kensington, London, S.W. [*Ruston, Lincoln.*]

1884. Rutherford, George, General Manager, Bute Shipbuilding Engineering and Dry Dock Co., Bute Dry Dock, Roath Basin, Cardiff. [*Caisson, Cardiff.*]

1885. Ryan, John, D.Sc., Professor of Physics and Engineering, University College, Bristol.

1883. Ryder, George, Turner Bridge Iron Works, Tong, near Bolton. [*Ryder, Machinist, Tong, Bolton.* 33A.]

1866. Ryland, Frederick, Messrs. A. Kenrick and Sons, Spon Lane, Westbromwich.

1866. Sacré, Alfred Louis, 60 Queen Victoria Street, London, E.C. [*Sextant, London.* 1668.]

1883. Sadoine, Baron Eugène, 13 Place de Bronckart, Liége, Belgium.

1864. Saïd, Colonel M., Pasha, Engineer, Turkish Service, Constantinople: (or care of J. C. Frank Lee, 9 Park Crescent, Portland Place, London, W.)

1892. Sainsbury, Francis Charles Barrett, Woodlands, Bradford-on-Avon.

1859. Salt, George, Sir Titus Salt, Bart., Sons and Co., Saltaire, near Bradford; and 23 St. Ermin's Mansions, Westminster, S.W.

1874. Sampson, James Lyons, Messrs. David Hart and Co., North London Iron Works, Wenlock Road, City Road, London, N. [*Bascule, London.* 6699.]

1865. Samuelson, Sir Bernhard, Bart., M.P., F.R.S., Britannia Iron Works, Banbury; 56 Prince's Gate, South Kensington, London, S.W.; and Lupton, Brixham, South Devon.

1881. Samuelson, Ernest, Messrs. Samuelson and Co., Britannia Iron Works, Banbury.

1890. Sandberg, Christer Peter, 19 Great George Street, Westminster, S.W.

1881. Sanders, Henry Conrad, Messrs. H. G. Sanders and Son, Victoria Works, Victoria Gardens, Notting Hill Gate, London, W.; and Elm Lodge, Southall.

1871. Sanders, Richard David, Hartfield House, Eastbourne.

1886. Sandford, Horatio, Messrs. E. A. and H. Sandford, Thames Iron Works, Gravesend.

1881. Sandiford, Charles, Locomotive and Carriage Superintendent, North Western Railway, Lahore, Punjaub, India.

1891. Sands, Harold, 3 Harcourt Buildings, Inner Temple, London, E.C.

1874. Sauvée, Albert, 22 Parliament Street, Westminster, S.W. [*Sovez, London.* 3133.]

1891. Savill, Arthur Slater, Exhaust Steam Injector Company, 4 St. Ann's Square, Manchester.

1880. Saxby, John, Messrs. Saxby and Farmer, Railway Signal Works, Canterbury Road, Kilburn, London, N.W. [*Signalmen, London.*]; and North Court, Hassocks, R.S.O., Sussex.

1869. Scarlett, James, Messrs. E. Green and Son, 14 St. Ann's Square, Manchester.

1890. Schofield, George Andrew, Chief Assistant Locomotive and Carriage Superintendent, Chilian State Railways, Casilla No. 193, Santiago, Chili; and care of I. D. Schofield, Oakfield, Alderley Edge, Cheshire.

1886. Scholes, William Henry, 1255 n/n Rivadavia, Buenos Aires, Argentine Republic.

1883. Schonheyder, William, 4 Rosebery Road, Brixton, London, S.W.

1880. Schram, Richard, 17A Great George Street, Westminster, S.W. [*Schram, London.*]

1890. Schroller, William, 16 Baldry Gardens, Streatham Common, London, S.W.

1886. Schurr, Albert Ebenezer, Messrs. Fry Miers and Co., Suffolk House, 5 Laurence Pountney Hill, London, E.C.; and Lyncot, Romford.

1885. Scorgie, James, Professor of Applied Mechanics, Civil Engineering College, Poona, India: Poona Villa, Beechgrove Terrace, Aberdeen: (or care of Messrs. W. Watson and Co., 27 Leadenhall Street, London, E.C.)

1891. Scott, Arthur Forbes, 69 Swan Arcade, Bradford.

1882. Scott, Charles Herbert, Messrs. Summers and Scott, High Orchard Iron Works, Gloucester.

1890. Scott, Frederick McClure, 89 Victoria Street, Liverpool.

1891. Scott, F. Walter, Messrs. George Scott and Son, 44 and 46 Christian Street, London, E. [*Thirty-four, London.* 4390.]

1875. Scott, Frederick Whitaker, Atlas Steel and Iron Wire Rope Works, Reddish, Stockport. [*Atlas, Reddish.*]

1881. Scott, George Innes, Three Indian King's Court, Quayside, Newcastle-on-Tyne.

1891. Scott, Henry John, Ellangowan, Headlands, Kettering.

1877. Scott, Irving M., Union Iron Works, San Francisco, California.

1881. Scott, James, General Manager, President Land and Exploration Co., Pretoria, Transvaal, South Africa: (or Douglasfield, Murthly, Perthshire.)

1886. Scott, James, Consett Iron Works, Consett, R.S.O., County Durham.

1885. Scott, Robert, 50 Cornwall Road, Westbourne Park, London, W.

1891. Scott, Robert Julian, School of Engineering, Canterbury College, Christchurch, New Zealand.

1861. Scott, Walter Henry, Locomotive Superintendent, Ferro Carril del Norte, San Eugenio, Uruguay: (or care of H. Eaton, 75 Tulse Hill, London, S.W.)

1884. Scott-Moncrieff, William Dundas, 14 Victoria Street, Westminster, S.W.

1868. Scriven, Charles, Whinfield Mount, Chapel Allerton, Leeds. [*Scriven, Leeds.*]

1882. Seabrook, Alfred William, Engineer Surveyor to the Port of Bombay, Port Office, Bombay.

1892. Seaman, Charles Joseph, Stockton Forge Works, Stockton-on-Tees. [*Forge, Stockton-on-Tees.*]

1882. Seaton, Albert Edward, Earle's Shipbuilding and Engineering Works, Hull.

1886. Seddon, Robert Barlow, Manager, Wigan Wagon Works, Wigan.

1891. Selby, Millin, 14 Rue de la Gare, Lille, France.

1882. Selfe, Norman, 279 George Street, Sydney, New South Wales.

1884. Sellers, Coleman, E.D., Professor of Engineering, Stevens Institute, and Franklin Institute; 3301 Baring Street, Philadelphia, Pennsylvania. United States.

1888. Sellers, George, Ings Foundry, Wakefield; and Holly Cottage, Wakefield.

1865. Sellers, William, Pennsylvania Avenue, Philadelphia, Pennsylvania, United States.

1891. Sellier, Alphonse Louis, 74 St. James Street, Sanfernando, Trinidad.

1883. Shackleford, Arthur Lewis, General Manager, Britannia Railway-Carriage and Wagon Works, Saltley, Birmingham.

1884. Shackleford, William Copley, Manager, Lancaster Wagon Works, Lancaster.

1872. Shanks, Arthur, Fairmile, Cobham, Surrey.

1884. Shanks, William, Messrs. Thomas Shanks and Co., Johnstone, near Glasgow. [*Shanks, Johnstone.*]

1881. Shapton, William, Sir William G. Armstrong Mitchell and Co., 8 Great George Street, Westminster, S.W.

1890. Shardlow, Ambrose, Ealing Works, Washford Road, Attercliffe, Sheffield.

1891. Sharp, Henry, 23 College Hill, London, E.C.; and Watersgreen, Brockenhurst, R.S.O., Hants.

1875. Sharp, Thomas Budworth, Consulting Engineer, Muntz Metal Works, Birmingham; and County Chambers A, Martineau Street, Birmingham. [*Budworth, Birmingham.*]

1881. Shaw, Joshua, Messrs. John Shaw and Sons, Wellington Street Works, Salford, Manchester.

1881. Shaw, William, Messrs. W. Shaw Kirtley and Co., Wellington Cast Steel Foundry, Middlesbrough.

1890. Sheldon, Harry Cecil, Messrs. W. P. Thompson and Boult, 63 Long Row, Nottingham.

1891. Shenton, James, Messrs. Tinker Shenton and Co., Hyde Boiler Works, Hyde, near Manchester.

1892. Shepherd, James, Messrs. Edward Ripley and Son, Bowling Dye Works, Bradford.

1861. Shepherd, John, Union Foundry, Hunslet Road, Leeds.

1875. Sheppard, Herbert Gurney, Chief Engineer, Assioot-Girgeh Railway, Assioot, Upper Egypt: (or 89 Westbourne Terrace, Hyde Park, London, W.)

1876. Shield, Henry, Messrs. Fawcett Preston and Co., Phœnix Foundry, 17 York Street, Liverpool.

1888. Shin, Tsuneta, Director, Ishikawajima Shipbuilding and Engineering Co. Tokyo, Japan.

1892. Shirlaw, Andrew, Suffolk Works, Oozells Street, Birmingham. [*Shirlaw, Birmingham.*]

1889. Shone, Isaac, Great George Street Chambers, Westminster, S.W.

1890. Shoosmith, Harry, 329 Norwood Road, London, S.E.

1890. Shore, Alfred Thomas, Government Inspector of Steam Boilers, Custom House, Bombay, India.

1885. Shuttleworth, Alfred, Messrs. Clayton and Shuttleworth, Stamp End Works, Lincoln. [*Claytons, Lincoln.*]

1885. Shuttleworth, Major Frank, Messrs. Clayton and Shuttleworth, Stamp End Works, Lincoln; and Old Warden Park, Biggleswade. [*Claytons Lincoln.*]

1891. Siemens, Alexander (*Life Member*), 12 Queen Anne's Gate, Westminster, S.W

1888. Siemens, Frederick, 10 Queen Anne's Gate, Westminster, S.W.

1888. Siemens, Dr. Werner von, Messrs. Siemens and Halske, 94 Markgrafen Strasse, Berlin.

1871. Simon, Henry, 20 Mount Street, Manchester. [*Reform, Manchester.*]

1877. Simonds, William Turner (*Life Member*), Messrs. J. C. Simonds and Son, Oil Mills, Boston.

1876. Simpson, Arthur Telford, Engineer, Chelsea Water Works, 38 Parliament Street, Westminster, S.W.

1883. Simpson, Charles Liddell, Messrs. Simpson and Co., Engine Works, 101 Grosvenor Road, Pimlico, London, S.W. [*Aquosity, London.*]

1885. Simpson, James Thomas, Executive Engineer, Public Works Department, Shwebo, Upper Burmah.

1882. Simpson, John Harwood, Manchester Ship Canal, 65 King Street, Manchester.

1889. Sinclair, Nisbet, The William Cramp and Sons Ship and Engine Building Co., Philadelphia, Pennsylvania, United States.

1847. Sinclair, Robert, care of Messrs. Sinclair Hamilton and Co., 17 St. Helen's Place, Bishopsgate Street, London, E.C. [*Sinclair, London.*]

1857. Sinclair, Robert Cooper, Chesham House, Cavendish Road, Balham, London, S.W.

1891. Sinclair, Russell, 81 Pitt Street, Sydney, New South Wales: (or care of Messrs. Potter Brothers, 112 Fenchurch Street, London, E.C.)

1881. Sisson, William, Quay Street Iron Works, Gloucester. [*Sisson, Gloucester.*]

1872. Slater, Alfred, Gloucester Wagon Works, Gloucester.

1892. Slight, George Henry, Sub-Director of Lighthouses, Valparaiso, Chile.

1885. Slight, William Hooper, care of B. N. Powell, Soerabaya, Java: (or care of G. H. Slight, 64 Cromwell Road, Fitzhugh, Southampton.)

1891. Sloan, Robert Alexander, Messrs. Sloan and Lloyd Barnes, Castle Chambers, 26 Castle Street, Liverpool. [*Technical, Liverpool.*]

1886. Small, James Miln, Messrs. Urquhart and Small, 17 Victoria Street, Westminster, S.W.

1889. Smelt, John Dann, Argentine Great Western Railway, 4 Finsbury Circus, London, E.C.

1879. Smith, Charles Hubert, Board of Trade Offices, Middlesbrough.

1860. Smith, Henry, Messrs. Hill and Smith, Brierley Hill Iron Works, Brierley Hill; and Summerhill, Kingswinford, near Dudley. [*Fencing, Brierley Hill.*]

1881. Smith, Henry, Messrs. Simpson and Co., 101 Grosvenor Road, Pimlico, London, S.W.

1860. Smith, Sir John, Parkfield, Duffield Road, Derby.

1876. Smith, John, Wintoun Terrace, Rochdale.

1883. Smith, John Bagnold, Newstead Colliery, near Nottingham.

1891. Smith, John Reney, 16 Seaton Buildings, 17 Water Street, Liverpool.

1890. Smith, John Windle, Messrs. Marshall Sons and Co., and Messrs. Walsh Lovett and Co., 308 Calle Peru, Buenos Aires, Argentine Republic: (or care of Edward Smith, The "Lock," Gainsborough.)

1857. Smith, Josiah Timmis, Hæmatite Iron and Steel Works, Barrow-in-Furness; and Rhine Hill, Stratford-on-Avon.

1870. Smith, Michael Holroyd, Royal Insurance Buildings, Crossley Street, Halifax; and 18 Abingdon Street, Westminster, S.W. [*Outfall, London.*]

1886. Smith, Reginald Arthur, Messrs. Dorman and Smith, Ordsal Station Electrical Works, Salford, Manchester.

1881. Smith, Robert Henry, Professor of Engineering, Mason Science College, Birmingham; and 124 Hagley Road, Edgbaston, Birmingham.

1885. Smith, Thomas, Steam Crane Works, Old Foundry, Rodley, near Leeds, [*Tomsmith, Leeds.*]

1890. Smith, Thomas Ridsdill, Earle's Shipbuilding and Engineering Works, Hull.

1881. Smith, Wasteneys, 59 Sandhill, Newcastle-on-Tyne. [*Wasteneys Smith, Newcastle-on-Tyne.* 429.]

1890. Smith, William, London and Manchester Plate Glass Co., Sutton, St. Helen's, Lancashire.

1863. Smith, William Ford, Messrs. Smith and Coventry, Gresley Iron Works, Ordsal Lane, Salford, Manchester.

1887. Smith, William Mark, District Locomotive Carriage and Wagon Superintendent, Great Southern and Western Railway, Cork.

1882. Smyth, James Josiah, Messrs. James Smyth and Sons, Peasenhall, Suffolk.

1884. Smyth, William Stopford, Engineer, Alexandra Docks, Newport, Monmouthshire.

1883. Snelus, George James, F.R.S., Ennerdale Hall, Frizington, near Carnforth.

1885. Snowdon, John Armstrong, Stanners Closes Steel Works, Wolsingham, near Darlington.

1887. Sorabji, Shapurji, Bombay Foundry and Engine Works, Khetwady Bombay: (or care of Messrs. S. and E. Ransome and Co., Billiter Buildings, 49 Leadenhall Street, London, E.C.)

1884. Soulsby, James Charlton, 17 Mount Stuart Square, Cardiff.

1889. Souter-Robertson, David, Assistant Superintendent, Government Canal Foundry and Workshops, Roorkee, North Western Provinces, India.

1885. Southwell, Frederick Charles, Messrs. Richard Hornsby and Sons, Spittlegate Iron Works, Grantham.

1877. Soyres, Francis Johnstone de, 4 Arlington Villas, Clifton, near Bristol.

1887. Spence, William, Cork Street Foundry and Engineering Works, Dublin.

1887. Spencer, Alexander, 77 Cannon Street, London, E.C.

1878. Spencer, Alfred G., Messrs. George Spencer and Co., 77 Cannon Street, London, E.C.

1892. Spencer, Henry Bath, British Steam Users' Insurance Society, Manchester; and 25 Brundrett's Road, Chorlton-cum-Hardy, Manchester.

1877. Spencer, John, Globe Tube Works, Wednesbury [*Tubes, Wednesbury.*]: and 3 Queen Street Place, Cannon Street, London, E.C. [*Tubes, London.*]

1867. Spencer, John W., Newburn Steel Works, Newcastle-on-Tyne. [*Newburn, Newcastle-on-Tyne.*]

1885. Spencer, Mountford, Messrs. Luke and Spencer, Ardwick, Manchester; and The Hill, Teignmouth.

1854. Spencer, Thomas, Newburn Steel Works, Newcastle-on-Tyne. [*Newburn, Newcastle-on-Tyne.*]

1891. Spencer, William, Messrs. James Spencer and Co., Chamber Iron Works, Hollinwood, near Manchester.

1885. Spooner, George Percival, Locomotive Superintendent, Bolan Railway, Hirokh, Beluchistan, India; and Whitehall Club, Parliament Street, Westminster, S.W.

1883. Spooner, Henry John, 309 Regent Street, London, W.

1869. Stabler, James, 13 Effra Road, Brixton, London, S.W.

1877. Stanger, George Hurst, Queen's Chambers, North Street, Wolverhampton.

1875. Stanger, William Harry, Chemical Laboratory and Testing Works, Broadway, Westminster, S.W. [3117.]

1888. Stanley, Harry Frank, Messrs. Pontifex and Wood, Farringdon Works, Shoe Lane, London, E.C.; and 84 Finsbury Park Road, London, N.

1888. Stannah, Joseph, 20 Southwark Bridge Road, London, S.E.

1884. Stanton, Frederic Barry, 18 Bishopsgate Street Within, London, E.C. [*Barry Stanton, London.* 4084.]

1874. Stephens, Michael, Chief Locomotive Superintendent, Cape Government Railways, Cape Town, Cape of Good Hope.

1868. Stephenson, George Robert, 9 Victoria Chambers, 17 Victoria Street, Westminster, S.W. [*Precursor, London.*]

1879. Stephenson, Joseph Gurdon Leycester, 6 Drapers' Gardens, London, E.C. [*Fluvius, London.*]

1888. Stephenson-Peach, William John, Askew Hill, Repton, Burton-on-Trent.

1876. Sterne, Louis, Messrs. L. Sterne and Co., Crown Iron Works, Glasgow [*Crown, Glasgow.*]; and 2 Victoria Mansions, 28 Victoria Street, Westminster, S.W. [*Elsterne, London.* 3066.]

1891. Stevens, James, 9 and 11 Fenchurch Avenue, London, E.C.

1887. Stevenson, David Alan, F.R.S.E., 84 George Street, Edinburgh.

1892. Stevinson, Thomas, Messrs. Hender and Stevinson, Nailsworth, near Stroud, Gloucestershire.

1877. Stewart, Alexander, 3 Southbrook Terrace, Bradford.

1887. Stewart, Andrew, 41 Oswald Street, Glasgow.

1878. Stewart, Duncan, Messrs. Duncan Stewart and Co., London Road Iron Works, Glasgow. [*Stewart, Glasgow.* 531.]

1851. Stewart, John, Blackwall Iron Works, Poplar, London, E. [*Steamships, London.*]; and 8 Stamford Avenue, Preston Park, Brighton.

1888. Stiff, William Charles, Smethwick, Birmingham.

1892. Still, William Henry, Aden, Arabia.

1880. Stirling, James, Locomotive Superintendent, South Eastern Railway, Ashford, Kent.

1885. Stirling, Matthew, Locomotive Superintendent, Hull Barnsley and West Riding Junction Railway and Dock Co., Hull.

1867. Stirling, Patrick, Locomotive Superintendent, Great Northern Railway, Doncaster.

1888. Stirling, Robert, Locomotive Department, North Eastern Railway, Gateshead.

1875. Stoker, Frederick William, Messrs. Easton and Anderson, Erith Iron Works, Erith, S.O., Kent.

1877. Stokes, Alfred Allen, Elmcote, Godalming.

1892. Stone, Edward Herbert, Deputy Consulting Engineer to the Government of India for State Railways, Simla, India.

1887. Stone, Frank Holmes, P.O. Box, Kingston, Jamaica.

1877. Stothert, George Kelson, Steam Ship Works, Bristol.

1888. Strachan, James, Stenton, near Prestonkirk.

1892. Strachan, John, 29 The Walk, Cardiff.

1888. Straker, Sidney, Messrs. Ewen and Straker, 37 Walbrook, London, E.C.; and 240 Stanstead Road, Forest Hill, London, S.E.

1891. Stringer, William, 20 Mount Street, Manchester.

1884. Stronge, Charles, Locomotive Department, Porto Alegre and New Hamburg Railway, São Leopoldo, Rio Grande do Sol, Brazil: (or 1 Albion Street, Hyde Park, London, W.)

1873. Strype, William George, 115 Grafton Street, Dublin. [*Strype, Dublin.*]

1889. Stuart-Hartland, Dare Arthur, 31 Dalhousie Square South, Calcutta, India.

1882. Sturgeon, John, Shrublands, Hoole Road, Chester.

1890. Stutzer, Waldemar, Koltchugin Brass and Copper Mill Co., Alexandrov Station, Jaroslav Railroad, Russia.

1882. Sugden, Thomas, Babcock and Wilcox Co., 114 Newgate Street, London, E.C.

1890. Sulzer, Jacob, Messrs. Sulzer Brothers, Winterthur, Switzerland.

1861. Sumner, William, 2 Brazennose Street, Manchester.

1875. Sutcliffe, Frederic John Ramsbottom, Engineer, Low Moor Iron Works, near Bradford.

1883. Sutton, Joseph Walker, 36 Bedford Street, Strand, London, W.C.

1880. Sutton, Thomas, Carriage and Wagon Superintendent, Furness Railway, Barrow-in-Furness.

1887. Suverkrop, John Peter, Chief Engineer, East Tennessee Land Co., Harriman, Tennessee, United States: (or 1 Kew Gardens Road, Kew, Surrey.)

1882. Swaine, John, Messrs. Wright Butler and Co., Panteg Steel Works, near Newport, Monmouthshire.

1884. Swan, Joseph Wilson, 57 Holborn Viaduct, London, E.C.; and Lauriston, Bromley, Kent.

1882. Swinburne, Mark William, Wallsend Brass Works, Newcastle-on-Tyne; and 117 Park Road, Newcastle-on-Tyne. [*Bronze, Wallsend.*]

1864. Swindell, James Swindell Evers, Homer Hill, Cradley, Staffordshire.

1890. Swinerd, Edward, Superintendent, Locomotive Carriage and Wagon Departments, Mogyana Railway, Campinas, Brazil: (or care of Messrs. Fry Miers and Co., Suffolk House, Laurence Pountney Hill, London, E.C.)

1890. Swinnerton, Robert Allen William, Executive Engineer, Public Works Department, Bolarum, Dekkan, India.

1878. Taite, John Charles, Messrs. Taite and Carlton, 63 Queen Victoria Street, London, E.C. [1618.]; and The Corner House, Shortlands, S.O., Kent.

1882. Tandy, John O'Brien, London and North Western Railway, Locomotive Department, Crewe; and 4 Wellington Villas, Wellington Square, Crewe.

1875. Tangye, George, Messrs. Tangyes, Cornwall Works, Soho, near Birmingham. [*Tangyes, Birmingham.*]

1861. Tangye, James, Messrs. Tangyes, Cornwall Works, Soho, near Birmingham; and Aviary Cottage, Illogan, near Redruth.

1879. Tartt, William, Maythorn, Blindley Heath, Godstone, near Red Hill.

1876. Taunton, Richard Hobbs, 10 Coleshill Street, Birmingham.

1882. Tayler, Alexander James Wallis, 77 Victoria Road, Kilburn, London, N.W.

1874. Taylor, Arthur, Manager, Sociedad Anglo-Vasca, Villanueva del Duque, Provincia de Cordoba, Spain: (or 21 Victoria Road, Kensington, London, W.)

1887. Taylor, James, Messrs. Buckley and Taylor, Castle Iron Works, Oldham.

1873. Taylor, John, Midland Foundry, Queen's Road, Nottingham.

1875. Taylor, Joseph Samuel, Messrs. Taylor and Challen, Derwent Foundry, 60 and 62 Constitution Hill, Birmingham. [*Derwent, Birmingham.*]

1874. Taylor, Percyvale, Messrs. Burthe and Taylor, Paris; and 21 Victoria Road, Kensington, London, W.

1882. Taylor, Robert Henry, 2 Collingwood Street, Newcastle-on-Tyne; and 2 Winchester Terrace, Newcastle-on-Tyne.

1882. Taylor, Thomas Albert Oakes, Messrs. Taylor Brothers and Co., Clarence Iron Works, Leeds.

1864. Tennant, Sir Charles, Bart. (*Life Member*), The Glen, Innerleithen, near Edinburgh.

1882. Terry, Stephen Harding, 17 Victoria Street, Westminster, S.W.

1891. Tetlow, Ernest, Messrs. Tetlow Brothers, Bottoms Iron Works, Hollinwood, near Manchester.

1877. Thom, William, Messrs. Yates and Thom, Canal Foundry, Blackburn.

1889. Thomas, James Donnithorne, 25A Old Broad Street, London, E.C. [*Kooringa, London.*]

1867. Thomas, Joseph Lee, 2 Hanover Terrace, Ladbroke Square, Notting Hill, London, W.

1888. Thomas, Philip Alexander, Cornwall Buildings, 35 Queen Victoria Street, London, E.C. [*Argument, London.*]

1864. Thomas, Thomas, 10 Richmond Road, Roath, Cardiff.

1874. Thomas, William Henry, 15 Parliament Street, Westminster, S.W.

1891. Thompson, James, Highfield Boiler Works, Ettingshall, Wolverhampton. [*Boiler, Wolverhampton.*]

1875. Thompson, John, Highfield Boiler Works, Ettingshall, Wolverhampton. [*Boiler, Wolverhampton.*]

1883. Thompson, Richard Charles, Messrs. Robert Thompson and Sons, Southwick Shipbuilding Yard, Sunderland.

1880. Thompson, Thomas William, Eastham Ferry Pier, near Birkenhead.

1887. Thompson, William Phillips, 6 Lord Street, Liverpool.

1875. Thomson, James McIntyre, Messrs. John and James Thomson, Finnieston Engine Works, 36 Finnieston Street, Glasgow. [*Engineering, Glasgow.*]

1868. Thomson, John, Messrs. John and James Thomson, Finnieston Engine Works, 36 Finnieston Street, Glasgow. [*Engineering, Glasgow.*]

1889. Thomson, Robert McNider, Kobe Engine Works, Kobe, Japan: (or care of William Hipwell, Hillside House, Sharnbrook, Bedford.)

1868. Thornewill, Robert, Messrs. Thornewill and Warham, Burton Iron Works, Burton-on-Trent.

1885. Thornley, George, Messrs. Buxton and Thornley, Waterloo Engineering Works, Burton-on-Trent.

1877. Thornton, Frederic William, care of The Hydraulic Engineering Co., Chester.

1882. Thornton, Hawthorn Robert, Lancashire and Yorkshire Railway, Horwich, near Bolton.

1888. Thornton, Robert Samuel, West's Patent Press Co., Etawah, North Western Provinces, India.

1876. Thornycroft, John Isaac, Messrs. John I. Thornycroft and Co., Steam Yacht and Launch Builders, Church Wharf, Chiswick, London, W. [*Thornycroft, London.*]

1882. Thow, William, Locomotive Engineer, New South Wales Government Railways, Eveleigh Workshops, Sydney, New South Wales: (or care of Joseph Meilbek, 7 Westminster Chambers, 13 Victoria Street, Westminster, S.W.)

1884. Thwaites, Arthur Hirst, Messrs. Thwaites Brothers, Vulcan Iron Works, Bradford. [*Thwaites, Bradford.* 325.]

1887. Thwaites, Edward Hirst, Messrs. Thwaites Brothers, Vulcan Iron Works, Bradford. [*Thwaites, Bradford.* 325.]

1891. Tilley, Albert, care of Bernard Dawson, York House, Malvern Link, Malvern.

1885. Timmermans, François, Managing Director, Société anonyme des Ateliers de la Meuse, Liége, Belgium. [*Société Meuse, Liége.*]

1884. Timmis, Illius Augustus, 2 Great George Street, Westminster, S.W. [*Timmis, London.*]

1886. Tipping, Henry, 38 Croom's Hill, Greenwich, London, S.E.

1890. Titley, Arthur, Messrs. Vaughton and Titley, 28 Lower Temple Street, Birmingham.

1888. Todd, Robert Ernest, Mechanical Engineer, Tucuman, Estacion Provincia, Argentine Republic: (or care of William H. Todd, County Buildings, Land of Green Ginger, Hull.)

1875. Tomkins, William Steele, Messrs. Sharp Stewart and Co., Atlas Works, Glasgow; and 2 Victoria Mansions, 28 Victoria Street, Westminster, S.W.

1857. Tomlinson, Joseph, 64 Priory Road, West Hampstead, London, N.W.

1888. Topple, Charles James, Machinery Department, Royal Arsenal, Woolwich.

1883. Tower, Beauchamp, 5 Queen Anne's Gate, Westminster, S.W.

1889. Towler, Alfred, Messrs. Hathorn Davey and Co., Sun Foundry, Leeds.

1886. Towne, Henry Robinson, Yale and Towne Manufacturing Co., Stamford, Connecticut, United States.

1890. Trail, John, Marine Superintendent, Knott's Prince Line of Steamers, Newcastle-on-Tyne.

1888. Travis, Henry, Machinery Department, Royal Arsenal, Woolwich.

1889. Trenery, William Penrose, 73 Via Milano, Genoa, Italy.

1883. Trentham, William Henry, Albert Villa, Nore Road, Portishead, S.O., Somersetshire.

1876. Trevithick, Richard Francis, Locomotive and Carriage Superintendent, Japanese Government Railways, Kobe, Japan: (or care of Mrs. Mary Trevithick, The Cliff, Penzance.)

1886. Trew, James Bradford, High Street, Watford, Herts.

1887. Trier, Frank, Messrs. Brunton and Trier, 19 Great George Street, Westminster, S.W.

1885. Trueman, Thomas Brynalyn, Ferro Carril Buenos Aires al Pacifico, Junin, Argentine Republic: (or care of Thomas R. Trueman, 3 The Barons, Twickenham.)

1887. Turnbull, Alexander, Messrs. Alexander Turnbull and Co., St. Mungo Works, Bishopbriggs, Glasgow. [*Valve, Glasgow.* 1270.]

1885. Turnbull, John, Jun., 255 Bath Street, Glasgow. [*Turbine, Glasgow.*]

1866. Turner, Frederick, Messrs. E. R. and F. Turner, St. Peter's Iron Works, Ipswich. [*Gippeswyk, Ipswich.*]

1886. Turner, George Reynolds, Vulcan Iron Works, Langley Mill, near Nottingham; and 81 Highgate Road, London, N.W.

1887. Turner, Joshua Alfred Alexander, Inspector, Commissariat Mills and Bakeries, Bombay Presidency, Poona, India.

1882. Turner, Thomas, Shelton Iron and Steel Works, Stoke-on-Trent.

1886. Turner, Tom Newsum, Vulcan Iron Works, Langley Mill, near Nottingham.

1876. Turney, Sir John, Messrs. Turney Brothers, Trent Bridge Leather Works, Nottingham. [*Turney, Nottingham.*]

1891. Tweddell, Ralph Hart, 14 Delahay Street, Westminster, S.W. [*Tweddell, Westminster, London.*]

1882. Tweedy, John, Messrs. Wigham Richardson and Co., Newcastle-on-Tyne.

1856. Tyler, Sir Henry Whatley, K.C.B., Pymmes Park, Edmonton, Middlesex.

1877. Tylor, Joseph John, 2 Newgate Street, London, E.C.

1889. Tyrrell, Joseph John, Messrs. Clayton and Shuttleworth, Stamp End Iron Works, Lincoln.

1878. Tyson, Isaac Oliver, Ousegate Iron Works, Selby.

1878. Unwin, William Cawthorne, F.R.S., Professor of Engineering, City and Guilds of London Central Institution, Exhibition Road, London, S.W.; and 7 Palace Gate Mansions, Kensington, London, W.

1875. Urquhart, Thomas, General Manager, Nevsky Engineering Works, St. Petersburg, Russia.

1880. Valon, William Andrew McIntosh, 140 and 141 Temple Chambers, Temple Avenue, London, E.C.; and Ramsgate. [*Valon, Ramsgate.*]

1885. Vaughan, William Henry, Royal Iron Works, West Gorton, Manchester. [*Vaunting, Manchester.* 5106.]

1862. Vavasseur, Josiah, 28 Gravel Lane, Southwark, London, S.E.; and Rothbury, Blackheath Park, London, S.E. [*Exemplar, London.*]

1889. Vesian, John Stuart Ellis de, 5 Crown Court, Cheapside, London, E.C. [*Biceps, London.*]

1891. Vicars, John, Messrs. T. and T. Vicars, Furnace Foundry, Earlestown, Newton-le-Willows, Lancashire.

1865. Vickers, Albert, Messrs. Vickers Sons and Co., River Don Works, Sheffield.

1861. Vickers, Thomas Edward, Messrs. Vickers Sons and Co., River Don Works, Sheffield.

1888. Voysey, Henry Wesley, 1 Fordwych Road, Brondesbury, London, N.W.

1883. Waddell, James, 9 Ashton Terrace, Dowanhill, Glasgow.

1856. Waddington, John, 35 King William Street, London Bridge, London, E.C.

1879. Wadia, The Hon. Nowrosjee Nesserwanjee, C.I.E., Manager, Manockjee Petit Manufacturing Co., Tardeo, Bombay: (or care of Messrs. Hick Hargreaves and Co., Soho Iron Works, Bolton.) [*Wadia, Tardeo, Bombay.*]

1882. Wailes, George Herbert, St. Andrews, Watford, Herts.

1875. Wailes, John William, South Shore, Gateshead-on-Tyne.

1884. Wailes, Thomas Waters, General Manager, Mountstuart Dry Dock and Engineering Works, Cardiff. [*Mountstuart, Cardiff.*]

1888. Waister, William Henry, Assistant Locomotive Superintendent, Great Western Railway, Stafford Road Works, Wolverhampton.

1881. Wake, Henry Hay, Engineer to the River Wear Commission Sunderland.

1882. Wakefield, William, Locomotive Superintendent, Dublin Wicklow and Wexford Railway, Grand Canal Street, Dublin.

1892. Waldron, Patrick Lawrence, Chief Engineer, Mail Steamer "Yomah," Irrawaddy Flotilla Co., Rangoon, Burma, India: (or 20 St. Joseph's Road, Aughrim Street, Dublin.)

1890. Walkeden, George Henry, Temperance Life Buildings, Swanston Street, Melbourne, Victoria.

1891. Walker, Arthur Tannett, Messrs. Tannett Walker and Co., Goodman Street Works, Hunslet, Leeds.

1875. Walker, George, 95 Leadenhall Street, London, E.C.

1890. Walker, Henry, 11 Oxford Terrace, Gateshead-on-Tyne.

1875. Walker, John Scarisbrick, Messrs. J. S. Walker and Brother, Pagefield Iron Works, Wigan; and 3 Alexandra Road, Southport. [*Pagefield, Wigan.*]

1884. Walker, Matthew, 16 London Street, Fenchurch Street, London, E.C.

1884. Walker, Sydney Ferris, Cardiff Electrical Works, Severn Road, Cardiff [*Dynamo, Cardiff.*]; and Hunter's Forge, New Bridge Street, Newcastle-on-Tyne. [*Dynamo, Newcastle-on-Tyne.*]

1876. Walker, Thomas Ferdinand, Ship's Log Manufacturer, 58 Oxford Street, Birmingham.

1878. Walker, William, Kaliemaas, Alleyne Park, West Dulwich, London, S.E. [*Bromo, London.*]

1890. Walker, William George, University College, Bristol.

1863. Walker, William Hugill, Messrs. Walker Eaton and Co., Wicker Iron Works, Sheffield.

1878. Walker, Zaccheus, Jun., Fox Hollies Hall, near Birmingham.

1881. Walkinshaw, Frank, Hartley Grange, Winchfield.

1884. Wallace, John, Backworth Collieries, near Newcastle-on-Tyne.

1884. Wallau, Frederick Peter, Messrs. Harland and Wolff, Belfast.

1868. Wallis, Herbert, Mechanical Superintendent, Grand Trunk Railway, Montreal, Canada.

1891. Walmsley, John, Queen's Mills, Huddersfield.

1865. Walpole, Thomas, Messrs. Ross and Walpole, North Wall Iron Works, Dublin. [*Iron, Dublin.* 311.]

1877. Walton, James, 28 Maryon Road, Charlton.

1881. Warburton, John Seaton, 19 Stanwick Road, West Kensington, London, W.

1882. Ward, Thomas Henry, 24 Church Lane, Smethwick, near Birmingham.

1876. Ward, William Meese, Newton Villa, Claremont Road, Handsworth, R.O., near Birmingham.

1864. Warden, Walter Evers, Phœnix Bolt and Nut Works, Handsworth, R.O., near Birmingham. [*Bolts, Birmingham.*]

1882. Wardle, Edwin, Messrs. Manning Wardle and Co., Boyne Engine Works, Hunslet, Leeds. [*Manning, Leeds.*]

1886. Warren, Frank Llewellyn, 73 Breakspears Road, St. John's, London, S.E.

1885. Warren, Henry John, Jun., Cornwall Boiler Works, Camborne.

1885. Warren, William, 5 Victoria Buildings, Hong Kong, China; care of Messrs. Jardine Matheson and Co., Hong Kong, China: (or care of Walter Ross, Hill Top, Blythe Hill, Catford, London, S.E.) [*Wakeful, London; or Hong Kong.*]

1882. Warsop, Henry, Clarendon Hotel, Nottingham.

1889. Warsop, Thomas, Coniston Copper Mines, Coniston, S.O., Lancashire.

1858. Waterhouse, Thomas (*Life Member*), Claremont Place, Sheffield.

1891. Waterous, Julius E., Waterous Engine Works Co., Brantford, Ontario, Canada.

1881. Watkins, Alfred, 121 Fenchurch Street, London, E.C.

1862. Watkins, Richard, 71 Blenheim Crescent, London, W.

1890. Watkinson, William Henry, Central Higher Board School, Orchard Lane, Sheffield.

1890. Watson, George Coghlan, Manganèse Bronze and Brass Co., St. George's Wharf, Deptford, London, S.E.; and 1 Farquharson Road, Croydon.

1882. Watson, Henry Burnett, Messrs. Henry Watson and Son, High Bridge Works, Newcastle-on-Tyne. [*Watsons, Newcastle-on-Tyne.* 439.]

1879. Watson, Sir William Renny, Messrs. Mirrlees Tait and Watson, Glasgow.

1891. Watt, Charles, Melbourne Cable Tramway, 418 Little Collins Street, Melbourne, Victoria.

1877. Watts, John, Broad Weir Engine Works, Bristol.

1886. Weatherburn, Robert, Locomotive Manager, Midland Railway Works, Kentish Town, London, N.W.

1884. Webb, Richard George, Messrs. Richardson and Cruddas, Byculla Iron Works, Bombay, India: (or care of Messrs. Richardson and Hewitt, 101 Leadenhall Street, London, E.C.)

1890. Webster, John James, 39 Victoria Street, Westminster, S.W.

1887. Webster, William, 6 Oxley Road, Singapore, Straits Settlements.

1883. Weck, Friedrich, Lilleshall Old Hall, near Newport, Shropshire.

1891. Weightman, Walter James, Engineer-in-Chief, Nilgiri Railway, Coonoor, Madras, India.

1888. Wellman, Samuel T., Wellman Iron and Steel Works. Thurlow, Pennsylvania, United States.

1882. West, Charles Dickinson, Professor of Mechanical Engineering, Imperial College of Engineering, Tokyo, Japan.

1876. West, Henry Hartley, Naval Architect and Engineer, 14 Castle Street, Liverpool. [*Referee, Liverpool.*]

1891. West, Leonard, Ravenhead Plate Glass Works, St. Helen's, Lancashire.

1874. West, Nicholas James, Messrs. Harvey and Co., 186 Gresham House, Old Broad Street, London, E.C.; and The Turret, West Heath Road, Hampstead, London, N.W.

1877. Western, Charles Robert, Broadway Chambers, Westminster, S.W. [*Donbowes, London.* 3199.]

1877. Western, Maximilian Richard, care of Bombay Burmah Trading Corporation, Bangkok, Siam: (or care of Messrs. Wallace Brothers, 8 Austin Friars, London, E.C.)

1862. Westmacott, Percy Graham Buchanan, Sir W. G. Armstrong Mitchell and Co., Elswick Engine Works, Newcastle-on-Tyne; and Benwell Hill, Newcastle-on-Tyne.

1880. Westmoreland, John William Hudson, Lecturer on Engineering, University College, Nottingham.

1867. Weston, Thomas Aldridge, Yale and Towne Manufacturing Co., 62 Reade Street, New York; and P.O. Box 230, Ridgewood, New Jersey, United States.

1880. Westwood, Joseph, Napier Yard, Millwall, London, E. [*Westwood, London.* 5065.]

1888. Weyman, James Edwardes, Messrs. Weyman and Johnson, Church Acre Iron Works, Guildford.

1884. Whieldon, John Henry, 75 Ivydale Road, Nunhead, London, S.E.

1882. White, Alfred Edward, Borough Engineer's Office, Town Hall, Hull.

1887. White, Alfred George, 11 Queen Victoria Street, London, E.C.

1874. White, Henry Watkins, 23 Leadenhall Street, London, E.C.; and 122 Lavender Hill, London, S.W.

1888. White, William Henry, C.B., F.R.S., Assistant Controller and Director of Naval Construction, Admiralty, Whitehall, London, S.W.

1890. Whitehouse, Edwin Edward Joseph, Monkbridge Iron Works, Leeds.

1876. Whiteley, William, Holly Mount, Edgerton, Huddersfield.

1891. Whittaker, John, Messrs. William Whittaker and Sons, Sun Iron Works, Oldham.

1869. Whittem, Thomas Sibley, Wyken Colliery, Coventry.

1888. Whittle, John, Union Railway Wagon Works, Chorley.

1878. Whytehead, Hugh Edward, North Staffordshire Tramways, Stoke-on-Trent.

1878. Wicks, Henry, Messrs. Burn and Co., Howrah Iron Works, Howrah, Bengal, India: (or care of John Spencer, 125 West Regent Street, Glasgow.)

1868. Wicksteed, Joseph Hartley, Messrs. Joshua Buckton and Co., Well House Foundry, Meadow Road, Leeds.

1891. Widdowson, John Henry, Britannia Works, Ordsal Lane, Salford, Manchester.

1878. Widmark, Harald Wilhelm, Helsingborgs Mekaniska Verkstad, Helsingborg, Sweden.

1889. Wigham, John Richardson, Messrs. J. Edmundson and Co., Stafford Works, 35 Capel Street, Dublin.

1881. Wigzell, Eustace Ernest, Billiter House, Billiter Street, London, E.C. [*Wigzell, London.* 1844.]

1890. Wild, John, Falcon Iron Works, Oldham. [*Falcon, Oldham.*]

1884. Wilder, John, Yield Hall Foundry, Reading.

1886. Wildridge, John, Messrs. Wildridge and Sinclair, 81 Pitt Street, Sydney, New South Wales.

1890. Wildy, William Lawrence, Messrs. Richard Hornsby and Sons, 84 Lombard Street, London, E.C.

1892. Wilkinson, Edward R., 71 Wood Street, Barnet.

1888. Willans, Peter William, Messrs. Willans and Robinson, Ferry Works, Thames Ditton, Surrey. [*Willans, Thamesditton.*]

1885. Willcox, Francis William, 45 West Sunniside, Sunderland.

1883. Williams, Edward Leader, Engineer, Manchester Ship Canal Co., 41 Spring Gardens, Manchester. [*Leader, Manchester.* 688.]

1884. Williams, John Begby, Messrs. William Gray and Co., Central Marine Engineering Works, West Hartlepool.

1885. Williams, Nicholas Thomas, New Morgan Gold Mining Co., Dolgelly.

1847. Williams, Richard, Brunswick House, Wednesbury.

1890. Williams, Thomas David, Egremont, Battle Road, Ore, near Hastings.

1881. Williams, William Freke Maxwell, 29 Great St. Helen's, London, E.C. [*Wabash, London.*]

1873. Williams, William Lawrence, 16 Victoria Street, Westminster, S.W. [*Snowdon, London.*]

1889. Williams, William Walton, Jun., 25 Oakley Street, Chelsea, London, S.W.

1883. Williamson, Richard, Messrs. Richard Williamson and Son, Iron Shipbuilding Yard, Workington.

1870. Willman, Charles, 26 Albert Road, Middlesbrough.

1878. Wilson, Alexander, Messrs. Charles Cammell and Co., Cyclops Steel and Iron Works, Sheffield.

1882. Wilson, Alexander Basil, Holywood, Belfast. [*Wilson, Holywood.* 201.]

1884. Wilson, James, Chief Engineer of the Daira Sanieh, Egypt: Cairo, Egypt.

1881. Wilson, John, Engineer, Great Eastern Railway, Liverpool Street Station, London, E.C. [*Wilson, Eastern, London.*]

1863. Wilson, John Charles, St. Werburgh's, Eversley Road, Bexhill-on-Sea.

1892. Wilson, John Charles Grant, Locomotive Superintendent, Manila Railway, Manila, Philippine Islands.

1879. Wilson, Joseph William, Principal of School of Practical Engineering, Crystal Palace, Sydenham, London, S.E.

1890. Wilson, Joseph William, Jun., Vice-Principal of School of Practical Engineering, Crystal Palace, Sydenham, London, S.E.

1880. Wilson, Robert, 10 St. Bride Street, London, E.C.; and 7 St. Andrew's Place, Regent's Park, London, N.W.

1883. Wilson, Robert, 7 Westminster Chambers, 13 Victoria Street, Westminster, S.W.

1890. Wilson, Robert James, 17 Kelvinhaugh Street, Glasgow.

1891. Wilson, Thomas, Morro Foundry, Iquique, Chile.

1873. Wilson, Thomas Sipling, Messrs. Holroyd Horsfield and Wilson, Larchfield Foundry, Hunslet Road, Leeds.

1888. Wilson, Walter Henry, Messrs. Harland and Wolff, Belfast.

1881. Wilson, Wesley William, Messrs. A. Guinness Son and Co., St. James' Gate Brewery, Dublin.

1891. Wimshurst, James Edgar, Messrs. William Esplen, Son, and Swainston, Billiter Buildings, 22 Billiter Street, London, E.C.

1890. Winder, Charles Aston, Messrs. Winder Brothers, Royds Works, Attercliffe, Sheffield.

1886. Windsor, Edwin Wells, 1 Rue du Hameau des Brouettes, Rouen, France.

1890. Wingfield, Digby Charles, Messrs. E. Beanes and Co., Falcon Works, Hackney Wick, London, N.E.

1887. Winmill, George, Locomotive and Carriage Superintendent, Oudh and Rohilkund Railway, Lucknow, India: (or Hare Street, Romford.)

1872. Winstanley, Robert, Mining Engineer, 28 Deansgate, Manchester.

1872. Wise, William Lloyd, 46 Lincoln's Inn Fields, London, W.C. [*Lloyd Wise, London.* 2766.]

1884. Withy, Henry, Messrs. Furness Withy and Co., Middleton Ship Yard, West Hartlepool. [*Withy, West Hartlepool.* 4.]

1878. Wolfe, John Edward, General Manager, Alagoas Railway, Maceio, Brazil: (or care of Rev. Prebendary Wolfe, Arthington, Torquay.)

1878. Wolfenden, Richard, 17 Dudley Street, Moss Side, Manchester.

1878. Wolfenden, Robert, Revenue Cutter "Ling Fêng," care of Commissioner of Customs, Shanghai, China: (or 17 Dudley Street, Moss Side, Manchester.)

1888. Wolff, Gustav William, M.P., Messrs. Harland and Wolff, Belfast.

1881. Wood, Edward Malcolm, 2 Westminster Chambers, 3 Victoria Street, Westminster, S.W.

1887. Wood, Henry, Messrs. John and Edward Wood, Victoria Foundry, Bolton.

1880. Wood, John Mackworth, Engineer's Department, New River Water Works, Clerkenwell, London, E.C.

1868. Wood, Lindsay, Mining Engineer, Southhill, near Chester-le-Street.

1885. Wood, Robert Henry, 15 Bainbrigge Road, Headingley, Leeds.

1884. Wood, Sidney Prescott, Semaphore Iron Works, Newport, Melbourne, Victoria: (or care of H. W. Little, Messrs. McKenzie and Holland, Vulcan Iron Works, Worcester.)

1890. Wood, Thomas Royle, Assistant Locomotive Superintendent, Sola Works, Ferro Carril del Sud, Buenos Aires, Argentine Republic.

1890. Wood, William, 4 Wolfington Road, West Norwood, London, S.E.

1882. Woodall, Corbet, Palace Chambers, 9 Bridge Street, Westminster, S.W.

1888. Woodford, Ethelbert George, Beira Railway, Port Beira, South Africa: (or care of George Pauling, 28 Victoria Street, Westminster, S.W.)

1887. Worger, Douglas Fitzgerald, Assistant Engineer, Southwark and Vauxhall Water Works, Southwark Bridge Road, London, S.E.

1874. Worsdell, Thomas William, Stonycroft, Arnside, near Carnforth.

1877. Worssam, Henry John, Messrs. G. J. Worssam and Son, Wenlock Road, City Road, London, N. [*Massrow, London.* 6656.]

1886. Worthington, Charles Campbell, Messrs. Henry R. Worthington, Hydraulic Works, 145 Broadway, New York, United States: (or care of the Worthington Pumping Engine Co., 153 Queen Victoria Street, London, E.C.)

1888. Worthington, Edgar, Messrs. Beyer Peacock and Co., Gorton Foundry, Manchester; and Mill Bank, Vicarage Lane, Bowdon, R.O., near Altrincham.

1860. Worthington, Samuel Barton, Consulting Engineer, 33 Princess Street, Manchester; and Mill Bank, Vicarage Lane, Bowdon, R.O., near Altrincham.

1866. Wren, Henry, Messrs. Henry Wren and Co., London Road Iron Works, Manchester. [*Wrens, Manchester.*]

1881. Wrench, John Mervyn, Chief Engineer, Indian Midland Railway, Jhansi, N.W. Provinces, India; and 39 Conduit Road, Bedford.

1876. Wright, James, Messrs. Ashmore Benson Pease and Co., Stockton-on-Tees. [*Wright, Gasholder, Stockton.* 12.]

1867. Wright, John Roper, Messrs. Wright Butler and Co., Elba Steel Works, Gower Road, near Swansea.

1859. Wright, Joseph, Metropolitan Railway-Carriage and Wagon Co., Saltley Works, Birmingham; and Arundel House, Lower Road, Richmond, Surrey.

1860. Wright, Joseph, 16 Great George Street, Westminster, S.W.; and Lawnswood, Alexandra Road, Upper Norwood, London, S.E.

1878. Wright, William Barton, Cambridge House, Dover.

1871. Wrightson, Thomas, M.P., Messrs. Head Wrightson and Co., Teesdale Iron Works, Stockton-on-Tees.

1891. Wroe, Joseph, 26 Park Avenue, Manchester, S.E.

1891. Wylde, Thomas, P.O. Box 455, Johannesburg, Transvaal, South Africa.

1886. Wylie, James, 83 Malvern Street, Stapenhill, Burton-on-Trent.

1865. Wyllie, Andrew, 1 Leicester Street, Southport.

1883. Wynne-Edwards, Thomas Alured, Agricultural Engineering Works, Denbigh. [*Foundry, Denbigh.*]

1877. Wyvill, Frederic Christopher, 19 East Parade, Leeds.

1889. Yarrow, Alfred Fernandez, Isle of Dogs, Poplar, London, E.

1878. Yates, Henry, Brantford, Ontario, Canada.

1882. Yates, Herbert Rushton, Assistant Engineer, Michigan Air Line Railway Extension, Pontiac, Michigan, United States: (or care of Henry Yates, Brantford, Ontario, Canada.)

1881. Yates, Louis Edmund Hasselts, District Locomotive and Carriage Superintendent, Eastern Bengal State Railway, Saidpore, Bengal, India: (or care of Rev. H. W. Yates, 98 Lansdowne Place, Brighton.)

1880. York, Francis Colin, Locomotive Superintendent, Buenos Aires and Pacific Railway, Junin, Buenos Aires, Argentine Republic: (or care of Messrs. Samuel York Sons and Co., Snow Hill, Wolverhampton.)

1889. Young, David, 11 and 12 Southampton Buildings, London, W.C.

1879. Young, George Scholey, Engineer, Thames Iron Works, Orchard Yard, Blackwall, London, E.

1874. Young, James, Managing Engineer, Lambton Engine and Iron Works, Fence Houses.

1879. Young, James, Salroyd, 21 Cambalt Road, Putney, London, S.W.

1892. Young, Robert, Superintending Engineer, Penang Steam Tramways, Penang, Straits Settlements.

1887. Young, William Andrew, Messrs. Lobnitz and Co., Renfrew, near Paisley. [*Lobnitz, Renfrew.* 57, *Paisley.*]

1881. Younger, Robert, Messrs. R. and W. Hawthorn Leslie and Co., St. Peter's Works, Newcastle-on-Tyne.

ASSOCIATES.

1880. Allen, William Edgar, Imperial Steel Works, Cross George Street, Sheffield.

1881. Barcroft, Henry, Bessbrook Spinning Works, County Armagh, Ireland; and The Glen, Newry, Ireland.

1889. Barr, John, Glenfield Engineering Works, East Shaw Street, Kilmarnock.

1886. Bennison, William Clyburn, Messrs. Samuel Osborn and Co., Clyde Steel and Iron Works, Sheffield.

1890. Birch, John Grant, 10 and 11 Queen Street Place, London, E.C.

1892. Bowman, Frederic Hungerford, D.Sc., F.R.S.E., Ash Leigh, Ashley Heath, Bowdon, R.O., near Altrincham.

1888. Brown, Harold, 2 Bond Court, Walbrook, London, E.C.

1890. Burt, John Mowlem, Messrs. John Mowlem and Co., 19 Grosvenor Road, Pimlico, London, S.W.

1892. Carpenter, Henry James, Messrs. George Kent and Co., 200 High Holborn, London, W.C.

1891. Carter, Frederick Heathcote, 9 Oxford Street, Manchester. [*Girder, Manchester.*]

1889. Castle, Frederick George, The People's Palace Technical Schools, Mile End Road, London, E.

1889. Chamberlain, John George, Messrs. Joseph Wright and Co., Neptune Forge, Tipton.

1888. Chrimes, Charles Edward, Messrs. Guest and Chrimes, Brass Works, Rotherham.

1887. Chubb, Edward George, Ironbridge Gas Works, Ironbridge, R.S.O., Shropshire.

1890. Chubb, Richard, Messrs. Gillison and Chadwick, 10 Tower Buildings, Liverpool.

1879. Clowes, Edward Arnott, Messrs. William Clowes and Sons, Duke Street, Stamford Street, London, S.E. [*Clowes, London.* 4558.]

1892. Cooper, Thomas Lancelot Reed, 267 Temple Chambers, Temple Avenue, London, E.C.

1891. Cornett, James Porteus, Ford Paper Works, near Sunderland. [*Ford, Hylton. Nor. Dis.* 135.]

1892. Cryer, Arthur, University College, Cardiff.

1892. Davis, George Brown, Palace Wharf, Stangate, London, S.E.

1890. Day, Arthur Godfrey, Director of Studies, Science Art and Technical Schools, Bath.

1892. Fauvel, Charles James, 15 George Street, Mansion House, London, E.C.

1891. Foster, George, Hecla Foundry Steel Works, Sheffield; and 40 Dixon Street, Rotherham.

1889. Golby, Frederick William, 36 Chancery Lane, London, W.C.

1889. Götz, Carl Johann Wilhelm, Messrs. John M. Sumner and Co., 2 Brazennose Street, Manchester.

1889. Gregory, George Francis, Boarzell, Hawkhurst.

1887. Hind, Enoch, Edgar Rise, Nottingham.

1891. Jackman, Joseph, Persberg Steel Works, Pothouse Road, Attercliffe, Sheffield. [*Persberg, Sheffield.* 94.]

1884. Jackson, Edward, Midland Railway-Carriage and Wagon Works, Birmingham. [*Wagon, Birmingham.*]

1882. Jackson, William, Kingston Cotton Mill, Hull. [*Cotton, Hull.*]

1891. Jennings, George Henry, Stangate, Lambeth, London, S.E. [*Jennings, London.* 4680.]

1890. Jennings, Sidney, Stangate, Lambeth, London, S.E. [*Jennings, London.* 4680.]

1884. Livesey, Joseph Montague, Stourton Hall, Horncastle.

1865. Longsdon, Alfred, 9 New Broad Street, London, E.C.

1881. Lowood, John Grayson, Gannister Works, Attercliffe Road, Sheffield. [*Lowood, Sheffield.* 131.]

1886. Mackenzie, Keith Ronald, Gillotts, Henley-on-Thames.

1868. Matthews, Thomas Bright, Messrs. Turton Brothers and Matthews, Phœnix Steel Works, Sheffield. [*Matthews, Sheffield.*]

1890. McGillivray, William, Messrs. Austin McGillivray and Co., Falcon Works, Sheffield. [*Austin, Sheffield.*]

1889. McKinnel, William, Messrs. Samuel Osborn and Co., Clyde Steel and Iron Works, Sheffield.

1891. McMeekin, Adam, Cogry Flax Spinning Mills, Doagh R.S.O., Co. Antrim, Ireland.

1890. Meggitt, Samuel Newton, Messrs. Ibbotson Brothers and Co., Globe Steel Works, Sheffield.

1889. Miles, William Henry, 23 Barnato Buildings, and P.O Box 1860, Johannesburg, Transvaal, South Africa.

1891. Monie, Hugh, Jun., Textile Section, Victoria Jubilee Technical Institute, Byculla, Bombay, India: (or care of Hugh Monie, Springfield, Belfast.)

1892. Morley, John, Sanitary Engineering Works, Palace Wharf, Stangate, London, S.E.

1889. Nasmith, Joseph, 4 Arcade Chambers, St. Mary's Gate, Manchester.

1887. Neville, Edward Hermann, Fabrica del Piles, Gijon, Spain.

1886. Newton, Henry Edward, 6 Bream's Buildings, Chancery Lane, London, E.C.

1874. Paget, Berkeley, Low Moor Iron Office, 2 Laurence Pountney Hill, Cannon Street, London, E.C. [*Gryphon, London.*]

1886. Peacock, William J. P., Wells Street, Oxford Street, London, W.; and 41 St. James' Street, London, S.W.

1888. Peake, Robert Cecil, Cumberland House, Redbourn, near St. Albans.

1887. Peech, Henry, Phœnix Bessemer Steel Works, near Sheffield.

1887. Peech, William Henry, Phœnix Bessemer Steel Works, near Sheffield.

1890. Perry, Edwin, Queen Street, Newcastle-on-Tyne.

1884. Phillips, Richard Morgan, 21 to 24 State Street, New York, United States. [*Sarita, New York.*]

1891. Pirrie, John Barbour, Barn Flax Spinning Mills, Carrickfergus, Co. Antrim, Ireland.

1891. Plant, George, Moseley Road School, Birmingham.

1891. Rankin, Thomas Thomson, Principal, Coatbridge Technical School and West of Scotland Mining College, Coatbridge.

1886. Raven, Henry Baldwin, Messrs. Hare and Co., Temple Chambers, Temple Avenue, London, E.C.

1892. Reed, Ernest Charles, Thames Paper Mills, Purfleet, Essex.

1882. Ridehalgh, George John Miller, Fell Foot, Newby Bridge, Ulverston.

1891. Rochfort, Bertram, Rua da Guitanda 55, Rio de Janeiro, Brazil.

1891. Rowcliffe, William Charles, 1 Bedford Row, London, W.C.

1888. Rowell, John Henry, New Brewery, High Street, Gateshead.

1883. Sandham, Henry, Keeper, Science and Art Department, South Kensington Museum, London, S.W.

1875. Schofield, Christopher J., Vitriol and Alkali Works, Clayton, near Manchester.

1890. Schofield, John William, Messrs. Gregory and Bramall, Soho Steel and File Works, Sheffield.

1887. Scott, Walter, Victoria Chambers, Grainger Street West, Newcastle-on-Tyne. [*Contractor, Newcastle-on-Tyne.*]

1891. Spencer, Francis Henry, Robinson Gold Mining Co., Johannesburg, Transvaal, South Africa.

1878. Stalbridge, The Right Hon. Lord, 12 Upper Brook Street, Grosvenor Square, London, W.

1892. Stead, John Edward, 5 Zetland Road, Middlesbrough.

1886. Stumore, Frederick, 34 Leadenhall Street, London, E.C.

1890. Taylor, John, 99 and 101 Fonthill Road, Finsbury Park, London, N.; and Stockport.

1884. Tilfourd, George, Messrs. Samuel Osborn and Co., Clyde Steel and Iron Works, Sheffield.

1887. Tozer, Edward Sanderson, Phœnix Bessemer Steel Works, near Sheffield.

1869. Varley, John, Leeds Forge, Leeds.

1878. Watson, Joseph, Patent Office, 25 Southampton Buildings, London, W.C.

1892. Whitehead, Richard David, Municipal Technical College, Green Hill, Derby.

1892. Widdows, Francis R., Messrs. Colman's Mustard Mills, Carrow Works, Norwich.

1883. Williamson, Robert S., Cannock and Rugeley Collieries, Hednesford, near Stafford.

1891. Wiseman, Edmund, Cheapside and John Street, Luton. [*Wiseman, Luton.*]

GRADUATES.

1884. Adam, Frank, Sir W. G. Armstrong Mitchell and Co., Elswick, Newcastle-on-Tyne; and 193 Albion Road, Stoke Newington, London, N.

1892. Adams, Sidney Rickman, London and South Western Railway, Nine Elms, London, S.W.

1885. Addis, Frederick Henry, District Locomotive Superintendent, Rajputana Malwa Railway, Mhow, India: (or care of Messrs. Grindlay and Co., 55 Parliament Street, London, S.W.)

1890. Alderson, George Alexander, 13 Cheviot Street, Lincoln.

1874. Allen, Francis, Messrs. Allen Alderson and Co., Gracechurch Street, Alexandria, Egypt: (or care of Messrs. Stafford Allen and Sons, 7 Cowper Street, Finsbury, London, E.C.)

1882. Anderson, William, Locomotive Department, North Eastern Railway, Dairycoates, Hull.

1878. Appleby, Charles, Jun., 89 Cannon Street, London, E.C.

1889. Ashford, John, Messrs. Carr and Ashford, Apollo Works, Blews Street, Newtown Row, Birmingham; and Lyndale, Hall Road, Handsworth, near Birmingham.

1890. Aubin, Percy Adrian, 29 St. James' Street, St. Helier's, Jersey.

1888. Bailey, Wilfred Daniel, India-rubber Gutta-percha and Telegraph Works, Casilla de Correo 1212, Buenos Aires, Argentine Republic.

1888. Barker, Eric Gordon, Locomotive Superintendent, Wirral Railway, Docks Station, Birkenhead; and Guyse House, Oxton, R.O., near Birkenhead.

1889. Barrow, Arthur Robert Maclean, Locomotive Department, North Western Railway, Sukkur, India: (or care of A. M. Barrow, 13 Upper Maze Hill, St. Leonard's-on-Sea.)

1882. Barstow, Thomas Hulme, Manager, Kaihu Railway, Dargaville, Auckland, New Zealand.

1888. Bell, Alexander Dirom, The Woll, Hawick.

1884. Bell, Robert Arthur, Locomotive Department, South Indian Railway, Negapatam, India: (or care of Mrs. Bell, 30 Brompton Crescent, London, S.W.)

1890. Bell, William Thomas, Messrs. Robey and Co., 10 Kaiser Wilhelmstrasse, Breslau, Germany.

1880. Birkett, Herbert, Calle Lima 1159, Buenos Aires, Argentine Republic: (or 62 Green Street, Grosvenor Square, London, W.)

1884. Bocquet, Harry, care of Arthur E. Shaw, Estacion Central, Buenos Aires, Argentine Republic: (or care of Mrs. Bocquet, Llanwye, Hampton Park, Hereford.)

1888. Bradley, Arthur Ashworth, Robinson Gold Mining Co., P.O. Box 787, Johannesburg, Transvaal, South Africa: (or care of Rev. Gilbert Bradley, St. Edmund's Vicarage, Dudley, Worcestershire.)

1887. Bremner, Bruce Laing, 21 Langworthy Road, Manchester: (or Streatham House, Canaan Lane, Edinburgh.)

1892. Bromly, Alfred Hammond, Castell Carn Dochan Gold Mine, Llanuwchllyn, near Bala, Merionethshire; and 359 Amhurst Road, Stoke Newington, London, N.

1878. Brooke, Arthur, Post Office, Otahuhu, Auckland, New Zealand: (or care of Miss Helen Brooke, Sunnymead, The Rise, Sidcup, S.O., Kent.)

1890. Brousson, Robert Percy, Messrs. Woodhouse and Rawson, Cadby Hall Works, Hammersmith, London, W.

1889. Brown, Arthur Selwyn, Hayes Street, Neutral Bay, Sydney, New South Wales.

1880. Buckle, William Harry Ray, 11 Billiter Buildings, 49 Leadenhall Street, London, E.C.

1886. Budenberg, Christian Frederick, Messrs. Schäffer and Budenberg, 1 Southgate, St. Mary's Street, Manchester; and Bowden Lane, Marple, Stockport. [*Manometer, Manchester*. 899.]

1892. Bulwer, Ernest Henry Earle, Messrs. George Fletcher and Co., Poplar Iron Works, King Street, Poplar, London, E.

1890. Burne, Edward Lancaster, 21 Great College Street, Westminster, S.W.

1879. Burnet, Lindsay, Moore Park Boiler Works, Govan, near Glasgow. [*Burnet, Glasgow*. 1513.]

1891. Butcher, Walter Edward, Locomotive Department, London Brighton and South Coast Railway, Brighton; and 4 Exeter Street, Brighton.

1891. Buttenshaw, George Eskholme, Messrs. Guest and Chrimes, Brass Works, Rotherham.

1889. Calastremé, John Carlos, Locomotive Department, Midland Railway, Kentish Town, London, N.W.

1891. Caswell, Charles Henry, Naval Construction and Armaments Works, Barrow-in-Furness.

1889. Challen, Walter Bernard, Messrs. Taylor and Challen, Derwent Foundry, 60 and 62 Constitution Hill, Birmingham.

1890. Chatwood, Arthur Brunel, Lancashire Safe and Lock Works, Bolton.

1891. Church, Harry, Ingeniero, para la Colonia San Gustavo, La Paz, Entre Rios, South America: (or care of George Church, Willington, Bedford.)

1890. Cleeves, John Frederick, Rotherham. [*Cleeves, Rotherham.*]

1892. Cleverly, William Bartholomew, Jun., Lambeth Water Works, Brixton, London, S.W.

1885. Clift, Leslie Everitt, 1 Holborn Place, High Holborn, London, W.C.

1883. Clinkskill, Alfred Alphonse Rouff, Messrs. James Clinkskill and Son, 1 Holland Place, St. Vincent Street, Glasgow.

1892. Collingridge, Harvey, Messrs. Fawkner Rogers and Co., 11 and 12 New Bridge Street, London, E.C.; and The Elms, Hornsey, London, N.

1889. Cook, George Norcliffe, Messrs. Thomas Firth and Sons, Norfolk Works, Sheffield.

1888. Cox, Herbert Henry, Hillside, Falmouth.

1887. Crosland, DelevanteWilliam, 22 Royal Crescent, Kensington, London, W.

1891. Cutler, Samuel, Jun., Messrs. Samuel Cutler and Sons, Providence Iron Works, Millwall, London, E.

1890. Davidson, Albert, Messrs. Hattersley and Davidson, Arundel Engineering Works, 14 and 16 Arundel Street, Sheffield.

1884. Dixon, John, Eastwood Villa, Lytham, near Preston, Lancashire.

1891. Douglass, Alfred Edwards, South Staffordshire Water Works, Paradise Street, Birmingham.

1891. Drummond, Hawtrey Marks, care of Mrs. Drummond, 145 Bruntsfield Place, Edinburgh.

1868. Dugard, William Henry, Messrs. Dugard Brothers, Vulcan Rolling Mills, Bridge Street West, Summer Lane, Birmingham. [*Vulcan, Birmingham.*]

1891. Duncan, Martin Gordon, Lexden, 1 Elmfield Road, Upper Tooting, London, S.W.

1892. Edgcome, James Edmund, Notting Hill Electric Light Co., Bulmer Place, Notting Hill Gate, London, W.

1891. Edwards, Herbert Francis, Messrs. Forster Brown and Rees, Guild Hall Chambers, Cardiff.

1885. Edwards, Walter Cleeve, Assistant Engineer, Midland Railway, Greymouth, New Zealand.

1887. England, William Henry, 53 Hillary Street, Leeds.

1892. Fletcher, Joseph Ernst, Messrs. Thomas Firth and Sons, Norfolk Steel Works, Sheffield.

1890. Garrett, Frank, Jun., Messrs. Richard Garrett and Sons, Leiston Works, Leiston, R.S.O., Suffolk.

1891. Gillatt, Thomas Stanley, North British Railway Works, Cowlairs, Glasgow.

1891. Gregory, Henry Hodges Mogg, Messrs. Sanders and Co., Rivington Works, Rivington Street, Great Eastern Street, London, E.C.

1890. Hatton, Thomas Reginald, Murdoch House, Kington, Herefordshire.

1889. Hayward, Robert Francis, Messrs. Crompton and Co., Arc Works, Chelmsford.

1877. Heaton, Arthur, Messrs. Heaton and Dugard, Metal and Wire Works, Shadwell Street, Birmingham. [*Heagard, Birmingham.*]

1874. Hedley, Thomas, P. O. Box 19, Portland, Oregon, United States.

1883. Hill, John Kershaw, Engineer and Manager, West Surrey Water Works, High Street, Walton-on-Thames.

1891. Hodgson, William James, Central Chemical Co., 182 London Road, Nottingham.

1887. Hogg, William, Craigmore, Blackrock, Dublin.

1867. Holland, George, Mechanical Department, Grand Trunk Railway, Montreal, Canada.

1884. Holt, Follett, Ferro Carril Buenos Aires y Rosario, La Banda, Argentine Republic: (or care of Robert Hallett Holt, Land Registry, Staple Inn, Holborn, London, W.C.)

1889. Hosgood, Thomas Watkin, Sketty, near Swansea.

1891. Hosgood, Walter James, Locomotive Department, Barry Dock and Railways, Barry, near Cardiff.

1889. Hosken, Arthur Fayrer, Messrs. Neilson and Co., Hyde Park Locomotive Works, Glasgow.

1889. Howard, Geoffrey, Britannia Iron Works, Bedford.

1883. Howard, Harry James, Messrs. Colman's Mustard Mills, Carrow Works, Norwich.

1891. Hughes, Edward Sinclair Bremner, care of Mrs. McCreery, 1 Kelvinside Terrace North, Glasgow; and Ericstane, Helensburgh.

1890. Jones, Arthur Dansey, Locomotive Works, Lancashire and Yorkshire Railway, Horwich, near Bolton.

1891. Jordan, Frederic William, 42 Wells Street, Mortimer Street, Cavendish Square, London, W.

1889. Joy, Basil Humbert, 9 Victoria Chambers, 17 Victoria Street, Westminster, S.W.; and Manor Road House, Beckenham.

1883. Lander, Philip Vincent, Lyndhurst, Hampton Wick, R.O., Kingston-on-Thames: (or care of W. W. Lander, Imperial Ottoman Bank, 26 Throgmorton Street, London, E.C.)

1881. Lawson, James Ibbs, Resident Engineer, New Zealand Railways, Invercargill, Otago, New Zealand.

1886. Lewis, William Thomas, Jun., Engineer's Office, Bute Docks, Cardiff; and Langholm, Dinas Powis, near Cardiff.

1881. Macdonald, Ranald Mackintosh, Messrs. Booth Macdonald and Co., Carlyle Engineering and Implement Works, Christchurch, New Zealand; and P.O. Box 267, Christchurch, New Zealand: (or care of Messrs. Redfern Alexander and Co., 3 Great Winchester Street, London, E.C.)

1883. Mackenzie, Thomas Brown, Messrs. J. Copeland and Co., Pulteney Street Engine Works, Glasgow; and 342 Duke Street, Glasgow.

1883. Malan, Ernest de Mérindol, Signal and Telegraph Department, Hull Barnsley and West Riding Junction Railway and Dock Co., Alexandra Dock, Hull. [*Engineer, Deepdock, Hull. Nat.* 106.]

1868. Mappin, Frank, Messrs. Thomas Turton and Sons, Sheaf Works, Sheffield.

1892. Marks, Alfred Pally, 155 Adelaide Road, London, N.W.

1889. Marshall, Frank Theodore, Messrs. R. and W. Hawthorn Leslie and Co., St. Peter's Works, Newcastle-on-Tyne.

1888. Marten, Hubert Bindon, Contractor's Office, Great Western Railway, Pangbourne, near Reading; and Pedmore, Stourbridge.

1886. Mattos, Alvaro Gomes de, 98 Rua da Sande, Rio de Janeiro, Brazil: (or care of Messrs. Fry Miers and Co., Suffolk House, 5 Laurence Pountney Hill, London, E.C.)

1887. May, Harold Milton, Lakemead, Totnes.

1892. Miles, Frederick Hudson, Metropolitan Railway Works, Neasden, London, N.W.

1891. Mills, Matthew William, Moss Foundry, Heywood, near Manchester.

1867. Mitchell, John, Swaithe Hall, Barnsley.

1868. Moor, William, Ocean House, Hartlepool.

1892. Murray, David James, 54 Park Street, Greenheys, Manchester.

1878. Newall, John Walker, Suffolk House, Laurence Pountney Hill, London, E.C.

1883. O'Connor, John Frederick, 16 and 18 Exchange Place, New York, United States.

1883. Osborn, William Fawcett, Messrs. Samuel Osborn and Co., Clyde Steel and Iron Works, Sheffield.

1892. Osmond, Frederick John, Whitworth Works, Rea Street South, Birmingham.

1881. Oswell, William St. John, 121 Calle Defensa, Buenos Aires, Argentine Republic.

1883. Palchoudhuri, Bipradas, Moheshgunj Factory, Krishnugher, Bengal.

1887. Paterson, John Edward, Chief Mechanical Engineer's Office, New South Wales Government Railways, Wilson Street, Eveleigh, Sydney, New South Wales.

1892. Payton, Frank John, Boiler Insurance and Steam Power Co., 67 King Street, Manchester.

1890. Philipson, John, Jun., Messrs. Atkinson and Philipson, 27 Pilgrim Street, Newcastle-on-Tyne.

1884. Philipson, William, Messrs. Atkinson and Philipson, 27 Pilgrim Street, Newcastle-on-Tyne. [*Carriage, Newcastle-on-Tyne.* 415.]

1890. Powell, Frederick, York House, Malvern Link, Malvern.

1892. Power, Arthur Cyril, 2 Rheinstein Villas, Church Road, Teddington, S.O., Middlesex.

1887. Price-Williams, John Morgan, 19 Elsworthy Road, Primrose Hill, London, N.W.

1892. Ransom, Herbert Byrom, Messrs. Manlove Alliott and Co., 57 Gracechurch Street, London, E.C.

1892. Redfern, Charles George, 122 Bethune Road, Stamford Hill, London, N.

1884. Reynolds, Thomas Blair, 28 Victoria Street, Westminster, S.W.

1892. Ridley, James Cartmell, Jun., 3 Summerhill Grove, Newcastle-on-Tyne.

1885. Ripley, Philip Edward, Messrs. Ransomes Sims and Jefferies, Orwell Works, Ipswich.

1889. Roope, Walter, Stisted, Badulla, Ceylon: (or care of Mrs. Roope, Hangerfield, Witley, Godalming.)

1884. Roux, Paul Louis, 54 Boulevard du Temple, Paris.

1888. Rümmele, Alfredo, 17 Via Principe Umberto, Milan, Italy.

1890. Sanders, Percy Henry, Messrs. H. G. Sanders and Son, Victoria Works, Victoria Gardens, Notting Hill Gate, London, W.

1890. Saxelby, Herbert Raffaelle, Messrs. J. Copeland and Co., Poultney Street Engine Works, Dobbies Loan, Glasgow.

1892. Scarfe, George Norman, Fursby House, Finchley, London, N.

1881. Scott, Ernest, Close Works, Newcastle-on-Tyne. [*Esco, Newcastle-on-Tyne.* 432.]

1892. Seymour, William Frederick Earl, care of Captain J. Seymour, Dunkeld, Newlands Park, Sydenham, London, S.E.

1892. Shepherd, James Horace, Great Western Railway, Swindon.

1886. Silcock, Charles Whitbread, Wood House, Ely.

1887. Simkins, Charles Wickens, Jun., Amguri Tea Estate, Amguri Post Office, Sibsagar, Assam, India: (or care of Charles W. Simkins, The Lodge, Lowdham, near Nottingham.)

1889. Smith, Henry Buckley Bingham, Messrs. Sharp Stewart and Co., Atlas Works, Glasgow.

1891. Smith, Joseph Philip Grace, Polytechnic School of Engineering, 309 Regent Street, London, W.

1891. Snell, John Francis Cleverton, Albert Electric Lighting Co., Albert Hall Estate, South Kensington, London, S.W.; and 85 Peterborough Road, Fulham, London, S.W.

1892. Stokes, Frank Torrens, Locomotive Department, North London Railway, Bow, London, E.

1883. Swale, Gerald, The Cedars, Wiarton, Ontario, Canada: (or care of H. J. Swale, Ingfield Hall, Settle.)

1887. Tabor, Edward Henry, Fennes, Braintree.

1889. Tangye, Harold Lincoln, Messrs. Tangyes, Cornwall Works, Soho, near Birmingham.

1885. Tangye, John Henry, Messrs. Tangyes, Cornwall Works, Soho, near Birmingham.

1884. Taylor, Joseph, 24 Hawthorn Grove, Heaton Moor, near Stockport.

1884. Taylor, Maurice, 39 Rue de Lisbonne, Paris.

1884. Templeton, Edwin Arthur Slade, 42 Boscombe Road, Shepherd's Bush, London, W.

1889. Treharne, Gwilym Alexander, Pontypridd; and Aberdare.

1891. Vaizey, John Leonard, Locomotive Works, Great Eastern Railway, Stratford, London, E.; and 6 Grove Crescent, Stratford, London, E.

1892. Vezey, Albert Edward, Electrical Department, London and North Western Railway Works, Crewe.

1878. Waddington, John, Jun., 35 King William Street, London Bridge, London, E.C.

1888. Waddington, Samuel Sugden, 35 King William Street, London Bridge, London, E.C.

1885. Wakefield, William Marsden, Buckingham House, Government Place, Calcutta, India.

1884. Walker, Ralph Teasdale, Fabrick Olean, Sitoebondo, Java: (or Kaliemaas, Alleyne Park, West Dulwich, London, S.E.)

1892. Wallis, William Wallace, Messrs. George Fletcher and Co., Masson and Atlas Works, Litchurch, Derby.

1888. Waring, Henry, Engineer, Dublin Laundry Co., Milltown, near Dublin; and 40 Frankfort Avenue, Rathgar, Dublin.

1892. Warton, Richard George Frank, Wallsend Pontoon Works, Bute Docks, Cardiff.

1886. Wesley, Joseph A., Clarke's Crank and Forge Works, Lincoln.

1883. Westmacott, Henry Armstrong, Sir W. G. Armstrong Mitchell and Co., Elswick Works, Newcastle-on-Tyne; and Benwell Hill, Newcastle-on-Tyne.

1880. Weymouth, Francis Marten, 3 Goldsworth Villas, Heathfield Road, Wandsworth, London, S.W.

1888. Whichello, Richard, Messrs. Max Nothmann and Co., Rio de Janeiro, Brazil: (or 44 Trumpington Street, Cambridge.)

1889. Wigham, John Cuthbert, Messrs. J. Edmundson and Co., Stafford Works, 35 Capel Street, Dublin.

1892. Williams, Arthur Edward, Messrs. Bramwell and Harris, 5 Great George Street, Westminster, S.W.

1889. Willis, Edward Turnley, 99 Shooter's Hill Road, Blackheath, London, S.E.

1890. Wilson, Alexander Cowan, Osgathorpe Hills, Sheffield.

1889. Winkfield, Richard Ernest, Electric Light Department, Great Western Railway, Paddington, London, W.

1890. Winmill, Hallett, 14 Hamfrith Road, Stratford, London, E.

1887. Wrench, John Henry Kirke, 80 Bismark Court, Chicago, United States: (or care of E. M. Wrench, Park Lodge, Baslow, Chesterfield.)

1889. Wright, Howard Theophilus, 16 Great George Street, Westminster, S.W.

1890. Wright, William Carthew, General Post Office, Melbourne, Victoria: (or care of Dr. Gaskoin Wright, 253 Eccles New Road, Salford, Manchester.)

1888. Yates, Edward, Wolverton Road, Stony Stratford.

1891. Yerbury, Frederick Augustus, 17 Victoria Street, Westminster, S.W.

THE INSTITUTION OF MECHANICAL ENGINEERS.

Memorandum of Association.

AUGUST 1878.

1st. The name of the Association is "THE INSTITUTION OF MECHANICAL ENGINEERS."

2nd. The Registered Office of the Association will be situate in England.

3rd. The objects for which the Association is established are:—

(A.) To promote the science and practice of Mechanical Engineering and all branches of mechanical construction, and to give an impulse to inventions likely to be useful to the Members of the Institution and to the community at large.

(B.) To enable Mechanical Engineers to meet and to correspond, and to facilitate the interchange of ideas respecting improvements in the various branches of mechanical science, and the publication and communication of information on such subjects.

(C.) To acquire and dispose of property for the purposes aforesaid.

(D.) To do all other things incidental or conducive to the attainment of the above objects or any of them.

4th. The income and property of the Association, from whatever source derived, shall be applied solely towards the promotion of the objects of the Association as set forth in this Memorandum of Association, and no portion thereof shall be paid or transferred directly or indirectly, by way of dividend, bonus, or otherwise howsoever, by way of profit to the persons who at any time are or have been Members of the Association, or to any of them, or to any person claiming through any of them: Provided that nothing herein contained shall prevent the payment in good faith of remuneration to any officers or servants of the Association, or to any Member of the Association, or other person, in return for any services rendered to the Association, or prevent the giving of privileges to the Members of the Association in attending the meetings of the Association, or prevent the borrowing of money (under such powers as the Association and the Council thereof may possess) from any Member of the Association, at a rate of interest not greater than five per cent. per annum.

5th. The fourth paragraph of this Memorandum is a condition on which a licence is granted by the Board of Trade to the Association in pursuance of Section 23 of the Companies Act 1867. For the purpose of preventing any evasion of the terms of the said fourth paragraph, the Board of Trade may from time to time, on the application of any Member of the Association, impose further conditions, which shall be duly observed by the Association.

6th. If the Association act in contravention of the fourth paragraph of this Memorandum, or of any such further conditions, the liability of every Member of the Council shall be unlimited; and the liability of every Member of the Association who has received any such dividend, bonus, or other profit as aforesaid, shall likewise be unlimited.

7th. Every Member of the Association undertakes to contribute to the Assets of the Association in the event of the same being wound up during the time that he is a Member, or within one year afterwards, for payment of the debts and liabilities of the Association contracted before the time at which he ceases to be a Member, and of the costs, charges, and expenses for winding up the same, and for the adjustment of the rights of the contributories amongst themselves, such amount as may be required not exceeding Five Shillings, or in case of his liability becoming unlimited such other amount as may be required in pursuance of the last preceding paragraph of this Memorandum.

8th. If upon the winding up or dissolution of the Association there remains, after the satisfaction of all its debts and liabilities, any property whatsoever, the same shall not be paid to or distributed among the Members of the Association, but shall be given or transferred to some other Institution or Institutions having objects similar to the objects of the Association, to be determined by the Members of the Association at or before the time of dissolution; or in default thereof, by such Judge of the High Court of Justice as may have or acquire jurisdiction in the matter.

Articles of Association.

AUGUST 1878.

INTRODUCTION.

Whereas an Association (hereinafter called "the existing Institution") called "The Institution of Mechanical Engineers" has long existed for objects similar to the objects expressed in the Memorandum of Association of the Association (hereinafter called "the Institution") to which these Articles apply, and the existing Institution consists of Members, Graduates, Associates, and Honorary Life Members, and is possessed of books, drawings, and property used for the objects aforesaid;

And whereas the Institution is formed for furthering and extending the objects of the existing Institution, by a registered Association, under the Companies Acts 1862 and 1867; and terms used in these Articles are intended to have the same respective meanings as they have when used in those Acts, and words implying the singular number are intended to include the plural number, and *vice versâ*;

Now THEREFORE IT IS HEREBY AGREED as follows:—

CONSTITUTION.

1. For the purpose of registration the number of Members of the Institution is unlimited.

MEMBERS.

2. The subscribers of the Memorandum of Association, and such other persons as shall be admitted in accordance with these Articles, and none others, shall be Members of the Institution, and be entered on the register as such.

3. Any person may become a Member of the Institution who, being a Member of the existing Institution, shall agree to transfer his membership of the existing Institution, and all rights and obligations incidental thereto, to the Institution, and to be registered as a Member of the Institution accordingly.

4. Any person may become a Member of the Institution who shall be qualified and elected as hereinafter mentioned, and shall agree to become such Member, and shall pay the entrance fee and first subscription accordingly.

5. The rights and privileges of every Member of the Institution shall be personal to himself, and shall not be transferable or transmissible by his own act or by operation of law.

QUALIFICATION AND ELECTION OF MEMBERS.

6. The qualification of Members shall be prescribed by the By-laws from time to time in force, as provided by the Articles.

7. The election of Members shall be conducted as prescribed by the By-laws from time to time in force, as provided by the Articles.

GRADUATES, ASSOCIATES, AND HONORARY LIFE MEMBERS.

8. Any person may become a Graduate, Associate, or Honorary Life Member of the Institution, who, being already a Graduate, Associate, or Honorary Life Member of the existing Institution, shall agree to transfer his interest in the existing Institution, and all rights and obligations incidental thereto, to the Institution.

9. The Institution may admit such other persons as may be hereafter qualified and elected in that behalf as Graduates, Associates, and Honorary Life Members respectively of the Institution, and may confer upon them such privileges as shall be prescribed by the By-laws from time to time in force, as provided by the Articles: Provided that no Graduate, Associate, or Honorary Life Member shall be deemed to be a Member within the meaning of the Articles.

10. The qualification and mode of election of Graduates, Associates, and Honorary Life Members, shall be prescribed by the By-laws from time to time in force, as provided by the Articles.

ENTRANCE FEES AND SUBSCRIPTIONS.

11. The Entrance Fees and Subscriptions of Members, Graduates, and Associates, shall be prescribed by the By-laws from time to time in force, as provided by the Articles: Provided that no Entrance Fee shall be payable by a Member, Graduate, or Associate of the existing Institution.

EXPULSION.

12. If any Member, Graduate, or Associate shall leave his subscription in arrear for two years, and shall fail to pay such arrears within three months after a written application has been sent to him by the Secretary, his name may be struck off the list of Members, Graduates, or Associates, as the case may be, by the Council, at any time afterwards, and he shall thereupon cease to

have any rights as a Member, Graduate, or Associate, but he shall nevertheless continue liable to pay the arrears of subscription due at the time of his name being so struck off: Provided always, that this regulation shall not be construed to compel the Council to remove any name if they shall be satisfied the same ought to be retained.

13. The Council may refuse to continue to receive the subscriptions of any person who shall have wilfully acted in contravention of the regulations of the Institution, or who shall in the opinion of the Council have been guilty of such conduct as shall have rendered him unfit to continue to belong to the Institution; and may remove his name from the list of Members, Graduates, or Associates (as the case may be), and such person shall thereupon cease to be a Member, Graduate, or Associate (as the case may be) of the Institution.

GENERAL MEETINGS.

14. The first General Meeting shall be held on such day, within four months of the registration of the Institution, as the Council shall determine. Subsequent General Meetings shall consist of the Ordinary Meetings, the Annual General Meeting, and of Special Meetings as hereinafter defined.

15. The Annual General Meeting shall take place in London in one of the first four months of every year. The Ordinary Meetings shall take place at such times and places as the Council shall determine.

16. A Special Meeting may be convened at any time by the Council, and shall be convened by them whenever a requisition signed by twenty Members of the Institution, specifying the object of the Meeting, is left with the Secretary. If for fourteen days after the delivery of such requisition a Meeting be not convened in accordance therewith, the Requisitionists or any twenty Members

of the Institution may convene a Special Meeting in accordance with the requisition. All Special Meetings shall be held in London.

17. Seven clear days' notice of every Meeting, specifying generally the nature of any special business to be transacted at any Meeting, shall be given to every Member of the Institution, and no other special business shall be transacted at such Meeting; but the non-receipt of such notice shall not invalidate the proceedings of such Meeting. No notice of the business to be transacted (other than such ballot lists as may be requisite in case of elections) shall be required in the absence of special business.

18. Special business shall include all business for transaction at a Special Meeting, and all business for transaction at every other Meeting, with the exception of the reading and confirmation of the Minutes of the previous Meeting, the election of Members, Graduates, and Associates, and the reading and discussion of communications as prescribed by the By-laws, or any regulations of the Council made in accordance with the By-laws.

PROCEEDINGS AT GENERAL MEETINGS.

19. Twenty Members shall constitute a quorum for the purpose of a Meeting other than a Special Meeting. Thirty Members shall constitute a quorum for the purposes of a Special Meeting.

20. If within thirty minutes after the time fixed for holding the Meeting a quorum is not present, the Meeting shall be dissolved, and all matters which might, if a quorum had been present, have been done at a Meeting (other than a Special Meeting) so dissolved, may forthwith be done on behalf of the Meeting by the Council.

21. The President shall be Chairman at every Meeting, and in his absence one of the Vice-Presidents; and in the absence of all Vice-Presidents a Member of Council shall take the chair; and if

no Member of Council be present and willing to take the chair, the Meeting shall elect a Chairman.

22. The decision of a General Meeting shall be ascertained by show of hands, unless, after the show of hands, a poll is forthwith demanded, and by a poll when a poll is thus demanded. The manner of taking a show of hands or a poll shall be in the discretion of the Chairman, and an entry in the Minutes, signed by the Chairman, shall be sufficient evidence of the decision of the General Meeting. Each Member shall have one vote and no more. In case of equality of votes the Chairman shall have a second or casting vote: Provided that this Article shall not interfere with the provisions of the By-laws as to election by ballot.

23. The acceptance or rejection of votes by the Chairman shall be conclusive for the purpose of the decision of the matter in respect of which the votes are tendered: Provided that the Chairman may review his decision at the same Meeting if any error be then pointed out to him.

BY-LAWS.

24. The By-laws set forth in the schedule to these Articles, and such altered and additional By-laws as shall be added or substituted as hereinafter mentioned, shall regulate all matters by the Articles left to be prescribed by the By-laws, and all matters which consistently with the Articles shall be made the subject of By-laws. Alterations in, and additions to, the By-laws, may be made only by resolution of the Members at an Annual General Meeting, after notice of the proposed alteration or addition announced at the previous Ordinary Meeting, and not otherwise.

COUNCIL.

25. The Council of the Institution shall be chosen from the Members only, and shall consist of one President, six Vice-Presidents, fifteen ordinary Members of Council, and of the Past-

Presidents; and the first Council (which shall include Past-Presidents of the existing Institution) shall be as follows:—

PRESIDENT.

JOHN ROBINSON	Manchester.

PAST-PRESIDENTS.

SIR WILLIAM G. ARMSTRONG, C.B., D.C.L., LL.D., F.R.S.	Newcastle-on-Tyne.
FREDERICK J. BRAMWELL, F.R.S.	London.
THOMAS HAWKSLEY	London.
JAMES KENNEDY	Liverpool.
JOHN PENN, F.R.S.	London.
JOHN RAMSBOTTOM	Manchester.
C. WILLIAM SIEMENS, D.C.L., F.R.S.	London.
SIR JOSEPH WHITWORTH, BART., D.C.L., LL.D., F.R.S.	Manchester.

VICE-PRESIDENTS.

I. LOWTHIAN BELL, M.P., F.R.S.	Northallerton.
CHARLES COCHRANE	Stourbridge.
EDWARD A. COWPER	London.
CHARLES P. STEWART	London.
FRANCIS W. WEBB	Crewe.
PERCY G. B. WESTMACOTT	Newcastle-on-Tyne.

COUNCIL.

DANIEL ADAMSON	Manchester.
JOHN ANDERSON, LL.D., F.R.S.E.	London.
HENRY BESSEMER	London.
HENRY CHAPMAN	London.
EDWARD EASTON	London.
DAVID GREIG	Leeds.
JEREMIAH HEAD	Middlesbrough.
THOMAS R. HETHERINGTON	Manchester.
HENRY H. LAIRD	Birkenhead.
WILLIAM MENELAUS	Dowlais.
ARTHUR PAGET	Loughborough.
JOHN PENN, JUN.	London.
GEORGE B. RENNIE	London.
WILLIAM RICHARDSON	Oldham.
JOHN C. WILSON	Bristol.

26. The first Council shall continue in office till the Annual General Meeting in the year 1879. The President, two Vice-Presidents, and five Members of the Council (other than Past-Presidents), shall retire at each succeeding Annual General Meeting, but shall be eligible for re-election. The Vice-Presidents and Members of Council to retire each year shall, unless the Council agree amongst themselves, be chosen from those who have been longest in office, and in cases of equal seniority shall be determined by ballot.

27. The election of a President, Vice-Presidents, and Members of the Council, to supply the place of those retiring at the Annual General Meeting, shall be conducted in such manner as shall be prescribed by the By-laws from time to time in force, as provided by the Articles.

28. The Council may supply any casual vacancy in the Council (including any casual vacancy in the office of President) which shall occur between one Annual General Meeting and another, and the President or Members of the Council so appointed by the Council shall retire at the succeeding Annual General Meeting. Vacancies not filled up at any such Meeting shall be deemed to be casual vacancies within the meaning of this Article.

OFFICERS.

29. The Treasurer, Secretary, and other employés of the Institution shall be appointed and removed in the manner prescribed by the By-laws from time to time in force, as provided by the Articles. Subject to the express provisions of the By-laws the officers and servants of the Institution shall be appointed and removed by the Council.

30. The powers and duties of the officers of the Institution shall (subject to any express provision in the By-laws) be determined by the Council.

POWERS AND PROCEDURE OF COUNCIL.

31. The Council may regulate their own procedure, and delegate any of their powers and discretions to any one or more of their body, and may determine their own quorum: if no other number is prescribed, three Members of Council shall form a quorum.

32. The Council shall acquire the property of the existing Institution, and shall manage the property, proceedings, and affairs of the Institution, in accordance with the By-laws from time to time in force.

33. The Treasurer may, with the consent of the Council, invest in the name of the Institution any moneys not immediately required for the purposes of the Institution in or upon any of the following investments (that is to say):—

(A.) The Public Funds, or Government Stocks of the United Kingdom, or of any Foreign or Colonial Government guaranteed by the Government of the United Kingdom.

(B.) Real or Leasehold Securities, or in the purchase of real or leasehold properties in Great Britain or Ireland.

(C.) Debentures, Debenture Stock, or Guaranteed or Preference Stock, of any Company incorporated by special Act of Parliament, the ordinary Shareholders whereof shall at the time of such investment be in actual receipt of half-yearly or yearly dividends.

(D.) Stocks, Shares, Debentures, or Debenture Stock of any Railway, Canal, or other Company, the undertaking whereof is leased to any Railway Company at a fixed or fixed minimum rent.

(E.) Stocks, Shares, or Debentures of any East Indian Railway or other Company, which shall receive a contribution from Her Majesty's East Indian Government of a fixed annual percentage on their capital, or be guaranteed a fixed annual dividend by the same Government.

(F.) The security of rates levied by any corporate body empowered to borrow money on the security of rates, where such borrowing has been duly authorised by Act of Parliament.

34. The Council may, with the authority of a resolution of the Members in General Meeting, borrow moneys for the purposes of the Institution on the security of the property of the Institution.

35. No act done by the Council, whether *ultra vires* or not, which shall receive the express or implied sanction of the Members of the Institution in General Meeting, shall be afterwards impeached by any Member of the Institution on any ground whatsoever, but shall be deemed to be an act of the Institution.

NOTICES.

36. A notice may be served by the Council of the Institution upon any Member, Graduate, Associate, or Honorary Life Member, either personally or by sending it through the post in a prepaid letter addressed to such Member, Graduate, Associate, or Honorary Life Member, at his registered place of abode.

37. Any notice, if served by post, shall be deemed to have been served at the time when the letter containing the same would be delivered in the ordinary course of the post, and in proving such service it shall be sufficient to prove that the letter containing the notice was properly addressed and put into the post office.

38. No Member, Graduate, Associate, or Honorary Life Member, not having a registered address within the United Kingdom shall be entitled to any notice; and all proceedings may be had and taken without notice to such Member in the same manner as if he had had due notice.

By-laws.

(Last Revision, January 1890.)

MEMBERSHIP.

1. Members, Graduates, Associates, and Honorary Life Members of the existing Institution, may, upon signing and forwarding to the Secretary of the Institution a claim according to Form D in the Appendix, become Members, Graduates, Associates, or Honorary Life Members respectively of the Institution without election or payment of entrance fees.

2. Candidates for admission as Members must be Engineers not under twenty-four years of age, who may be considered by the Council to be qualified for election.

3. Candidates for admission as Graduates must be Engineers holding subordinate situations and not under eighteen years of age; and they may afterwards be admitted as Members at the discretion of the Council.

4. Candidates for admission as Associates must be gentlemen not under twenty-four years of age, who from their scientific attainments or position in society may be considered eligible by the Council.

5. The Council shall have the power to nominate as Honorary Life Members gentlemen of eminent scientific acquirements, who in their opinion are eligible for that position.

6. The Members, Graduates, Associates, and Honorary Life Members shall have notice of and the privilege to attend all Meetings, but Members only shall be entitled to vote thereat.

7. The abbreviated distinctive Titles for indicating the connection with the Institution of Members, Graduates, Associates, or Honorary Life Members thereof, shall be the following: — for Members, M. I. Mech. E.; for Graduates, G. I. Mech. E.; for Associates, A. I. Mech. E.; for Honorary Life Members, Hon. M. I. Mech. E.

8. Subject to such regulations as the Council may from time to time prescribe, any Member, Graduate, or Associate may upon application to the Secretary obtain a Certificate of his membership or other connection with the Institution. Every such certificate shall remain the property of, and shall on demand be returned to, the Institution.

ENTRANCE FEES AND SUBSCRIPTIONS.

9. An Entrance Fee of £2 shall be paid by each Member, except Members of the existing Institution, who shall pay no Entrance Fee, and Graduates admitted as Members, who shall pay an Entrance Fee of £1. Each Member shall pay an Annual Subscription of £3.

10. An Entrance Fee of £1 shall be paid by each Graduate, except Graduates of the existing Institution, who shall pay no Entrance Fee. Each Graduate shall pay an Annual Subscription of £2.

11. An Entrance Fee of £2 shall be paid by each Associate, except Associates of the existing Institution, who shall pay no Entrance Fee. Each Associate shall pay an Annual Subscription of £3.

12. All Subscriptions shall be payable in advance, and shall become due on the 1st day of January in each year; and the first Subscription of Members, Graduates, and Associates, shall date from the 1st day of January in the year of their election.

13. In the case of Members, Graduates, or Associates, elected in the last three months of any year, the first subscription shall cover both the year of election and the succeeding year.

14. Any Member or Associate whose subscription is not in arrear may at any time compound for his subscription for the current and all future years by the payment of Fifty Pounds. All compositions shall be deemed to be capital moneys of the Institution.

15. The Council may at their discretion reduce or remit the Annual Subscription, or the arrears of Annual Subscription, of any Member who shall have been a subscribing Member of the Institution for twenty years, and shall have become unable to continue the Annual Subscription provided by these By-laws.

16. No Proceedings or Ballot Lists shall be sent to Members, Graduates, or Associates, who are in arrear with their subscriptions more than twelve months, and whose subscriptions shall not have been remitted by the Council as hereinbefore provided.

ELECTION OF MEMBERS, GRADUATES, AND ASSOCIATES.

17. A recommendation for admission according to Form A in the Appendix shall be forwarded to the Secretary, and by him be laid before the next Meeting of the Council. The recommendation must be signed by not less than five Members if the application be for admission as a Member or Associate, and by three Members if it be for a Graduate.

18. All Elections shall take place by ballot, three-fifths of the votes given being necessary for election.

19. All applications for admission shall be communicated by the Secretary to the Council for their approval previous to being inserted in the ballot list for election, and the approved ballot list shall be signed by the President and forwarded to the Members. The ballot list shall specify the name, occupation, and address of the Candidates, and also by whom proposed and seconded. The lists shall be opened only in the presence of the Council on the day of election, by a Committee to be appointed for that purpose.

20. The Elections shall take place at the General Meetings only.

21. When the proposed Candidate is elected, the Secretary shall give him notice thereof according to Form B; but his name shall not be added to the list of Members, Graduates, or Associates of the Institution until he shall have paid his Entrance Fee and first Annual Subscription, and signed the Form C in the Appendix.

22. In case of non-election, no mention thereof shall be made in the Minutes, nor any notice given to the unsuccessful Candidate.

23. A Graduate or Associate desirous of being transferred to the class of Members shall forward to the Secretary a recommendation according to Form E in the Appendix, signed by not less than five Members, which shall be laid before the next meeting of Council for their approval. On their approval being given, the Secretary shall notify the same to the Candidate according to Form F if an Associate, and according to Form G if a Graduate; but his name shall not be added to the list of Members until he shall have signed the Form H, and, if a Graduate, shall have paid £1 additional entrance fee, and £1 additional subscription for the current year.

ELECTION OF PRESIDENT, VICE-PRESIDENTS, AND MEMBERS OF COUNCIL.

24. Candidates shall be put in nomination at the General Meeting preceding the Annual General Meeting, when the Council are to present a list of their retiring Members who offer themselves for re-election; any Member shall then be entitled to add to the list of Candidates. The ballot list of the proposed names shall be forwarded to the Members. The ballot lists shall be opened only in the presence of the Council on the day of election, by a Committee to be appointed for that purpose.

APPOINTMENT AND DUTIES OF OFFICERS.

25. The Treasurer shall be a Banker, and shall hold the uninvested funds of the Institution, except the moneys in the hands

of the Secretary for current expenses. He shall be appointed by the Members at a General or Special Meeting, and shall hold office at the pleasure of the Council.

26. The Secretary of the Institution shall be appointed as and when a vacancy occurs by the Members at a General or Special Meeting, and shall be removable by the Council upon six months' notice from any day. The Secretary shall give the same notice. The Secretary shall devote the whole of his time to the work of the Institution, and shall not engage in any other business or profession.

27. It shall be the duty of the Secretary, under the direction of the Council, to conduct the correspondence of the Institution; to attend all meetings of the Institution, and of the Council, and of Committees; to take minutes of the proceedings of such meetings; to read the minutes of the preceding meetings, and all communications that he may be ordered to read; to superintend the publication of such papers as the Council may direct; to have the charge of the library; to direct the collection of the subscriptions, and the preparation of the account of expenditure of the funds; and to present all accounts to the Council for inspection and approval. He shall also engage (subject to the approval of the Council) and be responsible for all persons employed under him, and set them their portions of work and duties. He shall conduct the ordinary business of the Institution, in accordance with the Articles and By-laws and the directions of the President and Council; and shall refer to the President in any matters of difficulty or importance, requiring immediate decision.

MISCELLANEOUS.

28. All Papers shall be submitted to the Council for approval, and after their approval shall be read by the Secretary at the General Meetings, or by the Author with the consent of the Council.

29. All books, drawings, communications, &c., shall be accessible to the Members of the Institution at all reasonable times.

30. All communications to the Meetings shall be the property of the Institution; and be published only by the authority of the Council.

31. None of the property of the Institution—books, drawings, &c.—shall be taken out of the premises of the Institution without the consent of the Council.

32. All donations to the Institution shall be enumerated in the Annual Report of the Council presented to the Annual General Meeting.

33. The General Meetings shall be conducted as far as practicable in the following order :—

1st. The Chair to be taken at such hour as the Council may direct from time to time.

2nd. The Minutes of the previous Meeting to be read by the Secretary, and, after being approved as correct, to be signed by the Chairman.

3rd. The Ballot Lists, previously opened by the Council, to be presented to the Meeting, and the new Members, Graduates, and Associates elected to be announced.

4th. Papers approved by the Council to be read by the Secretary, or, with the consent of the Council, by the Author.

34. Each Member shall have the privilege of introducing one friend to any of the Meetings; but, during such portion of any meeting as may be devoted to any business connected with the management of the Institution, visitors shall be requested by the Chairman to withdraw, if any Member asks that this shall be done.

35. Every Member, Graduate, Associate, or Visitor, shall write his name and residence in a book to be kept for the purpose, on entering each Meeting.

36. The President shall ex officio be Member of all Committees of Council.

37. Seven clear days' notice at least shall be given of every meeting of the Council. Such notice shall specify generally the business to be transacted by the meeting. No business involving the expenditure of the funds of the Institution (except by way of payment of current salaries and accounts) shall be transacted at any Council meeting unless specified in the notice convening the meeting.

38. The Council shall present the yearly accounts to the Members at the Annual General Meeting, after being audited by a professional accountant, who shall be appointed annually by the Members at a General or a Special Meeting, at a remuneration to be then fixed by the Members.

39. Any Member wishing to have a copy of the Papers sent to him for consideration beforehand can do so by sending in his name once in each year to the Secretary; and a copy of all Papers shall then be forwarded to him as early as possible prior to the date of the Meeting at which they are intended to be read.

40. At any Meeting of the Institution any Member shall be at liberty to re-open the discussion upon any Paper which has been read or discussed at the preceding Meeting; provided that he signifies his intention to the Secretary at least one month previously to the Meeting, and that the Council decide to include it in the notice of the Meeting as part of the business to be transacted.

APPENDIX.

FORM A.

Mr.   being not under twenty-four years of age, and desirous of admission into the Institution of Mechanical Engineers, we the undersigned proposer and seconder from our personal knowledge, and we the three other signers from trustworthy information, propose and recommend him as a proper person to become a   thereof.

Witness our hands, this   day of

Members.

FORM B.

Sir,—I have to inform you that on the   you were elected a   of the Institution of Mechanical Engineers. In conformity with the rules, your election cannot be confirmed until the enclosed form be returned to me with your signature, and until your Entrance Fee and first Annual Subscription be paid, the amounts of which are   and   respectively. If these be not received within two months from the present date, the election will become void.

I am, Sir,

Your obedient servant,

Secretary.

FORM C.

I, the undersigned, being elected a   of the Institution of Mechanical Engineers, do hereby agree that I will be governed by the regulations of the said Institution, as they are now formed or as they may hereafter be altered; that I will advance the objects of the Institution as far as shall be in my power, and will attend the Meetings thereof as often as I conveniently can: provided that, whenever I shall signify in writing to the Secretary that I am desirous of withdrawing from the Institution, I shall (after the payment of any arrears which may be due by me at that period) be free from this obligation.

Witness my hand, this   day of

FORM D.

As a   of the Institution of Mechanical Engineers, I claim to become a   of the Association incorporated under the same name.

Please register me as a

FORM E.

Mr. being of the required age, and desirous of being transferred into the class of Members of the Institution, we, the undersigned, from our personal knowledge, recommend him as a proper person to become a Member of the Institution of Mechanical Engineers.

FORM F.

Sir,—I have to inform you that the Council have approved of your being transferred to the class of Members of the Institution of Mechanical Engineers. In conformity with the rules, your transference cannot be confirmed until the enclosed form be returned to me with your signature. If this be not received within two months from the present date, the transference will become void.

I am, Sir,

Your obedient servant,

Secretary.

FORM G.

Sir,—I have to inform you that the Council have approved of your being transferred to the class of Members of the Institution of Mechanical Engineers. In conformity with the rules, your transference cannot be confirmed until the enclosed form be returned to me with your signature, and until your additional Entrance Fee (£1) and additional Annual Subscription (£1) be paid for the current year. If these be not received within two months from the present date, the transference will become void.

I am, Sir,

Your obedient servant,

Secretary.

FORM H.

I, the undersigned, having been transferred to the class of Members of the Institution of Mechanical Engineers, do hereby agree that I will be governed by the regulations of the said Institution, as they now exist, or as they may hereafter be altered; that I will advance the objects of the Institution as far as shall be in my power, and will attend the Meetings thereof as often as I conveniently can: provided that, whenever I shall signify in writing to the Secretary that I am desirous of withdrawing from the Institution, I shall (after the payment of any arrears which may be due by me at that period) be free from this obligation.

Witness my hand, this day of

Institution of Mechanical Engineers.

PROCEEDINGS.

February 1892.

The Forty-Fifth Annual General Meeting of the Institution was held in the rooms of the Institution of Civil Engineers, London, on Thursday, 4th February 1892, at Half-past Seven o'clock p.m.; Joseph Tomlinson, Esq., Retiring President, in the chair, succeeded by Dr. William Anderson, F.R.S., President elected at the Meeting.

The Minutes of the previous Meeting were read, approved, and signed by the President.

The President announced that the Ballot Lists for the election of New Members, Associates, and Graduates, had been opened by a committee of the Council, and that the following thirty-three candidates were found to be duly elected:—

MEMBERS.

Capt. Francis Edward Dyke Acland,	London.
Edwin Ault,	London.
Peter Macleod Baxter,	Falmouth.
George Maximilian Borns,	Sydney.
Arthur F. Bowker,	Snodland.
William J. Bromiley,	Bolton.
James Fiddes Brown,	London.
John T. Buckley,	Preston.
David Sing Capper,	London.
William George Causer,	Birmingham.
Henry Aylwin Bevan Cole,	London.
Sherard Osborn Cowper-Coles,	London.
Archibald Campbell Elliott, D.Sc.,	Cardiff.

Hilary Sheldon Forrest, . . .	Bolton.
Henry Edwin Hodgson, . . .	Halifax.
Carl Albert Holmström, . . .	London.
James John Howitt, ...	Northwich.
William John Jenkins, . . .	Manchester.
Thomas Kennedy, ...	Kilmarnock.
Dr. Antonio Augusto Machado, . .	Bahia.
James Mactear, F.R.S.E., . . .	London.
John Fraser Masterton, . . .	Glasgow.
John McDonald, ...	Tokyo.
William Forbes McIntosh, . . .	Hong Kong.
Middleton Pratt, ...	Huddersfield.
James Shepherd, ...	Bradford.
William Henry Still, . . .	Aden.
Edward R. Wilkinson, . . .	Barnet.

ASSOCIATES.

George Brown Davis, ...	London.
John Edward Stead, ...	Middlesbrough.

GRADUATES.

Sidney Rickman Adams, . . .	London.
William Bartholomew Cleverly, Jun., .	London.
Joseph Ernst Fletcher, . . .	Sheffield.

The following Annual Report of the Council was then read:—

ANNUAL REPORT OF THE COUNCIL.

1892.

At this Forty-Fifth Annual General Meeting of the Institution, the Council have the pleasure of presenting the following as their Annual Report to the Members respecting the proceedings of the Institution during the past year.

At the end of last year the number of names of all classes on the roll of the Institution was 2077, as compared with 1943 at the end of the previous year, showing a net gain of 134. During 1891 there were added to the register 181 names; against which the loss by deceases was 19, and by resignation or removal 28.

The following eleven Transferences of Graduates to the class of Members have been made by the Council in 1891 :—

EWART CHARLES AMOS,	London.
FREDERICK EVELYN CAIRNES,	London.
SIDNEY WARD CONYERS,	Sydney.
HENRY STREATFEILD COTTON,	London.
Lieut. GILBERT HARWOOD HARRISON, R.E.,	Woolwich.
ARCHIBALD POTTER HEAD,	Middlesbrough.
RAY DOUGLAS THEODORE HEAP,	London.
WILLIAM WADE FITZHERBERT PULLEN,	Cardiff.
HORACE WYON ROGERS,	London.
THOMAS RIDSDILL SMITH,	Hull.
FRANK LLEWELLYN WARREN,	Shanghai.

The following twenty-five Deceases of Members of the Institution have occurred during the past year :—

AUGUSTE BERNARD ALBARET,	Liancourt.
ALEXANDER ALLAN,	Scarborough.
HENRY JOHN CARD ANDERSON,	Cairo.

Samuel Barratt, ...	Manchester.
Peter Borrie Blair, ...	Stockton-on-Tees.
Anthony Bower, ...	Liverpool.
Henry Brown, ...	Birmingham.
Archibald Douglas Bryce-Douglas, . . .	Barrow-in-Furness.
John Chapman, ...	Eton.
John William Cole, ...	Gawler, S. Australia.
Capt. Charles Fairholme, R.N. (Associate),	London.
William Fox, ...	Leeds.
David Greig, ...	Leeds.
Thomas Comings Hide, ...	London.
Samuel Osborn, ...	Sheffield.
Loftus Perkins, ...	London.
Francis Preston, ...	Huddersfield.
Robert James Ransome, ...	Ipswich.
John Seddon, ...	Wigan.
Richard Sennett, ...	London.
Charles Percy Bysshe Shelley, . . .	London.
Edward Slaughter, ...	Bristol.
Thomas Tucker (Associate).	Gateshead.
Benjamin Walker, ...	Leeds.
Joseph Whitley, ...	Leeds.

Of these Mr. Allan was one of the original Members of the Institution on its foundation in 1847, and for eleven years was also a Member of Council. Mr. Greig was for seven years a Member of Council, and afterwards for four years a Vice-President. Mr. Walker was for six years a Member of Council.

The following sixteen gentlemen have ceased to be Members of the Institution during the past year:—

Thomas Napier Armit, ...	Dundee.
William James Baker, ...	Scarborough.
Claude Carter, ...	Birmingham.
George Alexander Corder, ...	London.
George Crowe, ...	Cardiff.
St. John Vincent Day, F.R.S.E., . . .	Edinburgh.
Alfred George Hamilton, ...	London.
John Ferguson Haskins, ...	London.
Ashton Marler Heath (Graduate), . . .	London.

JOSEPH LEDGER, ...	Darlington.
EDWARD MCKILLOP NICHOLL, ...	Amritsar.
THOMAS HENRY OWEN, ...	Cardiff.
VINCENT RHODES, ...	Granville, N.S.W.
JUAN EMILIO SANCHEZ (Graduate), . . .	Buenos Aires.
ALLISON DALRYMPLE SMITH, ...	Newport, Victoria.
DAVID WALKER, ...	London.

In addition to these there have been twelve Resignations of membership.

The Accounts for the year ending 31 December 1891 are now submitted to the Members (*see* pages 10–13), after having been passed by the Finance Committee, and certified by Mr. Robert A. McLean, chartered accountant, the auditor appointed by the Members at the last Annual General Meeting. The receipts during the year were £7,212 5*s.* 10*d.*, while the expenditure, actual and estimated, was £5,097 10*s.* 2*d.*, leaving a balance of receipts over expenditure of £2,114 15*s.* 8*d.* The financial position of the Institution at the end of the year is shown by the balance sheet: the total investments and other assets amount to £36,401 7*s.* 3*d.*; and allowing £600 for accounts owing but not yet rendered, the capital of the Institution amounts to £35,801 7*s.* 3*d.*, of which the greater part, as seen from the balance sheet, is invested in Railway Debenture Stocks, registered in the name of the Institution. The certificates of the whole of the securities have been duly audited by the Finance Committee and the auditor.

The competitive designs sent in by Graduates of the Institution for the Certificate of Membership have been examined by the Committee appointed for the purpose; and the premium of five guineas offered for the accepted design has been awarded by the Council to Mr. William W. F. Pullen, Cardiff. The preparation of the certificates in conformity with this design has since been proceeded with; and their issue in accordance with by-law 8 will take place as early as possible.

The Research Committee on Friction, of which the President is the chairman, presented to the General Meeting in March the report on their experiments upon the friction of pivot bearings. Like their previous experiments these last were conducted in as general a manner as possible, in order that the results arrived at might be so far independent of any special circumstances affecting individual pivot bearings of peculiar form. The experimental apparatus constructed for this purpose, and described in the report, has since been lent by the Council to the Engineering Department of University College, London, at the request of Professor Beare, on the understanding that any further results obtained by the use of the machine, while so lent during the pleasure of the Council, be not published through other means until they have been communicated to the Institution, and until the Research Committee have had the opportunity of investigating them, and of publishing them, if so directed by the Council, in the Institution Proceedings.

Since the Research Committee on Marine-Engine Trials presented at the Spring Meeting their report upon the trials of the "Iona," a further trial has been conducted by their chairman, Professor Kennedy, upon the "Ville de Douvres," one of the Belgian paddle steamers performing the mail and passenger service between Dover and Ostend, having a pair of two-cylinder compound inclined engines working on two cranks. The results of this trial are now being worked out with a view to the early presentation of a report thereon for reading and discussion at a General Meeting, as in the case of the previous trials. The trials made by the Committee during the last four years have thus included fairly representative examples of the following kinds of modern marine engines:—triple-expansion compound, two-cylinder compound, and twin two-cylinder compound, all in screw-steamers; and two-cylinder compound inclined for a paddle-steamer. The types of vessel in which these engines have been tested have been as varied as the engines themselves. The Council are accordingly of opinion that the objects for which this Committee was appointed—namely, as defined by the instructions in the resolution appointing the Committee, to endeavour (1) to draw up a standard system for conducting marine-engine trials, and

(2) to arrange for the carrying out of such trials in accordance with this system—have now been realised to an extent as complete as was originally contemplated, and as the interests of the Institution seem to render desirable. While not looking forward therefore to any further trials being undertaken, they desire to call the attention of the Members to the practical importance and value of the results already arrived at, and also to the large amount of time and labour which has been so liberally devoted to this research by Professor Kennedy and the other members of the Committee who have so cordially co-operated with him in the work, as well as by the Members of the Institution and others who have formed the staff of observers on the various trials.

The time and labour involved in the Marine-Engine Trials have hitherto delayed the completion of the next report now in progress of preparation by the Research Committee on the Value of the Steam-Jacket, which however will be accelerated for presentation as soon as the one remaining report on the last marine-engine trial has been cleared off. Meanwhile under the direction of the chairman, Mr. Henry Davey, a large amount of time has already been devoted to the work of the Steam-Jacket research, as far as has been compatible with the due prosecution of the earlier subject.

The Alloys Research Committee, of which Dr. Anderson is the chairman, have had the gratification of presenting to the last General Meeting the first Report of Professor Roberts-Austen upon the investigations he is conducting in connection with this subject. The interest felt in this important research was evidenced by the nature of the discussion which followed the reading of the Report; and further valuable results are awaited, as soon as Professor Roberts-Austen has had time to extend his investigations further.

The additions that have been presented to the Library of the Institution during the past year are enumerated in pages 14–22. For these contributions the Council have much pleasure in recording their thanks to the several Donors. Members are invited to present original pamphlets on engineering subjects and records of experimental research, and other works valuable for reference, which are always welcome for permanent preservation in the Library.

The General Meetings in 1891 were the Annual General Meeting and two Spring Meetings, all of which were held in London; the Summer Meeting in Liverpool; and the Autumn Meeting in London. Altogether ten sittings were occupied in the reading and discussion of the following Papers, which are published in the Proceedings:—

On some different kinds of Gas Furnaces; by Mr. Bernard Dawson.

On the Mechanical Treatment of Moulding Sand; by Mr. Walter Bagshaw.

Fourth Report of the Research Committee on Friction: Experiments on the Friction of a Pivot Bearing.

On recent Trials of Rock Drills; by Mr. Edward H. Carbutt and Mr. Henry Davey.

Research Committee on Marine-Engine Trials: Report upon Trials of the S.S. "Iona;" by Professor Alexander B. W. Kennedy, F.R.S.

A Review of Marine Engineering during the past decade; by Mr. Alfred Blechynden.

Description of the Warehouse and Machinery for the Storage and Transit of Grain at the Alexandra Dock, Liverpool; by Mr. William Shapton.

On the Experimental Marine Engine and the Alternative-Centre Testing Machine in the Walker Engineering Laboratories of University College, Liverpool; by Professor H. S. Hele-Shaw.

On the Mechanical Appliances employed in the construction of the Manchester Ship Canal; by Mr. E. Leader Williams.

On some details in the Construction of modern Lancashire Boilers; by Mr. Samuel Boswell.

First Report to the Alloys Research Committee; by Professor W. C. Roberts-Austen, C.B., F.R.S.

The attendances during 1891 were as follows:—at the Annual General Meeting 96 Members and 44 Visitors; at the earlier Spring Meeting 52 Members and 32 Visitors; at the later Spring Meeting 80 Members and 56 Visitors; at the Summer Meeting 292 Members and 80 Visitors; and at the Autumn Meeting 108 Members and 61 Visitors.

The Summer Meeting of the Institution was held in Liverpool, after an interval of nineteen years since the previous Meeting held there in 1872. Under the auspices of the Mayor, Joseph B. Morgan, Esq., as chairman of the Reception Committee, aided by

Mr. George Heaton Daglish as honorary treasurer and Mr. Henry H. West as honorary secretary, a highly attractive programme was arranged, including visits to important local engineering works and establishments. Three of these were closely connected with papers read at the Meeting, namely the Alexandra Grain Warehouse, the Walker Engineering Laboratories, and the Manchester Ship Canal; while the review of Marine Engineering progress was exemplified by visits to some of the principal Atlantic liners then in port, and to various marine engineering works on both sides of the Mersey. In their visit to the Manchester Ship Canal the Members enjoyed the advantage of seeing the nearly completed portions of the work, before the part below water-level was rendered invisible through being submerged. Of the visits arranged to the Horwich Locomotive Works and the Vyrnwy Water Works, the former was highly gratifying to all who took part in it; while the latter had reluctantly to be abandoned at the last for lack of sufficient numbers, much to the regret of those who wished to avail themselves of the favourable opportunity for inspecting so magnificent an engineering work as Lake Vyrnwy.

Through the kind influence of the Mayor the Meetings were held in the Concert Room of St. George's Hall; and to his hospitality the Members were indebted for their enjoyment of the Conversazione to which they were invited by him in the Walker Art Gallery and the Liverpool Free Library and Museum. The indefatigable exertions of Mr. West, as Honorary Secretary, in maturing and carrying out the arrangements for the Meeting, have been recognised by the Council by a resolution conveying to him the cordial thanks of the Institution, accompanied by a time-piece and inkstand as memorials of his signal services, to which the President had the pleasure of drawing the attention of the Members at the Autumn Meeting.

In accordance with the Rules of the Institution, the President, two Vice-Presidents, and five Members of Council, retire from office this day. The result of the ballot for the election of the Council for the present year will be announced to the Meeting.

Dr. ACCOUNT OF EXPENDITURE AND RECEIPTS

Expenditure.	£	s.	d.	£	s.	d.
To Printing and Engraving Proceedings of 1891	904	12	5			
Less Authors' Copies of Papers, repaid	69	11	6	835	0	11
„ Printing Library Catalogue				81	4	8
„ Stationery and General Printing				206	14	3
„ Certificate of Membership				32	9	5
„ Binding				67	15	11
„ Rent				708	15	0
„ Salaries and Wages				1,942	19	0
„ Coal, Firewood, and Gas				44	9	2
„ Fittings and Repairs				38	17	4
„ Postages				315	16	1
„ Insurance				7	0	6
„ Petty Expenses				49	0	0
„ Meeting Expenses—						
Printing	174	1	9			
Reporting	75	17	4			
Diagrams, Screen, &c.	107	19	4			
Travelling and Incidental Expenses	189	9	7	547	8	0
„ Dinner Guests				54	17	4
„ Research				160	10	10
„ Books purchased				4	11	9
				5,097	10	2
Accounts owing, not yet rendered, say	600	0	0			
Less Reserve in previous year for accounts since paid	600	0	0	0	0	0
Balance, being excess of Receipts over Expenditure, carried down				2,114	15	8
				£7,212	5	10

	£	s.	d.
To Investment—			
£1,450 *Taff Vale Railway 3% Debenture Stock*	1,398	16	6
Cash Balance 31st December 1891	2,960	7	3
	£4,359	3	9

FOR THE YEAR ENDING 31ST DECEMBER 1891. *Cr.*

Receipts.	£	s.	d.	£	s.	d.
By Entrance Fees—						
147 *New Members at £2*	294	0	0			
13 *New Associates at £2*	26	0	0			
20 *New Graduates at £1*	20	0	0			
11 *Graduates transferred to Members at £1*	11	0	0	351	0	0
„ Subscriptions for 1891—						
1620 *Members at £3*	4,860	0	0			
62 *Associates at £3*	186	0	0			
135 *Graduates at £2*	270	0	0			
11 *Graduates transferred to Members at £1*	11	0	0	5,327	0	0
„ Subscriptions in arrear—						
80 *Members at £3*	240	0	0			
2 *Members, instalments*	2	4	0			
5 *Graduates at £2*	10	0	0	252	4	0
„ Subscriptions in advance—						
35 *Members at £3*				105	0	0
„ Interest—						
From Investments	667	15	4			
From Whitworth Bequest	334	2	6			
From Bank	48	10	0	1,050	7	10
„ Reports of Proceedings—						
Extra Copies sold				126	14	0
				£7,212	5	10

	£	s.	d.
By Balance brought down	2,114	15	8
By Life Compositions, 3 Members at £50	150	0	0
Cash Balance 31st December 1890	2,094	8	1
	£4,359	3	9

Dr. BALANCE SHEET

	£	s.	d.
To Sundry Creditors—			
Accounts owing, not yet rendered, say	600	0	0
Capital of the Institution at this date	35,801	7	3
	£36,401	7	3

Signed by the following members of the Finance Committee:—

JOSEPH TOMLINSON,
WILLIAM ANDERSON,
SIR JAMES N. DOUGLASS,
WILLIAM H. MAW.

AS AT 31ST DECEMBER 1891.

Cr.

	£ s. d.	£ s. d.	£ s. d.
By Cash—*In Union Bank, on Deposit*		2,100 0 0	
„ „ „ *on Current account* . .		360 7 3	
In Imperial Bank	483 12 3		
In hand	16 7 9	500 0 0	2,960 7 3
„ Investments—(*cost* £20,287 5*s.* 9*d.*)			
£ 3,178 *London and North Western Ry.* 4% *Debenture Stock*			
2,200 *North Eastern* „ „ „ „			
1,800 *Great Western* „ „ „ „			
1,270 *Great Eastern* „ „ „ „			
2,755 *Metropolitan* „ „ „ „			
2,325 „ „ 3½% „ „			
1,000 *Aire and Calder Navigation* „ „ „			
3,288 *Midland Railway* 3% „ „			
1,450 *Taff Vale* „ „ „ „			
700 *Sir J. Whitworth and Co., Ld.* 5% „ „			
Two hundred £10 shares Sir J. Whitworth and Co., Ld.			
The Market Value of these investments at 31st Dec. 1891 was about . .			26,948 0 0
„ Subscriptions in Arrear, probable value			290 0 0
„ Office Furniture and Fittings			343 0 0
„ Library			1,240 0 0
„ Proceedings, back numbers at cost			4,520 0 0
„ Drawings, Engravings, Models, Specimens, and Sculpture .			100 0 0
			£36,401 7 3

Audited and Certified by

ROBERT A. McLEAN, Chartered Accountant,
1 Queen Victoria Street, London, E.C.

LIST OF DONATIONS TO LIBRARY.

Mémoire sur le Viaduc de Garabit, by Gustave Eiffel; from the author.

Blast-Furnace Practice at Ormesby Iron Works, Middlesbrough, by Charles Cochrane; from the author.

Report of the Kew Observatory Committee, 1890; from the Committee.

Congreso Internacional de Ingenieria celebrado in Barcelona durante 1888; from the Commissioners.

Congrès International des Procédés de Construction, Comptes Rendus des Séances et Visites du Congrès; from the French Ministry of Commerce, Industry, and the Colonies.

Notes on the Construction of Ordnance; from the Ordnance Office, Washington, U.S.

Metropolis Water Supply, 1890 and 1891, by Richard Hassard and Arthur W. N. Tyrrell; from the authors.

Solutions to questions given at the extra first-class Engineers' Examinations, by Edward J. M. Davies; from the author.

Classified Lists and Distribution Returns of Establishment, Indian Public Works Department, to 31 Dec. 1890; ditto to 30 June 1891; from the Registrar.

Maps showing lines of Equal Magnetic Declination for 1 January 1891; from the editor of the "Colliery Guardian."

Civil Engineer's Pocket-Book, 1891, by John C. Trautwine; from Mr. John C. Trautwine, Jun.

Spons' Engineers' and Contractors' Diary and Reference Book 1891 and 1892; from the publishers.

"Practical Engineer" Pocket-Book and Diary 1891 and 1892; from the publishers.

Lockwood's Builder's and Contractor's Price Book, 1891; from Messrs. Crosby Lockwood and Co.

Recent Standard Locomotives (eight sheets of lithographed drawings); from Mr. Theodore West.

Catalogue of Incandescent Electric Lamps and Electrical Fittings and Instruments; from Messrs. Edison and Swan.

Steel, as applied to Armour-plates, by Charles Weston Smith; from the author.

Modern Cotton-Spinning Machinery, its principles and construction, by Joseph Nasmith; from the author.

Recent Experiments on the Flow of Water over Weirs, by M. Bazin; from Mr. John C. Trautwine, Jun.

Some new Telemeters or Range-finders for military and engineering use, by Professor Archibald Barr and William Stroud; from Professor Archibald Barr.

List of Chinese Lighthouses, Light Vessels, Buoys, and Beacons, 1891; from the Inspector-General of Chinese Customs.

Tests of Metals &c. at Watertown Arsenal, Massachusetts, 1888 and 1889; from the Ordnance Office, Washington, U.S.

Laws of Force and Motion, by John Harris; from the author.

Decimal Coinage, Weights and Measures popularly explained, by Sir Guilford Molesworth, K.C.I.E., and J. Emerson Dowson; from Mr. J. Emerson Dowson.

Decimal Coinage, Weights, and Measures, by J. Emerson Dowson; from the author.

The following from Mr. Bryan Donkin, Jun.:—Principes d'Hydraulique, by Du Buat; Principles of Hydrostatics, by S. Vince; Outlines of Natural Philosophy, by John Playfair; Mathematics (Hutton's), by Olinthus Gregory and T. S. Davies; Mathematics for practical men, by Olinthus Gregory; Treatise on Mechanics, by Olinthus Gregory; Traité de Mécanique, by S. D. Poisson; Political Economy of Railroads, by Henry Fairbairn; Mechanics (Ferguson's), by David Brewster; Tracts on Mathematical and Philosophical subjects, by Charles Hutton; Canals and Reservoirs, by John Sutcliffe; Mechanics' Magazine, vols. 21, 22, and 23; Mechanical Philosophy, by John Robison.

Forth Bridge, and other new Works, as they affect employment, wages, and trade, by Henry Barcroft; from the author.

Annual Report of the U.S. Chief of Ordnance, 1891; from the Ordnance Office, Washington, U.S.

Congrès International de Mécanique Appliquée, Paris, 1889; from the Committee.

Electric Lighting, by Professor Henry Robinson; from the author.

Bombay Water Works; from Mr. T. Bernard Hall.

M'Neil's Steel Embossed Man and Mud-hole Doors; from Mr. Charles M'Neil, Jun.

The following from Messrs. Whittaker and Co.:—Electricity in our Homes and Workshops, by S. F. Walker; Electric Light Installations, by Sir David Salomons, Bart.; Electric Transmission of Energy, by Gisbert Kapp; Metal Turning; Practical Iron Founding; Conversion of Heat into Work, by William Anderson; Hydraulic Motors, Turbines, and Pressure Engines, by G. R. Bodmer; Working and Management of an English Railway, by George Findlay (third and fourth editions); First book on Electricity and Magnetism, by W. Perren Maycock; Practical Telephone Handbook, by Joseph Poole.

Utilisation de la Force Motrice des Marées, by J. Diamant; from the author.

Rationnalisation des Expériences de Regnault sur la Vapeur, by J. Macfarlane Gray; from the Conservatoire des Arts et Métiers.

Application de la Théorie Éthéro-thermique à l'étude de la Vapeur d'eau, by J. Macfarlane Gray; from the Conservatoire des Arts et Métiers.

Presidential Address to the Liverpool Polytechnic Society, 1890, by Professor H. S. Hele-Shaw; from the author.

Education of an Engineer, by Professor H. S. Hele-Shaw; from the author.

Missing Link in Industrial Education, by Professor H. S. Hele-Shaw; from the author.

System of International Measure and Weight, by C. J. Haussen; from the author.

The following official publications from the Government of New South Wales:—

New South Wales, its History and Resources; Wealth and Progress of New South Wales 1889–90, by T. A. Coghlan; Statistical account of the Seven Colonies of Australasia, by T. A. Coghlan; Blue book for the year 1890; Third Annual Report of the Board of Water Supply and Sewerage; Annual Report of the Department of Mines, 1890; First Annual Report of Chief Engineer for Water Conservation. Also Reports on the following:— Proposed College for the training of teachers of public schools; Proposed extension of Kaima to Nowra Railway into the town of Nowra; Proposed Hospital Buildings, Macquarie Street, Sydney; Progress Report on the extension of the Railway into the City and the North Shore Bridge connection; Addendum to Progress Report on the extension of the Railway into the City, and Extension of the Railway to the Suburbs of Sydney.

Descriptive list of Instruments manufactured by the Cambridge Scientific Instrument Co.; from Mr. Horace Darwin.

Critique upon a report on the Drainage of Melbourne, by Isaac Shone; from the author.

Observations on various matters of interest to Engineers in the United States of America, by Jeremiah Head; from the author.

Presidential Address to the Asiatic Society of Bengal, 1891, by H. Beveridge; from the Society.

Report of a Test of the Steam Plant at the Washington Mills, Lawrence, Mass., U.S., by E. D. Leavitt and J. T. Henthorn; from Mr. E. D. Leavitt.

Wire, its Manufacture and Uses, by J. Bucknall Smith; from the author.

Register of the Institute of Chemistry of Great Britain and Ireland, 1891; from the Institute.

Divorce of Silver and Gold, by Sir Guilford L. Molesworth, K.C.I.E.; from the author.

Silver and Gold, the money of the world, by Sir Guilford L. Molesworth, K.C.I.E.; from the author.

Recenti Applicazioni Meccaniche usate nella Preparazione dell'Ossigeno, by C. Marzocchi; from Mr. E. J. Ristori.

Report on the section of Transportation and Engineering in the United States National Museum, 1888; from the Smithsonian Institution.

Papers read before the Engineering Society of the School of Practical Science, Toronto, 1887–88, 1889–90, 1890–91; from the University of Toronto.

Patents for Inventions, and how to procure them, by G. G. M. Hardingham; from the author.

Serbatoi Recipienti Economici in tele impermeabili, by Gius. Perelli-Minetti; from the author.

Refrigeranti Economici, by Gius. Perelli-Minetti; from the author.

Electric-Light Cables, Silvertown; from the India-rubber, Gutta-percha, and Telegraph Works Co.

British Iron Trade Report, 1890; from Mr. J. S. Jeans.

Steam-Boiler Construction, by Walter S. Hutton; from the author.

Board of Trade Reports on Boiler Explosions; from the Board of Trade.

Railway Safety Appliances; hydraulic working and interlocking of railway Points and Signals; from Messrs. Saxby and Farmer.

The Railway Problem, by A. B. Stickney; from Messrs. Brentano.

Artillery of the Future, and the new Powders, by J. A. Longridge; from the author.

Résistance des Matériaux, by A. Madamet; from the author.

Trial of an 11½-inch all-steel Armour-Plate at the Naval Ordnance Proving Ground, Annapolis, Md., 20 January 1891; from Lieut. W. H. Jaques.

Bethlehem Armour and recent Armour Experiments at the Naval Ordnance Proving Ground, Annapolis, Md.; from Lieut. W. H. Jaques.

Report of a visit to several Continental and English Technical Schools; from the Council of the Manchester Technical School.

Building Materials, by Professor James Scorgie; from the author.

Washington Bridge over the Harlem River, New York, by W. R. Hutton; from Mr. Charles Cochrane.

Fourth International Congress on Inland Navigation, Manchester, 1890; from Mr. E. Leader Williams.

Tables of Diameters and Thicknesses of wrought-iron and steel Tubes to resist hydraulic pressures of from 500 lbs. to 20,000 lbs. per square inch; from Mr. S. Earnshaw Howell.

Communication between Passengers, Guards, and Engine-drivers, by Thomas Urquhart; from the author.

Electric Lighting by Municipal Authorities, by Professor Henry Robinson; from the author.

Permanent Fortification for English Engineers, by Major J. R. Lewis, R.E.; from the Royal Engineers' Institute.

The following from Mr. Henry Chapman :—Moteur à Gaz Simplex, Expériences faites sur un moteur de cent chevaux indiqués, by A. Witz; Compteurs d'Électricité, systèmes Frager et Cauderay-Frager; Mémoire justificatif de la demande en concession déposée avec les pièces à l'appui au Ministère des Travaux Publics, by the Channel Bridge and Railway Company; Conférence sur la Transmission du Travail Mécanique par les Courants Électriques, by H. Tresca; Nouveau système de Ponts Portatifs, à montage et démontage rapides; Étude sur les améliorations de l'Estuaire de la Seine et de l'entrée du Port du Havre, by de Coene; Note sur l'emploi comparé des Huiles de Colza et des Huiles minérales de Péchelbronn et du Caucase pour le graissage du matériel roulant, by Louis Salomon; Fêtes du Centenaire de la Loi de 1791 sur Brevets d'Invention; Éclairage Électrique par le Moteur à Gaz, by P. Delahaye.

Experiments in Aerodynamics, by S. P. Langley; from the Smithsonian Institution.

Recherches Expérimentales Aérodynamiques, et données d'expérience; from Mr. S. P. Langley.

Geology of the Bridgwater Railway, by J. F. Mostyn Clarke; from the author.

Photo-lithograph of Parliament Building, Toronto; from the Department of Public Works, Ontario.

Peshawar City Water Supply project (with Plans), by Rai Bahadur Ganga Ram; from the author.

Glossary of Notes on Water-Works, by Rai Bahadur Ganga Ram; from the author.

Condensed Mechanics, by W. G. C. Hughes; from the author.

Memorial Medal of the late G. A. Hirn; from M. William Grosseteste.

Manifestation en l'honneur de G. A. Hirn; from M. William Grosseteste.

Catalogue of Science Library, South Kensington Museum; from the Science and Art Department.

Dosage du Grisou par les limites d'inflammabilité, by H. Le Chatelier; from the author.

Pyromètre optique, by H. Le Chatelier; from the author.

Catalogue of Boilers, Engines, and Machinery; from Messrs. Galloway.

Calendars for 1891-92 from the following Colleges :—Royal Technical High School, Berlin; Mason Science College, Birmingham; University College, Bristol; Yorkshire College, Leeds; King's College, London; City of London College; Sheffield Technical School, and Firth College, Sheffield; School of Practical Science, Toronto.

Cornell University Register, 1890-91; from the University.

Cornell University, general and technical courses; from Professor R. H. Thurston.

Calendar 1890-91 of McGill College and University, Montreal; from the University.

Announcements of the Faculties of Applied Science and of Arts, McGill University, 1891-92; from the University.

Construction of Platinum Thermometers, by H. L. Callendar; from the author.

Railway System of London, by Frederick McDermott; from the editor of "Railway News."

Notes on the Leicester and Swannington Railway, by Clement E. Stretton; from the author.

Ironmonger Diary, 1892; from the publisher.

Year-Book of Australia, 1886 and 1890; from the editor.

Report of the Hydraulic Engineer on the Water Supply of Queensland; from Mr. J. B. Henderson.

Note sur les Transmissions par Bielle et Manivelle, by J. Massau; from the Ghent Association of Engineers.

From the United States Geological Survey.

Ninth Annual Report of the United States Geological Survey, 1887-88, by J. W. Powell.

Bulletins of the United States Geological Survey, Nos. 58-61, 63, 64, and 66.

Mineral Resources of the United States, 1888.

The following Monograph of the Survey;—I. Lake Bonneville, by Grove Karl Gilbert.

The following Publications from the respective Societies and Authorities :—

Reports of the Academy of Science, France.

Annales des Ponts et Chaussées, Paris.

Proceedings of the French Institution of Civil Engineers.

Journal of the French Society for the Encouragement of National Industry.

Annales des Mines.

Journal of the Marseilles Scientific and Industrial Society.

Proceedings of the Industrial Society of St. Quentin et de l'Aisne.

Proceedings of the Industrial Society of the North of France.

Proceedings of the Industrial Society of Rouen.

Proceedings of the Industrial Society of Mulhouse.

Annals of the Association of Engineers of Ghent.

Proceedings of the German Society of Engineers.

Reports of the Royal Academy of Science, Belgium.

Reports of the Royal Institute of Engineers, Holland.
Proceedings of the Engineers' and Architects' Society of Canton Vaud.
Proceedings of the Engineers' and Architects' Society of Austria.
Proceedings of the Engineers' and Architects' Society of Prague.
Proceedings of the Architects' and Engineers' Society of Hannover.
Proceedings of the Italian Engineers' and Architects' Society.
Proceedings of the Engineers' and Architects' Society of Milan.
Proceedings of the Russian Imperial Institute of Engineers.
Proceedings of the Swedish Technical Society.
Journal of the Norwegian Technical Society.
Journal of the Franklin Institute.
Transactions of the American Society of Civil Engineers.
Transactions of the American Society of Mechanical Engineers.
Transactions of the American Institute of Mining Engineers.
School of Mines Quarterly, Columbia College, New York.
Report of the Smithsonian Institution.
Report of the Master Car-Builders' Association, New York.
Proceedings of the United States Naval Institute.
United States Patent Office Gazette.
Transactions of the Canadian Society of Civil Engineers.
Proceedings and Journal of the Asiatic Society of Bengal.
Proceedings of the Committee of Locomotive and Carriage Superintendents of India.
Proceedings of the Engineering Association of New South Wales.
Proceedings of the Institution of Civil Engineers.
Journal of the Iron and Steel Institute.
Transactions of the Society of Engineers.
Journal of the Institution of Electrical Engineers.
Transactions of the North of England Institute of Mining and Mechanical Engineers.
Proceedings of the South Wales Institute of Engineers.
Transactions of the Institution of Engineers and Shipbuilders in Scotland.
Transactions of the Chesterfield and Midland Counties Institution of Engineers.
Transactions of the Liverpool Engineering Society.
Transactions of the Midland Institute of Mining, Civil, and Mechanical Engineers.
Proceedings of the Cleveland Institution of Engineers.
Transactions of the Mining Institute of Scotland.
Transactions of the North-East Coast Institution of Engineers and Shipbuilders.
Transactions of the Hull and District Institution of Engineers and Naval Architects.
Proceedings of the South Staffordshire Institute of Iron and Steel Works Managers.

Proceedings of the Royal Society of London.
Proceedings of the Royal Society of Edinburgh.
Proceedings of the Royal Institution of Great Britain.
Transactions of the Surveyors' Institution.
Journal of the Royal United Service Institution.
Professional Papers of the Royal Engineers' Institute.
Journal of the Royal Agricultural Society of England.
Journal of the Royal Statistical Society.
Report of the British Association for the Advancement of Science.
Report of the Royal Cornwall Polytechnic Society.
Transactions of the Institution of Naval Architects.
Transactions and Journal of the Royal Institute of British Architects.
Transactions of the Incorporated Gas Institute.
Proceedings of the Physical Society of London.
Proceedings of the Literary and Philosophical Society of Manchester.
Transactions of the Manchester Geological Society.
Journal of the Royal Scottish Society of Arts.
Proceedings of the Philosophical Society of Glasgow.
Transactions and Proceedings of the Royal Irish Academy.
Transactions and Proceedings of the Royal Dublin Society.
Transactions of the Institute of Marine Engineers.
Journal of the Liverpool Polytechnic Society.
Journal of the Society of Arts.
Journal of the Society of Chemical Industry.
Proceedings of the Society of Architects.
Transactions of the Manchester Association of Engineers.
Report &c. of Tenth Session of the Junior Engineering Society.
Reports of the Manchester Steam Users' Association; from Mr. Lavington E. Fletcher.
Records of Boiler Explosions, vol. 2, 1880 to 1889, and Annual Record 1890; from Mr. Edward B. Marten.
Report of the National Boiler Insurance Company; from Mr. Henry Hiller.
Report of the Engine, Boiler, and Employers' Liability Insurance Company; from Mr. Michael Longridge.
Report of the London Association of Foremen Engineers and Draughtsmen.
Eighth Annual Report of the Barrow-in-Furness Free Public Library.
Thirty-eighth Annual Report of the Liverpool Free Public Library.
Catalogue of Additions to the Radcliffe Library, Oxford, during 1890.

The following Periodicals from the respective Editors :—

Revue générale des Chemins de fer.
Revue universelle des Mines.
Revue industrielle.
Stahl und Eisen.
Der Civil-Ingenieur.
Glaser's Annalen.
Giornale del Genio Civile.
Ingeniero y Ferretero Español y Sud Americano.
The American Engineer.
The American Manufacturer.
The Engineering and Mining Journal.
The National Car and Locomotive Builder.
The Railroad and Engineering Journal.
The Railway Master Mechanic.
The Indian Engineer.
The Engineer.
Engineering.
The Railway Engineer.
The Marine Engineer.
Iron.
The Iron and Coal Trades Review.
Ryland's Iron Trade Circular.
The Ironmonger.
Ironmongery.
The Mechanical World.
The Mining Journal.
The Colliery Guardian.
The Machinery Market.
The Builder.
The Electrician.
The Electrical Review.
The Chamber of Commerce Journal (from Mr. Henry Chapman).
The Contract Journal.
The Plumber and Decorator.
The Shipping World.
The Steamship.
The Fireman.
Industries.
Invention.
The Practical Engineer.
Electrical Plant.
Hardware Trade Journal.
Trade Mark Times.
The Railway Review.
Phillips' Monthly Machinery Register.
The Engineering Review.

The President moved that the Report of the Council with the statement of accounts be received and adopted.

Dr. William Anderson, President elect, had much pleasure in seconding the motion.

No Member offering any remarks, the motion was put to the Meeting, and agreed to.

The PRESIDENT announced that the Ballot Lists for the election of Officers for the present year had been opened by a committee of the Council, and that the following were found to be elected:—

PRESIDENT.

DR. WILLIAM ANDERSON, F.R.S., . . Woolwich.

VICE-PRESIDENTS.

SIR JAMES N. DOUGLASS, F.R.S., . . London.
EDWARD B. MARTEN, ... Stourbridge.

MEMBERS OF COUNCIL.

JOHN A. F. ASPINALL, . . . Horwich.
WILLIAM DEAN, ... Swindon.
FRANCIS C. MARSHALL, . . . Newcastle-on-Tyne.
HENRY D. MARSHALL, . . . Gainsborough.
J. HARTLEY WICKSTEED, . . . Leeds.

In consequence of the election of Dr. Anderson as President, the resulting vacancy in the list of Vice-Presidents had been supplied by the Council by the appointment of Sir DOUGLAS GALTON, K.C.B., D.C.L., F.R.S., as a Vice-President for the present year; and for supplying the consequent vacancy among the Members of Council they had also appointed as a Member of Council for the present year Mr. JOHN G. MAIR-RUMLEY, whose name stood next highest in the voting for the election at this Meeting.

The Council for the present year will therefore be as follows:—

PRESIDENT.

DR. WILLIAM ANDERSON, F.R.S., . . Woolwich.

PAST-PRESIDENTS.

THE RIGHT HON. LORD ARMSTRONG, C.B.,
D.C.L., LL.D., F.R.S., . . . Newcastle-on-Tyne.
SIR LOWTHIAN BELL, BART., F.R.S., . Northallerton.
SIR FREDERICK J. BRAMWELL, BART.,
D.C.L., F.R.S., ... London.

PAST-PRESIDENTS (*continued*).

EDWARD H. CARBUTT, . . .	London.
CHARLES COCHRANE, ...	Stourbridge.
THOMAS HAWKSLEY, F.R.S., . . .	London.
JEREMIAH HEAD, ...	Middlesbrough.
JOHN RAMSBOTTOM, ...	Alderley Edge.
JOHN ROBINSON, ...	Leek.
JOSEPH TOMLINSON, ...	London.
PERCY G. B. WESTMACOTT, . . .	Newcastle-on-Tyne.

VICE-PRESIDENTS.

SIR JAMES N. DOUGLASS, F.R.S., . .	London.
SIR DOUGLAS GALTON, K.C.B., D.C.L., F.R.S., ...	London.
ALEXANDER B. W. KENNEDY, F.R.S., .	London.
EDWARD B. MARTEN, ...	Stourbridge.
SIR JAMES RAMSDEN, ...	Barrow-in-Furness.
E. WINDSOR RICHARDS, . . .	Low Moor.

MEMBERS OF COUNCIL.

JOHN A. F. ASPINALL, . . .	Horwich.
WILLIAM DEAN, ...	Swindon.
JOHN HOPKINSON, JUN., D.Sc., F.R.S., .	London.
SAMUEL W. JOHNSON, . . .	Derby.
ARTHUR KEEN, ...	Birmingham.
WILLIAM LAIRD, ...	Birkenhead.
JOHN G. MAIR-RUMLEY, . . .	London.
FRANCIS C. MARSHALL, . . .	Newcastle-on-Tyne.
HENRY D. MARSHALL, . . .	Gainsborough.
EDWARD P. MARTIN, ...	Dowlais.
WILLIAM H. MAW, ...	London.
T. HURRY RICHES, ...	Cardiff.
WILLIAM H. WHITE, C.B., F.R.S., .	London.
J. HARTLEY WICKSTEED, . . .	Leeds.
THOMAS W. WORSDELL, . . .	Arnside.

The PRESIDENT said that, in relinquishing the presidential chair to his successor, Dr. Anderson, he wished to return his thanks to all the Council who had worked with him during the past two years, for their cordial support and co-operation and for their frequent attendances at the numerous Council and Committee meetings. During the whole of the two years he had presided over their deliberations not a single matter had ever come before them that had not been decided with the greatest unanimity. He now asked Dr. Anderson to take the chair.

Dr. WILLIAM ANDERSON, President elect, on taking the chair, said the first duty he had to perform was undoubtedly a most agreeable one, namely to tender the heartiest thanks of the Council, and he was sure of the Members also, to their late President, Mr. Tomlinson. Certainly no one who had taken an interest in the Institution could have failed to observe the admirable way in which he had fulfilled his duties. The Council, who had had the pleasure of working with him, all knew well with what ready tact and with what good humour and good nature he had always guided their discussions: so that the unanimity of which he had just spoken was really in a great measure due to himself.

He did not know in what terms sufficiently to express his appreciation of the honour that the Members had done him in electing him to fill the chair which had been occupied by so many distinguished men. He had the greatest possible diffidence in accepting the post, partly because he really felt a doubt as to his qualifications, and partly because, in his position at the head of the great government manufacturing establishments, he was really not master of his own time: so that, with the best possible disposition to attend to the interests of the Institution, he should probably be often unable to do so to the full extent which he desired, and which the Members had a right to expect. He must obey the summonses connected with his official duties, and attend meetings of all kinds at the War Office and elsewhere when required to do so. This very afternoon, for example, he had hoped to be able to attend the Council meeting of the Institution; but at the last moment a meeting of the Ordnance Committee had been fixed for the same time, and he had therefore been unable to attend the Council. To calls of that sort he was afraid he should be liable; but the Members might rest assured that he should do his best to repay the confidence they had placed in him, by letting nothing of a trivial nature stand in the way of attending to his duties in connection with the Institution. As ill luck would have it, he was obliged both this evening and tomorrow to ask their kind indulgence. This evening was to be repeated at the Royal Institution the highly important lecture by the distinguished American electrician, Mr. Tesla; and as a member of that Institution he felt it almost a point of honour to attend, more especially as in doing so he should now be able to represent the Institution of Mechanical Engineers. The Members therefore he hoped would forgive him if in the course of the meeting he asked their Vice-President, Sir James Douglass, to take the chair in his place. The reason why he should have to ask their indulgence in regard to a portion of tomorrow evening's meeting was that Professor Roberts-Austen was to lecture at the Royal Institution on steel at high temperatures. The lecture was in a great measure the outcome of the research which Professor Roberts-Austen was now conducting on behalf of this Institution; and he had long ago promised him to attend on the occasion of its delivery, which likewise he should now be able to do as the representative of this Institution.

He would now ask Mr. Carbutt to invite them to accord to Mr. Tomlinson the vote of thanks which he had so admirably earned. He hoped he might be able himself in some small measure to follow in Mr. Tomlinson's footsteps, and to discharge with equal fidelity the trust committed to him.

Mr. Edward H. Carbutt, Past-President, heartily concurred in the terms employed by the President in so appropriately proposing a vote of thanks to Mr. Tomlinson. That the election of the new President should have called Dr. Anderson now to occupy the chair was indeed a peculiar satisfaction to himself, inasmuch as he had for years exerted himself in the direction of getting a mechanical engineer placed at the head of the government manufacturing departments. The Government had done the Institution the honour of choosing one of the Vice-Presidents for that post; the Members were all delighted at the appointment, and felt that it was an honour which the Institution had fully deserved. They had now conferred on Dr. Anderson the further honour of electing him as their President; and he was quite sure that he would worthily fulfil the duties of the office.

In seconding the vote of thanks which had been proposed to the retiring President, he wished to say that he had known Mr. Tomlinson ever since he himself had commenced his own mechanical training. He had served his time on the Midland Railway, with which Mr. Tomlinson was then connected; and having followed his career ever since, he had seen how sound had been all the work he had done as a mechanical engineer, and how much he had at heart the well-being of the mechanical engineering of the country. In all that he had done, he had always endeavoured to make progress, to get on well with his men, and to make everything work satisfactorily for the railways and others for whom he was acting. He had now been the President of this Institution for two years, and the Members all knew how he had fulfilled his duties, and how he had striven in every way to promote the welfare of the Institution. In the first year of his presidency the Members who had attended the summer meeting held in Sheffield were all witnesses what a successful meeting it had proved. He had himself been at the Liverpool meeting last year, which he considered was one of the most successful the Institution had ever held. He hoped that, although Mr. Tomlinson now retired from the presidency, he would continue the work on which he had been for some years engaged, as chairman of the Research Committee on Friction. He should like to see a larger proportion of the surplus, which was being put by every year, spent on research. If their retiring President, in conjunction with the four locomotive superintendents who were now on the Council —Mr. Aspinall of the Lancashire and Yorkshire Railway, Mr. Dean of the Great Western, Mr. Johnson of the Midland, and Mr. Riches of the Taff Vale Railway—and others, would only form a committee for going into the question of the rolling friction of rolling stock, including both engines and carriages, he believed that they would materially help to reduce the cost of carriage, and in the end would do much more good to the railways and the country than the railway rates enquiry, which had been going on at so large a cost, and seemed to have settled nothing satisfactorily. If such a committee were formed by the Institution of Mechanical Engineers, and if a sufficient amount of money were placed at their disposal for carrying out the research in a proper manner, the Institution he considered would be benefited as well as the country at large. No one could be a better chairman of such a committee than their retiring President. He was sure he was speaking on behalf of every member of the Institution in heartily seconding the vote of thanks to Mr. Tomlinson for his conduct as President during the past two years, trusting that he would enjoy long life, health, and happiness, and would thereby be enabled to continue to serve the Institution in the future as effectively as he had done in the past.

Sir JAMES N. DOUGLASS, Vice-President, had great pleasure in supporting the resolution, because he had served under Mr. Tomlinson both on the Council and on committees, and as a London member of the Institution had enjoyed many opportunities which country members did not possess of witnessing the hard work that he had done for the Institution in his position as President during the last two years. He did not hesitate to say that no president could possibly have given more earnest attention to the duties of his office; and he was sure the Members would most heartily thank him for all he had done on their behalf.

The vote of thanks was passed with applause.

Mr. TOMLINSON said that, although he knew it had long been usual to pass a vote of thanks to the retiring President, he was none the less sensible of the value of the compliment conveyed in the vote of thanks which had just been passed. Two years ago, when he became President of the Institution, he had promised that he would do the best he could in that position; and throughout his term of office he had endeavoured to redeem the promise. If happily he had succeeded in doing so to the satisfaction of the Members, he was himself highly gratified. It was possible of course that the Members in general might not quite realise all the work which the members of the Council had to do. It was naturally upon the London members of the Council that the largest demands were made for their time and attention; but there were also many others not living in London, of whom he must say that during his presidency they had been as efficient as it was possible for men of business to be who had to come two or three hundred miles to attend the Council meetings. As a London member himself of course he had not begrudged taking his full share of the work; and he had received cordial support throughout from every one of his colleagues. Looking at the progress of the Institution from year to year, it would be observed that during the last two years the actual increase in the number of members, after allowing for deaths and resignations—and the loss by deceases had been rather heavy during the last two years—had been 10½ per cent.; which he thought spoke for itself, and showed that the Institution had been doing something which had been appreciated by engineers. During the same period between £4,000 and £5,000 additional capital had been accumulated, in regard to which he fully agreed with the observation of Mr. Carbutt, and should be glad if that money were utilised in some other way than in the purchase of debentures. He did not despair of being able to push forward the matter of a house for the Institution; but Rome was not built in a day, and in London it was impossible to buy a piece of ground where they would like, and at the price they would like. If they wanted to buy a piece of freehold land suitable for a building anything like that of the Institution of Civil Engineers, they would have to get rid of the whole of their savings simply to buy the land, and then they would have nothing left wherewith to build upon it. The matter therefore was one which required some little thought, and it seemed better to wait until they had more money at their disposal, rather than to begin prematurely and run into debt. At present they had no need to run into debt, and therefore possibly the suggestion made by Mr. Carbutt might be realised, as he hoped it would be in his time. It had been a great pleasure to himself to serve during the last twelve years as a Member of Council, as a Vice-President, and as President; but it was a sorrow to him to find, in looking through the present list of members, that of the 259 who were members in 1857—the year in which he joined the Institution—there were now only thirty remaining in the list whose membership was older than his own, including the President who had joined six months before himself; and of the original members at the formation of the Institution in 1847 there were now only eight remaining. He thanked the members for the cordial reception they had given to the motion proposed in regard to himself; and he assured them he should always continue to do his best for the Institution as far as labour was concerned, so long as his health and strength permitted.

The PRESIDENT reminded the Members that at the present meeting the appointment had to be made of an Auditor for the present year.

On the motion of Mr. HENRY CHAPMAN, seconded by Mr. DAVID JOY, it was unanimously resolved that Mr. Robert A. McLean, chartered accountant, 1 Queen Victoria Street, London, be re-appointed to audit the accounts of the Institution for the present year, at a remuneration of Ten Guineas, being the same as heretofore.

The PRESIDENT said he had the pleasure of announcing that the Summer Meeting of the Institution in the present year would be held in Portsmouth; and a cordial promise of important facilities for the occasion had been received from the Mayor of Portsmouth, T. Scott Foster, Esq. The arrangements would include also a visit to Southampton, for which a similar welcome promise had been received from the Mayor of Southampton, James Lemon, Esq. The Meeting would commence on Tuesday the 26th of July, and would last four days. He had no doubt the Admiralty would give special facilities for seeing the interesting works which were being carried on in the Royal Dockyard.

The following Paper was then read and discussed:—

"Notes on Mechanical Features of the Liverpool Water Works, and on the supply of Power by pressure from the public mains, and by other means;" by Mr. JOSEPH PARRY, Water Engineer, Liverpool.

At Twenty minutes to Ten o'clock the Meeting was adjourned till the following evening. The attendance was 99 Members and 55 Visitors.

The ADJOURNED MEETING of the Institution was held at the Institution of Civil Engineers, London, on Friday, 5th February 1892, at Half-past Seven o'clock p.m.; Dr. WILLIAM ANDERSON, F.R.S., President, in the chair.

The following Paper was read and discussed:—

"On the Disposal and Utilization of Blast-Furnace Slag;" by Mr. WILLIAM HAWDON, of Middlesbrough. Communicated through Mr. CHARLES COCHRANE, Past-President.

On the motion of the Chairman a vote of thanks was unanimously passed to the Institution of Civil Engineers for their kindness in granting the use of their rooms for the Meeting of this Institution.

The Meeting then terminated at Nine o'clock. The attendance was 33 Members and 35 Visitors.

NOTES ON MECHANICAL FEATURES OF THE LIVERPOOL WATER WORKS, AND ON THE SUPPLY OF POWER BY PRESSURE FROM THE PUBLIC MAINS, AND BY OTHER MEANS.

BY MR. JOSEPH PARRY, WATER ENGINEER.

When the Members of this Institution visited Liverpool in the summer of 1863, a paper on the Mechanical Features of the Liverpool Water Works was read by the late Mr. Thomas Duncan, who was at that time Water Engineer to the Corporation. Five of the pumping engines and the whole of the mechanical appliances on the Rivington pipe line then described are still in use. Three Cornish pumping engines, each capable of lifting about 1½ million gallons a day, were added between 1865 and 1868. A resolution of the City Council in 1874, to obtain a large additional supply from a distant gathering ground, diverted attention from the further development of the well system; and there have consequently been no recent additions to the pumping machinery. The extensive works for conveying water to Liverpool from the river Vyrnwy, which were commenced in 1880 and are now approaching completion, include many new and interesting mechanical appliances; but a description of these does not come within the scope of the present paper. Of the engines described by Mr. Duncan in 1863, there may still be seen on active duty, as examples of veteran pumping machinery, the Cornish engine and boilers erected at the Windsor station (Fig. 2, Plate 2) in 1840 by Messrs. Rigby and Co. of Hawarden; a crank engine made in 1837 by Messrs. Mather Dixon and Co., and now at the Green Lane station; also at the same station a Cornish engine and boilers made by Messrs. Harvey and Co. of Hayle in 1846, to the specification of the late Mr. James Simpson. This last engine is an excellent specimen of the workmanship of forty-five years ago. Its cylinder is 50 inches diameter, steam-jacketed, with 9 feet stroke; pump 17¼ inches diameter, 8 feet 9 inches stroke; total lift 278½ feet. It has three boilers, each 26 feet long by 5 feet 9 inches diameter. The average boiler pressure is 35 lbs. per square inch. Since it began its career it has lifted altogether 19,061 million foot-tons of water to the end of 1891. Its duty as ascertained on a recent trial was 55·7 millions of foot-lbs. per cwt. of coal. Indicator diagrams taken during the trial gave 86·66 indicated horse-power, and 71·2 pump horse-power. The coal consumption per indicated horse-power per hour was 3·29 lbs.; and the water evaporated per lb. of coal from and at 212° Fahr. was 8·541 lbs., with jacket in circulation.

The cost of pumping from the deep wells and at the high-level station in Aubrey Street (formerly called Audley Street) during the year ending 31st December 1890 was as follows, per thousand gallons lifted one hundred feet:—

Aubrey Street station	. . . . .	0·4203 penny
Green Lane „	. . . . . .	0·2560 „
Windsor „	. . . . . .	0·3129 „
Dudlow Lane „	. . . . . .	0·2101 „
Bootle „	. . . . . .	0·2647 „

Water Supply.—The compulsory district of supply includes 36 townships, parishes, or places, and embraces an area of 114 square miles, Fig. 1, Plate 1. Outside of the compulsory area there are two townships in Lancashire supplied with water in bulk. The total population now receiving a supply from the Liverpool works is about 800,000. The average quantity of water distributed per day is:— from deep wells and boreholes sunk in the new red sandstone in and around the city, 6½ million gallons; from the Rivington reservoirs, situated in Lancashire between Bolton and Blackburn, and having a gathering ground of 10,000 acres, 12 million gallons. The average rate per head per day for all purposes is about 24 gallons. There is also a small supply of salt water pumped from the Mersey Railway tunnel, for public swimming baths, street watering, and flushing. In the summer months this amounts to 2 or 3 million gallons per week.

The water is distributed on the constant-service system; but the constant supply has been occasionally interrupted in recent years in consequence of the inadequacy of the existing sources to meet the demands during periods of high consumption.

Motive Power.—The special subject dealt with in this paper is the supply of water for Power by direct pressure from the street mains, and its cost to consumers as compared with other kinds of power. Schemes for the distribution of power from a central station, by means of water pressure, compressed air, steam, gas, and electricity, have of late years occupied attention in Liverpool as elsewhere; but up to the present time only water pressure and gas have got beyond the experimental stage. It is probable that at no distant date a supply of power transmitted through a pipe or wire from a central station will be looked upon as a necessity of urban life, for domestic service as well as for the purposes of trade. Already there has been sufficient advance in this direction to claim for the subject the serious attention of engineers, and to invest with interest any facts which relate to it In towns where much machinery is employed for manufacturing or commercial purposes, and especially where power is required intermittently, a strong primâ facie case can be made out in favour of a system of distribution from a single station; but it is a matter of common experience among those who have been engaged in such schemes that there is great difficulty in obtaining accurate information with regard to the cost of doing work under the detached system as distinguished from a combined or centralized system; and this difficulty is due as much to the absence of exact records as to the reluctance of power-users to give details of their expenditure.

Most of the comparative figures here given were obtained in the course of an investigation made by the writer about two years and a half ago, for the purpose of reporting to the Corporation of Liverpool on a proposal of the Hydraulic Power Company to extend their mains outside of an area within which they had been allowed to supply power under an Act passed in 1884. Liverpool would appear to offer a particularly favourable field for the establishment of a system of power distribution, because there are so many large warehouses, where goods for import and export are stored, occupying a limited and well-defined area, in streets which are entirely devoted to trade. The weight of goods imported into and exported from the port of Liverpool in one year exceeds 14 million tons; and the weight of goods stored at one time in the Liverpool warehouses is estimated to be about $\frac{3}{4}$ million tons. There is therefore a large amount of lifting and lowering to be done, from and into ships, railway wagons, and warehouses, apart altogether from the trading and manufacturing needs of the city population. A great deal of this work too is of an intermittent character. Bales and boxes are carted to the warehouses as the ships arrive; and as the deliveries are irregular, the hoisting machinery is often idle, and is seldom in full work the whole of a day.

Use of Hydraulic Machinery.—Use was first made in Liverpool of pressure from the street mains for the working of hydraulic machinery about the year 1847, when Lord Armstrong, then Mr. W. G. Armstrong, erected a hydraulic crane at the Albert Dock for the late Mr. Rendel. This form of power does not appear to have been much in demand by Liverpool warehouse owners down to thirty years later; for in 1877 the number of hydraulic machines supplied from the public mains was only 89. The Mersey Docks and Harbour Board introduced a high-pressure system of their own, and have for many years had a very extensive and efficient hydraulic plant extending along the whole line of docks.

The number of hydraulic machines now worked by direct pressure from the street mains of the Corporation is 162. The following is a classification of the machines, with a statement of the quantity of water consumed by each class during the year ended on 31st December 1890:—

Number of Machines.	Use of Motors.	Consumption of Water. Gallons per annum.
25	Organ blowing	4,864,000
10	Passenger Lifts	21,536,000
114	Goods Lifts	96,896,000
7	Motors for ventilators &c.	1,352,000
3	Curtain Lifts for theatres	320,000
3	Hair Dressers	632,000
Total 162		125,600,000

Pressure of Water.—The pressure of water in the mains in the warehouse districts varies from about 50 to 80 lbs. per square inch. On the line of docks the pressure is seldom less than 75 lbs. per square inch; but notwithstanding that this excellent pressure is available, and that the water is sold at a lower price per thousand gallons than is generally charged for trade purposes, it has been used to only a very limited extent for power purposes in the warehouses near to the docks. Nearly the whole of the 114 goods lifts included in the above table are in general-produce warehouses and in bonded stores, at some distance from the docks, and at a higher level.

Other Powers.—The great bulk of the loading and unloading work in Liverpool is done by steam power. Gas engines have of recent years been fixed in many of the new warehouses, and have replaced steam in some of the old warehouses; the total number of gas engines now in use is about 650. Compressed-air machinery was tried in a few warehouses several years ago; but it has not been sufficiently successful to lead to its adoption on a large scale.

Cost of Water Power.—The average amount paid per annum for working a goods hoist from the Corporation mains is £13; and, as this represents an average payment of only 10*d.* per hoist per working day, the cost is very small for the convenience afforded. With an available pressure in the mains of 60 lbs. per square inch, properly utilised, and at the present rate of sevenpence per thousand gallons, the cost of lifting one ton to a height of 50 feet is one penny, equal to about 1*s.* 1½*d.* per horse-power per hour. There are many places in which large consumers of water might take advantage of pressure from the mains for mechanical purposes, and afterwards use the water for their boiler or other trade requirements; but so far as is known to the writer the only instance in which this is done is at the Exchange Station, Liverpool, where the Lancashire and Yorkshire Railway Company have two passenger and luggage lifts worked by pressure from the mains, and where the exhaust is utilized for feeding boilers and locomotives, and for the general supply of a large passenger station.

Cost of different Powers.—One of the principal objects of the writer's investigation was to obtain trustworthy data for enabling him to calculate the actual cost in practice of similar work done by different classes of machinery; and for this purpose observations were made and measurements taken by his assistants, with the co-operation of the owners and occupiers of warehouses and other buildings. In Table 4 (page 47) are contained particulars of cases which were selected as being of a representative character, and fair examples of average and relative costs. The figures given as the cost per indicated horse-power per hour are only estimates, based upon the best information that was available as to the average weight and number of packages lifted. The actual amount of work done could not be ascertained in these instances with the same accuracy and completeness as in the examples given in Table 5 (page 48), which were subsequently and specially obtained for this paper.

Low-pressure Hydraulic Hoists.—The following Table 1 gives the results of one day's working of three hydraulic hoists in a provision warehouse by direct pressure from the street mains :—

TABLE 1.

One Day's Working of three Low-pressure Hydraulic Hoists.

Hoist		A	B	C
Total Weight lifted	cwts.	493	94	69
Total number of lifts	no.	197	37	18
Average Weight per lift	cwts.	2·50	2·54	3·87
Height of lift	feet	1 to 24	2 to 35	4 to 20
Mean Speed*	feet per minute	115	140	110
Total Work	foot-tons	167·8	108·6	39·2
Cost of Water, total	pence	30·0	15·4	4·9
" " per foot-ton	penny	0·1788	0·1418	0·1250

* The speed varied from 69 to 200 feet per minute.

The following observations in Table 2 were made in separate warehouses :—

TABLE 2.

Working of Low-pressure Hydraulic Hoists.

Warehouse		D	E	F
Total Weight lifted	cwts.	144	126	108
Total number of lifts	no.	24	9	27
Height of lift	feet	26	15½	26½
Time occupied in lifting	minutes	23	35	35
Water Pressure in mains	lbs. per sq. inch	76	75	75
Water consumed	gallons	960	1,662	1,250
Cost per ton lifted 50 feet	pence	1·79	5·95	3·05

In judging of the economy of low-pressure hydraulic machinery, allowance must be made for the fluctuations of pressure in the street mains, as well as for variations of load. For intermittent purposes and for small loads this form of power is probably less costly than any other, when water is as cheap and the pressure as high as in Liverpool. The first cost of the machinery is moderate, and the expenditure of power ceases as soon as the duty is performed. No skilled attendance is required; the manipulation of the valves is simple, and to a large extent automatic; and there is freedom from risk of fire.

No counterbalance weights are attached to the hydraulic goods lifts; and consequently the weight of the platform and ram has to be added, in order to arrive at the total weight lifted by the water pressure. None of the hydraulic machines in Liverpool are fitted with devices for varying the expenditure of water according to the variation of load.

Steam Power.—Most of the mechanical work in the warehouse districts of Liverpool is done by steam power. To a large extent the warehouses have been built in blocks, each block under one ownership; and the hoists are worked by shafting driven from one central steam engine. In such cases the cost of skilled attendance per hoist is reduced; but where steam is employed for only one or two warehouses, the loss in getting up steam and in stoppages, added to the wages of an engine-driver and the expenditure for repairs, makes the total cost of steam considerably higher than that of gas or hydraulic power.

Two typical examples G and H of steam machinery in blocks of warehouses are given in Table 4 (page 47). In example G the engine and boiler are on the top floor of a fire-proof warehouse. The engine is horizontal, with single cylinder 12 inches diameter by 28 inches stroke, speed 60 revolutions per minute, and steam supplied at 60 lbs. per square inch from a tubular boiler. The estimated annual expenditure is as follows, and the approximate cost per indicated horse-power comes to 0·62*d.* per hour :—

	£	*s.*	*d.*
Coal and slack	40	0	0
Oil, cotton waste, &c.	7	6	0
Water supply	5	14	0
Engineman's wages	78	0	0
Repairs	4	0	0
Total working expenses . .	£135	0	0

In example H the hoists are driven by an engine with single cylinder 10½ inches diameter and 21 inches stroke, running at 90 revolutions per minute; vertical boiler with 45 lbs. pressure. The estimated annual working expenses are as follows, and the approximate cost per indicated horse-power is 1·00*d.* per hour :—

	£	*s.*	*d.*
Coal and slack	45	6	8
Oil, tallow, &c.	5	13	4
Water supply	5	0	0
Engineman's wages	78	0	0
Total working expenses . .	£134	0	0

Three additional examples N O P of the cost of working warehouse hoists by steam, which have been recently obtained for this paper, are summarised in Table 5 (page 48).

Example N.—Block of five warehouses, with six floors and basement. One warehouse contains two hoists, and four warehouses contain one hoist each. The six hoists are driven by one horizontal engine fixed in the jigger loft of the middle warehouse, and coupled direct on to the line shaft. Maximum load per hoist 7 cwts.; usual load 6 cwts. Single-cylinder engine, 9 inches diameter and 16 inches stroke, 120 revolutions per minute, allowed to run continuously during business hours, whether doing useful work or not. Vertical boiler, steam pressure 40 lbs. per square inch. Machinery erected about thirteen years ago. Working expenses for year 1891:—

	£	s.	d.
Slack	38	0	0
Oil and stores	5	10	0
Water	3	5	0
Wages	78	0	0
Repairs	4	10	0
Total working expenses . .	£129	5	0

Cost per indicated horse-power per hour 1·27*d.* Total number of lifts during the year 69,925. Total estimated foot-tons 450,155. Example of one day's work, 1,000 bales lifted an average height of 35 feet; average weight per bale 5½ cwts.; making 9,625 foot-tons.

Example O.—Block of three warehouses, with six floors and basement; one hoist in each warehouse. Maximum load per hoist 7½ cwts.; average load 4½ cwts. Speed about 120 feet per minute. Separate doors for receiving and delivering goods. Hand jiggers used for lowering. Horizontal engine, cylinder 8 inches diameter and 16 inches stroke, 98 revolutions per minute, said to be started and stopped to suit requirements of work. Vertical boiler 4 feet diameter and 8 feet high; ordinary steam pressure 40 lbs. per square inch. Engine fixed in basement and driving horizontal shafting in jigger loft through bevel wheels and about 70 feet of vertical shafting. Warehouseman attends to engine and boiler. Working expenses for year 1891:—

	£	s.	d.
Slack	21	9	6
Oil and Water	4	3	3
Proportion of wages	26	0	0
Repairs	5	10	0
Total working expenses . .	£57	2	9

Cost per indicated horse-power per hour 0·74*d.* Total number of lifts during the year 20,350. Total foot-tons about 151,110.

Example of one day's work, 344 packages lifted an average height of 27 feet; average weight of each 6½ cwts.; making 3,018 foot-tons.

Example P.—Two blocks of warehouses of five floors, separated by a space of 20 feet. Two hoists in each block, all driven by one engine fixed on top floor. Engine drives through bevel wheels an upright shaft and a line shaft crossing the open space between the two blocks; altogether 230 feet of 2½ inch shafting. Engine cylinder 11¼ inches diameter and 24 inches stroke, 90 revolutions per minute when running light; allowed to run continuously during business hours, unless work is very slack. Vertical boiler 5½ feet diameter and 10 feet high; steam pressure 40 to 45 lbs. per square inch. Maximum load on hoists 50 cwts., double purchase; usual load 7 to 8 cwts. Speed 150 to 160 feet per minute. First cost of machinery £1,500; erected fifteen years ago. Working expenses for year 1891:—

	£	s.	d.
Slack	25	10	0
Oil	4	14	6
Water	2	16	8
Wages	78	0	0
Repairs	6	0	0
Total working expenses . .	£117	1	2

Cost per indicated horse-power per hour 0·59*d*. Total number of lifts during the year 18,524. Total foot-tons about 147,755. Example of one day's work by two of the hoists, 677 bales lifted from 12½ to 47½ feet, equivalent to 5,611 foot-tons.

Gas Engines.—In Table 4 (page 47) are given two examples J and K of the cost of working warehouse hoists by gas engines.

A subsequent example Q in Table 5 (page 48) is taken from a block of four warehouses, with six floors and basement, having one hoist in each warehouse. Maximum load with single purchase 8 cwts., with double purchase 13 cwts. Speed about 140 feet per minute. Crossley engine 8 HP. nominal, attended to by warehouse keeper; fixed on concrete floor in jigger loft, midway in length of block. Fast and loose pulley on each end of crank shaft, driving up to line shafts which work the hoists. One belt drives shaft for Nos. 1 and 2 hoists, and the other drives shaft for Nos. 3 and 4; so that either two or four hoists can be worked, as desired. Engine in use seven years. Cost of working for the year 1891:—

	£	s.	d.
Gas	20	0	0
Oil, stores, and water	5	0	0
Proportion of wages	20	16	0
Repairs	3	10	0
Total working expenses	£49	6	0

Cost per ton lifted 50 feet 2·71*d.* Total number of lifts during the year 36,236. Total foot-tons 218,254. Example of one day's work, 400 lifts averaging 5½ cwts. and 23 feet height, equivalent to 2,530 foot-tons, costing 0·76*d.* per ton lifted 50 feet.

In Table 6 (page 49) the results are given of three days' observations of the working of two gas engines, R and S, each 8 HP. nominal, lifting bales of cotton into warehouses adjoining the docks. In connection with these observations it transpired that it was the usual practice of the attendants to start the engines in the morning, or when delivery of goods was first expected, and to let them run throughout the day, or so long as they were likely to be wanted, rather than take the trouble to stop and re-start them. Two men were required to start each engine. For comparison with the results obtained under the conditions dealt with in the table, the following two examples T and U are given, in which the consumption of gas also corresponded with the specific performance of the duty stated.

Example T.—70,000 lbs. lifted 30 feet high in 1¾ hour. Speed 120 feet per minute. Gas consumption 200 cubic feet, costing 6·4*d.*, or 0·34*d.* per ton raised 50 feet, or 1*d.* per nominal HP. of engine per hour.

Example U.—Gas engine of 3½ HP. nominal, working single hoist; 106 tons of grain in bags lifted 40 feet in nine hours. Gas consumption 1,025 cubic feet, costing 32·8*d.*, or 0·39*d.* per ton raised 50 feet.

Into another warehouse W, 605 bags, each of 200 lbs. average weight, in eleven wagon-loads, were lifted 10 feet 8 inches, at a speed of 80 feet per minute, by a 3½ HP. gas engine working a geared hoist by open and crossed belts. The delivery of the eleven loads extended over 4½ hours, and the consumption of gas as read at the beginning and end of the time was 1,300 cubic feet, the cost being 41·6*d.* The cost per foot-ton for gas was therefore 0·07*d.*, or 3·5*d.* per ton lifted 50 feet. In this instance it was found that in ordinary practice the engine was allowed to run all day, though hoisting was required only at intervals. If the engine had here been stopped and re-started for each wagon-load, the cost for gas would have been reduced by about one half.

In a large block of offices X, an 8 HP. gas engine is employed to pump water into an accumulator for working by water pressure two passenger and two goods lifts. The passenger lifts travel 55 feet, at a speed of 220 feet per minute; they run nine hours per day, during which they make about 400 journeys each; the average load per lift is 3·75 cwts., including weight of lift. The goods lifts travel 20 feet, at 100 feet per minute; the average load per lift is 8 cwts. The total working expenses per annum are as follows:—

	£	s.	d.
Gas	80	0	0
Oil, waste, tallow, &c.	50	0	0
Engine driver at 26*s.* per week . .	67	12	0
Repairs	2	4	0
Total working expenses . .	£199	16	0

Compressed Air.—Two examples L and M of compressed-air machinery are given in Table 4 (page 47). Example L comprises one steam cylinder and two compressors, all three being 12 inches diameter and 24 inches stroke, with an air pressure of 45 lbs. per square inch when all the nine hoists are working; and the cost of working comes to about 1·00*d.* per indicated horse-power per hour.

In example M there are two steam cylinders 12 × 22 inches, and two compressors 14 × 22 inches, with 45 lbs. air pressure for full work; the eighteen friction hoists are each worked by two air cylinders coupled direct to the friction-wheel shaft. The annual expense of this machinery is made up of the following items, and the cost per indicated horse-power comes to about 0·40*d.* per hour:—

	£	*s.*	*d.*
Slack, 106¼ tons	39	17	0
Oil, &c.	5	5	0
Water	5	0	0
Wages of attendant	78	0	0
Repairs	5	0	0
Total working expenses . .	£133	2	0

High-pressure Water.—The following are two examples Y and Z of warehouse hoists worked by high-pressure water from the Hydraulic Power mains.

Example Y.—20 bales or 10,000 lbs. lifted 20 feet in 7½ minutes. Speed 200 feet per minute. Water used 66 gallons, costing 3·16*d.*, or 1·75*d.* per ton lifted 50 feet.

Example Z.—339 bales, averaging 4½ cwts. each, lifted from 8½ to 50 feet. Water used 1,920 gallons, costing 0·045*d.* per foot-ton, or 2·25*d.* per ton lifted 50 feet. The returns kept in this warehouse showed 8,542 bales to have been lifted in six weeks at a cost for water of 0·24*d.* per bale.

Examples of work done by pressure from the Hydraulic Power mains were not sought to the same extent or investigated as fully as in the case of steam and gas, because with the assistance of the company's published scale of charges the cost of any given amount of work can be easily calculated.

The relative cost of high-pressure and low-pressure water for equivalent motive power is approximately shown by the following Table 3, in which the cost of water from the public mains at sevenpence per thousand gallons, with a pressure of 60 lbs. per square inch in the mains, and omitting friction of machinery, is compared with the cost of water from the Hydraulic Power mains at 700 lbs. per square inch. This comparison is also plotted graphically in Fig. 3, Plate 3.

TABLE 3.

Comparative Cost of High-pressure and Low-pressure Water.

See Fig 3, Plate 3.

Water Consumption per quarter.	HIGH-PRESSURE, at Hydraulic Power Co.'s charges.		LOW-PRESSURE, from public mains, for equivalent power, at 7*d.* per 1,000 galls.
	Rate per 1,000 gallons.	Total.	
Gallons.	*s.* *d.*	£ *s.* *d.*	£ *s.* *d.*
4,000	10 0	2 0 0	1 7 2
10,000	7 0	3 10 0	3 8 0
25,000	5 9¾	7 5 0	8 10 1
50,000	5 0	12 10 0	17 0 3
100,000	4 0	20 0 0	34 0 6
150,000	3 5	25 12 6	51 0 10
200,000	3 1½	31 5 0	68 1 0
300,000	2 10	42 10 0	102 1 8
Above 300,000	2 6		

The supply of power for hydraulic machines from the public water mains in Liverpool has been facilitated by the existence throughout the city of special fire mains, to which it has been practicable to make the connections without appreciably affecting the efficiency of the general distribution of water for trade and domestic purposes. It is obvious that ordinary water mains laid for domestic and manufacturing services can be utilized only to a very limited extent for power purposes, and that the relatively low pressures they carry cannot be economically applied as a motive power for driving large machinery. With a pressure of 700 lbs. per square inch, which is the pressure usually adopted for hydraulic work, a main of 6 inches diameter will carry about 100 HP. To obtain the same efficiency from a main conveying water at the pressure ordinarily available for town supplies, the diameter of pipe required would be about 21 inches. As seen from the foregoing Table 3, pressure from the mains of corporations and water companies cannot compete with high-pressure water from mains specially laid for hydraulic purposes when the power water is sold at rates not exceeding five shillings per thousand gallons.

Under the instructions of the Corporation the writer prepared a scheme for a system of high-pressure mains with pumping machinery and accumulators for the distribution of power water; but subsequently the Corporation entered into an agreement with the Hydraulic Power Co., under which the latter were permitted to extend their mains into all the business parts of the city.

In compliance with the wish expressed in the discussion (pages 57 and 60), Table 7 has been added so as to give in a tabulated form the costs per foot-ton, in so far as it has been practicable to reduce to a common standard the examples quoted in the paper. In applying the figures contained in this and other tables, it is necessary to remember the statement on page 35 as to the intermittent character of the work done; and also that, in consequence of the great diversity and irregularity in the size and weight of the packages lifted, there is often a considerable waste of power.

TABLE 4.

Cost of working Warehouse Hoists by Steam, Gas, and Compressed Air.

Example, and Motive Power employed.	Nominal Horse-Power.	Number of years at work.	Particulars of Hoists.						Total first Cost.	Cost of working.		
			Number.		Load per hoist.		Speed.	Height of lift.		Total per annum.	Per Hoist per annum.	Per I.H.P. per hour.
			Total.	Average working at one time.	Maximum.	Average.	Feet per minute.					
	H.P.	Years.	No.	No.	Cwts.	Cwts.	Feet.	Feet.	£	£ s. d.	£ s. d.	Penny.
G Steam	12	4	7	4	8	5	180	55	1,000	135 0 0	19 6 0	0·62
H Steam	8	10	5	3	7	4½	158	54	600	134 0 0	26 16 0	1·00
J Gas	8	10	4	3	6	5	120	69	600	44 0 0	11 0 0	0·33
K Gas	3½	3	1	1	2½ to 10	4	300 to 75	58	140	28 10 0	28 10 0	0·52
L Comp. Air	...	20	9	6	8 to 10	5	250	60	...	184 0 0	20 9 0	1·00
M Comp. Air	20	3	18	8	6	5	250	33*	2,700	133 0 0	7 8 0	0·40

Particulars of Machinery.

STEAM.—In Example G the hoists were driven by a horizontal engine, with single cylinder 12 inches diameter and 28 inches stroke, running at 60 revolutions per minute; boiler pressure 60 lbs. per square inch. In Example H they were driven by an engine with single cylinder 10½ inches diameter and 21 inches stroke, running at 90 revolutions per minute; vertical boiler with 45 lbs. pressure.

COMPRESSED AIR.—Example L, one steam cylinder 12 inches diameter and 24 inches stroke, two compressors 12 × 24 inches; boiler 5½ × 20 feet, with 45 lbs. steam pressure; air pressure 20 to 45 lbs. Example M, two steam cylinders 12 × 22 inches, two compressors 14 × 22 inches; boiler 6 × 24½ feet, with 40 lbs. steam pressure; air pressure 20 to 45 lbs.

* Average height of lift.

TABLE 5.

Cost of working Warehouse Hoists by Steam and by Gas.

Example.	Motive Power employed.	Number of Hoists.	Average Cost of Working.				Example of Cost of one day's working.	
			Per Hoist per annum.	Per I.H.P. per hour.	Per foot-ton.	Per lift.	Per foot-ton.	Per lift.
		No.	£ s. d.	Penny.	Penny.	Penny.	Penny.	Penny.
N	Steam	6	21 10 10	1·27	0·0689	0·443	0·0104	0·101
O	Steam	3	19 0 11	0·74	0·0907	0·673	0·0147	0·129
P	Steam	4	29 5 3	0·59	0·1902	1·510	0·0163	0·135
Q	Gas	4	12 6 6		0·0542	0·323	0·0152	0·096

TABLE 6.

Results of three days' working of Two Gas Engines, each 8 horse-power nominal, driving Hoists in Warehouses.

Example, and Duration of test.		Height of lift.		Load lifted.		Total number of lifts.	Average Load per lift.	Total Work.	Cost of Gas.					
				To each floor.	Total.				Total.	Per hour.	Per nominal horse-power per hour.	Per foot-ton.	Per fifty foot-tons.	Per lift.
	Hours.	Feet.	Ins.	Tons.	Tons.	No.	Cwts.	Ft.-Tons.	Pence.	Pence.	Penny.	Penny.	Penny.	Penny.
R	6	49 31	2 9	23·6 20 8	44·4	200	4·4	1824·7	35·2	5·86	0·73	0 0193	0·965	0·1760
R	6¼	49 31 40 10	2 9 6 0	97·2 41·5 26·5 4·4	169·6	720	4·7	7225·9	43·2	6·91	0·86	0·0059	0·295	0·0600
R	5½	40 12	9 0	171·5 4·5	176·0	775	4·5	7068·0	40·0	7·27	0·91	0·0056	0·285	0·0516
S	6	40 22	9 6	10·6 10·6	21·2	100	4·2	673·1	19·5	3·25	0·41	0·0290	1 450	0·1950
S	6½	31	7	28·3	28·3	184	3·1	897·1	25·9	3·98	0·50	0·0289	1·445	0·1407

The mean speed of lifting was 118 feet per minute in each test. There are four hoists worked by each engine.

TABLE 7.

Cost of Motive Power per foot-ton,
summarized from Examples quoted in paper.

Page.	Example.	Cost per foot-ton.	Motive Power, and conditions of measurement.
		Penny.	*Water Pressure from public mains.*
37	A	0·1788	Measurement of work done in one day. Cost of water only.
37	B	0·1418	
37	C	0·1250	
38	D	0·0358	
38	E	0·1190	
38	F	0·0610	
			Steam Engines.
40,48	N	0·0689	Total work done and total working expenses, during one year.
40,48	O	0·0907	
41,48	P	0·1902	
40,48	N	0·0104	Work done in one day, but cost of working calculated on average of annual expenditure.
40,48	O	0·0147	
41,48	P	0·0163	
			Gas Engines.
42,48	Q	0·0542	Total work done and total working expenses, during one year.
42,48	Q	0·0152	One day's work, and average of working expenses.
42,49	R	0·0193	Measurement of work done in one day. Cost of gas only.
42,49	R	0·0059	
42,49	R	0·0056	
42,49	S	0·0290	
42,49	S	0·0289	
42	T	0·0068	Measurement of work done and gas consumed in 1¾ hour.
42	U	0·0077	Do. do. 9 hours.
43	W	0·0700	Do. do. 4½ hours.
			Water Pressure from Hydraulic-Power mains.
44	Y	0·0350	Measurement of work done in 7½ minutes. Cost of water only.
44	Z	0·0450	Do. one day. Cost of water only.

Discussion.

Sir James N. Douglass, Vice-President, occupying the chair in the absence of the President, mentioned that this paper had been kindly prepared by the author for last year's Summer Meeting of the Institution in Liverpool, from which it had had to stand adjourned owing to want of time for the reading and discussion to take place on that occasion.

Mr. Parry said it might be useful to add to the information given in the paper the price of coal and slack in Liverpool, and also the price of gas. The present price paid by the Corporation under contract was 10*s.* 2*d.* per ton for coal, and 6*s.* 11*d.* per ton for slack, the mean of the two being 8*s.* 6½*d.* per ton. In 1890, in which year most of the figures in the paper had been obtained, the contract price for coal had been 10*s.* 9*d.*, and for slack 7*s.* 8*d.*, the mean being 9*s.* 2½*d.* per ton. The price of gas per thousand cubic feet was at present 3*s.*, and last year 1891 it had been 2*s.* 8*d.*

In regard to utilizing for station purposes in the manner described in page 36 the exhaust water from the lifts at the Exchange Station of the Lancashire and Yorkshire Railway, Liverpool, the credit of that application of the water supply was due to Mr. Aspinall, who he hoped would give some further information with regard to the result of his experiment, in order that it might be the means of inducing others to follow his excellent example. There were many large consumers who might apply the pressure from the public mains for the performance of useful work, before using the water for the purpose for which the supply was primarily intended.

Mr. John A. F. Aspinall, Member of Council, said that, when it had been determined to put the passenger and luggage lifts in the hotel at the new Exchange Station in Liverpool, it became necessary to decide what course should be adopted for working them: whether

(Mr. John A. F. Aspinall.)
hydraulic plant should be provided for the purpose or not. Seeing that a large quantity of corporation water was used for locomotive purposes, and that it could be supplied at the hotel at about 60 lbs. pressure per square inch, it was thought desirable to try to utilize this pressure for working the hotel lifts, and then to pass the water on into a tank to be used afterwards for the locomotives. By that means the power could be got out of the water, without wasting it in any way. In this particular instance the plan had some disadvantages, which no doubt the author would be able to obviate in any future erections of the kind, because they were simply due he thought to the street mains having been laid before any such use of the water had been contemplated. Hence the fact was that during the daytime the mains were subject to certain variations of pressure, which materially affected the speed of the lifts; while during the night, when visitors were not in such a hurry to get up and down stairs, there was an ample amount of pressure. With the exception of that difficulty he considered the plan had worked well. If it were not for the fact that every drop of water could be utilized afterwards, it would be an expensive plan; but at present it practically cost nothing to work the lifts.

Mr. EDWARD B. MARTEN, Vice-President, asked how the water was afterwards used, and what was the back-pressure thrown upon the lifts by raising their exhaust water to the height required for supplying the locomotives.

Mr. ASPINALL replied that the exhaust water from the lifts passed through a main from the hotel, which was at one end of the station, to the other end of the station, where there was a large tank containing some thousands of gallons; and from that tank the water flowed to the different water columns for supplying the engines. The height that the water had to rise from the exhaust of the lifts into the storage tank was about 14 or 15 feet; and this amount of head was consequently lost for working the lifts.

On the Lancashire and Yorkshire Railway there were in both counties a large number of warehouses, some of which were worked on the old plan by means of shafting driven by a steam engine, and working either cranes or overhead jiggers or capstans in the station yard, and driving those capstans continuously. The difficulty with that plan was of course that all the capstans were running at once. It was an old-fashioned system, but it undoubtedly had some advantages, because it was so economical in the first instance; it did not involve spending a large amount of capital in laying it down. But it had grave objections: all the machinery was running at once, and if a piece of shafting broke the whole yard was at once stopped. Moreover the shafting was run underground, and was thus subject to having dirt from the yard washed down upon the bearings, which gave a great deal of trouble in consequence. The hydraulic system, which was much more modern, might be said to be much more convenient in every respect. It was not hampered by having to go in straight lines, as was the case with shafting; and the pipes could be carried in and out of a building almost in any way. With regard to gas, there were a certain number of small warehouses that had been fitted originally with steam engines, which worked jiggers overhead and probably three or four capstans. It had been found that, as they were working intermittently early in the morning and perhaps late at night, it was probably more economical to pull out the steam-engine, and put in a gas-engine. Several instances of this kind had occurred in small warehouses; but it would be useless to attempt to get any economy in a large warehouse in this way, and any modern establishment was now fitted up throughout with high-pressure hydraulic machinery.

Mr. DAVID JOY said that thirty-five years ago he had been intimately connected with the utilization of water-pressure for the purposes of lifting and of organ-blowing (Proceedings 1857, page 184). For the latter purpose however he had no idea that it had become so extensively used as it appeared to be in Liverpool, namely to the extent of nearly five million gallons per annum. In one instance, which was somewhat parallel with that described by Mr. Aspinall, two or three hydraulic blowers had been applied to a large church organ, where the waterworks company were to charge a certain

(Mr. David Joy.)
rate per meter for the water used. As the water was supplied at a pressure of 90 lbs. per square inch, there was a superabundance of power; and it was decided to pass the exhaust water from the engines into the tanks used for supplying the schools. The churchwardens had previously been paying £40 a year to the water company for the supply for school and household purposes; but when they came to utilize the water from the organ blower, the payment was reduced to £26. Since the time when he began to apply hydraulic engines for organ blowing, a great improvement had taken place in the attitude of water companies towards such applications. In those early days, when he went to ask for water for blowing an organ in a church or a cathedral, he was met with the objection that the water could not be spared for such a use, as it might interfere with the general household supply for the town or city; and it was only after giving an exact calculation of the quantity required, and an idea of the times when it would be wanted, that he had at last succeeded in obtaining the desired supply. Then came the question of the size of the supply pipe. He wanted a 3-inch pipe, not with a view to the large quantity of water which a pipe of that size was calculated to supply, but in order to get the water to pass through it without loss of power from friction. At last he had succeeded in getting it, after having shown practically that even with large pipes of 3 or 4 or even 6 inches diameter the blowers did not use any more water. Indeed it was with a smaller pipe that more water would have had to be used, because so much more power would have been lost in friction. In this connection he had also tried some experiments in order to ascertain how much of the theoretical value of the water pressure could be obtained in practice; and he should like to ask the author if he could give any information as to the proper speed at which a piston should be run so as to get the full commercial value out of the water. In his own experiments he had found that with a rotating engine, when the piston speed did not much exceed 80 feet per minute, nearly 80 per cent. of the full power due to the head of water could be obtained. With the organ blowers running very slowly, frequently as much as 85 and even 90 per cent. of the true value of the water pressure was obtained. The original

difficulty which he had to contend with in introducing those machines was in the mechanical arrangement for getting from water pressure a reciprocating motion with no dead point and no shock; and he had found that the way of getting it was by having a small subsidiary valve or four-way cock for working the main slide-valve, which latter was moved by the water pressure: the piston of the machine gave the reciprocating motion, and in doing so moved the small valve, which in turn admitted the water pressure upon a piston at either end of the main slide-valve, thereby reversing the latter entirely, however slowly the piston of the machine might be moving. The reversal was thus effected without the assistance of momentum and without a dead point. This was the first form in which what was now known as the "servo-motor" had been introduced.

Mr. JEREMIAH HEAD, Past-President, said the interesting and instructive paper now read, containing such a large mass of figures, appeared to him strongly to confirm the notion that, while water-power for cranes and motors of various kinds was one of the best and most useful where they were intermittently employed, it could not compete with other kinds of power where there was anything like continuous employment. There must also be a great loss of power in the friction of pipes, when they were continued to any great length. For in page 33 he noticed that the cost of pumping was given "per thousand gallons lifted one hundred feet." Now a gallon of water weighed 10 lbs., and a thousand gallons 10,000 lbs.; and this multiplied by a hundred feet gave just one million foot-pounds. A horse-power was 33,000 foot-pounds per minute, or say two million foot-pounds per hour; so that one million foot-pounds, if continued for an hour, would be equal to half a horse-power. Again, omitting the Aubrey Street Station, the average of the other four stations in page 33 came to 0·2609 penny, or say one farthing, for one million foot-pounds. This meant that half a horse-power cost one farthing per hour—actual horse-power in the water pumped, not indicated power in the cylinder of the low-pressure pumping engines; and for one penny per hour two actual horse-power were obtained. The average cost of the coal used during

(Mr. Jeremiah Head.)
1890 had been given by the author at 9*s.* 2½*d.* per ton. Taking it at 9*s.* 4*d.* that would be just 1–20th of a penny per pound. A good modern compound pumping engine consumed about 2 lbs. of coal per indicated horse-power per hour. The value of such an indicated horse-power would therefore be 2–20ths or 1–10th of a penny per hour. Such a good engine would therefore yield for a penny ten indicated horse-power per hour, instead of only the two actual horse-power in the water delivered to whatever height it was wanted. The conclusion was that the remaining eight horse-power, representing the difference between what was exerted on the piston and what was got at the reservoir, must be lost in friction of the engine and in friction of the pipes, supposing the other conditions to be the same. But if instead of 2 lbs. of coal per indicated horse-power per hour 4 lbs. were taken as the consumption—because he imagined that the engines referred to in page 33 were not any of them modern engines, and presumably not compound—then there would be two actual horse-power in the reservoir against five indicated in the cylinder of the engine, and the loss in friction would be somewhere between three and eight horse-power. If five horse-power were taken as the loss, so that for every two of potential energy in the reservoirs there was a loss of five, this would mean that 2½ times as much power was lost as could be utilized. In page 36 the power actually obtained from the hydraulic hoists had been spoken of as costing 13½*d.* per horse-power per hour; so that there was a further and much greater loss from friction in the delivery pipe and hoists. As was to be expected, the further the water went through the pipes, the greater was the loss of power; until at length, instead of getting even so little as two horse-power per hour for a penny, only 1–13th of a horse-power was obtained at the hoists. It would further appear that gas was cheaper. If, as stated in page 42, it gave one horse-power per hour for a penny, it was certainly much more economical than the hydraulic hoists. The opinion he thought must therefore be confirmed, that, while the methods described might be exceedingly convenient for intermittent work, they could never compete with a good well-ordered steam engine exerting large power under favourable circumstances.

Mr. WILLIAM SCHÖNHEYDER could quite understand the difficulties that must have been experienced by the author in collecting all the information he had given as to the cost of different kinds of machinery for hoisting and other purposes; and it was doubtless in consequence of these difficulties that in some of the examples mentioned the real cost appeared not to be given in full, but only a portion of it. In page 33, for instance, the cost was given of pumping from deep wells at the various stations in Liverpool; but no mention was made of what the cost included. No doubt it included the cost of fuel, oil, engine-men's wages, and so on; but was anything allowed for interest, or depreciation, or the cost of buildings? because it was impossible to obtain power without a large expenditure in this last direction. In London the corporation some time ago dug a well in the East End, and got it completed after a large expenditure; and it had recently been stated that the annual cost of working would be £140. But if the interest on the original cost was added, the cost of working he believed would really be twice that amount. The same general remarks would apply to the cost of steam power as given in pages 39 to 41. It was not sufficient to give merely the cost of coal, oil, and engine-men's wages; the interest on the first cost should be added, in order to compare, for instance, with water: because when water was supplied either at low pressure from a town main or at high pressure from hydraulic pumping machinery, the cost of the water under pressure included all interest on machinery, buildings, and so on. In the use of either water or steam or gas, after the power itself had been paid for, something else still remained to be charged. For hydraulic power there had to be added the cost of rams, multiplying pulleys, and ropes; for steam power, the cost of ropes or bands, pulleys, and shafting; for gas engines similar costs should be added; and every statement of the cost of power ought to include interest and depreciation. In some of the examples in the paper the cost of power was reckoned per foot-ton; while in others, especially with steam power, it was given per horse-power. It would be an advantage if the statements could be made all to refer to one standard.

Mr. George Cochrane noticed that in the use of high-pressure hydraulic power the cost was put down in page 44 at $1\frac{3}{4}d.$ per 50 foot-tons in example Y, and at $2\frac{1}{4}d.$ in example Z, which appeared to him in both cases to be a great deal higher than it should be. In neither case however was the maximum power stated of the hoist, and he should like to know what it was: because it was quite possible the hoists might not have been working up to their full capacity when those figures were obtained. If the load were reduced one-half, the cost would be doubled. He had frequently met with cases in which the cost of working a hydraulic-power machine was more than doubled, simply from ignorance on the part of those using the machine that it was necessary to have the machine suited to the load or the load to the machine: that is, that a hydraulic machine should always lift as nearly the maximum load as possible. In other cases which he had often come across, the man in charge had imagined that by screwing down the inlet valve he reduced the quantity of water used, which was of course absurd. The correction of these mistakes had been attended with the result that the consumption of water had been reduced to something like half. It was sometimes found that cranes or hoists made to lift a ton were lifting only half or a quarter of a ton, so that 50 or 75 per cent. of the water was being thrown away.

Mr. Edward B. Marten, Vice-President, considered that the paper was exceedingly interesting in more ways than those already mentioned. The history of what had been done at Liverpool in the past was of great interest at any time, and just now especially so to himself, coming as he did from near Birmingham, where there was at present under consideration a proposal for going to so great a distance as Radnorshire for an additional water supply; whilst many still thought they might well be content to pump more extensively from the sandstone beneath their feet. From the present paper he gathered that Liverpool had gradually been driven away from the original plan of pumping water from the local sandstone, and had resorted first to Rivington and now to Vyrnwy. It had there been found necessary at whatever cost to get a gravitation supply; and he thought it would be the same at Birmingham. The remarks which had just been made with respect to the use of water for power rather confirmed this view. When he had first had to do with such applications—not many in number, but for various small purposes, such as hair-cutting and organ-blowing, as well as for lifting heavy goods—he had been in the position of leasing the water-works at Stourbridge, and was naturally anxious to make everything he could out of the water. It was a great pity, he then thought, that with a supply of water at 80 or 100 lbs. pressure per square inch a larger revenue should not be made, and that people were not induced to take the water for motive power. On going carefully into the question of how it could be made more available, he found that for small works, such as those described in the paper, where it was wanted intermittently, it did well; but in the case of large quantities it did not pay. Nor would it pay a water company to supply water at such a low pressure as 80 lbs., merely for the sake of using it in large quantities in lifts; and this he thought was the conclusion to be drawn from the paper. In regard to the price of water for such purposes as lifts, he had lately watched one of the hydraulic lifts in Stourbridge, and found it was constructed to lift one ton and was lifting to a height of 70 feet. The charge was a penny per lift, and he thought it took nearly a pennyworth of water. Of course if a man alone went up the lift he cost a penny, just as if a ton of goods went up; and therefore it was wasteful for the user, because the exhaust water was not used afterwards. That was why he had asked Mr. Aspinall (page 52) how he managed to use the exhaust water afterwards; and the explanation was that it was discharged into a tank with only 15 feet of back pressure on the lift. The primary object however of a water company was to supply water to houses in every direction and at a convenient pressure; and if the water were used for a machine and the pressure absorbed, it would be difficult to use it afterwards in houses, because it would not go upstairs, and it would therefore be wasted. In Liverpool he thought it would be found that the water, though brought by gravitation from so great a distance, would still have sufficient pressure for enabling it to be applied economically for working low-pressure lifts. Where

(Mr. Edward B. Marten.)
however all the pressure had to be put into it by pumping, it was not so economical for such a purpose. In Stourbridge, instead of creating pressure by pumping for the use of lifts, he had found it economical to reduce the pressure to that most suitable for domestic service, by dividing the pumping into two stages of about half the total height, so as not to have to pump against such a high pressure where it was not wanted. Where water had to be pumped, lifts could hardly be worked economically unless the water was used over again after the pressure had been taken out of it by the lifts. Where the water was supplied by gravitation, it would certainly be a large source of revenue, even if it were wasted by using a larger quantity for working low-pressure lifts.

Mr. H. PERCY BOULNOIS, City Engineer, Liverpool, considered the questions raised in the paper were of the greatest importance to engineers generally, who were anxiously desiring to obtain some unit of cost for the different motive powers now in use. An endeavour had been made by the author to reduce his calculations and investigations to something like a unit of cost for each power; and he hoped that other engineers would follow out the same idea in connection with any kind of motive power which they might have to employ, as any authentic facts would be extremely useful. So far as he gathered from the paper, water power did not come out exceedingly well; and it appeared to him that gas engines still held their own as an intermittent power. In the cases given by the author, notwithstanding that the gas engines had been running all day although used only intermittently, the working expenses had been much in favour of gas as a motive power. Nothing however had been said with regard to the cost of producing electrical energy; and information upon this subject would be acceptable for comparison with water, steam, and gas. Such a comparison as had been made by Mr. Head, in regard to the cost of producing a horse-power by different modes of using water power, was most valuable, and might be followed out for other motive powers with great advantage.

Mr. JAMES PLATT agreed with Mr. George Cochrane (page 58) in regard to the waste of water by using a one-ton lift to lift only a quarter of a ton. The remedy proposed he presumed would be to put four quarter-ton packages together and lift them all at once; for he did not know of any really good arrangement in general use for reducing the consumption of water under pressure for lighter loads. There was indeed Hastie's rotary engine with variable stroke (Proceedings 1879, page 484); but he did not know that it was largely used. Some device of that kind was certainly desirable for using high-pressure water. Fifteen or twenty years ago he had had to do with putting in a number of water-pressure engines in Liverpool to be worked by the pressure from the public mains, and they answered well until the gas engine came in; and then the demand fell off, because the gas engine worked so much more economically. The present paper dealt mainly with water pressure applied as motive power for the performance of work, and the comparison with the gas engine was rendered rather unfair on account of the gas engines having worked all the day (page 42) in consequence of the difficulty of starting them. The more modern gas engines however could be readily started according to the requirements of the work. One man could now start a 20 HP. gas engine; so that the former difficulty of requiring two or three men to start the engine had been overcome. The engines therefore could now be used only when they were wanted, stopping after they had done their work, and starting again whenever required. Hence the gas engine would now show to a greater advantage than it did in the paper. But if some good devices could be introduced for using a quantity of water in proportion to the load, the high-pressure hydraulic supply would be greatly enhanced in value. Being already most convenient to apply, and safe from fire, and requiring no attention, it would still be much more valuable if this improvement could be brought about; but the object was difficult to manage, and had not yet been satisfactorily accomplished.

Sir JAMES N. DOUGLASS, Vice-President, was sorry that the discussion had not gone further into the subject of the paper, because

(Sir James N. Douglass.)
he considered that it had not yet been thoroughly fathomed. A pertinent remark had been made by Mr. Boulnois (page 60) on the question of a satisfactory unit of cost. For himself he looked upon the question in some respects as an "oil" man, and considered that the price of gas was too high at 2*s.* 8*d.* per thousand cubic feet. No doubt the reason of so high a charge was that the gas was in the hands of a company; and private companies wanted of course to make large dividends, which they were able to do wherever they were not subject to competition. When going through the United States five years ago, he had remarked upon the high price of gas notwithstanding the fact that mineral oil was obtained at 6*d.* per gallon, which would be equivalent to gas at about 1*s.* 3*d.* per thousand. The Americans however were satisfied to pay as much as 2 dollars or 8*s.* per thousand cubic feet for coal gas; and on enquiring the reason he was told that it was simply because concessions had been made to companies who charged just what they liked. Electricity had no doubt already to some extent altered that state of things; and with gas at so high a price electricity certainly had greater advantages in America than it had in this country. In connection with the use of gas engines, which had been referred to by Mr. Platt (page 61), it was well to bear in mind that at the present time oil was at a fairly uniform price, the best oil for lighting being 6½*d.* per gallon, while for oil-gas engines it could be purchased at 4*d.* From investigations made by Professor Kennedy and other engineers he believed that, while the price of London gas was about 2*s.* 6*d.* or 2*s.* 8*d.*, the cost of oil for driving engines of the same power would be only about 1*s.* 4*d.* Of course he did not mean to recommend oil engines in all directions; but rapid strides were being made with them, and he believed they were coming to the front for electric-lighting purposes. In cost per candle-power electricity could not at present compete with mineral oil, except in the case of very high powers in arc lights, as in light-houses. A letter had been received from Mr. A. Bromley Holmes, the managing engineer of the Liverpool Electric Supply Company, who wrote, "I think it may be of interest to the Members to know that electricity is now supplied for working motors in Liverpool at a

cost of fivepence per horse-power per hour by this company, whose mains have been laid in all the principal streets. In one instance electric motors are used by a manufacturing company twelve hours daily." Probably most mechanical engineers would agree that, except as a matter of luxury, such a high price was prohibitive; and he understood from Mr. Aspinall that on this account electricity would not do for him just at present. It might answer for some small powers, as for dentists and hair-dressers; but it would not yet compare with the results furnished in Mr. Parry's paper from the employment of other motive powers.

Mr. PARRY said that, like Mr. Boulnois, he should have been glad if the discussion had elicited from electrical engineers some information with regard to what could now be done by means of electro-motors. He had also hoped that the paper might bring out in the discussion more information with reference to the distribution of power from a centre, as compared with the production of power in detached buildings; this was the question he had principally had in view in collecting the statistics given in the paper. His primary object had naturally been to get facts for the information of the Liverpool Corporation, in order to determine the actual cost of practical working with different kinds of motive power in Liverpool. Without any desire to advocate one power rather than another, he had simply endeavoured to ascertain what it was costing in warehouses and other buildings in Liverpool to obtain the motive power which was being used. The principal facts so compiled had been stated in the paper. There had necessarily been omissions, which could not in all cases be supplied; but the information obtained had been sufficient for his purpose, and the figures given were relatively correct. It was of course most desirable that all costs of work done should be reduced to a common unit; but it was not always practicable to do this, and those who had endeavoured to collect similar data would well understand the difficulties that had to be encountered. In the cases referred to by Mr. George Cochrane (page 58) the hoists were not working to their full capacity; and the difficulty of varying the expenditure of power according to the amount of work to be done

(Mr. Parry.)
was precisely the reason why hydraulic power often compared unfavourably with other kinds of power.

With reference to the cost of the high-pressure water in Liverpool, in page 45 were given the actual charges made per thousand gallons by the Hydraulic Power Co. for various quantities of water at the constant pressure of 700 lbs. per square inch; and from that table a calculation could easily be made as to the exact cost of obtaining any amount of power that might be required. The facts that had been given were simply examples from actual observation, showing the cost of work in those particular instances.

In regard to additional safety from fire by using water power, to which allusion had been made by Mr. Platt (page 61), at one time he had himself thought that there was a great deal more benefit to be gained from the use of water owing to that circumstance than he now found to be the case. The insurance companies he had understood would make a considerable or at all events a substantial reduction in the rates of insurance in buildings where water power was used instead of a steam or other heat engine; but he had now found that it was not as he had supposed. In Liverpool at all events no such distinction was made; the building regulations were there so excellent, and the care taken to enclose and isolate the steam engines was so effective, that there was no practical risk of fire, and no longer could any saving in insurance be obtained by adopting water power in preference to other powers.

The cost of pumping in page 33 was given on the usual basis followed in stating the expenditure incurred in lifting water; and he thought it was generally understood that the cost of pumping meant the cost of wages, repairs, coal, oil, and all such expenses of working and maintenance, apart from interest on the machinery and depreciation. At any rate that was how the cost had been given in the paper. It had however been worked out a little more fully per indicated horse-power, with the result that in three of the pumping stations it came to 0·35*d.* and 0·39*d.* and 0·40*d.* severally per indicated horse-power per hour, including fuel, wages, stores, repairs, and all expenses other than interest and depreciation. Of course the engines referred to could not now be regarded as being in any sense economical engines; as explained in the paper they were old engines, and they were referred to as matters of history and as of interest on that account. If allowance had been made for interest and depreciation, as suggested by Mr. Schönheyder (page 57), a speculative element would have been introduced, which would have altered the character of the paper. All such estimates were intentionally omitted.

With reference to the loss by friction, alluded to by Mr. Head (page 55), there was undoubtedly a large apparent loss if a comparison were instituted between the first cost of raising water from wells or rivers and the cost of its ultimate delivery and use as a motive power in any town. But the bulk of the water supplied to Liverpool was obtained and delivered by gravitation; and the loss of head by friction in the mains, between the service reservoirs and the warehouses in which the low-pressure hoists described in the paper were situated, was not more than from 15 to 25 feet, or from 8 to 13 per cent. of the hydrostatic head. The cost at the point of delivery included the cost of collection in large reservoirs and carriage through thirty miles of aqueduct, besides distribution, and establishment expenses.

With regard to the question of organ-blowing, referred to by Mr. Joy (page 53), the form of motor almost exclusively used for this purpose in Liverpool was a double-cylinder engine, designed as a water meter many years ago by the late Thomas Duncan. It was liked by organ builders because of its steady reciprocating motion and the simplicity of the connections between the pistons and the organ bellows. As to the efficiency obtained in practice, he had not himself made any very exact experiments; but the percentage of useful work was certainly not so high as that mentioned by Mr. Joy, nor had he found any organ-blowing engine to give such a high proportion of efficiency. A low rate of speed was obviously economical in such engines. The Duncan engines usually worked at an effective piston-speed of about 25 feet per minute. It would be seen from page 35 that the average cost of organ-blowing by water pressure in Liverpool did not amount to more than £5 14*s.* per engine per annum. For large church organs, to which such engines

(Mr. Parry.)

were chiefly applied, this was a moderate expenditure, and much below the expense of blowing by hand. The statement in page 35 did not include the organ in St. George's Hall, which was not blown by hydraulic power but by a steam engine.

With respect to the remark of Mr. Platt (page 61) as to the unfairness of comparing gas engines under the conditions given, it would be found that in page 42 examples were given of the cost of gas actually consumed in the performance of a definite amount of work. And as to recent devices for starting gas engines, he did not know of any arrangement that entirely prevented the loss to which attention was called in the paper. Notwithstanding the application of such devices, the engines would not and could not, in practice, be stopped whenever the hoists were idle, and be re-started only when lifting had to be done.

Sir JAMES N. DOUGLASS had much pleasure in proposing a hearty vote of thanks to Mr. Parry for his interesting paper, which he was sure would be found to be a valuable addition to the proceedings of the Institution.

Mr. E. B. ELLINGTON, being unable to be present at the meeting, wrote that it was somewhat difficult to arrive at any definite conclusion from the paper itself as to the relative cost of different descriptions of power when applied to lifting. This was due, not so much to the want of exact data, as to the insuperable difficulty of reducing the varying practical experience of different individuals to a common standard. The question may be regarded either from a theoretical or from a practical standpoint, but cannot be usefully treated by mixing up the two. Considering the question of the theoretical efficiency of lifting appliances worked by steam, gas, hydraulic, and electrical power, either high or low, it does not appear that there ought to be much difference between them. They all depend upon the combustion of coal; and it is seen at once that the real question at issue is the relative economy of different methods of distributing power from a common centre. Steam is out of the question altogether for application over a large area. Gas and electricity have much larger fields, and are highly economical methods of distributing power; but for lifting and intermittent work hydraulic power undoubtedly occupies the first place. The efficiency of a hydraulic pumping engine is 84 per cent., while the loss in distribution through properly constructed mains is not over 2 per cent. per mile; and the efficiency of the motor is often as high as 90 per cent.; so that the combined efficiency, or useful work done in proportion to indicated power expended, is as high as 74 per cent. within a mile from a supply station. An efficiency of this amount cannot be exceeded, though it may be approached, by other methods. It is well known however that no such results are realized in practice by users of lifting machinery; and the kind of power to be adopted really depends upon a number of practical considerations, which have nothing to do with theoretical efficiency at all.

What is required in lifting appliances for warehouses or offices is: (1) that the apparatus shall be always ready for work when wanted; (2) certainty, speed, and safety of action; (3) simplicity of construction, and absence of noise and of dirt and of fire risk; (4) small amount of space occupied; (5) little attention needed; (6) an initial outlay and a working cost of moderate account. It will be found on investigation that these requirements are best met by the use of public hydraulic power on the high-pressure system, as already carried out in London, Liverpool, Birmingham, and Hull, and as now being applied in Manchester. As to the question of high or low pressure for hydraulic power, low pressure from ordinary waterworks mains is unsuitable, firstly because of its being necessary to place the machinery in the basement, instead of in the roof where it is most wanted; secondly because of the large size of the apparatus needed, and the slow speed which is generally attained; and thirdly because of the great variations in pressure that occur in practice. The first cost of high-pressure hydraulic machines worked from public mains is less than that of machinery on other plans. It is these practical considerations

(Mr. E. B. Ellington.)

which have been mainly instrumental in determining the success of the high-pressure hydraulic-power system. In Liverpool there are already about 250 machines, either actually at work or in process of being connected to the hydraulic-power mains, although the system has only recently been started in operation there. In London during the past eight years over 1,600 machines have been put in operation; and in all places where hydraulic power in its integrity has been available it is found that for lifting and for some other intermittent purposes it is able not only to hold its own, but practically to supersede every other system. In all cases the power is supplied under a sliding scale of charge, which is so devised as to enable the consumer to secure the contingent benefits of the supply without any increase of expense; and in the great majority of cases there is a saving in actual annual charge of from 10 to 50 per cent., when all items of expense are taken into account.

Referring to the interesting particulars given by the author as to the actual amount of work done in several warehouses during a year, estimated in foot-tons, the hydraulic-power supply in Liverpool has an available pressure of 700 lbs. per square inch throughout the system. Taking 50 per cent. efficiency for the hydraulic cranes, which is commonly realized in practice, each gallon should do say 4 foot-tons of work in goods raised. In the examples given by the author the annual consumption of water from the hydraulic-power mains should therefore be as follows:—example N 112,000 gallons; O 38,000; P 37,000; and Q 54,000 gallons. The cost for power in each case should be about as follows:—

Example N	£38	for Hydraulic Power,	compared with	£129	for Steam.
„ O	£17	„	„	£57	„
„ P	£18	„	„	£117	„
„ Q	£22	„		£49	for Gas.

In this calculation it is assumed that the work is equally divided throughout the four quarters of the year, and meter rent has been added. The amount of work done is small in all these cases; three times as much work could be done at only twice the charge, and the great practical economy of the high-pressure hydraulic power would then be made still more apparent. It may also be mentioned that the block of offices X (page 43) with two passenger and two goods lifts, referred to by the author as being worked by an 8 HP. gas engine with pumps and accumulator, has for some time past been connected with the hydraulic-power mains, and the gas pumping machinery has been abandoned.

ON THE DISPOSAL AND UTILIZATION OF BLAST-FURNACE SLAG.

BY MR. WILLIAM HAWDON,
OF NEWPORT IRON WORKS, MIDDLESBROUGH.
COMMUNICATED THROUGH MR. CHARLES COCHRANE, PAST-PRESIDENT.

Production of Slag.—Whilst it is sometimes difficult to realize a profit in disposing of the pig-iron produced from the blast-furnace, it is almost always difficult to avoid a loss in getting rid of the Slag made in conjunction with the iron. In other words, it is only under exceptional circumstances that the slag can be sold as a useful commodity and a profit be made from it. As a rule, slag has at a certain cost for labour to be tipped over more or less valuable land, which is thereby covered up and wasted; or in some cases it has to be carried quite away from the locality of the blast-furnaces, at a further cost of carriage. Thus the disposal of slag from the blast-furnace has almost always entailed a cost which has had to be reckoned in the manufacture of pig-iron; and when it is borne in mind that about twelve million tons of slag are made yearly in Great Britain, it will at once be seen that the aggregate amount of this cost is enormous, and that the subject is well worthy of attentive consideration.

Uses of Slag.—In some few localities a profit is realized by selling slag as a material for road making. It is first of all run in its molten state into balls, formed in the usual boxes, and carried away on bogies, from which it is tipped; when cool it is broken into pieces of the requisite size, and sold at whatever profit can be obtained. At a few blast-furnaces bricks for paving are made, but only to a limited extent. Where the traffic is not too heavy they are found to stand fairly well, as in stables and stable-yards, and in the back streets of towns; in Edinburgh indeed some of the better streets are paved with them, notwithstanding that granite can be obtained near at hand. Slag is also cast in various forms, such as channels for water courses, kerbs for footpaths, &c. Glass has been made from it in small quantities, but the author believes not on a commercial scale. The whole however of the slag utilized for all other purposes than road metal amounts to only a very small percentage indeed of the total quantity of slag made in the country.

Previous modes of Disposal.—In the Cleveland district, where probably some $3\frac{3}{4}$ million tons of slag are made per annum, the question of its disposal has received somewhat serious consideration, and has been solved in one of two ways: either land on which it can be deposited has been purchased at a high price, from £400 to £1,000 per acre; or the slag has been sent out to sea in hopper barges, and there deposited in deep water. In the early days of iron manufacture, slag used to be run into holes dug in the sand of the pig bed, into which a stake of iron had been fixed; and when the slag was set, the lump was lifted out by a crane laying hold of the stake. Sometimes the slag was run into channels made in sand or ashes, and was afterwards broken up by hand. But with the subsequent larger makes from the blast-furnaces it was found impossible to clear it away by these means. Cast-iron boxes were then placed to form moulds for the slag to run into, and iron bogies were run underneath them to carry them away; or the sides of the boxes were themselves mounted on bogies which formed the bottoms of the boxes. The slag is thus conveyed away in balls of 2 to 4 tons; when it is cool, the frame or sides of the box are lifted off the bottom by a crane, the bogie is drawn away to the tip, and manual labour is employed to bar the balls off, in order either to level up uneven ground, or to build up a mountain, which is carried to a great height so as to cover as small a base as possible on the costly land.

When sent to sea, the slag had to be broken up by hand, and wheeled barrowful by barrowful into the hopper barge. But this was such costly work that means were devised for breaking up the slag balls by fixing in the centre of the bogie a hollow casting: the idea being that, when the mass of slag cooled and contracted round this casting, the ball would break up into pieces small enough to be tipped into the barge without much injury to the vessel. The hollow core was cast about one-third the width of the ball, and was fixed firmly on the top of the bogie; and when the box or frame was placed on the bogie, the slag was run into the space surrounding the core. It was found in practice however that this also was too costly a procedure. The central castings soon became burnt and cracked and broken, and repairs were found to be heavy. Moreover the slag balls did not break into small pieces, and occasionally into only two or three large heavy pieces, which wrought havoc to the plates of the barges, springing the joints so that the pumps had to be kept constantly going to keep the water out; and often a hole was smashed through the plates and the vessel disabled by a heavy mass of slag, falling as it did down a shoot placed at the height necessary for loading at high tide, and thus acquiring an additional fall of 14 or 16 feet when loading at low water. The idea of the central hollow core or casting for causing the slag to break up in cooling was no doubt a good one, and the plan was certainly an advance on the old method of breaking up the balls by hand.

In some few instances the American plan is adopted of running the slag in flushes into a "boat" or tank on wheels, lined throughout with fire-brick; or simply into a ladle placed on a truck. By this method there is a good deal of labour in cleaning out the skulls or crusts of slag which deposit on the sides of the vessel. In either case the slag is merely run or tipped out of the vessel containing it. If intended for further removal, it is run from the tank or ladle on to cast-iron plates, so as to form layers of slag, which can afterwards be removed by hand when cool.

The cast-iron core was first introduced by Mr. J. A. Birkbeck at the Acklam Iron Works, Middlesbrough. The hollow casting was made double pear-shaped in plan, and of the full height of the slag ball, with the finer ends or knife-edges towards the front and back of the bogie: so that, when the bogie was tilted up, the slag, which had already cooled and cracked through contraction round the core, was cleft in two or more pieces, and fell into the barge. But these pieces were too large and heavy to be thrown with safety into the vessel.

A vast amount of thought and money was expended by Messrs. Cochrane and Co. at the Ormesby Iron Works, Middlesbrough, in their endeavours to bring to a successful issue this method of breaking up and loading the slag. They introduced as a central core what they called a "hog-back" casting, shown in Figs. 10 to 12, Plates 7 and 8. By this plan the slag was broken up into much smaller pieces than by Mr. Birkbeck's. After the slag had been run and had solidified, the bogies were placed under a series of perforated pipes, from which sprays of water played on the top of the still hot mass; the sudden cooling assisted in cracking the balls, which contracted round the hog backs. But although not so costly as the old plan of breaking the balls by hand-labour and wheeling the slag into the barges, it was not satisfactory, the wear and tear, especially of the hog-back core, being very great. This method has now been abandoned, and replaced by the apparatus shown in Plate 4, about to be described.

At the Normanby Iron Works, Middlesbrough, where also the slag is sent to sea, Mr. Edwin Jones has built iron bogies with wrought-iron sides about 15 inches high, and widening outwards from the bottom. These are filled with slag, and when it is sufficiently cooled are run up to the slag shoot, and tipped up bodily so as to shoot the block of slag out, and let it fall upon knife-edged iron castings, by which it is broken up into somewhat large pieces; these then roll into the barge, or into a large hopper shoot fixed so as to store about 100 tons of slag. Still however the pieces are too large for the safety and wear of the barge.

Several mechanical contrivances have been designed to deal with molten slag, the earliest probably being that of Mr. Thomas Bell at the Walker Iron Works near Newcastle-on-Tyne. As there was little land adjoining the works, he designed about 1871 or 1872 a machine for running the slag into such small pieces as could be readily handled and tipped into barges at the wharf adjoining the works, and thence sent to sea. The slag was run into a series of cast-iron trays, each about 3 feet wide, which were fixed on an endless horizontal belt travelling a distance of about 12 feet. When the slag was sufficiently set or cooled, water was played on it to cool it further, and it was then tipped into trucks. For some reason unknown to the writer this apparatus was abandoned.

In 1871 designs were submitted to the Cleveland ironmasters by Mr. David Joy, who read a paper in 1873 to the Cleveland Institute of Engineers, describing his plans, none of which however, so far as the writer is aware, were put into practice. One idea was to have long endless belts running from the furnaces to the river for shipment of the slag. The belts were to consist of iron plates ¼ inch thick, fixed on steel-wire ropes; it is needless to say that these would have been too light to resist the action of molten slag run upon them. Another design was for disintegrating the slag by running it into water, out of which it was to be dredged by an arrangement of buckets. Other designs were also drawn out for running the slag into moulds fixed on the inner side of a horizontal wrought-iron drum, which revolved on small rollers: the idea was to assist the cooling of the slag by means of water jets, and to carry it in the moulds nearly up to the top of the circle described by the revolution of the drum, and then to tilt it out into a trough shooting it into trucks. An experimental machine constructed on this plan could not deal with such a large flow of slag as was discharged from the furnaces of that date; and it does not appear to have been put into practical use.

Utilization of Slag.—To the utilization of slag a good deal of attention has been given by Mr. Charles Wood of Middlesbrough, by whom several machines have been designed for dealing with it in this direction. In a paper read before the Society of Arts in London (Journal, 14 May 1880, page 576) Mr. Wood described several appliances which he had designed with this object. The first is a machine for making "slag shingle," and consists of a horizontal annular table about 16 feet diameter, which is formed of segments or cooling plates about 2 feet wide, made of cast-iron, and having wrought-iron pipes cast in them, through which water flows. While the table is revolving slowly, the molten slag is run upon it, and spreads out to a thickness of from ½ to ¾ inch. When it has travelled about 10 or 12 feet and has consolidated, water is run upon it to cool it further; and then a set of scrapers push the slag off into iron wagons, into which it falls in large flat pieces. When cold it is tipped from the wagons, and falls into small pieces. The slag shingle produced by this machine is largely used for making concrete.

Another machine is for making slag sand, and consists of a vertical wheel or short drum of wrought-iron, the outer rim of which is trough-shaped and about 14 feet diameter, revolving about five times per minute on a horizontal axle. Into the lower part of the trough-shaped rim, of which the opening faces inwards or towards the centre, water is run to a depth of from 18 to 24 inches; and into this water the molten slag is run, where it forms into a spongy sort of sand. Perforated screens or elevators are fixed radially across the trough at intervals all round the wheel, and in passing through the water carry away the sand and lift it to the top of the machine, where it drops into a spout and is shot out into wooden wagons. This sand has been extensively used for making concrete bricks, which are useful for structural purposes; some millions have been sent from the Tees to London for house building, and have sometimes been sold at a profit at as low a price as 10*s.* or 11*s.* per thousand. In many parts of the country cement mortar is made from slag in this condition, and is very tough and durable.

Another form in which blast-furnace slag has been usefully applied by Mr. Wood is that of slag wool or silicate cotton; and he appears to have been the first to manufacture it in this country on a commercial scale. The process is interesting, and extremely simple. According to his own description in the paper already referred to (page 585), a jet of steam is made to strike upon the stream of molten slag, as it flows from the usual spout from the furnace. The steam scatters it into shot, and as each shot leaves the molten stream it draws out a fine thread: just in the same way as, when treacle is touched lightly with the finger, it is drawn out into a fine thread. On losing its heat the fine thread of slag drawn out by each shot becomes set like glass. The shot itself being heavy drops to the ground; but the light thread is sucked into a large tube by an induced current of air caused by the steam jet, and is discharged into a large chamber. The finer qualities float about and settle near the outside of the chamber, whilst the heavier or larger fibres lie chiefly in the centre. The wool is then ready for use, and is packed in bags for sending away. It is principally used for covering boilers or steam pipes, for which purpose it is peculiarly adapted, being a splendid non-conductor of heat, and incombustible. As only one quarter of a cwt. is made from each ton of molten slag operated upon, it will be seen that the process is not a very rapid one.

Endless-Chain Slag Machine.—The disposal of slag by mechanical means engaged the writer's attention more particularly in the year 1885, when all the land available for slag tipping at the Newport Iron Works of Sir B. Samuelson and Co. at Middlesbrough had been pretty well filled up. He then designed the apparatus shown in Figs. 1 and 2, Plates 4 and 5, of which Fig. 1 is a longitudinal elevation, and Fig. 2 a longitudinal section to a larger scale. The endless chains are made of long steel or iron links, which are fastened together by pins or rivets. The primary driving shaft D is driven by a small engine, or if more convenient by a belt. The pair of cast-steel pulleys L, over which the endless chains pass, are driven from the shaft D by geared wheels, and cause the chains to travel in the direction shown by the arrows. The pans which carry the slag are fixed on the chains, and are shown in detail in Plates 6 and 7: in plan in Fig. 3, in side elevation in Fig. 4, in cross section in Fig. 6, and in longitudinal section in Fig. 7; while Fig. 5 is an inverted plan showing the chains CC. The pans are ninety in number, and are each made in three pieces, Fig. 6; by means of two lugs cast on the bottom they are bolted on the chains C. The slag is conveyed from the furnace by the trough T, Fig. 1, from which it flows in a molten state into the pans as they travel beneath it. The pans then pass through the water trough W, Fig. 2, after which the slag is still further cooled by being sprinkled with water from the perforated pipe P. Finally in passing over the pulleys U the slag is tipped out of the pans into a shoot discharging into wagons beneath. For taking up any wear on the chains, a worm wheel and screw fixed at S are connected by links to the shaft of the pulleys U, whereby the chains can at any time be tightened up as required.

The links of the chains are made alternately single and double, as shown in section in Figs. 8 and 9: the single are of ⊥ section, Fig. 9, while the double are of a flat-bottomed U section, Fig. 8, so that each has a broad bearing surface in contact with the pulleys. The links first adopted were simply wrought-iron bars with a hole punched in each end; but the surface in contact with the pulleys was not broad enough, and cut into the pulleys, which at that time were made of cast-iron. Cast-steel links were then adopted of the section above described, which ought to have stood well enough; but for some reason or other they were not reliable, some being brittle and others honey-combed, causing occasionally a break-down. Of late malleable cast links have been used, and these have stood better. Now however links are being made under a steam hammer out of rolled steel plates, and the eyes are bushed with cast-iron; this is being done because they are both cheaper to make and also stronger than the cast-steel links can be.

At the Newport Works the eight machines are each driven by a steam engine with single 5-inch cylinder; but only about half its power is really necessary for driving them. The engine works on the primary driving or crank shaft D, Fig. 1, on which is fixed a pinion $6\frac{1}{4}$ inches diameter, gearing into a wheel $34\frac{3}{8}$ inches diameter; and on the shaft of the latter is again fixed a pinion $6\frac{1}{4}$ inches diameter, gearing into a still larger wheel $57\frac{1}{4}$ inches diameter, which is fixed on the same shaft as the pair of chain pulleys L. This gearing is used, not to gain power, but in order to obtain a sufficiently slow rate of travel to suit the flow of slag into the pans. The chains are run at slightly varying speeds, according to the output of slag, the average rate being about 13 feet per minute. The eight machines together deal with 1,000 tons of slag per twenty-four hours; any one of them alone deals with 180 to 200 tons per twenty-four hours. In the water trough W, Plate 5, into which the pans dip down with the chains after passing the bearing pulley B, the water is kept at a level reaching about two-thirds up the sides of the pans; and being kept boiling by the heat, a certain portion of it usually splashes over into the pans and assists in cooling the slag; but this is not essential to the process. Two or three wagons are kept in reserve on a slight incline, so that when one wagon is full another is lowered into its place without stopping the machine. The wagons are made with bottom doors, or with side or end tip, to suit the particular requirements of the works where they are employed.

At casting time the slag which may follow the iron at the end of the cast is run into cast-iron troughs, and when cooled is broken up by the slagger and thrown into the wagons: so that bogies and boxes are dispensed with in the general working of the blast-furnace when these machines are employed. There is thus a considerable saving in labour, and in wear and tear of machinery and material employed in the disposal of slag from ordinary blast-furnaces. Burst balls of slag, which might burn up the sleepers and roads and cause labour, are unknown. It is now no longer necessary to bar the balls off the trucks at the tip; and the constant repair, renewal, and shifting of rails and sleepers on the tip, are now unnecessary. Two men per shift do the whole of the work required for the disposal of 6,000 to 7,000 tons of slag per week at the Newport Works, with one locomotive per shift, which is assisted on the day shift only for one half of the time by a second locomotive, thus averaging $1\frac{1}{4}$ locomotive per shift, in place of three locomotives when tipping on a mountain of slag. The enormous wear and tear of bogies and boxes due to the hot slag is now done away with; and the wear on locomotives is reduced to a minimum, as the dust and dirt due to the old method are dispensed with.

The slag is run into the pans about 1 to 2 inches thick, and breaks up into pieces from the size of a nut to a few pounds weight. It is largely used for road making, especially for the foundation of new roads. For concrete, being already small in size, it requires little further breaking to render it suitable; and for this purpose it has been found to be particularly adapted, some thousands of tons having been used for the walls of piers, wharves &c., and also for the walls of buildings.

Experiments are now being made on a large scale with the view of using it extensively in the manufacture of cement, and it would appear likely to be largely used in future for the purpose. Not that it is a new idea to use slag for cement making; but the form and consistency into which it is run and manipulated in this apparatus adapt it in a special manner for cement manufacture. The slag is annealed or tempered in passing through the water; and that made from hematite iron, for instance, which contains a high percentage of lime, does not fall away into dust, as it otherwise does when run into balls and exposed for a short time to the atmosphere. In some districts vast quantities of ballast are brought for miles, and obtained in the first instance at some cost, for packing the sleepers of railways, which actually run past the very furnaces where this slag can so easily be made useful for the purpose. On the North Eastern Railway, which forms a network throughout the Cleveland and Durham districts, ashes from the ironworks and coke ovens are universally used. This ballast retains a lot of rain and surface water, and tends to rot the sleepers; and where steel sleepers are used, nothing is more injurious and wasting. Were this slag used wholly or in part as ballast, a drier and more lasting road would be the result. On many railways the larger class of ballast is used entirely, and in some cases slag is the material chosen. There appears therefore to be a wide opening for the use of this material, which in most localities is at present a nuisance and a source of expense.

The apparatus now described is a simple contrivance, having probably no great merit as a mechanical device; but few things are more difficult to deal with than the forces of expansion and contraction in metals, especially when brought about rapidly, and with continuous alternations. This is the case at blast-furnaces in any mode of dealing with the slag, and particularly so when artificial cooling is resorted to; and though at last simplicity has now been arrived at, a good deal of time has been occupied and many devices tried, before the desired result has been reached. Wear and tear, consequent on the rough usage which such apparatus is necessarily subjected to, require to be met by special consideration in the design of the different parts. Cheapness of manipulation being a matter of vital importance in iron manufacture, any apparatus or any method of treating the materials which realizes this object lays claim to the best attention not only of the makers of iron, but also of engineers generally, by whom this metal is so largely employed in all their works.

Discussion.

Mr. HAWDON exhibited a collection of samples of slag as it came from the belt of pans and was tipped into the trucks, in which form it was at once useful for ballast on railways. Also two slag bricks, which had been sent to him by Mr. Charles Wood, and were made from slag-sand and cement; they had been taken out of the outer wall of a building in Middlesbrough where their outer side had been exposed to the weather for the last fourteen or fifteen years, and they were seen to be still in perfectly sound condition, their outer face being simply darkened by smoke and showing no signs of wear. The large collection exhibited of samples of slag-wool had been sent by Mr. Frederick H. M. Jones, of the Silicate Cotton Works in Kentish Town, London. The slag-wool or silicate cotton was a material which was generally used for covering boilers and pipes, and was a very good non-conductor of heat. Samples of bricks for paving roads were also exhibited. Two years ago, at a meeting of the Society of Arts (Journal 31 Jan. 1890, page 221), an interesting paper on the utilization of slag had been read by Mr. Gilbert R. Redgrave, dealing with it chiefly as a material for the manufacture of cement; and on that occasion it had been stated by Mr. Thomas C. Hutchinson (page 234) that if slag were allowed to cool before it was granulated or disintegrated it lost its cement-making property. This however he believed was not the case; for he had been credibly informed by cement manufacturers that the slag delivered from the pans of the machine described in the paper, if it was cooled by using a little more water, causing it to be honey-combed or spongy, was specially adapted for the manufacture of cement, and was indeed preferable in this form to any other for several reasons. It was not at all necessary to take it in the hot state; that is, it was quite as good for cement making when it was cool as when it was used in the hot state.

Mr. DAVID JOY said that in connection with the subject of the paper he had formerly had considerable experience, and had at one time done a great deal of work. He had therefore hunted up a few of the plans which had been placed before the Cleveland ironmasters in November 1871 (page 74), the fact having then come to his knowledge that many of the ironmasters were getting into a difficulty from want of sufficient ground for tipping their slag, and from the high prices charged by the Tees Conservancy for carrying the slag out to sea to make their breakwater. Several of the ironmasters were therefore ready enough to listen to any good proposal for the removal of their slag. Besides examining the various methods in which at that time slag was being treated at different blast-furnaces, he had also watched how coal was dealt with, both at the collieries and also at the principal shipping ports and in the large trans-shipping barges on the Thames; as well as how other materials were shifted, such as grain in the large grain warehouses in Liverpool. At a meeting of the Cleveland ironmasters, specially called to discuss the question, he had submitted various proposals for disintegrating the slag as it flowed from the blast-furnace. In the first plan a band of wrought-iron plates was to be carried on a couple of steel-wire ropes; and though the plates were to be only $\frac{1}{4}$ inch thick, he believed they would have answered the purpose, because the band was intended to run fast enough to take small quantities of slag quickly, so as not to have to deal with large masses containing so much heat as might be likely to damage such thin plates. By this plan blast-furnaces situated near the river would have had their slag delivered direct into the barges for carrying it away: probably the cheapest possible way of dealing with it. Another plan was to use a chain of buckets, instead of a plate belt. A third plan was to receive the slag into shallow troughs or moulds inside the conical rim of a large wheel, which

(Mr. David Joy.)
revolved on an axle that was inclined at 45° to the horizon. The angle of the cone was nearly a right angle, and the rim received the slag at the bottom; then in turning round it gradually raised and tilted the slag, until when the latter got nearly to the top of the wheel it dropped out of the shallow moulds into a hopper, from which it slid down into the wagon. One machine was made on that plan, and so far as it worked it proved an efficient contrivance; the only thing was that it was perhaps not quite big enough; but it would have been made larger if the plan had gone on. A somewhat similar wheel, but only very slightly coned, had also been under construction, intended to revolve on a horizontal instead of an inclined shaft; it received the slag at the bottom in shallow moulds inside the rim, carried it round nearly to the top, and there dropped it out. In both machines the slag while in the wheel was being treated with water, which made it boil up into a sort of scum or scoria, and in that form of course it broke up readily in its fall out of the wheel into the wagon. Another plan was to blow the slag away by a sort of air ejector. The intention was by applying a jet of air under the slag spout to carry away the slag while liquid, thereby causing it immediately to form itself into spheres or small drops; and so practically to drive it away in the form of sand. The slag did so form itself into sand, but it drew after it a residuum which was the terror of every one about the place: that was the little thread or tail which was afterwards found to be the right thing for making the slag wool or silicate cotton. The experiments in that direction were brought to rather a sudden termination, because these little tails being left free got down the back of every one near, and they were so sharp and irritating that it was impossible to stay more than an hour or two about the place. This discovery led however to what was afterwards utilized by Mr. Wood in his arrangement for making silicate cotton; in this the slag was blown into a closed receptacle, and the cotton being caught in meshes as it was going through the receptacle formed the highly valuable product of which specimens were now exhibited. Afterwards Messrs. Gilkes Wilson Pease and Co. took up that vertical wheel, and under Mr. Wood's management slag sand was made by that plan, and

the process became satisfactory. A committee of ironmasters having been formed for deciding which was the best of the plans proposed, they worked first the diagonal wheel, then two or three arrangements of vertical wheels; and then, after these had gone through a good summer's work in 1872, the end came. This was not that the plans tried were unsuccessful; but that the Tees Conservancy, finding that they required a much larger quantity of slag for making their breakwater, offered to take it in such enormous quantities and at such a low rate that the ironmasters found it to their advantage to abandon all idea of putting their capital into these contrivances, when they could get their slag taken away for half the money that they could get rid of it for by any other means possible. The whole of those plans therefore were rendered unnecessary for the time being; but the work was perhaps not all in vain, because a great deal had been learnt by it. Among other plans which had been tried was that of taking away the slag blocks unbroken and carrying them out to sea, and depositing them three or four miles out, permission to do so having been obtained from the Board of Trade. The blocks were simply taken solid from the common bogies into which the slag had been run. Mr. Wood and himself went out with them one day, and witnessed a miniature submarine volcano: some of the blocks were still so hot that when they were dropped into the sea they broke, and the heat of the interior of the mass caused such a violent ebullition as created a sort of volcanic agitation at the surface of the water. At Mr. Thomas Bell's invitation he had seen his machine at the Walker Iron Works (page 73), and it did not strike him as differing much from the author's design described in the paper. It had been a wonder to himself why Mr. Bell's apparatus had not been continued in use, because the Walker Iron Works were so close to the river Tyne that they had really almost no slag ground at all, and the machine so far as he could see was doing well. Still he had come then to the conclusion that one or other of the revolving wheels which he had designed was better, using a separate wheel for each furnace; because if any wheel went wrong, it could at once be taken away and another put in its place, each wheel being so constructed that it would go upon

Mr. David Joy.)
the existing slag bogie exactly where the slag box should go. Either the vertical or the diagonal wheel could be dealt with in this way, and had therefore seemed to him really the best machine to answer the purpose. A horizontal wheel on a vertical axle was the plan that had been used by Mr. Wood; but it took up more room than the diagonal or the vertical wheel.

Mr. JEREMIAH HEAD, Past-President, though coming from the land of slag, could not say that he had ever given any special attention to the subject; but he believed he was acquainted with all the schemes for its utilization which had yet been devised. The greatest consumption of slag for any useful purpose in the Cleveland district had certainly been for making breakwaters and piers. Large quantities had been taken away by the Tees Conservancy; and there was a noble pier on the south bank of the Tees, called the South Gare breakwater, which was made entirely of slag formed into concrete with cement. Large blocks had been made, for forming what was called the "toe" of the pier; they were made on shore and floated out by barges, many of them weighing he believed as much as 230 tons. Nevertheless in a heavy sea those huge blocks were occasionally found to have been slightly shifted. Since that breakwater had been completed it had been damaged once or twice; but the cracks and interstices had been filled up with cement, and he understood that it was now in a thoroughly good condition. On the north bank and nearly at the mouth of the Tees the construction was at present proceeding of another breakwater, which was to be about a mile long, and was being formed in precisely the same way. Lighters carrying slag he believed were now also going round to Hull, where it was being used in making embankments and other similar structures. As mentioned in the paper, a large quantity was now being taken in hopper barges by the Tees Conservancy to a distance of three or four miles outside the mouth of the Tees, where it was dropped in deep water. Some persons had thought that in time an island would begin to show its head above the water level, and that in this way additional surface of ground would gradually be secured. Such a hope however he was afraid would be blighted, because it was found that after a heavy sea the neighbouring shore was strewn with small pieces of slag, similar to the samples exhibited by Mr. Hawdon: showing that, although those heaps might rise to a considerable height under the sea, they were quickly lowered again and spread out as soon as a strong gale occurred.

With regard to the slag sand, he had been concerned some years ago in an attempt to turn the slag sand into bricks by mixing it with lime and pressing it in moulds. It was found that each little particle of slag sand was in reality a kind of hollow sphere containing water, and that there was enough water so contained in the sand to moisten the lime sufficiently for forming it into bricks: so that there was no occasion to add any water at all. Being mainly silica the slag particles were very sharp. On one occasion he remembered that a trial had been made to reduce them still smaller by passing them through a Carr's disintegrator, the beaters of which were made of hard steel; but the slag sand cut them to pieces in a few days. Although the bricks so made seemed to be practically successful, and got harder as they grew older, and looked very well, the manufacture was not a financial success, and it was ultimately abandoned.

The manufacture of the slag wool he believed was still carried on by Messrs. Wilsons Pease and Co., of which firm Mr. Wood was the practical manager. Although this substance looked so like wool, it really consisted of very fine threads of a silicious substance like glass. It was a good material for covering boilers he believed as long as it was comparatively new, its non-conducting property depending largely on its open nature. Some that he had seen taken off the tops of marine boilers after longer use had presented a curious appearance. The heat of the boiler seemed to have annealed the little filaments of slag, and they had settled down together into a closer and more compact mass, so that it looked more like a lump of glass. Whether in that state it still retained fully its non-conducting property he did not know, but he should rather doubt it.

(Mr. Jeremiah Head.)

The annealed bricks exhibited were rather an important industry. They were formed by running the molten slag into iron moulds, which before they got cooled were conveyed into a kiln, and there allowed to cool gradually during about ten or eleven hours. The effect of this annealing was to make the bricks exceedingly hard. If they were allowed to cool quickly in the open air, the bricks would fly to pieces. Where slag was run into large tubs, as was done in some places, and these tubs were allowed to stand for a time before being tipped on the slag heap, so that the slag balls did not break up immediately on being tipped out, it would be found, when they were subsequently broken up, that the heart or core was as hard and solid as could possibly be: showing that there was all the difference between quick cooling and slow annealing. These annealed bricks were now used in the north extensively; they came in well where a road crossed a footpath; and in by-streets in towns, where there was not a great deal of traffic, they were used for paving right across the roadway, and made a nice clean finish. Some of them he believed had been sent up to London, although he had never seen any of them actually laid down in the London streets. Their edges had a slight fin, and were so exceedingly hard and sharp that he had heard it remarked they would be rather bad for shoe-leather, which he thought was likely enough.

Attempts had several times been made to use slag sand for making Portland cement, inasmuch as slag that was composed of silica, lime, and alumina contained closely the same ingredients as Portland cement. It had therefore occurred to many that it ought to be capable of being utilized in that way; and small works had actually been put up in the Cleveland district more than once with that object, and apparently with good results. Cement made in that way had been used successfully he believed by Mr. Thomas C. Hutchinson, whose name had already been mentioned (page 80), and who was the managing director of the Skinningrove Iron Works, in the construction of a pier or sort of breakwater from his works out to sea. The adviser in that work had been the late Mr. John Fowler, engineer of the Tees Conservancy, who had a strong belief in cement made from slag sand. In March 1887 a paper had been read before the Cleveland Institution of Engineers by Mr. J. E. Stead, in which a description was given (page 113) of the manufacture of this cement by Mr. Edward Larsen, who was then introducing into this country a German plan of cement-making; but the manufacture did not seem to be going on commercially at the present time. Mr. Larsen's main idea had been that the slag should be ground to an exceedingly fine or almost impalpable powder; and that, if it could only be ground sufficiently fine, so as to enable the molecules to get close enough together, this was almost all that was required for ensuring its forming a hard-setting cement. His method was to put it into an iron cylinder with a number of spherical balls inside, and those balls rolling over and over among the stuff ground it to a fine powder almost impalpable. Still it was not a commercial manufacture at the present time.

Mr. Hawdon's apparatus he had seen at work, and he certainly thought it had been brought to a great state of perfection. Besides Messrs. Samuelson's works it was also in operation at several others, and it seemed to do its work remarkably well. It was always going on steadily taking the slag away as fast as it came from the furnace, and delivering it into the trucks in a form in which it could be conveyed away at once. Very few men were employed about it, and there did not seem to be any great wear and tear, or any great quantity of water used; in fact it seemed to give hardly any trouble at all. It was an invention which only wanted to be better known, and it would then he thought come into almost universal use.

Mr. EDWARD B. MARTEN, Vice-President, said there was an enormous quantity of blast-furnace slag to be dealt with in the South Staffordshire district, where it lay in great heaps, through which he had himself had to tunnel in a great many places for surface streams in connection with the Mines Drainage Commission. At first he had anticipated there would be great difficulty in doing so, because it was such an unkind material to tunnel through; but he had found that, when the men were used to it and knew exactly what to expect, they could really tunnel through it a great deal faster and better than through some other ground. The great use he had made

(Mr. Edward B. Marten.)

of it was for pitching the beds of the streams. When he first began to make the streams water-tight in order to protect the mines, it had been difficult to find puddle enough for the purpose, and it had to be strengthened by mixing it with gravel. The use of slag in small pieces was then suggested, because being so angular and rough these were more likely to keep the puddle in its place; and this was found to be an improvement upon gravel. Almost a hundred miles of pitching had been carried out entirely with broken slag, in whatever size the pieces could be obtained; it was used merely as surface pitching, and he had been surprised to find how exceedingly well it stood. The little interstices filled up with the mud of the stream; and the slag formed a much better pitching than many other materials. It was a great deal better than bricks, which perished badly in frost. There were plenty of places which had now been done some ten or twelve years, and were still as good as originally, although at the time the slag had looked so unpromising a material.

Remarks had been made in the paper about the difficulty of making slag blocks that would fall to pieces for convenience of road making. At some of the older ironworks in South Staffordshire, such as the Netherton Iron Works, the slag had been made into rectangular blocks, with which a number of walls had been built; and so long as they were not disturbed the walls remained sound; but if one block fell out, it generally tumbled into pieces not much bigger than a fist. The blocks appeared to have been made by running the slag in a very small stream from the furnace into a box about two feet square, so large that the small stream was continuously cooling as it filled up the box, and consequently the block contained a great many cooled faces which did not join. For such purposes as foundations therefore the blocks so made were nearly useless, because they fell to pieces so easily. Hence if it were wanted to get the blocks or balls to fall to pieces, that mode of running the slag from the furnace would insure their doing so. Lately he had been surprised to see that the London and North Western Railway, and he believed the Great Western also, had put up large plants at Coneygree in the neighbourhood of Dudley for breaking slag into small pieces for ballast, which was done at a great cost. He had also been surprised to see in the same neighbourhood enormous heaps of very small pieces, which it was said were useless; but certainly it was exceedingly useful for making roads or paths, because it formed for either a remarkably good and open bottom. If it was worth while to break the slag for ballasting, surely it must be worth while to run it from the furnace in such a manner as not to let it get set hard into large lumps which would have to be broken up again, but to take it from the furnace in the shallow pans described in the paper. It seemed to him that it ought to be most useful to provide such a material as this slag ready broken. The slag bricks, like those exhibited, although when used for paving they might be sound and lasting, would be terrible he thought for wearing horses' shoes and leather boots, and would make too rough a road or path. In the construction of the shallow pans in the author's machine he enquired what was the object of making each pan in three pieces, as described in the paper and shown in Fig. 6, Plate 7.

Mr. Arthur Keen, Member of Council, enquired what would be about the cost of the necessary machinery on the author's plan for dealing with the slag from a blast-furnace producing for instance 500 tons of iron per week, for the purpose of reducing the slag to a useful size for road making. Also what would be about the space that would be required for dealing with the slag from such a furnace with that object. It was already the case in some districts that, if slag could be produced of a size which would be useful for making roads, it became a saleable commodity. If the cost of the machinery and the space occupied were known, and also the cost of working the machinery for making the slag into a form suitable for road material, there would then be the means of forming an opinion as to the usefulness and value of the plan for other districts besides that of Middlesbrough, which alone was referred to in the paper.

Mr. Perry F. Nursey, having occasionally seen the large heaps of slag in various parts of the country, had been much impressed with the idea that they must betoken a great waste of valuable material. Having also seen some of the methods adopted for the

(Mr. Perry F. Nursey.)

utilization of slag, he supposed the most obvious was that originally employed of breaking it into small pieces and using it for road metal. Some years ago he remembered seeing slag applied in that way in Middlesbrough; but it was not at all successful, inasmuch as it was damaging both to the horses and to the vehicles going over it. Some walls which he had seen, built of square balls of slag, were beautiful to look upon, and appeared solid; but beyond those instances he believed not much more had been done in the way of using slag for the erection of walls. As regarded the use of slag for ballast on railways, he had seen in ironworks both in Germany and in France a simple method of disintegrating the slag by merely running it from the blast-furnace into a stream of water: it there became disintegrated, and by hand labour it was shovelled out into trucks. It had been and still was largely used for railway ballast, for which purpose it was particularly useful because it kept the permanent way so well drained. One objection which he had heard to its use was that in course of time it became too solid, and therefore rigid, otherwise it would be used for English railways. That objection might stand good with the high speeds usual in England; but probably it would not be of such importance with the lower speeds in vogue on the continent. Ten or twelve years ago he had gone over the slag works at Middlesbrough, where he had seen the interesting processes by which the slag was utilized for making bricks and other such purposes, and by which also the slag wool was made. The latter process was carried on inside a shed containing a number of screens, upon which the slag wool was blown. It had a beautiful appearance, causing the whole of the shed and the screens to look as if covered with freshly drifted snow. Though the formation of an island in the North Sea by the tipping of the slag from the Cleveland blast-furnaces was thwarted by the violence of the sea (page 85), yet he believed similar tipping had been carried out largely on the more sheltered coast at Barrow, without the slag getting washed away in storms; on the contrary a considerable extent of territory had already been added to the ironworks there by the tipping of their slag, which he believed was still going on. At the Clarence Iron Works, Middlesbrough, a large area of level ground had been gained behind the works by judiciously tipping the slag which had been previously broken. At the present time also at the new Dowlais Works just outside Cardiff the slag was being tipped, not into the sea for extending the premises, but over the surface of the ground for raising it to a higher level. It was most interesting to watch the development of any contrivance which, even though it could not reduce the existing slag heaps, could yet prevent their increase; and as far as he could judge from the description of the apparatus devised by the author it appeared to be one that was working in the right direction, and was deserving of the success which he had no doubt it would attain. One of the most valuable uses of slag which had lately become known he thought was that of basic slag from the Thomas-Gilchrist process for agricultural purposes. Its value in this direction was of course due to the fact that this particular slag contained chemical constituents which were absent from other slags.

Mr. Joseph Tomlinson, Past-President, mentioned that slag was already being used to a large extent for ballasting railways in this country. Nearly all the large railways were now coming to use it. All the railways in Wales that ran near the ironworks, such as the Taff Vale, were entirely ballasted with slag, which was laid about two feet thick all over the line; and it lasted so long that the railways could not themselves utilize all the slag which was being made.

Mr. Hilary Bauerman had been struck in reading the paper with the length of time it had taken to get away from so thoroughly bad a system as the original plan of casting the slag in big blocks, which it had been so great a trouble to supersede. There was no means of handling those big blocks without heavy expense, and they had been a great mistake. The true principle undoubtedly was that of the older and smaller furnaces, running their slag into much smaller masses, and getting it away as quickly as possible. This principle seemed to him to be carried out in a thoroughly effective manner in the author's machine now described, supposing always that the train

(Mr. Hilary Bauerman.)
of dish-shaped troughs ran at the proper rate for taking the slag in the right thickness, and that the slag was not too liable to run out at the joints. It was certainly wonderful to see one of the large Cleveland furnaces making such enormous quantities of slag; and he had been much interested in listening to the description of the machine by which it was now successfully dealt with. One of the best plans for disposing of the slag he believed was the American mode of "slopping," or running it over the ground in a thin layer. The great thing was not to have it in too great a thickness. Most of the American blast-furnaces made only a small quantity of slag, and on this account also the smallest quantity of land was damaged by that plan of dealing with it, because the slag then lay closer in filling the inequalities of the ground than when tipped in large solid masses. The method devised by the author appeared to be something of the same kind, getting rid of the slag as quickly as possible and in small thin pieces which could be easily handled. The American method had been introduced at the Dowlais furnaces at Cardiff, but with a modification which appeared to him to be a decided improvement on the original plan of taking such a large quantity as ten tons of slag at a time: a much smaller flush was taken in an ordinary ladle, such as was used for conveying the melted metal to a Bessemer converter. These smaller quantities of slag being then run out over the ground set quickly in a thin layer of not more than $\frac{1}{2}$ inch or $\frac{3}{4}$ inch thickness, which could then be easily broken up, and handled just as conveniently as gravel or ballast. It certainly seemed to him that the author's machine was based on a thoroughly good principle.

Mr. J. E. STEAD, being unable to attend the meeting, wrote that he had seen the author's machine in operation at Middlesbrough, and it appeared to do its work satisfactorily. At other furnaces besides the Newport Iron Works it had been adopted and continued in use, and he understood it was about to be applied to still more. There could be no doubt that it was a labour-saving machine; and information would be acceptable as to the relative cost of this process and of the plans previously used. One important element in slag utilization was the heat contained in the molten slag, which should receive most careful consideration; and it seemed strange that no effort appeared yet to have been made towards saving this large amount of heat. As the heat contained in 15·3 tons of molten slag was equal to that given out by 1 ton of coal burnt completely, it followed that the heat wasted in the twelve million tons of slag referred to in page 70 was equivalent to 784,000 tons of coal. Probably this waste had hitherto continued to pass unnoticed because of the silent way in which the heat was dissipated; but surely there was here a source of economy worthy of being followed up.

Mr. J. ADDISON BIRKBECK, regretting that he was unable to be present, wrote that, in connection with the reference made in the paper (page 72) to his own method of dealing with blast-furnace slag by using a hollow cast-iron core on the bogie, it was remarked that the slag ball so treated was cleft in two or more pieces, which were too large and heavy to be thrown into the barges. At the Acklam Iron Works however, where the core system was tried in exact accordance with his plan, it had been found most successful in breaking up the slag into pieces which as a rule were smaller than those broken by hand for filling the barges. In any case where the slag had been found to break into large pieces only, the cause must have been the adoption of too small a core, which did not follow the dimensions he had himself found to be best. The same cause would also account for the excessive wear and tear of the core. Had the walls of the slag ball itself been kept from exceeding a reasonable thickness, this difficulty would have been obviated, and the outer casing also would have lasted longer.

In regard to the endless-chain arrangement, first tried at the Walker Iron Works by Mr. Thomas Bell, he had been told by the late Mr. Thomas James, who had been thoroughly acquainted with that machine, that the reason of its abandonment there was its great cost in maintenance; the wear and tear of the links and pins had been a constant trouble and expense. This being the case, it would be interesting to learn the cost of keeping up Mr. Hawdon's, of which the belt he should imagine must be five or six times the length of

(Mr. J. Addison Birkbeck.)
that at Walker. He enquired also the cost of the adoption of the complete machine in the first instance, including water supply, special wagons for the conveyance of the slag from the machine, and renunciation of the bogies previously in use. His own plan carried out at Acklam had been designed specially with a view to the economy of making use of the existing appliances common to all the furnaces in Cleveland; and it was the only plan which dispensed with the necessity for fresh outlay, as the cost of the core on each bogie was so small a matter.

Mr. Hawdon thought that some of the designs which had been brought forwards twenty years ago by Mr. Joy (page 81) might have saved the unfortunate ironmasters many thousands of pounds if they had been then adopted. They appeared indeed to have been a little before their time. The thin wrought-iron plates however, only ¼ inch thick, of the band machine first described (page 81) he thought would have been burnt through directly. In Mr. Thomas Bell's machine, referred to both by Mr. Joy and by Mr. Birkbeck, it was true that there was a good deal of wear and tear of the pins and the other working parts, and that this was the reason why the machine had been abandoned. Apparently it was something similar in design to his own; but the precaution had not been taken of cooling the slag by the water tank or the sprinkling pipe that were used in the latter machine. Nor had the pans been made in several pieces, which was an essential feature in such machines; for each pan became considerably enlarged by expansion when receiving its fresh charge of molten slag, and then contracted afterwards in cooling; and unless it was made in several pieces it was sure to break up.

The manufacture of slag bricks, referred to by Mr. Head (page 85), he understood from Mr. Wood had been a great success both practically and financially, and that the reason why it had been abandoned was that the North Eastern Railway had bought the land on which it was being carried on.

The slag cement, mentioned by Mr. Head (page 86) as having been used by Mr. Hutchinson in building the pier at the Skinningrove

Iron Works, he understood had contained about 60 to 75 per cent. of Cleveland slag. The slag so used however had not been in the form of an almost impalpable powder, but in the form of sand. Cement manufacturers had told him that for their purposes they greatly preferred the slag made in the open honey-combed form of the specimens now shown, or even honey-combed a little more open, which was done by running it rather deeper into the water. It was then much more easily dealt with for cement manufacture than in the form of a fine sand, the grains of which were filled with water; whereas the open honey-combed slag was perfectly dry and easily ground. If the slag were run into balls, and these were allowed then to fall to pieces as was the case with slag balls run from hæmatite iron, the cement makers had a great deal of trouble with the dust.

He was glad to hear of the employment of slag in South Staffordshire for such useful purposes as those mentioned by Mr. Marten (page 88). There was no doubt an enormous waste in throwing slag away. Of the twelve million tons of slag made in this country per annum the great bulk was in many places entirely wasted. At a great number of works he thought it ought to be made in the form of the small pieces exhibited from his own machine, and to be used for ballasting railways, especially where they were laid with steel sleepers. The water drained away so readily that the road bed soon got perfectly dry after rain. In that form the slag was much better for ballasting than when it was reduced still smaller into dust or sand. In the latter case, as had been mentioned by Mr. Nursey (page 90), it set into a sort of cement far too hard; the water could not get away quickly, and the road was too hard for running, as there was no spring in it.

The cost of the eight machines at the Newport Iron Works was from £200 to £220 each. All the eight were of the same size, and together dealt with 7,000 tons of slag per week: while any one of them was capable of dealing with 1,300 or 1,400 tons of slag per week, which was the largest quantity made by a single furnace. The working cost of producing the slag, including the use of the locomotives, was about 2·8*d.* per ton of *iron*. This was the cost of

(Mr. Hawdon.)
making the slag in the form of the broken pieces shown, which he considered suitable for road making and for railway ballasting. Besides the use of the locomotives, this cost included wages, the wear and tear of the machine, and the coal consumption for locomotives. Without the locomotives, the working cost was exactly 2*d.* per ton of *iron*; or 1·3*d.* per ton of slag, as there was a ton and a half of slag to a ton of iron.

The method of using the slag at the Dowlais Works, mentioned by Mr. Bauerman (page 92), seemed an improvement on the plan generally adopted in America of running it into a large shallow tank or boat on wheels. The objection in the latter case was that the tanks or trucks soon got much choked with skulls or crusts of slag sticking so fast to their sides that there was great labour in cleaning them out. Every few days the trucks had to be stopped in order to clear the skulls out by hand labour; and this added greatly to the cost. The slag had also to be broken up after it had been run out into the thin layers, and then to be filled into other trucks; whereas by his own apparatus it was tipped direct into the trucks from the chain of pans.

Mr. Stead's idea of utilizing the heat from the slag (page 93) was a good one. It had often been thought of, but he did not see how it could be carried out; the hot slag could not be run under the boilers for raising steam. At present there was no doubt a great waste of heat.

The hollow cast-iron core introduced by Mr. Birkbeck had been adopted by Messrs. Cochrane and Co. at the Ormesby Iron Works, and they had then altered it to the "hog-back" casting described in the paper. Nevertheless the slag was still broken up into pieces too large for the height of fall into the barges. The rise of the tide was 14 feet; there was therefore a great drop at low water, and large pieces of slag, say half a ton in weight, falling that height were not a good thing for the barge plates. The plan had accordingly been abandoned at Ormesby, and the endless chain of pans had been substituted, the cost of which was from 2½*d.* to 3*d.* less per ton of *iron* than the old plan of slagging with bogies.

Professor ALEXANDER B. W. KENNEDY, Vice-President, occupying the chair at the request of the President, was sure the Members would agree in giving the author a cordial vote of thanks for his paper. It was obvious that Mr. Hawdon and his coadjutors and predecessors were anxious to do with iron what had been done with gas, and he believed with one or two other manufactured commodities: namely to make a material which could be sold for nothing, because so much profit was made out of the by-products. Although this point had not been quite reached yet in the iron manufacture, he remembered that one of the earliest lessons which he had learned under Dr. Percy was that, if it could be discovered how to get out of slag all that was in it, iron itself could then be made for nothing. This he supposed was the actual truth; and he hoped that some member of the Institution of Mechanical Engineers might in future show how to succeed in so great an achievement. In the meantime to every one who had made any practical step towards that end the whole of the engineering world was unquestionably indebted.

MEMOIRS.

HENRY BROWN was born at Slough on 18th October 1832, his father being a florist. After being educated at Edmonton he was articled to Mr. Murray, University Street, London. At an early age he entered the service of the London Brighton and South Coast Railway as a draughtsman; and in 1854 went to the railway carriage works of Messrs. Joseph Wright and Sons, Saltley, near Birmingham, where he rose from the position of draughtsman to that of general manager, which he continued to hold when the business became known in 1862 as the Metropolitan Railway Carriage Company. In 1871 he resigned that post in order to take the entire management of the Kingston Metal Works, Birmingham. Subsequently he entered into business on his own account in Birmingham as a railway stores contractor, and travelled extensively in connection therewith. His death took place at Perry Barr, Birmingham, on 10th October 1891, in the fifty-ninth year of his age. He became a Member of this Institution in 1863.

JOHN PIGGIN FEARFIELD was born at Stapleford, near Nottingham, on 23rd June 1850, and was educated at Manchester, at Belper, and in Germany. With a view to engaging in engineering business he received some practical training in a large engineering establishment in Manchester; but returned to Stapleford to assist in his father's lace-making business. On his father's death at the end of 1884, he succeeded to the business, and so improved it by his energy as to render it now one of the largest warp-lace manufactories in the world. He also fulfilled the duties of many local positions of responsibility and influence. His death took place from heart disease on 1st February 1892, in the forty-second year of his age. He became a Member of this Institution in 1884.

John William Miers was born at Villa Vicencio in the Andes on 2nd May 1819, and passed his childhood in Chili. He was educated in England, and, after a course of theoretical and practical instruction in civil and mechanical engineering, went out to Brazil in 1841. There he devoted his attention to the improvement of the rude appliances then used in the coffee and sugar estates; and in 1845 established at Rio de Janeiro, in conjunction with his brother, engineering works for the construction of machines, and the building and repairing of steamboats, &c. A plan of movable coffer-dam, which they adopted for the purpose of examining and removing the screw propeller of a ship while still floating was mentioned by Sir Frederick Bramwell in 1878 (Proceedings, page 179). They also carried out several important contracts for dredging, and for erecting iron structures, lighthouses, and other work for the Brazilian government. He returned to England in 1863. His death took place at Kensington on 28th January 1892, in the seventy-third year of his age, from pneumonia following influenza. He became a Member of this Institution in 1864.

Francis Preston was born at Higher Ardwick, Manchester, on 10th January 1823. For the greater part of his life he was actively engaged in engineering works of a varied character. He died suddenly of heart disease on 24th December 1891 at Blackpool, in the sixty-ninth year of his age. He became a Member of this Institution in 1856.

Institution of Mechanical Engineers.

PROCEEDINGS.

MAY 1892.

The SPRING MEETING of the Institution was held in the rooms of the Institution of Civil Engineers, London, on Thursday, 5th May 1892, at Half-past Seven o'clock p.m.; Dr. WILLIAM ANDERSON, F.R.S., President, in the chair.

The Minutes of the previous Meeting were read, approved, and signed by the President.

The PRESIDENT announced that His Royal Highness the Duke of Cambridge, who had been the Guest of the Institution at the Anniversary Dinner yesterday evening, had consented to become an Honorary Life Member of the Institution of Mechanical Engineers, and had today been elected by the Council.

The PRESIDENT announced that the Ballot Lists for the election of New Members had been opened by a committee of the Council, and that the following forty-two candidates were found to be duly elected :—

MEMBERS.

JOHN WEMYSS ANDERSON,	Liverpool.
JAMES MEREDITH AUSTIN,	London.
GEORGE HENRY BANISTER,	Woolwich.
THOMAS EDWIN BICKLE,	Plymouth.
JOHN BLECHYNDEN,	Kobe, Japan.
ROBERT GRUNDY BROOKE,	Manchester.
WILLIAM WALKER CAMPBELL,	Paisley.
GEORGE CHALMERS,	London.

Rupert Thomas Cooke, . . .	Stockton-on-Tees.
Philip Burne Corin, ...	Penzance.
William James Davy, . . .	London.
George D. Dennis, ...	London.
Richard Elihu Dickinson, . . .	Bradford.
James Fenwick, ...	Sydney, N.S.W.
Francis Haniel Harvey, . . .	Hayle.
John Basil Hope, ...	Whitby.
George Craig Lambe, . . .	Bilbao.
Richard John Lee, ...	Limerick.
Charles O'Keefe Mackay, . .	Horwich.
Thomas George Martin, . . .	London.
David Home Morton, . . .	Birmingham.
Arthur David Nelson, . . .	Sydney.
William Norris, ...	Lincoln.
William Heather Parratt, . .	Demerara.
Thomas Henry Parrott, . . .	Birmingham.
George Pirie, ...	Dum Dum, Calcutta.
Thomas Quentrall, ...	Kimberley.
John Rodgers, ...	Newcastle, N.S.W.
Henry Ronald, ...	Dum Dum, Calcutta.
George Henry Slight, . . .	Valparaiso.
Thomas Stevinson, ...	Nailsworth.
John Charles Grant Wilson, . .	Manila.
Robert Young, ...	Penang.

ASSOCIATES.

Henry James Carpenter, . . .	London.
Charles James Fauvel, . . .	London.

GRADUATES.

James Edmund Edgcome, . . .	London.
Alfred Pally Marks, . . .	London.
Frederick John Osmond, . . .	Birmingham.
Frank John Payton, . . .	Manchester.
Herbert Byrom Ransom, . . .	London.
George Norman Scarfe, . . .	London.
Frank Torrens Stokes, . . .	London.

The election of Mr. J. A. S. Levick, of Glasgow, whose name had been included in the Ballot List, was unfortunately rendered void by his subsequent decease.

The President then delivered his Inaugural Address: after which the following Paper was read and partly discussed:—

"Research Committee on Marine-Engine Trials: Report upon Trial of the Steamer Ville de Douvres"; by Professor Alexander B. W. Kennedy, F.R.S., *Chairman.*

Shortly before Ten o'clock the Meeting was adjourned to the following evening. The attendance was 119 Members and 82 Visitors.

The Adjourned Meeting was held at the Institution of Civil Engineers, London, on Friday, 6th May 1892, at Half-past Seven o'clock p.m.; Dr. William Anderson, F.R.S., President, in the chair.

The Discussion upon the Report on Marine-Engine Trials was resumed and concluded; and the following Paper was read and discussed:—

"On Condensation in Steam-Engine Cylinders during admission;" by Lt.-Colonel Thomas English, of Jarrow.

On the motion of the President a vote of thanks was unanimously passed to the Institution of Civil Engineers for their kindness in granting the use of their rooms for the Meeting of this Institution.

The Meeting then terminated at Twenty minutes past Nine o'clock. The attendance was 74 Members and 70 Visitors.

The ANNIVERSARY DINNER of the Institution was held at The Criterion, Piccadilly, on Wednesday evening, 4th May 1892, and was largely attended by the Members and their friends. The President occupied the chair; and the Dinner was honoured by the presence of Field Marshal His Royal Highness the Duke of Cambridge, K.G., K.T., K.P., Commander-in-Chief; and the following Guests accepted the invitations sent to them, though those marked with an asterisk * were unavoidably prevented at the last from being present. The Right Hon. Earl Brownlow,* Parliamentary Under-Secretary for War; Sir Ralph W. Thompson, K.C.B., Permanent Under-Secretary for War; General Robert J. Hay, C.B., R.A., Director of Artillery; Major-General Sir H. J. Alderson, K.C.B., R.A., President of the Ordnance Committee; Major-General Bateson; Captain Sir J. Sydney Webb, K.C.M.G.,* Deputy-Master, Trinity House; Mr. George Lawson, C.B., Director of Contracts, War Office; Mr. Albert J. Durston, Engineer-in-Chief, Admiralty; Commander Barrington H. Chevallier, R.N., Central Staff, Woolwich; Captain Frederic L. Nathan, R.A., Central Staff, Woolwich; Mr. George M. Tapp, Central Staff, Woolwich; Colonel Matthew T. Sale, C.M.G., R.E., Superintendent, Building Works Department, Woolwich; Commander F. C. Younghusband, R.N., Superintendent, Gun Factory, Woolwich; Lt.-Colonel W. McClintock, R.A., Superintendent, Royal Small Arms Factory, Birmingham; Colonel F. W. J. Barker, Gunpowder Factory, Waltham Abbey; Mr. John Rigby, Superintendent, Royal Small Arms Factory, Enfield Lock.

Mr. George Berkley, President of the Institution of Civil Engineers; Sir William T. Lewis,* President of the South Wales Institute of Engineers; Professor W. E. Ayrton, F.R.S., President of the Institution of Electrical Engineers; Mr. Charles F. Amos, President of the Hull and District Institution of Engineers and Naval Architects; Mr. James Forrest,* Secretary of the Institution of Civil Engineers; Mr. Julian C. Rogers, Secretary of the Surveyors' Institution.

Mr. Robert A. McLean, Auditor; Mr. Harry Lee Millar, Treasurer.

Professor T. Hudson Beare, F.R.S.E.,* University College, London; Professor Boulvin, Ghent University; Professor W. C. Roberts-Austen, C.B., F.R.S., Chemist to the Royal Mint; Monsieur J. Kraft,* Engineer-in-Chief, Société Cockerill; Monsieur Ortmans, Engineer, Société Cockerill; Mr. A. G. Ashcroft*; Mr. C. E. Stromeyer; Mr. Charles J. Wilson, F.I.C.

James Lemon, Esq., Mayor of Southampton; Mr. George C. Day, Southampton; Mr. William Forbes,* General Manager, London Chatham and Dover Railway; Mr. William Matthews, Waterworks Engineer, Southampton; Mr. John Morgan, Secretary, London Chatham and Dover Railway; Mr. Philip Murch,* Borough Engineer, Portsmouth; Mr. E. W. Verrinder, Traffic Superintendent, London and South Western Railway.

The President was supported by the following Officers of the Institution:—*Past-Presidents,* Mr. Edward H. Carbutt, Mr. Charles Cochrane, Mr. Thomas Hawksley, F.R.S., and Mr. Joseph Tomlinson. *Vice-Presidents,* Sir Douglas Galton, K.C.B., D.C.L., F.R.S., Professor Alexander B. W. Kennedy, F.R.S., and Mr. E. Windsor Richards. *Members of Council,* Mr. John A. F. Aspinall, Mr. William Dean, Dr. John Hopkinson, F.R.S., Mr. Samuel W. Johnson, Mr. John G. Mair-Rumley, and Mr. Henry D. Marshall.

After the usual loyal toasts, the President proposed "The Queen's Land and Sea Forces," which was acknowledged by His Royal Highness the Duke of Cambridge, K.G., K.T., K.P., and by Commander Barrington H. Chevallier, R.N. Sir Douglas Galton, K.C.B., D.C.L., F.R.S., Vice-President, proposed the toast of "Other Scientific Societies," which was acknowledged by Mr. George Berkley, President of the Institution of Civil Engineers. The concluding toast of "The Institution of Mechanical Engineers," proposed by General Robert J. Hay, C.B., R.A., Director of Artillery, was acknowledged by the President.

ADDRESS BY THE PRESIDENT,

WILLIAM ANDERSON, Esq., D.C.L., F.R.S.

Gentlemen,—It has, I think, now become the established practice for the newly elected President to address you, as soon as may be convenient after his election, on some topics that are of interest to the Members, or with which his own experience has been specially concerned. I shall endeavour tonight to fulfil this duty; and will commence by saying a few words about our Institution. But before doing so, I should like to refer to the severe losses which have been sustained through the death of many of our old and distinguished members.

I may specially allude to David Greig, for many years a Member of Council and Vice-President, a man who to very strong mechanical instincts added great powers of organization. To him certainly belongs the honour of having taken a leading part in working out to a practical issue the great problem of steam cultivation. It was my good fortune, as Engineer to the Royal Agricultural Society, to be associated with him on many occasions, and to have numerous opportunities of admiring his energy, his comprehensive knowledge, and his fertility of resource in the midst of difficulties.

Equally to be deplored is the loss of another Member of Council, Benjamin Walker, who has left a gap which will not easily be filled up. A first-rate mechanic and an enterprising manufacturer, he founded and managed with signal success, both from a mechanical and from a commercial point of view, one of the largest machinery shops in the kingdom; and he has left the indelible mark of his genius on the particular class of machinery to which he devoted his attention.

No less must we mourn the untimely death of Richard Sennett, who, having been brought up in H.M. service and having attained to the highest position therein, ended his days as managing director of one of our oldest and most successful marine engine works.

But let me turn from these sorrowful reflections to the more pleasing subject of the progress and present state of our Institution. To save time, and to make our past history and present condition visible at a glance, I have had a graphic representation prepared in Plates 9 and 10, of the figures contained in the accompanying Table (page 108), in which are recorded the total number of members, the numbers attending the meetings, the annual income, and the total invested capital, for each year since the founding of the Institution in 1847. The curves, you will notice, for all the items except the attendances, indicate that the increase has gone on roughly in proportion to the age of the Institution. The attendance at meetings, though showing a tendency to increase, has not progressed by any means to so great an extent as we should have liked; and this perhaps is natural, because a larger proportion of our members are now abroad or living in distant parts of this country, and in consequence regular attendance at meetings is not to be expected by us any more than by other scientific and technical societies, who are of necessity in the same predicament as ourselves. I am not aware that the steady increase in our prosperity is due to any special exertion on the part of the members. Engineers join our ranks spontaneously, because, I presume, they consider it their interest to do so, either, when living at a distance, because they value our transactions, or because they derive benefit and instruction from attending the meetings and joining in the discussions.

I trust that I have not appeared in the least vainglorious on your behalf; but lest we should become so, we, like St. Paul, have had a thorn in the flesh inflicted on us, in the shape of some tolerably sharp adverse criticism; and, as in the case of the great Apostle, there seems but small chance of our getting rid of it; nay, like him, we should welcome it, because it warns us of our shortcomings, and

Institution of Mechanical Engineers.

Record of Progress from commencement. See Plates 9 *and* 10.

Year.	Total Number of Members.	Annual Income.	Total Invested Capital.	Total number of Members and Visitors attending Meetings in the year.		
		£	£	Members.	Visitors.	Total.
1847	107	515				
1848	189	741				
1849	201	600				
1850	202	676				
1851	203	619				
1852	204	614				
1853	216	703				
1854	228	702				
1855	218	620				
Commencement of peripatetic Summer Meetings in 1856.						
1856	259	817				
1857	301	997				
1858	341	1,097				
1859	391	1,275				
1860	428	1,445				
1861	464	1,391				
1862	497	1,618				
1863	540	1,889				
1864	572	1,926				
1865	652	2,141				
1866	728	2,462				
1867	791	2,388				
1868	825	2,357	4,500			
1869	855	2,650	4,500			
1870	862	2,823	4,500			
1871	875	2,806	5,973			
1872	912	2,953	6,973			
1873	956	3,208	7,978			
1874	992	3,270	9,178	416	231	647
1875	1,018	3,471	10,188	434	203	637
1876	1,041	3,440	11,016	357	117	474
1877	1,075	3,641	11,869	521	168	689
Removal from Birmingham to London in 1877.						
1878	1,140	3,801	8,868	394	263	657
1879	1,178	3,782	8,868	441	314	755
1880	1,210	4,085	8,868	414	269	683
1881	1,276	4,248	9,217	485	279	764
1882	1,370	4,660	9,617	509	270	779
1883	1,440	4,690	10,617	397	189	586
1884	1,554	5,094	10,617	887	248	1,135
1885	1,640	5,330	11,967	607	362	969
1886	1,674	5,701	13,467	620	370	990
1887	1,741	5,753	15,469	554	260	814
1888	1,806	6,121	16,490	442	244	686
1889	1,858	6,260	17,489	418	246	664
1890	1,943	6,784	18,888	518	236	754
1891	2,077	7,212	20,287	628	273	901
1892			22,536			

should stimulate us to fresh exertion. The recent comments in some of the professional papers, and the correspondence to which they have given rise, have, I think, been prompted by a sincere desire for our wellbeing; and it behoves us to see whether we cannot profit by the advice and the warnings given, and so raise the Institution higher than ever in the estimation of our profession.

One of the complaints is that the papers are poor and few. But, gentlemen, is there a scientific society in existence in which the same shortcomings do not arise? and in which the responsible officers do not feel keenly the difficulty of procuring papers of merit and of interest? The very large number of societies constituted to foster particular branches of our profession—the Iron and Steel Institute, for instance, the Naval Architects, the Institution of Electrical Engineers, the numerous, and many of them excellent, provincial societies—require an enormous number of papers to feed them; while at the same time the day of original or startling mechanical developments has almost passed away, or at any rate we must admit such revelations to be, like angels' visits, few and far between. If we could only have the pick of the papers which are scattered through the kindred societies, how greatly more important our meetings would appear, how the records thereof would be prized, and how easily we should be able to satisfy our exacting critics. Besides, we have been spoiled and cloyed by the rapid progress of Mechanical Engineering: so that papers which are not revelations of something new are condemned as unworthy of the Institution. Is there any form of steam engine, for example, which it would be worth while now to describe, unless it be a monster of exceptional proportions, the details of which we should like to see in our engravings? Who would venture to read a paper on a bridge of even 800 feet span, and to illustrate it with all the pomp of type and plates which characterized the two accounts of the Britannia Bridge, when the Forth Bridge, a structure of more than double that opening, has recently become familiar to us? Nay, the creator of this later marvellous work has not even thought it worth while to present an authentic account of it to the parent Institution, of which he is one of the most distinguished members. I am afraid that, in consequence of the state at which we have arrived, and, in respect of originality, the untoward age in which we live, we must be content with many papers that may justly be termed poor so far as novelty alone is concerned. We must therefore rely for excellence on a more scientific treatment of our subjects, and on the care with which the details of construction are worked out and presented in the illustrative drawings. Our critics should remember also that originality is not our only quest, and that we are not all veterans, to whom design comes almost by instinct. We have a large body of younger and less experienced members, and to them I feel sure, from my own past experience, that our Proceedings offer practical examples and guidance which are appreciated all over the world; and the desire to possess these is, I take it, the main cause of the ever-increasing strength of the Institution.

The important work done by the Research Committees is unquestionably also a reason why membership with us is sought and appreciated. So far back as the presidency of Dr. Siemens in 1872–73, Mr. Arthur Paget, for so many years past an active member of our Council, started the idea of devoting a portion of our funds to the prosecution of independent Research; but it was not till 1878, when he himself became a member of the governing body, that his views were warmly adopted and carried into effect. The investigations indeed are of a nature which but few individuals would be either competent or able to undertake; while the high professional position of the men who have devoted their time and talents to the work, the care with which the experiments have been conducted, the manner in which details have been recorded, and the absence of motive for leaning to one view or another, inspire the most perfect confidence in the results obtained, and give a warrant that the information supplied may safely be acted on in our practice. I trust that we shall never slacken in using our funds, either in following up fresh lines of investigation ourselves, or in assisting those who may be so engaged.

There is another sphere of usefulness in which our abundant means would enable us to do good service: it is in the compilation of a brief reference index to all mechanical matters at home and abroad. Were we to establish a staff—and it might be a very modest one—whose duty it should be to index under proper heads every important article relating to mechanical science which comes out week by week, we should in time, and at moderate cost, form an invaluable record, from which an inquirer would be able to find in a few minutes where to look for complete information on any subject connected with our special branch of engineering. The Royal Society is doing a similar work for scientific papers generally; and in the United States Messrs. Haferkorn and Heise have compiled a most useful index of books printed in English relating to technical matters, but the work stops at 1888 and does not contain references to the isolated letters and papers, often of great interest, which appear in English and foreign journals. An index such as I describe should be published from time to time and distributed among our members; and any member by writing to the secretary should be able to obtain, by return of post, the latest as well as complete references connected with the subject he may be interested in. We must I think pay more and more attention to foreign technical literature. The Institution of Civil Engineers has rendered good service by publishing its foreign abstracts; but these require to be indexed in the convenient form which I suggest.

Naturally, in the criticism to which we have been subjected, the vexed question of the constitution of the Council has not escaped handling; and a good deal has been said of that famous and time-honoured remedy for all defects, the infusion of "new blood." A glance however at the Councils which have year by year been constituted reveals the fact that on the average fully two new men have been added at every election; and if we exclude the Past-Presidents, the President, and the Vice-Presidents, who form, if I may say so, the upper or permanent house, the change in the Council has amounted to nearly 14 per cent. per annum. Of the Council of 1882, elected only ten years ago, but one member now remains, the others having been removed either by death, by translation to the upper house, or by the ballot; and no one, I am confident, can question that the men whose names have been suggested by the Councils, and whom the Members, by a far wider suffrage than is usual in institutions like ours, have elected, are men who stand in the forefront of our profession, who have risen to positions of trust and importance by their proved capacity, and who on these accounts must be eminently qualified to conduct the affairs of the Institution in the best manner possible, while by the regularity of their attendance at the Council meetings they demonstrate the deep interest which they feel in the work. Of course our critics are free to assert that the renovation of the Council does not take place often enough. I wonder how many of them, if in charge of important factories, would advise frequent changes in their staff? I, for one, value the tried wisdom of experienced men above the restless aspirations of energetic new-comers.

But although I have thus lightly traversed the representations urged against the management of this Institution, I should be wanting in my duty were I not to admit that there is some foundation for them, and to exhort you all to do your very best to contribute towards the continuance of a prosperity which I think our keenest critics must admit we have a good right to be proud of.

The position which I have the honour to hold in the Royal Ordnance Factories will lead you naturally to expect some allusion to them; and it will probably be interesting to you if I say a few words respecting the manner in which new Warlike Appliances are introduced into Her Majesty's service, and about the process by which necessary changes are made in existing patterns. I am afraid that the bulk of the nation, nearly all those who are not responsible for the organization of the defences of the empire, consider that insurmountable inertia and red-tape stand in the way of enlightened progress, and tend to extinguish every effort of outside inventors to improve our armaments and our means of offence and defence. The critics who, with so much confidence, express their opinions in the public press, and denounce the shortcomings of the War Office, would change their tone very materially, if they were placed in a position to give effect to their views, and had to assume the heavy responsibility of the changes which they advocate.

The difficulties which the responsible Minister of State has to face when about to introduce a new armament are indeed enormous. In the first place, naval and military officers of great experience, like other men, frequently hold incompatible views: so that, at the outset, the Secretary of State has to discriminate between conflicting professional opinions. In the next place, when, rightly or wrongly, the chief military nations of the world have adopted a certain arm—such for example as the 110-ton gun, the small-bore magazine-rifle, or smokeless powder—and when it becomes evident that we must follow suit, the difficulty arises how to select the best weapon, or the most suitable explosive, on which to expend millions of public money. Inventors flock in from all sides, and proceed to bring pressure of every kind to bear, in order to secure the adoption of their particular inventions. It becomes imperative to examine a vast number of proposals; and the only method of doing this, in the public service, is by committees of officers or experts who have specially devoted their attention to similar subjects. But however long the inquiries may last, inventions still keep pouring in: so that a final decision seems to be unattainable, unless an arbitrary limit of time be set; and this leads at once to charges of want of appreciation of inventors. And at the best, experience shows that, when new weapons or stores have been most carefully and exhaustively proved by committees and by experienced officers at the various proof grounds, unlooked-for defects will appear when these appliances are issued for service to the troops. In the meantime every failure and defect which may come to light is laid hold of by disappointed inventors or by their friends, the newspapers teem with severe strictures on the Government, and the outcry is re-echoed to a certain extent in the House of Commons. The responsible Minister has however to pursue his course unmoved by the agitation around him, except so far as to inquire into each fresh charge, and urge his responsible advisers to spare no efforts to perfect the matter in hand with as little delay as possible. Furthermore the enormous extent of the British Empire, and the diversity of climate, render extreme caution necessary. Warlike material must be capable of being stored for years, and be always fit for use, amidst the snows and frosts of Canada, the parching heat of Aden and some parts of India, and the damp oppressive atmosphere of our tropical maritime possessions: their keeping qualities must be such that, when accumulated in magazines scattered all over the world, or in ships' holds, they shall remain serviceable for a reasonable number of years.

A case in point is presented by the smokeless explosive now being introduced. Had it been a question of use in England, or even in Europe only, many varieties of powder would have proved suitable, because the comparatively equable climate would not have taxed their keeping qualities severely, while the relatively short and safe transport would have made it possible to arrange for the rapid manufacture and supply of fresh powder in case of war. But the conditions of imperial supply are very different, and an explosive had to be sought, and I believe has been found, which will not undergo slow chemical change, but will maintain its shooting qualities when stored in our far-off dependencies, where the temperature ranges from $-40°$ F. up to $150°$ F., and where every degree of dampness and dryness may be met with.

The Ordnance Factories are often accused of being devoid of inventive powers, while endowed with great acquisitiveness for the inventions of others. Yet surely such an accusation, if true, would point only to a state of things which might be expected, seeing that the officers employed in the designing branches are relatively few in numbers, while the busy brains outside, eager at invention and prompted by the hope of substantial reward, are legion: so that, according to the doctrine of probabilities, more inventions must be expected to come from outside than from within, more especially as officers who show conspicuous ability are very soon lured away from the service by the manufacturers. But the accusation is a baseless one, for the number of inventions and improvements made in the factories is very large; but because they are restricted to the services and are not made public, they are not generally known. Besides, even in the case of outside inventions, it is found that many which claim to be original have been known and used for years in the royal workshops; and applications for reward have constantly to be resisted on this ground. The Government has a right to use any invention that may be of service to the State; but means are provided by which inventors have their claims to reward assessed, and suitable royalties granted.

Another difficulty which stands in the way of adopting new arms, or improvements in those already existing, is the necessity of incurring the enormous expense of changing the stores all over the world, and, what is almost as bad, the alteration of drill and tactics thereby involved: so that hesitation is only natural, and is even proper, until absolute necessity for change becomes apparent.

Warlike stores are now made as far as possible to uniform patterns, and the components are interchangeable not only in each service, but also, when practicable, between the various services: so that, for example, a ship might in the event of necessity be armed with fortress guns, or might fill up her ammunition from stores intended for the land service; and therefore all fittings and appliances must be made, as far as possible, common to all. This most important and desirable condition of things is well maintained by the apparently cumbersome process, according to which the design of naval as well as of military ordnance stores is placed under the control of the Director of Artillery, who is assisted by the Ordnance Committee, a council composed of distinguished officers, representing every branch of the sea and land forces, aided by two of our most eminent civil and mechanical engineers.

Patterns of stores once approved and sealed cannot be departed from without the sanction of the highest authority. Any necessary alterations are published from time to time in the "lists of changes," in accordance with which the sealed drawings and patterns are also formally amended.

The drawings which govern manufacture either by the Ordnance Factories or by the trade are very carefully dimensioned; and limits of toleration, usually confined to a few thousandths of an inch, are prescribed. In addition gauges of all kinds and in great numbers are provided for the guidance of the manufacturer, and for use by the inspection officers, who form a branch of the department of the Director of Artillery, and are entirely independent of the Ordnance Factories, whose work is examined in exactly the same way and with the same strictness as that of private contractors. The number of gauges is sometimes very large; thus in the magazine rifle, which is built up of 98 components, they number no less than 370. Every component is passed separately; and so accurate is the machine-work that in assembling the arms scarcely any handwork is required: as may be judged from the fact that one man will put together as many as twelve rifles in a day of nine hours, or a complete rifle every three-quarters of an hour.

Gun mountings or carriages, even for the heaviest ordnance required for land service, have not only to be critically measured and compared with the drawings in every detail, and examined for soundness of materials, but they are also fully erected and tested by firing five full charges either at Shoeburyness or at the proof butts at the Royal Arsenal. The guns themselves are proved by firing six rounds, four of which are proportioned so as to give from 20 to 25 per cent. pressure in the chamber in excess of the normal pressure for which the gun is constructed. After proof the interior of the gun is examined by means of impressions taken on warm gutta-percha, and by very careful gauging to the thousandth part of an inch, at regular distances along the bore, on the horizontal and vertical diameters. These gaugings are preserved for future reference, for the guns are all numbered and the history of each is carefully recorded: so that, in the event of accident, the source from which the materials were obtained, their chemical and mechanical properties, the treatment they received, the number of rounds fired by the gun, the nature of the projectiles, and the charges of powder employed, can all be ascertained at once.

In the case of other stores the same care is exercised. Shot, shell, fuzes, powder, small-arm ammunition, swords, bayonets—all have to pass the strictest gauging and the prescribed tests before they can be received into the service.

There is of necessity, I venture to think, so much of the commercial element in Mechanical Engineering that it is not likely to be considered inopportune if I dwell briefly on the financial arrangements of the Royal Ordnance Factories. It is one of the functions of the national workshops to afford to the Director of Contracts information respecting the actual cost of warlike stores, and so to prevent the formation of "rings," which have not been altogether unknown among the comparatively few firms who devote themselves to the manufacture of military appliances. It will therefore doubtless be of interest to describe the methods by which our cost prices are arrived at.

The Ordnance Factories have no floating capital, and no profit or loss entries in their accounts. The money required for the purchase of materials and for the payment of wages has to be found concurrently with the progress of manufacture, by our customers, who are the Army, the Navy, India, the Colonies, and to a very limited extent private firms, who occasionally require small alterations to be made in the supplies which they have already delivered, but which are found to be not quite up to pattern. The total expenditure is still further controlled by an annual vote in Parliament, founded upon an estimate of the probable expenditure in wages and materials for the approaching financial year.

The fixed capital is composed firstly of about £450,000, representing the value of the stocks of materials, which now include not only the materials currently used and mostly purchased to meet orders already given, but also what are termed "imperial stores." Some of the latter—such as sulphur, saltpetre, and acetone—being of foreign origin, have to be collected and stored in advance of orders, to meet the contingency of a war which might shut us out from our sources of supply; and the same is the case with others—such as native and foreign timber, and gunstocks—which require time to season, and have therefore to be kept for several years before they can be used. The second portion of the capital consists of buildings, taken at £572,000; and machinery, which stands at £764,000. These sums may not absolutely represent the actual value of our buildings and plant, for this has never been ascertained by direct valuation; but they answer the important purpose of enabling the Secretary of State to judge whether the capital account is shrinking or expanding; and they also serve as a basis for levying a rate of 5 per cent. depreciation on buildings and 10 per cent. on machinery. This rate is recovered by a charge against production, which has to bear in addition the cost of maintaining the buildings and plant in thorough repair.

The sum of money set aside for depreciation, amounting to about £100,000 per annum, is expended wholly or in part on various capital improvements: such for example as organizing the railway establishment, the electric lighting, the hydraulic power, the wood converting and other arrangements; all of which have grown up by degrees as departmental services, but which under the present organization it is found more economical and convenient to work on a general system in connection with the army and navy store departments and the inspection branch. Or again the fund is devoted to the covering over of certain premises in which operations have previously been carried out at great disadvantage in the open air; or it is expended in moderate additions, such as the cordite factory and the cartridge-strip rolling-mill, which have been rendered necessary by the introduction of smokeless powder, and by the adoption of solid-drawn cartridge-cases for rifles and quick-firing guns. The aim of the Secretary of State is to avoid increasing largely the productive power of the factories, while permitting such gradual expansion as is warranted by the continually increasing complexity of munitions of war: the object being to have a good surplus of orders to place with private manufacturers, who thus become valuable auxiliaries, and whose establishments co-operating with the Royal Factories will permit of almost indefinite increase of production in time of need.

A very large proportion of the work is done by the piece; and it will be readily understood that in establishments so ancient, and constantly employed on the same class of work, prices exist for nearly every operation which a workman can be called upon to perform. Naturally therefore the number of prices is very large; in the Royal Carriage Department alone, for example, they amount to nearly 300,000. Besides the ordinary single piece-work, there is a good deal of fellowship work, and working by shops. In both these cases the actual cost of the individual articles which are made is arrived at by computation from the piece-work prices of the components and from the actual cost of their materials; it is therefore to a considerable extent a matter of estimate, which however is checked by the known value of the total products of the fellowship or of the shop. No subletting of any kind is permitted; neither do the foremen ever receive bonuses on the output. The charges for machines and tools are levied as a uniform percentage of the wages, where the machines do not vary much in value among themselves. But in the Gun Factory, for example, where there are lathes and boring and rifling machines which range from those adapted for dealing with three-pounder quick-firing guns up to machines suited to produce 110-ton breech-loaders, the tools are grouped, and assessed at different rates. There is also a special shop rate, which covers incidental expenses not chargeable directly to workmarks. Finally a charge, also levied on the wages, and varying in rate in each department, but which may be taken at about 54 per cent., provides for the expenses of the local and central administration. These include all salaries, account keeping, pay of foremen, cost of storekeeping, coals, oil, gas, electric light, railways, rolling stock, canals, police, medical staff, hospital stores, sick and injury pay, schools, chapels, gratuities to men leaving the factories after a certain number of years' service, and various smaller expenses incidental to a large manufacturing establishment.

It sometimes happens that capital expenditure has to be incurred in the preparation of stores—such, for example, as the new magazine rifle—the manufacture of which must necessarily cease in the course of a few years. In such cases the capital expenditure incidental to preparing for manufacture may be wiped off gradually by an annual charge against production: the object being to avoid bringing out the store too dear in the first year, and much too cheap in succeeding years.

The percentages which I have indicated cannot be ascertained exactly till some time after the end of the financial year and after the annual stock-taking of the semi-manufactures or partly finished work; and as there are no profit or loss or suspense accounts, the cost ledgers are obliged to be kept open till the exact cost of each service is ascertained.

The stores are kept with great care. Commodious buildings are provided; and by means of tally boards and suitable books the quantity of any store in stock may be ascertained at any moment. No annual stock-taking is done; but a set of men is always employed in examining the remains, and applying in that way a random check on the accuracy of the store-keeper. Materials are issued at their actual invoice values, the cost of storage being part of the general indirect expenditure; they constitute about 44 per cent. of the value of the manufactures, and are all obtained by contract from the various trades.

From this account it will be seen that the Ordnance Factories' cost embraces every expenditure which would be incurred by a private manufacturer, with two important exceptions. The one is that no interest on capital is charged against production; and the other, that no expense has to be incurred in getting the business together, and no bad debts have to be allowed for. But on the other hand there is a heavier charge for medical attendance, police, sick and injury pay, and holidays, than is usual in private firms; and the rates of wages are not altogether governed by the laws of demand and supply. The Director of Contracts, in judging of trade prices, considers that they may be fairly taken at from 15 to 20 per cent. higher than the costs we furnish to him.

There has recently been a considerable outcry against the War Department: on the one hand, for appropriating all the government orders to the Ordnance Factories and so starving private manufacturers; and on the other hand, for reducing the output of the Ordnance Factories in order to give work to the trade at greatly enhanced prices. These two opposite complaints might be allowed to neutralise each other; but I have already pointed out that the Secretary of State acts upon the principle that it is in the interest of the nation that private firms should be employed even at a sacrifice to the taxpayer, in order that we may have enhanced powers of production in the event of war; and this is the more necessary, because, from the rapid changes which are constantly taking place in war material, it would be unwise to keep very large reserve stores.

The Ordnance Factories, like private firms, are subject to periods of great activity, such as the last two or three years, during which strenuous efforts had to be made to supply big guns and their ammunition to the navy and coaling stations, a new field and horse artillery equipment, and the magazine rifle with its ammunition. These services having been nearly completed, a slack time is coming on, or at any rate the activity will be diverted from one department to another, and a reduction or change of men will become necessary. In private firms reductions are made, and no one complains; but in our affairs the newspapers and members of parliament of all shades of opinion have a good deal to say, especially on the eve of a general election; and however much they may disagree in politics, it is wonderful how unanimous all are in abusing the War Office, although it is only trying to act in a rational manner, by meeting the wants of the services with due regard to the pockets of the taxpayers.

The total value of the warlike stores annually produced by the Ordnance Factories is about 2½ millions sterling; and the average number of men employed is 16,700. The following statistics of annual expenditure for the financial year 1890–91 may also be of interest:—

	£
Cost of Materials	1,124,207
Gross amount of Wages, including sick pay .	1,359,440
Total Value of Productions	2,610,162

Non-Manufacturing Expenses.	£
General Superintendence	6,920
Clerical staff	6,011
Mechanical staff	763
Medical establishment	2,791
Schools and Library	773
Divine Service	585
Police	9,787
Maintenance of Buildings, Machinery, Railways, Rolling Stock, Electric-lighting Plant, &c. . .	35,223
Carriage of Stores	490
Travelling expenses	921
Horse hire	1,099
Rents	821
Postage	112
Furniture	131
Insurance of Boilers	332
Rates	8,826
Official Quarters	855
Sick and Injury Pay	14,970
Water	1,456
Stationery	3,552
Gas	14,542
General Administration	23,057
Pensions	3,869
Gratuities	485
Hydraulic establishment	2,312
Canal staff	261
Inspection charges	316
Depreciation on Buildings &c.	116,712
Other Miscellaneous items	7,338
Total of Non-Manufacturing expenses .	£265,310

The non-manufacturing expenses are thus equivalent to 19½ per cent. on the gross amount of wages; or to about 27⅓ per cent. on the direct labour cost only. The total indirect expenditure, comprising materials, labour, and outside liabilities—excluding Waltham Abbey, at which factory the charges are abnormal—amounts to $52\frac{3}{4}$ per cent. on the direct labour expenditure.

Notwithstanding that there is but little scope for originality nowadays, some problems still remain to be solved; and among them is one which is of the greatest practical importance to Mechanical Engineers, while at the same time it is of extraordinary theoretical interest: I allude to the question of the nature and composition of Steel, and indeed of Alloys generally. Since the year 1879 this Institution has been engaged more or less continuously in trying to unravel the mystery which surrounds the behaviour of steel, in connection with its chemical and molecular composition, combined with changes of temperature. The inquiry languished for a while, because all our efforts seemed incompetent to raise the veil, though here and there a ray of light would struggle through. But the researches of Sir Frederick Abel, Dr. Sorby, M. Osmond, Mr. Hadfield, and Professor Roberts-Austen, aided by the Le Chatelier pyrometer, have given the inquiry new life; and I am in great hopes that the active measures taken by the Institution through the Alloys Research Committee will result at no very distant time in the solution of the enigma and in the establishment of definite laws.

The problem indeed is excessively involved: it amounts in fact to a consideration of the number of permutations or combinations possible among some ten variables, the relations of which to each other are also dependent not only on actual temperature, but also on the rate of its changes and on the uniformity of these changes throughout the mass. In the first place it seems almost certain that pure iron, which is the basis of steel, is allotropic, and can exist in at least two forms, one of which is hard and the other soft. Carbon again, which is an essential ingredient, also exists in steel in two forms: either in chemical union with the iron; or not merely in such union, but at the same time also in the form of detached particles of carbon suspended in the mass. There are other ingredients besides, even in the purest steel—if theoretical purity be considered a combination of iron and carbon only, as some authorities hold that it is—which, it is well known, even in very minute quantities exert a notable influence on the mechanical properties of the material; and these properties are further dependent on the temperature to which it is heated, and on the manner and rate of heating and cooling. In consequence of the changes through which iron, carbon, and possibly other constituents pass during changes of temperature, the chemist is impotent to pronounce from mere analysis what the quality of steel may be; and the ordinary mechanical tests are not of much avail, because the specimens are not and cannot be in the same condition of internal stress—on which again the molecular arrangement appears to depend—as the masses from which they are cut. Moreover specimens for mechanical testing cannot always be taken from the central parts of the huge forgings and castings now in use for many purposes. It certainly appears to me that the method of noting the rate of cooling by curves automatically traced, as now so well and so ingeniously worked out by Professor Roberts-Austen, affords the best promise of placing in the hands of the mechanic a means of judging at any rate of the uniformity in composition of the material, and even perhaps of its actual chemical nature, so far as his needs are concerned. It is besides no small advantage that the thermo-electric autographic apparatus is cheap, that it occupies but little space, that it can be employed in an ordinary room, and that the results sought can be obtained in a few minutes.

Some practical engineers, I have reason to think, look upon our investigations with a sort of good-natured contempt, as though we were frittering away the funds of the Institution in investigations much better fitted for the Royal Society than for practical men like ourselves. I can only say that such critics can never have been placed in positions of responsibility, where the safety of ships, the lives of their passengers and crews, the efficiency of armaments, and their own financial position were in question; they can never have looked at masses of steel with the view of deciding whether they were fitted for the purpose for which they had been produced; nor can they ever have felt the helplessness, and the want of reasonably secure guidance, which it is still the lot of the responsible judge to experience. May I venture to hope that the process I have alluded to will attract the attention of those of our members who are specially engaged in the manufacture of steel; and that they will start independent researches in the midst of their regular work, and communicate to us results which I feel convinced will prove of the utmost practical importance.

One more subject which is now attracting great attention, and which seems to open up a field for the inventive faculties of Mechanical Engineers, is the use of Petroleum or mineral oil. This substance has been in commercial use for many years as an illuminant, for which purpose its employment is yearly extending, especially among the poorer members of the community, because it has proved itself to be the cheapest and at the same time the most brilliant means of lighting at present available. Its use is also extending for the public lighting of moderate sized towns and villages, in which, owing to the comparatively small production of gas, and to the great length and cost of mains compared with the value of the services taken from them, the price of gas runs high. At Erith, for example, where I live, and where we pay 4*s.* 6*d.* per thousand cubic feet, the Local Board finds that oil lamps give quite as good a light, and are sensibly cheaper than gas, having besides the advantage of being capable of arrangement without reference to the situation of the gas mains. The oil lamps however, though very good and efficient, are by no means perfect, for they are apt to go out occasionally, especially in stormy weather. This I attribute partly to the somewhat defective make, quality having been sacrificed to cheapness; but there is undoubtedly room for invention, and I should welcome a paper in this Institution on the construction of out-door lamps intended for burning heavy mineral oils.

In lighthouses petroleum has now become the standard illuminant. Cautiously and step by step, as became the introduction of a substance in some of its forms not free from danger, it has gained ground; and thanks to the persevering ingenuity of Sir James Douglass, Mr. H. Defries, and others, it has now established itself as next in value to electricity, and as a perfectly safe, trustworthy, and economical illuminant. The lamps are all constructed on the principle of Argand, and have concentric wicks rising to as many as ten in number, the oil being maintained at a constant level between 2 and 3 inches below the flame; and so perfectly adapted to the oil are the structure and material of the wick, and the adjusting mechanism, that a lamp will burn some 500 hours consecutively without trimming. The unassisted candle-power of these lamps ranges from 20 to 2,215.

As a source of power, petroleum is, I think, rapidly gaining ground, especially where motors of moderate size are needed. The records of the Royal Agricultural Society show that for many years past efforts have been made to produce petroleum engines, but never, until quite recently, with any practical success: chiefly, I think, because oils of low flashing point, or petroleum spirit, were used. The dangerous nature of these would alone have condemned any engine, however efficient for general use: except indeed in the form advocated by Mr. Yarrow, in which petroleum spirit acts only as the working substance or agent for the conversion of heat into work, and is therefore not expended, except by way of leakage, so that the difficulty of supply does not arise. It was not till the show at Nottingham * in 1888 that Messrs. Priestman brought out their engine working with heavy oil having a high flashing temperature. That engine was tested by Lord Kelvin and by myself independently, and gave an efficiency of one brake horse-power to 1·73 lb. of oil per hour. At the next year's show the consumption fell to 1·42 lb.; at the next in 1890 to 1·243 lb.; and Professor Unwin this year reports † that a brake horse-power has been obtained by the combustion of 0·946 lb. per hour. It is proved by experience that these engines do not need any special attendants; neither boiler nor chimney is required; the fuel is much more cleanly; and the engine can be got to work in a few minutes; it is certain therefore that they will increase greatly in favour with the public, and will prove

* Journal of the Royal Agricultural Society, 1889, vol. xxv, page 96; and succeeding vols.

† Proceedings of the Institution of Civil Engineers, 1892, vol. cix, page 18.

formidable competitors to gas engines. Naturally also, Messrs. Priestman's success has stimulated the inventive spirit, and already more than one successful form of motor is in the field, the tendency being to simplify the details and to render them less delicate in adjustment. But much still remains to be done. The useful work on the brake is under 14 per cent. of the energy latent in the fuel; while the heat carried off by the water-jacket round the cylinder and by the exhaust is equivalent to 75 per cent. of the total thermal capacity of the oil. It will surely be within the resources of mechanical ingenuity to reduce this loss very materially. I think that probably a combination of the direct-combustion engine with the spirit engine of the Yarrow type will give the best results, especially if a more advantageous cycle than that of the Otto gas engine can be adopted.

As a lubricant also petroleum is taking a prominent place. The circumstance that it is devoid of fatty acids makes it peculiarly fitted for use with steam machinery, and for work which it is desired to protect from rust or verdigris. It can be obtained also of any degree of fluidity, from the most mobile of liquids to the consistency of jelly, while its cheapness serves to recommend it to every consumer.

There are probably few in this room who, having realised the rapid increase in the use of petroleum, have not asked themselves the question, whether the stores of petroleum in the bowels of the earth will long be able to stand the demand made upon them. I will not trouble you with statistics, because, when we come to talk of fifty or a hundred millions of barrels being annually consumed, the figures do not convey any clear idea, at any rate to my own mind, of the magnitude of the consumption; it is however already very great, and is increasing with extraordinary rapidity, doubling in about ten years. The statistical trade returns, besides, take no account of the enormous volume of natural gas evolved in some localities, nor of the waste which occurs when the fountains of petroleum get out of control.

It is commonly assumed, without any good reason however, that petroleum is of the nature of coal, and has been formed like it out of the debris of primeval forests or out of the remains of marine animals;

and that, like coal, the deposit will be exhausted in time. But it seems not unlikely, as the distinguished Russian chemist Dr. Mendeléeff has suggested, that petroleum is constantly being formed by the action of water on metallic deposits in the heated interior of the earth; and that there is good hope therefore, not only that rock oil can never be exhausted, but that it will be found in most parts of the earth if borings sufficiently deep be made; and it should be borne in mind that the depth of a boring adds very little to the cost of getting, because the oil usually rises naturally to the surface or very nearly to it.

Petroleum is an almost pure hydro-carbon, the American variety having a composition homologous with marsh gas or fire-damp, $C\ H_4$; that is, composed according to the general formula $C_n\ H_{2n+2}$, n ranging in value from 1 to 15. The Caucasian oil has the general formula $C_n\ H_{2n}$; and olefiant gas or ethylene, $C_2\ H_4$, appears to be the lowest of the series, n rising in value to 15. When exposed to heat—either in the ordinary process of distillation, or when, by working under pressure, the temperature is raised above that due to the atmospheric boiling point—the crude oil "cracks," as it is termed, and the vapours of different boiling points, but still preserving a homologous chemical composition, are given off in succession, and in varying proportions; indeed, in some districts rock oil issues from the ground in the form of gas, even at ordinary temperatures and pressures. Petroleum, in a form not to be distinguished from the natural product, has been produced artificially by the action of steam at high temperature and pressure upon the carbides of metals, more especially on those of iron: the water is decomposed, the oxygen combining with the metal, and the hydrogen, in part at least, with the carbon. This circumstance, among others, led Dr. Mendeléeff in 1877 to propound a theory, which I will sketch very briefly, because if correct—and I believe it to be gaining in general acceptance—it gives an assurance of inexhaustible supplies of oil, and also indicates the probability of its occurring in every part of the world quite irrespective of the age of geological formations; and so holds out motives to engineers to perfect the means of penetrating much deeper into the heart of the earth.

Laplace's theory of the origin of the planetary system is generally accepted as correct; and according to it the earth must be composed of the same materials as the sun. This view has in latter days received striking confirmation from the spectroscope, by means of which it has been demonstrated that there exist in the sun many of our metals, and especially iron, in the state of vapour: while meteoric stones, which belong to the same order of substances as the planets, have been found by actual analysis to be largely composed of iron and its carbides. The law of the diffusion of gases would lead us to expect that on the condensation of the metallic vapours the substances of higher specific gravity or greater atomic weight would collect chiefly nearer the centre of the future globe, while the lighter matters would tend to aggregate on the surface. The mean specific gravity of the earth is about 5, while that of its superficial deposits ranges from only 2½ to 4: so that it is evident that the interior of the globe must be composed of substances having high specific weights—such as iron, for example, which ranges between 7 and 8. Moreover it is certain that the rocks at a comparatively short distance down from the surface exist in a highly heated if not in a molten condition; and that the solid crust covering them is relatively thin and easily fissured, as is abundantly proved by the upheaval of the land in geological and even in modern times, and by the earthquake disturbances which prevail more or less over the whole world even now. Dr. Mendeléeff points out that the oil-bearing regions generally lie parallel to mountain ranges, such as the Caucasus in Russia, the Alleghanies in America, and the Andes in Peru; and that petroleum does not appear to belong to any particular geological formation, inasmuch as it occurs in Europe usually in rocks of the Tertiary period, while in the United States it is found in the Devonian and Silurian strata. He also points out that, on account of the volatile nature of rock oil, it could not have been borne from a distance on the surface, like many other deposits, but must have been formed near the spct where it is found. The fissuring of the earth's crust by the upheaval of mountain chains and by other disturbances allows surface-waters to penetrate into the heated internal portions of the earth; and there, coming in contact with the glowing metals and their carbides, they give rise to the chemical reactions which result in the formation of petroleum in the state of vapour, and in the evolution of steam. These vapours penetrate through the fissured crust into the upper and cooler regions, where they are either wholly or partially condensed, forming deposits of petroleum very commonly associated with water; and the gases which cannot be condensed by cold escape to the surface. The precise compounds which are formed depend upon the temperature and pressure met with; and hence we find associated every grade of product—gas, oil, mineral pitch, ozokerit, and other substances. The extraordinary average persistence of the oil wells leads to the conviction that the substance must be forming as fast almost as it is removed; and I have very little doubt that improved boring appliances, which it is your business, gentlemen, to contrive, will enable engineers to penetrate to depths not even dreamt of now: so that, by the time that our coal resources come to an end, from the exhaustion of the mineral, or from the condition of perpetual strike to which we seem tending, oil springs will be tapped which will have the priceless advantage of yielding their riches without the agency of underground labour. If my memory serves me rightly, we have not had a paper on deep boring at any rate within the last ten years; and the Institution would, I am sure, welcome a communication which would lay before us the present state of this most important art.*

There are two or three more subjects to which I should like to have alluded—such as the influence of bicycle building on machinery design, and on the mechanical details of dynamos and electric motors, which have been raised so lately from the region of mathematical instrument making to the workshop of the mechanical engineer. But I feel that I have trespassed too long already on your patience:—so long, that I cannot but thank you for

* A more extended account of Dr. Mendeléeff's theory will be found in the Report for 1889 of the British Association, President's Address to the Mechanical Section, pages 727–730.

the attention and good humour with which you have listened to much that must have appeared to you mere platitudes.

Sir FREDERICK BRAMWELL, Bart., Past-President, believed he was the senior Past-President present, and probably not a long way from being the senior member of the Institution. There were some advantages attaching to the post of seniority, one of them being the opportunity of expressing his feelings of gratitude to the President for such an address as had just been delivered. The members he thought would all agree that it was a model of accurate statement and of sound reasoning from beginning to end. It had touched upon a variety of most interesting subjects. Beginning with a mournful reference to the loss of many of their esteemed colleagues, it then went on to the growth of the Institution, which, let people criticise it as they might, was to his mind highly satisfactory. The Institution had been in existence forty-five years, and now numbered more than two thousand members. It was not the mere number of the members however, which in his opinion constituted the quality of an Institution; but it was the merits of the members themselves, and of the work that was brought before the Institution and published as the result of the year's labours. Any one who attended the meetings, which he himself had not been able to do much lately, or who read the Proceedings, which could be taken into the country and read there, would agree that the meetings were in themselves extremely instructive, and that the records were such as would be useful to every mechanical engineer in the world. Another point alluded to by the President was the effort made by the Institution in the direction of scientific research, which redounded greatly to its credit.

(Sir Frederick Bramwell, Bart.)

The address had then not unnaturally dealt at some length, but not at anything like the length demanded by the subject, with one particular matter of engineering—namely artillery and small arms and work connected therewith—which was of interest to every one in the nation, concerning as it did their safety and honour; and to mechanical engineers the subject was especially interesting in regard to the wonderful amount of detail and the ability shown in carrying out the work. It was a subject on which of course the President was of all persons the best qualified to speak, holding as he did the post of Director General of Ordnance Factories. It would be impossible to follow this part of the address in detail, without occupying more time than had been devoted to the address itself. But there could be no doubt that, leaving small arms out of account, the subject of making large cannon was one of great interest to every mechanical engineer who appreciated good work, and knew the difficulties it involved. It should be remembered that the pressures dealt with in modern ordnance were not as in boilers 150 lbs. or even 1,000 lbs. to the square inch, but were pressures reckoned in tons per square inch. A structure had to be made that would bear these high pressures, and for each firing the breech had to be opened to receive the charge. The opening alone it might be said was easy enough; but then it had to be closed again after the new charge had been put in, and it had to be so well closed as to make an absolutely gas-tight joint. Think of making in a few seconds a tight joint at the end of a gun of 16¼ inches bore, with a breech block of about 24 inches diameter, against a pressure of 20 tons per square inch; and compare that with an engineer screwing a cylinder cover on, and flattering himself that, with three or four men to help him, a good day's work was done if a joint was made by the end of an hour. A piece of work that produced with certainty the result attained in the breech-loading cannon was one which he thought would commend itself to every mechanical engineer. But he must not be betrayed into dwelling longer on this subject, although the interest he felt in it was so great.

The President had further touched upon the use of petroleum for motive power, and had held out the consoling prospect of saying good-bye to coal and to smoke and to strikes, and of enjoying the use of a clean trustworthy fuel which would be perennially renewed. He hoped the President was right. With regard to gas as a source of motive power, it had also been stated in the address that petroleum was likely to prove a great rival to it. There was certainly one reason for this anticipation, if there were no other. During the past winter there had been several days of severe fog; and when those who possessed gas engines, which they used for working dynamos for lighting their premises with electricity, went with great confidence to these engines, there was no gas to be had. In the case of the club to which he belonged, which was furnished throughout with the old gas burners and the modern electric fittings, no light could be obtained from either. There was no gas to light the burners, and no gas to work the engine; and the members had to stumble about until a few tallow candles and oil lamps could be raked up. Thereupon the electric-lighting committee considered naturally enough whether the gas engines could not be replaced by petroleum engines, because the petroleum would not fail in a fog any more than on a clear day. They determined however to apply to the electric company, whose mains passed their door, to make a connection there, which they could use when need arose, while continuing to employ ordinarily the gas engines.

After the survey of these various matters in the address, he was quite sure that, if the time had only admitted, the President could have dealt with many more subjects; for he believed there was nothing in the whole range of mechanical engineering with which the President was not thoroughly acquainted. He had great pleasure in moving that a hearty vote of thanks be given to him for his valuable and interesting address.

Mr. Edward H. Carbutt, Past-President, in seconding the vote of thanks as a younger Past-President than Sir Frederick Bramwell, felt that to follow so genial a speaker, and to endeavour to imitate his clear way of expressing his views, was rather a hard task. The address had raised many questions of great interest, which he was sure that many of the members would carefully consider at their

(Mr. Edward H. Carbutt.)
leisure; and he had no doubt that the suggestions which had been made by the President would be fruitful in after years. Those who had taken an interest in the Research Committees hoped that the Institution would go on with this kind of work, and would continue to find money for the purpose. He was glad the President had paid so well deserved a tribute to their former Vice-President, Mr. Paget; because, even had Mr. Paget done no other good work, he had succeeded in persuading the Institution to spend money on these researches, from which he believed a great deal of good had already arisen. One of the Committees, dealing with the subject of friction, was presided over by Mr. Tomlinson, and a good deal of information on the subject had been obtained. As he had already remarked at the previous meeting, he trusted these enquiries would be still further extended, and that the subject of rolling friction in railway trains would be more scientifically investigated, because he believed that a great deal of power might be saved in that direction. The Alloys Committee, presided over by the President and aided by the indefatigable zeal of Professor Roberts-Austen, had also done good work, which would be further stimulated by what the President had said on this subject in his address; and all would agree that the Institution could not do better than spend money in these directions. The Institution had now £22,000 in investments; and if they were not going to build, he would rather see a larger amount spent for the benefit of present engineers in doing good scientific work; for it seemed a pity to hoard up the money year after year, while not knowing what their successors would do with it. He therefore hoped that Dr. Anderson during his presidency would devise some further means of spending money advantageously.

While he was glad that the President had replied to some of the criticisms which had been showered upon the army in the newspapers, he was sure that he would not deprecate all criticism from the outside world; because it would be remembered that it was through the criticism of the outside world and the newspapers that the present Secretary of State for War, Mr. Stanhope, had been persuaded to put a mechanical engineer in charge of the manufacturing departments at Woolwich. Criticism had thus been of service in bringing about the appointment of the President to the position which all felt proud that he now occupied. He had been glad to hear the Duke of Cambridge at the dinner yesterday evening acknowledge how well Dr. Anderson filled his position; and also to hear from the President himself that the military and naval men had received him with the utmost cordiality, and had done all they could to co-operate with him for the benefit of the country: both military and mechanical engineers doing their best to put its armaments in a proper condition. That was the spirit which should animate all Englishmen. He was glad to have the opportunity of seconding the vote of thanks to the President for his address; and was sure that all who had listened to it would feel every confidence that he would do good work in the responsible position to which he had been appointed by the Government of the country.

The vote of thanks having been passed with applause was acknowledged by the President.

RESEARCH COMMITTEE ON MARINE-ENGINE TRIALS.

REPORT UPON TRIAL OF THE P.S. "VILLE DE DOUVRES."

BY PROFESSOR ALEXANDER B. W. KENNEDY, F.R.S., *Chairman.*

Since their Report of April 1891, the Research Committee on Marine-Engine Trials have been enabled through the kindness of the Belgian Government to carry out a trial on a large paddle-steamer, which in many respects differs considerably from the vessels tested previously.

Steamer.—The steamer tested was the "Ville de Douvres," a paddle-steamer owned by the Belgian Government, and employed by them on the postal service between Ostend and Dover. The Committee are very deeply indebted to the Government and to the officers of their Service du Génie Maritime, especially to M. Allo, Directeur Général de la Marine, M. Delcourt, Ingénieur-en-chef, Directeur, and M. Lecointe, Ingénieur de première classe; also to M Kraft, Ingénieur-en-chef, and M. Ortmans, Ingénieur, of the Société Cockerill; for the many facilities which they afforded throughout for carrying on the trial, and for the arrangements they kindly made for the accommodation and comfort of all on board. The Committee have also to thank Professor Boulvin of Ghent University, for the trouble he took in obtaining permission from the Belgian Government to test one of their steamers, and for the assistance he rendered during the trial, which originated in a conversation between Professor Boulvin and Mr. Donkin at Ghent.

The "Ville de Douvres," which is driven by feathering paddles, was built and engined by the Société Cockerill, Seraing, Belgium, and was delivered by them to the Belgian Government in February 1890. Between that time and the date of the trial, she made 220 voyages from Ostend to Dover and back. She is a vessel of 271 feet length between perpendiculars, 29 feet breadth, and 15·5 feet depth moulded. Her registered tonnage is—gross 855 tons, under deck 776 tons, and net 495 tons. Her draft during the trial was 8 feet 6½ inches forward, and 9 feet 7 inches aft, or 9 feet 0¾ inch mean; corresponding with a displacement of 1,090 tons.

The trial was made on Tuesday 8th September 1891, upon a run from Ostend over a course round the three light-vessels Nord-Hinder, Outer-Gabbard, and Corton, in the North Sea. Previous to the trial, the vessel had been lying in Ostend harbour, where her machinery had been overhauled. She left her moorings about half-past seven o'clock on the morning of the 8th September, and observations were begun to be taken at 8.0 a.m., but owing to various causes, the trial was not started until 10.0 a.m. It was continued until 7.0 p.m. on the same day, thus lasting for nine hours. The quantity of fuel burnt and other conditions did not allow of its further prolongation. The weather was fine throughout, with smooth sea, so that no difficulty was experienced in taking any of the usual observations. The speed of the engines was fairly constant throughout the trial, with the exception of a period of fourteen minutes from 1.10 to 1.24 p.m., when the engines slowed down a little, owing to temporary failure of the centrifugal circulating-pump, and consequent loss of vacuum in the condenser. The reading of the counter was noted several times during this period, in addition to the readings taken in ordinary course. The valve-gear remained unaltered throughout the trial.

Engines.—The "Ville de Douvres" is fitted with compound inclined surface-condensing engines, working on two cranks at right angles, the high-pressure crank leading. Sections of the cylinders are shown in Plates 16 and 17; and the general arrangement of the engines and boilers in Plate 19. The high-pressure cylinder is situated on the port side of the vessel. The diameters of the cylinders are 50·12 inches and 97·12 inches, by gauges. The piston

"Ville de Douvres."

rods are both 8·27 inches diameter, and there are no tail-rods. The stroke of both cylinders is 72 inches. Neither of the cylinders is specially jacketed; the intermediate receiver encircles the high-pressure cylinder. The clearance volumes of the cylinders as given by the makers are, high-pressure 15 per cent., and low-pressure 12 per cent., of the volumes swept through by their respective pistons. The volume of the receiver is 164 cubic feet, or 1·76 times the volume of the high-pressure cylinder and clearance. The total area of the interior surface of the receiver is 379 square feet.

The high-pressure cylinder is provided with a pair of piston-valves, and the low-pressure cylinder with a single one. The valve-gear is the Allan link-motion with a variable cut-off of from 50 to 75 per cent. of the stroke. There are two feed-pumps, two air-pumps, and two bilge-pumps, all vertical and single-acting, one of each driven from each of the two cross-heads by bell-crank levers L, Plate 19; and there is one centrifugal circulating-pump C, driven by a separate double-acting vertical engine, taking steam from the main boilers and exhausting into the condenser S. The feed-pumps and bilge-pumps are all 6·69 inches diameter, and the air-pumps are each 32·68 inches diameter. The stroke of all these six pumps is 24·61 inches. The circulating-pump is 34 inches diameter over the blades, and the suction and delivery pipes are each 14·2 inches bore. The circulating-pump engine has one cylinder 10 inches diameter and 10 inches stroke.* The donkey pump D takes steam from the main boilers, and exhausts through a three-way cock either into the atmosphere or into the condenser. During the fourteen minutes when, owing to the failure of the circulating-pump, the donkey pump was in use during the trial, it was made to exhaust into the condenser. The surface condenser, Plates 20 and 21, contains 6,540 square feet of tube surface, in 6,026 brass tubes of 0·79 inch external diameter and 5 feet 3 inches length between tube-plates. There are four tube-plates T, the tubes being arranged in

* With an estimated mean effective pressure of 50 lbs. per square inch in the cylinder, and at 240 revolutions per minute, which was its mean speed throughout the trial, the circulating-pump engine would develop about 47 indicated horse-power.

two sets; and the circulating water, entering at W, is divided between them by means of deflecting plates P. In this way the circulating water is made to pass three times through the condenser, entering at the top so that the coldest water meets the hottest steam. There is no feed-heater. The base of the condenser is 13 inches above the foot-valve of the air-pump, Plate 20, and 3·15 inches above the upper surface of the air-pump piston when at the bottom of its stroke. The bottom edge of the aperture by which the water escapes from the hot-well is 17·7 inches above the upper surface of the air-pump piston when at the top of its stroke.

Paddle-wheels.—The two paddle-wheels are fitted with curved feathering floats of steel. Their diameters are 22 feet 10 inches over the floats, and 14 feet 2 inches inside the floats, so that their mean diameter is 18 feet 6 inches. In each wheel there are nine floats, each 10 feet broad and 4 feet 4 inches deep, pivoted round a circle of 19 feet 2 inches diameter. When the vessel has a mean draft of 9 feet 0¾ inch, which was her mean draft during the trial, the maximum immersion of the inner edges of the submerged floats is about 17 inches, or one-third of the depth of the floats.

Boilers.—Steam is supplied by four single-ended boilers, Plate 18, designed for a working pressure of about 118 lbs. per square inch above the atmosphere. The furnaces, combustion chambers, tube-plates, and lower portions of the fronts of the boilers are of iron, and the remainder is of Siemens-Martin steel. The boilers are 13 feet mean diameter and 10 feet long, with three plain furnaces in each, or twelve furnaces in all. The total grate area is 236 square feet, each grate being 6 feet 6¾ inches long and 3 feet broad. There are 66 wrought-iron fire-bars of the ordinary type in each fire, arranged in three sets of 22 each. The total heating surface in the four boilers is 7,340 square feet, of which 6,280 square feet is tube surface. The total heating surface is therefore 31·1 times, and the tube surface 26·6 times the grate area. There are in all 1,356 tubes, of which 312 are stay-tubes. The tubes are all 2·48 inches external diameter and 7 feet 1⅝ inch long between plates, their length being therefore 34·5 times their external diameter. The internal diameter of the ordinary tubes is 2·24 inches, and of the stay-tubes 1·93 inch. There are two funnels, one for each stoke-hold; their internal diameter is 5 feet 3 inches, and the total height of each is 49 feet 8 inches above the centre of the lowest furnaces. The total cross-sectional area through the tubes is 35·0 square feet, and the total area across the two funnels is 43·3 square feet. The volume of the boilers, including the space between each pair, is 6,210 cubic feet.

The boilers are worked under forced draught on the closed stoke-hold system. The two stoke-holds are situated one forward and one aft of the engine-room, and communicate with it by air-locks. On the stoke-hold side of each of the two bulkheads which separate the stoke-holds from the engine-room is placed a centrifugal fan F, Plate 19, 60 inches diameter and 20¼ inches wide, driven direct by a three-cylinder Brotherhood engine in the engine-room; these maintain an air-pressure equivalent to from 0·7 to 1·0 inch of water in the stoke-holds. The two fan engines have each three cylinders, 5 inches diameter, and 5 inches stroke.* They are supplied with steam from the main boilers and can be exhausted into the atmosphere, the receiver, or the condenser; during the trial they both exhausted into the atmosphere. The effect of this arrangement of forced draught with the block fuel used during the trial was in this case to fill the whole atmosphere in the stoke-hold with a fine dust, the blocks being broken up with hammers instead of being cut with hatchets.

Weights.—The total weight of the engines and boilers and all mountings, including water in condenser, pipes, and boilers, is about 361 tons, exclusive of the two paddle-wheels. The following are the weights of the various parts of the machinery, boilers, and accessories, as given by the makers :—

* Mr. Brotherhood states that each of these fan engines would develop about 14 indicated horse-power at a speed of 404 revolutions per minute, which was the mean speed during the trial. The corresponding mean effective pressure in the cylinders would be about 47 lbs. per square inch.

	Tons.
Main engines alone	145·27
Paddle-wheels	45·21
Circulating pump	2·20
Donkey pump	1·74
Pipes, valves, &c.	10·37
Water in condenser, pipes, and pumps	7·85
Boilers, &c.	132·35
Water in boilers	61·51
Total	406·50

Duration of Trial.—The duration of the trial from start to finish was exactly 9 hours, or 540 minutes.

Fuel Measurement.—Block fuel was used throughout the trial. It had been hoped that the total quantity of fuel used might be arrived at by counting the blocks and weighing only a portion of them; but they were found to vary so much in size and weight that this was impossible, and the whole of the fuel used throughout the trial was weighed in baskets by means of spring balances in the stoke-holds, as in the former trials (Proceedings 1889, page 237; 1890, pages 205, 214, 226; and 1891, page 204). About 150 lbs. at a time was put down on the floor beside each boiler. The trial was started with clean floors, and as far as possible the time of first stoking from each weighed lot of fuel was noted, as well as the time when the whole of the weighed fuel was finished. Owing to the rapidity with which the firing had occasionally to be carried on, it was not always found possible to note these times, the whole attention of the observers being sometimes devoted to the weighing and registering of the quantity of fuel used, and to seeing that no unweighed fuel was thrown on the fires. Therefore although the total quantity of fuel used during the trial, as well as the quantity used by each of the four boilers, may be taken as correct within the ordinary limits of error, the intermediate points on the line of fuel consumption cannot be vouched for to the same degree of accuracy.

The fires were cleaned once during the trial, and the clinker and ashes were measured after the trial was over, the total amount being

2,367 lbs. in the forward stoke-hold, and 2,393 lbs. in the after stoke-hold, or a total of 4,760 lbs. in all, which is about 7·2 per cent. of the whole fuel used.

The block fuel used throughout the trial was from the manufactories of M. Dehaynin of Marcinelle, and M. Henin of Chatelineau. Samples were taken from the weighed lots of fuel frequently during the trial; and after a thorough mixture of all these samples, the final analyses, which have again been kindly made by Mr. C. J. Wilson, are as follows:—

	Fuel as used.	Dry Fuel.
Carbon	84·65 per cent.	86·74 per cent.
Hydrogen	3·98 " "	4·08 " "
Moisture	2·41 " "	0·00 " "
Ash	5·30 " "	5·43 " "
Nitrogen, Sulphur, Oxygen, &c., by difference	3·66 " "	3·75 " "
	100·00	100·00

A sample of the ashes was also collected, and has been analysed by Mr. Wilson as follows:—

Loss on ignition (= Carbon)	42·86 per cent.
Mineral matter	57·14 " "
	100·00

The calculated calorific value of the fuel is 14,390 thermal* units per lb., which corresponds with the evaporation of 14·90 lbs. of water from and at 212° Fahr., and to an equivalent carbon-value of 0·99 lb. per lb. of fuel. The total fuel used was as follows:—

After stoke-hold,	port side	15,583 lbs.
" "	starboard side	16,742 lbs.
Forward "	port side	16,737 lbs.
" "	starboard side	17,358 lbs.
		66,420 lbs.

This amounts to 7,380 lbs. per hour, or 123 lbs. per minute.

* For values of constants, see Proceedings 1891, page 287.

Furnace Gases.—Eighteen samples of the furnace gases were obtained, a sample from each funnel being collected over mercury by Mr. Wilson each hour as the trial went on. The analyses of these samples are given in Table 13, pages 144 and 145, the samples collected from each funnel being grouped together and their mean analyses given separately, as was done in the case of the s.s. "Colchester" (Proceedings 1890, Table 4, pages 216 and 217), which also was provided with a pair of funnels. The mean analyses for both funnels are also given; and these have been used in calculating the heat account, and the quantity of air required for combustion. In the present instance the volumetric analysis of each sample was not determined separately, but in several cases three or four samples from the same funnel were mixed together in equal volumes, and the analysis of each of these mixtures was determined, as indicated in Table 13. Allowance has been made for this in calculating the mean analyses for each funnel, and for both.

Two mercury thermometers of the same description as those used in former trials (Proceedings 1890, page 218) were employed to read the temperatures of the gases in the two chimneys. Both of these thermometers read up to 860° Fahr.; but unfortunately it was found that in every case, with the exception of four, the temperature considerably exceeded the maximum possible reading of the thermometer. The mean temperature in the two chimneys is therefore unknown; but from observations made of the motion of the mercury as it approached the end of the thermometer tube, it has been thought proper for the purpose of working out the heat account to assume that an excess of 50° Fahr. above the maximum possible reading of the thermometer may be taken as representing this temperature within reasonable limits of accuracy. For the purposes of the calculations which follow, 910° Fahr. is therefore assumed to be the mean temperature of the chimney gases. The mean air temperature on deck was 64° Fahr.

The mean vacuum in the chimneys was measured by a U gauge in each funnel, each of which gave a mean reading of 0·22 inch of water. The mean chimney-draught during the trial corresponded

TABLE 13 (*continued on opposite page*).

VILLE DE DOUVRES TRIAL.

Analyses of Funnel Gases by Volume.

	No. of Sample.	Carbonic Acid.	Carbonic Oxide.	Oxygen.	Nitrogen.
		Per cent.	Per cent.	Per cent.	Per cent.
FORWARD FUNNEL.	2	9·99	0·00	9·52	80·49
	4	10·19	0·00	9·22	80·59
	6, 8, 10 *	12·66	0·00	6·69	80·65
	12, 14, 16, 18 *	9·78	0·00	9·92	80·30
	Mean	**10·81**	**0·00**	**8·72**	**80·47**
AFTER FUNNEL.	1	12·61	0·00	6·58	80·81
	3	14·06	0·00	5·23	80·71
	5, 7, 9 *	12·21	0·00	7·32	80·47
	11, 13, 15, 17 *	11·84	0·00	7·72	80·44
	Mean	**12·30**	**0·00**	**7·18**	**80·52**
Mean of Both Funnels		**11·55**	**0·00**	**7·95**	**80·50**

* Mixed in equal volumes,

(*continued from opposite page*) TABLE 13.

VILLE DE DOUVRES TRIAL.

Analyses of Funnel Gases by Weight.

No. of Sample.	Carbonic Acid.	Carbonic Oxide.	Oxygen.	Nitrogen.	Time of Collecting.
	Per cent.	Per cent.	Per cent.	Per cent.	
2	14·66	0·00	10·16	75·18	10.45 a.m.
4	14·95	0·00	9·83	75·22	11.45 „
6, 8, 10 *	18·39	0·00	7·07	74·54	12.45 p.m., 1.45 „, 2.45 „
12, 14, 16, 18 *	14·36	0·00	10·60	75·04	3.45 „, 4.45 „, 5.45 „, 6.45 „
Mean	**15·80**	**0·00**	**9·29**	**74·91**	
1	18·32	0·00	6·95	74·73	10.15 a.m.
3	20·31	0·00	5·49	74·20	11.15 „
5, 7, 9 *	17·76	0·00	7·74	74·50	12.10 p.m., 1.20 „, 2.15 „
11, 13, 15, 17 *	17·25	0·00	8·18	74·57	3.15 „, 4.15 „, 5.15 „, 6.15 „
Mean	**17·88**	**0·00**	**7·60**	**74·52**	
Mean of Both Funnels	**16·84**	**0·00**	**8·44**	**74·72**	

and the mean analysis taken.*

therefore with a pressure of from 0·92 to 1·22 inch of water (page 140).

Feed-Water Measurement.—Owing to the large quantity of water required by the engines (over 1,100 lbs. per minute), it was found impracticable to measure the feed in tanks as in former trials; and it was ultimately decided to employ water-meters for this purpose. The arrangement of the meters and their connections during the trial is shown in Plate 11. The meters used were two 4-inch Kennedy's positive piston water-meters, kindly lent to the Committee by the makers, the Kennedy's Water-Meter Co., Kilmarnock, by whom they were specially constructed for use on this trial. The Committee are greatly indebted to the Company and to their managing director, Mr. Thomas Kennedy, for the generous way in which they volunteered to provide the meters and their fittings for use on the trial. The two meters, each of which was capable of measuring 32,000 gallons per hour, were placed between the feed-pumps and the boilers, and were arranged in parallel circuit on a two-branch pipe fitted with stop-cocks, as shown in Figs. 1 and 3, so that the whole of the feed-water could be made to pass through either of the meters or could be divided between the two. Only one of the meters however was in use during the trial, as it was found to do its work admirably and to require very little care or attention on the part of the observers. The meters were tested at Kilmarnock before the trial under different conditions and at various speeds, and were found to give an average excess reading amounting to three-quarters of one per cent., which has been allowed for in the totals. After the trial was over, the meter which had been used to measure the feed-water was again tested, and its correction was found not to differ sensibly from that obtained during the tests made before the trial. The whole of the steam used by the main engines and all the auxiliary engines is supplied by the main boilers; so that the feed-water for all these engines passed through the meter, and is included in the totals.

The reading of the meter index was noted every quarter of an hour throughout the trial. The total amount of water used by all the engines was 595,620 lbs., during a total time of 9 hours, or

540 minutes. This is equivalent to 66,180 lbs. per hour, or 1,103 lbs. per minute. The rate at which the feed-water was supplied and the variations of its temperature are shown in Plate 12.

Both at the beginning and at the end of the trial the boiler pressure was rising, as will be observed from the diagram, Fig. 6, Plate 13; but the water-level in the boilers averaged 2·82 inches lower at the finish than at the commencement. This corresponds with a total of 6,220 lbs. of water, or a little over one per cent. of the total feed; this amount has been included in the figures given above. It may be of interest to note that an error of one inch in reading the level of one of the boilers, which is certainly a greater error than could have occurred, would make a difference of only about 550 lbs., or less than 0·1 per cent. of the total feed-water.

The temperatures of the circulating water and of the discharge water were measured at about half-hourly intervals throughout the trial. Calculating from these observations, the mean temperature of the circulating water was 61·7° Fahr., and of the discharge water 85·0° Fahr. The ratio of condensing water to condensed steam, calculated from the respective quantities of heat taken up per lb. of condensing water, and given out per lb. of condensed steam, was 43·1 to 1.

Feed-Water to Auxiliary Engines.—Separate runs of the auxiliary engines, to determine the quantity of feed-water used by them individually, were made on 9th September, the day following the trial. Each of these runs lasted for two hours, and from the results obtained the feed-water used by the auxiliary engines throughout the trial has been computed. In each run, the speed of the auxiliary engine and the boiler pressure were maintained as far as possible the same as during the trial. No water was supplied to the boilers during the run, the water-level in the boilers being allowed to fall gradually; and at the end of two hours the engine was stopped, and a sufficient quantity of water pumped through the meter into the boilers to raise the water-level in them just to the position in which it stood at the commencement of the run. This measured quantity of water has been taken to represent with sufficient accuracy the consumption of the engine during its two hours' run.

The water used by one of the two three-cylinder Brotherhood engines driving the fans was found to be 2,120 lbs. in the two hours' run, at a mean speed of 376 revolutions per minute. Correcting for the mean speed of these engines during the trial, namely 404 revolutions per minute, this is equivalent to 1,140 lbs. per hour for each fan engine, or 20,520 lbs. for both during the nine hours' trial.

The circulating-pump engine consumed 4,140 lbs. of feed-water at a mean speed of 240 revolutions per minute, which was also the mean speed during the trial. This is equivalent to 2,070 lbs. per hour, or 18,630 lbs. during the nine hours' trial.

On the assumption that these figures hold good for the nine hours' trial, the total feed-water used by the three auxiliary engines during the trial was 39,150 lbs., which is 6·57 per cent. of the whole feed-water consumed. The following table shows the relative consumptions of feed-water by the main engines and by the three auxiliary engines on this assumption.

Feed-Water consumption.	Total.	By Main engines.	By Auxiliary engines.
	Lbs.	Lbs.	Lbs.
Throughout trial (9 hours) .	595,620	556,470	39,150
Per hour	66,180	61,830	4,350
Per minute	1,103	1,030·5	72·5
Per revolution of main shaft .	29·96	27·99	1·97

It was found impossible to arrange for the separate measurement of the supplementary feed-water. All that was used passed through the meter, and is included in the totals given above.

Priming-Water Tests.—Three times, at about equal intervals during the trial, samples of the boiler water were taken, and at the same times quantities of steam from the main steam-pipe were collected and condensed. These samples were afterwards analysed by Mr. Wilson to determine the quantity of salt present in each, and gave the following results:—

Sample.	Salt present.
No. 1 from boilers	88·8 grains per gallon.
No. 2 " "	115·3 " " "
No. 3 " "	113·0 " " "
No. 1 from steam-pipe	Each contained less than 0·16 grain of salt per gallon.
No. 2 " "	
No. 3 " "	

These analyses show that in every one of the three cases, the condensed steam contained considerably less than one-fifth of one per cent. of the boiler water, or that there was practically no priming during the trial.

The method of employing this test for the purpose of ascertaining the amount of priming at any given time is as follows. A quantity of steam from the main steam-pipe is condensed in a special surface-condensing apparatus, and collected, and at the same time a sample of water is taken separately from the boilers. Both of these samples are carefully analysed to determine the quantity of salt present in each. As the whole of the salt found in the sample from the steam-pipe must have come over from the boiler in conjunction with priming water, and not with steam, a simple calculation will show how much boiler water corresponds with the quantity of salt, if any, found in the steam-pipe sample. From this it is easy to determine what percentage of the whole feed-water has passed over from the boilers in the form of water, or in other words, what percentage there is of priming. The chemical determination for salt is a very simple one, and is capable of being carried with ease to an exceptional degree of certainty.

In order to test the accuracy of the above method, Mr. Wilson took a measured quantity of each sample of boiler water, and after mixing it with a measured quantity of the corresponding sample of condensed water from the steam-pipe, he determined, both by analysis of the mixture and by calculation from the quantities taken, the apparent priming shown by each of the mixtures. The following table gives the results of these experiments :—

No. of Mixture.	Apparent Priming.		Difference between Analysed and Calculated percentages.
	by Analysis.	by Calculation.	
	per cent.	per cent.	
No. 1	0·92	0·85	0·07
No. 2	0·86	0·85	0·01
No. 3	1·30	1·24	0·06

These last experiments show that the actual method of testing for priming water may be taken as giving results correct to within eight per cent. of the real amount of priming, although the latter did not itself exceed 1·3 per cent. It will be observed that in every one of the three cases, actual analysis gave a somewhat higher result than that calculated from the quantities in the mixture.

A considerable amount of water came through the indicator cocks on both cylinders, during the trial. This fact, taken in conjunction with the results of the priming tests given above, shows that water appearing at the indicator cocks, even although in considerable quantities, is no criterion as to priming.

Power Measurements.—Indicator diagrams were taken at half-hourly intervals throughout the greater part of the trial; but towards the end the number was increased by their being taken at intervals of a quarter of an hour. In this way 21 sets of diagrams, or 84 single diagrams, were taken in all. Owing to temporary failure of two of the indicators however, a few of these diagrams were found to be imperfect and were accordingly rejected, the total number of diagrams available for purposes of calculation being thereby reduced to 78. The indicators used were all of the Richards type, made by the Crosby Steam Gage and Valve Co., and kindly lent to the Committee by the makers for use on the trial. The high-pressure diagrams were taken with springs having a scale of 60 lbs. per inch, and the low-pressure diagrams with springs of 20 lbs. per inch. The following are the mean effective pressures in the two cylinders in lbs. per square inch :—

Cylinder.	Front.	Back.	Mean.
High-pressure . .	57·90	53·08	55·49
Low-pressure . .	15·63	15·45	15·54

These pressures correspond with the following indicated horse-powers :—

High-pressure cylinder	1,444
Low-pressure cylinder	1,533
Total Indicated Horse-Power	2,977

The maximum indicated horse-power given by any one set of diagrams was 3,229, which occurred at 2.30 p.m., with 37·0 revolutions per minute, and boiler pressure of 108·8 lbs. per square inch. The minimum indicated horse-power by any one set of diagrams when not slowed down was 2,757 at 6.45 p.m., with 36·6 revolutions per minute, and boiler pressure of 97·3 lbs. per square inch. In Plate 14 is given the set of diagrams nearest to the mean, which was taken at 5.45 p.m., with 37·7 revolutions per minute, and boiler pressure of 100·8 lbs. per square inch. From the 78 diagrams which were taken during the trial a mean diagram has been plotted for each of the two cylinders, and these are given in Plate 15. The continuous variations of boiler pressure are shown in Fig. 6, Plate 13; of cylinder mean effective pressures in Fig. 8; of horse-power in Fig. 5; and of speed in Fig. 7.

If to the total indicated horse-power of the main engines, given above, were added the estimated indicated horse-powers of the three auxiliary engines in use during the trial, the mean total indicated horse-power of the main engines and the three auxiliary engines during the trial would be roughly about 3,050.

Speed.—The counter was read every half hour throughout the greater part of the trial, and every quarter of an hour towards the end, alternately with the times for taking indicator diagrams. The total number of revolutions made was 19,885 in 9 hours, giving an average speed of 36·82 revolutions per minute throughout the trial. The maximum number of revolutions per minute for any half-hour during the trial was 38·1; and the minimum, except during the time when slowed down, was 36·4 revolutions. The mean speed for the fourteen minutes from 1.10 to 1.24 p.m., when the engines slowed down owing to failure of the vacuum, was 20·2 revolutions per minute. The continuous increase of total revolutions from the beginning of the trial is shown in Plate 12.

Pressures.—As in the "Iona" trial, a standard gauge was used to check the pressure gauges, and a correction was made where necessary. The mean barometric pressure during the trial was 30·21 inches of mercury, or 14·84 lbs. per square inch. The mean boiler pressure was 105·8 lbs. per square inch above the atmosphere, averaged from the four pressure gauges on the boilers. The mean pressures in the high-pressure valve-chest and the intermediate receiver were respectively 99·4 and 12·0 lbs. per square inch above the atmosphere. The mean vacuum in the condenser was 20·6 inches of mercury by gauge, which corresponds with a mean absolute back-pressure in the condenser of 4·72 lbs. per square inch. The mean initial pressure in the high-pressure cylinder, as measured from the diagrams, was 93·2 lbs. per square inch above the atmosphere; and the mean pressure during admission, obtained in the same way, was 89·2 lbs. per square inch above the atmosphere. The mean vacuum in the low-pressure cylinder, as measured from the diagrams, was 8·78 lbs. per square inch below the atmosphere, which corresponds with a mean absolute back-pressure of 6·06 lbs. per square inch in the cylinder. The poor vacuum thus indicated was the subject of considerable discussion before the commencement of the trial; and it was then understood that this was the vacuum commonly obtained when the engines were working at full power.

Boiler Efficiencies.—The mean rate of combustion in the furnaces was 31·3 lbs. of fuel per square foot of grate area per hour, or 1·01 lb. per square foot of total heating surface per hour. The total amount of feed-water pumped into the boilers was at the rate of 8·97 lbs. per lb. of fuel. The average temperature of the feed-water throughout the trial was 158° Fahr., and the temperature corresponding with the mean boiler-pressure was 342° Fahr., so that each lb. of steam must have taken up 1,060 thermal units. The equivalent evaporation from and at 212° Fahr. was therefore 9·84 lbs. of water per lb. of fuel, or 9·94 lbs. per lb. of carbon-value in the fuel. The equivalent amount of heat utilised per lb. of fuel was 9,509 thermal units, or 66·1 per cent. of the whole thermal value of the fuel. This percentage therefore represents the actual boiler efficiency.

The total calculated calorific value of the fuel burnt per minute was 1,770,000 thermal units.

The weight of dry air per lb. of fuel, calculated from the series of analyses of chimney gases given in Table 13, page 145, works out to 17·9 lbs.; so that the total weight of furnace gases per lb. of fuel would be about 18·8 lbs. Assuming the estimate of the mean chimney temperature to be correct, as given in a former paragraph, page 143, these gases were raised in temperature from 64° Fahr., the temperature of the outer air, to 910° Fahr., the estimated mean chimney temperature. Taking the mean specific heat of the gases as 0·243, this corresponds with a loss of 3,853 thermal units per lb. of fuel, or 26·8 per cent. of the whole thermal value of the fuel. In estimating the weight and specific heat of the chimney gases, it has been assumed that the whole of the hydrogen in the fuel was burnt. This assumption may not be absolutely correct, but the error due to it is exceedingly small. The loss due to the evaporation of the moisture in the fuel was so small that it may be neglected; and none of the eighteen gas samples collected during the trial was found on analysis to contain carbonic oxide. The two percentages mentioned above, 66·1 and 26·8, add up to 92·9 per cent., so that the quantity of heat unaccounted for was 7·1 per cent. of the whole thermal value of the fuel. This is mainly covered by radiation, but would also include any losses due to unburnt carbon passing up the chimney, imperfect combustion of the hydro-carbons, and other causes.

The weight of air theoretically necessary for the combustion of 1 lb. of the fuel is 11·1 lbs. The air actually used, 17·9 lbs. per lb. of fuel, is therefore about 1·6 times the amount theoretically required.

The weight of water evaporated from and at 212° Fahr. per square foot of total heating surface was 9·90 lbs. per hour.

The average rate of transmission of heat through the material of the boiler was 9,560 thermal units per square foot of total heating surface per hour.

Fuel Consumption.—The total fuel burnt was equivalent to 7,380 lbs. per hour. This would amount to a gross consumption of 2·48 lbs. of total fuel per indicated horse-power of the main engines per hour; but assuming that 6·57 per cent. of the total amount was consumed in making steam for the three auxiliary engines, as already explained (page 148), this would reduce the net consumption of fuel on behalf of the main engines alone to 6,900 lbs. per hour, or 2·32 lbs. per indicated horse-power per hour, which is equivalent to 2·30 lbs. of carbon-value per indicated horse-power per hour.

Engine Efficiencies.—The total indicated horse-power of the main engines was 2,977; and the total gross feed-water used per hour by the main engines and the three auxiliary engines which were at work throughout the trial was 66,180 lbs. This would amount to 22·23 lbs. of gross feed-water per indicated horse-power of the main engines per hour; but, as was shown in an earlier paragraph (page 148), 6·57 per cent. or 4,350 lbs. per hour of the total feed-water was probably used by the three auxiliary engines: so that the net feed-water used by the main engines alone would be 61,830 lbs. per hour, which is equivalent to 20·77 lbs. of net feed-water per indicated horse-power per hour. The total actual heat received by the gross feed-water per minute was 1,169,000 thermal units, which is 66·1 per cent. of the whole heat of combustion. Similarly the actual heat received by the net feed-water used by the main engines would be 93·43 per cent. of this amount, or 1,092,000 thermal units per minute. The absolute engine efficiency of the main engines alone, or ratio of the heat turned into work by them to the heat given to the net feed-water, was 11·7 per cent.

Total Efficiency.—The combined efficiency of the boilers and engines, or ratio of the heat turned into work to the total heat of combustion of the fuel, allowing for the 6·57 per cent. of the feed-water which was consumed by the three auxiliary engines, was 0·661 × 0·117, which is equivalent to 7·7 per cent.

Steam from Indicator Diagrams.—The following are the results of measurements made upon the indicator diagrams taken, to ascertain the proportion of steam accounted for by them. The actual weight of net feed-water used by the main engines per revolution was 27·99 lbs.:—

Proportion of Steam accounted for by indicator diagrams.	Lbs. per Revolution.	Percentage of Total Feed.	Percentage present in cylinder as water.
	Lbs.	Per cent.	Per cent.
Steam present in high-pressure cylinder after cut-off, when the pressure was 81·2 lbs. per square inch above the atmosphere	22·57	80·6	19·4
Steam present in low-pressure cylinder near end of expansion, when the pressure was 0·8 lb. per square inch below the atmosphere . . .	20·29	72·5	27·5

"Ville de Douvres."

Speed of Vessel.—The following extract from the ship's log has been kindly furnished by M. Lecointe :—

Time.	Observations.	Distance measured from chart.
8 September 1891.		Nautical Miles.
7.30 a.m.	Left Ostend Harbour	0
10.30 „	Passed Nord-Hinder Light-vessel . .	26
12.14 p.m.	Passed Outer-Gabbard Light-vessel. . (speed reduced for fourteen minutes)	57
2.20 „	Passed Corton Light-vessel . . .	90
4.18 „	Repassed Outer-Gabbard Light-vessel .	125
6. 4 „	Repassed Nord-Hinder Light-vessel .	156
7.37 „	Returned to Ostend Harbour . . .	182

The mean speed between the times of passing Nord-Hinder light-vessel on the outward journey and arriving in Ostend Harbour, which includes nearly the whole of the trial, was therefore 17·1 knots. The mean speed over the trial, excluding the passage from Outer-Gabbard light-vessel to Corton light-vessel, during which the fourteen minutes of slowing down occurred, was 17·5 knots.

There is added to this report Table 14, pages 158–163, showing the leading results of the trial of the "Ville de Douvres," and beside these are placed the corresponding figures for the six trials which have been reported on previously, namely those of the "Meteor," "Fusi Yama," "Colchester," "Tartar," "Iona," and the subsequent trial of the "Iona" carried out by the Chief Engineer of the vessel. There is also added an appendix giving a list of the Members of this Institution and others who took part in the trial of the "Ville de Douvres."

APPENDIX.

Staff of Observers.

* Members of the Institution are marked with an asterisk.

One watch was taken by Mr. H. R. J. Burstall,* along with Mr. Bryan Donkin, Jun.,* Professor T. Hudson Beare, and Messrs. A. G. Ashcroft, W. Defries,* P. T. J. Estler, H. F. Friedrichs, L. A. Legros,* and H. A. Page. The other watch was taken by Professor Kennedy,* along with Professor D. S. Capper,* and Messrs. H. F. W. Burstall, J. T. Ewen, S. B. B. Hebb, C. E. Stromeyer, A. G. Way, R. C. B. Willis, and R. H. Willis. Mr. C. J. Wilson collected the gas samples throughout the whole of the trial. M. Kraft, Ingénieur-en-chef, M. Ortmans, and M. Nolet, of the Société Cockerill, and Professor Boulvin, of Ghent University, were on board the vessel during the trial, and the Committee are greatly indebted to them, as well as to M. Lecointe, Ingénieur de première classe, and to M. Pierrard, Ingénieur de troisième classe, of the Belgian Service du Génie Maritime, and to M. Davin, Mécanicien-chef of the "Ville de Douvres," and his staff, for the valuable assistance they rendered towards the successful carrying out of the trial. The Committee have also to express their thanks to M. le Commandant Romyn, of the "Ville de Douvres," and his officers, for their kindness and courtesy throughout the whole of the operations.

The thanks of the Committee are also due to the Chairman and Directors of the London Chatham and Dover Railway, for their kindness in providing the whole of the staff of observers with free passes from London to Dover and back, for the purpose of conducting the trial; and to M. Allo, Directeur Général de la Marine, for kindly arranging, on behalf of the Belgian Government, for the free conveyance of the staff between Dover and Ostend.

TABLE 14 (*continued to page* 163).

Comparative Results of the Trials of Five Screw-Steamers, and One Paddle-Steamer,

1	Name of Vessel	Meteor.	Fusi Yama.
2	Date of Trial	1888 24 June	1888 14 & 15 Nov.
3	Duration of Trial . . . hours	17·15	13·95
4	Type of Engines	Triple	Compound
5	Cylinder diameter, high-pressure inches	J 29·37	N 27·35
6	„ „ intermediate . inches	J 44·03	...
7	„ „ low-pressure . inches	J 70·12	N 50·3
8	Stroke, length . . . . inches	47·94	33
9	Boilers, number of main boilers . . .	2	1
10	„ single-ended or double-ended . .	double	single
11	Furnaces, total number	12	3
12	Heating surface, total . . square feet	6,648	2,257
13	„ „ tubes . . square feet	5,760	1,689
14	Grate area . . . . . square feet	208	52
15	Total heating surface to grate area ratio	32·0	43·4
16	Tube surface to grate area . . ratio	27·7	32·5
17	Grate area to flue area through tubes ratio	—	4·0
18	„ „ to area through funnel ratio	5·0	3·2
	Mean Pressures.	Lbs. per square inch.	
19	Mean boiler-pressure above atmosphere .	145·2	56·84
20	„ admission „ high-pressure cylinder above atm.	134·4	50·3
21	„ effective „ high-pressure cylinder	58·46	30·74
22	„ „ „ intermediate „	19·50	...
23	„ „ „ low-pressure „	12·38	10·87
24	„ „ „ total reduced to low-pressure cyl.	29·9	19·9
25	„ exhaust „ low-pressure cylinder below atm.	11·6	10·9
26	„ vacuum in condenser below atm.	12·17	12·48
26A	„ back-pressure in condenser, absolute	2·73	2·32

J = Jacketed. N = Not jacketed.

(*continued on next page*) TABLE 14.

S.S. "METEOR," "FUSI YAMA," "COLCHESTER," "TARTAR," "IONA"; P.S. "VILLE DE DOUVRES."

Colchester.	Tartar.	Iona. (Committee.)	Iona. (Chief Eng.)	Ville de Douvres.	1
1889 9 Nov.	1889 27 Nov.	1890 13 & 14 July	1890 15 August	1891 8 September	2
10·88	10·08	16	10·88	9	3
Twin Compound	Triple	Triple	Triple	Compound Inclined	4
(two) N 30	N 26·03	J 21·88	J 21·88	N 50·12	5
...	J 42·03	N 34·02	N 34·02	...	6
(two) N 57	J 68·95	N 56·95	N 56·95	N 97·12	7
36	42	39	39	72	8
2	2	2	2	4	9
double	double	single	single	single	10
12	8	4	4	12	11
5,820	5,226	3,160	3,160	7,340	12
4,770	4,366	2,590	2,590	6,280	13
220	161	42	42	236	14
26·5	32·5	75·2	75·2	31·1	15
21·7	27·1	61·7	61·7	26·6	16
5·5	4·5	2·3	2·3	6·7	17
4·8	4·2	1·4	1·4	5·5	18
Lbs. per square inch.					
80·5	143·6	165·0	162·0	105·8	19
64·3 / 59·4	121·4	142·5	147·8	89·2	20
45·65 / 42·07	36·89	46·65	47·33	55·49	21
...	20·07	20·44	20·21	...	22
13·42 / 12·42	7·18	7·16	7·96	15·54	23
24·8	19·8	21·13	21·95	30·17	24
10·6 / 10·5	10·5	12·74	12·8	8·78	25
12·49	12·9	13·88	13·76	10·12	26
2·51	1·70	0·70	1·04	4·72	26A

J = Jacketed. N = Not jacketed.

TABLE 14 (*continued from preceding page*).

Comparative Results of the Trials of Five Screw-Steamers, and One Paddle-Steamer,

	Name of Vessel		Meteor.	Fusi Yama.
27	Revolutions per minute, mean .	. revs.	71·78	55·59
28	Piston Constant, high-pressure cyl.	. H.P.	11·31	5·36
29	„ „ intermediate „	. H.P.	26·00	...
30	„ „ low-pressure „	. H.P.	66·65	18·32
31	Indicated horse-power, high-p. „	I.H.P.	662	168·2
32	„ „ inter. „	I.H.P.	507	...
33	„ low-p. „	I.H.P.	825	203·1
34	„ mean total	I.H.P.	1,994	371·3
35	Fuel burnt per minute . . .	lbs.	66·75	16·45
36	„ „ per hour	lbs.	4,005	987
37	„ „ per square foot of . . grate area per hour	lbs.	19·25	18·98
38	„ „ per square foot of total heating surface per hour	lbs.	0·602	0·437
39	„ „ per indicated horse-power per hour	lbs.	2·01	2·66
40	Carbon-value of 1 lb. of fuel as used .	lbs.	0·878	0·878
41	„ „ equivalent per indicated horse-power per hour	lbs.	1·76	2·33
42	Feed-water per minute . . .	lbs.	497·7	131
43	„ „ per hour	lbs.	29,860	7,860
44	„ „ per square foot of total heating surface per hour	lbs.	4·49	3·48
45	„ „ per lb. of fuel . . .	lbs.	7·46	7·96
46	„ „ per lb. of fuel from and at 212° F.	lbs.	8·21	8·87
47	„ „ per lb. of carbon-value from and at 212° F.	lbs.	9·62	10·10
48	„ „ per indicated horse-power per hour	lbs.	14·98	21·17

(*continued on next page*) TABLE 14.

S.S. "METEOR," "FUSI YAMA," "COLCHESTER," "TARTAR," "IONA"; P.S. "VILLE DE DOUVRES."

Colchester.	Tartar.	Iona. (Committee.)	Iona. (Chief Eng.)	Ville de Douvres.	
86·0 87·1	70·0	61·1	63·9	36·82	27
10·71 10·84	7·73	4·41	4·61	25·99	28
...	20·42	10·82	11·32	...	29
39·55 40·06	55·27	30·54	31·93	98·56	30
490·3 457·9	283·7	205·6	221·0	1,444	31
...	408·5	221·2	231·7	...	32
532·2 499·3	395·2	218·6	255·0	1,533	33
1,022·5 957·2	1,087·4	645·4	707·7	2,977	34
95·7	32·0	15·7	16·3	123	35
5,742	1,920	942	978	7,380	36
26·1	11·93	22·4	23·3	31·3	37
0·987	0·367	0·298	0·310	1·01	38
2·90	1·77	1·46	1·38	2·48 2·32*	39
0·913	1·031	1·02	—	0·99	40
2·65	1·82	1·49	—	2·46 2·30*	41
717	359·4	143·6	—	1,103	42
43,020	21,564	8,616	—	66,180	43
7·39	4·13	2·73	—	9·02	44
7·49	[11·23]†	9·15	—	8·97	45
8·53	[13·06]†	10·63	—	9·84	46
9·34	[12·67]†	10·42	—	9·94	47
21·73	[19·83]†	13·35	—	22·23 20·77*	48

* Allowing for the three auxiliary engines, as on pages 154 and 155.

† Proceedings 1890, page 233.

TABLE 14 (*continued from preceding page*).

Comparative Results of the Trials of Five Screw-Steamers, and One Paddle-Steamer,

	Name of Vessel	Meteor.	Fusi Yama.
49	Calorific value of 1 lb. of fuel as used Th.U.	12,770	12,760
50	Percentage of line 49 taken up by feed-water per cent.	62·0	67·2
51	„ „ „ „ carried away by furnace gases p. cent.	21·9	23·5
52	„ „ „ „ lost by imperfect combustion per cent.	3·6	0·0
53	„ „ „ „ used in evaporating moisture in coal p.c.	1·2	0·9
54	„ „ „ „ unaccounted for p.c.	11·3	8·4
55	Heat taken up by feed-water per min. Th. U.	528,600	141,100
56	„ turned into work per minute Th. U.	85,240	15,870
57	„ taken up by feed-water per indicated horse-power per minute Th. U.	265·6	380·0
58	Efficiency of boiler (line 50) . per cent.	62·0	67·2
59	„ of engine (line 56 ÷ line 55) p. c.	16·1	11·2
60	„ of engine and boiler combined (line 58 × line 59) per cent.	10·0	7·6
61	Mean velocity of steam through water-surface in boilers per minute feet	—	6·28
62	Space occupied by boilers per I.H.P., cub. ft.	2·72	4·53
63	Weight of engines, boilers, &c., with water, per I.H.P. ton	0·20	0·27
64	Clearance volume, high-p. cyl. per cent.	12·4	8·5
65	„ „ inter. „ per cent.	9·3	...
66	„ „ low-p. „ per cent.	8·02	5·0
67	Clearance surface, high-p. cyl. square feet	—	17·3
68	„ „ inter. „ square feet	—	...
69	„ „ low-p. „ square feet	—	42·5
70	Speed of vessel, mean, during trial knots	14·6	—
71	Mean temperature of outer air. . Fahr.	—	55°
72	„ „ „ chimney gases. Fahr.	791°	578°
73	„ „ „ feed-water . Fahr.	163°	129·5°
74	„ „ „ boiler steam . Fahr.	363°	304°
75	„ „ „ circulating water Fahr.	—	—
76	„ „ „ discharge water Fahr.	—	—
77	„ „ „ condensed steam Fahr.	138°	132°
78	Condensing surface, total square feet	3,200	—
79	Condensing surface ÷ feed-water per minute	6·43	—
80	Condensing water to condensed steam ratio	—	—

(concluded from page 158) TABLE 14.

S.S. "Meteor," "Fusi Yama," "Colchester," "Tartar," "Iona"; P.S. "Ville de Douvres."

Colchester.	Tartar.	Iona. (Committee.)	Iona. (Chief Eng.)	Ville de Douvres.	
13,280	14,995	14,830	—	14,390	49
62·0	—	69·2	—	66·1	50
28·0	22·1	16·2	—	26·8	51
1·3	0·0	0·0	—	0·0	52
0·4	0·0	0·0	—	0·0	53
8·3	—	14·6	—	7·1	54
788,700	[403,600]†	161,100	—	1,169,000 1,092,000*	55
84,630	46,490	27,590	30,250	127,300	56
398·4	[371·2]†	249·6	—	392·7 366·8*	57
62·0	—	69·2	—	66·1	58
10·7	[11·5]†	17·1	—	11·7*	59
6·6	9·7	11·8	—	7·7*	60
8·60	3·43	1·61	—	8·61	61
2·52	4·33	4·15	4·15	2·09	62
0·20	0·27	0·31	0·31	0·12	63
(two) 9·39	14·51	12·41	12·41	15·0	64
...	9·25	10·11	10·11	...	65
(two) 6·23	5·10	7·64	7·64	12·0	66
(two) 22·6	25·4	12·75	12·75	84·73	67
...	40·1	29·14	29·14	...	68
(two) 56·9	79·6	49·5	49·5	211·46	69
14·4	—	8·6	9·0	17·1	70
55°	55°	62°	91°	64°	71
835°	477°	452°	—	910°	72
113°	101°	106°	157°	158°	73
324°	362°	373°	371°	342°	74
—	—	55·8°	78·6°	61·7°	75
—	—	75·5°	94·4°	85·0°	76
135°	120°	90°	103°	160°	77
3,000	2,250	1,360	1,360	6,540	78
4·18	6·26	9·47	—	5·93	79
—	—	52·5	62·5	43·1	80

* † See footnotes to page 161.

Discussion.

The PRESIDENT said this report upon the trial of the "Ville de Douvres" was just an illustration of what he had ventured to say in his address as to the sort of facts which the Research Committee were bringing forward. The report contained an account of everything the observers saw; and the facts were given exactly as they were observed, for each person to draw his own deductions therefrom. The first wish of the Members he was sure would be to pass a cordial vote of thanks to Professor Kennedy, and to the observers who had been associated with him in carrying out this latest trial; and to Mr. Charles J. Wilson for the analytical work done in his laboratory. Also to the Belgian Government, and the officers of their Service du Génie Maritime: especially to Monsieur Allo, Directeur-Général de la Marine, Monsieur Delcourt, Ingénieur-en-chef, Directeur, and Monsieur Lecointe, Ingénieur de première classe, for the opportunity afforded for carrying out a trial on one of their steamers, and for all the facilities rendered by them in connection therewith. Also to the Société Cockerill and their officers: especially to Monsieur Greiner, Directeur-Général, Monsieur Kraft, Ingénieur-en-chef, and Monsieur Ortmans, Ingénieur; and to Professor Boulvin of Ghent University; for the assistance rendered by them for the successful carrying out of the trial, and for the valuable information supplied in connection with the preparation of the report. Also to Kennedy's Water Meter Co., Kilmarnock, and to their managing director, Mr. Thomas Kennedy, for specially constructing two of their 4-inch water meters for use on this trial, and for lending them to the Committee free of charge. And to the Chairman and Directors of the London Chatham and Dover Railway, for providing the whole of the staff of observers with free passes from London to Dover and back, for the purpose of conducting the trial; and to Monsieur Allo, Directeur-Général de la Marine, for arranging on behalf of the Belgian Government for the free conveyance of the staff between Dover and Ostend.

The vote of thanks was heartily passed.

The President wished, before the discussion commenced, to remind the meeting that the subject chiefly to be criticised was not the construction of the engines which had been experimented with. The Research Committee had been most generously allowed the use of the steamer, and had been furnished liberally with all the details of the engines; and therefore it would be a little ungenerous to criticise them in any disparaging sense. But any comment on points connected with the engines, which bore upon the results obtained, would of course be perfectly legitimate. What was wanted was criticism upon the observations made and upon the results obtained, and not upon the mechanical construction of the engines.

Sir Frederick Bramwell, Bart., Past-President, had but little to say in any event, but after the caution just received from the President that little had become still less, owing to the great difficulty of separating comments upon the results reported and remarks on the engines from which those results had been obtained. The Committee appeared to consider it right to separate the fuel and the water that were employed in driving the auxiliary engines from the consumption for the main engines. In this view he could hardly agree. The main engines could not work without circulating water for the condenser; and if the circulating water was supplied by a separate engine, as it was in this case, that separate engine was only a substitute for a circulating pump worked from the main engines themselves. Again the main engines could not work without feed-water; and if the feed-water was put in by an independent engine, the latter was only a substitute for a feed-pump attached to the main engines. Further, as regarded the fuel used in working the engines for the forced draught, this was an expedient by which the steam consumed by the now obsolete steam-jet was saved, or, looked at in another way, was an expedient by which an equally good draught could be obtained with a lower temperature of escaping gases; and therefore again it seemed to him that the fuel needed for the forced-draught engines was properly chargeable to the main engines. With respect to the temperature of the escaping gases, it struck him

(Sir Frederick Bramwell, Bart.)
however that in the present instance, notwithstanding the forced-draught, the gases did escape at a somewhat high temperature.

Mr. CHARLES COCHRANE, Past-President, wished to say how heartily he concurred in the vote of thanks to Professor Kennedy for the way in which he and his able colleagues had worked out this subject; and how continuously increasing had been the value of their work at every one of the six successive stages through which it had advanced, from the commencement with the "Meteor" right up to the present trial of the "Ville de Douvres." Much valuable information was in his opinion to be gathered from this latest trial, with advantage not only to the Institution, but also to the Belgian Government, who he was sure would be pleased to have candid criticisms on any fault which might be found out through the medium of the trial.

On the point of condensing surface, he had been puzzled to know why the "Ville de Douvres" should have given such a poor vacuum compared with the "Iona" and others, with only a slightly different ratio of condensing surface to steam condensed. On looking into the different conditions however he found that it was due mainly to the fact that the cold condensing water was introduced into the condenser at the opposite end to that at which it should be introduced; and he thought it would be a simple matter to alter that arrangement in the condenser, and so obtain higher results than those given in the report. The mean absolute back-pressure in the "Ville de Douvres" condenser was stated at 4·72 lbs. per square inch (page 152); while as a matter of course the condenser would work best if the cold water were admitted at the place where there was the least elastic force of steam, in order if possible to kill the last breath of steam by the coldest water that could be brought to bear upon it.

Another point was that between the vacuum in the low-pressure cylinder and the vacuum in the condenser there was a further considerable loss, amounting to 1·34 lb. per square inch. This seemed to point to the desirability of a larger exhaust pipe for conveying the steam from the cylinder to the condenser: which also was a matter of easy remedy if his surmise was correct.

Another highly interesting point was that, on comparing the various trials, it was found that the triple-expansion engine took the lead throughout, and in the most marked way; there was no question about the superiority of the triple throughout all these experiments. In the case of the "Iona," where there was a comparison between the Committee's trial and that of the Chief Engineer, he had been anxious to see what was the ratio of condensing water to condensed steam. In the Committee's trial of the "Iona" he found this ratio was 52·5 to 1; but in the Chief Engineer's it appeared to be 62·5 to 1, in order to obtain practically the same vacuum in the condenser. The difference arose from the simple fact that in the Mediterranean, with condensing water having an initial temperature of 78·6° and rising only 16° to 94·4° Fahr., a larger quantity had to be used, in order to maintain the same vacuum, than when the initial temperature was only 55·8°, rising 20° to 75·5°. In the "Ville de Douvres" the ratio was reduced to 43·1 to 1.

Another feature in the "Ville de Douvres" was that there was no feed-heater. In the "Iona," in the Chief Engineer's trial, the feed-water was sent into the boiler at 157° Fahr., whereas in the Committee's it was only at 106°, or about the temperature due to the pressure in the condenser. There must evidently have been some means of supplying extra live steam to the water in the Chief Engineer's trial, to have raised the temperature of the feed-water from 106° up to 157°. This he thought was probably worth notice.

Another point which had rather puzzled him was as to the waste of water that took place through the leakage of joints and so on. This he had in vain endeavoured to ascertain from the figures given in the report; and he now understood that it had not been observed in the trial of the "Ville de Douvres." It was however a matter which might well be inquired into in any further observations.

In connection with the previous report on the "Iona" trial (Proceedings 1891, page 287) he noticed that a higher calorific value had been given for hydrogen than had formerly been adopted. For the purpose of checking the results in the Tables, it was important to know what was now taken for the value of a unit of hydrogen burnt into water, as compared with what it formerly stood

(Mr. Charles Cochrane.)

at, in order to see how the thermal units had been arrived at in the report. A reference therefore to the calorific value adopted was desirable for making the calculations perfectly clear. Similarly in estimating at 11·7 per cent. (page 155) the absolute engine efficiency of the main engines alone, or the ratio of the heat turned into work by them to the heat given to the net feed-water, he should be glad if it were expressly stated that the calculation was made on the basis of Joule's equivalent, namely 772 foot-lbs. per heat unit. This would enable the calculations to be more readily followed.

M. KRAFT pointed out that the Ostend mail boats had been built under special conditions to perform a special service. Carrying as they did the English and continental mails, the first conditions they had to fulfil were high speed and safety. The entrance of the harbour at Ostend limited the draught of water to about 8¾ feet, thus giving predominance to the question of weight of the machinery. As the "Ville de Douvres" took only 3 to 3¼ hours for the passage from Ostend to Dover, consumption of fuel became so far a secondary consideration as hardly to concern the practical efficiency of the engines. The limit of weight allowed did not permit the use of highly economical engines; and forced draught was indispensable, not with the aim of economy of fuel, but for a sufficient production of steam. The contract speed required by the Belgian Government was 17·5 knots; and in the severe official trial, which was made not at a measured mile but over a much longer measured distance, a mean speed of 18·955 knots was realised, and the builders received a speed premium of £4,000. In this trial the main engines developed 4,431 indicated horse-power. As stated in the report (page 140), the total weight of the engines and boilers and all mountings, including water in condenser, pipes, and boilers, was about 361 tons, exclusive of the two paddle-wheels; and from the comparison given in line 63 of Table 14 it was seen that the weight per indicated horse-power, amounting to only 0·12 ton in the "Ville de Douvres," was less than in any of the other steamers tried, in which it ranged from 0·20 up to 0·31 ton per I.H.P. The high speed of so large a vessel, and the small weight of such powerful engines, were certainly worth noting.

As to the circulation of the water in the surface condenser, he entirely agreed with Mr. Cochrane (page 166) in the principle that every heating or cooling apparatus should be arranged with counter currents, instead of with both currents running in the same direction. But a calculation of the transmission of heat in this particular instance would show that the difference between counter current and similar current was inconsiderable. The principal reason of the indifferent vacuum did not lie there. It was due to the imperfect subdivision of the water current passing through the inside of the condenser tubes. There were corners of the water chambers in which the flow of water was not satisfactory, and in which deposits even of mussels had been found. Owing also to the predominant consideration of weight, the circulating pump was rather small for a condenser of that size. It was doubtless well to divide the current of condensing water between the two halves of the condenser, as was here done, so as to give it a better chance of penetrating everywhere; and it would perhaps be better still to have two circulating pumps, one to each half of the condenser. Why the vacuum had been so indifferent in the Committee's trial he could not say; in regular work it was much better, rising generally to 22½ or 23 inches of mercury and even higher.

The Belgian Government had now ordered two new fast mail boats, one from Messrs. Denny of Dumbarton, and the other from the Société Cockerill. These boats would be ready to start running next year, and were intended to have a speed of 21½ knots with about 8,000 indicated horse-power. At the proper time he should be very glad to place before the Institution the results obtained at the trials.

Professor BOULVIN considered that, as the object of the Research Committee on marine-engine trials was to collect as extensive data as possible on the behaviour and efficiency of the different kinds of marine engines now in use, they could not have done better than include among their trials one set of engines having such special features as those of a mail steamer making short runs at high speeds. The results of the present trial showed that there was room for some improvements in the condenser

(Professor Boulvin.)
of the "Ville de Douvres," and indicated clearly the nature of the improvements needed; but the defects pointed out belonged only to these particular engines themselves, and not to the class of which they were representatives. It might be of more interest to direct attention to the leading difference between the "Ville de Douvres" and the other engines tried previously, which lay in the much lighter weight of her machinery and boilers, as shown by the comparison in line 63 of Table 14. This weight, including the water in the boilers, amounted in the "Ville de Douvres" to only 0·12 ton per indicated horse-power. Taking for comparison, not the "Iona" with so much less power, but the "Meteor" with her weight of 0·20 ton per I.H.P., it would be seen that, for developing the 3,000 I.H.P. of the "Ville de Douvres," the engines of the "Meteor" would be some 240 tons heavier than those of the "Ville de Douvres." This comparison was made on the assumption that the figures given for the weights included corresponding portions of machinery; and it was to some extent to the disadvantage of the paddle-wheel engines, in which the piston speed was about 25 per cent. slower, thus rendering them so much heavier than they would otherwise have been.

If the consumption of fuel in the "Ville de Douvres," instead of being 2·46 lbs. of carbon-value per indicated horse-power per hour, had been the same as in the "Iona," or only 1·49 lbs., the saving in fuel would have been about 1 lb. per I.H.P. per hour, or 1·34 ton per hour for the 3,000 I.H.P. Assuming that the bunkers had to be filled with a supply sufficient for ten hours' steaming, corresponding with a double run in bad weather, the saving in weight of fuel would be 13 to 14 tons. As the cost of the extra fuel was of minor importance in this case, while the problem was to obtain the maximum speed with a steamer of given dimensions, the object to be aimed at was to reduce to a minimum, not the weight of fuel alone, but the sum of the weights of the fuel and the machinery. In long voyages this sum was always minimised by the more economical engines, because the weight of the fuel amounted to so large a percentage of the whole weight. This was the reason why it had been so long before compound engines had been applied to fast channel-steamers, and why low-pressure engines, some even oscillating, had been in use as late as ten years ago for this class of steamers. The saving of 13 tons in weight of fuel would no doubt be advantageous, if it could be realised without increasing by an equal amount the weight of machinery; but in order to obtain a reduction in fuel consumption, steam-jackets would be needed on both cylinders, as well as a greater extent of boiler surface to take up more heat from the gases. It might be doubted whether there was room for these additions to the weight of machinery, within so low a margin of saving as only 13 tons in weight of fuel.

It was thus clear that engine perfection was a different thing in its relation to land engines, locomotives, war steamers, trading vessels, and mail boats; and the Research Committee had therefore taken the best course in trying marine engines of different kinds and designed for different purposes. While the construction of the vessels themselves was a matter of naval architecture, shipbuilders would none the less feel much indebted to this Institution for the valuable series of reports which had been presented by the Committee on the six trials already made. The Belgian marine department could not have done anything more advantageous to their own interests than by availing themselves of the opportunity which had been afforded them of having one of their steamers tried by the Committee; and their warmest thanks were due to the Institution for the useful information thus obtained.

Dr. Alexander C. Kirk was sure the work that had been done by the Research Committee was highly appreciated by every one; and they had to thank the Chairman for the remarkable care with which he had worked out and recorded the results. The great thing done was that the performance of the boiler had here been separated from that of the engine. Ordinarily, by the help of the indicator, and by measuring the coal, sometimes more and sometimes less accurately, it was only the gross efficiency of the machine that was ascertained; but the separation of the boiler efficiency from that of the engine was a most important point. If some arrangement could be adopted whereby a record of the boiler performance could be obtained throughout a voyage, it would be a great help in maintaining the efficiency of the machinery and in judging what was the best

(Dr. Alexander C. Kirk.)

arrangement of boiler. The engine itself he thought was now known pretty thoroughly, and also what was required to make it efficient: but what was required with a view solely to efficiency was not always what was most suitable for the ultimate object aimed at. Consequently the shipowner sometimes did not think it worth while fitting his vessels with the most efficient engines; and in other cases, especially in high-speed steamers which came into port at frequent intervals after short runs, looking at the ship and engines as one whole machine, it did not pay to make the engines themselves of the highest possible efficiency. For instance the proportion of machinery in the "Iona" would be altogether unsuitable in a transatlantic steamer. Sometimes machinery was not made to work so economically as it should do, occasionally through sheer ignorance; while at other times it was quite necessary that it should be so made, in order that the ship might do its work with the utmost economy. Not only were all much indebted to the Committee for their trials, so far as these had gone, but he wished also to express the appreciation that was felt of the way in which some shipowners had put their machinery into the hands of the Committee for trial; and it would therefore be unbecoming that any criticism whatever should be made upon the greater or less relative efficiency of the different machines, except as to the cause of their efficiency or deficiency as the case might be.

Mr. BRYAN DONKIN, JUN., observed that, while the water meter had been so carefully tested both before and after the trial, nothing was said in the report with regard to the indicator springs having been tested; and he should be glad to know whether they had been tested or not. In his own experience he had found they were generally wrong.

Mr. H. GRAHAM HARRIS asked how the quantity of coal on the fire had been measured at the starting of the trial, both in the present and in the previous trials; and how what was left on the fire had been measured when the trial was finished. In the case of the "Ville de Douvres" there was a grate area of 236 square feet, on which with a fire 6 inches thick there might be roughly 120 cubic feet of fuel. In some of the many experiments which he had carried out with Sir Frederick Bramwell, he had first tried to guess, as had always been done previously, the depth of the fire upon the grate; and had then raked it all out and allowed it to cool, and had afterwards weighed it. In one case, with a vertical boiler having 16 square feet of grate area, he had found that there was a difference of something like 70 per cent. in the weight of the fuel in different trials, although the thickness upon the bars was practically the same, as well as it could be measured. Neither in the present trial nor in those previously reported did the Committee appear to have entered at all into the question of the quantity of fuel upon the grate, either at the commencement or at the end of the trial. In the "Ville de Douvres" burning 7,380 lbs. of coal per hour, if the fires were 6 inches thick and there were an error of the kind he had suggested amounting to only 50 per cent., this would represent about 60 cubic feet of fuel, amounting say to 3,500 lbs., which would be equivalent to about half an hour's run of the engines. As there were twelve furnaces, it might be said that the weight of the fuel would average itself throughout the twelve, and probably it did so; but he had mentioned this point because there was nothing said upon it in the report.

The President did not quite agree in the reasons that had been given by Mr. Cochrane for the imperfect vacuum, because the temperature at which the condensing water was discharged from the condenser was only 85° Fahr., whereas a good vacuum could be got with water leaving at 100° and more: so that, even at the part of the condenser which ought to have been coldest, the water was cold enough to produce a good vacuum, for it was discharged in the present trial not warmer than it often was in the Red Sea, where he understood there was no difficulty in getting a good vacuum.

Mr. Cochrane suggested that, for the further elucidation of the efficiency of the condenser, it would be desirable to add in Table 14, from the data furnished in the successive reports, the mean

(Mr. Charles Cochrane.)
temperature of the condensed steam corresponding with the observed pressure existing in the condenser, and the ratio of the condensing water to the condensed steam; also the actual area of the condensing surface, and the ratio of the condensing surface to the feed-water per minute. (See lines 26A and 77–80 accordingly added in Table 14.) When these additions were made, it would be seen that the temperature of the condensed steam at the pressure existing in the condenser of the "Ville de Douvres" was no less than 160°, which would of course militate seriously against a good vacuum, if it was really a matter of fact, as he thought must be the case.

With regard to fuel burnt per square foot of heating surface per hour, it would appear—from the comparison with the previous trials in line 38 of Table 14, and from the possibility of burning as much as 1·01 lb. of fuel per square foot of total heating surface per hour, as was done in the "Ville de Douvres," without the appearance of priming—that the "Iona" and the "Tartar," which ran on counter lines with only about one-third of that consumption, had an excessive amount of heating surface, so far as priming was concerned: that is, they need not have had so large an extent of heating surface in order to avoid priming. But this inference was immediately confronted with the counterbalancing disadvantage that, for want of larger heating surface in the "Colchester" and the "Ville de Douvres," the temperature of the chimney gases had risen to 835° in the former and 910° in the latter. Corresponding with these temperatures the heat wasted through being carried away by the chimney gases amounted to 28·0 per cent. and 26·8 per cent. respectively; while the temperature of the chimney gases in the "Iona" was only 452°, with a waste of 16·2 per cent., and in the "Tartar" it was only 477°, with a waste of 22·1 per cent. Between these two extremes there might of course be some intermediate proportion of heating surface which would secure the most economical results with a fair recovery of heat from the chimney gases.

With a view to arriving at a clearer appreciation of the conditions affecting the efficiency of condensation in the several trials already made, he had collected in Table 15 the principal figures bearing upon this portion of the subject. In this comparison it would be

TABLE 15.—*Comparative Efficiency of Condensation.*

Name of Vessel.	Condensing Surface.		Condensing Water per lb. of [illegible]d Steam.	Back-Pressure in Condenser. Lbs. per sq. inch [illegible]e.	Corresponding [illegible]re of Condensed Steam.
	Total.	Per lb. of Feed-Water per minute.			
	Sq. Ft.	Sq. Ft.	Lbs.	Lbs.	Fahr.
Meteor . .	3,200	6·43	—	2·73	138°
Fusi Yama . .	—	—	—	2·32	132°
Colchester . .	3,000	4·18	—	2·51	135°
Tartar . .	2,250	6·26	—	1·70	120°
Iona (Committee) .	1,360	9·47	52·5	0·70	90°
Iona (Chief Eng.) .	1,360	—	62·5	1·04	103°
Ville de Douvres .	6,540	5·93	43·1	4·72	160°

noticed how, although the "Tartar" and the "Ville de Douvres" had nearly the same ratio of condensing surface to feed-water, the vacuum was only 10·12 lbs. below atmosphere in the latter, against 12·90 lbs. in the former, or 4·72 lbs. per square inch absolute pressure in the "Ville de Douvres" against 1·70 lbs. in the "Tartar." To the reason of the poorer vacuum he had already adverted (page 166); but the explanation he had attempted was modified by what Mr. Kraft had since stated in regard to the vacuum ordinarily obtained in the "Ville de Douvres," in which the area of condensing surface per lb. of feed-water per minute was 5·93 square feet. In the "Colchester," where the corresponding area was only 4·18 square feet, it would not fail to be noticed how, with this still lower ratio of condensing surface to feed-water, the excellent vacuum was nevertheless obtained of 12·49 lbs. below atmosphere, or 2·51 lbs. absolute pressure. The highest vacuums obtained were in the two trials of the "Iona," namely 13·88 and 13·76 lbs. below atmosphere, or 0·70 and 1·04 lbs. absolute

(Mr. Charles Cochrane.)
respectively, but at the expense of no less than between 9 and 10 square feet of condensing surface per lb. of feed-water per minute. The question arose whether the combined economies secured in the "Iona," in low temperature of chimney gases and in high vacuum, were real economies from a pecuniary point of view.

Some explanation he hoped would be furnished with regard to the substitution made in the "Ville de Douvres" for a steam-jacket round the high-pressure cylinder, by surrounding it with the intermediate receiver into which the steam exhausted from the high-pressure cylinder before it passed into the low-pressure. It seemed to him a prominent feature in the construction of this engine that the high-pressure cylinder was encircled by a large concentric jacket, which received the steam at the lowest pressure at which it escaped from the high-pressure cylinder. In this particular case he gathered there was a difference of 80° between the temperature at which the steam was admitted into the high-pressure cylinder, and that at which it escaped from the high-pressure cylinder to go into the low-pressure cylinder at about 15 lbs. per square inch above the atmosphere, or about 30 lbs. absolute pressure. This low-pressure steam in the receiver must exercise a cooling influence on the steam in the high-pressure cylinder all through the stroke; and from the indicator diagrams there was evidence of a slight condensation taking place in the high-pressure cylinder right through the stroke, certainly after the steam was cut off, which might naturally be expected when a difference existed of 80° of temperature between the outside surface of the cylinder and the inside surface after the cut-off had taken place. He hoped that Professor Kennedy might have something to say upon this point, in regard to which it was so desirable that the truth should be realised. Already a large degree of truth had been attained on the subject generally, owing to the valuable data and calculations which had been furnished by the Research Committee; and he regarded all that had been done as a stepping stone to the establishment of fixed data, on which not marine engineers only but land engineers also might decide what was the best ratio for heating surface, boiler capacity, condensing surface, and every other element in steam engineering. Further experiments also he hoped would be made on quadruple expansion, which was a problem certain to come forward soon for consideration.

Mr. James Rowan considered the ultimate effect of these investigations would be to produce faster and more economical steamers. With regard to the present trial, he looked upon the "Ville de Douvres" as essentially a passenger steamer; and if this was the case it might safely be assumed that speed stood first, as had been pointed out by Mr. Kraft (page 168); and if the speed of the steamer could now be materially increased in consequence of the present investigations, he was sure the Committee would feel in some measure rewarded for their labours, and the result would also be an incentive to other owners to place steamers at their disposal for trial.

With the object of increasing the speed of the "Ville de Douvres" at a trifling expense and without increasing the coal consumption, he would in the first place suggest that a considerable increase of power would be obtained, were the low-pressure slide-valve made to cut off somewhat earlier; that is to say, more work would be realised per lb. of steam passing through the engines. The effect of this earlier cut-off would be to increase the mean effective pressure in the low-pressure cylinder, and diminish it in the high-pressure cylinder; and accordingly the indicated horse-power would be increased in the low-pressure cylinder and diminished in the high-pressure, but the net result would be a gain of perhaps 5 per cent. in the total horse-power. If it would be of interest to the Committee he could have indicator diagrams taken from the engines of either of two new fast paddle-steamers—the "Neptune" and the "Mercury," which had just been completed by his firm for the Clyde passenger traffic of the Glasgow and South Western Railway—first with the eccentric rod and slide-valve spindle in line, and secondly with the links linked up. These would show the increase in power consequent upon linking up in those engines, which indicated from 1,800 to 1,900 horse-power. Objection might be taken to this suggestion in so far that, if the horse-powers taken from each cylinder were too unequal, they might produce that jumping forward motion

(Mr. James Rowan.)
which all travellers on the Clyde had experienced in the river steamers fitted with the single-cylinder diagonal engines, until recently so popular there. [See pages 196–7.]

Secondly, it was mentioned in the report that there was no feed-heating apparatus. This however could be arranged with but little trouble and small expense, and in the most complete manner possible. All that was necessary was to lead a pipe of 1 inch or 1¼ inch diameter from the receiver to a heating coil in the hot-well, with a non-return valve upon it. When the engine was at rest the atmospheric pressure kept the valve shut; and when the engine was at work, whenever the receiver pressure rose above that of the atmosphere, then this feed-heating apparatus came into action automatically. Steam was certainly taken away thereby from the low-pressure cylinder; but by the adoption of this simple apparatus about 4 to 6 per cent. less coal would be required to produce a certain indicated horse-power than without it. It was always a most important matter that hot water and not cold water should be put into a boiler; for nothing was so severe on boilers as pumping cold water into them.

From his own experience of paddle-boats and their engines, he anticipated that, if these two proposals were adopted, at least one more revolution per minute would be got, which should propel the vessel at least a quarter of a knot faster than at present. As a marine engineer he wished to express his great admiration of the manner in which these investigations had been conducted by the Research Committee.

Mr. C. E. STROMEYER, having been one of the staff of observers, said there was one point which struck him as remarkable in these trials, namely the great regularity in the efficiency of the boilers. From line 58 in Table 14 it was seen that the lowest efficiency was 62 per cent. and the highest 69·2 per cent. In trying to find out whether there might not be some unexplained reason for this, he had paid particular attention to the percentage unaccounted for (line 54) of the calorific value of the fuel as used; and this item could at once be reduced a little, by taking into account the carbon contained in the ashes which were found in the furnaces. In the "Ville de Douvres" for example, there was 7·2 per cent. of ashes remaining after the trial (page 142), and, as far as he remembered, this was the total for twelve hours' firing, not for nine hours' only; so that the actual ashes for the nine hours' trial would be only 5·4 per cent. The mineral matter in the ashes was stated to be 57·14 per cent.; it followed that 3·08 per cent. of mineral matter remained in the stoke-hold. Moreover 42·86 per cent. of the ashes consisted of carbon (page 142), which would reduce the loss unaccounted for by 2·32 per cent. The analysis of the fuel showed 5·30 per cent. of ash; subtracting 3·08 per cent., the difference, or 2·22 per cent. of mineral matter, must therefore have been blown up the funnel, carrying carbon with it; and he had tried to find out how much carbon was lost in that way. The sample now exhibited of deck-sweepings had been sent him by Mr. Lecointe; as could be seen, it was very black and consisted chiefly of carbon, the mineral matter amounting to only about 20 per cent.; so that it might fairly be argued that, as 2·22 per cent. of mineral matter had gone up the funnel, it must have been accompanied by four times as much carbon or 8·9 per cent. The balance account of the fuel would then be as follows:—66·1 per cent. of its thermal value was taken up by the feed-water (page 153) while being converted into steam; 26·8 per cent. was wasted with the escaping gases, a rough estimate of course and not absolutely correct; 2·2 per cent. of carbon was found in the ashes in the stoke-hold; and 8·9 per cent. was blown away up the funnel. These items made altogether 104 per cent., showing that the efficiency of the fuel was greater in the boiler than in the calorimeter. Adding 3 per cent. for radiation of the boilers, or about 1 lb. of steam condensed per square foot of external boiler surface per hour, which he thought was not an excessive estimate, this would bring the total up to about 107 per cent., as against 93 per cent. in the report. Such an excess no doubt appeared amusing; but as regarded accuracy of results he could not see that there was much to choose between a deficit and a surplus. Even these results were highly satisfactory, and Professor Kennedy and

(Mr. C. E. Stromeyer.)
Mr. Wilson might be congratulated on having got so near to the true result; it meant indeed that, halving the surplus of 7 per cent., the boiler trial was correct within 3½ per cent., and so was Mr. Wilson's calorific estimate. A study of the experiments made on the calorific value of various fuels had shown him that no great certainty could be claimed for their results; unexpected differences occurred now and then, amounting he believed to as much as 10 or 15 per cent. The present difference was therefore comparatively small. Working out the results of the "Colchester" trial, he found that the total amounted to exactly 100 per cent., without allowing for radiation. The estimate given as regarded the carbon that had been blown away he had not the least doubt erred on the excess side, and the above percentages might safely be reduced; for, besides the pieces of coal which fell on deck, there would be much mineral dust blown away altogether. In forced-draught boilers the ends of the tubes were often coated with a sort of clinker, showing that a considerable amount of pure mineral dust was blown about, though none of this could be found in the sample exhibited: so that it would be better to assume that, instead of carrying away four times its own weight of carbon, the mineral matter was perhaps associated with only an equal amount of combustible. This estimate would reduce the efficiency of the fuel from 107 per cent. to 100·3 per cent.

In connection with the priming of the boilers, great care he thought should be exercised in collecting the boiler water. While he was in the stoke-hold he had noticed that one of the samples was taken from a gauge-glass connection; and he remembered pointing out that it ought to be blown out first, because there was a possibility of steam having become condensed in the upper part of the glass, which would find its way into the measuring pot and might produce error in the result. In fact this might account for the difference that existed in the three experiments. In the first sample of boiler water he noticed that there was only 89 grains of salt per gallon (page 149); whereas in the second and third the amounts were 115 and 113 grains respectively. It was possible that the difference of 25 grains might be due to the above cause; but if it was not, and supposing for argument's sake that all the samples were taken in the same way, it would show that during the first interval of three hours, between samples Nos. 1 and 2, the saltness had increased by 26 grains per gallon. Now as sea water contained 1,850 grains of salt per gallon, this increase meant that supplementary feed from the sea to the extent of more than one per cent. had been added to the boilers. Without desiring to attach any importance to these exact figures, he only wished to point out how in this way the priming of a locomotive or other land boiler could be measured; it could be done, for instance, by filling a boiler with salt water, and measuring its saltness before starting and subsequently. Unfortunately he had not yet been able to induce any locomotive superintendents to put salt into their boilers, even though, as these experiments showed, only a little was required; but he hoped that the experiment would yet be tried, as it offered a ready means of showing whether there was any priming going on in locomotive boilers or not. Mr. Wilson's valuable test for estimating slight variations in saltness might also be utilized for determining to what extent surface-condensers leaked. Considerable trouble was often experienced at sea from the ever increasing saltness of the boiler water; and it would be of great importance to be able to know whether it was due to a leaky condenser. All that would be necessary would be to test the saltness of the water in the hot well or in the feed pumps, and every two grains per gallon found there would mean that one-tenth of one per cent. of salt water was being constantly added to the feed.

One other matter to which he wished to refer was the great difference in the efficiency of the engines, which ranged, as shown in line 59 of Table 14, from 10·7 up to 17·1 per cent. Part of the difference was of course due to the various steam-pressures and the various grades of expansion; but there were other discrepancies which he thought could be accounted for only by restricted steam passages, and perhaps by the valve-motions: in fact by the friction of the steam. He would therefore further suggest that the trials would be still more interesting and valuable if in each case continuous indicator diagrams were to be constructed, showing the cylinder pressures one above the other at each point in their correct

(Mr. C. E. Stromeyer.)
relative positions, as well as a diagram showing the amount of openings of the slide-valves. This would assist in explaining the frictional resistance of the steam in its passage from one cylinder to another.

Mr. P. W. WILLANS said it was a great pleasure to read a report like the present, which however was a most difficult one to criticise, because there appeared to be nothing omitted from it, and almost every suggestion that had been made in connection with the previous engine trials appeared to have been acted upon in this. There was one novelty in the present trial, which he thought it would be a good thing to see carried further, namely the introduction of water meters. It was a simple matter with almost every boiler to use a water meter; and the information which would thereby be obtained on a long voyage as to water consumption, and the relation it bore to coal consumption, would be highly valuable, as distinguished from what was obtained on a twelve hours' run under trial. It would be most valuable to have a continuous record of the water used. In his own works he had now a meter continually in use on the boiler that was employed for testing engines, and he thus kept a continuous record both of the coal burnt and of the water consumed. Until lately meters had not been so accurate that they could be absolutely depended upon; but they were now made more accurate. One of Mr. Schönheyder's meters that he was using was found to be correct within 1 per cent., which was near enough for any ordinary trials, and gave a satisfactory record.

With regard to the mode of determining the priming, it might be of interest to mention that there was another simple means of doing so, provided the priming was not excessive. A little steam was allowed to flow from the steam pipe into a small vessel attached to it, the steam being taken as far as possible of average quality. The following three observations were then made: the pressure in the steam pipe close to the vessel; and the pressure and also the temperature of the steam in the vessel. The percentage of priming water in the sample could then be calculated from the excess of the observed temperature above that of saturated steam at the reduced pressure in the vessel. Of course in any method of sampling for priming—not only in the one he was now speaking of, but also in the method of determining the priming by salt—it was only a certain sample of steam from which the priming was determined; and that was not exactly the same thing as taking all the steam and condensing it in the calorimeter; both methods had their disadvantages. The specific heat of superheated steam was not accurately known; it was liable to an amount of doubt, which could not at present be determined, and this might affect the result to the extent of 2 or 3 per cent. of the total priming. Having lately used this method several times, he had found the priming to be under 1 per cent., a result which agreed with previous calorimetric tests that he had made under similar conditions.

With regard to the condenser in the "Ville de Douvres," he thought there was no doubt Mr. Kraft was right (page 169) as to the water probably not flowing through the whole of the condenser tubes: the figures given of the trial pointed to this conclusion. There was hot feed-water, and at the same time the temperature of the discharge water was low. It looked as though the bulk of the condensing water was flowing through a certain portion only of the tubes, and that these tubes had reached the limit at which they could transmit heat. Supposing for instance that 90 per cent. of the tubes had stagnant hot water in them, while the remaining 10 per cent. had a large quantity of cold water flowing through them, the effect would be that the limit of the capacity of the majority of the tubes to transmit heat from the steam to the water would be reached, and notwithstanding a comparatively cold circulating water the condensed steam would still be hot. This appeared to be the explanation in the present case. Whether the effect of the comparatively poor vacuum was serious was another question altogether. For some time past he had been trying to find what was the best vacuum to work with under certain conditions. It seemed to vary much; and there was no doubt that an economic limit could be reached. Owing partly to the fact that the compression of steam at a very low tension must fail to fill the clearances and to warm the clearance surface at the beginning of

(Mr. P. W. Willans.)
the stroke, and owing partly to the general reduction of temperature in the cylinder when the pressure was very low, there must be a degree of vacuum which it was not worth while to exceed. Mr. Kraft had apologised for the performance of the engines in regard to economy; but it was rather curious to observe from line 59 in Table 14 that—although the steamer was built entirely for speed and there was no great amount of expansion in the cylinders, and the expansion, as pointed out by Mr. Rowan (page 177), was not so well distributed as theoretically it might be—still the performance of the engines was practically as good as that of the other two-cylinder compound engines in the "Fusi Yama" and the "Colchester": pointing to the conclusion that expansion might often be carried too far, and that a high vacuum might not be an entire gain. In the "Ville de Douvres" the condenser appeared to have rather more than two square feet of surface per indicated horse-power; whereas with only one square foot he had frequently got a better vacuum. This showed what might be done by properly distributing the circulating water, so as to ensure its passing through every part of the condenser equally.

Lt.-Colonel English asked whether the capacity of the intermediate receiver could be stated, and also the extent of its surface. [This has since been added accordingly in page 138.] It was evident from the low-pressure diagrams that it had largely affected the working of the engine.

Mr. Frederick Edwards corroborated the statement in the report (page 141) with regard to the fuel consumption, that the amount could not be arrived at either by counting the blocks or by measuring. Every portion of it he considered ought to be weighed; and he was very glad that in the present case the whole of it had been weighed. The priming test he regarded as highly instructive; and this he believed was the first trial in which there had been practically no priming. In this connection therefore it would be of great interest and use to the members if fuller particulars were supplied of the boilers, of which only the general dimensions appeared to be given:

especially the actual capacity of the steam space, and how and where the steam was taken from the boiler; and also the position and dimensions of the internal steam-pipe, if there was one. In the former trial of the "Tartar" the amount of feed-water that had been measured had been found to be more than the coal would evaporate. A portion had therefore been attributed to priming; and he thought it would be of great use if a similar test were made on the "Tartar" to see what amount of priming went on in her boilers. His impression was that it would not take long to do; the ship often came to London, and it would be useful to know how much her boilers did really prime.

With regard to what Mr. Cochrane had said (page 166) about the coldest water meeting the hottest steam in the condenser, as was the case in the "Ville de Douvres," he had had several ships arranged in that way, including one of his best and largest boats; and it gave good results. His impression was indeed that he had obtained quite as good results in that way as by the reverse method of a counter-current. He greatly regretted that owing to ill-health he had been prevented from continuing to take an active part in the more recent trials. In any future trials however he should be happy to offer the assistance of some members of his staff.

The President hoped that Mr. Edwards had carried out his intention (Proceedings 1889, page 290) of fitting all his vessels with permanent feed-measuring apparatus.

Mr. Edwards said he had fitted one ship, and he hoped to have been able to have a trial with her, and that the Committee would have been able to test her; but she had not been to London, and circumstances had not led to his being able to arrange the matter otherwise. Whenever he was in a position to do so, he should be happy to inform the Committee.

Mr. J. Macfarlane Gray considered that the present report, like all the others of the Research Committee, was a most careful and valuable one. In the discussion upon the "Iona" report he

(Mr. J. Macfarlane Gray.)
had since observed he was said to have stated (Proceedings 1891, page 260) that, if the efficiency of the "Iona" engines had been calculated on the method introduced by Mr. Willans, it would have been raised from 69·2 per cent. to about 72 per cent. What he meant to say was that the figures of the Committee's trial would have given 69·2 per cent.; and if the evaporation were taken at the same rate for the coal consumption recorded in the Chief Engineer's subsequent trial, the engine efficiency on Mr. Willans' plan would have been about 72 per cent. Another 69·2 per cent. occurred in the report; but the two numbers were not related, and the equality was wholly accidental.

The results obtained in the "Iona" and in the "Ville de Douvres" were very different; but the engines were designed for quite different conditions. The latter was a fast steamer running short voyages; the former a slow cargo steamer making long voyages. The "Ville de Douvres" had only two-thirds of the initial pressure in the "Iona"; yet the "Iona" gave only two-thirds of the mean effective pressure in the "Ville de Douvres," referring all to the low-pressure cylinder. The piston area per indicated horse-power was more than 50 per cent. greater in the "Iona" than in the "Ville de Douvres." That a high initial pressure with a low back-pressure should be more economical than low initial pressure and high back-pressure was only what would be expected. It was not for demonstrating such obvious conclusions that the reports of the Research Committee were chiefly valuable. They enabled these obvious differences to be separated from the enquiry, and the theoretical duty of the steam working within the given limits of pressure to be compared with the result actually obtained; and then the more or less occult causes to be enquired into, which might have produced a defect in efficiency. When that method was applied to the "Ville de Douvres," the margin of difference was much contracted. Taking as the limits the pressures bounding the mean indicator diagrams, not quite the same as he had taken before for the "Iona," the method he had explained in Proceedings 1889, page 449, gave the following results:—

	"Iona."	"Ville de Douvres."
Initial pressure absolute, lbs. per sq. inch	164	109
Terminal " " "	5	14
Back " " "	1⅔	6
Pounds of steam per I.H.P. theoretically	9·36	13·84
" " " actually	13·35	20·77
Efficiency of engine per cent.	70·1	66·6
Relative efficiencies	100	95

To account for this 5 per cent. difference, the "Iona" had her high-pressure cylinder jacketed with boiler steam, whereas the high-pressure cylinder of the "Ville de Douvres" was wholly surrounded by its own exhaust steam.

With reference to the efficiency of the boiler, he had given in the discussion on Mr. Blechynden's paper last year at Liverpool (Proceedings 1891, page 355) a sort of rule of thumb for the loss of calorific effect due to the heat escaping up the funnel, which he had stated was always equal to 4 per cent. of the elevation of temperature in degrees Fahr.:—that is, 4 per cent. of the difference between the temperature of the atmosphere and the temperature of the escaping gases. In the case of the "Ville de Douvres" the temperatures reported were 64° and 910° respectively; whence the loss would be 33·84 per cent., with the result that the boiler efficiency ought to be 66·16 per cent. The efficiency reported (page 153) as the result of the trial, after a somewhat elaborate calculation, was 66·1 per cent. This result might perhaps induce some of the members, when they had no better way of guessing at the boiler efficiency, to apply the rule of thumb.

Mr. CHARLES J. WILSON, referring to the calorimetric method of measuring priming mentioned by Mr. Willans (page 182), considered an objection to that plan was that it applied only to cases in which the priming was somewhat large. The great advantage in employing the salt method was that it could always be shown thereby, in cases where there was no priming, that the priming was below a certain extremely small amount, less for instance than one-tenth of one per cent. In nearly all cases where he had himself had to investigate

(Mr. Charles J. Wilson.)
priming, there had either been none at all, or only a very minute quantity; and in his own experience he had met with only one instance in which the priming was sufficiently large to be measured at all accurately by the calorimetric method.

The elaborate calculations made by Mr. Stromeyer (page 179) with regard to the small percentage of heat unaccounted for seemed to him to be hardly necessary. It was an unfortunate feature of the recent trial that the two thermometers employed were incapable of measuring the temperatures of the exit gases; and there was therefore obviously a certain error in the calculation. Under these circumstances he thought it was not worth while to endeavour to draw any deduction from deck-sweepings; and in any case he should be unable to recognise deck-sweepings as a sample of anything in particular. From the mere fact that the deck-sweepings were black, it would be an error to assume that they were entirely carbon. Nor as to the extent of priming could he follow the curious inference which seemed to be drawn (page 181) from the small increase in the amount of salt in the boiler; any such increase he believed had no bearing upon the matter of priming.

Mr. G. S. Young enquired how the samples of steam were taken from the steam-pipe in the test for priming. It might seem at first sight quite as simple as to draw off a sample from a cask of liquor; but it was not so simple to draw off a true sample of mixed water and steam travelling at 100 feet per second along a steam-pipe. If the opening was simply in the side of the pipe, the momentum of the priming water would be so great at 100 feet per second that it would shoot right past the orifice, and the pure steam alone would pass out into the condensing apparatus. If on the other hand there was an elbow projecting inside the steam-pipe, and the elbow was turned to face the current of mixed water and steam, even then, unless the velocity within that small elbow coincided exactly with the velocity at which the mixture of steam and water was passing along the main steam-pipe, there would not be a true sample taken; because the momentum of the water being greater than that of the steam, the sample would contain a larger quantity of water and a smaller quantity of steam than represented their true proportion in the steam-pipe. Under those circumstances he thought it would be difficult to draw any absolute conclusion from such observations. They would indeed be comparable one with another, provided the method of taking was equal in all cases; but he thought they did not represent in any case the actual amount of priming. A comparison of the analysis of the water in the hot-well with that in the boiler, as suggested by himself at a previous meeting (Proceedings 1890, page 272), he was still of opinion was the best method of obtaining an actual measure of the quantity of priming.

With regard to the condenser, there seemed no reason to doubt that the circulating water, being admitted into the top of the central chamber between the two nests of tubes, flowed outwards along the top compartment in either half of the condenser, and then turned round at the outer end and flowed inwards through the middle compartment of the same half; but on reaching the central chamber it seemed to him probable that it would not turn round again and flow outwards through the bottom compartment of the same half, but would shoot across the central chamber and flow outwards through the bottom compartment of the other half of the condenser. The consequence would be that the current through one half of the condenser would interfere with that through the other half, and the condensing surface in one half or the other would not be doing its proper share of work. From the experience he had had with many surface condensers, he had found that if two things were properly attended to there was not much fear of a bad vacuum, assuming that there was sufficient condensing surface, a sufficient quantity of circulating water, and a sufficient capacity of air-pump. The first point to be attended to was to distribute the steam horizontally over the whole surface in the upper part of the condenser, and to allow it to fall by gravitation to the bottom: not using any of the momentum given to the steam in its passage into the condenser to drive it down to the bottom, but using its whole momentum to spread it horizontally and uniformly over the whole of the top condensing surfaces. Any one who had noticed a feather falling in a vacuum within the exhausted bell-glass of an air-pump

(Mr. G. S. Young.)
would understand how the steam would drop to the bottom of the condenser readily enough by its own weight. But if the steam was blown down to the bottom of the condenser without in its passage having been in contact with sufficient cooling surface to condense it, there would be a vapour of undesirable weight lying at the bottom, which would not be condensed until it was lifted up into contact with other parts of the cooling surface by a fresh volume of incoming steam; and there would thus be a higher pressure in the condenser. The other point to be attended to was the distribution of the circulating water so that it should travel uniformly over the whole of the condensing surface. It seemed obvious that, if the steam was properly distributed over the whole of the outside surface, and the water over the whole of the inside surface, nothing could be done in either respect for rendering the condenser more efficient. Both of these conditions could easily be obtained by fitting guide plates to direct the flow of both the steam and the water. Looking at the two ribs projecting downwards from the inside of the top of each half of the condenser in Fig. 18, Plate 21, it appeared to him they would have the effect of preventing the steam from spreading laterally over the whole surface of the tubes. With such an arrangement of condenser it seemed to him it would naturally be expected to find that there was a large amount of surface inoperative, partly on account of the water not being uniformly circulated, and partly also on account of the steam being directed downwards and reaching the bottom of the condenser before it was condensed. The steam would indeed have to be buoyed up before it would reach the other portions of the condensing surface.

Mr. Robert Bruce noticed in the diagram of boiler pressure, Fig. 6, Plate 13, a drop of about 27 lbs. between one and two o'clock, corresponding with the time when the engine was slowed down. It was not explained in the report whether or not that drop was due to the fire-doors being opened, and the cold air rushing right through from the fans and cooling down the boilers. In dealing with manœuvring ships it was an important point, as was well known, to have a complete control of the evaporation; and it was evident that if, as he supposed, the doors were opened when the boilers cooled down, there was no possibility of having such control. The temperature of the escaping gases had been referred to, and the value of the heat carried away by them could be realized. If the air going into the furnace could be heated up to perhaps 200° Fahr. above atmospheric temperature, he calculated that it would be equivalent to a recovery of 870 heat-units per lb. of fuel. The heating of the air he had alluded to in the discussion upon Mr. Blechynden's paper read at Liverpool (Proceedings 1891, page 369); and in connection with the two new steamers now mentioned (page 169) as having recently been ordered, he would strongly recommend a study of the examples of forced draught given in that paper (1891, page 336), which he thought would be found of considerable advantage in working the new steamers.

Professor KENNEDY, Vice-President, said the point raised by Sir Frederick Bramwell (page 165)—whether the steam consumption of the auxiliary engines ought not to be included with that of the main engines, because the latter could not get on without the former—was a matter about which there might perhaps be a difference of opinion. There was no doubt that the steam consumption for the auxiliary engines was quite essential, and from that point of view the whole steam consumption had been given in the report; but as the auxiliary engines used about four times as much steam per indicated horse-power as the main engines, it was perhaps hardly fair to the latter, regarded as machines in themselves, to debit them with that large amount of steam: so that he thought the Committee were justified in giving the figures both ways, as they stood in the report.

Although Mr. Cochrane had not attended the meetings of the Committee, it should be mentioned that he had rendered them great assistance by the care with which he had examined the successive reports before they were issued to the members, and had worked out the calculations independently, and had thereby found out several arithmetical and other discrepancies, which had consequently been rectified.

(Professor Kennedy.)

The brief remarks (page 171) of such a veteran in marine work as Dr. Kirk were particularly welcome, as showing how cordially he concurred in the importance of separating the efficiency of the boiler from that of the engine. It had taken a long time to convince some marine engineers that these were two different things; but surely there should not now remain any doubts upon this point.

An interesting point had been raised by Mr. Harris (page 173) about the quantity of coal on the fires at the beginning and end of the trial. It was of course physically altogether out of the question to draw the fires at the beginning and end of a trial of this kind. The best way therefore to get rid of the error was naturally to make the trial as long as possible; and all these trials had been so long that he thought, even if they had been carelessly made, the error must have been practically eliminated. In the present instance the trial lasted only nine hours, and therefore was not so long as the previous trials had been. On a former occasion (Proceedings 1889, page 252) he had given the method followed for endeavouring to secure the same value of the fires at the starting and at the end of the trial. The straightness or otherwise of the line representing the coal consumption—like that marked "total fuel consumption" in Fig. 4, Plate 12—was really the best measure of whether there was any difference between the beginning and the end of the trial. As a matter of fact in this particular trial the fuel was put on the fire as hard as ever it could be from the beginning to the end; and if there had been any slackness of firing the boiler pressure would have gone down. Very generally in similar trials the line representing the coal consumption was something like that sketched in Fig. 24, Plate 24, the time being plotted horizontally and the consumption vertically, in the same way as was done in Fig. 4. It was found that after starting the trial, the steam probably having been allowed to drop and the fires to get a little low, the coal consumption line rose at a rather steep inclination from A to B, Fig. 24, and then from B it remained practically straight and less steep. At C, towards the end of the trial, if the plan were followed which he thought most engineers adopted of allowing the boiler pressure to fall, a nearly horizontal line was obtained from C to the very end at D. In such cases he had over and over again found that, if a straight line were drawn from the starting point at A to the end at D —that is, if the whole coal consumption were taken over the whole time—the mean line AD was parallel to the straight portion BC of the actual line: showing that the average rate of consumption over the whole trial was the same as that which existed through the straight part BC of the actual line. By this mode of procedure therefore at the beginning and at the end he thought that in a carefully conducted trial of sufficient duration it was possible practically to eliminate errors due to the difference between the amount of fuel on the fires at the beginning and at the end. Certainly the error was less than would be due to the unavoidable irregularities consequent on any attempt to draw fires at start and finish in the stoke-holds of a steamer. But he fully recognised the fact pointed out by Mr. Harris that a running start and finish were liable on a short trial to lead to considerable errors; and in any case the greatest care was of course essential.

In connection with Mr. Rowan's remarks (page 177), he had further understood from him that, by the alteration of cut-off which he recommended, not merely had he obtained a greater speed and a greater power, but also the increase of power had been in a larger ratio than the increase of speed. This was a matter of such interest that he hoped Mr. Rowan would be able to give the figures of the actual increase both in speed and in power, and also to explain the alteration of cut-off by which so important a result was obtained. [See pages 196–7.]

The balancing of the heat account by aid of the deck-sweepings (page 179) he supposed was hardly intended by Mr. Stromeyer to be taken seriously. In certain other trials which he had himself conducted with the greatest care, he had found that he had accounted for everything to within about 3 per cent. But from this very fact he had felt great doubts whether he was right. At any rate in an ordinary marine boiler he was sure that the losses by radiation were much more than 3 per cent.—more nearly three times as much in ordinary cases. Certainly therefore the heat account ought not to balance up to 100 per cent. without allowance for radiation &c.; and

(Professor Kennedy.)
the fact that Mr. Stromeyer's approximations came out in such a fashion seemed to him to carry with it a proof that they must be erroneous.

The important question of priming had very properly been raised by Mr. Willans (page 182) and subsequent speakers. This was really a subject which he suggested might well be taken up by the Research Committee, and its investigation would probably not occupy a long time. On a former occasion it had been suggested by Mr. Young (Proceedings 1890, page 272) that the saltness of the water in the hot-well might be assumed to represent fairly that of the condensed steam; and it was possible that the extent of priming might be determined by that means. He was very sorry that he had omitted to avail himself of that suggestion in the present trial. The question had also been raised, how far it was possible to sample the steam correctly. Besides the Barrus calorimeter, which he presumed was the apparatus alluded to by Mr. Willans (page 182) for dealing with small amounts of superheating, there were also various other contrivances for estimating the extent of priming; but all of them involved certain assumptions, and most of them involved the particular assumption of the accuracy of a sample obtained of the steam or of the water, which to begin with was a pretty big assumption. By means however of comparing several methods with one another, all being applied to the same boiler at the same time, he thought a result might be arrived at of some value; and he hoped the Committee might be able to undertake these experiments.

In reply to Mr. Young's enquiry (page 188) as to how the samples of steam were collected in the test for priming, he had to say that it was by an elbow pipe projecting into the steam pipe and facing the current of steam.

In reference to Mr. Cochrane's remark as to the calorific value of hydrogen (page 167), the value he had used in the present trial was the same that he had given in the reply to the discussion upon the previous trial (Proceedings 1891, page 287); and the value of Joule's equivalent had been taken throughout as 772 foot-lbs. per heat unit.

In the "Iona" it would be remembered that the jacket of the high-pressure cylinder was itself surrounded by steam of lower pressure in a receiver, so that the jacket served more or less to transmit heat both to inside and to outside; and he had made some remarks on this in the discussion upon the "Iona" trial (Proceedings 1891, page 284). In the "Ville de Douvres" however the condition was so far different that the high-pressure cylinder itself without a jacket was here enclosed in receiver steam (page 176); and he could not but think that this must detract considerably from the steam economy. The use of jackets he understood had been given up because of the additional weight which they entailed.

The reduction of steam pressure referred to by Mr. Bruce (page 190) was caused by the forced draught being stopped for a few minutes, and not by the opening of the fire-doors.

He desired to express the regret of every one who had taken any part in the trials that Mr. Edwards' health had not allowed him to be with them in the last two trials. Any regret felt by Mr. Edwards himself that he had not been present was not half so great as the regret of every one who had had the chance of working with him in any of the trials.

The members of the Institution, he desired to add, were greatly indebted to the Belgian Government, to Mr. Kraft, and to the Société Cockerill, not only for placing at their disposal the steamer and all its machinery, but for the cordial and kind way in which they had helped the trial. In the same connection there were also two other names that ought to be mentioned, which had not been specially referred to. The work that had been so frequently done in earlier trials by Mr. Ashcroft, in seeing that everything was got ready and into proper order before the trial came off, had on the present occasion been taken up by his friend Mr. H. R. J. Burstall, a member of the Institution. It was no less due to Mr. Ewen to say that a great deal of trouble and care had been taken by him in the collation and preparation and arithmetical working out of the large mass of figures contained in the report; and the fact that he happened to be one of the staff of the Institution was no reason why they should not recognise the extreme care he had bestowed upon the work. In

(Professor Kennedy.)
conclusion he thanked the Members for the reception given to the work of the Committee; and he hoped it might turn out that engineers in general, and shipowners and marine engineers in particular, would follow it up by making trials on their own account, now that they knew that the trials were quite practicable and would yield results of such great practical value.

The PRESIDENT said that, although he had already invited the Members to pass a formal vote of thanks to all to whom they were specially indebted, he thought they ought not to lose the opportunity afforded by Mr. Kraft's presence for presenting to him a cordial and unanimous vote of thanks for the help he had given in the course of this investigation. He hoped that Mr. Kraft would understand how greatly his kindness was appreciated; and he trusted that no remarks which had been made respecting the construction of the engines would be otherwise than agreeable to him.

Mr. JAMES ROWAN sent the indicator diagrams shown in Figs. 19 to 23, Plates 22 to 24, which he had subsequently taken as offered (page 177), in illustration and confirmation of the power gained by earlier cut-off in the low-pressure cylinder. These diagrams were taken from the engines of the paddle steamer "Mercury," in the course of one of her regular trips between Prince's Pier, Greenock, and Rothesay on the Firth of Clyde. The cylinders were 33 inches and 62 inches diameter with 60 inches stroke. The diagrams were taken on the afternoon of Tuesday 5th July: first with the low-pressure valve-gear linked up to the extent of $1\frac{5}{8}$ inch distance of the link block in the slot from the position of full gear, causing the low-pressure valve to cut off at 37·8 inches or 63 per cent. of the stroke; and ten minutes later with the low-pressure link in full gear, cutting off at 41·64 inches or 69·4 per cent. of the stroke. On both occasions the high-pressure link was kept in full gear, cutting off at 44·7 inches or $74\frac{1}{2}$ per cent. of the stroke. Under both conditions the revolutions were counted for five minutes. The boiler pressure was 110 lbs. per square inch above atmosphere, and the condenser vacuum 26 inches of mercury or 12·8 lbs. per square inch below atmosphere. The throttle-valve was kept full open all the time. The following were the different results obtained in the two modes of working:—

	Full Gear.	Linked up.
Receiver Pressure by gauge, lbs. per square inch above atmosphere	15½	20
Revolutions per minute, mean of five minutes	51·8	53·6
Indicated horse-power, high-pressure cylinder	782	748
" " low-pressure "	918	1,069
" " total	1,700	1,817
Weight in lbs. of Steam used per revolution, measured at pressure of 71 lbs. per square inch above atmosphere	9·74	9·84
Work done per lb. weight of steam, in ft.-lbs.	111,130	113,700

Hence the increased work done per lb. weight of steam amounted to 2,570 ft.-lbs.; or 2·31 per cent. increase of work was obtained by linking up the low-pressure valve-gear. The great gain was the increase in revolutions, which increased the speed of the vessel from 17 up to 17½ knots; and the net increase in total horse-power obtained was 6·88 per cent.

ON CONDENSATION IN STEAM-ENGINE CYLINDERS DURING ADMISSION.

BY LT.-COLONEL THOMAS ENGLISH, OF JARROW.

In 1887 and again in 1889 the author had the pleasure of laying before this Institution the results of some experiments which he had had the opportunity of making on steam engines, with the object of determining the amount of loss by Condensation in the Cylinder, as distinguished from the liquefaction necessarily caused by the performance of work. These experimental results, he is glad to think, were accepted as trustworthy by engineers on whose opinions he would most implicitly depend; but the conclusions which he drew from them, and embodied in the shape of approximate formulæ for determining the condensation under given conditions, did not meet with the same favour. It was objected that they practically left out of account several factors, notably the range of temperature in the cylinder, which must be the principal agents concerned in producing the observed amounts of cylinder condensation; and that therefore the formulæ must be quite empirical, and applicable only to engines of the type experimented on. Probably a good deal of this criticism arose from the formulæ not being presented in a shape which could be readily compared with the observed result generally recorded in steam-engine experiments, namely the proportion of water present at cut-off; and the expression then obtained for the weight of steam condensed at cut-off (Proceedings 1889, page 652) has therefore now been reduced into one for the proportion of steam condensed to steam uncondensed at cut-off. When this is done, it becomes more apparent that the ratio of expansion, and therefore indirectly the range of temperature in the cylinder, is really a factor in the formula.

In jacketed cylinders the author considered (Proceedings 1889, page 652) that the weight of steam condensed per stroke and not re-evaporated at cut-off is represented by the expression

$$\frac{56}{\sqrt{\text{revs. per second}}} \times \frac{(S_c - S_1)}{L} \rho_1$$

where S_c is the unjacketed clearance surface in square feet, S_1 the fresh surface exposed during admission up to cut-off, ρ_1 the initial density of the steam in pounds per cubic foot, and L the latent heat of evaporation in thermal units. If d be the diameter of the cylinder in feet, l the length of stroke in feet, m the proportion of stroke up to cut-off, $\mu = \frac{S_c}{2 \times \text{area of cylinder}}$, and N the number of revolutions per minute: then S_c = unjacketed clearance surface $= \frac{\mu\pi d^2}{2}$; $S_1 = \pi dml$; $\sqrt{\text{revs. per second}} = \frac{\sqrt{N}}{7 \cdot 75}$; and the foregoing expression may be written

$$\text{Weight condensed} = \frac{56 \times 7 \cdot 75}{L \times \sqrt{N}} \left(\frac{\mu\pi d^2}{2} - \pi dml \right) \rho_1$$

$$= \frac{868}{L \times \sqrt{N}} \left(\frac{\mu}{ml} - \frac{2}{d} \right) \frac{\pi d^2 ml}{4} \rho_1$$

But $\frac{\pi d^2 ml}{4} \rho_1$ is the weight of steam per stroke uncondensed at cut-off, and 868 may be taken as an approximate value for L; therefore for jacketed cylinders $y = \frac{\text{weight condensed}}{\text{weight uncondensed}} = \frac{1}{\sqrt{N}} \left(\frac{\mu}{ml} - \frac{2}{d} \right)$.

For unjacketed cylinders a similar approximate expression is $y = \frac{1 \cdot 5}{\sqrt{N}} \left(\frac{\mu}{ml} - \frac{2}{d} \right)$.

In order that the applicability of these formulæ to the varying conditions met with in practice may be readily tested, the author has collected in the accompanying Tables 14 to 21, for comparison with results calculated from these formulæ, the observed percentages of condensation deduced from all the experiments to which he has had means of reference. These range between initial pressures of 12 and 190 lbs. per square inch, with ratios of expansion from 1·1 to 10·0, and revolutions from 14·6 to 422 per minute; and they were carried out in engines with cylinders of from 5 inches to 3 feet 9 inches diameter, and from 6 inches to 6 feet length of stroke.

As the amount of condensation must in all cases necessarily be arrived at by subtracting in some form or other the number of thermal units otherwise accounted for from the total number of thermal units supplied, the best standard of comparison is obtained by using the values of $1-x$, which represents the ratio of the weight of water present at cut-off to the weight of saturated steam supplied per stroke, or in other words, the percentage of steam supply that is condensed at cut-off: x denoting the fraction (by weight) of saturated steam in the working mixture of steam and water contained in the cylinder. This gives the same relative value to accidental errors of observation, whether the condensation be large or small. Tables 14 to 21 are accordingly arranged on this plan, to show in parallel columns the observed values of $1-x$, and those calculated according to the author's formulæ.

A diagram is also given in Fig. 1, Plate 25, showing the same thing graphically for all the experiments, 210 in number, by the vertical distances of dots representing the observed results from the diagonal line representing calculation.

It should be observed that where, as in cases of wire-drawing, there is a large difference between the initial pressure and the pressure at cut-off, the point of cut-off should for calculation be taken at where it would have been without wire-drawing.

Where the compression is slight or wanting, the value of m, that is, the proportion of stroke up to cut-off, should be taken to include clearance.

The value of μ, that is, the ratio of the unjacketed clearance surface to twice the area of the cylinder, where it cannot be found from the experimental data, may be taken, according to the author's experience, as equal to $2\sqrt{\frac{l}{d}}$ for slide-valves of ordinary proportion.

(*continued on page* 212.)

TABLE 14.

Initial Condensation in Single Engines Jacketed.

Authority.	Index reference.	Cylinder.		Cut-off. Percentage of Stroke.	Pressure at cut-off. Lbs. per sq. inch.	Revolutions per minute.	Ratio of Clearance Surface to twice Cylinder area.	Ratio of Steam Condensed to Uncondensed.	Percentage of Steam supply condensed at cut-off.			Remarks.
		Diam.	Stroke.						Calculated.	Observed.	Difference.	
		d	l	m		N	μ	y	$1-x$	$1-x$		*
MAIR.		Feet.	Feet.	P. c.	Lbs.	Revs.	Ratio.	Ratio.	P. c.	P. c.	P. c.	
Inst. C.E. 1882	A	3·75	5·5	13·1	33·0	14·62	1·90	0·56	36	39·8	+ 4	C P 4%
vols. 70	L	2·67	5·5	22·5	45·9	20·31	2·86	0·37	27	28·3	+ 1	P 1%
and 79. 1885	N	2·25	6·0	30·0	44·4	20·0	3·26	0·20	17	21·0	+ 4	P 1%
	O	2·25	6·0	30·0	53·0	20·0	3·26	0·20	17	13·8	− 3	W P 1%
ENGLISH.	IV	0·83	1·17	10·0	58·1	121	2·12	1·44	59	50	− 9	O S
	VI	”	”	20·8	63·1	120	”	0·58	36	42	+ 6	O
	VIII	”	”	10·4	50·5	119	”	1·38	58	44	−14	O S
Proceedings	XII	”	”	13·7	59·9	118	”	1·00	50	46	− 4	O
	XIII	”	”	15·4	63·5	115	”	0·88	47	46	− 1	O
Inst. Mech. E.	XIV	”	”	19·3	86·4	121	”	0·64	39	33	− 6	O
	XV	”	”	17·8	61·4	120·5	”	0·71	41	41	0	O
October 1889.	XVI	”	”	14·7	75·9	122	”	0·90	47	38	− 9	O
	XVII	”	”	10·5	44·8	120	”	1·37	58	45	−13	O S
	XVIII	”	”	16·4	78·0	120	”	0·79	44	39	− 5	O

* C = Cylinder Cover jacketed. P = Priming, to the extent of the percentage denoted by the figure annexed.

W = Wire-drawing. O = No compression. S = Superheating.

TABLE 15.

Initial Condensation in Single Engines Not *jacketed.*

Authority.	Index reference.	Cylinder. Diam. d	Cylinder. Stroke. l	Cut-off. Percentage of Stroke. m	Pressure at cut-off. Lbs. per sq. inch.	Revolutions per minute. N	Ratio of Clearance Surface to twice Cylinder area. μ	Ratio of Steam Condensed to Uncondensed. y	Percentage of Steam supply condensed at cut-off. Calculated. $1-x$	Observed. $1-x$	Difference.	Remarks. *
		Feet.	Feet.	P. c.	Lbs.	Revs.	Ratio.	Ratio.	P. c.	P. c.	P. c.	
MAIR. Inst. C.E. 1885 vol. 79.	M	2·67	5·5	24·9	45·7	20·26	2·86	0·45	31	36	+ 5	P 1%
ENGLISH.	Plate 88	1·33	1·5	28·8	71·3	40·2	1·92	0·70	41	53	+12	O
	89	,,	,,	29·0	72·6	42·2	,,	0·69	41	53	+12	O
	90	,,	,,	29·8	83·7	40·4	,,	0·68	40	47	+ 7	O
	91	,,	,,	29·8	87·6	39·8	,,	0·68	40	45	+ 5	O
	92	,,	,,	17·2	89·0	39·5	,,	1·43	59	58	− 1	O
	93	,,	,,	17·0	82·6	39·7	,,	1·43	59	62	+ 3	O
Proceedings	94	,,	,,	17·2	83·3	40·4	,,	1·43	59	64	+ 5	O
	95	,,	,,	17·3	86·5	40·3	,,	1·41	59	62	+ 3	O
Inst. Mech. E.	96	,,	,,	17·0	83·3	40·4	,,	1·45	59	62	+ 3	O
	97	,,	,,	17·4	90·9	40·1	,,	1·40	58	54	− 4	O
September 1887.	98	,,	,,	17·5	91·6	40·3	,,	1·40	58	55	− 3	O
	99	,,	,,	15·8	87·4	40·9	,,	1·55	61	62	+ 1	O
	100	,,	,,	14·9	86·8	41·2	,,	1·67	62	62	0	O
	101	,,	,,	14·8	88·1	40·1	,,	1·69	63	62	− 1	O
	102	,,	,,	14·6	87·8	39·3	,,	1·74	63	62	− 1	O
	103	,,	,,	14·6	89·3	39·5	,,	1·74	63	62	− 1	O

WILLANS.	Table I S 40	1·17	0·5	60·4	40·9	393·5	0·92	0·10	9	11·7	+ 3	* Cylinder Cover taken as jacketed.
	I S 50	,,	,,	43·7	50·6	408·4	,,	0·18	15	19·3	+ 4	
	I S 70	,,	,,	33·9	68·7	409·1	,,	0·27	21	26·5	+ 5	
	I S 80	,,	,,	29·6	78·7	403·2	,,	0·33	25	23·7	− 1	
	I S 90	,,	,,	26·4	92·6	400·9	,,	0·38	28	24·8	− 3	
Proceedings	I S 100	,,	,,	23·8	98·1	397·7	,,	0·44	31	31·2	0	
	I S 110	,,	,,	21·6	106·3	406·2	,,	0·49	33	29·6	− 3	
Inst. C.E. 1888	V S 50	,,	,,	43·7	49·5	200·6	,,	0·27	21	23·9	+ 3	
	V S 50	,,	,,	43·7	49·0	110·5	,,	0·35	26	34·5	+ 8	
vol. 93.	V S 70	,,	,,	33·9	71·1	205·2	,,	0·39	28	34·4	+ 6	
	V S 70	,,	,,	33·9	69·1	112·7	,,	0·51	34	45·8	+12	
	V S 90	,,	,,	26·4	88·5	223·0	,,	0·52	34	24·7	− 9	
	V S 90	,,	,,	26·4	89·4	122·8	,,	0·71	41	42·5	+ 1	
	V S 110	,,	,,	21·6	109·0	223·7	,,	0·67	40	42·3	+ 2	
	V S 110	,,	,,	21·6	108·7	138·0	,,	0·87	46	44·5	− 1	
ENGLISH.	I	0·83	1·17	11·8	48·9	118	2·12	1·77	64	53	−11	O
	II	,,	,,	18·2	67·9	115·5	,,	1·06	52	46	− 6	O
	III	,,	,,	17·6	60·3	120	,,	1·09	53	53	0	O
Proceedings	V	,,	,,	15·2	73·7	122	,,	1·31	57	49	− 8	O
Inst. Mech. E.	VII	,,	,,	12·5	59·7	119·5	,,	1·66	63	59	− 4	O
October 1889.	IX	,,	,,	16·1	77·0	119	,,	1·22	55	48	− 7	O
	X	,,	,,	20·0	61·7	119	,,	0·92	48	46	− 2	O
	XI	,,	,,	20·4	85·3	121	,,	0·89	47	36	−11	O
COTTERILL.	G K	1·5	3·5	21·0	62	68	3·05	0·51	34	38	+ 4	
Steam-Engine	D	3·0	2·5	21·0	47	57	1·84	0·57	36	29	− 7	
1890 p. 334.	H H	2·0	5·6	21·0	55	30	3·35	0·52	34	30	− 4	

* P = Priming, to the extent of the percentage denoted by the figure annexed. O = No compression.

Initial Condensation in Woolf Engines Jacketed.

Authority.	Index reference.	Cylinder.		Cut-off. Percentage of Stroke.	Pressure at cut-off. Lbs. per sq. inch.	Revolutions per minute.	Ratio of Clearance Surface to twice Cylinder area.	Ratio of Steam Condensed to Uncondensed.	Percentage of Steam supply condensed at cut-off.			Remarks. *
		Diam. d	Stroke. l	m		N	μ	y	Calculated. $1-x$	Observed. $1-x$	Difference.	
		Feet.	Feet.	P. c.	Lbs.	Revs.	Ratio.	Ratio.	P.c.	P.c.	P.c.	
MAIR. Inst. C.E. vols. 70 and 79. 1882	C	1·83	3·58	21	60	19·62	4·5	1·11	53	38·5	−14	C S P 2¼%
	D	2·04	3·42	36·5	70	33·73	5	0·52	34	35	+ 1	C S P 2½%
	E	”	”	37	66·5	34·22	5	0·51	34	35	+ 1	C S P 2½%
	K	2·42	5·42	32	58	17·78	3·5	0·29	22	22	0	C P 3%
1885	R	”	”	37	58	17·78	3·5	0·22	18	21·3	+ 3	C P 3%
	S	1·83	4·59	22	63	28·30	3·5	0·45	31	30	− 1	C P 1%
	T	”	”	28	65	28·16	3·5	0·31	24	29	+ 5	C P 1%
	U	”	”	28	65	27·48	3·5	0·30	23	28	+ 5	C P 1%
KENNEDY.	83	0·5	1·0	77	48·3	96·2	8	0·65	39	33	− 6	H E
	54	”	”	58	57·8	101·0	4	0·29	22	31	+ 9	H
Proceedings	38	”	”	26	78·0	98·2	4	1·14	53	54	+ 1	H
	44	”	”	44·5	87·8	99·3	4	0·50	33	49	+16	H
Inst. Mech. E.	19	”	”	19	82·1	99·0	4	1·70	63	57	− 6	H
	75	”	”	76	85·0	96·3	8	0·67	40	39	− 1	H E
October 1889	50	”	”	27·5	87·7	96·3	4	1·08	52	57	+ 5	H
	39	”	”	39	88·3	99·3	4	0·63	39	41	+ 2	H
pages 742–5.	91	”	”	27·5	94·0	97·2	4	1·08	52	43	− 9	H
	103	”	”	31·5	94·6	97·3	4	0·89	47	54	+ 7	H
	105	”	”	30·5	95·0	97·2	4	0·94	48	52	+ 4	H

* C = Cylinder Cover jacketed. S = Long Slide-valve.

P = Priming, to the extent of the percentage denoted by the figure annexed.

H = Clearance in High-pressure cylinder taken as 8 per cent. E = Exhaust port of high-pressure cylinder taken as open.

TABLE 17.

Initial Condensation in Woolf Engines Not *jacketed.*

Authority.	Index reference.	Cylinder.		Cut-off. Percentage of Stroke.	Pressure at cut-off. Lbs. per sq. inch.	Revolutions per minute.	Ratio of Clearance Surface to twice Cylinder area.	Ratio of Steam Condensed to Uncondensed.	Percentage of Steam supply condensed at cut-off.			Remarks. *
		Diam.	Stroke.						Calculated.	Observed.	Difference.	
		d	l	m		N	μ	y	$1-x$	$1-x$		
		Feet.	Feet.	P.c.	Lbs.	Revs.	Ratio.	Ratio.	P.c.	P.c.	P.c.	
MAIR.	B	1·83	3·58	38	55	17·84	4·5	0·93	48	42·5	− 5	S P 8½%
Proceedings	F	2·04	3·42	47	68	34·52	5	0·54	35	31·5	− 3	S P 2½%
Inst. C.E. 1882	G	1·31	4·25	23·5	43·5	80·45	5	0·59	37	38·6	+ 2	S P 4%
vol. 70.	H	,,	,,	23	52	81·5	5	0·60	37	39·6	+ 3	S P 4%
KENNEDY.	100	0·5	1·0	85	39·4	96·1	8	0·83	45	48	+ 3	H E
	98	,,	,,	87	59·6	97·0	8	0·79	44	37	− 7	H E
Proceedings	18	,,	,,	26·5	77·4	99·1	4	1·68	63	62	− 1	H
	17	,,	,,	43	76·5	97·3	4	0·78	44	54	+10	H
Inst. Mech. E.	16	,,	,,	25·5	80·0	75·3	4	2·03	68	69	+ 1	H
	73	,,	,,	92	87·7	95·7	8	0·72	42	40	− 2	H E
October 1889	35	,,	,,	26	89·7	95·5	4	1·74	64	66	+ 2	H
	62	,,	,,	37	89·3	98·7	4	1·02	51	55	+ 4	H
pages 742–5.	80	,,	,,	28·5	94·6	97·0	4	1·53	61	63	+ 2	H
	99	,,	,,	27	95·0	96·1	4	1·64	63	63	0	H

* S = Long Slide-valve. P = Priming, to the extent of the percentage denoted by the figure annexed.
H = Clearance in High-pressure cylinder taken as 8 per cent. E = Exhaust port of high-pressure cylinder taken as open.

TABLE 18.

Initial Condensation in Compound Engines Jacketed.

Authority.	Index reference.	Cylinder.		Cut-off.	Pressure at cut-off. Lbs. per sq. inch.	Revolutions per minute.	Ratio of Clearance Surface to twice Cylinder area.	Ratio of Steam Condensed to Uncondensed.	Percentage of Steam supply condensed at cut-off.			Remarks.
		Diam. d	Stroke. l	Percentage of Stroke. m		N	μ	y	Calculated. $1-x$	Observed. $1-x$	Difference.	*
		Feet.	Feet.	P.c.	Lbs.	Revs.	Ratio.	Ratio.	P. c.	P. c.	P. c.	
MAIR. Inst. C.E. 1882 vol. 70.	J	1·75	5·5	20·5	64	23·98	3·52	0·40	29	30	+ 1	P 4%
WILLANS.	Table II C 80	0·83	0·5	60	78·6	400·0	1·28	0·10	8	5·2	− 3	Cylinder and Cover taken as jacketed.
	II C 90	”	”	60	87·5	401·1	”	0·10	8	5·0	− 3	
	II C 90	”	”	52·3	90·0	397·6	”	0·13	11	7·6	− 3	
	II C 100	”	”	52·3	98·8	401·5	”	0·13	11	6·23	− 5	
	II C 100	”	”	47	97·7	405·3	”	0·15	13	10·2	− 3	
Proceedings	II C 110	”	”	47	109·3	402·9	”	0·15	13	9·5	− 3	
	II C 110	”	”	42·7	109·0	402·7	”	0·18	15	11·3	− 4	
Inst. C.E. 1888	II C 120	”	”	42·7	121·3	402·7	”	0·18	15	10·5	− 4	
	II C 120	”	”	39·2	119·9	404·1	”	0·21	17	12·5	− 4	
vol. 93.	II C 130	”	”	39·2	130·6	405·8	”	0·21	17	11·7	− 5	
	II C 130	”	”	36·2	129·9	401·9	”	0·23	19	14·2	− 5	
	II C 140	”	”	36·2	139·7	398·7	”	0·23	19	13·9	− 5	
	II C 140	”	”	33·6	141·8	405·1	”	0·26	21	14·7	− 6	
	II C 150	”	”	33·6	149·8	404·0	”	0·26	21	15·1	− 6	
	II C 150	”	”	31·3	?	402·1	”	0·28	22	18·8	− 3	
	II C 160	”	”	31 3	158·5	401·2	”	0·28	22	17·0	− 5	

WILLANS.	Tablo											* Cylinder and Cover taken as jacketed.
	IV C 130	,,	,,	47·0	128·8	406·8	,,	0·15	13	8·9	− 4	
	IV C 130	,,	,,	42·7	129·2	405·0	,,	0·18	15	10·2	− 5	
	IV C 130	,,	,,	33·6	131·1	402·6	,,	0·26	20	14·3	− 6	
	IV C 130	,,	,,	31·3	130·1	400·0	,,	0·29	22	18·4	− 4	
	IV C 130	,,	,,	23·3	128·1	404·4	,,	0·43	30	25·0	− 5	
	VI C 90	,,	,,	60·4	90·4	210·8	,,	0·13	11	12·6	+ 2	
	VI C 90	,,	,,	60·4	90·8	122·0	,,	0·17	14	20·2	+ 6	
	VI C 110	,,	,,	47	109·1	212·0	,,	0·21	17	20·3	+ 3	
Proceedings	VI C 110	,,	,,	47	110·1	123·8	,,	0·27	21	25·2	+ 4	
	VI C 130	,,	,,	39	128·8	216·4	,,	0·28	22	19·1	− 3	
Inst. C.E. 1888	VI C 130	,,	,,	39	128·8	130·9	,,	0·36	27	29·7	+ 3	
	VII C 60	,,	,,	47	61·0	399·9	,,	0·15	13	17·5	+ 4	
vol. 93.	VII C 70	,,	,,	47	72·6	413·1	,,	0·15	13	15·2	+ 2	
	VII C 80	,,	,,	47	81·1	399·8	,,	0·15	13	10·1	− 3	
	VII C 90	,,	,,	47	89·8	405·7	,,	0·15	13	13·2	0	
	VII C 120	,,	,,	47	120·6	409·6	,,	0·15	13	12·2	− 1	
	VII C 130	,,	,,	47	128·8	406·8	,,	0·15	13	8·9	− 4	
	VII C 60	,,	,,	47	60·9	400·3	,,	0·15	13	15·4	+ 2	
	VIII C 160	,,	,,	36·2	155·4	421·7	,,	0·23	18	13·2	− 5	
	VIII C 160	,,	,,	33·6	157·3	411·3	,,	0·26	20	14·6	− 5	
	IX C 130	,,	,,	33·6	131·1	402·6	,,	0·26	20	14·3	− 6	
Inst. Mech. E.		1·5	4·0	13	63·3	58·7	5·15	1·12	53	46·3	− 7	L 6%
Steam-Jacket		2·67	,,	25·5	12·0	,,	3·46	0·34	25	22	− 3	
Committee,		1·5	,,	22·5	55·3	63·6	5·15	0·54	35	35	0	L 5%
and English.		2·67	,,	20	16·2	,,	3·46	0·45	31	28	− 3	
		1·5	,,	26	58·9	65·7	5·15	0·445	31	43	+12	L 4%
		2·67	,,	16·5	21·2	,,	3·46	0·55	36	37	+ 1	

* P = Priming, } to the extent of the percentage denoted by the figure annexed.
L = Liquefaction, }

TABLE 19.

Initial Condensation in Compound Engines Not *jacketed.*

Authority.	Index reference.	Cylinder. Diam. d	Cylinder. Stroke. l	Cut-off. Percentage of Stroke. m	Pressure at cut-off. Lbs. per sq. inch.	Revolutions per minute. N	Ratio of Clearance Surface to twice Cylinder area. μ	Ratio of Steam Condensed to Uncondensed. y	Percentage of Steam supply condensed at cut-off. Calculated. $1-x$	Observed. $1-x$	Difference.	Remarks. *
	Table	Feet.	Feet.	P. c.	Lbs.	Revs.	Ratio.	Ratio.	P.c.	P.c.	P.c.	Cylinder Cover taken as jacketed.
WILLANS.	II C 80	1·17	0·5	60·4	45	400·0	0·92	0·10	9	11·7	+ 3	L 3·5%
	II C 90	"	"	"	51	401·1	"	0·10	9	11·7	+ 3	L 3·5%
	II C 90	"	"	"	45	397·6	"	0·10	9	12·9	+ 4	L 4·5%
	II C 100	"	"	"	47	401·5	"	0·10	9	16·7	+ 8	L 4·5%
	II C 100	"	"	"	45	405·3	"	0·10	9	11·4	+ 2	L 5·0%
Proceedings	II C 110	"	"	"	49	402·9	"	0·10	9	11·2	+ 2	L 5·0%
	II C 110	"	"	"	45	402·7	"	0·10	9	7·8	− 1	L 5·5%
Inst. C.E. 1888	II C 120	"	"	"	45	402·7	"	0·10	9	13·6	+ 5	L 6·0%
	II C 120	"	"	"	45	404·1	"	0·10	9	13·4	+ 4	L 5·5%
vol. 93.	II C 130	"	"	"	47	405·8	"	0·10	9	13·1	+ 4	L 6·0%
	II C 130	"	"	"	43	401·9	"	0·10	9	11·0	+ 2	L 6·0%
	II C 140	"	"	"	45	398·7	"	0·10	9	13·9	+ 5	L 6·0%
	II C 140	"	"	"	45	405·1	"	0·10	9	13·3	+ 4	L 6·5%
	II C 150	"	"	"	47	404·0	"	0·10	9	14·6	+ 6	L 6·0%
	II C 160	"	"	"	51	401·2	"	0·10	9	14·8	+ 6	L 6·5%

WILLANS.	IV C 130	,,	,,	,,	51	406·8	,,	0·10	9	14·6	+ 6	* L 5·5% Cylinder Cover taken as jacketed.
	IV C 130	,,	,,	,,	51	405·0	,,	0·10	9	13·8	+ 5	L 6·0%
	IV C 130	,,	,,	,,	47	402·6	,,	0·10	9	10·0	+ 1	L 6·0%
	IV C 130	,,	,,	,,	43	400·0	,,	0·10	9	12·0	+ 3	L 6·5%
	IV C 130	,,	,,	,,	40	404·4	,,	0·10	9	12·4	+ 3	L 6·0%
	VI C 90	,,	,,	,,	45	210·8	,,	0·14	12	20·3	+ 8	L 4·5%
	VI C 90	,,	,,	,,	50	122·0	,,	0·18	15	27·2	+12	L 4·0%
	VI C 110	,,	,,	,,	45	212·0	,,	0·14	12	21·7	+10	L 5·0%
Proceedings	VI C 110	,,	,,	,,	48	123·8	,,	0·18	15	26·9	+12	L 4·5%
	VI C 130	,,	,,	,,	47	216·4	,,	0·14	12	20·4	+ 8	L 6·0%
	VI C 130	,,	,,	,,	47	130·9	,,	0·18	15	27·1	+12	L 5·0%
Inst. C.E. 1888	VII C 60	,,	,,	,,	31	399·9	,,	0·10	9	10·3	+ 1	L 4·0%
	VII C 70	,,	,,	,,	35	413·1	,,	0·10	9	8·3	− 1	L 4·0%
vol. 93.	VII C 80	,,	,,	,,	37	399·8	,,	0·10	9	11·8	+ 3	L 5·0%
	VII C 90	,,	,,	,,	40	405·7	,,	0·10	9	10·9	+ 2	L 5·0%
	VII C 120	,,	,,	,,	50	409·6	,,	0·10	9	16·1	+ 7	L 5·0%
	VII C 130	,,	,,	,,	53	406·8	,,	0·10	9	15·2	+ 6	L 5·5%
	VII C 60	,,	,,	,,	30	400·3	,,	0·10	9	4·3	− 5	L 4·0%
	VIII C 160	,,	,,	,,	51	421·7	,,	0·10	9	15·4	+ 6	L 7·0%
	VIII C 160	,,	,,	,,	51	411·3	,,	0·10	9	13·7	+ 5	L 7·0%
	IX C 130	,,	,,	,,	45	402·6	,,	0·10	9	9·5	+ 0	L 6·5%
Ins. Mech. E. Steam-Jacket Committee, and English.		1·5	4·0	28	56·7	57	5·15	0·66	40	33	− 7	
		2·7	4·0	17	16·4	57	3·46	0·89	47	50	+ 3	L 5·0%
Ins. Mech. E. Marine-Engine Trials 8.	Fusi-Yama	2·28	2·75	45	65	55·6	2·12	0·17	15	16·9	+ 2	Piston-valves
	Colchester	2·5	3·0	50	77	86·5	2·94	0·19	16	28·0	+12	

* L = Liquefaction, to the extent of the percentage denoted by the figure annexed.

TABLE 20.

Initial Condensation in Triple Engines Jacketed.

Authority.	Index reference.	Cylinder.		Cut-off.	Pressure at cut-off. Lbs. per sq. inch.	Revolutions per minute.	Ratio of Clearance Surface to twice Cylinder area.	Ratio of Steam Condensed to Uncondensed.	Percentage of Steam supply condensed at cut-off.			Remarks.
		Diam.	Stroke.	Percentage of Stroke.					Calculated.	Observed.	Difference.	
		d	l	m		N	μ	y	$1-x$	$1-x$		*
WILLANS.	Table	Feet.	Feet.	P. c.	Lbs.	Revs.	Ratio.	Ratio.	P. c.	P. c.	P. c.	Cylinder and Cover taken as jacketed.
	III T 150	0·583	0·5	69·4	151·9	405·6	1·72	0·077	7	5·33	− 2	
	III T 150	”	”	64·7	149·6	409·0	”	0·094	8·5	5·46	− 3	
	III T 160	”	”	64·7	159·5	401·2	”	0·095	8·5	4·43	− 4	
	III T 160	”	”	60·6	158·1	408·4	”	0·110	10·0	6·84	− 3	
Proceedings	III T 170	”	”	60·6	?	414·8	”	0·110	10·0	10·32	0	
	III T 170	”	”	60·6	172·5	400·4	”	0·112	10·0	5·01	− 5	
Inst. C.E. 1888	† T 170	”	”	48·4	168	402·6	”	0·183	15·5	8·8	− 7	
	† T 170	”	”	48·4	164·5	406·2	”	0·183	15·5	7·2	− 8	
vol. 93.	III T 150	0·83	”	60·4	105	405·6	1·28	0·092	8·5	11·8	+ 3	L 3%
	III T 150	”	”	60·4	100	409·0	”	0·092	8·5	8·9	0	L 3%
	III T 160	”	”	60·4	100	401·2	”	0·093	8·5	12·8	+ 4	L 3·5%
	III T 160	”	”	60·4	100	408·4	”	0·092	8·5	9·0	0	L 3%
† vol. 96	III T 170	”	”	60·4	100	414·8	”	0·091	8·5	8·2	0	L 3%
page 255.	III T 170	”	”	60·4	100	400·4	”	0·093	8·5	11·9	+ 3	L 4%
	† T 170	”	”	60·4	?	402·6	”	0·093	8·5	12·3	+ 4	L 5%
	† T 170	”	”	60·4	?	406·2	”	0·093	8·5	10·9	+ 2	L 5%
Inst. Mech. E. Marine-Engine Trials 1889 and 1891.	Meteor	2·45	4·0	45	147	71·8	4·0	0·17	15	22·9	+ 8	M
	Iona	1·82	3·25	21	163	61·1	2·46	0·32	25	28·8	+ 4	I

* L = Liquefaction, to the extent of the percentage denoted by the figure annexed.
M = "Meteor," having piston valves; and observed condensation includes jackets.
I = "Iona" having 7·8% priming and jacket water.

TABLE 21.

Initial Condensation in Triple Engines Not *jacketed.*

Authority.	Index reference.	Cylinder.		Cut-off.	Pressure at cut-off. Lbs. per sq. inch.	Revolutions per minute.	Ratio of Clearance Surface to twice Cylinder area.	Ratio of Steam Condensed to Uncondensed.	Percentage of Steam supply condensed at cut-off.			Remarks.
		Diam. d	Stroke. l	Percentage of Stroke. m		N	μ	y	Calculated. $1-x$	Observed. $1-x$	Difference.	*
WILLANS.	Table	Feet.	Feet.	P. c.	Lbs.	Revs.	Ratio.	Ratio.	P. c.	P. c.	P. c.	Cylinder Cover taken as jacketed.
	III T 150	1·17	0·5	60·4	45	405·6	0·92	0·10	9	15·1	+ 6	L 7%
	III T 150	"	"	"	42	409·0	"	0·10	9	13·5	+ 4	L 7%
	III T 160	"	"	"	42	401·2	"	0·10	9	15·9	+ 7	L 7·5%
Proceedings	III T 160	"	"	"	40	408·4	"	0·10	9	15·1	+ 6	L 7%
Inst. C.E. 1888	III T 170	"	"	"	40	414·8	"	0·10	9	13·1	+ 4	L 7·5%
vol. 93.	III T 170	"	"	"	42	400·4	"	0·10	9	16·2	+ 7	L 8%
† vol. 96	† T 170	"	"	"	?	402·6	"	0·10	9	24·8	+16	L 9·5%
page 255.	† T 170	"	"	"	?	406·2	"	0·10	9	11·1	+ 2	L 9%
REYNOLDS.	I 41	0·42	0·83	30·4	190	146	2·82	0·65	40	40	0	
	I 35	"	"	40·0	187	229	"	0·38	28	29	+ 1	
	I 40	"	"	47·0	189	322	"	0·19	16	22	+ 6	
Proceedings	II 41	0·67	0·83	37·0	66	127	2·24	0·60	38	37	− 1	L 4%
Inst. C.E. 1889	II 35	"	"	37·0	70	215	"	0·43	30	34	+ 4	L 4%
vol. 99.	II 40	"	"	40·0	75	320	"	0·325	25	26	+ 1	L 4%
	III 41	1·0	1·25	32·0	20	109	2·24	0·52	34	43	+ 9	L 8%
	III 35	"	"	27·0	22	184	"	0·51	34	40	+ 6	L 8%
	III 40	"	"	34·0	21	276	"	0·30	23	23	+ 0	L 9%
Inst. Mech. E. Marine-Engine Trials 1890.	Tartar	2·17	3·5	30·0	136	70	4·05	0·52	34	38	+ 4	Priming 17%. Piston-valves.

* L = Liquefaction, to the extent of the percentage denoted by the figure annexed.

(*continued from page* 200.)

For piston valves and many other descriptions of valve gear, the ratio μ is largely in excess of the figure thus found, and a special value must be assigned in each case. The greatest value of the ratio μ which the author has measured is 5·15; but it seems probable that many engines, especially those of Woolf type fitted with long slide-valves, would give values nearly as great.

In the column of observed values of $1-x$ in Tables 14 to 21 the estimated percentage of liquefaction due to work already performed before the steam enters the cylinder has been deducted. Any priming given in the data has also been deducted from the observed percentage of condensation.

Assuming all the observations to be of equal weight, the average of their results would indicate about 3·6 per cent. of the total steam supply as the probable error in predicting from the formulæ the amount of condensation; and the author therefore trusts that he has sufficiently established the possibility of employing formulæ of this description as useful checks, both in new designs and in determining the most economical conditions for the working of existing engines.

Discussion.

Mr. P. W. Willans said he had made many thousand engine-trials within the last few years, but had so far resisted the temptation to construct any formula, because as soon as a formula was attempted something was sure to crop up showing that some other cause was at work. The formulæ now given in page 199 seemed to him to have only casual connection with the subject. One of the main factors in the first expression given he observed was $S_c - S_1$, where S_c was the unjacketed clearance surface, and S_1 was the fresh surface exposed during admission up to cut-off. One of the chief objects he supposed of a formula of that kind was to help in designing an engine which should have as little initial condensation as possible; and if he understood the above expression rightly, it would appear that, if the unjacketed clearance surface were equal to the fresh surface exposed during admission up to cut-off, there would be no condensation at all. It was not a difficult matter to design an engine in which those two surfaces should be equal; and absence of condensation was such a desirable result that he thought by this time some engine would have been found in which that equality of the two surfaces existed. Why the fresh surface exposed during admission up to cut-off should be deducted from the unjacketed clearance surface he did not understand. There were also several conditions in ordinary steam engines, with which the formula did not seem to have anything to do. One was the extent of compression after the closing of the exhaust; because the effect of the unjacketed clearance surface must largely depend upon its temperature at the beginning of the stroke. If its temperature had been raised by compression, the clearance surface must have a different effect on the initial condensation from what it would have if its temperature had not been so raised. But beyond all this there was the yet larger question of the state of the surface; and the further he had gone into this matter, the more had he found that the different conditions of dampness or dryness of surface were the main factors in determining the amount of the initial condensation. If the clearance surfaces were damp during the exhaust, their action was quite different from what it was if they were dry. In the high-pressure diagram of the "Ville de Douvres," Fig. 11, Plate 15, there was seen to be a great drop in the pressure at the end of the expansion in the high-pressure cylinder; and attention had been called by Mr. Rowan (page 177) to the fact that the drop could be reduced by cutting off earlier in the low-pressure cylinder. This however he thought would not be an entire gain, because he considered that the drop in pressure at the end of the steam stroke had the same effect in drying the surface and reducing the condensation as the compression at the end of the exhaust stroke had. The high-pressure cylinder in the recent "Ville de Douvres" trial showed less condensation he thought than had been met with in any of the marine-engine trials previously reported to the Institution by the Research Committee.

The PRESIDENT said the members of the Institution knew that Colonel English had devoted a vast amount of time and attention to this subject, and had made a great number of experiments upon it, as evidenced by the diagram accompanying the present paper. To his own knowledge these experiments had been going on for many years. Some valuable practical results he had no doubt would follow from the formula that the author had constructed; although, as Mr. Willans had shown, it might not bear being pushed to extremes. He enquired whether in any of the experiments it had been tried how far the condensation was affected by the presence of grease on the clearance surfaces. And where $1-x$ was said in the paper to represent a certain ratio, he presumed it was meant that the ratio of $1-x$ to 1 represented the ratio referred to.

Lt.-Colonel ENGLISH replied that in one or two experiments the clearance surfaces had been coated with grease, but he had not been able to find any difference on that account in the extent of condensation. It was the ratio of $1-x$ to 1 which represented the percentage of steam supply that was condensed at cut-off (page 200). The expression $1-x$ for representing this percentage had originated with Professor Cotterill, he believed, the symbol x denoting the fraction (by weight) of saturated steam in the working mixture of steam and water contained in the cylinder: so that $1-x$ denoted the fraction (by weight) of water in the working mixture. (See Proceedings 1887, page 491.)

In connection with the expression given in page 199 for the initial condensation, it had been remarked by Mr. Willans (page 213) that, if the fresh surface exposed during admission up to cut-off was equal to the unjacketed clearance surface, there would according to this formula be no condensation at cut-off. That was certainly the case, and he thought Mr. Willans himself had reached that very result as nearly as possible in practice. For in the experiments on his triple engine, recapitulated in Table 20 (page 210), it would be seen that in the third line, marked III T 160, the actual observed condensation was only 4·43 per cent.; and it would be difficult he thought to come much nearer than this to having no condensation at all. As far as the experiments which he himself had analysed had gone, it appeared that, when the extent was increased of the fresh surface exposed during admission up to cut-off, the difference between the initial condensation and the subsequent re-evaporation up to cut-off was diminished; and in that particular example the two had nearly reached a balance.

With regard to the effect of compression upon the temperature of the unjacketed clearance surface (page 213), in no experiment with which he was acquainted had the compression ever reached the initial pressure in the case of an unjacketed engine; there was always a deficiency of pressure to be made up by the fresh incoming steam.

Mr. Willans pointed out that in most of his experiments the compression did reach the initial pressure.

Lt.-Colonel English replied that his last remark applied expressly to unjacketed engines; and all through his calculations he had dealt with Mr. Willans' engine as being jacketed. In any engine, whether jacketed or not, it would be impracticable he thought to ascertain from the indicator diagrams what condensation was going on during the period of compression, or to separate this from condensation on admission. The state of the surface (page 213) he considered was an exceedingly difficult question, for it was practically impossible to see what was the state of the interior surface of a cylinder while the engine was working; this he admitted was a weak point in his investigations.

The President was sure the members would be glad to pass a vote of thanks to Colonel English for his present paper, which with the two previous ones represented a vast amount of conscientious work, and formed a store-house wherefrom he was confident that a great deal of useful and practical information could be derived by those who were connected with the construction and working of steam engines.

Mr. W. H. NORTHCOTT, being unable to attend the meeting, wrote that the physical basis of the author's formulæ was not quite clear from his present paper; but from his previous paper (Proceedings 1889, pages 647–8) it appeared that the four following assumptions were made:—(1) "that the initial condensation varies directly as the density of the incoming steam;" (2) "that the number of thermal units of heat transferred must vary directly as the area of that portion S_c of the clearance surface which is colder at the moment than the entering steam, whether the surface be that of a film of water or of the actual metal;" (3) that the heat transferred "varies, at any rate approximately, as the square root of the time of exposure, or as $\frac{1}{\sqrt{N}}$, where N is the number of revolutions per second;" (4) that, when the flow of heat to the outside is prevented, "the effect of range of temperature will become negligible."

In regard to the first of these assumptions, if the initial condensation varies literally as the initial density of the steam, there would be the same condensation with steam of say 100 lbs. pressure, worked without expansion and expelled against a back-pressure of 99 lbs., as with steam of the same density expanded down to 1 lb. absolute and expelled at this low pressure. With a given terminal pressure or with a given back-pressure, it might with some show of reason be thought that the condensation during admission would increase directly with the initial pressure or initial density; but without some such limitation in regard to the extent of expansion the author's formulæ surely seem to be erroneous.

With regard to the second assumption, if the condensation increases directly as the colder clearance surface, then by increasing sufficiently the area of this surface all the steam admitted could be condensed, and the cylinder would consequently be full of water at the point of cut-off. Now a pound of steam of 100 lbs. absolute pressure would carry into the cylinder 1,182 units of heat; while at 1 lb. absolute pressure it needs only 1,113 units to be fully saturated. Therefore, as under the assumed conditions no actual work would be performed, the steam on leaving the cylinder would itself be superheated by about 150° Fahr. This state of things is not inconceivable by any means. An initial condensation of 62 per cent. has been measured by the author himself in an engine, of which the working was spoken of in the discussion upon his first paper as "very economical" (Proceedings 1887, page 501). In a less perfect engine, what might not the condensation become?

It would be interesting to ascertain experimentally the precise effect of surface upon cylinder condensation; and this could be done by constructing a cylinder cover with a series of thin iron discs inside, as sketched in Fig. 2, Plate 25. These discs need not be more than 1-8th inch thick; and if a few holes were drilled in each, to give the steam free access to their entire surface, a space of 1-8th inch between the discs would probably suffice. A steam cylinder 14 inches diameter would have ordinarily about 3 square feet of initial condensing surface. For this cylinder the two faces of each disc would together expose an area of just 2 square feet: so that three such discs would treble the condensing surface, and with twelve such discs there would be a total of 27 square feet of condensing surface, instead of the usual 3 square feet. If the condensation does increase directly with the surface, the initial condensation would be nine times greater with the twelve-disc experimental cover on, than when the ordinary cover was on. By varying the number of the discs, the variation of condensation with surface would be ascertained with certainty. As seen in Table 15, page 202, the author has experimented with an engine of about this size, and he has there given the observed and calculated initial condensations obtained by him. For half a dozen of the experiments given it will be seen that the observed and calculated quantities are practically identical, while the steam condensed was about 1·7 times greater than the steam uncondensed. Had the suggested cover been applied to that engine and the surface increased nine times, other conditions remaining the same, would the author anticipate an initial condensation of $9 \times 1 \cdot 7 = 15 \cdot 3$ pounds per pound of working steam?

So many different actions are involved in bringing about cylinder condensation, that in the writer's opinion no accurate physical formula, simple enough to be of use to engineers, seems likely to

(Mr. W. H. Northcott.)
be devised. In the pistons of the "Ville de Douvres," for instance, shown in Plate 16, it will be seen that, apart altogether from the absorption and emission of heat by the clearance surfaces, an appreciable quantity of heat would probably flow through the piston itself, from the hot steam during admission and expansion to the colder steam on the other side of the piston, whence it would then be expelled to the condenser. This flow of heat too would be in opposite directions for each single stroke. So also when the same ports and passages are used for admission and release, the contact of damp exhaust steam with the walls of the passages must cool their surfaces down; and as they are heated up again on the admission of the steam, a portion of the steam admitted will be condensed from this cause. In Plate 16 it will be seen that the damp cool exhaust steam is passing out into the condenser in forcible contact with one side of metallic walls, whilst hotter steam is being admitted on the other side of them. This must cause a transfer of heat, and the transfer must produce initial condensation. The author's formulæ deal only with the heat absorbed and emitted by the clearance surfaces; but these surfaces do not occasion all the mischief by any means. In many cases there is more re-evaporation during exhaust than during expansion; most of which would appear to be caused by the use of the same passages for admission and release. The condensation arising from the foregoing conditions will mainly depend, in the writer's opinion, upon the wetness of the steam at the moment of release. Wet steam is known to absorb heat readily and rapidly; whereas dry steam can be superheated only by passing it in contact with a large area of heating surface having a considerably higher temperature on its other side. It may therefore be concluded that, when steam is dry at the moment of release, the use of the same ports and passages for admission and exhaust leads to comparatively little loss; and indeed, so far as the writer knows, no experiments have shown any appreciable superheating during exhaust.

It appears to be taken for granted by the author and others who have dealt with initial condensation, that the quantity of heat available for absorption by the clearance walls is unlimited, and that the actual quantity absorbed and emitted is limited only by the power of the metallic walls to absorb and emit heat in the given time. This seems to the writer to be an entire fallacy; and he believes that there is a maximum possible quantity of heat for each case, and that this maximum quantity of heat limits the possible initial condensation. Furthermore, when this maximum is reached, the area of clearance surface may be increased, or the engine may be run slower, without appreciably increasing the rate of initial condensation. On the other hand however, the condensation may be reduced by running the engine faster or by reducing the area of condensing surface.

All the heat that goes into an engine must come out again: that is to say, the heat admitted is equal to the sum of the heat rejected and the heat converted into work. If then the steam goes to the condenser dry but not superheated, as apparently it always does with high rates of expansion, the maximum initial condensation is easily calculated for any given instance. Taking the case of an unjacketed cylinder, and assuming that internal engine friction balances radiation, then, if the admission pressure is 100 lbs. absolute, and the condenser pressure 2 lbs. absolute, each pound weight of steam brings into the cylinder 1,182 units of heat, and carries away 1,120 units. Therefore each pound weight of steam can do work equivalent to $1{,}182 - 1{,}120 = 62$ thermal units, and no more. If each pound of steam appearing in the indicator diagrams does that amount of work and no more, the writer thinks there will be no initial condensation, however large may be the clearance area, or however slowly the engine may run; and the quantity of steam as measured from the diagrams will be the quantity actually used. But if by expansive working or by any other means work equivalent to twice 62 thermal units is performed by each pound weight of steam appearing in the diagrams, then he believes it will be found that the steam appearing in the diagrams is just half the quantity actually used: in other words the initial condensation will now be 50 per cent. The heat equation for the first case is $1{,}182 = 1{,}120 + 62$; while for the second case it will be $1{,}182 \times 2 = 1{,}120 \times 2 + 62 \times 2$. Under these conditions —that is, with an unjacketed cylinder, and with the loss by radiation

(Mr. W. H. Northcott.)
balanced by the gain from cylinder friction, and with the steam leaving the cylinder dry but not superheated—the weight of steam in pounds per indicated horse-power per hour will be $\frac{33,000 \times 60}{772} \div (H_a - H_b) = 2,565 \div (H_a - H_b)$, where H_a and H_b are the total heats of steam of admission and exhaust pressure respectively. For the two above cases the steam consumption will therefore be $2,565 \div 62 = 41 \cdot 37$ lbs. The initial condensation per pound weight of steam appearing in the diagrams will be $C = W \div (H_a - H_b) - 1$, where W is the heat-value of the work performed by a pound weight of working steam, or steam appearing in the diagram at cut-off. The quantity $1 + C = W \div (H_a - H_b)$ is the total steam used per pound weight of steam appearing in the diagram at cut-off. These simple expressions cover the action not only of the clearance surfaces, but also of the other condensing tendencies already mentioned, under the conditions assumed; but they give maximum results. The condensation and steam consumption may be less; but they cannot be greater, so far as the writer can see.

Reasoning in another way, it may be considered certain, firstly that, when steam is admitted into a cylinder having cold metallic walls, the walls will be warmed up to the temperature of the steam by heat abstracted from it: so that, unless the steam be superheated, a portion of it must become condensed. Secondly, when the walls are once raised to the temperature of the admitted steam, they will cause no further condensation so long as they remain at that temperature. Thirdly, if heat be abstracted from the metallic walls, they will then be enabled to abstract heat from the hotter steam next brought into contact with them. Fourthly, in any series of alternate actions of this kind the quantity of heat taken away from the walls must be equal to the quantity of heat taken away from the steam; otherwise the cylinder would continue to become hotter and hotter or colder and colder. Fifthly, in order therefore to ascertain what initial condensation will take place in any given instance, it must be ascertained what heat can be previously abstracted from the surfaces with which the steam is brought into contact. Neglecting radiation, friction, and jacketing, and assuming the steam to be dry on reaching the engine, the maximum quantity of heat that can be abstracted from any cylinder by the total steam passing through it will be $H_m = W - (H_a - H_b) + H_bC$. But as H_m is made up of two quantities, one of which is H_bC, or the heat required to re-evaporate the condensed steam against the pressure of the condenser, for H_m may be written $H_y + H_bC$. Then $H_y + H_bC = W - (H_a - H_b) + H_bC$; and therefore $H_y = W - (H_a - H_b)$ now represents the maximum quantity of heat abstracted from any steam cylinder per pound weight of steam appearing in the diagrams. This means the same thing as the previous expressions. To cover the effect of jacketing, friction, superheating, and radiation—putting J and F and S for the gain of heat by jacketing, by friction, and by superheating respectively, and R for the loss by radiation, all in heat-units per pound weight of working steam—the equation becomes $H_y = W - (H_a - H_b) - J - F - S + R$; and the pounds weight of steam consumed per indicated horse-power per hour, exclusive of jacket steam, will be $2,565 \div (H_a - H_b + J + F + S - R)$. In the case of compound engines each cylinder has to be considered by itself.

Instead therefore of calculating what heat the clearance surfaces are capable of absorbing and emitting, it is more necessary to calculate what heat there is for them to absorb and emit. But as under certain circumstances they do not absorb and emit the full quantity available, it is still useful to ascertain the conditions which limit their absorption and emission. What these conditions are is still uncertain, although Colonel English has done a good deal towards ascertaining them. A rough formula for practical use may easily be devised; but physical accuracy is another matter altogether. For Corliss and other such engines, where the exhaust passages are underneath the cylinder and independent of the admission passages, the case is simplified; but the use of passages in common for admission and exhaust complicates the problem. Apart from this also it has to be remembered that, although the area of the clearance surface is generally an important factor, it is nevertheless tolerably certain that, with this area reduced to nothing, and indeed even assuming the use of a non-conducting cylinder, initial condensation might still attain its maximum.

Lt.-Colonel ENGLISH wrote that Mr. Northcott's criticism of two only out of the four assumptions, which he had specified as forming the bases of the author's formulæ, rendered it somewhat difficult to make a complete reply; but with regard to the first point raised—whether there would be the same condensation with steam of 100 lbs. pressure worked without expansion, as with steam of the same pressure expanded down to 1 lb.—the author thinks his experiments taken as a whole show that up to the point of cut-off the weight of steam condensed would be approximately the same in each case.

With regard to the second assumption, all the experiments quoted by the author show a progressive increase in the proportion of steam condensed as the clearance surface is increased, up to a condensation of two pounds out of every three admitted; and he is therefore quite inclined to the belief that, with a sufficient condensing surface, and a slow enough speed of revolution, all the steam admitted could be condensed, provided that, as in a surface condenser, the air and condensed water were continuously removed.

It has been assumed by Mr. Northcott that a cylinder 14 inches diameter would have ordinarily about 3 square feet of initial condensing surface; but in the author's opinion this is much too small an area, and 4½ square feet would be nearer the usual proportion.

It is not apparent to the author how condensation could be lessened by the use of separate exhaust-ports. The entering steam cannot be prevented from coming into contact with them, and the clearance surface would be increased rather than diminished.

The principal difference between Mr. Northcott's views and the author's appears to arise from the latter, as stated in Proceedings 1889, page 699, being based on the assumption that, whatever the action of initial condensation may be, it is practically instantaneous compared with any present means of measuring the time involved; and that, re-evaporation at release being equally sudden, the cylinder and clearance surfaces must be practically dry and inoperative during the whole of the exhaust.

MEMOIRS.

Richard Birtwistle was born at Haslingden, near Manchester, on 27th March 1834. After serving his apprenticeship to engineering, he entered into partnership with Mr. Samuel S. Stott, with whom he purchased in 1866 the engineering and mill-wright business of his late father-in-law, Mr. John Lindsay, at Laneside Foundry, Haslingden. At that time the place was of small dimensions, but through his ability and energy it rose to be ranked among the leading engineering establishments in the country, finding employment for a large number of workpeople. On the death of his partner more than two years ago, he became the sole proprietor of the business. Having been for several years a sufferer from bronchitis, he died at Southport on 18th April 1892, in the fifty-ninth year of his age. He became a Member of this Institution in 1888.

Sir James Brunlees was born at Kelso, Roxburghshire, on 5th January 1816. After leaving school he was put by his father to gardening and farm work, with the intention of his being trained as a landscape gardener; but having a desire for higher culture, and having himself earned enough to carry out his wish, he went to Edinburgh University for two sessions, and there became known to Mr. Alexander Adie, by whom he was engaged in 1838 as his assistant on the Bolton and Preston Railway, one of the earliest lines constructed in this country. Subsequently he joined the staff of Messrs Locke and Errington, and took an important part in the laying out of the Caledonian line from Beattock to Carstairs, with branches to Edinburgh and Glasgow. On the completion of this work he was occupied on the Lancashire and Yorkshire Railway as acting engineer, under Sir John Hawkshaw, engineer-in-chief. In 1850 he went to Ireland to engage in constructing the Londonderry and Coleraine Railway. Two years later he undertook the difficult work of the Ulverston and Lancaster Railway across Morecambe Bay, which involved heavy enbankments and several iron bridges. Subsequently he became actively engaged in engineering work of all kinds, both at home and abroad, in addition to acting as arbitrator in the settlement of disputed contracts and other railway matters. Among the principal works he carried out were the Solway Junction Railway, which included a viaduct of a mile and a quarter in length over the Solway Firth; the Clifton Extension Railway; the Mersey Tunnel Railway between Liverpool and Birkenhead, in connection with which he received in May 1886 the honour of knighthood; the Mont Cenis Summit Railway, involving very steep gradients; the Avonmouth, King's Lynn, and Whitehaven docks; the Southport, New Brighton, and Llandudno piers, and the new pier and pavilion at Southend; the Central Uruguay, San Paulo, Bolivar, and other railways in South America and elsewhere. He was also joint engineer with Sir John Hawkshaw for the Channel Tunnel scheme. He became a Member of this Institution in 1870. He was a Past-President of the Institution of Civil Engineers, and a Fellow of the Royal Society of Edinburgh. His death took place at Wimbledon on 2nd June 1892, in the seventy-seventh year of his age.

PETER WILLIAM WILLANS was born at Leeds on 8th November 1851, and received a classical education at the grammar school of that town. His father, who owned large cloth mills in Leeds, intended that he should be brought up to that trade; but he preferred to become an engineer, and served his apprenticeship from 1867 to 1872 with Messrs. Carrett and Marshall, Sun Foundry, Leeds. He then went to London, where, after working two years in the drawing office of Messrs. John Penn and Sons, Greenwich, he designed his first high-speed engine. On leaving them he acted for a short time as a consulting engineer in the firm of Messrs. Willans and Ward. From 1875 to 1880 he superintended the manufacture of his engine at the works of Messrs. Hunter and English, Bow. At that time the use of the three-cylinder single-acting high-speed

Willans engine was almost entirely confined to launches; and as it was difficult to make launch builders believe in high-speed engines, he started works at Thames Ditton in partnership with Mr. Mark H. Robinson, for building not only the engines but the launches as well. From small beginnings these works grew in importance, at first gradually and then at a more rapid rate, and in 1888 the firm was converted into a company, in the management of which Mr. Willans was actively engaged, both as chairman and as chief engineer. He devised and elaborated a scheme of profit-sharing, in order to identify the men's interest with that of the company, and in this he entirely succeeded. Besides his three-cylinder engine, he invented also later the "central-valve" engine, much used for the direct driving of dynamos; an electrical governor, of which the latest form was completed only a short time before his death; and a magnetic coupling for the transmission of power without mechanical contact between the parts. He was well known as a scientific experimenter of the highest order, as shown by his valuable researches on the action of steam in engines, and his extensive experimental investigations into the economical performances of steam engines working under various conditions. His death took place on 23rd May 1892, in the forty-first year of his age, as the result of an accident in which he was thrown out of his dog-cart and fell upon his head. He became a Member of this Institution in 1888, and took part frequently in the discussion of subjects connected with his own researches.

Institution of Mechanical Engineers.

PROCEEDINGS.

JULY 1892.

The SUMMER MEETING of the Institution was held in Portsmouth, commencing on Tuesday, 26th July 1892, at Ten o'clock a.m.; Dr. WILLIAM ANDERSON, F.R.S., President, in the chair.

The President, Council, and Members were received in the Town Hall by the Mayor of Portsmouth, T. SCOTT FOSTER, Esq., J.P.

The MAYOR said it gave him great pleasure to welcome the Members of the Institution to Portsmouth, and he was very glad to place the Town Hall at their disposal for the occasion of their meeting here. The Dockyard alone he was sure they would consider amply sufficient to repay their visit, even were they to see nothing else; and besides this there were also the Clarence Victualling Yard, the Eastney Sewage Works, and the Gas Works, which would well compensate for the lack of other manufacturing works in any great variety or on any extensive scale. One of the objects of the meetings of the Institution he understood was to facilitate the interchange of ideas respecting improvements in the various branches of mechanical science; and he had no doubt that so excellent an object would be largely realised by their present meeting, and by the opportunity they would have of inspecting the places he had mentioned, in which would be found many things calculated to enlarge the range of individual ideas. This result he thought was likely to be further promoted by the papers announced for reading and discussion, some of which he observed had special reference to the most important works in Portsmouth and the neighbourhood. He trusted the Members would enjoy the advantage of fine weather

(The Mayor of Portsmouth.)
during their stay, and that their visit would afford them recreation in addition to engineering interest.

The PRESIDENT was sure the Members would feel with himself how much indebted they all were to the Mayor for his kind words of welcome, none the less because they were few and directly to the purpose; and all would join in heartily thanking his Worship, and those associated with him in the administration of the affairs of this important municipality, for the great kindness which they had shown in connection with the arrangements so obligingly made for the comfort and convenience of all attending the present meeting. No better evidence of this could be asked than was afforded by the magnificent Hall they were now assembled in, which had been so handsomely placed at their disposal by the Mayor, together with the very convenient accommodation which he had also provided in adjoining rooms for the secretaries' office and for luncheon. From these auguries it was certain that the Members would enjoy a most pleasant and profitable meeting; in view of which he was sure they would all join him in thanking his Worship in the warmest manner for what he had done for them.

The Minutes of the previous Meeting were read, approved, and signed by the President.

The PRESIDENT announced that the Ballot Lists for the election of New Members had been opened by a committee of the Council, and the following thirty-eight candidates were found to be duly elected:—

MEMBERS.

JAMES ATKINSON,	London.
JOHN WALTER BROOKE,	Lowestoft.
JOHN BRUNLEES,	London.
FRANCIS FURLONG BYRNE,	Dublin.
DAVID CARNEGIE,	Woolwich.
OSBERT CHADWICK, C.M.G.,	London.
CHARLES BUTLER CLAY,	Sunderland.

Robert Henry Collen, . . .	Northfleet.
James Dunlop, ...	Bombay.
Thomas Carline Eastwood, . . .	Derby.
Frederic Macdonnell Evanson, . .	London.
Percy Alexander Forbes, . . .	Walsall.
George Edward Hall, . . .	Manchester.
Robert Blackwell Hansell, . .	Glasgow.
Abraham Wyke Harrison, . . .	Abergavenny.
Edward Cartwright Harvey, . .	Johannesburg.
Alfred Herbert, ...	Coventry.
Arthur Hullah, ...	Bombay.
Thomas Irons, ...	Granville, N.S.W.
Richard Henry Lea, ...	Coventry.
Peter McGregor, ...	Hong Kong.
Charles Ralph Pinder, . . .	Johannesburg.
Alfred Lee Pogson, ...	Madras.
Harry Alfred Richardson, . . .	Bolton.
Andrew Shirlaw, . . .	Birmingham.
Henry Bath Spencer, ...	Manchester.
Edward Herbert Stone, . .	Simla.
John Strachan, ...	Cardiff.

ASSOCIATES.

Frederic Hungerford Bowman, D.Sc., F.R.S.E., ...	Halifax.
Arthur Cryer, ...	Cardiff.
Alfred Edwin Stove, ...	London.
Arthur Turner, ...	London.

GRADUATES.

Ernest Henry Earle Bulwer, . .	Derby.
Charles George Redfern, . . .	Derby.
James Cartmell Ridley, Jun., . .	Newcastle-on-Tyne.
James Horace Shepherd, . . .	Swindon.
Albert Edward Vezey, . . .	Crewe.
William Wallace Wallis, . . .	Derby.

The following Papers were then read and discussed:—

"On Shipbuilding in Portsmouth Dockyard"; by Mr. WILLIAM H. WHITE, C.B., F.R.S., Director of Naval Construction and Assistant Controller of the Navy.

"On the Applications of Electricity in the Royal Dockyards and Navy"; by Mr. HENRY E. DEADMAN, Chief Constructor at the Admiralty, late of Portsmouth Dockyard.

"Description of the Lifting and Hauling Appliances in Portsmouth Dockyard"; by Mr. JOHN T. CORNER, R.N., Chief Engineer, Portsmouth Dockyard.

Shortly after One o'clock the Meeting was adjourned to the following morning.

The ADJOURNED MEETING was held in the Town Hall, Portsmouth, on Wednesday, 27th July 1892, at Ten o'clock a.m.; Dr. WILLIAM ANDERSON, F.R.S., President, in the chair.

The following Papers were read and discussed:—

"Description of the new Royal Pier at Southampton"; by JAMES LEMON, Esq., J.P., Mayor of Southampton.

"Description of the Portsmouth Sewage Outfall Works"; by Sir FREDERICK BRAMWELL, Bart., D.C.L., LL.D., F.R.S., Past-President.

"Description of the new Floating Bridge between Portsmouth and Gosport"; by Mr. H. GRAHAM HARRIS, of London.

"Description of the Southampton Sewage Precipitation Works and Refuse Destructor"; by Mr. WILLIAM B. G. BENNETT, Borough Engineer and Surveyor.

The remaining Papers announced for reading and discussion were adjourned to a subsequent meeting.

The President proposed the following Votes of Thanks, which were passed with applause :—

To the Mayor of Portsmouth, T. Scott Foster, Esq., J.P., for his kindness in granting the use of the Town Hall for the present Meeting, and for the numerous facilities accorded in connection therewith, for welcoming the Members to Portsmouth.

To the Admiral Superintendent of the Royal Dockyard, the Superintendent of the Royal Clarence Victualling Yard, and the Authorities of the other Works visited by the Members in Portsmouth and the neighbourhood, for the efficient arrangements made for the occasion by their kindness.

To the Mayor of Southampton, James Lemon, Esq., J.P., for his welcome invitation to Southampton, and for his obliging aid in arranging the visit.

To the Chairman and Authorities of the Southampton Docks and of the Union Steamship Company, the Director General of the Ordnance Survey Office, and the Proprietors of the various Works in Southampton and the neighbourhood, who have kindly opened their establishments to the visit of the Members.

To the Directors of the London and South Western Railway, and of the London Brighton and South Coast Railway, for their kindness in providing special railway and steamboat facilities for the Excursions in connection with this Meeting.

To the Honorary Secretaries, Mr. Philip Murch and Mr. George C. Day, for their advice and aid in maturing the arrangements for the Meeting in Portsmouth and the Visit to Southampton.

The Meeting then terminated at One o'clock. The attendance was 187 Members and 107 Visitors.

ON SHIPBUILDING IN PORTSMOUTH DOCKYARD.

BY MR. WILLIAM H. WHITE, C.B., F.R.S.,
DIRECTOR OF NAVAL CONSTRUCTION
AND ASSISTANT CONTROLLER OF THE NAVY.

Portsmouth Dockyard.—The Members of the Institution of Mechanical Engineers will find much to interest them in the Shipbuilding and Engineering departments of Portsmouth Dockyard. At the same time they cannot fail to be impressed by the fact that the establishment exists, and has its scale and character determined, by the circumstance that it has been for centuries, and still remains, our principal naval arsenal. For nearly seven centuries Portsmouth is known to have been one of the head-quarters for the national fleet. In 1212 the sheriff of the county of Southampton was ordered to enclose the King's Docks by a strong wall, and to provide suitable storehouses. A dockyard properly so-called was founded there by Henry VIII, coming next after that at Woolwich, which was closed in 1869: so that Portsmouth yard is now the oldest as well as the most important in existence. (See plans of Dockyard in Plates 32 and 33.)

Growth of Dockyard.—Like the Royal Navy itself, the yard had a modest beginning and a slow development. In 1540 the total area is said to have been about 8 acres. Until nearly the end of last century there was no basin in which ships could lie while completing or undergoing repairs; and the docking accommodation was poor. In the face of much opposition, and largely on the initiative of General Bentham, what was thought a great extension was carried out in the closing years of last century. A basin about 2½ acres in area, and six dry docks adjoining it, were constructed; these are still in useful work. Dwarfed though they are by subsequent extensions, these additions to the dockyard were admirably conceived and executed. Only a few years ago the "Victory," Nelson's flag-ship, was undergoing repairs in the yard, and her presence in the basin and dock gave an object-lesson on the sufficiency of the scale on which the work was done for the utmost demands thought possible a hundred years ago. To Bentham also was largely due the introduction into the dockyards of steam power and mechanical improvements of various kinds, including saw mills, metal mills, pumping appliances, &c. Other able men, amongst them the elder Brunel, aided in this movement; and the famous block-making machinery devised by Brunel dates from this period. At the end of last century the yard had an area of 90 acres, and was capable of building and repairing the largest classes of war ships. For forty years there was little change.

The development of steam propulsion for war ships necessitated the next great change during the period 1843–50. A steam factory with all proper adjuncts, including engine and boiler shops, foundries, &c., was then added to the yard. Another basin 7 acres in area, and four docks capable of receiving the largest ships of the period, were constructed. By 1850 the area of the dockyard had reached 115 acres: an increase of rather more than one hundred acres in three hundred years. So far the development of war-ship building had left the materials and structural arrangements practically the same as early in the century; and the five building slips on the harbour front were still capable of use in the construction of ships of all classes and of the largest size. In the latest of the screw line-of-battle ships a full sail-equipment was retained, and the character and arrangement of the armament were little changed from the days of sailing ships.

With the introduction of armour plating and iron hulls a new era of rapid development began. Sail power was diminished or abandoned. Greater steam power and higher speeds were introduced. The old smooth-bore guns and their simple carriages gave place to rifled guns of greater weight and power, with mountings of novel design. Ships became larger in dimensions, more complicated in structure and equipment. Auxiliary machinery of all kinds—steam, hydraulic, and electrical—was introduced, either to perform operations formerly done by manual power, or to meet new requirements. All these and many other changes, to which reference cannot here be made, have necessitated corresponding changes in the plant and workshops of the dockyards, and have involved large expenditure.

There was a short period, from 1859 to 1869, during which Portsmouth fell behind Chatham and Pembroke in its equipment for iron shipbuilding, and in the scale of its operations. In 1869 however orders were given to lay down at Portsmouth the "Devastation" turret ship, which was the first iron-hulled armoured vessel built here. Since that date the old standing of the yard as a first-class shipbuilding establishment has been fully restored. A complete and extensive plant has been installed in convenient positions near the slips and docks where ships are built. New workshops and stores have been erected for electrical work, torpedoes and torpedo apparatus, gun mountings, &c. In short the dockyard has been made capable of meeting all requirements in the construction and equipment of ships of all sizes and classes.

Shipbuilding in Docks.—In one particular, present conditions compare unfavourably with past. Formerly there were five building slips, three of which were capable of receiving the largest class of ships then built. Now there is only one large slip suitable for building first-class armoured ships, and another suitable for ships of moderate size. On the other hand, docks are available, and are made use of for building the largest ships. The "Centurion" will be seen now building in the dock from which the "Royal Sovereign" was floated out in the presence of Her Majesty the Queen. Experience shows that, when docks are available, it is advantageous in many ways to use them for building ships. The risks and expense of launching are avoided; the work of construction is simplified, especially in dealing with heavy weights, such as armour plates; and economy is rendered possible in various directions. It is recognised of course that a plan such as this, while reasonable enough in a dockyard with ample dock accommodation, is not applicable in private establishments generally. One eminent shipbuilding firm however works on similar lines, having the good fortune to possess docks. Alone among the contractors for the first-class battle-ships now building, Messrs. Laird of Birkenhead are constructing the "Royal Oak" in a dry dock, and will not float her out until she is far advanced towards completion, with hull armour fixed and engines erected on board.

Latest Extension.—The latest and grandest extension of the dockyard to meet the new conditions was commenced rather more than twenty-five years ago, and is now nearly completed. It includes three enclosed basins, having a total area of 50 acres, in which the largest ships can lie afloat; a tidal basin of 10 acres; five large docks, three of which are in use and two are being completed; and two entrance locks, available also as dry docks, Plate 33. The total area of the dockyard has thus been increased to nearly 300 acres. The fitting-out basin is still incomplete. Adjoining it, on Coaling Point, special machinery has been erected for the supply of coal to the fleet. Except at the period when the naval manœuvres are taking place, there may be seen in the basins a large number of vessels, constituting the fleet reserve, ready for service at short notice. Besides these vessels, there may always be seen new ships in various stages of progress; older ships undergoing repairs; and others lying in reserve, until they can be taken in hand. Moored in the harbour are many other vessels, some in commission, others in reserve. The squadron of royal yachts; the armoured ship carrying the flag of the Commander-in-Chief; the numerous gunboats, torpedo boats, and small craft, attached to the gunnery and torpedo schools; as well as the transport and store-ships, are all certain to attract attention. Amongst the reserve ships are many famous vessels, but none more beautiful or more worthy of admiration than the first sea-going ironclad, the "Warrior." The "Victory" remains to show what line-of-battle ships were a hundred years ago. Specimens of the screw three-deckers may also be seen, shorn of their masts and spars, but even thus left impressive by their height and bulk. In fact nowhere else can be seen such an epitome of the history of shipbuilding for the navy, as is constantly on view at Portsmouth.

New Construction.—This brief glance at the history of the dockyard and the range of its work has been taken, in order that a fairer appreciation may be gained of the conditions under which new construction must necessarily be carried on here. Important as this section of the work undoubtedly is at all times, and specially important as it has become in the last three or four years, it still has to be subordinated to the requirements of the fleet. Sudden and exceptional demands from the fleet in commission have to be met even in time of peace: more serious demands must be made when war is being waged. These are the conditions which necessarily govern the scale and character of the establishment: influencing largely the arrangements of workshops, plant, basins, and docks. Great as are the resources of the yard for shipbuilding, and large as are the provisions for engineering and manufactures of special kinds, these are necessarily overshadowed to some extent by arrangements growing out of fleet requirements. Unavoidable difficulties have therefore to be faced in connection with new construction, as compared with the facilities existing in private establishments of recent date, created solely for the construction of ships and machinery. When inspecting the dockyard, Members of the Institution will not fail to remember these essential differences; and they will make due allowance for its antiquity, for its gradual development, and for the enormously rapid changes in naval material and equipment during the last forty years, which have revolutionised the shipbuilding and engineering departments.

New construction falls chiefly upon the shipbuilding department of the yard, of which the Chief Constructor is the principal officer. In recent years the manufacture of new machinery and boilers for ships of the navy has been undertaken to a limited extent in the Chief Engineer's department. New engines and boilers have been made for the "Rupert" now undergoing a thorough repair, their aggregate horse-power with natural draught being 4,500. The engines and boilers for the second-class cruiser "Fox" now building at Portsmouth are also being made in the yard; they will develop 7,000 H.P. with natural draught, and 9,000 H.P. with moderate forced draught. This new course has been taken by the Admiralty for special reasons, which have led also to a certain amount of new machinery and boilers being made at Devonport, Chatham, and Sheerness. The steam factories at the dockyards however are chiefly engaged in the repairing of machinery or boilers, and the making of new boilers. As a rule the machinery and boilers for new ships have been obtained from the leading marine engineers of the country; and this policy is still acted upon.

The engineering departments at Portsmouth and the other yards do a large amount of new work in manufacturing torpedo tubes, and special apparatus of various kinds, which in the public interest it is considered preferable not to have made by contract. Such apparatus is necessarily experimental in the earlier stages; and the bulk of this experimental work, in connection with gunnery as well as with torpedo fittings, has fallen upon Portsmouth, this port being the headquarters for the gunnery and torpedo schools of the fleet. The first electrical workshop under the Admiralty was organised in the Chief Constructor's department of Portsmouth yard, and has now grown into a large establishment of increasing importance; and all the other yards have now at work electrical departments, the staffs of which have mostly been trained at Portsmouth.

Expenditure.—Under the Naval Defence Act of 1889 the expenditure on new construction at Portsmouth, as well as at most of the other home dockyards, has been relatively increased. Large sums necessarily have to be expended also on the repairs and maintenance of the fleet, and on other services. The magnitude of the operations at Portsmouth is indicated by the following figures taken from the Navy Estimates for the current financial year 1892–93.

Estimated Expenditure.

	Labour.	Materials.	Total.
	£	£	£
New construction	255,450	184,380	439,830
Reconstruction, repairs &c. of fleet	134,150	65,750	199,900
Miscellaneous services, manufactures, work for other departments &c.	149,110	65,390	214,500
	£538,710	£315,520	£854,230

It will be seen that nearly half the estimated expenditure on labour and more than half the expenditure on materials are to be devoted to new construction. At present about 7,500 men are employed in Portsmouth dockyard. About 60 per cent. of these are in the shipbuilding department under the Chief Constructor; 25 per cent. are employed under the Chief Engineer; and the remainder are distributed in other departments.

Past Shipbuilding.—In this paper it is impossible to give even a summary of past shipbuilding operations at Portsmouth. The appended Table 1 has therefore been prepared simply to illustrate what has been done during the last thirty-three years in the construction of typical ships; and it is by no means complete. A few explanatory remarks must consequently suffice.

Following the list given in Table 1, the first three vessels, "Prince of Wales," "Duncan," and "Victoria," were splendid examples of the screw line-of-battle ships, with unarmoured wood hulls, which formed the main strength of the navy when the ironclad reconstruction began. The "Victoria" was the largest three-decker launched at Portsmouth. It may be interesting to add that the cost of such a ship, including machinery, but excluding armament and ordnance stores, was about £230,000. This was about three times the cost of a 100-gun line-of-battle ship at the beginning of the century, and about twice that of the 110-gun sailing line-of-battle ships of 1840. On the other hand it is less than one-third the cost of a first-class battle-ship of the present day.

The "Royal Sovereign" of 1864 was launched in 1857 as an unarmoured line-of-battle ship. In 1862 her conversion into an armoured turret-ship of low freeboard was begun on the plans of Captain Cowper Coles, R.N. The upper works were cut down, and the weight saved in this way and by other changes was devoted to armour-plating the hull and the turrets. She was the first armoured ship produced at Portsmouth, and was tried in 1864.

The "Royal Alfred" was another converted ironclad, armed on the broadside principle, with high freeboard and good sail-power.

She was one of a considerable number of such vessels built when a resolute effort was being made to overtake the French in the iron-clad reconstruction.

The "Devastation" laid down in 1869 was the first iron-hulled armoured ship built at Portsmouth. She has other claims to notice. Originally designed as a breastwork monitor with low freeboard to the upper deck, she was altered during construction by adding light superstructures; and as completed was the pioneer of the numerous class of sea-going turret-ships with moderate freeboard and a few heavy guns, which are now on service in the navy. Sail-power was abolished; duplicate engines and twin-screws were trusted to entirely for propulsion. After nearly twenty years' service the "Devastation" may now be seen in hand at Portsmouth for a complete refit, which includes the removal of the old horizontal engines, and the boilers carrying only 30 lbs. steam-pressure, and the substitution of vertical triple-expansion engines, and boilers working at 145 lbs. pressure. The turret armament will be altered also to 10-inch 29-ton breech-loaders. In this way the steaming capability and fighting efficiency of the ship will be much increased. Her iron hull is as sound and strong as ever. The first cost of the "Devastation," excluding guns, ammunition, stores, and incidental charges, was about £360,000.

The "Shah" and "Boadicea" are unarmoured cruisers, with iron hulls wood-sheathed and coppered for service on distant stations where docking facilities do not exist. At the time of their construction they were notable for high speed, powerful armaments, and good sail-power. The "Boadicea" is still on service as flag-ship in the East Indies. The "Shah" is now at Bermuda, employed as a stationary ship. She will long be remembered for her fight with the Peruvian turret-ship "Huascar."

The "Inflexible," commenced in 1874, was the largest ironclad constructed for the navy up to that date; and her displacement was not again equalled until the construction of the "Trafalgar" in 1886–90. She is a turret-ship of moderate freeboard, and was the first of the central-citadel type. Her design grew out of the report of the Committee on Designs appointed by the Admiralty in 1871, with Lord Dufferin as chairman; and great controversy arose respecting the arrangement of the armour. She is remarkable not only for her size, but for other qualities. Her hull carries the thickest armour of any ship in the navy; it is of iron in two thicknesses, which together make up as much as 24 inches. Her turrets carry steel-faced iron armour in two thicknesses amounting together to 17 inches; they contain four 80-ton guns, the heaviest muzzle-loaders afloat. She was the first of the battle-ships to be fitted with internal electric lighting, and with submerged tubes for discharging torpedoes. Many changes and additions were made during her construction, and she was about 7½ years in hand, costing about £812,000, exclusive of armament and establishment charges. She played a prominent part in the bombardment of Alexandria. Having been thoroughly refitted, she is now on service in the Mediterranean.

The "Colossus" is a smaller central-citadel turret-ship, carrying modern breech-loading guns, with steel hull and steel-faced iron armour. Her capabilities as a steamer are much superior to those of the "Inflexible;" but in thickness of armour and weight of armament she is inferior to the larger ship. She is now in the Mediterranean on active service. Her first cost on the same basis as for other ships was about £662,000.

The "Impérieuse" is officially classed as an armoured cruiser, although she compares favourably with many vessels ranking as battle ships in foreign fleets. She possesses high speed, and is an excellent steamer. Her bottom is sheathed with wood and copper. Her armament and armour are moderate in character. At present she is serving as flag-ship on the China station. Her first cost was £544,000.

The "Camperdown" is a barbette ship of the "Admiral" class, carrying a powerful armament, with armour of considerable thickness, but limited in area on the hull. She possesses high speed and a large coal-supply. Respecting the defensive powers of this class also there has been much debate. Apart from this feature, the class has found much favour. Three years' working in the Channel squadron has established their reputation as good steamers and excellent sea-boats, while the arrangement of the armaments is generally approved. The heavy guns, mounted in barbettes, are carried considerably higher above water than in most turret and broadside ships. Although Portsmouth has built only the "Camperdown" of this class, three other vessels built at Pembroke—the "Collingwood," "Howe," and "Anson"—came to Portsmouth in an incomplete state, and were here made ready for service. This fact is mentioned because it illustrates a section of work on new ships, of which a large share has fallen upon Portsmouth, Pembroke yard not being equipped for finishing large ships for sea. The first cost of the "Camperdown" was about £677,000.

The "Trafalgar" and her sister ship the "Nile"—the latter having been built at Pembroke and completed at Portsmouth—are the two most powerful turret-ships afloat. They are of moderate freeboard, like the "Inflexible," but are armoured on a different principle and exceptionally well protected. The armament is very powerful, and includes not only four 67-ton guns carried in two turrets, but also six 4·7-inch quick-firing guns mounted on the broadside in a protected central battery. The "Trafalgar" has an immense advantage over preceding turret-ships in this arrangement of the auxiliary armament. She is an excellent steamer. Her keel was laid in January 1886, she was launched in September 1887, and commissioned for service in the Mediterranean in April 1890. The time occupied in her construction was much less than that spent in building the battle-ships which preceded her. The cost of construction was relatively less; excluding armament and incidental charges it amounted to about £740,000.

"*Royal Sovereign.*"—The "Royal Sovereign" of 1892 is a very different ship in all respects from her predecessor of the same name completed in 1864. So much has of late been said and written respecting the vessel that further description of the leading features in her design is here unnecessary.* It will not be out of place

* Reference may be made to papers by the author "On the Designs for the new Battle Ships," published in the Transactions of the Institution of Naval Architects, 1889, vol. xxx, page 150; and "On the new Programme for Shipbuilding," published in the Proceedings of the Institution of Civil Engineers, 1889, vol. xcviii, page 375.

however to give a few particulars of the remarkable speed with which the work of building was carried on. Instructions to prepare to build were received on 1st May 1889. Full drawings and specifications were furnished on 3rd August. The keel was laid on 30th September 1889, by which time a large amount of preparatory work on framing &c. had been done. On 26th February 1891 the ship was floated out of the dock where she had been built, fully armoured and weighing at the time considerably over 7,000 tons. The engines and boilers were ready, and were at once put on board. By 19th April 1892 she was practically complete, and passed successively through her steam, gunnery, torpedo, and electrical trials, from that date to 22nd April. A month later she was passed into the fleet reserve as ready for commission; on 31st May she was commissioned, and on 8th June she left for Portland, carrying the flag of Vice-Admiral Fairfax in command of the Channel squadron. From the date of laying the keel to the date of commissioning, only two years and eight months elapsed. No similarly rapid construction of a modern first-class battle-ship has been achieved; and the Admiralty officially expressed their satisfaction with the conduct of all concerned.

The magnitude of the work will be better appreciated from the following figures. In the first three months after laying the keel, over 1,000 tons of structural steel &c. were worked into the ship. At the end of six months this had been increased to 2,100 tons. Twelve months from the start about 3,800 tons had been worked into the hull, exclusive of armour which alone represented about 1,400 tons; so that, inclusive of armour, a weekly addition of 100 tons weight had been averaged. Six months later, or a month after the ship was floated, the hull weights exceeded 4,600 tons, and the armour approached 2,900 tons. Two years from laying the keel these weights had become 5,300 tons in the hull, and 2,950 tons in the armour; and when the ship was finished the total weight worked in had risen to 9,800 tons, inclusive of about 3,000 tons of armour.

This rapid progress demanded the most thorough organization and strenuous effort on the part of the officers at Portsmouth. It was rendered possible by the steps taken at the Admiralty in placing orders for materials—especially armour, propelling and other machinery, guns and gun mountings, and hydraulic apparatus—so that there should be no check to progress. Private firms entrusted with this work also deserve much credit for the manner in which, under considerable difficulties, they executed their contracts. The armour plates were ordered before any portion of the ship was to be seen. Within 4½ months from laying the keel, and before the ship was ready to receive armour, the first plate was delivered. When she was floated out of dock nearly all the armour was in place. Messrs. Cammell spared no exertion in effecting this delivery. The machinery contractors, Messrs. Humphrys and Tennant, began delivery of the engines and boilers in November 1890, anticipating considerably their contract dates, so that their work was waiting for the ship when she was floated out. Sir W. G. Armstrong, Mitchell and Co., the designers and makers of the hydraulic gun-mountings &c. for the 67-ton guns, and of the 6-inch quick-firing guns and their mountings, gave equally early delivery, and by so doing contributed greatly to rapid completion. In short, this unprecedented feat in war-ship building was the result of careful forethought and organization, as well as of unremitting and harmonious effort on the part of all concerned.

The "Royal Sovereign" has been built with no important variation from the original design. The intentions of the design have been fulfilled or surpassed in every particular. Rapid construction moreover has been conducive to economy, and the vessel has been completed for about £35,000 within the estimate for dockyard work. The original estimate of cost, excluding guns and ammunition as well as incidental charges, was £780,000, which was subsequently raised to £790,000; and the actual cost will be under £760,000. Guns, ammunition and reserves, and the consumable stores carried, will bring the total value of the ship as she goes to sea up to a million sterling.

Ships now in progress.—The vessels now building or completing at Portsmouth include specimens of four types introduced into the navy under the Naval Defence Act of 1889. Three of these are exemplified by vessels building in the yard; the fourth is represented by the second-class cruiser "Intrepid," built on the Clyde and now completing here for service. In Table 2 appended the principal particulars of these vessels are given, together with corresponding information for the torpedo-depôt ship "Vulcan," which was also built at Portsmouth.

"*Centurion.*"—The most important ship is the "Centurion" battleship. She was laid down at the end of March 1891, and is now well advanced towards the condition for floating out, with hull-armour practically complete. Figs. 1, 2, and 7, Plates 26, 27, and 30, illustrate the general arrangements of armour, armament, &c. Her principal armament consists of four 10-inch 29-ton guns, mounted in pairs in two armoured barbettes, with large arcs of horizontal training. All these guns can be fought on either broadside. Manual power can be employed for all the operations of loading or working these guns. Steam power is also provided for training the guns and hoisting ammunition. The mountings are being made by Sir Joseph Whitworth and Co. Between the barbettes are mounted ten 4·7-inch quick-firing guns; and seventeen smaller guns, 6-pounders and 3-pounders, are also carried. There are seven stations for ejecting torpedoes; two of these are under water, four on the broadside behind light armour, and one at the stern.

A belt of armour, having a maximum thickness of 12 inches, protects the water-line region for about 200 feet of the length. This belt armour is associated with a protective deck 2 to 2½ inches in thickness; and is completed at the extremities by armoured bulkheads crossing the ship. A strong protective under-water deck runs from these bulkheads to the bow and stern respectively. Above the belt armour the broadside is protected by light armour, equivalent to a total thickness of 4 inches of steel, up to the height of the main deck, which is about 9¼ feet above water. Protection is also given to the quick-firing guns and their crews, by revolving shields or fixed casemates.

The hull is of steel, built on the system usually followed in the navy, and is minutely subdivided. Steel armour made by Messrs.

Vickers is fitted on the hull and the barbettes. Wood and copper sheathing are fitted on the bottom, to prevent fouling and consequent loss of speed after long periods afloat. The ship is thus specially adapted for service on foreign stations. She is of light draught as compared with most battle-ships of her size; and will be able to pass through the Suez Canal with a large quantity of coal on board, being in all other respects fully laden.

Relatively high speed is aimed at in the design; and larger and heavier boilers than have hitherto been fitted in battle-ships have been provided, so as to secure the maintenance of speed on long-distance steaming. The Greenock Foundry Co. are the contractors for the propelling machinery.

The estimated cost of the "Centurion," exclusive of armament and incidental charges, is about £550,000. She will probably be completed in the autumn of next year.

"Royal Arthur" and "Crescent."—The "Royal Arthur" and "Crescent" are two of the nine first-class cruisers provided for under the Naval Defence Act. In nearly all respects these vessels are similar to the "Edgar," of which the steam trials have been made most successfully, the ship realizing a maximum speed of nearly 21 knots. Like the "Centurion," these cruisers will be wood-sheathed and coppered. The "Royal Arthur" was laid down in January 1890, and is to be completed this year. The "Crescent" was begun in October 1890, and will be completed in the next financial year.

Figs. 3, 4, and 8, Plates 27, 28, and 30, illustrate the general arrangements of the "Royal Arthur." She belongs to the protective-deck type of cruiser. As indicated in the cross section, Fig. 8, the hold-spaces containing engines, boilers, magazines, &c., are covered in by a curved steel deck, having a maximum thickness of 5 inches on the slopes, and $2\frac{1}{2}$ inches on the horizontal or nearly horizontal portion. This deck runs from stem to stern, and is supplemented by coal-bunker protection. Vertical engines are fitted to drive the twin-screws; and in order to secure good working conditions for continuous steaming, the cylinders are allowed to come above the level of the protective deck, and are furnished with special armour protection.

The armament is exceedingly powerful. It includes one 9·2-inch 22-ton gun, mounted aft on the upper deck, where it can be fought on either broadside or as a stern-chaser; and twelve 6-inch quick-firing guns, of which four are carried on the main deck, six on the upper deck, and two on the high forecastle, which stretches back more than 100 feet from the bow. Besides these there are seventeen smaller quick-firing guns, 6-pounders and 3-pounders, mounted on the decks and other stations. All these guns and their crews have protection given them either by steel shields revolving with the guns or by fixed armour.

The "Royal Arthur" and "Crescent" differ from the other first-class cruisers in the possession of the high forecastle with two 6-inch guns mounted thereon. In the other vessels there is no forecastle, the freeboard forward is about 7½ feet less, and a single 9·2-inch gun forms the bow armament.

The estimated cost of the "Royal Arthur," exclusive of armament and incidental charges, is rather less than £350,000.

"*Fox.*"—Another type of protected cruiser is presented in the "Fox," which is as yet in an early stage of construction. Figs. 5, 6, and 9, Plates 29 and 30, illustrate her general arrangements. In size, armament, and protection, this cruiser is much inferior to the "Royal Arthur"; the speed is about the same. The estimated cost, exclusive as in the "Royal Arthur" of armament and incidental charges, is rather more than £200,000. These vessels are of high freeboard throughout their entire length, and carry all their guns on the upper deck, at a great height above water. These features will add largely to their habitability, fighting efficiency, and power of maintaining speed in rough water. There are eight vessels of this type in hand, all to be wood-sheathed and coppered.

"*Intrepid.*"—The "Intrepid" is a somewhat smaller type of cruiser than the "Fox," costing about 90 per cent. of the latter. In protection the two classes are identical: in armament the "Fox" is somewhat superior. In speed the difference will be small. The most important distinction is that in the "Intrepid" a long poop and forecastle are fitted, with an upper deck or waist between them at a moderate height above water, instead of the flush upper deck of the "Fox." Twenty-one sister ships to the "Intrepid" have been built under the Naval Defence Act, and most of them are now complete. The "Latona," the leading vessel of the class, has been tried in an experimental cruise to the Mediterranean, as well as in the manœuvres of 1891, and has been most favourably reported on.

"*Vulcan*."—The "Vulcan" is a vessel quite unique in character. Her special mission is to carry eight torpedo-boats of the second class; and for this purpose she is fitted with powerful hydraulic cranes capable of lifting boats weighing 20 tons. She is equipped with a workshop containing a large amount of machinery for use in repairing torpedo-boats, torpedoes, &c. A very complete laboratory is provided on board, for use in connection with torpedo and mining operations. The ship also carries a full equipment of mines and stores; and is fitted with specimens of torpedo-ejecting apparatus both above and below water. Besides all this special equipment, she carries a fair armament of guns, including eight 4·7-inch quick-firers and twelve 3-pounders. Her vitals are defended by a strong steel deck, having a maximum thickness of 5 inches. She is in fact a lightly armed protective-deck cruiser of high speed, capable of defending herself against all but the most powerful cruisers or battle-ships; and in addition she is a torpedo-boat carrier, a floating factory, a torpedo-practice ship, and a depôt for mining stores. Her trials at sea have been deferred in consequence of difficulties which arose in connection with leakages of boiler tubes. Experiments are now in progress in other ships, which it is hoped will enable these difficulties to be surmounted. From the steam trials already made it is certain that, if this hope is realized, the estimated maximum speed of 20 knots will be attained with ease. The cost of the "Vulcan" complete with all her special machinery, but excluding guns and incidental charges, is rather less than £340,000.

Reconstruction and Repairs.—The work of this kind now in hand at Portsmouth, which there is an opportunity of inspecting at the present time, may be illustrated by the "Devastation," respecting which particulars have already been given, and also by the "Hercules," the "Sultan," and the "Rupert." The "Hercules" is now approaching completion, having been fitted with vertical triple-expansion engines and high-pressure boilers, while her sails have been removed and rig altered. Her main armament remains unchanged; but quick-firing guns have been added, and many other changes made to increase her fighting efficiency. The "Sultan," which is a very similar ship to the "Hercules," ran aground at Malta, and afterwards foundered; but having been raised she was brought to England. Her repairs are to be commenced in the present year, and her refit will probably follow lines similar in many respects to those of the "Hercules." For the "Rupert" turret-ship, built with special reference to her use as a ram, new vertical triple-expansion engines have been made in the dockyard; and in the turrets breech-loading guns are to be substituted for muzzle-loaders. These and similar works of reconstruction are bringing once more into line, with greater fighting powers, ships which have already done good service.

Although this paper has extended to considerable length, it is even now a most incomplete sketch of the work and capabilities of Portsmouth dockyard. It may however serve a useful purpose if it facilitates the inspection of the establishment on the occasion of the present meeting of the Institution. In the published Proceedings also it may have some permanent interest as a record, however imperfect, of the shipbuilding operations carried out in England's greatest naval arsenal.

TABLES 1 AND 2.

TABLE 1. *Particulars of typical Ships built*

Name of Ship.	Dates of Commencement and Completion.	Material of Hull.	Class.	Length.		Breadth.		Displacement.
				Ft.	Ins.	Ft.	Ins.	Tons.
Prince of Wales	— / 4 Dec. 61	Wood	1st-rate screw Battle-ship (3 decker)	252	0	60	2	6,900
Duncan	— / 1 Feb. 64	Wood	2nd-rate screw Battle-ship (2 decker)	252	0	58	1	4,555
Victoria	— / 24 Oct. 64	Wood	1st-rate screw Battle-ship (3 decker)	260	0	60	1	6,960
Royal Sovereign	— / Aug. 64	Wood, iron-cased	4th-rate screw Turret-ship	240	7	62	2	4,965
Royal Alfred	1 Dec. 59 / 22 Mar. 67	Wood, iron-cased	3rd-rate screw Battle-ship.	273	0	58	7	6,720
Devastation	12 Nov. 69 / 19 Apl. 73	Iron	Twin-screw armoured 2nd-class Battle-ship.	285	0	62	3	9,330
Shah	8 Aug. 70 / 8 Dec. 76	Iron, wood-sheathed	Screw 2nd-class Cruiser	334	8	52	0	6,250
Boadicea	20 Jan. 73 / 22 May 77	Do.	Do.	280	0	45	0	4,140
Inflexible	24 Feb. 74 / 18 Oct. 81	Iron	Twin-screw armoured 2nd-class Battle-ship	320	0	75	0	11,880
Colossus	5 June 79 / 31 Oct. 85	Steel	Do.	325	0	68	0	9,420
Impérieuse	10 Aug. 81 / 7 Sep. 86	Steel	Twin-screw armoured 1st-class Cruiser.	315	0	62	1	8,400
Camperdown	18 Dec. 82 / 2 Aug. 89	Steel	Twin-screw armoured 1st-class Battle-ship	330	0	68	6	10,600
Trafalgar	18 Jan. 86 / 26 Apl. 90	Steel	Do.	345	0	73	0	11,940
Royal Sovereign	30 Sep. 89 / 31 May 92	Steel	Do.	380	0	75	0	14,150

at Portsmouth Dockyard 1859 *to* 1892. TABLE 1.

Indicated Horse-Power.	Speed.	ARMAMENT. Number and Description of Guns. Pr.=Pounder. Q.F.=Quick-firing. M.L.R.=Muzzle-loading, rifled. B.L.R.=Breech-loading, rifled.		ARMOUR. Maximum Thickness. Hull.	Turrets or Barbettes.
I.H.P.	Knots.	Number.	Description.	Inches.	Inches.
3,352	12	115 { 1 10 16 88	110-pr. Armstrong 82 cwts. 40-pr. " 35 " 8-inch 65 cwts. M.L. 32-pr. 58 " M.L.	Nil.	Nil.
2,826	12½	81 { 1 12 34 34	110-pr. Armstrong 82 cwts. 40-pr. " 35 " 8-inch 65 cwts. M.L. 32-pr. 58 " M L.	Do.	Do.
4,200	12¼	102 { 2 10 60 30	110-pr. Armstrong 82 cwts. 40-pr. " 35 " 8-inch 65 cwts. M.L. 32-pr. 58 " M.L.	Do.	Do.
2,460	11	5	9-inch 12 tons M.L R. in four turrets	5½ inches iron	10 inches iron in 2 thicknesses
3,430	12¼	18 { 10 8	9-inch 12 tons M.L R. 7-inch 6½ " M.L.R.	6 inches iron	Nil.
10,000	13¾	4	12-inch 35 tons M.L.R.	12 inches iron	14 inches iron in 2 thicknesses
7,480	16½	26 { 2 16 8	9-inch 12 tons M.L.R. 7-inch 6½ " M.L.R. 64-pr. 71 cwts. M.L.R.	Nil.	Nil.
5,130	14⅞	14 { 12 2	7-inch 6½ tons M.L.R. 64-pr. 71 cwts. M.L.R.	Do.	Do.
8,010	13¾	12 { 4 8	16-inch 80 tons M.L.R. 4-inch 13 cwts. B.L.R.	24 inches iron in 2 thicknesses	9 inches outside, steel-faced; 8 ins. inside, iron
7,500	15½	9 { 4 5	12-inch 43 tons B.L.R. 6-inch 5 tons B.L.R.	18 inches, steel-faced	16 inches, steel-faced
10,000	17	14 { 4 10	9·2-inch 24 tons B.L.R. 6-inch 5 tons B.L.R.	10 inches, steel-faced	8 inches, steel-faced
11,500	16¾	10 { 4 6	13·5-inch 67 tons B.L.R. 6-inch 5 tons B.L.R.	18 inches, steel-faced	14 inches, steel-faced
12,000	16¾	10 { 4 6	13·5-inch 67 tons B.L.R. 4 7-inch 2 tons B.L.R.,Q.F.	20 inches, steel-faced	18 inches, steel-faced
13,000	18	14 { 4 10	13·5-inch 67 tons B.L.R. 6-inch 6½ tons B.L.R.,Q.F.	18 inches, steel-faced	17 inches, steel-faced

TABLE 2. *Particulars of Ships building or*

Name of Ship.	Date of Commencement.	Material of Hull.	Class.	Length.		Breadth.		Displacement.
				Ft.	Ins.	Ft.	Ins.	Tons.
Centurion	30 Mar. 91	Steel, wood-sheathed	Twin-screw armoured 1st-class Battle-ship	360	0	70	0	10,500
Royal Arthur	20 Jan. 90	Steel, wood-sheathed	Twin-screw 1st-class Cruisers	360	0	61	0	7,700
Crescent	30 Oct. 90							
Fox	11 Jan. 92	Steel, wood-sheathed	Twin-screw 2nd-class Cruiser	320	0	49	6	4,360
Vulcan	18 Jan. 88	Steel	Twin-screw Torpedo-depôt ship	350	0	58	0	6,620
Intrepid	6 Sep. 89	Steel, wood-sheathed	Twin-screw 2nd-class Cruisers	300	0	43	8	3,600
Indefatigable	6 Sep. 89							

completing at Portsmouth in June 1892. TABLE

Indicated Horse-Power.	Speed.	ARMAMENT. Number and Description of Guns. Q.F. = Quick-firing. B.L.R. = Breech-loading, rifled.		ARM Maximum Hull.	Tu Ba
I.H.P.	Knots.	Number.	Description.	Inches.	Inches.
13,000	18¼	14 { 4 10	10-inch 29 tons B.L.R. 4·7 inch 2 „ B.L.R.,Q.F.	12 inches steel	9 inches st
12,000	19½	13 { 1 12	9·2-inch 22 tons B.L.R. 6-inch 6½ tons B.L.R.,Q.F.	6 ins. protecting cylinders, steel-faced; 5 ins. deck, steel	Nil.
9,000	19½	10 { 2 8	6-inch 6½ tons B.L.R.,Q.F. 4·7-inch 2 „ B.L.R.,Q.F.	5 ins. protecting cylinders, steel-faced; 2 ins. deck, steel	Do.
12,000	20	8	4·7-inch 2 tons B.L.R.,Q.F.	6 ins. protecting cylinders, steel-faced; 5 ins. deck, steel	Do.
9,000	19¾	8 { 2 6	6-inch 5 tons B.L.R. 4·7-inch 2 tons B.L.R.,Q.F.	5 ins. protecting cylinders, steel-faced; 2 ins. deck, steel	Do.

Discussion.

Mr. White said the purpose of the paper was expressed in its concluding paragraph. He had feared lest the members of the Institution in passing through the dockyard, unless they had something in the nature of a guide, might fail to realise the great diversity and extent of the operations there carried on, because of the very magnificence of the establishment. It had also occurred to him that many members who could not be present might be glad to have some account of what would be seen in the visit to the dockyard this afternoon under the guidance of Admiral Fane and the officers of the establishment. At the present time the ships of the fleet had gone from Portsmouth to take part in the naval manœuvres; otherwise the members would have been much more strongly impressed by their presence with the enormous reserve of power which England now possessed in ships completed and ready for service at the shortest notice, and of which the fleet reserve in Portsmouth dockyard was only one illustration. In the three great naval ports of the country there existed a numerous and powerful reserve of the most modern ships, ready for service, in addition to the large fleet which was in permanent commission. While specimens might be seen of all or nearly all the classes of vessels building under the Naval Defence Act, there would at the present time be no specimen visible in the dockyard of the class of which the "Royal Sovereign" was the first completed ship. That vessel he was sure would have been an interesting sight if she had been in port; but Portsmouth dockyard had proceeded with such extreme rapidity in her construction that, at a time when according to all precedent she would have still been here and open to inspection, she was already at sea, flying the flag of the Admiral commanding the Channel Squadron.

The President considered that the members were greatly indebted to Mr. White for the trouble he had taken to compile this paper; and he was sure that its interest would extend much further than the body of Mechanical Engineers now assembled in Portsmouth.

It afforded a proof that, with the best possible appliances, and with what he might call a marvellous combination of talent and good will, it had taken nearly three years to build the latest ironclad. It appeared to him that wars were getting shorter, and that the time occupied in preparing warlike materials was getting longer. The consequence was that it had become indispensable for a nation which desired peace to have not only an abundant reserve of warlike stores, but also vast powers of producing them; and it must be content to pay for the security thus obtained. In Nelson's despatches, written during the year and three-quarters 1803–5 when he was watching Toulon in the hope that the French fleet would come out and fight, he mentioned that frigates were sent from time to time to look into the harbour and observe what was doing, and they reported that various war vessels were in process of building and repairing; and by the time the fleet did at last come out, the French had managed to build and equip some new ships, and to send out a force superior to his. That was a thing which could not be done now-a-days. In the first place, no blockade was going to last so long as nearly two years; and in the next, two years would not suffice to produce a single ironclad or a single powerful gun. The only time therefore for building ships and guns was during a time of peace: so that, if ever war should break out, Britain might have a reserve such as no other nation or combination of nations could bring against her. The present paper would enable the members to understand what they would see this afternoon under the guidance of the officers who had kindly undertaken to conduct them over the dockyard; and he was sure they would all desire to accord a hearty vote of thanks to Mr. White for the preparatory information which he had given them.

ON THE APPLICATIONS OF ELECTRICITY IN THE ROYAL DOCKYARDS AND NAVY.

By Mr. HENRY E. DEADMAN,
Chief Constructor at the Admiralty, late of Portsmouth Dockyard.

This is an age of Electricity; and the fact is well exemplified on board a modern battle-ship of the Royal Navy. Some of the applications of electricity in the naval service do not come within the scope of ordinary mercantile practice; and in the case of those which do, the naval experience is of a special kind, and may be of some value to those outside the service. A statement therefore of naval methods and practice up to date may not be without interest and instruction to the members of this Institution. At the present time visitors to Portsmouth dockyard will probably be surprised, if not disappointed, at the somewhat primitive character of the workshops devoted to the electrical testing and construction. This branch of work was commenced in a very small way about seventeen years ago, and found a temporary home in a small shed which had been previously used for a different purpose. As the work from time to time has grown, this shed has been enlarged as far as available space would permit. Designs have been prepared for a more substantial and suitable building, in the hope that in time the Admiralty will be able to devote sufficient money to providing a proper habitation for this important branch of dockyard work.

The larger proportion of the plant required for electric installations in the navy is not made in the dockyard. Formerly all dynamos and their engines were obtained entirely by contract; but recently five sets of these have been designed and constructed in the dockyard. After patterns of the smaller fittings have been fixed, the supply of these also in bulk has generally been obtained by contract. There is however a great deal of work left to be done in the dockyard, in testing and installing these fittings in the ships, in carrying out repairs, and in devising and preparing patterns of new fittings to meet the constantly growing requirements of the navy. There is also some work, which for obvious reasons is confined to the royal workshops.

The applications of electricity in the navy may be dealt with under the following heads:—(1) search lights; (2) internal lighting of ships, including temporary installations in ships building and repairing; (3) torpedo and gun circuits; (4) electric communication; (5) other applications.

1.—SEARCH LIGHTS.

The introduction of the search light, without which no modern war-ship or torpedo boat would be considered complete, dates from 1876; and the first vessel in the navy fitted with a search-light apparatus was the "Minotaur." Some experiments had been carried out in the previous year by Messrs. Wilde and Co., of Manchester, on board the gunboat "Comet;" and these proved so far satisfactory that a complete plant was ordered and fitted on board the "Minotaur." The dynamo employed was one of the alternating-current type with 32 magnets, and it was driven at about 400 revolutions by a belt from an auxiliary pumping engine. The projector was of a primitive type, and pedestals were fixed in three different places, from any one of which the same projector could be used. It was fitted with a parabolic reflector and with dioptric and diverging lenses. A diaphragm was also provided for enabling flashing signals to be made. The lamp employed was Wilde's, and was a vertical one. The carbon rods were square in section, and their holders were made to slide on two pillars, and were moved up and down by a central pillar with a screw-thread cut in it. The lamp was hand regulated, and one lead was put to earth. The "Téméraire" in the same year was next fitted in a similar manner, with the exception that a Mangin projector was introduced, fitted with Wilde's lamp, lens &c. In the next year 1877 the "Dreadnought," "Neptune," and several other vessels were fitted with the same class of apparatus.

In 1878 direct driving was first introduced. In this year Messrs. Wilde and Co. coupled their machines to engines made by Brotherhood and by Chadwick of Manchester. The dynamo was also improved, so as to be able to maintain two arc search-lights at one time. In the same year also the "Triumph" was fitted by Messrs. Siemens Brothers with a search-light installation. The dynamos were four in number, of horizontal type, arranged in two pairs, and each connected in parallel to one circuit. A switchboard was fitted, for enabling any two of these dynamos to be coupled up together on any circuit. It consisted of a wood base with two sets of bars at right angles to each other, one set on the top of the base, and the other underneath. One set of bars was connected to the dynamos and the other to the circuits. At their intersections these bars could be connected as required, by means of suitable plugs. The projector used was a Siemens holophote, which was heavy and clumsy, being made largely of cast-iron. It was fitted with diverging and dioptric lenses; of these the latter was composed of concentric glass rings of triangular section, held in a metal frame. It had also a flashing arrangement. The lamp used was one of the Siemens self-regulating type, and had a small mirror fixed to it. It was complicated, and frequently got out of order. The carbons were apt to stick together, and the lamp was sensitive to any slight variation in speed of dynamos. This lamp did not come into general use.

Subsequently the Gramme dynamo was introduced for search lighting by Messrs. Sautter Lemonnier of Paris; and in 1881 the "Inflexible" was fitted with this dynamo. In this installation the Mangin projector was used, with a Mangin mirror instead of the dioptric lens. The lamp was hand regulated and inclined. The Gramme dynamo has since been superseded by others of later make; but the Mangin mirror and the inclined hand-lamp survived, and became the standard service fittings.

Naval-service Projector.—Out of these earlier attempts the modern naval-service projector has gradually been evolved, which has now become familiar to most persons. It consists of a cylindrical lantern made of very thin steel, with a silvered glass parabolic mirror at the back end, in the focus of which an arc light is produced between two carbon sticks, held in what is known as an inclined hand-lamp, the carbons standing at an angle of about 70° to the axis of the mirror. The feeding of the carbons is not automatic, but is accomplished by hand; and the lantern is so suspended that it has motion on its pedestal through the whole circle in azimuth and through about 60° in altitude. During its motion the electrical connections are kept up by suitable rubbing contacts. A switch is fitted in the pedestal for switching the current on and off. The circuit wires, main and return, are brought from the main switchboard to the switch on the pedestal, thence to the rubbing contacts, and finally pass close together up one of the hollow arms of the lantern support, and through the trunnion-bearing to two springs at the bottom of the lamp box. The lamp when put in makes contact with these springs, and completes the circuit through the carbons. Two sizes of these projectors are used in the naval service, a larger with a 24-inch mirror for ships, and a smaller with a 20-inch mirror for torpedo boats. The parabolic mirror in these projectors converges the rays into a powerful and penetrating cylindrical beam of light. At the front end of the lantern, diverging lenses can be attached, for spreading the beam over a larger surface when desired; or a flashing screen can be affixed for signalling purposes.

The two most important parts of these projectors for the production of a good and steady beam are of course the reflecting mirror and the carbons. For the supply of these articles the naval service was until recently dependent entirely upon French manufacturing firms. The Admiralty, fully alive to the dilemma in which this country might under certain circumstances be placed, have sought to induce English manufacturers to enter into competition with the French firms for the supply of these articles; and everything of this kind which has been offered with the slightest promise of success has been fully and carefully tested at Portsmouth. The English manufacturers have no doubt had no easy task to accomplish; for, disregarding cost, the French mirrors and carbons left nothing to be desired, and are still the standards to which all others are referred.

Mirrors.—As regards mirrors the qualities required are three:—first, that they shall project a cylindrical, sharply defined, and homogeneous beam of light of great intensity and penetrative power; second, that when in operation they shall not be liable to crack by contact with water, in the form either of rain or of sea-water spray, or by a blast of cold air; third, that they shall resist the concussion caused by the discharge of heavy guns. These qualities should be combined at a moderate cost if possible, although efficiency is the primary consideration. Thus far mirrors have been submitted for trial by six different English makers, and by others specimens are being prepared for future trial. From among these at least one successful English specimen has been obtained, which compares in all points favourably with the French production; and one half of the supply for the navy will this year be ordered from the firm producing this specimen.

Carbons.—With regard to English carbons, the results were for a long time disheartening; even after several attempts by the same firm, none of the specimens submitted for test approached in efficiency those of French make. The three qualities required in good carbons are that they shall maintain a steady arc without flaming or excessive hissing; that they shall be perfectly pure and homogeneous in structure, and shall preserve a well formed crater on the positive carbon and a well formed point on the negative; and that the waste shall be steady and uniform, without cracking or crumbling. A certain amount of success has already crowned the persevering efforts of English manufacturers. Although the English carbons are not even yet fully equal to those previously obtained from France, it is hoped they will soon come sufficiently near that standard to make them acceptable for use in the navy.

Projector requirements.—The present service projector requires at least one man close to it, to direct the beam of light in any desired direction, and also to feed and adjust the carbons as necessary. This is a disadvantage, because the position of a search light is evidently one affording a good mark for an enemy's fire, and is a bad position for observing the object illuminated; and it would be an advantage if the projector could be entirely manipulated from a protected position at some distance—for instance from the ordinary conning tower. For this purpose two requisites are necessary: first, a good automatic lamp; and second, an efficient motor for giving the necessary movements to the projector. It is to these two improvements that attention is now being given.

A good practical automatic lamp is the first necessity; without this a man must be stationed at the projector to feed the carbons, and being there he can also direct the beam of light. Several automatic lamps have been tested, but none have as yet been adopted for general use. One lamp, in which the carbons are regulated by a small electric motor in a shunt circuit, is promising; but in its final form it has not yet been returned from the makers. The field however is still open for invention in this direction.

Experiments have also been made with an electric motor attached to the service projector, by which the movements of the latter can be directed and controlled from a distance. The results are highly satisfactory, but the invention is not likely to be brought largely into use until it can be combined with a satisfactory automatic lamp. Recently a wheeled carriage has been introduced for using a ship's projector on shore.

2.—Internal Lighting of Ships.

Electricity is at the present day largely resorted to for the purpose of lighting internally the vessels of the navy; all the larger vessels, such as battle-ships and first and second-class cruisers, are now so lighted. Moreover this mode of lighting is not restricted to the habitable portions of the ship, but is carried out to the complete exclusion of other modes, although the latter may be fitted as a reserve. Thus the electric lighting is extended to machinery spaces, coal bunkers, magazines, shell rooms, store rooms, gun quarters, &c., as well as to the illumination of compass cards, telegraph dials, bow and mast-head and signal lanterns, semaphores, &c. Clusters of glow-lamps beneath an enamelled metal reflector are also employed for lighting the upper deck, when coaling or other operations are being carried on at night. In a large battle-ship like the "Royal Sovereign" there would be about 800 of these glow-lamps, necessitating for this system alone about 8 miles of electric leads, which are equivalent to something like 155 miles of copper wire of varying sizes, principally No. 20 legal standard wire-gauge or 0·036 inch diameter.

Progress.—The first installation of internal lighting in the navy was carried out on board the "Inflexible" in 1881 by the Anglo-American Brush Co. It was a combined system of arc and glow-lamps. The dynamos used were of the Brush type, of which the first specimen brought to this country was purchased by the Admiralty, and is still in use in Portsmouth dockyard. Each machine was capable of maintaining sixteen Brush arc-lamps of 2,000 candle-power each. The lamps had double sets of carbons, each pair burning eight hours. These arc-lamps were switched on and off by a switch opening and closing a shunt circuit of small resistance. With this plan it is evident a lamp could not be safely handled while the current was on, even if the light was switched off, because the lamp itself still formed a part of the circuit. A switch was accordingly devised in the dockyard, by means of which the lamp could be cut completely out of the circuit. The Swan glow-lamps were fitted in sets of eighteen in series, each set being placed in parallel between the main circuits of the arc-lamps. Each glow-lamp was fitted with an automatic cut-out, bringing into the circuit a resistance equal to about that of the lamp, in case the latter failed. This system now no longer exists in the "Inflexible," an installation on modern methods having been substituted for it.

The five Indian troopships were next fitted throughout with glow-lamps by the Edison Company, who used the Edison Hopkinson type of dynamo giving a current of 180 ampères at 110 volts. The "Polyphemus" was also entirely fitted by Messrs. Siemens, and the "Colossus" by the Brush Co., the dynamos in the latter being of the Victoria Brush type. The "Polyphemus" was the only ship in the navy fitted on the single-wire system. Afterwards the adoption of internal lighting became general; all new vessels except the smaller cruisers were fitted, as were also the previous vessels on the first opportunity of their coming in for repairs and refit. There are now in the navy about 300 vessels either fitted or being fitted with electric lighting, for search lights or internal lighting or both.

Difficulties.—The earlier installations, although quite satisfactory so far as the illuminating effect was concerned, were certainly not satisfactory in regard to the endurance of the plant; in fact they were a constant source of trouble, both the dynamos and the leads requiring frequent repairs or renewals. These troubles arose partly from faulty construction, and partly from the conditions necessarily existing on board ships of war, such as the necessity for placing the dynamos well down in the ship, and under protection from shot: which relegated them either to the engine room with its high temperature and moist atmosphere, or to some other confined and ill ventilated spot. Under these conditions the insulation soon suffered, and the machine quickly got out of action. The leads throughout the ship suffered too from access of salt water, which caused much trouble by destroying the insulation, and producing short circuiting, thereby frequently setting on fire the wood casing in which the leads were laid, or any other inflammable material in the neighbourhood.

Remedies.—Nearly all these difficulties have in the course of time been surmounted. While it is still necessary to place dynamos under protection for use in action, it has now become the practice to fit an additional dynamo, capable of running all the lights generally required at one time, and to place it with its motor in an open space between decks, or on the upper deck suitably sheltered. This is known as the peace or daylight dynamo. The introduction of lead-coated wires marked an epoch in ship-lighting; and these wires are entirely used in all the later installations, and are taking the place of the old leads as fast as the latter require renewal. These cables dispense entirely with the wood casings previously used, resulting in greater simplicity of fitting as well as increased efficiency. The wood casings are still used in America even with the lead-coated cables; but in the writer's opinion, while adding largely to the cost of the installation, they are likely to detract from its efficiency, as they form such convenient receptacles for water to lodge in.

Dynamos.—The Admiralty have not adhered to any one particular kind of dynamo; for although those of the Siemens type predominate in the navy at present, nearly all the well known makers are represented—such as Latimer Clark, Muirhead & Co., the Electric Construction Corporation, the Silvertown India-rubber Co., Goolden & Co., Crompton & Co., and Parsons & Co. The experience gained in the testing and use of these machines of different types was of immense benefit in enabling those points to be discovered in which dynamos are most likely to fail, and in suggesting the direction which improvements should take. The weak points in most of the machines supplied were imperfect insulation, owing to want of sufficient care in building; and the excessive internal heating of the armature. In some cases the section of copper used was too small, resulting in the heating of the dynamo and the rapid destruction of the insulation. By direction of the Admiralty a design was prepared by Mr. Lane, electrician of Portsmouth dockyard, for a machine to give an output of 400 ampères at 80 volts, which should embody the experience already gained, and should be free from the faults previously mentioned. Five machines have been built from this design, two of them for the "Rupert," a vessel now under reconstruction. No opportunity has yet occurred for testing them under the conditions of sea service; but two have been used for about a year under full load, as temporary lighting plant for ships building, &c. At times too they have been worked for long periods above full load, from 420 to 430 ampères; and they have always given perfect satisfaction. They are driven at about 330 revolutions a minute by open vertical compound engines of 56 I.H.P., designed and constructed in the Chief Engineer's department of the dockyard. The total weight of the dynamo and engine with bedplate is about 5½ tons.

Other Fittings.—As regards the other fittings required to complete a successful installation—such as switches, fusible cut-outs, lamp-holders, shades and their holders, &c.—patterns have been adopted or devised in accordance with accumulated experience, to meet the special requirements of the navy: these are simplicity, combined with strength to withstand the rough usage necessarily experienced on board a fighting ship. The guiding principle in the present installations is to make every part of the circuit inaccessible to water, by employing lead-coated cables, and by enclosing in metallic water-tight boxes all the principal switches and cut-outs, and by effecting in all cases in the interior of water-tight distributing boxes the subdivision of the mains, which themselves pass into the boxes, and their subdivisions out of the boxes, through water-tight stuffing-glands.

Switchboard.—With several dynamos and a combined installation of incandescent and search lighting, a good switchboard is a necessity. The one generally used in the navy at present is susceptible of improvement; and a new switchboard has just been designed and made by Mr. Lane, the dockyard electrician, which promises to fulfil all the requirements.

Temporary Installations. — The employment of temporary installations of electric lighting in ships building and repairing has been a marked feature of recent practice in Portsmouth dockyard. Probably more has been done here in this respect than has been attempted in any other royal dockyard or private shipbuilding yard in the country, although these are now finding it to their advantage to follow the lead. So highly satisfactory have been the results of the experiments in this direction, that, with the exception only of the smallest ships, the electric light is now installed as a matter of course on board every new vessel at an early stage of its construction, and also in all vessels where repairs of any magnitude are to be carried out. The character of these temporary installations varies with the circumstances of each case. Where vessels are building in the docks adjacent to the electric shop, it is of course more convenient and economical to supply the current from a dynamo in the shop itself, the leads being carried to the ship on temporary wooden poles put up for the purpose. Vessels building on the slips, which are too far from the electric shop to convey the current to them in an economical manner, are served by a dynamo and motor placed at some convenient spot near the ship. After a vessel is launched, and generally for all vessels afloat, the most satisfactory method is to place a dynamo and engine with a steam boiler in the ship itself, generally in a temporary wooden house on the upper deck. By this means the light is maintained without interruption during the many movements of the vessel about the basins. The leads, lamps, &c., are fitted roughly. The cables are simply tied to clips fastened to the beams, &c. The switches are hung from the cables. The lamps with their holders are secured to wooden bases, and protected by wire guards without any glass globes.

Cost.—Although no exact comparison can be given between the cost of this mode of illumination and that of the old plan of lighting by candles, it is estimated that in a first-class cruiser like the "Royal Arthur" the total cost of the electric lighting during the whole period of building, including depreciation of plant, would be about £1,200. This would probably not much exceed, if at all, the cost of candles for the same period; but the vastly superior illumination obtained by the electric light—enabling work to be done better, more quickly, and under better supervision, to say nothing of the advantages in health and comfort to the workmen—would justify the continuance of this system, even though the actual cost should be much greater than that of lighting by candles. It is hardly an exaggeration to say that the excellent results as regards celerity and cheapness of construction, which have lately been attained in Portsmouth dockyard, could scarcely have been realised without the aid of the admirable illumination afforded by the system of incandescent lighting.

Testing of Dynamos.—Before acceptance from the makers, all the dynamos purchased by the Admiralty are tested on receipt at a dockyard; and as no less than 150 have been tried at Portsmouth during the last two years 1890 and 1891, the experience gained has been considerable. At the north end of the electric shop a space is fitted with cast-iron holding-down plates, of sufficient area to take several sets of plant at one time; and also with an overhead traveller for lifting. Steam at 100 lbs. per square inch, or other pressure required, is supplied by three boilers of locomotive type in the adjoining boiler-house. Adjacent to the holding-down bed is a separate enclosed room, containing all the various electrical testing instruments; and into this room the leads from the dynamos under test are taken. For enabling a dynamo on any part of the holding-down bed to be rapidly connected up with the instrument room, two bare copper rods extending overhead the whole length of the bed are fixed to the roof trusses and insulated by slate. From these are suspended two hanging leads, the lower ends of which can be connected to the terminals of any dynamo under trial. The overhead bars are coupled to leads, which pass through resistances sufficient to give the full load, and then into the instrument room.

Nearly all the dynamos now purchased, except those for torpedo boats, are of a capacity of 400 ampères at 80 volts, the latter being the Admiralty standard voltage for all dynamos and electrical fittings; and they are all direct-current and compound-wound. The dynamos are coupled direct to open vertical compound engines, working generally with 100 lbs. steam pressure at a speed of about 330 revolutions a minute. The dynamo and engine are required to stand first a continuous trial of six hours' duration with full load. Every half hour the steam pressure, current, and electromotive force are recorded, as are also the temperatures of the testing room, field magnets, and armature.

It is stipulated that one minute after the end of the trial the temperature of any accessible part of the machine must not exceed that of the testing room by more than 30° Fahr.; and the maximum temperature of the armature at end of trial after stopping must not exceed that of the room by more than 70° Fahr. If these limits are exceeded by more than 10° the machine is liable to rejection. These tests for temperature are considered to be of great importance; for if unsatisfactory the dynamo is not likely to remain long efficient. The spare armature which is supplied with each machine is required to undergo a similar test for two hours. The current produced by any dynamo is measured by means of a Siemens ammeter; and as a check upon this the current can also be switched on to a Siemens dynamometer. There are several of these ammeters, and three dynamometers, of different sizes to suit the current to be measured. The leads pass through sliding resistance-boxes, so that the strength of the current can be varied as required for the compounding trials. The voltmeter leads taken from the dynamo terminals are also led into the instrument room, and the electromotive force is ascertained by means of three Evershed marine voltmeters. A switch is interposed, of such a nature that either any one or any two or all three of these voltmeters may be in the circuit, thus checking one another's accuracy. A Cardew voltmeter and also a Siemens voltmeter can be switched in, if desired. The Cardew voltmeter and the Siemens ammeter are the instruments actually used on board ships of the navy. Very delicate instruments, such as electro dynamometers, are found not to be suitable for use under such circumstances. As additional tests the engines and dynamos are expected to stand without injury the sudden removal of the full load by the breaking of the circuit; and under such circumstances the maximum increase in revolutions or in voltage must not exceed 25 per cent. The dynamos are required also to be compounded so as to give a constant electromotive force of 80 volts when the current is varied from 400 to 10 ampères, the speed being fairly constant during the time; and they are fully tested to ascertain that these conditions are fulfilled.

Engines.—As regards the engines, great importance is attached to their economy in the consumption of steam. The makers are required to state the consumption of water per electrical horse-power per hour which they will guarantee shall not be exceeded. A fine is incurred for every lb. of water per electrical horse-power per hour in excess of the maximum guaranteed; and if the excess amounts to more than 10 lbs. per hour, the machine is liable to rejection on this account. The water used is carefully measured, and every endeavour is made to avoid losses by leakage of steam, &c. Importance is also rightly attached to the governing arrangements, and it is stipulated that the increase of speed when the load is gradually removed must not exceed 5 per cent.

Resistances.—Before the six hours' trial is started, and immediately after its conclusion, the resistances are observed of the dynamo complete, of the armature, and of the shunt and series windings of the field magnets; the resistances are thus obtained both when the circuits are cold and also when they are hot. For this purpose the leads from the terminals of the part whose resistance is required are taken into the testing room, and the resistances are obtained by a bridge made by the Silvertown Co., by a Thomson's marine galvanometer, and by an ordinary scale, for which the light is given by an incandescent lamp behind. The ordinary Daniell cells are used, and a key is introduced, so that one, two, three, or four cells may be switched in at once without altering the leads.

The dynamo and engine are further tried when fitted on board ship, and when the whole installation is complete.

3.—Torpedo and Gun Circuits.

Torpedo Circuits.—When the Whitehead or fish torpedo was first adopted as an element in the offensive armament of ships of war, it was ejected from its carriage by means of an impulse-rod or piston actuated by compressed air. The ejection was effected by opening a communication valve between the impulse cylinder and a reservoir of compressed air. As it was necessary that the discharge of the torpedo should be effected from an observing and directing station remote from the torpedo itself, electricity was early pressed into service for effecting the communication between the observer and the firing valve; and in 1879 an electrical arrangement was first fitted for enabling the firing valve to be opened from a distance. The method of ejecting torpedoes from a carriage by means of an impulse-rod was soon abandoned however, and the air-gun principle took its place. By this method the torpedo is itself placed in a tube or gun, into which compressed air is admitted behind the torpedo; and the ejection is effected by the expansive force of the air acting directly upon the rear of the torpedo, exactly as the explosive force of gunpowder drives a shot out of a gun. When worked from a distant point the firing valve was still opened by electricity, but on an improved plan.

In the earlier ships the batteries were placed below the water-line, and the circuits were complete with main and return wires. Where not protected by armour or not below the water-line, the wires were duplicated and kept apart from each other, as greater security against both circuits being destroyed at one time from the same cause. The circuits now used are earth circuits, and are duplicated where exposed, as for instance near the torpedo tubes. The standing wires of the modern circuits are lead coated. The conductor is formed of a strand of seven wires of No. 22 legal standard wire-gauge or 0·028 inch diameter. It is insulated with india-rubber, which again is covered with three servings of cotton tape prepared with india-rubber. The enclosing lead tube is covered with a strong tape saturated with a water-proof mixture. Where flexible leads are required they are formed of thirty-six wires, each of No. 30 legal standard wire-gauge or 0·012 inch diameter, insulated as before with india-rubber and cotton tape, and protected by a binding of hemp. The whole is saturated with Hooper's composition.

Within the last few years powder impulse has been introduced for discharging torpedoes. For this purpose a cartridge containing a small charge of gunpowder is placed in a powder chamber prepared for it at the rear of the torpedo tube. The cartridge is fired by means of an electric current generated in a local circuit, that is, by a battery secured to the torpedo tube itself. This local circuit is momentarily completed by the action of the firing circuit from the observing station.

Gun Circuits.—Electricity is further employed in the navy for effecting the discharge of guns, either individually at the gun positions or simultaneously in groups, from some protected observing station. So far as the writer is aware, this system was first practically employed about 1874, and the current was obtained by means of pile batteries. These were formed of about 160 elements, consisting of alternate copper and zinc plates, separated by fear-nought (a kind of flannel) dipped in diluted vinegar. With these batteries high-tension fuses were used, which were not only dangerous, but frequently failed owing to the difficulty of keeping the circuits free from moisture. From 1874 down to about 1881 gun circuits were fitted as complete wire circuits. Now earth return circuits are adopted, except in some cases of auxiliary circuits which are still complete wire circuits.

Safety Arrangements.—These are highly necessary in the ordinary working circuit, in view of the possibility of the gun being fired prematurely with perhaps disastrous results. For instance the gun might by mischance be fired before the breech block was properly replaced and locked; and to avoid this danger the circuit is broken automatically by the act of unlocking the breech, and the connection is again made automatically in the act of relocking after loading. Again the gun might be fired after the breech block was secured, but before the gun was run out; in which case the energy of recoil would be expended in smashing up the gun mounting. To avoid this the circuit is automatically broken at the front end of the carriage by the action of recoil, and is remade automatically just as the gun again reaches its run-out position. The making and breaking of the circuits are effected by means of rubbing contacts; and while these give assurance against an accidental discharge of the guns, they are not an unmixed good, as they sometimes cause a misfire; and a continuous protected circuit without any of these breaks would be more certain and reliable.

In the earlier turret ships, especially in the rigged ones, the guns were masked by portions of the ship's structure at certain angles of training. In such cases devices were introduced for automatically breaking the firing circuit at these particular angles. In all the later turret and barbette ships, the turrets or barbettes are placed near the extremities of the ship; and while having a limited arc of training, they are unobstructed in their fire throughout the whole of this arc when the ship is cleared for action. Consequently there is now no necessity for any safety arrangements. In the modern as in earlier ships the full depression of the guns cannot be obtained throughout the whole horizontal arc of training; and the danger of firing into the deck is obviated by an ingenious automatic arrangement, which is in connection with the hydraulic or other mechanism for working the guns, and has nothing to do with the electric circuits.

4.—ELECTRIC COMMUNICATION.

The efficiency and safety of a modern war-ship are largely dependent upon the means of communication between the various working parts of the vessel. The officer directing from the conning bridge the movements of the vessel must have the means of instantly transmitting to the engine-room a few pre-arranged signals with regard to the starting, stopping, &c., of the engines ; and he likes to have the signal transmitted back to him, to assure him that his order has been correctly received, and is about to be carried out. When manœuvring in squadrons, in which his ship has to keep her correct station in relation to the others, he must be able to communicate to the engine-room his orders regarding small variations of speed, say one revolution more or less over a considerable range of speed. He also likes to be able to ascertain for himself on the bridge the actual revolutions of the engines; and he prefers to do this at a glance, without having to take the trouble to use a watch and count the revolutions say for a minute. Where the steering-wheel is immediately under his own eye, as it generally is when steam or other mechanical power is applied for steering the ship, he likes to have an indication on the bridge that the rudder is properly responding to the motions of the steering wheel. Should either of the alternative steering positions below have to be resorted to, he requires a certain means of communicating his orders to the steersman at that position. In addition to the means of transmitting a few pre-arranged signals, he also requires a means of conversing from the conning bridge with the engine-room, with the steering positions, and with a few other places. In action, when the vessel's movements may be directed from a protected conning station, he needs, in addition to the foregoing requirements, to be put in communication with the officers at the various gun quarters, including turrets or barbettes, and also with those at torpedo-discharging positions, &c. In fact he requires to be placed either directly or indirectly in communication with every part of the ship where officers and men are stationed. Where only a few pre-arranged signals are required, these intercommunications are effected by means of transmitting and receiving instruments, either mechanical or electrical; while for communications of a general character voice pipes are employed.

About six years ago many complaints were received from ships as to the unsatisfactory nature of the means of signalling and intercommunication; and as many new instruments, chiefly electrical, were then being proposed for use in the navy, a committee was appointed, consisting of naval officers, naval constructors, and marine engineers, of which the writer was a member, to enquire into and report upon the whole subject. In the course of the enquiry the views of a large number of representative naval officers were obtained, in order to determine firstly what intercommunication was necessary and best for a ship of war. On this primary question naval officers were not wholly agreed; but there was sufficient accord to enable a standard method to be laid down. In order to arrive at the best means for effecting these intercommunications, the committee studied all the methods that had been previously tried, including those in use in the mercantile marine. They carefully examined and tested many new instruments that were submitted to their notice, and carried out a number of independent experiments.

Voice Pipes and Call Bells.—Among other means of improvement, the committee recommended that all voice pipes connecting stations where there is noise or vibration, and all others if they are over 100 feet in length, should be increased from 1¼ inch internal diameter, the size then in use, to 2 inches internal diameter; and that with these larger pipes a system of electrical call-bells should be fitted. This plan is now generally carried out in the navy, and has resulted in greatly improving the means of communication. The call-bell arrangement consists of an ordinary push-board at the sending station, with the well-known annunciator or drop-shutter instrument at the receiving station; the dropping shutter indicates the station from which the call has come, and therefore the particular pipe to be spoken through. It is found that with voice pipes communication to places where there is much vibration or noise is not very satisfactory. Many devices have been and are still being tried with the object of improvement.

The committee were much impressed with the immense facilities offered by an electrical system of communication; and amongst other things, telephones of different forms were tried in substitution for voice pipes. The results were not such as to lead to their introduction into the ships of the navy.

Telegraphs.—Although for ship work a mechanical system has some great advantages over an electrical, yet as a means of establishing communication between instruments at distant stations the electrical system is far and away the best. The many disadvantages of a system of shafting and bevel wheels, led through a considerable portion of the length of a ship, scarcely need to be stated; while the great facilities offered by a simple insulated wire are equally self-evident. The committee therefore recommended that the one mechanical telegraph for communicating orders from the bridge or conning tower to the engine-room should be retained, but that the telegraph for orders respecting speed of revolution should be electrical. Many of these electrical telegraphs have been fitted in the navy; but the reports upon them have not been uniformly favourable, although in some instances they have been highly so. Whether the grounds for the unfavourable reports have always been good and sufficient is a matter of opinion; but at the present time electrical telegraphs in the navy appear to be somewhat under a cloud, so that the field is still open for invention in this direction. The telegraphs already in use are worked by primary or secondary batteries; while a new form has been proposed, to be worked direct from the ship's dynamo.

5.—Other applications of Electricity.

Submarine Mines. — There are some other applications of electricity in the naval service, which may be referred to without any detailed descriptions. One important application is that to submarine mining, whereby stationary torpedoes or mines are exploded by an electrical current from an observing station on shore, after the position of an enemy's ship in relation to the mines has been ascertained by means of range-finding instruments in two observing stations at a considerable distance apart but electrically connected.

An electrical apparatus has also been designed and experimented with for enabling a boat, without any person in it, to be sent from a safe distance into an enemy's mine-field, to explode and thus render harmless the mines laid there. By this apparatus the boat can be steered, the engines stopped or started, and the countermines dropped where desired, and exploded. In this way an otherwise exceedingly risky operation can be performed without danger to life; and the worst casualty that can happen is the loss of the boat only.

Night Signalling.—Electricity is also applied to night signalling. The ordinary semaphore instrument as used by day is illuminated at night by means of incandescent lamps in front of a reflector, the whole being contained in a wooden box placed a short distance in front of the semaphore arms, which are painted white in order to make them more distinctly visible by night in the rays of the electric lamps. Another system of night signalling has lately been introduced, consisting of four flashing lamps placed in a vertical line at topmast head. The lamps are incandescent with special filaments; and one lamp or more can be switched off at will by an arrangement of switches in the chart house, where there are also four small lamps which follow the movement of the lamps at masthead. By a preconcerted code the combinations of the lamps are interpreted. This arrangement is intended principally for squadron sailing at night, to convey orders as to altering course, speed, &c.

Gun Sights.—The sights for guns are now adapted for night firing by employing electricity for their illumination. This is effected by means of small incandescent lamps, for which the current is obtained from a primary battery consisting of three Leclanché cells in a box secured to some part of the gun mounting. A switch is also fitted for each sight, having a resistance connected with it, so that the current can be varied at will, and thus the lights can be made more or less brilliant to suit circumstances.

Examination of Gun Bores.—An apparatus is now supplied to ships for examining the bores of guns. It consists of an inclined mirror supported in a metal frame, which is made to fit the bore, and can be pushed into any part of the bore by a long rod screwed into it. According to the size of the gun the frame carries from one to four 100-candle-power lamps, which illuminate the bore; and by the aid of the inclined mirror a close examination can be made for defects.

Motors.—The application of electricity to motors has not yet gone very far, not so far in fact as in some other nations it has already gone or is now going. Of the success of its application to the working of guns, supply of ammunition &c., the writer is not very sanguine; but if results prove satisfactory, the navy will no doubt quickly profit by the experience of other nations. Electrical science is as yet in its infancy, and it promises enormous developments in the near future: so that in a few years operations may be performed by electricity which are not dreamt of now. In any case the British Navy will not fail to press into its service every development of science which will increase the efficiency of the nation's first line of defence.

Discussion.

The PRESIDENT said that Mr. Deadman, having very recently been promoted from Portsmouth to Whitehall, was prevented by his official duties from being present at the reading and discussion of his paper; but he had kindly arranged for it to be illustrated by the extensive collection of specimens now exhibited, to which Mr. Lane had been so good as to direct attention when they were severally alluded to in the course of the paper. Would Mr. Lane state what was about the weight of water now consumed per electric horse-power?

Mr. D. W. LANE, Electrician of Portsmouth Dockyard, replied that the consumption of water per electric horse-power varied from 32 to 35 lbs. per hour for high-pressure engines.

Mr. W. H. Allen mentioned that, in the system at present employed in Portsmouth dockyard for testing the steam engines used for driving dynamos, the steam was supplied from a large locomotive boiler, five or six times larger than was required for the engines, and measuring tanks of the ordinary form were employed for measuring the water pumped into the boiler. It was a combined system of testing the boilers as well as the engines; whereas it was desirable to be able to test the engines and dynamos alone, independently of the boiler. For such small engines as were employed for driving dynamos, he thought it was impossible to get a true record from so large a boiler, having regard to the fact that the measurement had to be taken from the water line indicated by a movable index or by the gauge-glass on the boiler itself, and the slightest error in observation, extending over so large a surface of water, would largely affect the correctness of the result. Moreover on that plan it occupied a long time to get the record, as much as six hours, making the trial a costly affair. When commencing at his own works the manufacture of dynamos and engines, he had at first followed the line of the Admiralty testing, and had then found that he could not get any trustworthy results in less than six hours, and at a cost of about £5, which was a serious item to add to the prime cost. It had then occurred to him whether some cheaper and readier plan might not be devised for testing the machines in his own shop; and he had been led to adopt the apparatus shown in Plate 36, which was suitable for an engine of 100 HP. The present regulation dynamo of the Admiralty was for a current of 400 ampères and 80 volts, and required an engine of only 60 HP., for which therefore a somewhat smaller apparatus would suffice. At the top was a horizontal surface-condenser C of ordinary construction, the circulating water passing through the inside of the tubes, while the exhaust steam from the engine circulated around them. The water from the condensed steam was drained into the collecting tank T below, whence it passed by the overflow pipe P and cock K into one or other of the two measuring tanks M, which formed the hollow columns for supporting the tank T and condenser C. As these columns were of small diameter in comparison with their height, the

(Mr. W. H. Allen.)
weight of the water condensed was shown with great accuracy by an external gauge-glass and a scale which had previously been carefully graduated by actual measurement. In order to make sure of the condenser tubes being kept always fully charged with water, the circulating water entered through the inlet pipe I at the bottom of the condenser, and passed away through the outlet pipe O at the top, following the course indicated by the arrows. Either one of the two measuring tanks M was sufficient for about fifteen minutes' working of a 100 HP. engine. By means of the tumbler cock K between the two measuring tanks M the test could be extended over a period of any length desired. The apparatus shown was intended for engines exhausting into the atmosphere; but it could equally be applied for condensing engines by inserting a valve to shut off the flow from the condenser C into the collecting tank T, and by connecting the condenser with the inlet to the air pump in the ordinary way, and the collecting tank T with the outlet from the air pump. Having tried this apparatus a great number of times, testing engines for as long as six hours and also for only ten minutes, he had found that the accuracy of the shorter tests was equal to that of the longer. The cost of a satisfactory trial with this apparatus was only four or five shillings, instead of five pounds for the longer trial. It had been suggested that, in testing with this apparatus, the leakage of the glands on the engine cylinders would not be accounted for; but he thought that in new engines there would be practically no leakage, or if any it would be so small as to make no appreciable difference in the result. It had also been suggested that the circulating water might possibly leak through from the condenser tubes into the measuring tanks, and thus produce an error in the record. The apparatus however lent itself readily to a method of testing in order to detect any leakage of this kind; by closing the outlet cock and turning on the circulating water, it would at once be seen whether the water was leaking from the tubes, as any leakage would be indicated by the gauge attached to the measuring tanks. This test was made before the commencement of every engine trial. In all new engines numerous adjustments had to be made before the consumption could be accurately determined; and each of these

adjustments often occasioned considerable difference in the quantity of steam used. It would therefore be understood that a single test would be of scarcely any avail; and in order to comply with the strict requirements of the Admiralty it often became necessary to test an engine something like twenty times. With the present apparatus a series of five accurate tests could be made in an hour at a cost not exceeding twenty shillings altogether.

As a manufacturer himself of dynamos he gave credit to the Admiralty for their work in connection with these machines. Had it not been for their strict requirements for testing, his own work would have been far behind its present position. Three and a half years ago the Admiralty regulation dynamo weighed nearly 8 tons; but at present, owing to their strict requirements, the weight was under 5 tons. Although he had originally been disposed to remonstrate against the requirements when they were first announced, he was now satisfied that they were really always for the good of the manufacturers.

Mr. R. E. B. CROMPTON thought electrical engineers might thank the author for having brought this subject forward, if only because it might lead to criticism. The chief remark that he had himself to make was that he was afraid the Admiralty was much behind the times; and that the best electrical apparatus now on board ship was to be met with only in foreign navies. This state of things he hoped would be amended in future.

With regard to the projector now exhibited of the Admiralty pattern (page 259), he had himself made a large number of these, and knew well their merits and demerits. This pattern was practically the projector of ten years ago, having hardly been altered in any respect. No doubt it had done good service. Eight or ten years ago, as a manufacturer of projectors of that kind for the Admiralty, and wishing like all other manufacturers to adhere to one pattern as closely as possible and to turn out as many as he could, he had found that he could sell projectors of the same pattern to other nations also. But that held good no longer; other nations had gone so far ahead that they would not now use a

(Mr. R. E. B. Crompton.)
projector of that pattern. The Germans had gone as far ahead of the French as the latter had of the English. Any one who had visited the Frankfort Exhibition last year had no doubt seen the beautiful projectors exhibited by Schuckert of Nuremberg. It was to be regretted that the English were at present so far behindhand, and he hoped that they would do something to recover their laurels. A projector which he had brought here for examination was about half the weight of that exhibited by the author; it possessed all the latter's merits, and he hoped none of its defects. In this lighter instrument it would be seen that advantage had been taken of all recent advances in mechanical engineering. The gun-metal arms, which were weak, clumsy, and heavy, were no longer used; but a hint had been taken from the work of the bicycle makers, and strong thin light steel tubes had been substituted. The long-focus mirror, still adhered to by the Admiralty, had been replaced by the short-focus mirror, which was due to the inventive genius of Mr. Charles R. Parsons. The hand-fed lamp, which involved the necessity that the man using the projector should be always under fire, had long been given up; and an automatic lamp had been produced, costing about one-fourth of the Admiralty hand-fed lamp, and having one-fourth of the number of parts in it. The automatic lamp exhibited had nothing but a right-and-left-handed screw in two pieces, connected by a sleeve so that they must turn together, but one part was capable of longitudinal movement so as to strike the arc; the striking was done by the simplest form of magnet, and the feed movement was given by causing the screws to revolve by means of a fine-toothed ratchet-wheel worked by a shunt magnet and paul. This improved form of projector he was now supplying to other governments; and he hoped the explanation he had offered of its principal features might be useful for leading to the introduction of improvements in the projectors used in the navy of this country.

In the use of search lights great advantage had been obtained from the beautiful invention by Sir James Douglass of fluted carbons, which had enabled much steadier lights to be produced than before. It was a curious thing that these carbons, although invented in this country, had always been manufactured abroad; there seemed to be something in this country that prevented the carbons from being manufactured here. But he wished to correct the impression which might be derived from the paper, that the French still held pre-eminence in the manufacture of carbons. That was no longer the case; the Germans had already surpassed the French, and the Belgians were now coming into the field; in fact they were rather ahead of the other two, having lately produced carbons at about half the price of any previously made, and sufficient for all useful purposes.

No reference had been made to an engineer who had done as much as any one, or even more, for ship-lighting, particularly in the ships of the navy: namely the late Mr. Willans, whose engines were the first high-speed engines used to any considerable extent for the lighting of ships. He was quite aware of the great value of Mr. Brotherhood's exertions in producing high-speed engines; but these were used principally for search-light purposes. Mr. Willans had worked hard at the subject, and had produced his magnificent engines, which were now almost exclusively used by electrical engineers. Nearly three-fourths of the horse-power now in use in England in central electric-lighting stations he believed was supplied by Willans engines. All were aware what a heavy loss the engineering profession had sustained, in consequence of the accident through which Mr. Willans had been killed a couple of months ago. Had he been present he would certainly have complained, like the previous speaker, of the totally inadequate means at the disposal of the dockyard officers for the proper testing of steam dynamos; indeed he had himself heard Mr. Willans make the very same objection to the large size of the boilers used; and he had further pointed out that, although premiums were offered for small differences in percentage of efficiency, there were yet no means of ascertaining where such differences in efficiency were obtained within a degree of accuracy anywhere approaching what the premiums had been offered for.

The whole state of electrical appliances on board ship he considered was at present not at all satisfactory. The great subject of electrical distribution and electrical energy for motive power as well as for light was so well established, that he thought any one making a

(Mr. R. E. B. Crompton.)

design, thoroughly well considered from first to last, for a new war-ship—which was really nothing but a great box of power intended for many and various purposes—ought to make it his aim to solve the problem of producing the lights and transmitting power to any part of the ship in the simplest way, and in such a manner that the practical working should be little liable to be crippled by the disturbance of any part of the apparatus during action with an enemy. Electrical distribution lent itself extremely well to that aim, he thought; and with present experience it would be easy enough to lay down all over a ship a network of mains for the distribution of light and power, which practically could never be totally severed from the source generating the power, that source being placed as at present down below the water line and away from the reach of shot. Instead of any of the other methods of dealing with the subject—by hydraulic, compressed-air, or steam power—he should prefer to confine himself to a well-considered electrical distribution, and to carry out every movement by a system of motors. As an electrical engineer he naturally advocated electricity; and he thought that this was the direction in which mechanical engineering was tending. In all large workshops and establishments, where power was required not continuously but in various directions and subdivided into small amounts, this he thought was what engineers would come to; and he was sure that any one of the most recent ships in the navy might now be considered in the same light as a large mechanical workshop, for it was practically nothing more nor less. He hoped to see the time when some of the officers of Portsmouth dockyard would begin to take up the subject thoroughly, and when the question of electrical distribution on board ship would be considered as a whole, and not merely as a means of producing electric light.

Mr. G. B. OUGHTERSON said no one could have a greater respect than himself for the late Mr. Willans as an engineer and as a man. But as a matter of fact, with reference to the adoption of high-speed engines for search lights, the first Brotherhood engine applied to search lights was two years before the date mentioned in page 258 of the paper. It was applied in 1876 by Messrs. Sautter Lemonnier to the French war-ship "Richelieu"; and upwards of two hundred Brotherhood engines had been applied to electric lighting for search lights before Mr. Willans had supplied any for that purpose. With regard to the method of engine testing adopted at Portsmouth, if it was defective the defects applied to a great many other trials besides those made in the dockyard. For the past thirty-five years he had had considerable experience in engine testing. The method he had generally adopted might be crude; it was that of measuring the feed-water pumped from a measuring tank into the boiler. About fifty engines had recently been tested in that way at Mr. Brotherhood's works, and re-tested at Portsmouth dockyard; and the differences of results obtained in the two tests did not exceed 2 per cent., which, with an average consumption of from 33 to 35 lbs. of water per electrical horse-power per hour, was not much. This showed that, however crude the testing might be at Portsmouth dockyard, it was equally crude at the works; and nevertheless the results obtained were practically identical. One of the Brotherhood high-speed engines working a sixteen-light Brush dynamo for a search light on board one of the ships of the navy had run three weeks at a speed of 800 revolutions per minute without the steam being once turned off.

Mr. JOHN T. CORNER, R.N., Chief Engineer of Portsmouth Dockyard, mentioned that the engine-testing apparatus at the dockyard was fully known beforehand to all persons tendering for the supply of engines; it was a condition of the contract that the test should be made with the apparatus there in use. Moreover, although the steam consumption was limited to a certain weight, say 32 lbs. per electrical horse-power per hour, another pound over was always allowed to make up for any little errors or differences in the measurements. That amounted to about 3 per cent.; and, as Mr. Oughterson had pointed out (page 283), the difference between the dockyard tests and his own amounted to only about 2 per cent. Nearly the whole of the makers supplying the electrical engines had expressed entire satisfaction with the apparatus in use and the means employed for testing their engines.

Mr. SYDNEY F. WALKER, having had the advantage of serving on board a man-of-war for eight years before devoting his attention to electrical work, was gratified to find from the paper and from the apparatus exhibited that the naval service was still keeping its place as well as it had always done. No doubt there were still difficulties, as there used to be in former times, from the Admiralty officials at Whitehall moving slowly; and they had the reputation in naval circles of not being kind to inventors; it would therefore not be surprising if the results obtained were a little behind the times. Although he had not had the honour of doing any work for the Admiralty, yet from his own experience of work in the mercantile marine and elsewhere, and as far as he could see from the paper and from an examination of the apparatus exhibited, the dockyard practice was certainly not behindhand.

In page 263 it was mentioned that lead-covered cables were now used entirely in all the later installations, and that their introduction marked an epoch in ship-lighting; and it seemed to be implied that all troubles from the cables were about to cease. This he was afraid would be found not to be the case. By coating the cables with lead, the difficulty was only put one degree back, if indeed it was put back at all. The reason why the uncoated cables had failed, as they undoubtedly had done formerly, was that the cable itself was badly insulated, and outside it there was a poor conductor. With the old wooden casing, the two cables having only a small amount of insulation were separated only by a small piece of wood, which was capable of absorbing a large amount of moisture. The result was similar to what had been witnessed at the recent electrical exhibition at the Crystal Palace, where in the high-tension experiments a piece of stick, which for most practical purposes could be used as an insulator, was subjected to a current having a tension of 20,000 volts, and in a few minutes was seen to be converted into a conductor and then gradually to become charred. There was no tension as high as 20,000 volts on board ship; but time would produce the same effect on board ship, or in any other place where there were similar conditions, even with the low tensions there used. With the old form of cable having bad insulation and wood casings, there was a small current going on across the insulation; and as long as the dynamo was at work and the cables were connected, that small current was doing its work, altering the nature of the insulation of the cable, and altering the nature of the wood casing itself, gradually converting it into a conductor. By and by, usually at a joint, it was enabled to throw a bridge across, a succession of sparks passed, and then the cable was destroyed. Practically the same thing would now occur with lead-coated cables, only perhaps a little later than with the wood casing. The occurrence might be delayed by increasing the insulation between the conductor and the lead casing; but it was still only a question of time, especially in the present iron ships. Sometimes there was a braiding outside the lead-coated cables; but this did not last long, and then practically the current would go across from one cable to the other by way of the lead, and the cable would be gradually broken down, just as it was with the wooden casing; the sparks would be thrown across at a favourable spot, at some favourable moment, say when the lights were turned out suddenly; and the cable would be wholly or partially destroyed. For this reason he considered it should be laid down as an absolute rule that no conductor should ever be placed outside the insulating envelope of the cable, if it could possibly be avoided. In drawing on a lead tube outside the insulating envelope, or wrapping it round with a coil of iron wires, there could never be any certainty that the insulator itself had not been damaged thereby; and although the damage might be only partial, the cable was deprived of some of its durability at the damaged place, and it was at that place that it would give way later on under the action of the minute current. Moreover the ship herself was constantly at work; and when she was in a heavy sea with the main engines working, there was probably no part that was still. If there was too rigid a casing outside the insulator, it offered resistance to that movement; and the insulation itself being the weakest part must get damaged, no matter whether it was protected by a lead covering or by a coil of iron wires. He had himself seen instances of this, and had preserved a piece of a wire-covered cable which had set fire to a cotton ship in

(Mr. Sydney F. Walker.)
this way; and also some pieces of a lead-covered cable which had broken down under similar conditions, and had given a great deal of trouble, though not actually setting fire to the ship. Having had to face this question some years ago in connection with mining work, he had had a well insulated cable covered with as much spun yarn as he could get the manufacturers to put on. Spun yarn was itself a good insulator, and was flexible and strong, and would keep the water out; and he had never had any trouble with a cable so covered.

With regard to the automatic cut-outs, switches, and safety devices for gun circuits (page 271), he doubted the advantage of resorting to these, and thought the eye of a properly trained captain of a gun was better than any automatic safety arrangements. In the conditions in which these were worked, it was necessary to use grease, and there was also a great deal of dirt about; and a little of either, settling upon the rubbing surfaces, would sometimes foul one of the automatic cut-outs, and then the contact was gone and the gun could not be fired when needed.

With regard to search lights, he suggested that their usefulness was by no means confined to the discovery of an enemy; and he was surprised they had not been more used to prevent shipwrecks. On a dark stormy night, when a captain knew his ship must be getting pretty near a certain coast, a search light showing some distance ahead would be a valuable help, and might be the means of saving the ship.

As to the use of telephones on board ship, he should expect to find they would be too delicate. It ought however to be quite practicable to use them if their connecting parts could only be made strong enough. It was not necessary that the working parts should be made so heavy that the power of the voice would not be enough to lift them. One source of trouble with nearly every electrical apparatus on board ship was keeping connections clean. The damp atmosphere charged with salt was too likely to eat away the metal surfaces and destroy the connection. The parts where connection was made were wanted to be big enough to stand cleaning; and then, if they were cleaned frequently, he thought the trouble would be got over. For holding communication between one ship and another he had been working out a plan for telephoning through the water; and had found that only 1-2,000th of a volt would be required to give a click in the telephone. By means of a single generator on board ship with a low resistance and a low voltage he believed it would be possible to send through the water a current from one ship to be picked up by another. Whether speech would be heard or not might be a matter of doubt; but the click of the Morse alphabet he believed could be taken up at a distance of about half a mile.

Mr. WILLIAM H. WHITE, C.B., Member of Council, speaking as one of the Whitehall officials who were considered to move so slowly, said that, if moving slowly was to be measured by being responsible for an annual expenditure on new construction of £4,000,000 and for the designing of 120 ships in 6½ years, then the Admiralty had been moving slowly; and if they were to be measured in their work at Whitehall by the scrutiny of any one of the many details which went to make up the equipment of a modern man-of-war, it was possible that they did move slowly. In the matter however of the application of electricity on board ship, the charge of moving slowly had not in his opinion been made out. Indirect evidence that this was not the case had been furnished by the remarks of previous speakers. When a member like Mr. Allen, who had very properly pointed out what he conceived to be possible improvements in the methods of testing engines, had ended his remarks by saying (page 279) that the Admiralty system of testing had led to substantial improvement in practice on the part of electrical engineers, there was not much fault to be found. It was now only about ten or eleven years since the application of electricity to the internal lighting of ships began to take practical shape in the navy. He had himself personally had to do with the first work of that kind; and ever since that time, allowing for the interval when he was not in the Admiralty service, but when his interest in electrical matters had by no means lessened, it had been his endeavour to know what was being done, and as far as possible to utilise whatever advances might be made in any direction.

(Mr. William H. White, C.B.)
But there was a much larger consideration to be borne in mind, namely that throughout this period the Admiralty had never failed to utilise private enterprise and invention in this country; they had never failed to encourage national industry and manufacture by all means in their power, and had never gone abroad for what was wanted unless compelled to do so. It had been admitted even by Mr. Crompton (page 280) that in some respects the manufacturers of this country had not kept the position which all had hoped they would have retained. No one desired more than the officials of the Admiralty that all the apparatus required should be procured in this country; and, as the writer of the paper had truly stated (page 259), everything which had in it the slightest promise of success had been carefully tested at Portsmouth, in the hope that it might prove successful and might enable work now obtained from abroad to be supplied at home. But while this was true, it was also true that in the naval service there were special requirements. All naval electrical apparatus had not merely to be provided for in the design of the ships, but it had also to stand the test of practice and to obtain the approval of the users of the apparatus, namely the officers who served in the ships, and who by the test of experience were sometimes persuaded that arrangements proposed by inventors or manufacturers were not the best for general adoption. Various plans which had offered promise of success beforehand had had to be set aside after the actual test of sea service. The process of testing and selection was always going on; and in the meantime ships had to be built and equipped. All therefore that the naval authorities could be expected to do at any time was to use the best appliances that they knew of at that time. It seemed natural enough that Mr. Crompton should have fully persuaded himself that the electrical distribution of power was the thing to be adopted; he started with that assumption, and admitted that he was influenced by the fact that he was himself an electrical engineer. There were many persons however, whose opinion was also entitled to respect, who did not agree with Mr. Crompton. In the use of heavy guns, their mounting and working, he found that the majority of those competent to give an opinion at the present time preferred hydraulic power to

electrical. Having himself had an opportunity of seeing probably the largest installation of electrical power ever yet fitted on board ship, which included working, loading, and all the other operations for heavy guns—operations ordinarily performed by steam or hydraulic power—he had been greatly struck with the possibilities of that system of distributing power; and he thought it was quite possible that in future many things would be done by electricity which were now done by other means. The time might come when the design of a ship would embody from the first the necessary arrangements for producing and distributing electrical power for a great variety of purposes; but at present that time had not been reached. With all respect to electrical engineers he would add that some of those officers in the naval service who had had the largest experience of electrical power, in connection particularly with the torpedo practice, were those who professed the greatest distrust of its large extension. There were indeed many things on board ship which involved special difficulties, and which necessitated great precautions for obtaining a trustworthy use and distribution of power. While the Admiralty were always desirous of securing for the service of the navy the full benefit of the manufactures or inventions of this country, they were bold enough to think that in details at least the experience and capability of their own officers enabled them to produce arrangements which were specially fitted for their own service; and he thought a careful examination of the extensive collection of specimens now exhibited would reveal many things which were at least suggestive, and which had been originated and worked out in the Admiralty and in the dockyards. It could hardly indeed be otherwise, where a succession of men had devoted themselves to that branch of mechanical science, some of whom certainly had not failed to make their mark outside the Admiralty service when they had chosen to leave it.

One great governing condition in regard to the use of electricity on board ship was the importance of economy of coal. The Admiralty system of engine testing was largely based upon that consideration. Even in the largest ship there were limits to the weight of coal that could be carried; and the introduction of

(Mr. William H. White, C.B.)
electric lighting, which was continually in use, made considerable demands upon the coal supply of a ship. It was therefore necessary to have an exact system of testing, and to regulate the orders given for machines in view of the results obtained. The matter had been put on the true footing in Mr. Corner's remarks about the system of testing (page 283): namely that, while this system of testing might be susceptible of improvement, it was for all practical purposes a good one, and at all events it was thoroughly understood by those who had to give guarantees with their machines.

Another thing which had been done by the Admiralty in connection with electrical work during the last ten years had been the organization and training of a competent body of electricians; and Mr. Lane, to whom they owed so much, was an example of what the Admiralty service contained. There had also been at the Admiralty men like Mr. Farquharson, whose name was so well known to electrical engineers; and his own present assistant, Mr. Richards, had succeeded Mr. Farquharson; while in the dockyards the assistant constructors, and the officers of junior grades in the ship-building department, were put through a course of practical training in electrical matters. There was also a competent staff of electrical workmen. Many matters therefore were now dealt with in the dockyards, which had originally been dealt with by private firms; and the change he conceived was attended with great advantage to the public service. The fitting up, for example, of internal electric lighting of ships was done almost without exception by the workmen in the dockyards. It was not a question of cost only; but when electric lighting had to be fitted into a ship it was evidently advantageous to have all the work done under one control; and it had been done well, as far as results indicated.

In this paper the author, whose unavoidable absence all regretted, had simply intended to give a statement of naval methods and practice up to date. There was no assertion that the electrical fitting of the ships in the navy was the best possible. The paper traced the progress made by gradual steps in advancing from one point to another in electrical fittings, until the stage had been reached which was illustrated by the specimens now exhibited.

One feature of great importance in relation to dockyard work was the use of temporary installations in ships that were building; and in page 265 of the paper reference had been made to Portsmouth having taken the lead in this respect. As a matter of fact however he believed that the first temporary installation used and made specially for that purpose had been procured by himself from Mr. Parsons, and used on board the "Victoria" at Elswick, at all events before he himself knew of any electrical work being done at Portsmouth. But the electrical work now being done at Portsmouth was much in excess of what could be done in any private yard. At present in every dockyard in which any large war-ship was being built a portable electric-light installation was in use; and, as pointed out in the paper (page 266), it could not be treated simply as a matter of economy in the production of light, but in working in those complicated and minutely subdivided ships it was an enormous advantage to have a good light, by which the men could work, and also a pure atmosphere; the two together constituted an advantage which it was difficult indeed to over-estimate. While regretting that Mr. Deadman was unable to be present to reply to some of the observations which had been made, he thought the meeting would agree with him that the information contained in his paper would be of great value to many members of the Institution.

Sir Frederick Bramwell, Bart., Past-President, thought the remarks made by Mr. Allen (page 277) about the method of testing engines in the dockyard, and about the method that he himself preferred, were consistent with both methods being used. Mr. Allen had objected to the test being made by measuring the feed-water supplied to the boiler, instead of by measuring the steam used by the engine after it had been condensed and collected; and he had pointed out that if the measurement were taken from the fluctuation in level of the large surface of water in the boiler, nothing short of a six hours' run could give a satisfactory result. But apart from the engine, in the act of testing the dynamo it appeared from the paper (page 267) and was well known by engineers that a six hours' run at least would have to be made in order to find out whether the dynamo

(Sir Frederick Bramwell, Bart.)
heated. There was therefore no hardship upon the manufacturer in using for the engine a method of testing which required six hours' working. In his own establishment naturally enough the manufacturer himself would prefer to have an apparatus like that described by Mr. Allen, by which the engine test could be made in a quarter of an hour as accurately as in six hours, or indeed even somewhat more accurately. But he thought both methods might well go together; and from some considerable experience in testing he could well understand how undesirable it was to depart from a system which had been in use for years, which was well known to all who had to submit to it, and which therefore rendered modern results capable of comparison with those which had been obtained in by-gone times. In the Royal Agricultural Society there had been the same desire to retain well-known methods of testing, in order to have the means of comparison.

The PRESIDENT said that, as the application of electricity was still in its infancy, it would be unreasonable to expect the same kind of perfection that might be looked for in a system brought to maturity by time and experience; and the divergence of opinion which had been expressed in the discussion was only natural. In listening to the paper he had been much struck with the progress which had been made in the application of electricity to ships of the navy, and with the close analogy between the history of its introduction and that of most other new warlike stores, for electricity might itself be called a warlike store. In every instance the introduction of important appliances had to go on by degrees. Inventors and manufacturers came with the utmost confidence, asserting that they could furnish exactly the thing that was wanted. Experiments were made in the dockyards and arsenals, committees of experienced officers were appointed, plans were investigated, and finally a design was approved and carried out. But in a short time, after the new appliances had been at work in various parts of the empire, exposed to the kind of treatment which was inseparable from service conditions, reports came home declaring that all was not satisfactory. It was one of the most difficult things, he should say an almost impossible task, to determine beforehand in connection with warlike stores what would work practically and what would not. It was only by experience, gained in the manner in which it was being acquired in the present instance and which applied equally to guns and ammunition, that trustworthy stores could be selected and adopted into the service. It therefore seemed to him that the process now going on in naval electrical apparatus showed that the Admiralty were proceeding with the caution which was essential in the interests of the public, when not only large sums of money depended upon the conclusions arrived at, but the efficiency of the service had to be maintained during the transition periods; because it must always be remembered that at any moment war might break out, and the defensive forces must therefore always be ready prepared with the best apparatus and the best means that the existing state of knowledge could supply. He had great pleasure in moving a vote of thanks to Mr. Deadman for the admirable paper, which had led to so interesting a discussion; and also to Mr. Lane for his kindness in explaining the extensive collection of apparatus that had been exhibited in its illustration.

Mr. Deadman, having been prevented from being present at the meeting, wrote that he had since seen the Crompton projector, and the following ascertained facts relating to this and to the naval service projector might be of interest:—

	Naval Service.	Crompton.
Total weight of projector complete, including mirror, lamp, &c.	5·25 cwts.	4·00 cwts.
Height	5 ft. 8 ins.	4 ft. 5 ins.
Focal distance of mirror	41 cm. = 16·14 ins.	27½ cm. = 10·83 ins.

The Crompton projector was thus only 24 per cent. and not 50 per cent. lighter, and the diminution of weight was almost wholly accounted for, firstly by the use of a lower pedestal, and secondly by the adoption of a mirror of shorter focal distance, thus diminishing the length of the lantern. To use the lower projector under

(Mr. Deadman.)

service conditions, it would have to be raised to the same height as the service projector; and this additional weight of pedestal had therefore to be added to it in the comparison. As regarded the use of a mirror with short focal distance, experiments had been carried out at Portsmouth in 1886 by the naval and dockyard officers with a Mangin mirror of 32 centimetres = 12·60 inches focal distance in comparison with one of 41 centimetres = 16·14 inches. It was found that the former became so hot that the hand could not be borne upon it; and on throwing a small quantity of water upon it, to represent what might take place if sea spray or heavy rain fell upon the projector while in use, the mirror at once fractured. The conclusion arrived at, as the result of the experiment, was that there was little to gain and much to risk by the adoption of a mirror with less focal distance than 41 centimetres or 16·14 inches. In the Crompton projector he noticed that an attempt had been made to ventilate the back of the lantern; but whether this would get over the difficulty above referred to could be ascertained only by exhaustive trials under service conditions. There were several points in which the lighter projector exhibited was inferior to the Admiralty projector for naval service, the arrangements in the lighter being such as had been fully tried and long since discontinued as unsuitable.

DESCRIPTION OF THE LIFTING AND HAULING APPLIANCES IN PORTSMOUTH DOCKYARD.

By Mr. JOHN T. CORNER, R.N.,
Chief Engineer, Portsmouth Dockyard.

In addition to the hand-power appliances which are common to all such large establishments, there are in use in Portsmouth Dockyard, for the cranes, capstans, &c., three separate systems of Mechanical Power:—namely Steam, Hydraulic, and Pneumatic. The first of these, besides being employed for working several machines direct, is also used for producing the pressures required for the two other systems. There are in the dockyard, in almost constant use for various purposes, no less than ninety-six boilers, with an aggregate of 784 square feet of grate surface, burning about 10,000 tons of coal per annum, and developing an average of 3,150 indicated horse-power during working hours. Most of them are fitted with feed-water heaters, by which some of the waste heat from uptakes is utilized, and the feed-water is supplied to the boilers at a temperature of about 130° Fahr. The majority of the boilers are worked at pressures of from 60 to 80 lbs. per square inch.

Hydraulic Power.—There are three separate pumping stations for obtaining the pressure for the general service of the yard, that is to say for the ordinary cranes, capstans, &c. The pipes, of which the arrangement is shown in the plan, Plate 33, are connected, so that any of the pumps can be used on the system. Water being here cheap, there are no return pipes, but the exhaust water runs away into the drains.

The first hydraulic pumping station is at B, Plate 33, near the main entrance to the yard, and includes two separate sets of engines, each with a pair of pumps. The larger set has cylinders 26¾ and 38 inches diameter with 2 feet stroke, the diameter of the pump plungers being $5\frac{7}{8}$ inches; and the indicated horse-power of the set is 128 I.H.P. The smaller set of engines is of about 40 horse-power, and is used chiefly as a stand-by. It is seldom required to keep more than one of the two sets of engines running for working all the hydraulic machines in connection with this station. These comprise eight cranes, with lifting powers varying from $1\frac{1}{2}$ to 10 tons, nine capstans, and nine lifts. The engines also supply pressure for working the chain-cable testing machines, as well as water for testing boiler and other tubes.

The second hydraulic pumping station is at the iron foundry V, where there are two sets of pumps worked from the shop shafting. These are primarily for the use of the foundry, in which there are three lifts; but being connected with the main pressure service pipes, the pumps can be used for the general service of the dockyard if necessary.

The third hydraulic pumping station is at the head of No. 8 dock, Plate 33, and the engine here is more modern than either of the others; it is compound, and capable of developing about 112 indicated horse-power.

The total length of pressure pipes is about 10,000 feet, or nearly two miles. The pipes vary from $1\frac{1}{2}$ to 4 inches diameter, and are placed entirely underground. There is therefore no serious trouble from frost during cold weather; all that is found necessary is to light small gas-jets in the various engine-boxes at the cranes and capstans.

In addition to this main hydraulic system, there are certain complete self-contained local systems. In the smithy K is placed one of these, comprising a 20 horse-power steam engine and accumulator, working two 40-ton cranes with travellers. In the boiler shop J there is another set of hydraulic plant for working the riveters, hydraulic presses, &c. Another application of hydraulic power is for stretching boiler tubes, and a machine for this purpose is now at work in the boiler shop. A commencement was made by stretching brass tubes which had been removed from boilers for cleaning; and new good ends were thus obtained for fitting in the tube plates. After having been extended to iron tubes, the practice has now been further extended to steel tubes, which above a certain limited size, say 2½ inches diameter, require heating before being stretched. Tube-stretching machines with hydraulic jacks have since been supplied to all the foreign dockyards of H.M. Government, so that tubes of boilers which have been removed for cleaning can now be stretched and put back again, instead of new tubes having to be used as formerly.

An additional complete hydraulic system with boilers and engines is now being fitted by Messrs. Tannett Walker and Co. at Coaling Point, at the north-western corner of the tidal basin, Plate 33. It is intended for coaling the fleet either direct from the shore or by means of lighters; and, besides ten 30-cwt. travelling cranes, it will comprise three 10-ton coal tips, with the necessary capstans, weighbridges, &c. This hydraulic system is fitted with return water-pipes.

Compressed Air.—In the most modern part of the dockyard the lifting and hauling appliances are worked chiefly by compressed air; the only exceptions are the heavy cranes and sheers, which are worked by steam-power direct. The air is compressed to 60 lbs. pressure per square inch into eight wrought-iron receivers, having a total capacity of 18,000 cubic feet. The compressing is done by one or other of two separate sets of pumps. One set consists of two pairs of compressing pumps worked through gearing, either separately or together, by a pair of simple engines of 90 indicated horse-power. The other set of pumps is worked by a pair of compound beam-engines of 200 indicated horse-power. This machinery is situated at the main pumping station R about the centre of the yard, Plate 33; and besides the air-compressing machinery the same building contains the main dry-dock pumping machinery of 1,000 indicated horse-power, and two pairs of 120 horse-power engines for general fire and dock-drainage purposes. The larger set of air-compressing pumps will fill the eight receivers to 60 lbs. pressure in one hour. No case of a receiver bursting has occurred here; and there is no record of a pipe having been replaced during the last two years. There is also less trouble with air joints than with steam and hydraulic joints.

The air pipes, which have a total length of 14,000 feet or about $2\frac{2}{3}$ miles, and vary from 3 to 12 inches diameter, extend round the large basins, as shown in the plan, Plate 33, and are connected to forty 7-ton capstans, to five 20-ton cranes, and to the machinery for working seven caissons Z and numerous penstocks, besides driving a small workshop engine. The air pressure is also used occasionally for driving small engines for carrying out machine-work on board ships building; and it has also been connected with the auxiliary steam-pipes of some of the larger battle-ships, so that air-pressure could be used instead of steam for driving the hydraulic pumping engines on board the ships for working the gun gear for drill purposes, and also for driving the electric-light engines and other auxiliary machinery on board. This obviates the necessity of getting up steam in the ships' boilers, thus admitting of their being kept systematically in a certain condition, either closed and dry, or quite full of water: which would be impracticable if they were being used at irregular intervals and at short notice.

Steam Power.—The direct use of steam for the lifting appliances is almost entirely confined to the heavy cranes and sheers, to the portable machines, and to a few cases where there is steam power conveniently near. The large tripod sheers U at the east end of the repairing basin, by Messrs. James Taylor and Co. of Birkenhead, have a working capacity of 80 tons and have been tested to 120 tons; their height is 140 feet, and they can overhang the basin wall 40 feet, the back leg being worked in and out by means of a screw. Two 50-ton steam cranes by Messrs. Cowans Sheldon and Co. are also situated at this part of the yard, one on the south side of the rigging basin and the other on the south side of the tidal basin. At the steam basin there are a 40-ton steam crane by Messrs. Fairbairn and Co., and a pair of 40-ton sheers worked by steam, the legs of the latter being shifted in and out by guys moved by steam crabs; also a 15-ton steam crane and two hand cranes, one for 15 tons and the other for $7\frac{1}{2}$ tons. On the harbour side at X, there is a pair of 50-ton sheers with fixed legs; and opposite to these, on the side of the ship basin, another pair of 25 tons. In addition to the above, there are several smaller steam and hand cranes and capstans, spread about the yard; also various travellers in the shops, worked by ropes, by shafting, and by hand.

The travellers in the iron foundry offer a good example of the saving of labour due to the application of power. These 24-ton travellers were originally manual, and required eight men to work them during the pouring of the metal for a moderately large casting. Since power has been applied to them, all the operations can now be performed by two men at the most, and with much greater certainty.

Comparative Advantages.—Steam power, as already stated, is almost entirely confined to single machines and central stations. If it could be used without being condensed or cooled in the pipes, it would of course be more economical than either air or water, because it would save all the efficiency lost in the air compressor or in the hydraulic pump. But it is quite out of the question for any extended general service over such a distance as a couple of miles of pipes, and cannot therefore be considered in comparison with the two other systems in regard to general convenience and economy. Comparing however the various advantages possessed by the hydraulic and pneumatic systems, the dockyard experience is greatly in favour of the latter. For a number of machines working with a certain amount of regularity, perhaps the hydraulic system would be the better as regards economy; but with work like that in the dockyard, which includes the use of a large number of machines spread about at considerable distances from the source of power, and with extensive variations in the demands made on the power, it is found that the requirements are best and most economically met by the compressed-air system. By means of the air receivers, sufficient power can be stored up to work about half the machines connected therewith for about two hours; the air-compressing engines can therefore be worked at regular speeds, and stopped if necessary for repairs, without fear of the various cranes and capstans being suddenly brought to a standstill for want of pressure; and it is hardly ever necessary for the pumps to be driven at their utmost speed, even when considerably more than ordinary demands are made upon the pressure. Consequently the necessity of working the boiler fires harder for only a few minutes' spurt is avoided, together with the waste of fuel due to steam blowing off at intervals. By pumping the receivers up in the evening before the yard closes, power is available for opening a caisson or penstock or working a capstan during the night, without having to start the compressing engines at all. Experiments made here show that the relative economical values of the air and hydraulic systems when applied to capstans are as 13 to 11 respectively, while the great balance of convenience is also in favour of the former: in addition to which it is found that the wear is much less in the air machinery than in the hydraulic, and moreover that the engines worked by air are less easily disarranged or put out of working order than those worked by water.

Machinery for working the Sliding Caissons.—In Figs. 1 to 4, Plates 34 and 35, is shown one of the sliding caissons, which consists of a number of horizontal compartments. Of these the lowest is an air chamber, containing some pig-iron ballast; and the next above is a water space, open at the ends. The third compartment is an air chamber, fitted with a water-ballast tank T at each end, and pumps P midway between, by which either tank can be pumped out or filled independently of the other; and the compartment above this is an air chamber A, in which are the cranks &c. for working the pumps. The next two compartments above are open at the ends, and together form a strongly bracketed structure for carrying a floor, to which is transmitted all the weight that passes over the upper platform F. The chamber or recess R into which the caisson floats when opening is covered over and forms part of the roadway; the upper platform of the caisson with its stanchions and hand-rails is therefore arranged so as to lower down sufficiently to pass under the road. The engines, which are situated at E and are worked by compressed air, have cylinders of 16 inches diameter and 17 inches stroke, and are arranged so as to open or close the penstocks of the docks or basins adjacent, to drive the shafting S for lowering the upper platform of the caisson, and for working the pumps P for pumping out the ballast tanks, besides hauling the caisson itself by means of chains. For lowering and raising the platform F, there are in the uppermost chamber a pair of carriages CC drawn together or moved apart by a shaft with right-and-left-handed screws, Fig. 4; and these carriages are connected to the platform above by sixteen pairs of links L. When the two carriages are furthest apart, the platform is up, and all the links are vertical, thus forming a rigid roadway for the heaviest traffic; but when by means of the traversing screws the carriages are brought together, the links become inclined and the platform is lowered. The stanchions and hand-rails are all jointed, and so connected to the caisson that, as the platform is lowered, they stow horizontally so as to clear. To prevent undue stress on the raising and lowering gear, the moving platform F is counterbalanced by weights W below.

In Figs. 1 and 2, Plate 34, is shown the lead of the chains which move the caisson. The chains on the lower barrels B draw the caisson into the recess, while those on the upper barrels U pass over sheaves or fair-leads secured to the masonry, thus forming loops, which, being connected below the fair-leads to projecting arms on the caisson, haul it outwards across the dock entrance. While the lower chains are winding on their barrels, the upper are unwinding, and *vice versâ*. The caisson while being moved is of course kept floating, the water-ballast tanks affording a ready means of regulating its displacement according to the height of the tide.

Dock Pumps.—Of main dock-pumps there are four sets. The oldest set, situated at G, Plate 33, comprises five chain-pumps, working in 24-inch barrels and driven by a 40-H.P. engine. These, though old-fashioned, still do good service in pumping out the docks in the old part of the yard. They are good for 1,600 tons per hour from an average depth of 15 feet; and besides pumping out docks they are used also for pumping up the old ship basin when it is necessary to have an extra depth of water over the dock sills.

The second set of dock pumps is situated at the back of the factory J, and consists of six single-acting lift pumps, with barrels 3 feet diameter and having a stroke of 2¼ feet; the power is obtained by means of gearing from a set of 300 I.H.P. compound engines. The aggregate capacity of these pumps is equal to 2,500 tons per hour.

The largest and most important dock pumps are the third set, situated in the main pumping station R, Plate 33. The engines, by Messrs. James Watt and Co., are of 1,000 I.H.P.; and the pumps have been made and fitted by Messrs. Easton and Anderson. The steam cylinders are 40 and 64 inches diameter by 5 feet stroke; boiler pressure 80 lbs. The pumps are single-acting, 8 feet diameter and 6 feet stroke. They were originally double-acting pumps, 6 feet diameter and 6 feet stroke, and were capable of pumping out one of the largest docks in 3¼ hours, which gave a rate of 20,500 tons lifted per hour from a mean depth of 23 feet. The wear and tear on the pumps was very great, and they were eventually changed for the present single-acting pumps, and the speed was slowed down, so that it now takes about 4 hours to pump out a dock, giving a rate of about 16,500 tons per hour.

Close at hand at M is the fourth set, comprising a pair of centrifugal pumps, 6 feet diameter, driven by a compound engine, with 27 and 50-inch cylinders and 3 feet stroke, working 31 revolutions per minute and developing about 400 I.H.P., and driving both pumps at 130 revolutions. These also are by Messrs. James Watt and Co. The pumps together lift at the rate of 7,500 tons per hour from the mean depth of 23 feet. In designing these centrifugal pumps it was intended that the two should work together until about 20 feet was pumped out of the dock, and then they were to be worked in series, the delivery from one pump becoming the suction for the other. But on the trials it was found that much better work could be done by letting the two pumps both go on drawing direct from the dock; and though the extreme depth from suction to delivery is 40 feet, yet the pumps continue to work efficiently to the end.

Discussion.

Mr. WALTER HUNTER was sure any one who had had an opportunity of visiting Portsmouth dockyard must have been struck with the complete nature of the appliances for lifting and hauling throughout the works, as described in the paper; and the same conclusion must be evident also from the fact mentioned in Mr. White's paper (page 242), that the "Royal Sovereign" had been built, completed, and commissioned within so short a time as only two years and eight months from the laying of her keel. It would be impossible for a ship to be so speedily and economically constructed unless the mechanical appliances available were of a high class. At the same time there was one particular in which it appeared to him that the naval dockyards were not up to the practice of the present day in regard to their lifting appliances, as compared with large mercantile docks and harbours. In the docks of London and Liverpool floating cranes were now in constant use for shipping and unshipping heavy weights, having practically superseded the fixed sheers formerly employed for this purpose; and he thought that any one experienced in lifting in docks would come to the conclusion that floating cranes gave the greatest facilities for loading and unloading in the shortest space of time. Two floating cranes which he had supplied to the St. Katharine's Docks and the East and West India Docks in London were of 50 and 30 tons lifting capacity. The 50-ton crane "Leviathan" was capable of masting the largest ships coming into the port of London; it was similar to the floating crane seen by the Members in their visit to the Tilbury Docks in August 1886 (Proceedings, page 451). It had lifted three boilers, weighing 55 tons each, on board H.M.S. "Blenheim" in one day, between nine o'clock in the morning and five in the afternoon, which he thought was a good day's work. He was informed by Mr. Robert Carr, chief engineer of the London and India Docks Joint Committee, that in 1891 in twelve months the "Leviathan" floating crane had lifted 3,740 tons, and the 30-ton floating crane, the "Titan," had lifted 3,112 tons; while in the first six months of the present year 1892 the larger crane had lifted 2,772 tons and the smaller 1,445 tons.

(Mr. Walter Hunter.)

The advantages of floating cranes were numerous. They took up no quay space themselves, and obviated the necessity for removing a ship from her berth to the sheer legs when heavy weights had to be taken off or put on board; the removal of the ship for such a purpose was always attended with some risk and considerable expense, besides entailing the stoppage of all unloading and loading of cargo while the ship was away from her berth. A floating crane could steam alongside a ship, take out or put in a crank-shaft, boiler, gun, or armour-plate, and move any other heavy weights, while the ordinary discharging or loading of cargo was being carried on from the quay, and without shifting the ship from her moorings. The hulls of the floating cranes were so constructed that guns or boilers or other heavy loads could be placed upon their decks, and conveyed to or from a ship, wherever she was lying. The crane could also lift weights out of a ship, and deposit them in a barge on the other side of its own hull, without moving from alongside the ship. It could be fitted with single-purchase gear when desired, for lifting lighter weights at a quicker rate, and so facilitating the ordinary unloading or loading of cargo or stores. The cranes could be fitted on a sea-going vessel to steam to any port, and the vessel could be arranged for carrying heavy guns out to a blockading fleet, so as to prevent a war-ship with damaged armament from having to come home in order to refit; the crane hull could further be fitted with machinery for executing repairs for the fleet. In times of war, if damaged ships came into harbour for repairs, a floating crane would be of the greatest service for enabling the work to be quickly performed; because there would be no absolute necessity for the ships under repair to lie alongside the quay, and they would not have to wait their turn to go under the sheer legs for getting heavy weights moved. It had already been remarked by the President (page 255) that in the present age wars were shorter, while the preparation of the means of defence was longer; and it was therefore of greater importance that the dockyards should be equipped with the best appliances for saving time when ships came in for repairs and equipment. With a view therefore to the efficiency of the navy, upon which the safety of the country was mainly dependent, it seemed to him most desirable that floating cranes should form part of the equipment of every naval station.

Mr. CHARLES COCHRANE, Past-President, enquired what was found to be the loss of pressure in the working of the pneumatic cranes. It appeared from page 297 that the air was compressed to 60 lbs. per square inch in the receivers, and was conveyed through pipes varying from 3 to 12 inches diameter; and he had been given to understand that it was so conveyed in some cases to a distance of as much as half or three-quarters of a mile from the receivers. It was an important matter therefore to know what loss of pressure there was in going such a distance.

Sir FREDERICK BRAMWELL, Bart., Past-President, had no doubt it would be generally agreed that one of the most important matters which mechanical engineers had to deal with at the present day was the transmission of power to a distance. A description had been given in Mr. Corner's paper of two modes of accomplishing this which were well known, namely by compressed air and by hydraulic pressure: both of which were undoubtedly serviceable modes. There was also an allusion to the direct use of steam, which naturally brought to mind the attempt that had been made in New York to transmit power in that way. He did not know the present condition of the enterprise, having heard nothing of it for the last two years; but when he was there eight years ago it was already coming into use, and he knew that since that time it had been largely extended. Some interesting results had been obtained with regard to the rate of condensation in pipes buried under the surface of the ground. There was also another mode which had been much thought of, but which the paper had not dealt with, being confined to methods actually in use at the dockyard: namely the transmission of power electrically.

In reference to the pneumatic transmission of power he should be glad to know what difficulty, if any, was experienced with the engines worked by compressed air, from moisture freezing in their outlet passages, and how it was got over; or whether it was a fact that, as the engines were crane engines or capstan engines, and were therefore

(Sir Frederick Bramwell, Bart.)
working only a few minutes at a time, there was not sufficient time for any formation of ice to take place. The great success of the pneumatic system at the dockyard he had not the least doubt was due to the enlarged views which had led to putting up the eight storage receivers having such an ample capacity as 18,000 cubic feet. This provision no doubt enabled the air-compressing engines to work continuously into the receivers, and also provided large storage for meeting any peculiar demands upon them. It had been a most wise course he thought to incur the considerable expense which must have been incurred in providing such large storage capacity. He should be glad if the writer could give any information as to the absolute power obtained from an engine worked by compressed air, compared with the power exerted for the compression of the air; and also the same information with respect to the hydraulic method, although the latter was pretty well known. Repeating the calculation which he had made at the York meeting of the British Association in 1881 (Report, page 504) about the transmission of power hydraulically, he found that a 10-inch pipe one mile long with the usual pressure of 700 lbs. per square inch would convey 300 horse-power with a loss of head of only 35 feet, or about 14 or 15 lbs. per square inch, being only 2 per cent. of the initial pressure. As he had already mentioned on more than one occasion, in earlier days he had been enabled to see the mode of transmitting power which was invented by Mr. John Hague, to whom he was then apprenticed; but it did not come into extensive use, because he was before his time. The principle was the converse of pressure, namely the exhaustion of air, maintaining a vacuum of 10 lbs. per square inch in the mains, and working the engines at the further end by the pressure of the atmosphere passing through them and going into the mains. It worked for a distance of over a mile and answered extremely well. One use made of it was for engines driving gunpowder machinery, the exhausting steam-engines being away from the building in which the pneumatic engines were placed; that application also worked very well. The whole subject of power transmission was highly important, and it was seldom that a record was obtained of an undertaking carried out on so large a

scale as that at Portsmouth dockyard, and described by a gentleman so thoroughly competent to explain it as the writer of this paper.

Mr. Jeremiah Head, Past-President, said that the important question raised by Mr. Cochrane (page 305), as to the loss of power when conveyed to a great distance either by compressed air or by water, recalled the discussion that had taken place at a recent meeting on the supply of motive power by water pressure in Liverpool (Proceedings 1892, pages 55–6). The great cost of power conveyed by water through a line of pipes to a distance, and there used intermittently and in small quantities, he thought was scarcely appreciated by engineers generally. In Liverpool it appeared that, whereas the cost of a horse-power in pumping into the supply reservoirs was only from a halfpenny to a penny per hour, yet when the water reached the warehouse cranes in the lower parts of Liverpool its cost amounted to between 13*d.* and 14*d.* per hour per horse-power actually used at the cranes. This showed the great loss that must be allowed for in conveying hydraulic power through a long line of pipes for intermittent use.

The President asked for a little more information about the process of stretching boiler tubes by hydraulic power, as mentioned in the paper (page 296). It was quite new to himself, and appeared to be a most interesting operation. There were various questions that might be asked, as to whether the tubes were much thinned by the stretching, and whether their diameter was reduced, particularly at the ends; and in regard to other practical details of the process further information would be acceptable.

Mr. Corner said his paper had not been written with a view of putting forward any novelties, but simply to give a general account of some of the useful appliances at work in the dockyard, so that the members in their visit to the establishment might know where to find these, and might there make any enquiries they wished about them.

(Mr. Corner.)

With regard to the recommendation to use floating cranes (page 303), these had already under a different form been used in the navy for many years. Before the construction of the present large basins and wharves, nearly all the lifting had been done by means of cranes mounted upon old sheer-hulks. The masts used to be floated down a slipway into the water, and towed off, and lifted into their places by these cranes. If they could only have lifted a crank shaft in the same way, they would have been still more serviceable; but as crank shafts were not so manageable in the water, it was preferred to bring the ships to the stationary crane or sheer legs on the wharf; and this plan he thought would be found to be the most economical and convenient in a place like a dockyard. No doubt if there was a large dock area with limited lifting power, it was an advantage to be able to move the crane about and take it from ship to ship; but in naval yards he thought there would not be found much use for the floating crane. As mentioned in Mr. White's paper (page 247), there was at present to be seen in the dockyard the "Vulcan," fitted with two 20-ton hydraulic cranes and two 10-ton hydraulic derricks. One intention of this ship was to take the place of a floating factory in the fleet, and also to serve as a floating crane. Moreover every battle-ship and first-class cruiser was fitted with a derrick capable of lifting nearly 20 tons. It would therefore be seen that there was now plenty of lifting power afloat, without going back to the old sheer-hulk.

With regard to the loss of pressure in the air pipes and also the increase of temperature due to compression, he had found that, in pumping up the air receivers from 40 lbs. to 60 lbs. pressure with a temperature of 75° Fahr. in the engine-room, the maximum temperature of the lubricating water drawn off from the bottom of the receiver, which approximately represented the temperature of the compressed air, was 118°; and the highest temperature that could be measured by fixing the thermometer directly upon the shell of the receiver was 106°: so that there was not any considerable or dangerous increase of temperature in the compression. The loss of temperature at a distance was found to be very small. The pipes themselves were good, and were made with flanged joints, faced and bolted together, which were found to keep tight; there was no record of any trouble having been experienced with the pipes during the last two years. With these tight joints and while all the machines were standing still, the pressure all over the system was practically uniform; but with a small draft from the pipes, perhaps for two or three capstans, there was a loss of from 3 lbs. to 5 lbs. of pressure. When there was a larger demand for air, the pressure went down to 20 lbs. or 35 lbs. less than that in the receivers.

With regard to formation of ice (page 305), the engines worked by air had not been troubled to any extent in this respect, as they did not work usually at any high rate of expansion. When they were linked up for a light load, as was sometimes the case to the extent of cutting off at one-sixth of the stroke, ice formed at the exhaust ports, especially in damp cold weather; but the exhaust ports had never become choked up, because they opened direct to the atmosphere; the engine, after two or three strokes, coughed and cleared itself, and then no more ice or snow came out for some time.

As to the relation between the power stored in the receivers and that given out by the compressing engine and the air-driven engine, he found that the compound air-compressing engines, in pumping up four receivers from 0 to 60 lbs. pressure, gave out 122½ horse-power in their steam cylinders, and did the work in half an hour. The air-compressing pumps of these engines developed 80 per cent. of the engine power, which was further reduced to 71 per cent. by friction in the pipes leading to the receivers, and probably also by the circulating water cooling the air a little: so that there was 71 per cent. of the engine power stored in the receivers. During experiments the work developed at the capstans in each case was measured by a friction dynamometer, consisting of a strap passed round the drum on the capstan head, and connected by long levers to the dynamometer spring; and allowance was made for the friction of the levers, as well as for the ropes not being exactly at right angles to them in all positions. The percentage of the air-compressing engine power that was obtained at the capstan engines at a distance of two miles from the receivers, with the heaviest load on the capstan that the dynamometer would admit of, was about

(Mr. Corner.)
20 per cent., with 60 lbs. pressure in the receivers. With 30 lbs. pressure in the receivers and the load on the capstan diminished to about two-thirds of the previous load, the efficiency was increased to about 30 per cent. With hydraulic engines and capstans, on the other hand, the percentage of efficiency was greater as the work increased: for at full capstan-power the percentage of main-engine power at the capstan engines was 40 per cent. with water pressure, against 20 per cent. with compressed air; but when the load was reduced to one-third of the full load, the hydraulic efficiency fell to 22 per cent.

Throttling the water to diminish the speed of hydraulic capstan-engines diminished the efficiency; while with air capstan-engines the efficiency remained practically the same with loads much varied. If half the actual maximum load used on the trial of these capstans were taken as the average work required, the air capstan cylinders gave 24 per cent. of the main-engine power, as against 29 per cent. for the hydraulic capstan cylinders. But comparing the work actually done by the capstans, it had to be noted that the air capstans were not working up to more than one-fifth of the power of which the capstan-engines were capable. By calculations made from the friction diagrams taken from air capstan-engines, it was found that with the full power the capstan would be doing at least 55 per cent. of the capstan-engine power. The efficiency of the hydraulic capstan with maximum load was actually 43 per cent. of the capstan-engine power. Hence the total efficiency of the air capstan would be $24 \times 55 \div 100 = 13$ per cent. of the main-engine power, against $29 \times 43 \div 100 = 12\frac{1}{2}$ per cent. of the main-engine power for the hydraulic capstan. In this calculation the work was assumed to be continuous; but as a matter of fact it was very irregular, and on this account the pneumatic system with its large receivers had a decided advantage over the water system. It was therefore necessary to modify the above percentage for water power, inasmuch as coal consumption trials showed that the hydraulic system suffered to the extent of at least 12 per cent. from irregular working, whilst the pneumatic lost none of its efficiency on that account. The total efficiency of the hydraulic capstan was thus reduced to $12\frac{1}{2} \times 88$

÷ 100 = 11 per cent. of the main-engine power. The relative efficiencies of the pneumatic and the hydraulic capstans were accordingly as 13 to 11, as given in the paper.

The tube-stretching machine employed in the dockyard had been used at first for stretching brass tubes. In the navy there were a number of boilers that came in for cleaning, and it had been found cheaper to take the tubes out than to send a man into the boiler to clean down between the tubes. But the number of times the tubes could be taken out for cleaning was limited by the great expense of putting new tubes in; and this was now got over in a measure by stretching the tubes by hydraulic power, making fast a clamp at one end and fixing a hydraulic jack to the other end, and stretching the tube by pumping water into the jack. A mandril was inserted in each end, in order to keep the ends of the tubes up to the full diameter. The tubes were not stretched much; a tube 6 feet long was stretched about 2 inches or 1–36th. The reduction in the thickness of the tube was therefore very little indeed; even the reduction in diameter was hardly perceptible, and in thickness it was still less. By drifting the ends as before, after cutting off the ragged or burnt extremities, a good tube was obtained, long enough to be refixed in the tube-plates for use over again. In this way it had been found that the brass tubes could be stretched cold five or six times. Iron tubes had then been tried, and it was found that these also could be stretched cold as well as the brass tubes. With steel tubes the pressure was not enough to stretch them cold, so that they were obliged to be warmed up to nearly a red heat; and by cooling the ends and driving a mandril into them the ends were kept sufficiently near their original size to fit into the tube-plates again properly. This process had had its origin in Portsmouth dockyard, and had been developed to such an extent that many of the ships now drew their own tubes for cleaning, and by means of one of the small hydraulic jacks stretched them, and then put them in again, which formerly was almost impracticable. The tubes required careful drawing, in order to admit of their being put back again into the boiler with their old ends. This process had proved a highly valuable one for the navy, and also very useful for engineers in general.

The PRESIDENT was sure the members would wish to join him in a vote of thanks to Mr. Corner for his most interesting paper, giving an insight into one of the many important features of Portsmouth dockyard.

DESCRIPTION OF THE NEW ROYAL PIER AT SOUTHAMPTON.

By Mr. JAMES LEMON, Mayor of Southampton.

Old Pier.—The Old Pier at Southampton was opened by Her Majesty the Queen, when Princess Victoria, on 8th July 1833. Since that time the increasing trade and prosperity of the town have rendered it necessary to erect for pleasure traffic a new pier adjoining, reserving the old pier for commercial purposes; but during the progress of the works it became necessary to reconstruct the old pier, so that the whole is now an entirely new structure.

New Pier.—The New Pier just completed is the largest in the South of England, covering an area of 3½ acres. From the plan shown in Fig. 1, Plate 37, it will be seen that it provides landing stages for ten steamers: so that ample provision is made for the ordinary passenger traffic to and from the Isle of Wight, and for the excursion traffic to and from Southsea, Bournemouth, Weymouth, and other places. The commercial traffic is also well provided for, so that goods, cattle, and sheep can be landed at all times of the tide.

General Description.—The pier is constructed upon cast-iron screw-piles, Fig. 7, Plate 40, on which are bolted main and cross girders of rolled iron: the object being to obtain lightness, strength, and durability, and not to impede the water way. The entrance buildings E, Fig. 1, are of white brick and stone, affording accommodation for a large traffic of passengers and visitors. The entrance gates G for the horse and carriage traffic are on the east side, and the roadway is extended to the pontoon bridge. The cattle and sheep entrances C are still further to the east, and are kept entirely distinct. Immediately inside the main entrance E is a paved quadrangle Q 80 feet by 50 feet, in the centre of which is an iron column carrying the electric light. Then follows the promenade or approach to the head of the pier, which is 780 feet long and 20 feet wide, having alcoves on the west side fitted with seats and plate-glass screens. At the south end of the promenade is the head H of the pier, 250 feet long by 165 feet wide; in the centre is a large and lofty octagonal band-stand, 26 feet diameter, enclosed by covered seats, the whole occupying an area of 5,800 square feet, and seating 264 persons. From each side of the pier head branches out an arm A, 185 feet long with an average width of 33 feet, for enabling passengers to land from the steamers or to embark therein. Ordinarily passengers will land upon the deck level; but spacious landing places are provided at two lower levels for enabling them to land both at ordinary and at extraordinary low tides. On the west side of the pier is provided a commodious landing place for passengers from yachts and from small boats, serviceable at all tides. The pier head commands a full view of the Southampton Water; it is fitted with seats for upwards of 400 persons, and accommodates also a club house of the Royal Southampton Yacht Club. The commercial traffic is confined to the site of the old pier. Goods, cattle, and sheep will be landed at the pontoon, and will pass along the carriage road to the town; and sheep and cattle for embarkation will be penned in a special area S provided for the purpose at the pier entrance. Through communication between Southampton and the Isle of Wight has been amply provided for by the South Western Railway Company and the Isle of Wight Steam Packet Company, passengers being brought upon the pier by the railway without changing carriages; and a spacious station for arrival and departure has been erected on the pier. Electric lighting has been substituted for gas over the whole of the pier.

Timber Piles.—The old pier having been constructed entirely of timber, it became necessary to reinstate portions thereof from time to time; and with these repairs the pier has lasted sixty years, showing that timber is more lasting for such work than is sometimes imagined. It was not possible to identify with certainty the original timber, but some of the piles were remarkably sound in the portions below the mud line.

Cast-iron Screw-Piles.—In the new pier the piles, Plates 40 and 41, are of cast-iron, 8 inches diameter externally and one inch thickness of metal, except those for the promenade, which are only 7 inches diameter and $\frac{3}{4}$ inch thickness. The average length measured from the top of the head to the point of the screw is 40 feet. They were cast in four lengths, Fig. 7, including the making-up piece at the top. The head was cast separate, Figs. 10 to 16, so that it could be slipped over the pile, and thereby assist in meeting the variation of length which must necessarily occur in putting down screw-piles. The joints to the piles are flanged, Figs. 17 and 18, planed and turned at their bearings, and secured at each joint by six bolts of one inch diameter. The screws to the 8-inch piles are 2 feet 9 inches diameter and 5 inches pitch; those to the 7-inch piles are 2 feet 6 inches diameter and the same pitch. It is always a difficult matter to determine the best form of screw, because so much depends upon the nature of the soil; and in the present case the result has been highly successful, enabling the piles to be got down satisfactorily in every way. It was decided that they should be screwed at least 5 feet deep into the solid ground after passing through the mud. This depth was considerably exceeded in many cases, the practice being to keep on screwing until the piles would not move any further.

Bracing.—The bracing to the piles is of two kinds, horizontal and vertical-diagonal. All the piles in the main body of the pier and in the arms are so braced, but the promenade and widening are braced with the vertical-diagonal bracing only. The horizontal bracings are of channel iron weighing 9 lbs. per foot run, kept about $1\frac{1}{4}$ inch apart, Figs. 19 to 21, Plate 42, by seven cast-iron distance pieces fixed by $\frac{1}{2}$-inch bolts; these bracings are secured to the piles by wrought-iron straps 3 inches by $\frac{1}{2}$ inch, and bolted together by $\frac{3}{4}$-inch bolts. The horizontal bracings are stiffened by diagonal ties of channel iron of the same section, Fig. 21, bolted to the web of the horizontal bracings. The total length of these horizontal bracings is

4 miles. The vertical bracing is of 1½-inch round iron, Figs. 7, 16, and 20: one end of each bar is forged flat, and bolted by a 1¼-inch bolt to the wrought-iron strap round the pile head, Figs. 7 and 16, and to the horizontal channel-iron that forms the horizontal bracing, Fig. 20; the other end is a screw with two hexagon nuts, passed through a tension ring of cast steel 9 inches inside clear diameter, Fig. 9. The total length of these vertical bracings is 2½ miles. They are applied only transversely, and there is no longitudinal bracing vertically.

Decking.—The main girders run longitudinally and form a continuous tie from end to end. They are supported on the pile heads by cast-iron chairs or stiffening pieces, Figs. 10 to 16, Plate 41, and are bolted thereto by 1-inch bolts, the stiffening pieces being also bolted to the pile heads by four 1-inch bolts to each head. The span is 20 feet from centre to centre of pile, Fig. 3, this being considered the most economical. The main girders are of rolled I iron 18 inches deep by 7 inches wide, and weigh 83 lbs. to the foot run. The outside girders of the pier head are only 12 × 6 inches, weighing 56 lbs. per foot run; and those to the arms are 12 × 5 inches, weighing 42 lbs. per foot run; the inside main girders of the arms are 12 × 6 inches. The whole of the girders are level on the top flange, the difference of depth being made up by pile-caps cast specially, Figs. 10 to 13. The floor girders run transversely, Figs. 7, 22, and 23, except for the last bay at the extreme end of the pier; their section is 9½ × 4½ inches, weighing 29 lbs. to the foot run. They are bolted to the main girders, being placed 5 feet from centre to centre; this distance was adopted in consideration of the quantity required for covering such a large area as upwards of two acres. The total length of floor girders is 3⅓ miles.

Upon each of the cross girders a wood camber piece is bolted to the top flange, Figs. 7, 22, and 23, Plate 42, rising to six inches in the centre on the head of the pier. On this is laid the flooring of pitch pine 2½ inches in thickness, laid close with iron tongues at the head of the pier, Fig. 23, and with open joints for the promenade and other portions. The moving load allowed for was 140 lbs. per square foot upon the main portion, and 112 lbs. upon the promenade.

Pier Arms.—In the construction of the arms of the pier the principle adopted is to use the cast-iron piles as supports for the decking and landings, and to place all the wood piles and bollards entirely independent of the iron structure, Figs. 4 to 6, Plate 39, so that the concussion of the steamers is taken by these wood piles and not by the iron structure. The wood piles are braced to one another both horizontally and vertically at angles of about 45°. The ends of the pier arms are constructed of greenheart in lieu of iron, the better to receive the concussion of vessels. The timbers are also of greenheart or jarrah wood, with the exception of the fender piles which are of Memel creosoted. The rubbing pieces to the fender piles are of American elm, Figs. 6 to 8, and the skidding irons thereto are of convex bar rolled iron, 6 inches wide and $\frac{3}{4}$ inch thick.

Landings.—The landings are placed at two levels, Fig. 3, Plate 38, to meet the variations of tide. They consist of cast-iron gratings $1\frac{1}{2}$ inch thick, cast in sections 3 feet 4 inches square, and supported on rolled iron girders. The total area is about 2,000 square feet to each arm, or 4,000 square feet to the two, so that ample space is provided for a large traffic.

Carriage road.—The carriage road is constructed with Lindsay's steel-trough decking, weighing 16 lbs. per square foot, supported by steel girders $15\frac{3}{4}$ inches in depth by 6 inches in width and weighing 62 lbs. per foot run. Steel girders were adopted on account of the increased moving load here allowed, amounting to 2 cwts. per square foot. The span of the girders is 20 feet, and of the trough decking 14 feet. The decking is covered with a coating of tar and pitch on the upper side, and is painted underneath. The roadway is formed of gravel laid upon Portland cement concrete, with a fall from the centre to the sides to facilitate under drainage.

Weight and Cost.—The total weight of the ironwork, including both cast and wrought iron and steel, is about 1,600 tons. The total cost of the work will be about £40,000. The new pier was opened on 2nd June 1892 by His Royal Highness the Duke of Connaught.

The engineers of this work are the author and Mr. E. Cooper Poole; and the contractor is Mr. Henry I. Sanders. The electrical engineer is Mr. J. G. W. Aldridge, and the contractor Mr. F. Shalders. On the visit of the members to the pier, the author will have much pleasure in affording whatever further information may be desired upon any other points of detail that may not have been alluded to in this paper.

The President said the members were greatly indebted to the Mayor of Southampton, who had done them the honour of preparing the paper just read, in anticipation of their visit to Southampton on the following day, when he had also kindly offered to conduct them over this interesting work, and to give on the spot any further explanations that might be desired.

Mr. Lemon said that, as Mayor of the neighbouring town of Southampton, he could not allow the present meeting of the Institution to take place in Portsmouth without doing what he could to promote its success. There might not be so many objects of engineering interest in the south of England as in the north; and everything that there was to be seen on the new pier he should be glad to show to the members. Happily the tide would be favourable at the time of their visit for enabling him to point out certain portions of the structure which were perhaps a little novel: such as the wood bollard piles of the pier arms, which were fixed entirely independent of the iron structure, and were strutted both horizontally and vertically, so that a vessel bumping against the pier would do no harm to the ironwork.

DESCRIPTION OF THE PORTSMOUTH SEWAGE OUTFALL WORKS.

By Sir FREDERICK BRAMWELL, Bart., D.C.L., LL.D., F.R.S., Past-President.

There is so little of novelty, or of matter calling for attention, in the Portsmouth Sewage Outfall Works, that the author feels bound to apologise to the Institution for presenting a paper upon them. He offers however these three facts in excuse:—the first, that the works are in the town where the Institution is holding its Summer Meeting, and can therefore be readily visited; the second, that they were carried out by one of your Past-Presidents; and the third, that the system depends entirely upon the employment of steam engines and pumps, and is therefore essentially one of mechanical engineering. This paper does not pretend to deal with the sewerage system of the district generally, with the original designing and laying out of which the writer had nothing to do. He found that system in existence; and, except that, prior to the commencement of the outfall works, there was undertaken the task of clearing out accumulations of sludge from the sewers, and that during the execution of the outfall works some portions of the system were renewed with better materials and rearranged so as to give better falls, he wishes it to be understood that the title of the paper practically expresses the limit of the work for which he is responsible.

District.—The district to be sewered being of considerable extent, Fig. 1, Plate 43, and as a rule flat and low-lying—indeed but slightly above the sea level—those who had laid out the works of sewerage had found it impossible to obtain, or even to approximate to, a proper inclination of the sewers, without causing their point of delivery into the sea to be at some depth below ordinary high-water; and therefore the original outfall plan, which was in operation immediately preceding the time when the writer undertook the work, was one that involved the pumping of some portion of the sewage.

Sewerage system and Pumping arrangements.—The system of sewers and the arrangements for pumping were generally as follows. For Portsmouth and for the outlying neighbourhood there were, and still are, two main arterial lines of sewers, following the course indicated in Fig. 1, Plate 43, which is a rough outline map of the area to be drained and of its surroundings: one of the lines is at a high level H, and the other at a low level L, each line passing through a different portion of the district, and each being provided with its tributary sewers, drains, and house connections. These two lines converged towards the Henderson Road at Eastney, and were there brought into a high-level sewer superposed on the low-level; these sewers delivered into sumps at the pumping station P, situated at the eastern end of the Henderson Road. From these sumps the sewage was lifted by engines, constructed by Clayton of Preston, into an outfall brick sewer R, which, proceeding from the pumping station, terminated on the sea-shore at Fort Cumberland, and was continued by two cast-iron pipes of 3 feet bore, extending well into the sea.

Outfall.—It need hardly be said that commonly the greatest difficulty, in fact *the* difficulty, in town drainage is to find an outfall which will admit of the delivery of the sewage without setting up a nuisance: a nuisance which either may be one recoiling upon those who have delivered the sewage, or may be, and not infrequently is, one that is inflicted upon another locality. In the case of sea-side towns, such as Portsmouth, the simplest and the most ready way of dealing with sewage, when collected, is to deliver it into the sea, choosing if possible some point where its delivery will not be a nuisance, and also if possible some point where it can be caused to flow into a tidal or other current, which will with certainty carry it away from the neighbouring shores. In this particular respect Portsmouth is well placed; for, as will be seen from Figs. 1 and 2,

Plates 43 and 44, there is situated about a mile to the east of Southsea the large natural reservoir or body of tidal water, called Langstone Harbour. The main channel to this harbour from the open sea passes near Fort Cumberland, and is there only some four hundred yards wide ; and as the greater portion of the water of Langstone Harbour flows in and out through this channel at each tide, the maximum current at spring tides has a velocity of sometimes as much as five to six miles an hour. It is obvious that, in order to obtain the advantages afforded by the proximity of Langstone Harbour to Portsmouth, the delivery of sewage should never be made during a flood tide, as that would simply mean carrying the matter into the harbour, to be there deposited and to accumulate. It is further obvious that the delivery should be confined to so much of the ebb tide as will cause the sewage to be carried to such a distance that it shall not return during the later portion of the flood, and thus shall not reach the harbour or the adjacent foreshores.

In the scheme of outfall which preceded the works carried out by the writer, the intention was that the pumping engines should go to work at about one hour after high water, continuing for the whole of the early portion of the ebb: in fact until the tide had fallen so low that the outfall sewer would deliver naturally. This arrangement however was attended by the difficulty of the cessation of the delivery of sewage during a considerable portion of the flood tide. And further—the production of sewage being continuous, although not uniform, while the period of delivery must vary in its time from day to day according to the tide—it followed that every fortnight there occurred times when the hours of greatest production of sewage were the very hours during which the sewers could not deliver naturally, and when also the pumps should not work. At such times the sewage had to accumulate in the sewers, the accumulation beginning from the immediate neighbourhood of the pumps, and extending backwards up the sewers to such distance inland as sufficed to contain it. It need hardly be said that this was a system where, owing to the cessation of all current in the sewers, subsidence of the more solid portions of the sewage readily took place; and a great part of this subsidence was not cleared out by the renewal of the current in the sewers when pumping was recommenced. Under these circumstances it was not surprising to find that, in the effort to make a compromise between the conflicting difficulties, the outflow was continued not only during the whole of the later portion of the ebb tide, but frequently also for some portion of the flood tide; and that, notwithstanding this improper extension of the pumping time, serious accumulations of material took place in the sewers.

Tidal Experiments.—Before proceeding to develop an outfall system which should overcome these difficulties, the writer thought it necessary to obtain practical and trustworthy information as to the behaviour of the currents after they had flowed out from Langstone Harbour, under all the changes of tides from springs to neaps, and as far as possible under the variations due to the direction and force of the wind. During the winter and spring of 1883–4 a long series of experiments were carried out in all sorts of weather, and in all states of the tide, by means of the usual vertical float, made of a piece of scaffold pole, six inches diameter and five feet long, and weighted at its lower end, so that in still salt water about four inches only of the upper part was above the surface. The floats were in all cases put into the sea as nearly as possible at the intended point of delivery of the sewage, and were followed by a boat, in which were the observers, who at regular intervals took the bearings of certain fixed points on the shore, by means of an azimuth compass, or of a theodolite. A compass used in these experiments, and of a convenient form, is exhibited. In all, something like thirty experiments were made. The records were afterwards plotted, and the results are shown roughly in Fig. 2, Plate 44. When put into the sea at J between $\frac{3}{4}$ and $1\frac{1}{2}$ hour after high water, the floats followed courses bounded by the lines AA; when put in between 2 and 3 hours after, the courses BB; and when between $3\frac{1}{2}$ and $4\frac{1}{2}$ hours after high water, their courses were limited to CC. From this it will be seen that, whether the tides were springs or neaps, and whatever might be the direction of the wind, the floats that were immersed shortly after high water were carried out to sea along the courses AA, and had no tendency to re-approach the shore.

These experiments confirmed the anticipation and hope entertained, namely that, if the delivery of the sewage could be confined to a brief period, beginning shortly after the commencement of the ebb tide, it would be possible to fulfil the condition of getting rid, once and for all, of the sewage matter of Portsmouth, without the production of nuisance either to Portsmouth itself or to any other locality.

With respect to the other part of the problem—that of preserving a continuous flow in the sewers—the solution clearly was to keep the pumping power constantly at work, and to provide it of such capacity that it should be able to deal with the maximum amount of sewage (not only for the present time but for fifteen years to come), and also with that portion of the rain water which would inevitably find its way into the sewers along with the house sewage, the ordinary rainfall being partly disposed of by a separate system.

Storage Tank.—These two conditions—of the pumping power being continuously at work, and of the delivery of the sewage being restricted to two brief periods in the twenty-four hours, namely shortly after each high water—involved the provision of a storage tank of sufficient capacity to contain all that would be pumped during the time the outfall was closed: namely the maximum sewage then being brought to the existing pumping engines, and the sewage of Stamshaw, Plate 43, an outlying part of the borough to the north, and of the Government establishments, dockyard, barracks, &c., not at that time connected with the system; and also a certain portion of the rain water. Further, an estimate was made, based on the increase of the population in the previous decennial period 1871–81, of the probable population to be dealt with at the end of the next fifteen years. Bearing these circumstances in mind, it was determined to give the tank a capacity of 4½ million gallons.

Having regard to the necessity for very rapid discharge from the tank into the sea, a discharge in fact restricted to an hour, it was obviously desirable that the site of the tank should be as near to the point of delivery as possible, so as to keep down the length of the very large outlet channels required. Fortunately the Corporation of Portsmouth, by agreement with the military authorities, were permitted to make use of the glacis of Fort Cumberland for the site of the tank; and were thereby enabled to construct it at T, Plate 43, on the very margin of the channel from Langstone Harbour.

As it was necessary to keep the upper surface of the tank at such a level that its cover, forming once more the glacis of the fort, should not be raised so as to interfere with the line of fire, and as it was also necessary that the floor of the tank should be but little below ordinary high water, in order to enable it to be entirely emptied when the tide had fallen but a small distance, the effective depth of the tank had to be limited to some 11½ feet, thus involving an area (including the outer walls) of some 1¾ to 2 acres.

The tank is divided in plan into three compartments A B C, as shown in Fig. 3, Plate 45, which are distinct and are practically equal in area. Any one of these can, if necessary, be thrown out of use for repair; and advantage has been taken of this division to enable any compartment, by means of the flushing culvert F extending across the east or sea end, to be flushed for cleansing purposes, by passing through it to the sea the contents of either or both of the other compartments. Further, although the tank is filled by pumping, and not by gravitation, and therefore there should not be pumped into it any sewage in excess of its capacity, yet each compartment of the tank is provided with an overflow W direct into the outlet culvert D, as shown in the longitudinal section of the tank in Fig. 5, Plate 46.

Ventilation.—As the tank occupies the glacis of Fort Cumberland, the military authorities, as one of the conditions upon which the use of the land was granted, insisted upon some artificial mode of ventilation being provided, because fears were entertained by them that possible nuisance might be caused to the garrison of the fort. Advantage was taken of the overflow and outlet culvert to ventilate the tank by connecting it through a flue U, Fig. 5, Plate 46, to a chimney having its outlet at a height of 90 feet above the ground level, the inlets A for fresh air being through the roof of the tank, each provided with a self-acting flap-valve, hung so as to open only inwards, and placed in a row across the sea end of the tank, the part most remote from the fort. A furnace is provided at the base of the chimney; and all the air supply for this furnace is, as a rule, drawn from the tank, and so through the fire, thus providing for the combustion of deleterious gases.

Construction.—The tank was constructed with Portland cement concrete walls, with an internal brick facing set in cement, and with a vertical water-tight collar-joint, also in cement, between the concrete and the brick lining. On the brickwork of the lining walls, in which at intervals brick piers or pilasters are built, and on cruciform brick piers founded on the concrete bottom of the tank, the brick arches forming the roof are carried, Fig. 4, Plate 45; the haunches of these arches are filled in with rough concrete, laid with a fall towards small drains which are carried on the cóncrete in the centres between the arches. On the completion of the arched roof the shingle and the mould were replaced, and the surface was sown with grass and clover seed.

The tank roof had to be of sufficient strength to carry field artillery, which in any military operations might possibly have to traverse it. Experiments were made to find the minimum number of half-brick rings necessary to support, with arches of this character, the required weights. A photograph is exhibited of an arch of two half-brick rings under a test load of 11¾ tons, and Fig. 7, Plate 47, is drawn from this photograph. After the photograph had been taken, additions were made to the test load, until at 16 tons a slight fracture occurred at the springing on one side; at 18½ tons this fracture increased, and a crack appeared near the centre of the arch; no further additions to the load were made. A series of photographs taken of the tank in various stages are also shown.

As a result of the tidal experiments, it was found desirable to provide for the whole of the sewage being discharged in a period of one hour, this period commencing from an hour to an hour and a half after high water. That is to say, with the very slight head provided by the tank, the whole of its contents, namely 4½ million gallons, had to be got rid of in sixty minutes. The outlet culvert D, Fig. 5, Plate 46, which extends crossways of the tank along the end farthest away from the sea, was therefore made as large as $6\frac{1}{2}$ feet by 5 feet, or roughly more than 30 square feet in area. This culvert delivers into a small chamber H, Fig. 3, from which three cast-iron outlet pipes P, each 3 feet 6 inches diameter of bore, were laid across the fore-shore to below low-water mark.

Discharge Valves.—In order to avoid the necessity of employing very large valves, needing several men to work them, it was found imperative to provide each of the compartments of the tank with as many as three outlet valves V, Fig. 3, Plate 45: so that, to deal with all three compartments, nine valves had to be worked. As the maximum time during which discharge should take place was one hour, it was obvious that only a few minutes could be afforded for the opening of these valves; for, allowing only the very modest time of five minutes for the opening of one valve, the whole of the nine would have required three-quarters of an hour; or, having regard to the shifting from valve to valve, more probably an hour: so that the last valve would not have been opened until the whole time at disposal for the emptying had elapsed.

Turbines.—After the first of these nine valves had been finished and tried at the factory, it became evident that some kind of motive power would be necessary for opening them. Many suggestions were made and considered. Finally a plan was proposed by Mr. Harris, now the writer's partner, which solved the whole difficulty. The principle upon which Mr. Harris's plan is based is that of making the sewage itself open the outlet valves. This is effected by gearing up each of the main valves to a small turbine, as shown in Fig. 9, Plate 48. Each turbine T is provided with a sluice 6 inches square, opened and closed by a spindle S, which is carried up above the roof of the tank and terminates in a square for a hand wheel to fit on. All that the tank attendant has to do, when outlet time arrives, is to open in succession these nine turbine sluices. As soon as the sluice is full open and the turbine running full speed, the further turning of the sluice spindle S has the effect, by means of a screw upon it acting through a lever L and an Addyman clutch C, of putting the turbine spindle into gear with the spindle of the main valve V; and the turbine continuing to work rotates through wheel-gearing the spindle of the main valve, thereby lifting the valve. As the valve rises, it carries up with it a horizontal projecting plate P, which, on reaching the top, closes the outlet from the bottom of the turbine, as shown by the dotted lines, thus causing the turbine to cease work, and bringing it and the main valve quietly to rest. The result of this arrangement is that one man is able to open the whole of the nine valves in succession in something less than ten minutes, each of the valves being about 4 feet by 3½ feet, equal to an area of about 14 square feet. Some doubts were expressed as to whether this plan, which seems complicated in description, would work satisfactorily, having regard to the fact that sewage was the operating liquid for the turbines; but the writer is glad to be able to say that from the time of opening the works until now no trouble at all has been experienced. Each turbine is surrounded by a galvanised iron screen N, so as to prevent large "flotsam and jetsam" from interfering with the working of the gear.

By means of these nine main valves, the outlet culvert, and the pipes, it is possible to empty the whole of the compartments of their contents in something like fifty minutes, or in less time than that shown by the tidal experiments to be necessary; and to do this at no greater cost than the wages of one attendant. For shutting down the valves there is ample time after the sewage has been discharged; the attendant readily closes them, which is a fairly easy operation, as their weight assists him.

Electrical Indicators.—The authorities desired to have the means of assuring themselves from day to day as to the time of tide when the discharge from the different compartments of the tank was commenced: and also as to the duration of the discharge. With the object of affording this information to them, and also to the men in charge of the pumping engines, electrical indicators were fixed in the engine house, which, combined with clockwork, show at every six inches depth the rate at which each compartment of the tank is being filled, and then show the time at which the discharge is commenced, and the duration of that discharge. Some of the actual diagrams thus obtained are exhibited, and one of them is represented in Fig. 8, Plate 47. From these records it will be seen how satisfactory has been the rate of emptying.

Outlet Pipes.—The fixing of the three cast-iron outlet pipes P, Fig. 3, Plate 45, laid across the fore-shore for the discharge of the sewage, was a work of great difficulty, and of some danger, owing to the swiftness of the tidal current, and to the exposed nature of the shore at this point. As will be seen from Fig. 10, Plate 49, the mode adopted for securing the pipes is an extremely substantial one, consisting of four rows of screwed piles, carrying cross iron supports. The outer ends of the pipes are protected by a dolphin, which is constructed of strong wooden piles driven into the fore-shore, and is surmounted by a beacon; the latter is used for the purposes of navigation of the entrance channel. Prior to the erection of this dolphin, some fears had been expressed, based upon what had happened to previous dolphins at this place, as to the possibility of any structure being erected capable of withstanding the severe effects of winter storms; but the writer is glad to say that the six years which have elapsed since the dolphin was erected have passed without damage to it of any kind.

Groynes.—The banks of the channel being exposed to very heavy weather, and to the scour produced by the rapid flow of the tide, observation was made as to the effect upon the fore-shore in times past; and it was found that in the neighbourhood there had been very considerable movements of shingle, sometimes by way of accretion, but also sometimes by way of removal, involving the destruction of a certain sea-wall work that had been carried out there. It was therefore necessary to resort to the use of groynes. These were simply constructed, consisting merely of rows of rough piles and planks: the piles in each row being about 6 feet apart, while the row is placed at such an angle to the shore as was deemed most advisable. To these piles rough 3-inch deals were spiked, only one row in height being fixed at a time, and this was left until the shingle had collected behind it, when another row was added, and so on: and in this way thousands of tons of shingle have been collected at the back of the groynes to protect the works. It may be mentioned that the collection of shingle is an operation which cannot be hurried, and that any attempt to complete a groyne to the full height at once, on this coast at all events, frustrates the object; it must be done little by little.

Main from Pumping Station.—The height at which the storage tank was placed involved the connection to it from the pumping station being under pressure; and this connection was therefore made by a cast-iron pipe 3 feet 6 inches diameter, laid to follow practically the surface line of the ground through which it passed, and provided at the high parts with automatic air-outlet valves. This main is commanded by a stand-pipe in the engine-house premises. It is practically parallel with the old brick outfall sewer R, Fig. 1, Plate 43, for the first 500 yards of its length; it then bends away to the north-east, passes round the moat of Fort Cumberland, and, approaching the tank at the north-west corner, is carried along its west wall, and is connected to the tank by three branches 2 feet 6 inches diameter, one at I in each compartment, Fig. 3, Plate 45, each connection being governed by a screw-down sluice-valve.

Pumping Engines.—The new pumping engines consist of two similar compound-cylinder beam-engines, manufactured by Messrs. James Watt and Co. of Soho, each competent to exert 150 indicated horse-power, with a boiler pressure of 80 lbs. on the square inch. The sewage pumps are arranged to receive the sewage from both the high and the low-level sewers, thus taking advantage of a portion of the sewage being delivered to the pumps at a higher level. The boilers are Lancashire boilers, also constructed by Messrs. James Watt and Co. The two separate Clayton engines, which had done all the work for the twenty years prior to 1887, have, since the new engines were put to work, been changed into a compound pair. Ordinarily one of the new engines in conjunction with the pair of old engines, or the two new engines by themselves, will suffice for the maximum work: thus giving a 50 per cent. stand-by of engine power, which can be employed when needed. The boilers of the old engines have also been thoroughly overhauled, repaired, and renewed where necessary. They have been connected to the new boilers, and the steam and other piping have been so arranged that either engine, or pair of engines, can obtain its steam from any or from all of the boilers. One of the engines at least is kept at work night and day from year's end to year's end, thus continuously extracting the sewage from the sewers, and getting rid of the necessity of allowing it to back up in the town sewers, as it had previously done for hours together. In Fig. 11, Plate 50, is shown a plan of the engine-house premises, containing the two engine and boiler houses and chimneys, condensing pond, workmen's cottages, coal store, &c.

The leading dimensions of the new engines are as follows:—

	Diameter.	Stroke.	
	Inches.	Ft.	Ins.
High-pressure cylinder . . .	20	4	2
Low-pressure cylinder . . .	30	6	0
Sewage Pumps, two in number to each engine and double-acting . .	30	3	3

According to the contract, the working speed of each engine was not to exceed 24 revolutions per minute; and at this speed, and with steam in the boilers at only 50 lbs. pressure per square inch above atmosphere, and when cutting off at half stroke in the high-pressure cylinder, each engine was to develop not less than 125 gross indicated horse-power.

Each of the four sewage pumps is of sufficient capacity and strength to lift 250,000 gallons of sewage per hour, against a total head of 40 feet. In order that they should lift this quantity, even when they are somewhat worn, they were to be of such dimensions that, if there were no waste or leakage past the piston or valves at all, each pump should lift $7\frac{1}{2}$ per cent. more, or 268,750 gallons per hour.

There are four Lancashire boilers, each 27 feet 1 inch long by 7 feet diameter, and each having two fire-flues 2 feet 8 inches diameter. Their working pressure is 60 lbs. per square inch above atmosphere.

Reconstruction of Sewers.—In addition to the outfall works, the sewers of the district generally were overhauled and repaired; and such modifications as were consistent with the system were made, with the object of increasing the rapidity of flow of the sewage, and of preventing flooding in times of heavy rain. Probably however the only matter of interest to the Institution in connection with this re-arrangement of sewers is shown in Figs. 12 to 16, Plate 51. A portion of the district of Southsea is upon bog land; there is no doubt that it was originally covered by the sea, in fact that a large portion of Southsea is merely reclaimed land. Many houses have had to be abandoned and left uninhabited, owing to settlement. The sewers which existed in the streets in this area were found on examination to be broken-backed and out of level, and altogether in an extremely unsatisfactory state. It was felt that some means should be adopted for preventing, as far as the sewers were concerned, a recurrence of these evils; and the plan shown in Figs. 15 and 16 was carried out. Rods were driven down through the peat, and it was found that, at depths varying from 10 to 25 feet below the road surface, there was solid gravel or shingle. A trench was sunk to the level of the intended underside of the concrete upon which the sewer was to be carried; and wooden piles were driven down in this trench until they were well into the gravel. The heads of the piles were cut off level with the bottom of the trench, and rolled iron joists were bedded upon them, thus bridging the distances from pile to pile. Upon the joists poling boards were laid transversely; and upon these poling boards was placed the concrete, upon which the sewer was to be bedded. The sewer was then put in upon the concrete, the trench filled in, and the road surface made good. This construction was no doubt somewhat expensive; but time has proved that the expense was fully justified. The sewers, on a recent examination, were found to be perfect in line and joint throughout; and this mode of construction has since been adopted for other portions of the town, where similar difficulties had to be contended with.

The whole of the new works were carried out without disturbance of those already in existence; and these with the system of sewage discharge in use prior to 1887 were maintained until the present outfall works were put into operation. They were completed in 1887, Alderman Sir William King, the then mayor, presiding over the opening ceremony, which took place on the 9th May in that year.

Discussion.

Sir FREDERICK BRAMWELL, Bart., referring to the statement in page 323 of the paper that the works had been designed not merely to suit the extent of population at the time of their construction but also with a view to an estimated increase of population during the next fifteen years, did not know whether he ought to be glad or sorry to say that the estimated increase had already been largely exceeded. This result was owing he presumed to the amenities of Southsea, and also perhaps to the improving health of the town generally. There had been no enlargement of district he understood to account for such a result. It appeared from the census returns that there was a population in 1871 of 113,500, and in 1881 of 128,000, being an increase of 13 per cent.; but between 1881 and 1891 the population had increased to 159,000, which was an increase of nearly 25 per cent.; so that at the present moment the works had to deal with a greater population than it had been expected at the time when they were projected that they would have to deal with in fifteen years. As mentioned at the outset of the paper, he did not pretend in the description of these works to have brought forward anything of novelty or of special interest, except indeed the extremely ingenious arrangement by Mr. Harris for making the sewage act as the motor for its own discharge.

Mr. Charles Cochrane, Past-President, considered the subject of this paper, as affecting the whole south coast of England, a most important one to be brought before the Institution; and he congratulated Portsmouth in having dealt with so vital a matter through an engineer of such ability, who by careful observations made beforehand had solved the problem, so far as Portsmouth was concerned, of avoiding the reflux of any nuisance to the town itself, or the discharge of that nuisance elsewhere as was too frequently the case. In Eastbourne prior to 1886 the same objection had existed of the sewage accumulating back in the sewers between high and low water; and it had been impossible to turn a corner of the street without perceiving an offensive smell from some of the sewer grates in consequence of that action, which he believed had since been corrected. At Bournemouth at the present time the discharge was being made right ahead of the pier by a sewer laid in the bed of the sea, and the mass of floating sewage was visible on the surface of the water, on which were busily engaged a number of sea-gulls, facetiously called the scavengers of Bournemouth. There were bathers there; and of course when there was an inshore wind the foul water flowed inshore, the gulls also came inshore, and the rest might be imagined. Those were nuisances that would have to be removed at some future time. It was astonishing to him that communities should go on making such mistakes, when there must surely be some other means of dealing with the sewage. It might be a question as to whether it was wise to send sewage out to sea at all: whether it should be allowed to pollute the sea, any more than theoretically it was allowed to pollute the rivers. But granting that there was no other method—although indeed there was the method of spreading the discharged sewage over the land for the benefit of the soil—it was a matter of congratulation that at Portsmouth there was the means of carrying the sewage out to sea right away, never to return. At

(Mr. Charles Cochrane.)
Folkestone there was an objectionable feature: the sewage was discharged within a few yards of the wall of the harbour, and it must interfere with the health of the place. At Portsmouth Sir Frederick Bramwell's work had had the effect of reducing the death rate; and in this respect he considered it was a most important subject to be brought before the Institution. At Torquay the sewage had been carried away, as at Portsmouth, to a point at which it was no nuisance to any one; but there were many places on the south coast, besides those he had referred to, which still needed to be dealt with. The storm water was a matter that needed to be dealt with even at Portsmouth. It occurred to himself, as probably it had already occurred to the Borough Engineer, that the storm water could be pretty well provided for, as at Torquay, by a side channel to carry off the water to the sea at any state of the tide. So long as it was only a few feet above the tide, it could be passed off before it reached the pumps, and in that way it might be disposed of; and the sewage being then so highly diluted, there was not the same objection to its flowing into the sea close inshore that there was to the ordinary mass of sewage doing so. Towards the end of the paper (page 331) he was glad to meet with the explanation of a condition of things which had perfectly astonished him on arriving in Portsmouth. In Clarendon Road, Southsea, he had noticed a house in which a large portion of the walls had ceased to be a support to the roof, and the roof itself was in a wavy condition, the whole tenement being uninhabitable. The explanation was given in the character of the ground beneath. It was a warning to architects to see what foundations they had to build upon.

Sir WILLIAM PINK was intimately acquainted with the sewage works at Portsmouth, having been chairman of the Drainage and Sanitary Committee of the Town Council more than twenty years; and he desired to thank Sir Frederick Bramwell for his valuable paper, which gave so minute and so correct an account of the works carried out, that it was highly interesting to all who listened to it. In dealing with the sewage of a large town he thought the members would agree with him that no back flow should ever be allowed, nor any accumulation of sewage in the sewers themselves. When formerly there had been no storage tank and the sewage could only be allowed to flow direct from the old outfall into the sea, it had been found that an accumulation of sewage took place in the sewers; and after storms, when the washings of the streets were carried into the sewers, the accumulation had become so great that it had been necessary on one occasion to take out 1,200 or 1,300 tons of silt and sewage from the bottom of the sewers by manual labour. The accumulation had been going on for years, and had practically filled some of the sewers, and caused the sewage to flow into the houses. It was also found that where sewage was stagnant it began to evolve gases. If the sewers were filled up, the gases would have to go somewhere; and no doubt they were driven backwards into the house connections, causing typhoid fever and other dangerous illness in Portsmouth, as doubtless had been the case elsewhere. It had therefore been considered necessary to deal with the sewage in a way that would be satisfactory both to the town and to the government authorities. General Crease, when he was Commandant of the Royal Marine Artillery, had done yeoman service for obtaining the present works. After various schemes had been proposed which were not considered satisfactory, the present scheme had been propounded by Sir Frederick Bramwell, in connection with the Committee; and during the whole time it had been in operation it had proved eminently successful. The healthiness of the town he supposed had tended to increase the population; and possibly other causes also had conduced to the same result. At any rate the population had certainly increased to a marvellous extent; and instead of so many weakly children and invalids being seen here as was the case a few years ago, there was now a healthy and he thought a generally contented and happy population. Care had of course been taken to let the public know what a comparatively salubrious place was Southsea, and to induce them to come and see the advantages it offered. All these circumstances taken together had brought so large a population to the place, that he supposed it would be necessary within the next few years to make some further provision for the sewage of Portsmouth.

(Sir William Pink.)

It was an important matter to separate the storm water from the sewage proper. A rainfall amounting sometimes to an inch in a few hours was very difficult to deal with in a flat district like that of Portsmouth. And if the washings of the streets, which were macadamised, were allowed to go into the sewers, there was a great deal of grit that would go into them; and if this flowed forward into the sump from which the pumps drew, it would get into the pumps and score their barrels, which was objectionable in other respects, besides being expensive.

In regard to the difference between the disposal of sewage by throwing it crude into the sea, and by precipitation and other methods, he thought it would be found that the advantage of throwing it into the sea was equal to a saving of about 3*d.* or 4*d.* in the pound in the assessment of the town. When therefore the deep sea fisheries could be encouraged, and the inhabitants supplied with cheap fish, by sending the sewage into the sea, he thought it would be admitted that this was the best possible mode of dealing with it.

Mr. Bryan Donkin, Jun., enquired what sort of turbines they were by which the discharge valves were opened; also what was their diameter, and the head upon them, and their speed.

Mr. Jeremiah Head, Past-President, enquired about the ventilation of the sewers. One of the first things a prudent householder would insist upon was that the outlet into the sewer in the street should always be efficiently trapped. The efficient trapping of the house drains had of course the effect of bottling up the sewer gas in the sewer itself, and driving it all the more to those who had not taken such precautions. It therefore became of the greatest importance in any town that the sewer gas should be released, by ventilating the sewers as completely as possible; otherwise, if there was any pressure in them, the gas was pretty sure to find its way into the houses where there was not efficient trapping. The ventilation of the sewers in towns he believed was really a great difficulty. If it were proposed to put chimneys to the sewers at different places, no one liked to have them near his own house, and there was great resistance in all localities to anything of the sort being done. If on the other hand the plan was employed, as it was in some towns, of making the ordinary gratings in the streets the means of ventilation, there was often an offensive smell in passing. In the present case, where the pumping engines were put to pump the sewage from the sumps into the storage tank, it was not quite clear to him whether in so doing they merely pumped the liquid, or whether there would be any of the sewer gas also following the pump pistons. If the latter was the case, then it would seem that any sewer gas so pumped would be delivered into the storage tank, together with any sewer gas which might further be formed between the sumps and tho storage tank; and the whole would find its way up the ventilating chimney of the tank. Otherwise it did not seem quite clear what really became of the sewer gas.

Mr. PHILIP MURCH, Borough Engineer, replied that the ventilation of the sewers, as far as it could be carried out, was by shafts. There were of course the usual open gratings in the street, but these were very objectionable as a means of ventilation. Where permission could be obtained, ventilating shafts 6 inches square inside were put up, which were found to answer the purpose admirably. A trial had also been made of putting trays of carbolic acid in the man-holes under the ventilators; and the sewer gas passing through the carbolic acid was thereby rendered comparatively harmless. This he thought was one of the best modes of ventilation that could be adopted.

Sir WILLIAM PINK added that practically there had not been any accumulation of gas in the sewers since the continuous flow of the sewage had been secured. Various plans had been proposed for incineration of the gases and for other modes of dealing with them; and he should advise all who had to do with sewage works to be extremely careful before they went to any great expense in adopting one or other of the different plans offered for getting rid of the gas, for they were not only expensive, but often by no means useful. There was no certainly sure method of ventilation that he was aware of. If the wind happened to be just in a particular direction in

(Sir William Pink.)
which it would carry a current of air into the sewers, it was beneficial; but if it was in the opposite direction the result was the other way. At present in his opinion there was no mode of ventilation so good as by shafts, which was the plan used in Portsmouth; and probably the addition of trays of carbolic acid was a further advantage. No doubt it was beneficial to have as many outlets as possible from the sewers, so that any sewer gas should escape readily in that way. Practically he should say that, since the present mode of discharging the sewage had been adopted, the quantity of sewer gas had been but slight, for the simple reason that, where there was a continuous current of sewage and water passing along the sewers, this naturally became mixed with the air, and the fermentation which occurred with stagnant sewage, and which was the cause of the evolution of the gas, had not time to take place. Portsmouth also enjoyed the advantage of having a magnificent water supply, which could be used not only for drinking and other purposes, but also for flushing the drains. A trial was about to be made of flushing the drains with salt water, and he believed that this would prove highly advantageous. If a capacious tank were erected on a high tower, to give a sufficiently large volume of water and velocity enough, the result would be satisfactory. The district was so flat that unless a large volume of water was used in flushing it was difficult to accomplish the desired object. There were baths in the town which held a large volume of water, and this was already used for flushing the sewers; but within half a mile of the inlet the water flowed so slowly, in consequence of the dead level, that practically very little benefit was found from it; but by and by, if the same plan could be carried out at several different places in the town and the sewers be flushed from their dead ends, he believed they would be able to be kept perfectly clear.

Mr. EDWARD B. MARTEN, Vice-President, referring to the elaborate way in which the reconstructed sewers were described as having been supported (page 331), enquired whether there had been any further experience of how they had behaved when the street sunk beside them, and how they were to be dealt with, and whether it was enough simply to raise the street. He wished to add his own thanks to those already offered to Sir Frederick Bramwell for this extremely clear paper. The utility of the plan adopted for getting rid of the sewage depended largely upon the extreme care with which the conditions to be dealt with had been ascertained. It was observed that Langstone Harbour formed a sort of ejector, which carried the sewage all away to a great distance, if only the discharge took place at the proper time. Upon this depended, as it seemed to him, the success of the plan at that particular place. The same conditions probably did not exist at some of the places mentioned by Mr. Cochrane (pages 333–4). Many who, like himself, had had to deal with large quantities of sewage had wished that they had the sea to discharge it into, and an ejector like the large volume of water in Langstone Harbour to carry it right away from the place of discharge.

Mr. JAMES LEMON, Mayor of Southampton, as one who had been engaged on similar works to those now described, desired also to thank Sir Frederick Bramwell for his careful paper. As to the discharge of crude sewage into the sea, it had been already pointed out that this was not advisable in all cases; but in the present special case he thought it was the right thing to do. It had been dealt with in an able manner by Sir Frederick Bramwell, who had gone to the root of the whole matter, inasmuch as from previous careful tidal experiments he had wisely determined to discharge the whole of the sewage in one hour. This he thought was the key to the whole question. The whole of the sewage was got rid of on the ebb tide, and he believed the inhabitants of Portsmouth never knew anything of it afterwards. At Southampton he had himself carried out a somewhat similar scheme twenty years ago. There his idea had been to discharge the crude sewage on the ebb tide in two hours, and the plan had answered very well. But there was a small low-lying district where it was discharged at all times of the tide, and this had had by no means a beneficial effect in the Southampton Water. He wanted to pump it up; but the local authorities having at that time a great horror of pumps, the proposal was never

(Mr. James Lemon.)
carried out. Subsequently he was happy to say the sewage had now for some years past been precipitated, as described in the paper about to be read by Mr. Bennett (page 354); and the result was that a comparatively pure effluent was discharged, which was a decided improvement upon the former discharge of crude sewage.

In regard to sewer gas, he considered the preferable course was to try not to have any, rather than to contrive the means of getting rid of it. The difficulty he thought had been solved in a great measure in Portsmouth by Sir Frederick Bramwell, because he had now ensured a continuous flow in the sewers; and so long as this could be maintained he thought there would not be much sewer gas. When smells were perceived from the ventilators, he regarded it as positive evidence that the sewers were in bad order, or in want of flushing, or otherwise needed looking after. Yet further, if there was any large quantity of sewer gas in a town, he considered it was evidence that the sewers were badly designed in the first instance, badly constructed, and badly looked after. It was preferable to go to the root of the evil, and to put the drainage of the town in a proper sanitary condition, as had now been done in Portsmouth, instead of resorting to tinkering arrangements for ventilation, as had been done in some towns. On this point he believed there was no difference of opinion amongst drainage engineers at the present time. Constant flow was what was now provided for in every good system of drainage.

With reference to the rainfall, he should like to know what quantity of rainfall had been provided for in laying out the present scheme. It had already been mentioned by Sir Frederick Bramwell how his careful calculations had been upset by the population increasing at such an abnormal rate at Portsmouth. For the quantity of rainfall however no such allowance would have to be made for probable increase.

Mr. H. J. BUTTER mentioned that a plan was being carried out in various parts of the country, and particularly in some of the suburbs of London, for the destruction of the injurious ingredients of the sewer gas by means of a furnace. There was a chimney to cause the ventilation of the sewer, and at the base of the chimney was a small gas furnace for the purpose of destroying the noxious germs, which was kept continually going by a small gas burner maintained at a trifling expense. There were two or three modifications of the plan in operation, he believed; and from the accounts received from chemists who had examined the condition of the air discharged from the chimney it appeared that the noxious germs were destroyed. The sewer was thus ventilated in the most effectual way. With the ordinary kind of ventilation, where such a furnace was not resorted to, the noxious germs drawn from the sewer must get into the atmosphere; but by the use of the furnace they were destroyed, and the air discharged into the atmosphere was nearly as pure as air which had not passed through any sewer at all. Under no circumstances was it possible, he thought, to make sure that there should be no accumulation of sewer gas, because in certain heavy conditions of the atmosphere the ventilation would not go on properly, and the gases would accumulate and afterwards discharge themselves in large quantities to the injury of health.

Mr. JAMES BUTLER mentioned that in Halifax there had been a great deal of trouble with sewer gas; and in order to ventilate the sewers cheaply and effectively, arrangements had been made with a number of manufacturers to allow vents from the drains to be connected under the furnaces of their boilers. In that way the sewer gases were able to be discharged and destroyed almost perfectly, and at a small cost. Along the course of every main sewer there were a few manufacturers who allowed these connections to be made; and thorough ventilation was thereby secured.

Mr. H. GRAHAM HARRIS said that the total head provided by the sewage for working the turbines was roughly 8 or 9 feet. The turbines were made by Messrs. Gilbert Gilkes and Co. of Kendal, and had fixed guide-blades; they were designed to develop one horse-power each with 3 feet head when running 95 revolutions per minute, and each required at that speed and head 235 cubic feet of sewage per minute. There had never been any trouble with them,

(Mr. H. Graham Harris.)
and they had worked satisfactorily ever since they were put in. With reference to the suggestion just made that there should be provision for ventilation of the sewers by connecting the gas outlets with a furnace or with some other means of burning the noxious germs, it was mentioned in the paper that this plan had been adopted for the storage tank on the recommendation of the military authorities, in order that noxious germs, if there were any, might be consumed in their passage through a furnace. But with regard to the sewers in the non-manufacturing part of a town, was it suggested that the furnaces should be placed at every fifty or hundred yards, in order to ventilate the length of sewer between them? It appeared to him that to make such a plan perfect there should be alternately an inlet opening for admission of air into the sewer and an outlet leading to one of the furnaces through which the gases should be compelled to pass; it was clear such a suggestion was not feasible in practice. The proper thing however, he agreed with Mr. Lemon and Sir William Pink, was not to have any sewer gas at all: to keep the sewage constantly flowing, as Sir Frederick Bramwell had tried to do; and to prevent any accumulation of gas, by having as many ventilating shafts as possible.

Sir Frederick Bramwell, Bart., regretted that the meeting had not heard anything from Mr. Burt, by whose firm the work had been carried out in a most admirable manner, and who had the satisfaction of knowing that everything had been well done.

With regard to the subject of the paper itself, he became more and more convinced, the older he grew, that there were two great difficulties in this world: the first was to earn one's living, which meant to get one's food; and the second, which was almost as great, was to get rid of the results. At the present time he did not know any engineering problem so difficult or so serious as that of getting rid of the refuse of a large town. Mr. Cochrane had spoken of the difficulties at other seaside towns; and mention had been made in the paper (page 321) of the advantage enjoyed by Portsmouth from the proximity of Langstone Harbour. The discharge of the sewage at that particular point had not been suggested by himself; the old sewage had been discharged there; and General Crease had recommended that the tank should be placed in this position. In fact no one could undertake the sewage outfall of Portsmouth without seeing that Langstone Harbour was the place which had been naturally provided for the purpose. There was an enormous body of water inside the harbour, a rapid current in and out, and, as was hoped, a current which in flowing out, if it were made use of at the right time of tide, would carry everything with it; and this had turned out to be the case. The outfall scheme at Portsmouth therefore was a remarkably easy one; and this was one of the reasons why he had apologised at the outset for presenting the paper to the consideration of the Institution.

With respect to Mr. Marten's enquiry (page 338) as to whether those sewers carried on piles and girders were found to remain true in line and level, although the roadway had varied, he had intended the remark in the paper (page 331), about recent examination having shown their perfect condition, to be the answer to this very question.

In regard to the rainfall (page 340), the provision that he had made for so much of the drainage area as was not provided with separate sewers was the same provision that had been made in London: namely, arrangements for dealing with a fall of a quarter of an inch distributed over the twenty-four hours. It was improper in his opinion to attempt to saddle any community with the cost of a pumping system which would deal with a heavy rainfall. To do so would mean a large cost at the outset; while on the other hand, when there was a heavy rainfall, it produced, as had been stated, such a dilution of the sewage that the whole might well be discharged direct at the storm overflow, as was the case in London and elsewhere. That was the way in which he had dealt with the question; for it had occurred to him that what was good enough for London was good enough for Portsmouth.

The PRESIDENT presented the thanks of the Institution to Sir Frederick Bramwell for his interesting paper.

DESCRIPTION OF THE NEW FLOATING BRIDGE BETWEEN PORTSMOUTH AND GOSPORT.

BY MR. H. GRAHAM HARRIS, OF LONDON.

Crossing Rivers.—Many arrangements have been devised for crossing rivers or arms of the sea, in those positions where it is necessary not only to facilitate the crossing, but also to keep the water-way open, so as to allow of the passage of large ships.

Opening Bridges.—One of the most notable of these, at present in course of construction, is the Tower Bridge over the Thames in London. Here a very heavy expenditure will be incurred to build a bridge, connecting the north and south sides of the river, between the Tower on the one side and Horselydown on the other. Fig. 5, Plate 54, is a diagrammatic elevation of the bridge, showing the means which have been adopted for keeping the water-way open for vessels with lofty masts, whilst still maintaining a possibility of traffic from side to side of the river. The central span, which is 200 feet in width, is an opening span; that is to say, the halves of the span can be lifted, being hinged on the piers at their inner ends, so as to leave the span clear. These piers are carried upwards, and at a clear height of 135 feet above ordinary high water are connected by a fixed bridge for foot traffic, enabling that traffic to be carried on even when the central span is open for the passage of large vessels. The piers are provided with staircases and with hydraulic lifts, the opening and closing of the central span being also performed hydraulically.

Ferries.—Another mode of obtaining the above end is that adopted in the case of the Thames ferries at Blackwall, or in the Birkenhead ferries for crossing the Mersey between Liverpool and

Birkenhead. In these cases there are one or more steamers, which ply from side to side, not being confined however to one particular line of course; being therefore affected by wind and tide, and requiring to be navigated; also requiring considerable power in the engines to enable these to propel the vessel at a sufficient speed to give steerage way when contending with strong tides and heavy winds. With ferries the difficulties of difference of level between the vessel and the shore, due to rise and fall of the tide, have to be provided for; this is usually done by a floating landing stage connected to the shore by a hinged prow or gangway.

Tunnels, &c.—A tunnel is of course an obvious mode of enabling a crossing to be made while maintaining the water-way open. Notable instances of such tunnels in Great Britain are the Severn tunnel, the Mersey tunnel between Liverpool and Birkenhead, and the new tunnel between Glasgow and Govan. With tunnels, as with a high-level bridge, the great difficulty which has invariably to be surmounted is that of the approaches. In order to allow of a workable incline for these, if the water to be crossed is deep and the tunnel level consequently some distance below the surface, the approaches must be of great length, or else hydraulic or other power lifts must be resorted to. Similar troubles arise with a high-level bridge. In the Mersey tunnel the traffic, which is a passenger traffic, is taken up and down by hydraulic lifts, thus avoiding any length of approach. In the new tunnel in Glasgow, hydraulic lifts will also be used, and will there be employed to deal with wheeled traffic, as well as with the ordinary foot traffic.

Floating Bridge.—No mode of attaining the desired end is so cheap, not only in first cost but also in working, and none is so satisfactory in many respects, as a floating bridge. Such a structure consists of a barge or vessel, which has passing through it one or more chains, supported on chain-wheels which are carried on the barge, the chain-wheels being made to revolve by power of some kind, carried by the bridge itself. The chains are laid from side to side of the water which has to be crossed, their ends being anchored on the shores; and the crossing is effected by the bridge hauling itself in either direction from one side to the other, lifting up the chains from the bottom, passing them through the bridge and over the chain-wheels, and dropping them behind it as it advances. The barge or vessel is made very shallow, so that it can approach close to the shore; and it carries at each end a projecting prow or landing stage, which has its inner end hinged upon the bridge, while the outer end is capable of being raised or lowered by power carried within the bridge.

Advantages and Disadvantage.—It will be seen that the result of such an arrangement as this is that with ordinary foot or wheeled traffic very little difficulty indeed is experienced with the approaches, nor are the differences of level of tide ordinarily of serious importance. Further, in its passage to and fro, the bridge keeps to a regular and fixed course, no matter what may be the state of the wind or tide. This is of great advantage to the traffic up and down the river, or in and out of the harbour, as the case may be. In addition the bridge requires no navigating, and can be readily stopped at any moment, being held or anchored by the chains in any position in which it is brought to rest. Most important of all, the power required to move it is small, and need not be in excess of that ordinarily required, because a structure worked in this way is not much affected by winds or by currents. These then are the advantages to be claimed for a floating bridge. The disadvantage, which is a very serious one, is common also to steam ferries, or to any mode of crossing other than by a fixed bridge: it is that any traffic desiring to cross, which arrives at the bridge end just after the bridge has started for the other side, has to wait until the bridge has crossed and has returned, before it can continue its journey. This is a disadvantage which has been felt between Portsmouth and Gosport, and has affected the communication there.

Torpoint Floating Bridge.—The Torpoint bridge at Devonport was designed by the late Mr. James Meadows Rendel, and was put to work in April 1834. A sectional elevation of this is shown in

Fig. 6, Plate 54. It is built of wood, and is roughly 55 feet in length, exclusive of the hinged prows or gangways at each end, and 45 feet in width, with a draught of water when fully loaded of 2 feet 6 inches. There are two condensing beam-engines, each with cylinder 19 inches diameter and 30 inches stroke, the revolutions being 35 per minute, and the boiler pressure 3½ lbs. per square inch. The width of the river at the crossing is 850 yards at high water, or nearly half a mile.

Nile Floating Bridge.—Over the Nile, about midway between Alexandria and Cairo, there is a form of floating bridge for conveying railway trains across the river, which is here about 370 yards wide. This is shown in Fig. 7, Plate 55. It was designed by Mr. Robert Stephenson, and was erected and put to work in 1856. It is 80 feet long, 60 feet wide, and 60 feet high; and consists of a flat-bottomed barge of shallow draught, carrying a vertical iron structure, in which is a platform extending the whole length of the barge, and sufficiently wide to carry two lines of rails of the ordinary gauge. The platform is capable of being raised and lowered by gearing worked by manual labour in the iron structure which supports it. An allowance of 27 feet between high and low Nile during the year has to be provided for; this variation occurs at the rate of not more than a few inches per day, and the platform is daily adjusted to the requisite height in the structure. The nicer adjustment, required for transferring the railway carriages from the barge rails to those of the timber jetties on each side of the river, is made by short lengths of hinged rails. The bridge is moved from side to side of the river by two high-pressure engines of 15 horse-power each, carried in the bridge, in a manner similar to that in which the Torpoint bridge is moved, and similar to that in which the floating bridge forming the subject of this paper is moved.

St. Malo Traversing Bridge.—In Fig. 8, Plate 55, is illustrated another method of crossing a tideway, by means of a traversing bridge, as employed across the harbour entrance between St. Malo and St. Servan. The distance to be traversed is about 110 yards, the spring tides here rising to a height of 34 feet above the bottom of the entrance, though at low water of these tides the passage is dry for some hours. The bridge consists of a platform supported from the bottom of the harbour entrance at a height of 36 feet on four braced wrought-iron columns. These rise from a framework carried on four wheels, which run on rails laid across the bed of the harbour entrance. Motive power is supplied to the bridge by means of hauling chains laid across the bottom, which are worked by a 10 horse-power winding engine on the St. Servan side, thus hauling the bridge backwards and forwards.

Portsmouth Floating Bridge.—Prior to the year 1840 the connection between Portsmouth and Gosport, then towns of considerable size—the population of Portsmouth being 50,000, and of Gosport including Alverstoke upwards of 14,000—was by means of the historical wherry. All readers of Marryat remember the Portsmouth wherry, which was practically similar to the boats now plying in the harbour and used by the watermen. It is an extremely useful and handy boat for the purposes for which it is intended.

An iron floating bridge to ply between the beach at Gosport and The Point at Portsmouth, Fig. 4, Plate 54, was designed by Mr. Rendel, and was put to work on 4th May 1840. This bridge was practically similar in principle to the one at Torpoint; it was 84 feet long and 60 feet wide, with 4 feet 6 inches draught of water. It was also practically similar to the new bridge, which was only launched on the 26th of this month, and has now to be towed round from the yard of the builders, Messrs. Allsup of Preston.

New Floating Bridge.—Except in small details the new floating bridge is so much like the original bridge that a description of the new one will be all that it is necessary to give. The new bridge is shown in Figs. 1 to 3, Plates 52 and 53. In plan it is a rectangle, Plate 53, 100 feet in length of hull by 62 feet in width, and draws 3 feet 3 inches of water. At each end there are two prows; those on each side and at opposite ends are coupled to each other by chains and wire-ropes passing over pulley wheels on the bridge, so that they counterbalance. The inner ends of the prows are hinged to the structure of the bridge; and either of them can have its outer end lowered so that it may allow of connection with the shore, at the same time raising its fellow at the other end of the bridge, Plate 52. The fellow prows are connected by a deck at a level of about 3 feet above the water; this deck forms the standing place for the wheeled traffic. The longitudinal centre portion of the bridge is occupied by cabins, and engine and boiler room, with an upper or promenade deck approached by a flight of steps from either end.

Gearing and Engines.—In the centre of the length of the vessel there is a cross shaft, carrying a mortice wheel M, Plate 53, with wooden cogs, 12 feet 4 inches diameter to the pitch line; and on each end of this shaft there is a chain-wheel W 10 feet 6 inches diameter, over which the hauling chains pass. The chains travel through the vessel in water-tight iron troughs, which form part of the structure and are employed to assist in bracing it; the chains enter and leave through holes considerably above the water line, and are supported in the troughs upon pulleys or rollers. The mortice wheel is geared up about 5 to 1 with a shrouded pinion in the centre of the length of the crank shaft of an overhead compound surface-condensing marine engine of ordinary type. The steam cylinders of the engine are 20 and 38 inches diameter, with 30 inches stroke. The speed is 60 revolutions per minute. The engine draws its steam from two cylindrical boilers, each 6 feet 6 inches diameter of shell and 13 feet long, and each having a single flue 3 feet 3 inches internal diameter, terminating in a combustion chamber, from which there are 92 tubes of $2\frac{3}{4}$ inches external diameter to the back end of the boiler; thence the products of combustion pass into the uptake and to the chimney and away.

Chains.—The old chains will be used for the new bridge. They are anchored at the shore ends by being connected in vertical pits to heavy weights, which are assisted by strong springs, thus allowing of a certain amount of elasticity at the anchorage. The chains pass over pulleys at the top of the pits. The links are of mild steel $1\frac{5}{8}$ inch diameter, and are of the ordinary shape and without studs. The chains are of excessive strength for the work required of them; but this is one of those cases where the mere weight of metal is of advantage, for this weight allows of the chains being at a very steep angle with the horizontal when the bridge is crossing, thus permitting vessels drawing a great depth of water to approach nearer to the bridge. The angle at which the chains enter the bridge at the forward or hauling end is practically one of 45° to 50°, so that at some few feet from the front of the bridge there is from 20 to 30 feet depth of water over the chains. Further than this, the chains are the portion of the work which most readily wears out, and which requires most frequent renewal. Their wear is largely due to their rubbing upon the gravel bottom of the harbour; and careful investigation has shown that it is desirable to make the chains as heavy as possible, with the object of preventing the necessity for their frequent replacement. The chain-wheels W on the bridge, by which the hauling is done, are made with fitted cast-steel whelps or cogs, bolted into the hollow rim of the cast-iron wheel, and shaped to suit the links of the chain. These are capablė of ready renewal when worn.

Electric Lighting.—There is nothing of especial note to comment upon in the engines or machinery of the bridge; and practically the main difference between the new bridge and that which it will displace is that the new one is of steel, and will be lit electrically, and that the engines and boilers are more modern in design and construction; whilst the old bridge was of iron and was lit by gas, which was filled into a gasholder carried in the bridge, by a flexible pipe connected to the shore at the Portsmouth end at those times when the bridge was waiting there for traffic. The electric-light plant of the new bridge consists of two vertical high-pressure engines, each coupled by a belt to a Paterson and Cooper dynamo. There are in all on the bridge lights equal to thirty-nine lamps of 16 candle-power; and in addition to these sufficient power has been allowed in the engines for two arc deck lights of 1,200 candle-power each. One of the engines and dynamos will do the whole of the work, the other being kept as a stand-by. Advantage has been taken of the fact of having these engines to make arrangements by which, in case of accident to the main engine, or if for any reason this cannot be worked, the bridge could be slowly brought to land by means of gearing working the main cross shaft through its mortice wheel. This gearing can also be worked by hand if desired.

The hull of the bridge is divided by transverse water-tight bulkheads into five compartments ; and there are two continuous longitudinal girders, one on each side of the bridge, formed by the sides of the cabins carried down to the skin of the vessel to strengthen it longitudinally. A journey is made by the present bridge, and will be made by the new one, in about five or six minutes, the distance being about 660 yards or 3-8ths of a mile. The advertised time for the starting of the bridge from the Portsmouth side is at the half hour and at the hour, and from the Gosport side at the quarter to and at the quarter past the hour. Under these circumstances it will be seen that it may be necessary, should a wheeled vehicle desiring to cross arrive just as the bridge has started to the other side, for it to wait for nearly half an hour before it can proceed on its journey. This is a disadvantage which has to be borne by the wheeled traffic; but in the case of the ordinary foot traffic the bridge company have provided in addition to the bridge itself a service of steam launches crossing in the intervals between the times of the bridge starting: so that as a fact it is possible for a passenger to cross without having to wait for more than some seven to ten minutes at the utmost. It is only within the last few years that these launches have been established. Previously it was necessary to wait for the bridge, or to take a wherry, there being a regular waterman's ferry plying for fares between these points. The bridge will take across as a maximum 1,200 foot passengers, or say one full regiment of soldiers.

Discussion.

Mr. HARRIS explained that, owing to unforeseen delays, the floating bridge, which was to have been launched a fortnight ago, had not been launched until yesterday. The members would however be able to see the old bridge, which, as mentioned in the paper, was very like the new one. There was one important difference however, in addition to that mentioned in the paper: in the new bridge the pair of projecting prows hinged on each end were made hollow, so that, in case they were struck off by a collision, they would not sink but would float, and there would thus be an opportunity of recovering them. For the purpose of bringing the new bridge round from Preston to Portsmouth, tugs would be employed; and rudders had been fitted to the bridge, which were coupled together, and were worked from a steam steering gear, whose engines obtained their steam from the boilers of the bridge itself; the steam steering gear was to be fixed in the engine room, and to be controlled by the captain on the upper deck. Sir Frederick Bramwell had suggested that the proper way was to tow her cornerwise, like a harrow, towing her from one corner and putting the rudders on the corner diagonally opposite. Probably that would have been a good plan; but the contractors had preferred to adopt the plan already mentioned. They had themselves suggested that they should put a false bow on as well; but he thought they had not done so.

Sir WILLIAM PINK, being interested in the passage between Portsmouth and Gosport, asked whether anything was going to be done with regard to the approaches, which at present were not only difficult but sometimes dangerous. A better approach was wanted on each side, but particularly on the Portsmouth side, which was pitched with granite blocks, and was sometimes so slippery that horses with heavy loads slipped in going down the steep descent, and accidents not unfrequently happened. If it was possible to make the approaches easier and better for the horses, it would be a great advantage, for which he was sure the inhabitants of Portsmouth and Gosport would all be most grateful.

Mr. HARRIS replied that all he had had to do had been to superintend the construction of the new bridge. The old chains were to be used, and he believed the old approaches. If the necessary expense could be incurred for improving the approaches, he had no doubt it could be done.

The PRESIDENT conveyed the thanks of the members to Mr. Harris for preparing this paper for their guidance in the visit which they would have made to the floating bridge, had it arrived at the time expected.

DESCRIPTION OF THE SOUTHAMPTON SEWAGE PRECIPITATION WORKS AND REFUSE DESTRUCTOR.

BY MR. WILLIAM B. G. BENNETT,
BOROUGH ENGINEER AND SURVEYOR.

Early in 1885 the Corporation of Southampton considered it expedient to introduce a more efficient system for the collection and disposal of house Refuse; and about the same time they found it desirable also to clarify by precipitation the Sewage of a particular district of the town, which was being discharged in its crude state direct into the Southampton Water at the Town Quay. Having been instructed to devise a scheme for accomplishing these objects, the author proposed the adoption of Messrs. Manlove, Alliott, Fryer and Co.'s refuse destructor, for destroying the ash-bin contents and garbage of the town; and also that the sewage-sludge should be transmitted to the destructor from the two existing reservoirs in which it was deposited in the process of clarification. These reservoirs are each 100 feet long and 60 feet wide, and at the lowest end 10 feet deep, Plate 60. Formerly the sewage of a district of the town, amounting to 500,000 gallons in twenty-four hours, from a population of about 13,000, for the most part flowed by gravitation into these reservoirs, whence it was discharged into the tide-way at low water; whilst a small portion, coming from a low-level sewer, passed direct into the tide-way through iron pipes laid under the reservoirs. The reservoirs act alternately, one being left still for precipitation of the sewage, whilst the other is being filled. In Plate 56 is given a plan of a portion of Southampton, showing the situation of the destructor works on the Chapel Wharf; the sewage precipitation tanks on the Town Quay, about one mile distant from the Chapel Wharf; the ejector stations; and the compressed-air and sludge mains.

Pneumatic Ejectors.—In order to render independent of the tide the discharge of the clarified effluent from the reservoirs, and to raise the low-level sewage into the reservoirs for treatment with the rest, two of Shone's pneumatic ejectors were erected, Plate 63, both of which are worked by power obtained from the destructor: the smaller of 360 gallons capacity is placed in the east reservoir, below the invert of the low-level sewer, and serves for transmitting the sludge from the reservoirs to the destructor, as well as for raising the low-level sewage; and the larger of 700 gallons capacity is placed in the east reservoir for discharging the clarified effluent into the Southampton Water. There is also a third ejector of 360 gallons capacity at E, Plate 56, which deals with the sewage of another district of the town near the destructor works, and is likewise worked by power obtained from the destructor; with an air-pressure of 12 lbs. per square inch it raises the sewage about 18 feet from a low-level sewer to a higher one. This ejector was formerly worked by an independent steam engine, costing for coals about £120 per annum, which is now saved since the adoption of the destructor.

A vertical section of one of the ejectors is shown in Plate 63, and its action is as follows. The sewage gravitates from the sewers through the inlet pipe I into the ejector, and gradually rises therein until it reaches the underside of the bell B. The air at atmospheric pressure inside this bell is thus enclosed; and the sewage continuing to rise outside above the rim of the bell compresses the enclosed air sufficiently to lift the bell, the spindle of which then opens the compressed-air admission valve A. The compressed air thus automatically admitted into the ejector presses on the surface of the sewage, driving the whole of the contents before it through the bell-mouthed opening at the bottom and through the outlet pipe D into the iron rising-main, or into the high-level gravitating sewer, as the case may be. The sewage can escape from the ejector by the outlet pipe only; because the instant the air pressure is admitted upon the surface of the sewage, the non-return flap-valve on the inlet pipe I falls on its seat, and prevents the sewage from escaping in that direction. As the sewage flows out of the ejector, its level therein falls to that of the cup C; and still continuing to lower, it leaves the cup full, until the weight of the stuff in the portion of cup thus exposed and unsupported by the surrounding sewage is sufficient to pull down the bell B and spindle, thereby shutting off the admission of compressed air to the ejector. The compressed air remaining within the ejector then exhausts through an air escape-valve in the top, which is opened by the fall of the cup C and spindle; and the sewage outlet non-return flap-valve falls on its seat, retaining the sewage in the rising main D. The sewage then flows once more into the ejector through the inlet I, driving the free air before it through the air escape-valve as the sewage rises; and so the action goes on as long as there is sewage to flow. The position of the bell and cup floats is so adjusted that the compressed air is not admitted until the ejector is full of sewage; and is not allowed to exhaust until the ejector is emptied down to the discharge level.

Reservoirs.—In each reservoir there is a floating sewage inlet P, Plates 61 and 62, consisting of a pipe hinged to the larger or effluent ejector, and shackled to a buoy; the latter causes the free end of the pipe to rise and fall with the level of the liquid, keeping its mouth, which is covered with perforated plate, a few inches below the surface of the liquid, in order to prevent the entrance of any floating solid matter. Directly the clarification by precipitation has been effected to a certain depth, a valve is opened, admitting the liquid into the larger ejector, whence it is at once discharged into the tide-way. A supplementary sewage outlet is also provided in each reservoir for discharging the liquid by gravitation when the tide is low enough. When the whole of the liquid has been thus drawn off, the buoy, resting now upon the floor of the reservoir, keeps the mouth of the inlet pipe P high enough to prevent the entrance of any sludge into the larger or effluent ejector; and by opening a valve the sludge is then admitted into the smaller or sludge ejector situated at the lower level, and is transmitted by air-pressure through a line of 4-inch cast-iron pipes, about a mile in length, to the destructor erected on the Chapel Wharf, Plate 56. An air-pressure of 40 lbs. per square inch is required for working the sludge ejector, and of 10 lbs. for the effluent ejector.

Precipitation.—Ferrozone supplied by the International Water and Sewage Purification Company is used for precipitating the sludge; it is mixed with just enough clean water to make the whole into a stiff paste, which is led through a shoot into a box with perforated sides, Fig. 3, Plate 57, placed in the sewer at B, Fig. 2. The sewage flowing past washes the ferrozone gradually out of the box, and is thoroughly mixed with it by the time it discharges into the reservoirs at a manhole 150 feet distant from the box. A small stream of water falling upon the ferrozone prevents it from consolidating. The box is filled three times in twenty-four hours; and this method of dosing the sewage has proved quite efficient and satisfactory.

Manure mixing.—On arriving at the destructor the sludge is delivered into a cell C, Fig. 4, Plate 58, from which it is drawn as required through a valve-pipe; and after mixture with road sweepings or sorted house-refuse it is turned out as a good manure, which from the commencement has all been readily bought up by agriculturists at 2*s.* per load delivered at the works. On an average 67 cartloads of ash-bin contents are daily collected and disposed of, the ascertained weight of the load in each cart averaging a little under 17 cwts. The road sweepings are never burnt; but to keep pace with the demand, the sludge is run into bays made of the road sweepings, and is also filled in with them; the quantity of road sweepings thus utilized amounts in twenty-four hours to about 8 tons. Arrangements were provided at first for burning the sludge, for which purpose it was discharged into a tank on the floor of the destructor, and drawn out through ports in the front, opposite the feed openings of the firing chambers, where its moisture was absorbed by the ash-bin contents, which were backed up against the ports with this object; and the mixture was then raked into the fires. Large quantities of sludge were thus destroyed; but the process has since been discontinued, owing to the ready sale of sludge when prepared for manure.

Destructor.—The refuse destructor, shown in Figs. 2 to 7, Plates 57 to 59, has six chambers or furnaces F, each capable of burning 8 to 11 tons of garbage per day. The products of combustion pass through a 30-H.P. multitubular steel boiler in the main flue M, Fig. 4, into a furnace shaft, which is of circular brickwork, 160 feet in height from the ground line, 6 feet inside diameter at top, and 7 feet at bottom, Fig. 2. The shaft is constructed upon a pedestal 14½ feet square and 24 feet high, of brickwork 3 feet thick; and thence upwards in four sections, of which the first is 27 inches thick and 30 feet high, the second 22½ inches thick and 30 feet high, the third 18 inches thick and 38 feet high, and the fourth 14 inches thick and 38 feet high. The first 30 feet height is lined with fire-brick, and behind the lining is a cavity 4½ inches wide, which is ventilated by apertures to the outside of the shaft. The foundation is loamy clay, upon which is laid a bed of concrete 30 feet square and 10 feet thick. The footings commence at 23 feet 2 inches square, and step off in regular courses upwards to 15 feet square at a height of 6 feet. The concrete was filled in continuously until completion. The pedestal was then run up and allowed to remain for nearly three months during the winter; after which the work was proceeded with until completion, occupying about six months more. The cap is white brick in cement, with a string course about 20 feet below the top. Foot irons are built inside in a winding lead up to the top. The shaft is provided with a copper-tape lightning-conductor, with iron rod and crow's foot 7 feet above the cap; the tape is about 215 feet long, the bottom end being carried into a well. In August 1888 the shaft was damaged by lightning, but was easily repaired, owing to the provision of the foot irons built inside it. At that time the shaft was plumbed, and was found to be quite vertical. The fires were only damped down during the repairs, which occupied about eight days. With the exception of this interval they have been constantly burning for nearly six years. The repairs have been almost nil. There is also a by-pass from the destructor to the shaft, for enabling the burning process to be continued when the boiler in the main flue is not required or during cleaning and repairs. No obnoxious fumes from the combustion have been perceived.

Steam Power.—The steam generated in the boiler is employed for driving a pair of engines E of 31½ indicated horse-power, which compress air into two large receivers RR at Chapel Wharf, Fig. 4, Plate 58, whence it passes through a 5-inch main to the Town Quay, where it is automatically supplied to the ejectors when required for working them. The air also serves for driving the precipitated sludge through the 4-inch main from the reservoirs to the destructor; for which purpose the air is led by a pipe from the receiver at Chapel Wharf to the head of the main at the Town Quay. A 6 horse-power engine, used in connection with the machinery for the preparation of fodder for forty horses at the corporation stables, is also driven by steam from the same boiler that supplies the air-compressing engines.

Utilization of Sludge and Refuse.—All obnoxious matters are collected throughout the borough in covered iron tumbler-carts of 2 cubic yards capacity, which go up the inclined roadway approach W to the destructor, Fig. 4, Plate 58, and discharge their contents into the firing chambers F. The road sweepings are frequently discharged into a hopper over an incorporator N, driven by a small engine, and are mixed with the sludge as required; this is generally done in wet weather. The residue from the continuous day and night combustion consists of about 20 per cent. of good hard clinkers and sharp fine ashes. The clinkers are used for the foundation of roadways and the manufacture of paving slabs; the latter have already been used in paving several footpaths of the town, and the new public baths, at a cost of 2*s.* 6*d.* per square yard. The fine ashes are also employed for making mortar, with which the stables and swimming baths have been erected; and for many other purposes. The mortar is also sold to builders at 7*s*. 6*d*. per cubic yard.

Electric Lighting.—The waste heat from the destructor is utilized for producing electricity. The air-compressing engines drive a dynamo of 150 volts. At the present time the works are lighted with two arc-lamps of 3,000 candle-power each, and twelve incandescent lamps of 16 candle-power each; and four streets in the vicinity of the works have been lighted experimentally for the information of the Corporation, who, from the successful results obtained, resolved to extend the installation to the municipal offices, a town clock, the Hartley Institution, and the Town Hall at the Bar Gate. For this purpose it was proposed to place accumulators in the basement of the municipal building, and charge them through a cable from the works. Circumstances having led to the abandonment of the street lighting, the public became financially the losers, and a private company is now supplying consumers.

Other Uses.—The destructor is also employed in lending a helping hand to a neighbouring authority, by supplying to the local board of Shirley and Freemantle, Plate 56, about 2½ miles from the works, sufficient compressed air to work ejectors which they have erected in connection with the disposal of their precipitated sewage sludge from a population of 15,000. The compressed air is conveyed through a 4-inch main from the destructor works to their precipitation reservoirs, thus saving them the cost of a pumping station, and bringing to the corporation a return of £200 a year, which is received for the compressed air. Thus the destructor works are now dealing with the sludge of nearly 30,000 inhabitants.

Cost.—The initial cost of the complete destructor—including engine house, inclined roadway, chimney shaft, boiler, and ironwork —was £3,723; and the sewage disposal works on the Town Quay cost about £3,000. This is exclusive of the Shirley and Freemantle works, which consist of three reservoirs very similar in construction to those at the Town Quay.

The annual expense for burning refuse is as follows:—

	£	s.	d.
Two stokers, one by day and one by night, at 25s. .	130	0	0
Two feeders, „ „ at 23s. 4d.	121	6	8
Total per annum	£251	6	8

Value of Refuse as Fuel.—The quantity of refuse burnt per day of 24 hours is a little over 50 tons, so that the cost of burning is about 3½d. per ton. The minimum quantity burnt per day of 24 hours is about 25 tons, which has been sufficient to maintain steam for the engines of 31½ indicated horse-power. This is equivalent to 16 cwts. of refuse per I.H.P. for 24 hours, or 75 lbs. of refuse per I.H.P. per hour.

The annual expenditure for the sewage clarification and disposal is as follows:—

	£	s.	d.
Precipitating material for 365 days, averaging about 5s. per day, say	90	0	0
Engine-driver, and labourers at reservoirs . .	128	0	0
Two men at wharf, mixing manure . . .	104	0	0
Total per annum	£322	0	0

Revenue.—The amount realised from the sale of manure and for the supply of compressed air during last year 1891 was £600. The products from the destructor—including concrete slabs, clinkers used for concrete foundations, and fine ashes for mortar and for foundations of footwalks—represent about £300. To these may also be added the saving of the coal which was required for working the engines previously to the establishment of the destructor.

Discussion.

Mr. BENNETT anticipated the pleasure of receiving the members of the Institution on the following day at Southampton, when he should be happy to give them any further information and to take them over the works.

Mr. CHARLES COCHRANE, Past-President, asked whether any attempt had been made in connection with the slaughter-houses to destroy the refuse from the slaughtered beasts. In Chicago in 1872 he had seen in operation a plan which worked admirably, of burning the refuse on the floor of the open hearth of a reverberatory furnace. He enquired also whether the occasion of the destructor chimney being struck by lightning (page 358) was before or after the lightning conductor had been provided; and, if it happened afterwards, whether the conductor was at fault, or where the fault lay; also what extent of damage had been done to the chimney by the lightning.

Sir WILLIAM PINK had for some time past been considering the advisability of getting a destructor established in Portsmouth. It seemed at first sight an easy method of getting rid of the refuse; and it appeared that at Southampton the refuse, along with the sewage sludge and road sweepings and other material, realised a considerable amount of money as manure. He should be glad however to know what advantage the farmers found from using this manure; because he had himself practically tried the corporation manure of Portsmouth, and other corporation manures, and had not found them worth the cost of carriage from the buildings to his farm, which was nine miles away. Possibly the sewage sludge at Southampton had produced something better, which might influence Portsmouth in deciding whether to go to the expense of building a destructor or not. He should like also to know whether there was any nuisance from the fumes escaping from the top of the chimney of the destructor. At some other places it had been stated that such fumes were extremely noxious. With regard to the clinkers obtained from the destructor, it appeared that in Southampton there was the means of utilising them for useful and valuable purposes; but in travelling upon the South Western Railway he had noticed not far from Nine Elms a destructor bearing the announcement "ashes and clinkers given away." If that was the case in the metropolis, he should be glad to know whether it was likely that in Portsmouth such clinkers could be disposed of permanently for anything like the amount realised at Southampton.

Mr. BENNETT replied that the offal from the slaughter-houses had frequently been burnt in the destructor without causing any nuisance. The lightning conductor had been provided on the chimney previously to the accident; and since the accident, during the time when the chimney was being repaired, a second conductor had been added. What advantage the farmers found from the use of the manure, he did not know; but at any rate they were glad to fetch it from the wharf, and pay 2*s.* a load for it. He had seen the destructor near Nine Elms, where clinkers and ashes were given away; but experience had led him to think that it was not necessary to give them away at Southampton, because they could readily be sold, or made into artificial paving slabs, or ground for making mortar. No complaints had been received of any noxious smells or any nuisance arising from the destructor works since they had been established.

Mr. COCHRANE enquired whether the lightning conductor had been discovered to be out of order after the accident, or whether the height of the rod was sufficient to protect the large diameter of the chimney top. He understood it would protect only a radius equal to its own height; and fancied the reason why the chimney was struck might be that the rod was not high enough.

Mr. BENNETT replied that the top of the lightning conductor was 7 feet above the summit of the chimney, and it was in good order. Before attaching it, the advice of good electricians had been obtained. The topmost however of the series of foot irons, which were built inside the chimney in spiral fashion up to the top, terminated a few

(Mr. Bennett.)
inches below the chimney top on the opposite side to that on which the lightning conductor was fixed. The lightning flash he presumed was thus shared between the conductor proper and the iron step, from which latter it sought the earliest escape it could make, rupturing the brickwork in its passage to the real conductor. Had the conductor proper been placed on the same side of the chimney as the top step, it would have conveyed the whole flash safely to earth, and would have rendered superfluous the subsequent addition of a second conductor and copper connecting belt. The damage caused was the displacement of about 2 feet depth of brickwork round the top of the chimney, costing about £20 for repairs.

The President asked Mr. Bennett to accept the thanks of the meeting for his interesting paper.

EXCURSIONS.†

On TUESDAY AFTERNOON, 26th July, after luncheon in the Mayor's Dining Room, a Visit was made to the Royal Dockyard (pages 370–375), by permission of the Lords of the Admiralty, and under the guidance of Admiral Fane, Admiral Superintendent.

In the evening the Institution Dinner was held at the Esplanade Hotel, Southsea. The President occupied the chair; and the following guests accepted the invitations sent to them, though those marked with an asterisk * were unavoidably prevented at the last from being present.

The Worshipful the Mayor of Portsmouth, T. Scott Foster, Esq., J.P. The Worshipful the Mayor of Southampton, James Lemon, Esq., J.P. Mr. Philip Murch, and Mr. George C. Day, *Honorary Secretaries.*

Authors of Papers.—Mr. Henry E. Deadman,* Chief Constructor at the Admiralty, late of Portsmouth Dockyard; Mr. John T. Corner, R.N., Chief Engineer, Portsmouth Dockyard; Mr. R. Edmund Froude, Admiralty Experiment Works, Haslar; Mr. H. Graham Harris; Mr. William B. G. Bennett, Borough Engineer and Surveyor, Southampton; Mr. William Matthews, Waterworks Engineer, Southampton.

Major-General Geary, C.B., R.A., Commanding Royal Artillery, Portsmouth; Mr. William Scott, Superintendent, Royal Clarence Victualling Yard, Gosport; Mr. George H. Stainer, Civil Assistant, Portsmouth Dockyard; Mr. L. G. Davies, Chief Constructor, Portsmouth Dockyard; Mr. D. W. Lane, Electrician, Portsmouth Dockyard; Sir William D. King, D.L., J.P., and Mr. John Baker, M.P., Directors, Floating Bridge, Portsmouth; Sir William Pink, J.P., Chairman, Drainage and Sanitary Committee, Portsmouth; Mr. A. W. White, J.P.,* General Manager, Portsmouth Tramways and Piers; Colonel Charles Mumby, J.P., Chairman, Gosport Local Board; Mr. W. Panter, Superintendent, Carriage and Wagon

---

† The notices here given of the various Works &c. visited in connection with the meeting were kindly supplied for the information of the Members by the respective proprietors or authorities.

Department, London and South Western Railway, Eastleigh; Mr. Charles Du Sautoy,* Superintendent Engineer, Union Steamship Company, Southampton; Mr. Robert Brown, Superintendent, Joint Railway Companies' Steam Packet Service, Portsmouth; Mr. Arthur J. Vinall, Postmaster, Portsmouth; Mr. Samuel B. Darwin,* Manager, Portsmouth Gas Works; Mr. Joseph H. Judd, Head Master, Technical School, Brighton.

The President was supported by the following officers of the Institution:—*Past-Presidents*, Sir Frederick Bramwell, Bart., D.C.L., LL.D., F.R.S., Mr. Charles Cochrane, Mr. Jeremiah Head, and Mr. Joseph Tomlinson; *Vice-Presidents*, Sir Douglas Galton, K.C.B., D.C.L., F.R.S., and Mr. E. Windsor Richards; *Members of Council*, Mr. Arthur Keen, Mr. John G. Mair-Rumley, Mr. Edward P. Martin, Mr. William H. Maw, Mr. William H. White, C.B., F.R.S., and Mr. J. Hartley Wicksteed.

After the usual loyal toasts, the President proposed "The Worshipful the Mayor of Portsmouth," which was acknowledged by his Worship, T. Scott Foster, Esq., J.P. The toast of "The Worshipful the Mayor of Southampton," proposed by Mr. Joseph Tomlinson, Past-President, was acknowledged by his Worship, James Lemon, Esq., J.P. Sir Frederick Bramwell, Bart., Past-President, proposed the toast of "The Army, Navy, and Auxiliary Forces," which was acknowledged by Major-General Geary, C.B., R.A., Commanding Royal Artillery, and Mr. John H. Heffernan, C.B., R.N., Chief Inspector of Machinery, Dockyard Reserve, Portsmouth. The toast of "The Authorities and Proprietors of the Works visited by the Members," proposed by Sir Douglas Galton, K.C.B., Vice-President, was acknowledged by Mr. John T. Corner, R.N., Chief Engineer, Portsmouth Dockyard; Mr. William Scott, Superintendent, Royal Clarence Victualling Yard, Gosport; Sir William Pink, J.P., Chairman, Drainage and Sanitary Committee, Portsmouth; and Sir William D. King, D.L., J.P., Director, Portsmouth Floating Bridge. The toast of "The Honorary Secretaries" was proposed by Mr. E. Windsor Richards, Vice-President; and was acknowledged by Mr. Philip Murch and Mr. George C. Day. The Mayor of Portsmouth, T. Scott Foster, Esq., J.P., proposed the toast of "The President," by whom it was acknowledged. The toast of "The Secretary" was proposed by Mr. Joseph Tomlinson, Past-President, and acknowledged by Mr. Alfred Bache.

On Wednesday Afternoon, 27th July, after luncheon in the Mayor's Dining Room, the Members were invited by his Worship to inspect the ancient Charters and Plate of Portsmouth, preserved in the Town Hall; and also to hear the fine Organ in the Hall, built by Messrs. Gray and Davison, which was played at his desire by Mr. Walter A. Griesbach, Organist of St. Thomas' Church.

A visit was then made in special brakes to the Eastney Sewage Works (pages 319–343), under the guidance of Sir William Pink, J.P., Chairman of the Drainage and Sanitary Committee, Sir Frederick Bramwell, Bart., Mr. H. Graham Harris, and Mr. Philip Murch, Borough Engineer. Before inspecting the storage tank, situated beneath the glacis of Fort Cumberland, the Members were invited by Major-General Geary, C.B., R.A., Commanding Royal Artillery, to witness in the fort the drill and firing of a 6-inch Hydro-Pneumatic Disappearing Gun, from which three rounds of 100-lbs. shot were fired at a floating target in tow of a steamer. The first round was with a half charge of powder for warming the gun and testing the mechanism; the next two rounds with full service charges of 48 lbs. of powder struck close to the target, and would have hit a war-vessel of moderate size. The range was given by a Watkin range-finder, and the gun laid with straight-edged sights, the word to fire being given as the bow of the imaginary war-vessel came on the edge of the sights. On arriving at the tank the Members witnessed the starting of the turbines, opening the outlet valves for the discharge of the sewage; and inspected one of the three compartments of the tank, which had been emptied and cleaned, and was lit up with candles for the occasion. In returning from the tank and the pumping station, the brakes passed through the drive of Eastney Barracks, by the invitation of Colonel Commandant Suther, A.A.G., who had also invited the Members to alight and inspect the barracks under his guidance.

Leaving South Parade Pier, Southsea, in special steam launches, the Members proceeded to visit the Royal Clarence Victualling Yard, Gosport (pages 375–376), by permission of the Lords of the Admiralty, and under the guidance of Mr. William Scott, Superintendent. They had been invited also to inspect the new Floating Bridge between Portsmouth and Gosport (pages 344–353), under the guidance of Sir William D. King, D.L., J.P., and Mr. John Baker, M.P., Directors; Sir Frederick Bramwell, Bart., and Mr. H. Graham Harris. But the new bridge not having been launched in time, the old one was seen in passing; and on the suggestion of his Worship the Mayor a trip was made on leaving the Victualling Yard to the head of Portsmouth Harbour, to visit the ruins of Porchester Castle, the old Roman fortress which formerly commanded the harbour.

A visit was also made to Whale Island, on the invitation of Admiral Fane, Admiral Superintendent of Portsmouth Dockyard, to witness the gunnery training there carried on.

On THURSDAY, 28th July, an Excursion was made to Southampton by special steamer. The Members were received at the new Royal Pier (pages 313–318) by the Mayor of Southampton, James Lemon, Esq., J.P., and John Miller, Esq., J.P., Chairman of the Harbour Board, and William Bone, Esq., J.P., Deputy Chairman; and inspected the structure under the guidance of his Worship and Mr. E. Cooper Poole, Engineers, and Mr. Henry I. Sanders, Contractor.

Visits were then made to the new Empress Dock (pages 377–378), under the guidance of Mr. Philip Hedger, Secretary and Superintendent, and Mr. A. Brydges Giles representing his brother Mr. George F. L. Giles, Assistant Engineer; and to the Union Steamship Co.'s new Repairing Shops (pages 378–379), under the guidance of Mr. Charles Du Sautoy, Superintendent Engineer. Sir Steuart Macnaghten, J.P., Chairman of the Southampton Dock Co., and Mr. Alfred Giles, Chairman of the Union Steamship Co. and Engineer of the Docks, were both of them prevented from being present.

Alternative Visits were made in special brakes from the Docks to the Ordnance Survey Office (pages 380–385), under arrangements made by Colonel Sir Charles W. Wilson, K.C.B., K.C.M.G., Director General; and to the following Works in Northam (pages 385–389):—Messrs. Day, Summers and Co., Northam Iron Works, Princes Street; Messrs. Driver and Co., Timber and Creosoting Works, Old Northam; Messrs. J. G. Fay and Co., Yacht Building Works, Millstone Point, Northam; Messrs. Summers and Payne, Yacht Building Works, Belvidere Road; and to the Gas Works, Marine Parade, under the guidance of Mr. Samuel W. Durkin, Manager. Also to the Refuse Destructor, Chapel Wharf, and the Sewage Precipitation Reservoirs and Ejectors, Town Quay (pages 354–364), under the guidance of Mr. William B. G. Bennett, Borough Engineer and Surveyor.

After luncheon at the Royal Victoria Assembly Rooms, the Members returned by brakes to the Docks Station, passing alongside the old Walls, through the West Gate, along the Town Quay, and past the South Tower. By special free train of the London and South Western Railway they were taken to Otterbourne, where the Pumping Engines and Water Softening Machinery of the Southampton Corporation Water Works were shown by Mr. William Matthews, Waterworks Engineer; and also to Eastleigh, where the new Carriage and Wagon Works of the London and South Western Railway (pages 390–391) were visited under the guidance of Mr. W. Panter, Superintendent of the Carriage and Wagon Department. Thence they returned to Portsmouth in the special train, the carriages of which had been newly turned out from the Eastleigh Works.

On Friday, 29th July, an Excursion was made round the Isle of Wight (pages 391–397) in the steamer "Heather Bell," specially offered free by the Joint Railway Companies. The party were accompanied by Mr. Robert Brown, Superintendent of the Joint Steam Packet Service, and Mr. David Greenwood, Superintendent of the Joint Railway Station, Portsmouth.

On SATURDAY, 30th July, a visit was made to the Locomotive and Carriage Works of the London Brighton and South Coast Railway at Brighton (pages 401–402), under the guidance of Mr. Robert J. Billinton, Locomotive Superintendent; and to the Technical School, Brighton (pages 402–403), under the guidance of Mr. Joseph H. Judd, Head Master. For this visit special facilities were offered by the London Brighton and South Coast Railway.

The following Works were also open to the visit of the Members during the Meeting:—St. Catherine's Lighthouse, Isle of Wight (pages 397–401), of which the Engineer in charge is Mr. H. C. Millett; Portsmouth Gas Works, Flathouse Quay, Landport, Portsmouth, of which Mr. Samuel B. Darwin is the Manager; and Portsmouth Water Works, situated in the parishes of Bedhampton and Farlington, near Havant, of which the Engineer is Mr. Henry Robert Smith.

During their stay in Portsmouth, the Members of the Institution were made Honorary Members of the Royal Naval Engineers' Club, Lion Terrace, Portsea.

H. M. DOCKYARD, PORTSMOUTH.

On entering the Dockyard by the main gates A, Plate 33, from what is known as the Common Hard, Portsea, on the left just within the gates are the mast-houses. Prior to the days of steam navigation, and to the manufacture of iron and steel masts, the work carried on within these buildings afforded unfailing interest to all who were capable of appreciating the construction of the masts and spars for the old battle-ships. At the end of the mast-houses, and a short distance to the west, is a building B containing the principal engines for the hydraulic service of the dockyard. From this point are seen two long rows of brick buildings, one surmounted about the middle by the clock tower C, and the other by the semaphore D. The clock-tower buildings are the storehouses, in which cordage, sails, light metal, and other goods are kept for supplying to the ships of the navy. The semaphore building contains the sail loft, the colours or flags loft, and the rigging house. Before the days of the electric telegraph, information was signalled from this semaphore to similar apparatus on the hills at the north of the town, and thence principally to Whitehall; and so complete was the system for the transmission of messages from semaphore to semaphore, that it is said any signal made from a ship arriving at Spithead was in the hands of the Admiralty in half an hour. As fogs sometimes interfered with the communication to the next station, horses were kept saddled and bridled ready to convey a message. Daily at one o'clock the fall of the time ball at Greenwich was communicated to Portsmouth; and as the semaphores were ready for the message, it commonly passed from London to Portsmouth in less than half a minute. The last message transmitted by this route and system was on 31st December 1847. From this elevated position a splendid view is obtained of the dockyard and harbour, the Solent and Spithead, and the whole of the adjacent towns.

Almost opposite to the clock tower C is Anchor Lane, which is filled from end to end with a double row of anchors piled closely together. It runs between a row of storehouses on the south side, built in 1777, and the old rope house E on the north side; the latter also is now used as a storehouse, the manufacture of ropes having been discontinued at Portsmouth. A little further north is the Admiral Superintendent's office F, fronting No. 2 dock, in which Nelson's ship, the "Victory," was recently repaired, and thereby obtained a fresh lease of life.

The bridge of No. 2 dock is near the centre of an almost quadrilateral area of just over 8 acres, which formed the original dockyard in 1540 during the reign of Henry VIII, Plate 32. During the Commonwealth 2 acres were added in 1658 on the east side; and during the reign of Charles II there were added 18 acres, 8 in 1663 on the south side, and 10 in 1677 on the east. William III added 10½ acres on the north side, beginning what was apparently the first reclamation of mud land, the northern boundary of the previous area having been the shore line. The reclamation then begun appears to have been completed in 1710 during the reign of Queen Anne. In 1723 George I extended the reclamation in the same northerly direction, adding 19 acres. In 1790 George III completed the northern portion of the yard by a further addition of 5½ acres. On the south and south-east sides of the dockyard a line of fortifications appears to have existed, which were afterwards demolished; and in this direction land was from time to time added to the yard, until the southern part extended up to the existing boundary wall built in 1711, in which the main gates are situated. From 1793 to 1843 there appears to have been no alteration in the boundary of the dockyard; but on the introduction of steam navigation increased accommodation was required, and in 1843 a purchase was made of 5½ acres of property, and about 12½ acres of mud land were reclaimed; this part of the yard was completed in 1848. In 1864 was commenced the work now known as the Dockyard Extension, which included the absorption of about 84 acres of land occupied chiefly by old fortifications, and the reclamation of about 93 acres of mud. This extension gives the dockyard at the present time an area of altogether about 293 acres.

From the bridge of No. 2 dock the old ship basin of 2½ acres, Plate 33, and the docks in connection therewith, are all within sight, the dock gates and bridges affording evidence of a past era. Two of the docks, Nos. 3 and 4, have been modernised by being fitted with caissons, and have also been lengthened. The masonry at the entrance to the basin bears the inscription "Britannia, 1801." At that date, under the direction of General Bentham, the entrance to the basin was reconstructed and fitted with a caisson, the first which had been made in England; and it is stated that at its first use a large number of people assembled, mostly incredulous of its success. It is assumed from the inscription that a ship named "Britannia" was the first vessel to pass through.

At the end of the bridge of No. 5 dock is situated at G the oldest dock-pumping machinery in the yard. Adjoining are the blockmills and sawmills H, where is to be found the complete set of block-making machinery designed by the elder Brunel in 1801, and manufactured by Messrs. Maudslay. The value of this machinery at the time it was erected may be judged from the fact that the price agreed to be paid for the design was the amount of the savings on one year's requirements of blocks for the navy; and this was estimated to amount to £16,621. Brunel was also paid £1,000 for his models, and a retaining fee to inspect the manufacture of the machines, making a total of £20,000. The dock pumps were renewed at about the same time, and the first steam engine was erected in the dockyard; this improvement was carried out by General Bentham, who also started the sawmills. Northwards are Nos. 7 and 10 docks, which by the removal of a caisson can be made into one dock 648 feet long. This is the longest dock in the establishment; and it appears from old plans to have been a natural creek, once known as the North Camber.

Northwards of No. 10 dock are the ship-building slips, of which there were five. In the days of wooden ships and less dock accommodation, all five were required for building purposes; but at the present period of iron shipbuilding, it is found convenient to utilize some of them as machine shops, &c. On the northernmost slip, known as No. 5, is being built the "Fox," intended for a cruiser of 4,360 tons displacement and 9,000 I.H.P. with a speed of 19½ knots. One of the last of the old wooden line-of-battle ships, the "Victoria" of 121 guns, was launched from this slip by Her Majesty in 1859, at which date each of the five slips had a ship building upon it: on No. 4 was the "Duncan," a two-decked ship; on No. 3 the "Prince of Wales," a three-decker; on No. 2 the "Frederick William," a two-decker; and on No. 1 a frigate.

The next point of interest is the west steam factory, Plate 33. In the boiler shop J at the northern end the boilers for the "Fox" are at present under construction. At the southern end of this shop are the turnery and the erecting shop, where are in course of construction parts of the engines for the "Fox," besides several submerged discharging tubes for Whitehead torpedoes. The upper floor over the whole 600 feet length of the building is used for light fitting work. Close at hand is the smithy K, with its furnaces, steam-hammers, forging machines, &c.; this building was partly constructed of materials from the Exhibition of 1851. Crossing the caisson of No. 7 dock, and passing along the south side of the steam basin, on the right are the offices L of the Chief Constructor and the Chief Engineer. Continuing round the steam basin, which covers an area of seven acres, on the east side are the 40-ton sheers and the 40-ton steam crane. At the head of No. 11 dock, which opens out of the steam basin, is situated the centrifugal dock-pumping station M; south of this is the shipbuilding shop N, eastwards of which at a little distance is the gun-mounting store and workshop P. Northwards again, at the south-west corner of the repairing basin, are the boiler house Q and the engine house R of the central pumping station, which last contains also the air-compressing machinery for the yard service. On the south side of the repairing basin is the electric fitting shop S; and further east No. 13 dock, in which is building the second-class battle-ship "Centurion," of 10,500 tons displacement, 13,000 I.H.P., and speed 18¼ knots. From the stern of this vessel the torpedo store T is about 300 yards to the south-east. On the east side of the repairing basin are the 80-ton sheers U. Outside the north wall of the fitting-out basin are three of the old wooden line-of-battle ships, "Asia," "Duke of Wellington," and "Marlborough," now used as barracks for seamen. At the western extremity of the fitting-out basin is "Coaling Point," now being completed and fitted with hydraulic coaling machinery by Messrs. Tannett Walker and Co.

Passing over the sliding caissons Z which connect the north and south locks and the deep dock with the tidal basin, a good view of the four basins and their appliances can be obtained. The greatest depth of water over the dock sills is 41½ feet, and the length of the docks here is 415 feet. Of the three enclosed basins one covers an area of 22 acres, and the other two of 14 acres each; the tidal basin covers 10 acres. The whole of the gear for the capstans, penstocks, and some of the cranes belonging to these docks and basins, and also the sliding caissons, are worked by compressed air.

For returning hence towards the main gates the route leads past the foundry V, the pattern shop, the joiners' shops, and also the residences of the principal officers and of the Admiral Superintendent.

At the end of the terrace of houses on the east is the fire-engine house W, surmounted by a tank containing 770 tons of water, which is used for supplying injection to the condensers of the various engines, as well as for fire-service purposes. Opposite is the chemical laboratory, and further on are continuations of the storehouses on each side. After passing through an arch, a little to the east is the dockyard church; beyond which, along the main road, is Admiralty House on the east side, the residence of the Naval Commander-in-Chief, and on the west the dockyard lawn with officers' residences. The latter originally formed the naval technical school presided over by the late Dr. Woolley. Many of the buildings in this part of the yard are said to have been erected under the supervision of Telford, who was employed as foreman of works at Portsmouth from 1784 to 1786; and he relates that while here "he had frequent opportunities of observing the various operations necessary in the foundations and construction of the docks, wharf-walls, and such like." After passing the Commander-in-Chief's house, on the east is the Royal Naval College, and the naval savings bank with the cashier's offices; and on the west are the boat-houses, opposite to the mast-houses, and close to the main gates of the dockyard.

ROYAL CLARENCE VICTUALLING YARD, GOSPORT.

This establishment, of which a plan is shown in Plate 64, exists for the supply of provisions, victualling stores, and clothing to the seamen of the Royal Navy on the Portsmouth station, and for the supply of provisions and victualling stores required by troops voyaging on Indian and Imperial services; that is to say, the supplies here dealt with are those which relate to the personal sustenance and comfort of the crews, as distinguished from those which form the stores for the equipment of the ships themselves apart from their crews. For the supplies of fresh provisions, such as fresh beef, mutton, corned pork, &c., ample accommodation has been provided in the shape of extensive lairs and abattoirs, constructed and arranged on the principles approved by the Agricultural Department, one of the main objects being to secure the inspection of every animal in a live state, as a preliminary evidence of health and quality. The animals are then stored for not less than a minimum period, in order that they may become tranquillized before slaughter, as this condition is also important for the healthfulness and keeping properties of the meat. The quantities issued of fresh provisions are large, as in this respect the Victualling Yard provides not only for the seamen afloat, but also for the troops and marines in garrison, representing daily rations in the aggregate for some 15,000 men in ordinary times, and for much larger numbers in times of pressure, such as reviews, mobilisations, expeditions, &c. There is also an extensive granary, a flour mill, and a biscuit bakery. Wheat is purchased under local contracts, and converted into flour and biscuit according to requirements. For some years past there has been a progressive diminution in the demands for biscuit, owing to the shorter periods during which the ships are now at sea, and to the introduction of soft bread baking on board troop ships and depôt ships &c.; hence the bakery machinery, which works economically and efficiently, remains practically what it was on its first introduction more than forty years ago, pending improvements which experience of the precise wants may ultimately show to be most suitable. Casks and cooperage utensils of all descriptions are manufactured on the premises by hand. There are also extensive storehouses for every description of provisions and clothing, water tanks, officers' and seamen's mess traps, miscellaneous implements for use on board ship, &c. The establishment further supplies water from its reservoirs for victualling purposes on board ships, to which it is delivered by the steam water-tanks belonging to the yard. A private railway station in this yard has been used by Her Majesty for many years on her journeys to and from Osborne.

SOUTHAMPTON DOCKS.

The Southampton Docks, Plate 65, are 78 miles west-south-west from London by the London and South Western Railway. This port enjoys the advantage of double tides, a peculiarity met with at no other port. In addition to the usual high water, a second high water occurs here about two hours after the first, in consequence of the Isle of Wight being situated across the entrance to Southampton Water. In the progress of the great tidal wave up the Channel, the portion flowing up the Needles passage into the Solent becomes separated from the main body of the wave, and reaching Southampton causes the first tide here at about the same time that the main flow arrives at Dunnose Head, the south-easternmost point of the island. This first tide on beginning to ebb from Southampton is stopped and driven up again by the main stream coming round through Spithead, and thus producing the second tide at Southampton about two hours later and six inches higher than the first. Low water is about three and a half hours after the second high water, or about five and a half hours after the first; and the flood tide lasts about six and a half hours. To ships using the port it is obviously of great advantage to have the water thus remaining nearly stationary at high tide for more than two hours, thereby allowing vessels to come out of the graving docks without risk, and others to enter without losing a tide. The docks are situated in a remarkably safe position within the land-locked port, and are accessible at all states of the tide and in all weathers.

The southernmost dock is the Empress Dock, which was inaugurated and named by Her Majesty the Queen on 26th July 1890. Its extent is 18½ acres, with a minimum depth of 26 feet at low water of spring tides, and an entrance 175 feet wide. It is provided with 3,750 feet of quays, and extensive cargo sheds of the most complete description. Owing to the peculiar rhomboidal or lozenge shape of the dock, vessels can be berthed alongside any of the quays without being turned, and with the greatest amount of ease. It has also been possible to lay the rails round the quays so that there is no curve of less radius than six chains or 400 feet, which has been found a great convenience for passenger traffic, as railway carriages can thus be taken to any part of the dock.

Further north is a tidal basin of 16 acres, with an entrance 150 feet wide, and a depth of nearly 31 feet at high water of spring tides. Adjoining the basin on the west side is a close dock of 10 acres, with 56 feet width of entrance, and a depth over the sill of 29 feet at high water of spring tides, and 25 feet at neaps. The length of the quays exceeds 7,500 feet; in addition to the extension quay 1,820 feet long, with 20 feet of water at low tide, which forms the eastern arm of the Empress Dock.

The four dry docks are capable of taking vessels of 8,000 tons and upwards. The following are their dimensions in feet:—

Dry Dock.	Length.	Width.	Depth of Water over blocks.
No. 1	400	66	21
„ 2	250	51	15
„ 3	500	80	25
„ 4	450	56	25

No. 2 dock was the first dry dock ever built of brick, previous dry docks having invariably been built of stone. There are three sets of sheers, worked by steam power, for lifting up to 100 tons.

UNION STEAMSHIP COMPANY'S NEW REPAIRING SHOPS, SOUTHAMPTON DOCKS.

The buildings are arranged in two parallel blocks, Plate 66, each $239\frac{3}{4}$ feet long by 49 feet wide. The north block is $33\frac{3}{4}$ feet clear height from ground level to under side of principals; and is well lighted from large side windows and skylights extending the whole length of the roof. It is divided into fitting and machine shops, engine and boiler house, and boiler-makers' shop. A 20-ton overhead hand-travelling crane, arranged to traverse the whole length of this block, is erected on piers 26 feet above floor-level, thus giving ample head-room for lifts. The main shafting, for driving lathes, radial drilling machines, punching and shearing machines, plate-bending rolls, etc., is carried on brackets secured to piers on one side of the building, and the counter-shafting is carried on double steel beams underneath of H section. The engine and boiler house contains two marine boilers, one compound engine for driving the machinery, a Cameron donkey-pump, and one Haslam cold-air machine delivering 20,000 cubic feet of cold air per hour. This machine is arranged to work two cold chambers, of 2,233 and 2,194 cubic feet capacity, which are built outside of the north block. The temperature of each chamber is continuously indicated upon a thermometer scale fixed in the superintendent's offices. In the boiler-makers' shop is a large plate-heating and annealing furnace, capable of admitting a plate 12 feet by 6 feet. In this shop are also placed a Root's blower for supplying blast to smiths' fires, a heavy punching and shearing machine, plate-bending rolls, etc.

The south block is only 20 feet high. It comprises pattern, tinning, and tinsmiths' shops, brass foundry, coppersmiths' and blacksmiths' shops, and stores. In the smiths' shop there are two steam-hammers of 12 and 4 cwts.

At the east end of the north block are placed the time offices and general stores. At the west end are the superintendent's offices on the ground floor, above which is a spacious drawing office; and above this again a photographing room, fitted with a novel arrangement for launching out frames for sun exposure.

In the yard of 30 feet width between the two blocks, rails are laid in connection with the dock service, having two large turntables to facilitate getting heavy pieces of machinery into the fitting and boiler-makers' and smiths' shops; when brought under the overhead traveller in the north block the pieces can be manipulated at will. Throughout the yard ample provision is made for extinguishing fire, several hydrants being conveniently arranged, and fitted with hose in cases, ready at any moment for use. An important feature in this factory is the telephone arrangement from the superintendent's offices to all the various departments, whereby full communication with foremen is obtained without loss of time.

ORDNANCE SURVEY OFFICE, SOUTHAMPTON.

The manuscript plans completed at any of the ten Ordnance Survey Offices throughout the country are sent to Southampton for publication on the following scales:—10·56 feet per mile, or 1-500th, for maps of towns of which the population exceeds 4,000; 5 feet per mile, or 1-1,056th, for maps of certain towns; 25·344 inches per mile, or 1-2,500th, for parish maps; 6 inches per mile, or 1-10,560th, for county maps; and 1 inch per mile, or 1-63,360th for a general map of the country. Town maps were originally engraved on copper, but are now published by photozincography, as are also the parish maps. Many of the six-inch county maps have been engraved on copper, and the process is still used for some counties. The manuscript plan, scale 1-2,500th, having been reduced by photography to the scale of six inches per mile, the photograph is traced on special tracing paper with a composition of lamp-black and gum and water, and the tracing is rubbed down upon a copper-plate, covered with a thin film of wax. The work is then engraved on the copper-plate, and printed for publication.

From about 1881 to 1890, for the cadastral survey, the six-inch maps were produced direct from the 1-2,500th manuscript plans by photozincography. To suit this process, all the detail in the manuscript plan is drawn to scale; but the names and ornament are sufficiently enlarged, so that when reduced they will be of suitable size. Moreover any names or ornament that would crowd the detail too much on the reduced scale are drawn on the manuscript plan in cobalt blue, and hence are not reproduced by photography. After the photographic transfer has been laid down on the zinc plate, the parks and also the mud along rivers, sea coasts, &c., are inserted by a tint. This is done by applying an etching solution wherever the tint is not required; after which, finely ruled lines can be transferred to the unetched portions of the zinc plate to produce the tint required. The advantages of the process are, firstly that the work is more accurate, any possibility of error in reduction being eliminated; secondly the operation is so much quicker that a plan which took two years to engrave can be produced in two days; thirdly there is a great saving in cost, estimated to exceed £100,000 on the total work.

In the case of the plans of Lancashire and Yorkshire now being published, the 1-2,500th manuscript plans are not specially drawn for reduction to six-inch scale. In this case impressions from zinc printed in cobalt blue are inked by pen in black, with such thickness of outline and size of lettering that when reduced they will produce the size required for the six-inch scale. Everything tending to overcrowd the detail on the reduced scale is left in cobalt, and is consequently not reproduced by photography. The four 1-2,500th impressions that are to form the six-inch quarter-sheet are then placed together, and the common edges made to fit accurately, and the whole is reduced by photography.

A reduction to the one-inch scale is made from the six-inch map, only so much of the detail being reduced as can be shown on the one-inch map without impairing its distinctness. A tracing of this is laid down on copper and engraved. The old process of producing the one-inch map was by a pentagraph reduction from the six-inch sheets; but the pentagraph has now been dispensed with, and such of the detail as is intended to appear on the one-inch scale is inked by pen in black, on blue impressions of the six-inch sheets; the remaining detail being left in blue is not reproduced by photography. The impressions are reduced by photography and transferred to zinc, from which a print is taken in transfer ink for rubbing down on copper for engraving. As soon as the engraving is completed for the outline edition, which shows contours and all other detail except hill-shading, a matrix is taken of the plate, and a duplicate plate made from it, and used for printing copies for sale. The hill features are engraved as a separate plate, partly by a graver and partly by etching, and a matrix and duplicate are then made. The main difficulty in engraving the hill features is to keep the different ranges of hills in proper subordination. The copper-plate printing press is a powerful one, being arranged to give a pressure of 40 tons on the cylinder, and the driving power is obtained from a small high-speed Willans engine, geared direct by means of a worm wheel. A second steam printing press has recently been erected.

The electrotype process, which is eminently suited for reproducing engraved copper-plates, was adopted on the Survey first at Dublin in 1840, one year after its discovery; and then at Southampton in 1847. By this means a matrix, in which all the detail is raised above the surface of the copper, is obtained from the original engraved copper-plate, and from the matrix is made a facsimile or duplicate of the original. This duplicate is used for printing from, and it can be reproduced from the matrix when signs of wear are apparent, usually after about 700 impressions have been pulled. The saving effected is great, for whereas an original engraved six-inch plate costs on an average £300, the cost of reproduction by electrotyping is only £6 or £7. The process is also used for taking copies of an engraved copper-plate map at different stages, for making corrections; and also for reproducing the stamps of trees, woods, &c., used in stamping the ornament on the manuscript plans.

The electrical current is obtained from a shunt dynamo by Messrs. Crompton and Co., and driven direct by a Willans three tandem compound high-speed engine, the current being kept constant at 35 ampères by means of a Willans electrical governor. The dynamo is connected to a number of electrolytic tanks by means of copper-wire leads. These tanks contain a solution of sulphate of copper slightly acidulated with sulphuric acid, and they are arranged in series, that is to say the whole current from the dynamo passes through the first tank, then through the second, and so on through all the tanks, 23 in number; and then through the solenoid of the electrical governor, finally returning to the dynamo. The connections are so arranged that any of the tanks can be cut out of the circuit at pleasure. A rocking motion is given to the tanks to keep the solution at a uniform density. At night the current is supplied from accumulators made by Messrs. Elwell and Parker, which are charged in the daytime by the dynamo; the rocking motion is caused by an electric motor driven by five of the storage cells. The engraved plate, after being chemically cleaned, is washed over with cyanide of silver, and afterwards with a solution of iodine in alcohol, and is then exposed to the sun for a few hours, the object being to prevent any of the deposited copper from adhering to the plate. It is then secured to a board, and the edges are covered with a composition of beeswax and tallow to prevent the copper from growing over the edges. A sheet of commercially pure copper, called the dissolving plate, is placed in one of the tanks, and the engraved plate is placed over it, being separated from it and kept parallel to it by means of wooden blocks. Electrical connection is then made to the leads that carry the current, by means of two short leads of twisted copper wire, soldered respectively to the dissolving sheet and to the engraved plate. As soon as the connection is made, electrolytic action commences, which causes copper to dissolve from the dissolving sheet, and to be deposited in equal quantity on the engraved plate. On a full-sized engraved plate it is deposited at the rate of one-tenth of a pound per hour. The smaller plates are electrotyped in a vertical tank. As soon as a matrix has been produced of 1-12th inch thickness, the engraved plate is removed from the trough, and the matrix separated from it. When a duplicate is required to be electrotyped, the matrix takes the place of the engraved plate, and the above process is gone through: except that, as the duplicate is required to be 3-16ths inch thick in order to stand the pressure of the printing press, the matrix is kept in the trough more than twice as long as before. The weight of a six-inch matrix is 24 lbs. and of its duplicate 40 lbs.; that of a one-inch new-series matrix is 8 lbs., and of its duplicate 20 lbs. Nickel-plating and steel-facing are also carried out in this department for various survey uses.

Photography was first applied to map reduction at Southampton in 1855, thus obviating the use of the pentagraph. Photozincography, or the art of printing a line photograph in permanent ink from a zinc plate, was discovered at the same place in 1859. This process was first used for the publication of facsimiles of ancient national manuscripts; but in 1881 it was utilized for the publication of the six-inch map, and in 1889 was adopted for reproducing the larger scale maps of the Ordnance Survey. On the focussing screen of the camera a rectangle is marked of the exact size of the required copy, the plan is fixed vertically on the plan board, and the camera is adjusted so that the rectangle on the focussing screen will exactly correspond with the sheet lines of the plan on the board. The glass plate having been coated with collodion, the latter is sensitized by being immersed in a silver nitrate bath, and then placed in a dark slide, which takes the place of the focussing screen in the camera. The length of exposure varies with the amount of light and temperature. The film is developed in the iron developer, fixed in cyanide of potassium, intensified in bromide of copper and nitrate of silver, and varnished with gum dammar. The glass plates used for copying the 1-500th and 1-2,500th scale plans measure 45 inches by 30 inches, are $\frac{1}{4}$ inch thick, and weigh 33 lbs. A negative is proved by a print on ferro-prussiate paper or by the platinotype hot-bath process. On the print the defects are detected by comparison with the manuscript plan; and they are corrected on the negatives by painting them out with lamp-black dissolved in benzole-varnish, and recutting the work in an accurate form with a finely tempered needle-point. A photo-transfer is made on Evans' thin paper coated with gelatine, sensitized by potassium bichromate. This prepared paper is exposed under the negative to the light, for two or three minutes in bright sunlight, or twenty minutes if electric light is used. The print is then laid face downwards on a zinc plate which has been covered with a thin coating of transfer ink, and is passed once or twice through the zincographic press so as to obtain a uniform coating of ink on the print. The print is now floated on hot water, to allow the gelatine unacted upon by light to swell, after which the operator sponges the surface of the print with tepid water, and then allows it to remain in a bath of hot water for about an hour, when the whole of the gelatine not acted upon by light will be washed away, leaving the lines on the paper covered with a fine coating of transfer ink. The print is then dried, and defects made good with Abney's transfer ink; and it is then ready for transfer to zinc. The arc light is used for photographic printing in bad weather.

Zincography is a process similar to lithography, and is used in preference at the Ordnance Survey Office, owing to the portability of a zinc plate, and the small space it occupies in storage; these are advantages of great importance, because many of the plates have to be preserved, and also because small editions are printed. Until recently the whole of the work was printed in hand-presses, from each of which only about 20 copies per hour can be turned out; but a steam printing machine by Messrs. Furnival and Co. has been introduced, by which 800 per hour can be printed. Owing to the small editions printed, it was necessary to make such arrangements that a zinc plate could be fixed in the press and removed with great rapidity. For this purpose a special cast-iron bed was designed, and a zinc plate can be fixed and removed again in one minute. The edges of the zinc plate have to be bevelled to fit in this bed, and this is done by the milling machine in the electro department. It has been found that the plates are preserved in good condition for much larger numbers of impressions in this machine than in the hand-presses.

Workshops containing a small quantity of machinery are maintained, in which are made several of the articles used on the survey, such as steel surveying chains, type-palettes for holding type for stamping, steel stamps, offset staves, &c. Repairs to machinery are also effected on a considerable scale.

The different standards of length used in the surveys of the chief countries in Europe were sent to Southampton for careful comparison for the purpose of measuring an arc of longitude. The delicate apparatus by means of which they were compared is still occasionally used for the comparison of various standards.

NORTHAM IRON WORKS, SOUTHAMPTON.

These works, carried on under the name of Day, Summers and Co., belong to Messrs. Day Brothers, and cover 10 acres of ground. The firm was founded in 1834 by the late father of the present members, in conjunction with Mr. William Summers and Mr. William Baldock, under the title of Summers, Day, and Baldock; later the name of the firm became Summers and Day, and then C. A. Day and Co., until in 1871 the present title was adopted.

Since 1834 there have been built at these works 103 vessels, including 27 steam yachts, and machinery of over 100,000 indicated horse-power. Among the most notable ships built here may be mentioned the "Hindostan" of 3,086 tons and 600 horse-power, and the "Surat" of 3,057 tons and 600 horse-power, both for the Peninsular and Oriental Steam Navigation Co.; the former was for many years one of the fastest of their fleet, and a favourite passenger vessel. Some years ago also the "Allemannia" of 2,500 tons and 400 horse-power was built here for the Hamburg-American Co.; on her first voyage to America she beat the record, and for some time was one of the fastest steamers crossing the Atlantic. The most recent work turned out has included the tripling of the Royal Mail Steamship Co.'s "Para" of 5,000 I.H.P., which attained a speed of 16¾ knots; new engines and boilers of 2,500 I.H.P. for the Union Steamship Co.'s "Nubian"; also the s.s. "Tyrian" of 1,500 tons and 1,400 I.H.P., and the s.s. "Norseman" of 950 tons and 850 I.H.P., which are running for the Union Steamship Co.'s colonial and continental traffic; and new engines and boilers of 1,600 I.H.P. with forced draught for the London and South Western Railway Co.'s steamer "Diana."

One speciality made here consists of traversing sheer legs, originally introduced in 1862, of which sixty-six sets have now been made for lifting weights varying from 20 to 150 tons; among these are several sets supplied to the British, foreign, and colonial governments, notably one of 150 tons for Cronstadt for the Russian Government. Another speciality is non-fleeting hauling-up slip-machinery, in which the hauling is done by a wire rope; two examples are in use in the works on different principles.

On entering the works the block of buildings on the right contains the general offices, with drawing office overhead; on the left is the pattern shop. Adjoining the offices are the main store, the turnery, and the fitting and erecting shops, which are at present in course of alteration to meet modern requirements. Across the tram lines is the central block of buildings, containing the iron foundry, smiths' shop, boiler shop, and brass foundry. The boiler shop has recently been entirely rebuilt and fitted with all new machinery of the most modern kind, including three hydraulic riveters, of which the largest is capable of exerting a pressure of 100 tons per square inch. Outside the boiler shop come the plate furnaces and boiler smithy, behind which is the copper shop. A pair of fifty-ton sheers overhang the river and the fitting-out wharves. Behind these are two building slips and the shipyard, in which are two hauling-up slips; the larger is capable of taking up vessels of 1,500 tons, and the smaller of 500 tons. Next to the smaller slip are three blocks of yacht stores, and facing them are berths for yachts. Nearer to the entrance is the timber yard, with its saw mills, at the end of which is a large block of buildings containing spar sheds, rigging sheds, &c.; and the joiners' shops overhead are fitted with some of the most modern wood-working machinery.

NORTHAM STEAM SAW MILLS AND WHARF, SOUTHAMPTON.

These works, the property of Messrs. Driver and Co., are situated on the land formerly occupied by the shipbuilding yard of Messrs. Money Wigram and Co. Considerable alterations however have been made since then. By carrying a wharf along the river front, the present proprietors, besides increasing the extent of ground, have also enabled their shipments to be landed direct from the vessels which have loaded them abroad. Steamers carrying over 1,800 tons measurement are now able to discharge at the wharf.

The machinery comprises a new sawing and planing mill, completed this year, and furnished throughout with machines of the latest type by Messrs. Thomas Robinson and Sons, of Rochdale. In addition there are other sawing and planing machines, a creosoting plant, and machinery for rendering timber fire-proof by the injection of material manufactured by the Pyrodene Fire-proofing Co. also a drying chamber for seasoning timber, with the necessary machinery. The power for working the various machines is supplied by a vertical compound condensing engine, built by Messrs. Robinson and Sons, and erected this year.

In addition to the works at Northam, the firm have also a yard with steam sawing and planing and moulding mills and joinery works, at Threefield Lane, St. Mary's, Southampton.

NORTHAM YACHT AND SHIPBUILDING YARD, SOUTHAMPTON.

This yard, belonging to Messrs. J. G. Fay and Co., and situated at Millstone Point, Northam, has been established now over twenty years, and three years ago was turned into a company, under the management of Mr. Joseph M. Soper. Of late years many of the fastest and best known English racing yachts of the larger classes have been built here, including "Irex," "Iverna," "Valkyrie," "Lethe," "Reverie," "Castanet," "Challenge," "Stranger," and "Columbine." The yard is also used for repairing and laying up yachts, having ample accommodation for hauling up; and every winter many of the racing fleet are hauled up in winter quarters. In the laying-up season the mud berths are always full with yachts of every description, from the smallest size up to the most luxurious steam-yachts afloat. There are two gridirons, on which yachts can lie ashore for scrubbing. By the aid of extensive machinery, work is carried out with the utmost despatch in every department from keel to truck.

YACHT BUILDING WORKS, BELVIDERE ROAD, SOUTHAMPTON.

These works, belonging to Messrs. Summers and Payne, were started in 1845 by the late Mr. Alfred Payne. Since that time they have built upwards of a hundred sailing yachts ranging from 2 to 120 tons, besides several steam yachts ranging from 8 to 271 tons. During the last three winters the number built has averaged fourteen, all of which have been designed by Mr. Arthur E. Payne. The busy time in the works is in the winter and spring; by the end of May all the yachts have left the yard for the season, and in the summer there is little to be seen beyond the machinery, which consists of wood-working machines of modern construction.

SOUTHAMPTON GAS WORKS.

These works appear to have been originally established by two private individuals in 1819, when the town had a population of 14,000 to 15,000. In 1823 they were taken over by a company, which was incorporated by act of parliament in 1848. The original works have long since disappeared, as have also those which succeeded them. To meet the growing requirements they were enlarged in 1853 under the then engineer and manager, Mr. James Sharp; and in 1866–67 a general reconstruction became necessary, which was carried out under the company's consulting engineer, Mr. J. Birch Paddon, under whom acted the present manager, Mr. Samuel W. Durkin.

The present works include a wharf, with viaduct and steam cranes, for discharging coals from screw colliers on the river Itchen. The coals are distributed into three retort houses, which are jointly capable of producing about 2,000,000 cubic feet of gas per 24 hours. In the rear of the retort houses are the condensers, washer, rotary scrubber, tower scrubber, purifiers, engine room, meter and governor house, and a laboratory. There are four gasholders, of the united capacity of 1,800,000 cubic feet. On the wharf has recently been erected a sulphate of ammonia plant by Messrs. R. and J. Dempster of Manchester. The town has now a population of between 60,000 and 70,000; and with villages and townships around, the works supply a large straggling district of about 80,000 inhabitants. The mains extend northwards to Eastleigh, more than five miles from the gas works, where they supply the railway station and the recently erected carriage and wagon works of the London and South Western Railway.

LONDON AND SOUTH WESTERN RAILWAY CARRIAGE AND WAGON WORKS, EASTLEIGH.

These works were opened in 1891 for the building and repairing of the carriage and wagon stock. They are situated on the eastern side of the main line from London to Southampton and Bournemouth, and cover an area of 42 acres including yard sidings, Plate 67. The floor area inside the shop walls is 456,000 square feet, and all the shops are on the ground floor. The timber is transferred over sidings on the Portsmouth line to the stacking ground and timber drying sheds, and thence is taken into the saw mill, where it is converted; it is then conveyed into the carriage building shop, where it is erected; and the carriages built are passed on over a traverser into the paint shop. The traverser, made by Messrs. Butler and Co. of Leeds, is capable of conveying a carriage of 45 feet wheelbase, which is the longest on this railway; it is driven by an engine fixed upon it. The shops have been built apart from one another, as a precaution in case of fire. The wagon building and repairing shop is at the side of the saw mill, and is connected up by a light tramway upon which timber and other stores are conveyed; it is supplied direct by the shunting engine from the sidings at the south end of the shop. The smithy and machine shops are in one block, placed between the wagon building shop and the carriage lifting shop. The steam hammers are supplied with steam from a nest of three boilers of locomotive pattern, situated at one end of the smithy; and also from two others, which are fed with the waste heat from the reverberating furnaces. These two boilers also supply the vertical engine in the machine shop. The blast for the smiths' fires and spring furnace is obtained from two of Root's blowers. The saw-mill engine is of the standard inside-cylinder locomotive type used on this railway, with cylinders 18 inches diameter and 26 inches stroke; it is supplied with steam from a nest of three boilers, one of which is fed by the sawdust and shavings from the mill. There are two lines of shafting, placed underground 116 feet apart, and driven by a 10-foot fly-wheel with ten white cotton ropes of $1\frac{3}{4}$ inch diameter. The paint shop, 420 feet long by 200 feet wide, is heated with steam pipes 3 inches diameter, which with the fittings complete were supplied by Messrs. Herring and Son of Chertsey; they are fed with steam from two boilers, one on each side of the shop. This warming apparatus has proved sufficient to keep the temperature in the shop up to 60° Fahr. even in the coldest days of winter. At one end of the shop, space is allotted for the trimmers. The other main buildings are 300 feet by 200 feet, 20 feet high from floor to roof plate, and 33 feet high in centre. The number of persons employed here is 1,150; and for the convenience of those living at a distance a dining hall has been erected at the entrance to the works, with ample accommodation for 600 men. The offices and stores are immediately facing the public road, and in close proximity to the entrance gates.

VOYAGE ROUND THE ISLE OF WIGHT.

The mouth of Portsmouth Harbour, Plates 43 and 54, between the Round Tower in Portsmouth and Blockhouse Fort on the Haslar or Gosport side, is only about one-eighth of a mile wide, yet has sufficient depth of water for a man-of-war to enter at any time of tide. Outside the harbour mouth a bank called the Spit runs out from its western side, extending nearly three miles to the south-east, Plate 31; whence the name Spithead is given to the roadstead or eastern arm of the sea which bounds the Isle of Wight on the north, up to the entrance into Southampton Water.

The Spithead Forts are four in number, situated on Spit Bank, Horse Sand, Noman's Land, and off St. Helen's, Plate 31. The foundations of the first three are built up from the bottom of the sea; they are circular forts, constructed of granite, and faced with heavy iron armour-plates. Spit fort, the nearest to the mainland and to the mouth of Portsmouth Harbour, has casemates seawards for nine large guns, and for several smaller on the side towards the harbour. Horse and Noman's are the two largest forts, each having casemates for forty-nine guns, arranged in a lower and an upper tier. The foundations for the forts were designed in 1861 by Sir John Hawkshaw, and were carried out under his superintendence from 1861 to 1868.

About midway between Southsea Castle and Ryde, a red buoy marks the place where on 29th August 1782 the "Royal George," carrying a hundred guns, sank so rapidly that of nearly 900 persons on board almost all were drowned, including Admiral Kempenfelt. For repairing the bottom on the starboard side the ship had been careened over too far; and the water rushed into the port holes on the lower gun deck too fast for the calamity to be averted. The wreck lay here until 1839, when it was blown up with gunpowder by Colonel Pasley.

Ryde Pier is about half a mile long. Alongside the steam railway, which starts from the pier head, runs also an electric railway constructed in 1885 by Messrs. Siemens.

The principal places round the coast of the island are seen in the following order, when starting eastwards from Ryde:—

Sea View, nearly due south of Southsea Castle. About two miles out eastwards lies the Warner Lightship, which has one mast carrying the lantern with a red ball at the mast-head; the light is white and revolving, of one minute period; a gong is sounded in thick or foggy weather. Then comes the ruined beacon-tower of old St. Helen's church; and just beyond is Brading Harbour, overlooked on the north side by St. Helen's, and from the south by Bembridge. From the estuary, which formerly extended a couple of miles inland up the eastern Yar to Brading, 700 acres of land have now been reclaimed, an unsuccessful attempt to accomplish the same object having been made 250 years ago by Sir Hugh Myddelton, who constructed the New River Water Works in London.

Off Foreland or Bembridge Point, which is the eastern extremity of the island, lies the Nab Lightship, nearly four miles out, having one mast carrying the lantern with a red ball above. It has a double-flashing light, giving two white flashes of 2½ seconds' duration every 45 seconds; a gong is sounded in thick or foggy weather. It is moored 1¾ mile eastwards of the Nab rock.

Whitecliff Bay is separated by Culver Cliff from Sandown Bay, where are seen Sandown, Shanklin and Shanklin Chine, Luccombe Chine, and Dunnose Head, off which the training ship "Eurydice" was capsized in a squall on Sunday afternoon 24th March 1878, when returning home from the West Indies.

Beyond Dunnose lie Bonchurch and Ventnor, backed by St. Boniface Down, 764 feet. At Bonchurch was born Admiral Sir Thomas Hobson, the hero in 1702 of Vigo Bay. Beyond Ventnor a conspicuous building is Steephill Castle, erected in 1835 ; and next is the Royal National Hospital for Consumption, commenced in 1869 and still extending westwards.

The Undercliff extends about seven miles from Bonchurch to Black Gang. The strata overlie one another in the following descending order:—chalk marl, chalk, green sandstone, blue marl, and red ferruginous sand. The blue marl, being readily acted upon by the land springs, softens into a yielding mud, locally called "blue slipper," which slowly oozes out seawards, and leaves the upper strata unsupported.

In Puckaster Cove stood thirty-five years ago the Royal Victoria Hotel, built on insecure ground, and wholly swept away by the sea, leaving only a trifling vestige of its foundation walls.

St. Catherine's Point Lighthouse, erected in 1838–40 on the extreme south point of the island, rises to 134 feet above sea-level, so that the light is visible 17 miles. Adjoining is the electric-light establishment, by which the lighthouse is now illuminated. A red sector of light is shown towards the Needles between the bearings of N 53½° W and N 43½° W from the lighthouse; this light is fixed red, varied by flashes every thirty seconds. (See pages 397–401.)

On the top of St. Catherine's Hill, the highest point in the island, 781 feet above the sea, are the ruins of a lighthouse commenced in 1785 by the Trinity House, but never completed on account of the mists whereby the light would have been so frequently obscured. Close by is the repaired octagonal tower 35 feet high of a hermitage erected towards the end of the thirteenth century, and endowed in 1323 by Walter de Godyton, for maintaining by night a signal-lamp for the guidance of mariners.

Rocken End, half a mile west of the lighthouse, is distinguished seawards by a "race," which in bad weather becomes dangerous to vessels approaching too near.

At Black Gang Chine the old coach road ran across the chine about half way up; and traces still remain, enabling its course to be followed along the cliffs, till it turns inland west of the chine towards Chale. Walpen chine, Ladder chine, and Whale chine succeed Black Gang westwards.

From Atherfield Point runs out far into the sea Atherfield Ledge, partly uncovered at low water. On this ledge the North German Lloyd steamer "Eider" from New York stranded in a fog on Sunday night 31st January in this year, believing herself to be then in the Solent; and becoming deeply embedded on the "blue slipper" or stiff sticky slaty clay which forms the ledge, she was not got off again till Tuesday 29th March, after having been lightened of the greater part of her cargo. The Royal National Life-Boat Institution established a station here last year, and the crew of the life-boat were successful in saving a large number of lives from the stranded steamer. Irrespective of fog, St. Catherine's light is not visible from Atherfield, being intercepted by the intervening ground at Rocken End.

Brighstone or Brixton Bay and several more chines intervene between Atherfield and Brook, where at low water can be seen the few remains of a submerged copse; and pieces of fossil wood are constantly being washed up here by the sea.

At Afton Down, on which are numerous barrows of the early Britons, the low red cliffs of the wealden formation are succeeded by the chalk of Freshwater cliff. Here terminates the military road, which skirts the coast from Chale; the Freshwater end has fallen into the sea at the spot marked by the two isolated masses of chalk called the Stag Rock and the Arched Rock.

Freshwater Gate, overlooked by a small fort, owes its latter name to the fact that it is the only opening down to sea-level through the range of cliffs and downs along the south coast of the island. The spring of fresh water, to which its first name is due, issues from the foot of Afton Down, a few hundred yards inland and below high-water level. Instead of finding its way here into the sea so close at hand, this western Yar flows northwards three miles to Yarmouth, widening out into an estuary on passing after the first mile Freshwater village, where a non-return flap-sluice prevents the tide flowing up the estuary from flooding the land which has thus been reclaimed between that point and the spring at Freshwater Gate. The High Down cliffs, rising nearly 480 feet, succeed Freshwater Gate westwards.

The Needles are three isolated rocks at the western extremity of the island. The lighthouse on the outermost was erected in 1859 by Mr. James Walker, and the light is 80 feet above high water. The light, which is a one-minute occulting, consists of a first-order lens for a fixed light, with a six-wick Trinity House Douglass mineral-oil burner in focus, showing white, red, and green sectors, to suit the requirements of navigation. A red sector, strengthened by vertical condensing prisms, is shown in the direction of St. Catherine's lighthouse between the bearings of SE ¼ S and SE from the lighthouse, in order to warn vessels approaching the Atherfield ledge.

Alum Bay is distinguished by its vertical strata, sharply defined and remarkably variegated in colour: purple, red, blue, yellow, gray, and black. The alum found in the bay is no longer collected for commercial purposes; the white sand at the north end of the bay is utilized for glass-making. Headon Hill, 400 feet high, surmounted by a fort, separates Alum Bay from Totland Bay and Colwell Bay; inland at a short distance from the latter is Golden Hill Fort on high ground.

Hurst Castle and its high and low lighthouses on the mainland, and Cliff End Fort at the north end of Colwell Bay, mark the entrance to the Solent, here only about a mile wide, across which the telegraph cable is laid.

Sconce Point was formerly marked by Carey's Sconce, a blockhouse erected in Elizabeth's reign by Sir George Carey, on the site of which now stands Victoria Fort.

Yarmouth, at the mouth of the western Yar which rises at Freshwater Gate, contains the ancient mansion of Admiral Sir

Robert Holmes, who was governor of the island 1667–92; the house is now the George hotel. The castle or round fort, erected in 1539 by Henry VIII, is a small semicircular battery commanding the entrance to the Yar estuary. The embankment of the marshes behind Yarmouth was carried out under Sir Robert Holmes' direction. A diminutive side-chapel of the church contains one of the finest pieces of sculpture in the island, a life-size statue in white marble of Sir Robert Holmes, beneath an arched canopy supported on solid columns of porphyry. Between Yarmouth and Lymington plies a steamboat in connection with the London and South Western Railway.

Beyond Bouldnor cliff a navigable creek with many branches runs inland up to Newtown, and nearly up to Shalfleet. Along its shores are numerous salterns, or shallow evaporating pans for obtaining salt from sea-water, which, as well as the neighbouring brickworks, no longer do so large a trade as formerly.

Egypt Point, the northern extremity of the island, facing the entrance to Southampton Water, is close to West Cowes at the mouth of the river Medina, which is navigable up to Newport, a distance of 4½ miles. At West Cowes were built many of the old line-of-battle ships, including Nelson's "Vanguard," some of them carrying as many as sixty-four guns. Here are now the yacht-building yards of Messrs. White, which were originated in 1815. Since 1856 the castle has served as the house of the Royal Yacht Club, established in 1812, from which earlier year dates the popularity of West Cowes as a yachting resort. Along the shore northwards from the castle extends the Green, laid out and presented to the town by Mr. George Robert Stephenson.

East Cowes lies opposite West Cowes. On Old Castle Point, Henry VIII constructed East Cowes Castle, of which no remains are now left. Above the point is Norris Castle, built of stone so prepared as to give an idea of greater antiquity.

Osborne Palace, of which the best view is obtained from the water, was designed and built by Mr. Thomas Cubitt.

Just beyond Wootton Creek are the ruins of Quarr Abbey, now converted into farm buildings. On the opposite coast is Stokes Bay, where is the measured mile for speed trials of steamships; at the eastern end of the bay is Gilkicker Point, protected by Gilkicker and Monckton Forts.

ST. CATHERINE'S LIGHTHOUSE.

Situated on the southernmost point of the Isle of Wight, near Niton, about 4½ miles west of Ventnor, this lighthouse was erected by Mr. James Walker, and the light was exhibited on 1st March 1840. The illuminating apparatus then consisted of a first-order dioptric lens for a fixed light, with a Fresnel four-wick oil-burner in focus, the focal plane being 178 feet above high-water and visible 19 miles. The approximate intensity of the light through the lens, as subsequently improved, was about 6,500 candles. Owing to the treacherous nature of the ground—consisting of upper greensand upon gault, acted upon by springs at the back which produce constant subsidences of the land—and owing also to the fact that the light was frequently capped by fog, it was deemed advisable in 1875 to reduce the height of the tower 44 feet, and at the same time to improve the lantern and illuminating apparatus by substituting a dioptric mirror in lieu of the four metallic reflectors originally fitted for utilizing the rear light; and also a six-wick Trinity House burner in place of the former four-wick. The intensity of the light thus improved was about 8,000 candles; and the focal plane is 134 feet above high-water, with a range of 17 miles.

On 1st May 1888 the electric light was installed. For this purpose additional dwellings were erected for the accommodation of the increased staff, an engine-house was built, and other additions were made to the establishment.

Electrical Apparatus.—This consists of two De Meritens magneto-electric machines, producing in the arc a light of about 22,000 candles when working singly, with an electromotive force of about 50 volts and a current of 240 ampères; or about 38,000 candles when working together as they do in hazy or foggy weather, with an electromotive force of 47 volts and a current of 470 ampères. Previous to their erection at St. Catherine's, these machines were used in 1885 at the South Foreland in the experiments on lighthouse illuminants. Each machine is composed of sixty steel permanent magnets, arranged in five rings of twelve magnets each, with twenty-four coils on each ring of the revolving armature. The coil consists of four layers of copper wire, 27 mm. or 1·06 inch deep, and about 100 mm. or 3·94 inches wide, which are connected in parallel circuit when all the brushes are in contact. The brushes are made in three sections, so that 20 or 40 or 60 per cent. or the whole current of a machine can be used in the arc. For the fine-weather intensity, 60 per cent. of the current of one machine is usually employed: that is, a current of 137 ampères with an electromotive force of about 50 volts. The machines make 600 revolutions per minute, and are driven direct by a Robey non-condensing compound engine; they are fully described in the Trinity House report of 23rd October 1885, issued as a parliamentary paper C 4551. The power absorbed by one machine is 12·6 net I.H.P., and by the two machines 25 net I.H.P. The distance of the magneto-electric machines from the lantern is about 375 feet, the current being sent through cables laid underground in glazed earthenware pipes; the pipes have lids or covers jointed with clay, so that the cables may be readily inspected or repaired.

Illuminating Apparatus.—This consists of a second-order sixteen-sided lens, of 700 mm. or 27·56 inches focal distance, giving flashes of five seconds' duration, at periods of thirty seconds. Each lens panel consists of a central lens and portions of fifteen refracting rings, that is, fifteen segments above and fifteen segments below, embracing a vertical angle of 97°. The maximum intensity of the flashes is estimated at from 6,000,000 to 7,000,000 candles. The electric lamps are of the Serrin Berjot type, and the carbons are Sir James Douglass' fluted craterless carbons of 50 and 60 mm. diameter, or 1·97 and 2·36 inches, made with a core of graphite; the smaller, having a sectional area of 1·6 square inch, are used in clear weather; and the larger, having a sectional area of 2·26 square inches, are used for the maximum intensity or when the clearness of the atmosphere is impaired. The consumption of carbon for each size is at the rate of about 7-8ths inch length per hour.

The lens is rotated by a miniature engine driven by compressed air. The lamp table is arranged for carrying two electric lamps, which are readily changed. A three-wick portable Trinity House Douglass mineral-oil lamp is also provided, and kept in readiness for immediate use, in case of interruption to the electric light. The subsidiary red light is produced by intercepting the rear light by means of a holophote of 187·5 mm. or 7·38 inches focal distance, embracing a vertical angle of 85°; and by reflecting the light by means of inclined mirrors, it is projected over the main apparatus in the direction of the Needles, for the purpose of warning vessels standing into danger. The red sector, which subtends an angle of 10°, is fixed and is varied by flashes, the fixed light being produced by the subsidiary apparatus, and the flashes every thirty seconds by the main apparatus.

Fog Signal.—This consists of four steel air-receivers, having a collective capacity of 480 cubic feet; and a double-noted automatic siren, controlled by a Slight centrifugal governor, and sounded by compressed air at a pressure of 30 lbs. per square inch, giving high and low notes every minute. The blasts and intervals are regulated by an admission valve, which is actuated by clockwork. The siren is cylindrical, 7½ inches diameter, made of gun-metal, carried on a vertical spindle, and rotating automatically at a speed of 600 revolutions per minute, within an outer fixed cylindrical casing. For the high-note siren the upper portion of the instrument has 32 ports round its circumference through both inner and outer cylinders, giving 320 vibrations per second. For the low-note siren the lower portion of the cylinders has 18 ports, giving 180 vibrations per second. The ports are cut obliquely to the radius and at opposite angles through the two cylinders, which causes the inner cylinder to revolve within the outer. The collective port-area is 20 square inches for the high note, and 18 square inches for the low. The entire instrument is enclosed in a cast-iron chest, having two separate chambers, one above the other, corresponding with the two sets of ports; into these chambers the compressed air is alternately admitted. The mouth of the trumpet is immediately above the siren, and when sounding is always turned to windward, except when the wind is directly off the shore. To render the signal instantaneous in action, three of the air-receivers are charged to a pressure of 170 lbs. per square inch; by this means the signal can be sounded and maintained by the stored air during the time occupied in getting an engine to work. The fog-signal apparatus is about 300 feet distant from the engine-house, from which the compressed air is conveyed along a 3½ inch steel pipe, laid underground.

Engines.—There are three Robey semi-portable compound non-condensing engines, working at a pressure of 150 lbs. per square inch, the cylinders being 6½ and 11½ inches diameter by 14 inches stroke; they make about 170 revolutions per minute, equal to a piston speed of about 400 feet per minute. The boilers are of locomotive type, constructed of mild steel, and fitted with copper fire-boxes, each boiler having 56 brass tubes 1¾ inch external diameter by 7 ft. 8 ins. long. The grate surface in each boiler is 7 square feet; and the heating surface is 192 square feet in the tubes and 41 square feet in the fire-box, or 233 square feet total. Each engine is fitted with a Johnson's double-cylinder single-acting air-compressor, driven direct by the piston rods. The air cylinders are 8½ and 11 inches diameter by 14 inches stroke, and are water-jacketed. When compressing air for storage in the air-receivers, both of the air-compressors are worked together up to 50 lbs., then the larger alone up to 80 lbs., and afterwards the smaller alone up to 170 lbs., more or less. Both cylinders of one air-compressor are used for sounding the fog signal. The power absorbed by the fog signal is 14½ net indicated horse-power. The engines, with the compressors disconnected, are of 49 indicated horse-power or 36 brake horse-power; and the consumption of fuel is 3·2 lbs. of gas coke per indicated horse-power per hour.

Water Supply.—The water supply is derived from a spring, distant about half a mile from the lighthouse. It is delivered into tanks of special construction, where it is treated by Clark's softening process, and the softened water is stored in underground tanks having a total capacity of 25,000 gallons. The water before treatment has 16·7 degrees of hardness on Dr. Clark's scale, which is reduced to 4 degrees by the above process, thus keeping the boilers free from incrustation. (Each degree of hardness represents one grain of calcium carbonate per gallon.) In the event of any interruption to the constant supply, the underground storage tanks would keep one engine working for twelve hours a day for about twenty-one days. The feed water on its way to the boilers is forced through a heater, in which it is heated by the exhaust steam to a temperature of 80° Fahr.

Fire Service.—A Merryweather double-acting steam fire-pump is fixed in the engine room, taking steam from either boiler, and is capable of throwing water over the lantern gallery at the rate of 300 gallons per minute. The steam cylinder is 8 inches diameter with 6 inches stroke, and the pump barrel is 6 inches diameter with the same stroke. The maximum speed of working is 260 revolutions per minute. The pump is also used for washing out the boilers.

Cost.—The total cost of this installation of the electric light and the fog signal, including purchase of land, buildings, machinery, and illuminating apparatus, was £12,449. The staff consists of an engineer-in-charge, and four assistant lightkeepers. The light and fog signal have worked satisfactorily and without interruption since the opening on 1st May 1888.

The whole of the work was carried out from the designs and under the direction of Sir James N. Douglass, F.R.S., Engineer-in-Chief to the Trinity House.

LONDON BRIGHTON AND SOUTH COAST RAILWAY LOCOMOTIVE AND CARRIAGE WORKS, BRIGHTON.

These works are situated on the triangular piece of ground bounded by the main line approaching the Brighton central station on the west side, and by the goods line and yard on the east and south, Plate 68. They have been built and enlarged at various times; and owing to their gradual increase, a portion has had to be built on the opposite side of the main line. On entering the lodge gates, which face the north end of the eastern arrival platform, the iron foundry is to the south, and next are the saw-mills for the carriage and wagon work. On the east of these are the Willans and Robinson high-speed electric-light engines, which drive three 50-arc Brush dynamos for supplying the station and goods yard. The carriage-building shop and the boiler and erecting shops are at present undergoing extensive alterations, so as to increase the capacity more especially of the boiler shop, with a view of putting down modern plant for the manufacture of boilers. The works are supplied throughout with a small tramway for the conveyance of material from one shop to another. There are the usual shops for repairs of locomotives and carriages, such as the machine and tool shop, wheel-turning shop, coppersmiths' and brass shop; and on the opposite side of the line are the trimming and painting shops for carriages. The supply of water for the works and running shed is pumped from deep wells on the premises; the quantity used is about one and a half million gallons per week. Locomotive, carriage, wagon, hydraulic, and general work, together with repairs to steamboats, is carried out more or less at these works.

BRIGHTON TECHNICAL SCHOOL.

The Technical School situated in York Place, Brighton, was erected in 1888 by private contributions, and was primarily intended as a day-class centre for manual instruction in connection with the public day-schools of the town, and for the development of science classes in the evening. Day and evening classes in mechanical and electrical engineering and building work have been successfully carried on. Courses of instruction in applied science, which extend over a period of three years, are provided for both day and evening students. The accommodation having become quite inadequate, it is intended shortly to erect a building in which the work can be extended to meet the increasing requirements. The mechanical engineering section has been supplied from private sources with the necessary appliances for the education of mechanical engineers. The machines comprise a 4½ H.P. Otto gas-engine, a 2 H.P. Tangye vertical engine and boiler, an 8-inch screw-cutting lathe, a 6½-inch sliding surfacing and screw-cutting lathe, a 5-inch lathe, and two small hand-turning lathes; a planing machine by Muir, a universal milling machine by G. Lister and Co. of Keighley, a forge and necessary equipment, small bench drill, power drill, and Siemens dynamo. In the wood-working section accommodation is provided for thirty pupils. In the class room are a number of wooden sectional models of engines and machine details, including a working model one-third full size of a set of triple compound engines, supplied for the s.s. "Ivy" by Westgarth, English and Co. of Middlesbrough. This model was made in the school, and is driven by electricity; being completely in section it is a valuable aid for class instruction. Several other branches of instruction, including plumbing, carpentry, and joinery, etc., are provided for evening students. Mr. Joseph Henry Judd is the head-master of the school.

MEMOIRS.

THOMAS EDWIN CRAVEN was born in Leeds on 21st December 1821, and served his apprenticeship from 1835 to 1841 at the Union Foundry, Leeds, under Messrs. Hay, Walton and Hobson, at fitting and turning for machinery and tools. From 1844 to 1856 he was engaged by Messrs. Smith, Beacock and Tannett, at the Victoria Foundry, Leeds, in designing and preparing drawings of engineers' tools. From 1856 to 1874 he was chief draughtsman with Messrs. Greenwood and Batley, Albion Works, Leeds, where he not only designed, but also acted as examiner of engineers' tools, and special tools for the manufacture of ordnance, small-arms, cartridges, &c. From 1875 he practised as a consulting engineer &c. in Leeds to the time of his death, which occurred at his residence on 26th June 1892, in the seventy-first year of his age. He became a Member of this Institution in 1890.

WILLIAM FIRTH was born at Leeds on 27th August 1835. On leaving school he joined his father in his business of iron and steel merchant, which had been started about 1823 and gradually increased, chiefly in connection with local engineering firms and with collieries in various parts of England. In the latter he did much in recent years to introduce the use of rolled steel girders in the place of timber supports for the roof of mines; and also a pit prop formed from a steel girder. He was also managing director of the Thornhill Iron and Steel Co., Dewsbury. His death took place at Leeds on 12th November 1892, in the fifty-eighth year of his age. He became a Member of this Institution in 1887.

AMANDUS EDMUND KEYDELL was born at St. Andreasburg, Hanover, Germany, on 16th July 1841. After being educated at Clausthal Grammar School, he commenced in 1860 his apprenticeship at the Ducal Engineering Works, Zange, and subsequently for several years studied at the Polytechnic Academy in Hanover.

In 1868 he entered the service of the North German Lloyd Steamship Co. at Bremen, as junior sea-going engineer, and passed through the various grades till he became chief engineer. In 1873 he superintended at Hull and Greenock the building and alteration of several vessels for the above company; and on the completion of the work in 1875 he started as a consulting engineer at Hull, and was soon afterwards elected a surveyor to Lloyd's Register of Shipping. In 1879 he became an exclusive surveyor to Lloyd's at Cardiff until 1887, when he was removed to Dundee, where he died on 22nd May 1892, in the fifty-first year of his age, from heart disease, after a lingering illness of several months. He became a Member of this Institution in 1885.

Dr. Alexander Carnegie Kirk was born on 16th July 1830 at the Manse of Barry, Forfarshire, of which parish his father was the minister. After being educated at the Burgh School at Arbroath, and subsequently at Edinburgh University, he served his apprenticeship as an engineer in the Vulcan Foundry of Robert Napier, Washington Street, Glasgow. On the completion of his apprenticeship he went to London, where he rose to the position of chief draughtsman to the firm of Messrs. Maudslay, Sons and Field. Some years later he returned to Scotland, and became engineer at the Bathgate paraffin oil works of Messrs. Young, Meldrum and Binney; on the dissolution of the firm he designed and erected for the senior partner the still greater establishment at West Calder. Here, after a long series of experiments on the use of atmospheric air, instead of the sulphuric ether then employed for cooling the paraffin oil, he produced an engine somewhat similar in principle to Stirling's air engine, with a regenerator. The results were so satisfactory that the ether engine was superseded. At the age of thirty-five he succeeded Mr. Rowan in the management of Messrs. James Aitken and Co.'s Engine Works at Cranstonhill, Glasgow, where he remained for five years, engaged in designing and constructing a large variety of marine, blowing, and winding engines, and oil machinery. After spending a short period in Glasgow as a consulting engineer, he became manager of the

Centre Street and Fairfield Engine Works of Messrs. Elder and Co., shortly after the death of Mr. John Elder; and for the remainder of his life devoted himself almost entirely to marine engineering. While with Messrs. Elder he carried into effect the idea of a triple-expansion type of marine engine, and ultimately succeeded in rendering triple-expansion marine engines commercially profitable by applying them to ocean-going vessels. In 1874 he designed and fitted on board the screw-steamer "Propontis" triple-expansion engines in which a high pressure of steam was employed; these engines were quite successful. In 1877, a year after the death of Mr. Robert Napier, he joined Messrs. John and James Hamilton in the firm of Messrs. Robert Napier and Sons, of which he was the senior partner. In 1881, after having endeavoured for some years to get the triple marine engine adopted, he constructed engines on this plan for the s.s. "Aberdeen," the first steamer for the celebrated line of clipper ships of Messrs. George Thompson and Co., who deserved great credit for adopting the principle. These engines, which were essentially of the same design as those of the "Propontis," used steam at 125 lbs. pressure in a boiler of the ordinary marine kind, and the economy in coal consumption was about 30 per cent. Having devoted much time to experimental work, especially on the strength and durability of metals, he was a frequent contributor on this and other engineering subjects to the proceedings of professional institutions; and in recognition of his scientific attainments and practical work he received from the University of Glasgow the honorary degree of LL.D. Having been in indifferent health for some time, he died from failure of the action of the heart on 5th October 1892, at the age of sixty-two. He became a Member of this Institution in 1872, and was connected also with many institutions and societies, having been a vice-president of the Institution of Naval Architects, and President in 1887 of the Institution of Engineers and Shipbuilders in Scotland.

HERMANN LUDWIG LANGE was born on 10th May 1837 at Plauen in Saxony, where he was educated. At the age of seventeen he went to Berlin, and there served an apprenticeship of three and a half years with a firm of engineers engaged in the construction of stationary engines, turbines, and water-wheels; during which time he was frequently employed in delivering and erecting the machinery. He next spent two years in a course of study in civil and mechanical engineering at the Polytechnic School at Carlsruhe. In 1861 he came to England at the invitation of the late Mr. Beyer, who was a townsman of his, and entered the engineering works of Messrs. Beyer Peacock and Co., locomotive and machine-tool makers, Gorton Foundry, Manchester, where he was first employed for about a year in the workshops. Afterwards he was engaged in the drawing office, first chiefly on machine-tool designs under the personal superintendence of Mr. Beyer, and subsequently on locomotive designs. In 1864 he was appointed to the position of head draughtsman, which he held until Mr. Beyer's death in 1876, when he became chief engineer and co-manager; and in 1888, the business having been converted into a company in 1883, he became one of the directors. His sound judgment in mechanical subjects and true appreciation of the fitness of details were particularly exemplified in carrying out designs for tramway engines and rack locomotives for steep inclines, which on account of the special conditions these two kinds of engines have to fulfil presented more difficulties of detail than usually occur in locomotives. He devoted himself so indefatigably to the business that his health gradually declined; and he died suddenly from paralysis on 14th January 1892, in the fifty-fifth year of his age. He became a Member of this Institution in 1877.

ALFRED PERRY was born at Taunton on 3rd June 1834. In early life he was engaged in the engineering works and drawing offices of Messrs. Ashbury, and Messrs. Ormerod, Grierson and Co., Manchester, and of the Birmingham Wagon Works, Smethwick. About 1861 he entered the service of Messrs. Chance Brothers and Co., and became the head of the drawing office in their Lighthouse Works, Smethwick, where up to the date of his last illness he assisted in the design and execution of the mechanical portions of dioptric apparatus and of iron towers, in association successively with Mr. James Chance, Dr. John Hopkinson, and Mr. J. Kenward.

He was untiring in perfecting whatever he undertook; and his long experience in lighthouse construction has been of considerable service to lighthouse authorities in many parts of the world. His death took place at Smethwick from heart disease, after two months' illness, on 12th December 1892, in the fifty-ninth year of his age. He became a Member of this Institution in 1882, and was also an Associate Member of the Institution of Civil Engineers.

WILLIAM RICHARDSON was born at Horbury, near Wakefield, on 11th August 1811; and in the following year his parents went to reside at Cooper Bridge, near Huddersfield, his father having obtained an appointment with the Calder and Hebble Navigation Company. At an early age he was sent to the village school, and remained there until he was eight years old, when he was set to work to assist in the domestic hand-loom weaving industry. When eleven years old he was sent to the cotton mill of Mr. Thomas Haigh, Colne Bridge, where suitable employment in connection with the various machines used for the preparing and spinning of cotton was found for him until 1823, when, having been bound apprentice, he entered the mechanics' shop connected with the mill, and remained there until 1833. During this period he diligently applied himself to remedy the defects of his scanty education, and to make himself practically acquainted with every machine in the mill, as well as with the water-wheel and its connections, by which the machinery of the mill was driven. On leaving Cooper Bridge, he removed to Marsden, and entered the service of Messrs. Taylor Brothers, engineers and boiler makers. In 1834 he entered the service of Messrs. Hibbert and Platt, Hartford Works, Oldham, makers of textile machines. In 1837, trade in Oldham being slack, he went to London, in order to gain further experience and knowledge, and succeeded in obtaining employment with Messrs. Beal and Henderbury, East Greenwich, where he learned the use of gauges and templates, and obtained an insight into the principle of interchangeability and accurate workmanship, which he never afterwards forgot. After working there about six months, he returned to Oldham to Messrs. Hibbert and Platt, who in the meantime had become busy. Here he applied himself with such diligence and success to the re-modelling of the scutcher department, that his employers advanced him to a leading position in the management of the business. From this time his career becomes part of the history of Hartford Works. The building of the Hartford New Works, Werneth, and the consequent transfer of several important departments, left him free to devote himself to the remainder of the business carried on at the old works, with which he had been more particularly connected. After the building of the new works, the still further growth of the business led to the establishment of what may be called the outside branches, such as the forge for the manufacture of wrought-iron, the brick works for the manufacture of dry-clay bricks (Proceedings 1859, page 42), and the collieries and coke ovens. With all of these he was specially associated, taking a leading part in their establishment and development; and on all these additions to the business he has left the impress of his judgment and ability. Although he was a prolific inventor, it was more in the character of constructor and organizer that his strength lay. A skilful handicraftsman, capable of enduring the most exhausting fatigue, well informed on all matters connected with his business, possessed of sound judgment, indomitable resolution, perseverance, and love of his calling, in cases of difficulty he was full of resources and a tower of strength. He took great interest in the progress of his adopted town, of which he was a borough magistrate and for a time a town councillor, promoting its educational interests, especially in the direction of technical education, having been President both of the Oldham Lyceum and of the Oldham School of Science and Art; to the Oldham Infirmary he was a generous benefactor. He became a Member of this Institution in 1859, and was a Member of Council from 1877 to 1884. He was a member of the Iron and Steel Institute from its foundation in 1869; and was also a member of other scientific and technical institutions. In 1854 he was admitted into the partnership of Messrs. Platt Brothers and Co.; and when in 1868 the business was transformed into a limited company, he became vice-chairman, and had the satisfaction of seeing his elder son a director at the same board as himself. He continued to hold the position of vice-chairman to the time of his death, which took place at his residence, Werneth, Oldham, on 16th December 1892, at the age of eighty-one.

George John Miller Ridehalgh was born at Prestwich, near Manchester, on 24th February 1835. As a young man he was educated as a mechanical engineer; but succeeding to an ample property he never practised in his profession. As a county gentleman he resided for many years at his mansion at Fell Foot, on the banks of Lake Windermere. He was colonel of the 2nd Westmorland Volunteer Battalion Border Regiment. He was the leading spirit of the Windermere regattas, and himself owned several sailing yachts, and a magnificent steam yacht, the "Britannia"; the machinery of the latter was constructed under his own personal superintendence, and to keep it in the most perfect order was his great interest. His death took place at his residence at Fell Foot on 16th October 1892, in the fifty-eighth year of his age. He became an Associate of this Institution in 1882.

John Varley was born in Leeds on 7th August 1828. At the age of fourteen he entered the works of Messrs. Maclea and March, Union Foundry, Holbeck, and rose to be assistant manager. On leaving them he acted as traveller for Messrs. J. Whitham and Sons, Perseverance Iron Works, Leeds, for three years; at the expiration of which he went to a firm in London as a draughtsman and engineer, and afterwards was engaged by Messrs. Charles Cammell and Co., Sheffield, for whom he acted for many years as their chief representative. He was next associated with the Farnley Iron Co., near Leeds, for twelve years, until mild steel superseded Yorkshire iron; from there he went to the Leeds Forge Co., Armley, where he was for upwards of ten years. His death took place in Liverpool on 17th November 1892, of syncope, at the age of sixty-four. He became an Associate of this Institution in 1869.

William Hugill Walker was born on the site of the present works of his firm in Sheffield on 11th August 1828, and in 1849 succeeded to his father's business of a steel manufacturer and merchant, having served his apprenticeship as an engineer at the Perran Foundry in Cornwall. On coming of age, he entered into partnership with Mr. John Eaton, who had been manager to his late father; and the new firm then went into the engineering and ironfounding trade, discontinuing the manufacture of steel. On the death of Mr. Eaton, he acquired the entire business as his own, and took the sole management. He was principally engaged in the manufacture of rolling-mill plant, including the engines and all the necessary machinery in connection therewith; and had put down most of the large mills in Sheffield. He was one of the first to bring to a successful issue the manufacture of chilled rolls direct from the cupola, instead of casting them from the air furnace as had previously been the custom; and he manufactured extensively both chilled and grain rolls of all sizes, from the smallest up to the largest necessary for rolling armour-plates. He carried on the business successfully for a period of forty-three years, continuing to take an active part in the management till less than eighteen months before his death, which took place at his residence, Sheffield, on 28th May 1892, in the sixty-fourth year of his age. He became a Member of this Institution in 1863.

Institution of Mechanical Engineers.

PROCEEDINGS.

October 1892.

The Autumn Meeting of the Institution was held in the rooms of the Institution of Civil Engineers, London, on Wednesday, 26th October 1892, at Half-past Seven o'clock p.m.; Dr. William Anderson, F.R.S., President, in the chair.

The Minutes of the previous Meeting were read, approved, and signed by the President.

The President called attention to the fact that since the last Meeting of the Institution Her Majesty had conferred the honour of a baronetcy upon Mr. Edward H. Carbutt, Past-President, and of knighthood upon Mr. William Renny Watson, Member. Letters of congratulation from the Council on behalf of the Institution had been addressed to Sir Edward H. Carbutt, Bart., and Sir William Renny Watson.

The President announced that the Ballot Lists for the election of New Members, Associates, and Graduates, had been opened by a committee of the Council, and that the following thirty-six candidates were found to be duly elected:—

MEMBERS.

Thomas Andrew,	Johannesburg.
John William Booth,	Leeds.
David Jollie Brand,	North Queensland.
William Brown,	Woolwich.
Henry William Butler,	Birmingham.
Charles Carrack,	Nottingham.

Richard Henry Dickinson, . .	Birmingham.
Joseph S. Ellis, ...	Chepstow.
Charles Frederick Focken, . .	Calcutta.
Harry Edward Gresham, . . .	Manchester.
Rupert Skelton Hawkins, . .	London.
Daniel Horsnell, ...	London.
Francis Gassiot Houghton, . .	Crossness.
Frederick Kensington, . . .	London.
John James Macbean, . . .	Singapore.
George Arthur Mitcheson, . .	Longton, Staffs.
William Ramsay, ...	Hong Kong.
Leslie Stephen Robinson, . . .	London.
Almond Rowe, ...	Singapore.
Francis Charles Barrett Sainsbury, .	Bradford-on-Avon.
Peter Scattergood, Jun., . . .	Stapleford, Notts.
Charles Joseph Seaman, . . .	Stockton-on-Tees.
Patrick Lawrence Waldron, . .	Rangoon.

ASSOCIATES.

Thomas Lancelot Reed Cooper, . .	London.
John Morley, ...	London.
Ernest Charles Reed, . . .	Purfleet.
Richard David Whitehead, . .	Derby.
Francis R. Widdows, . . .	Norwich.

GRADUATES.

Alfred Hammond Bromly, . . .	London.
Harvey Collingridge, . . .	London.
Frederick Hudson Miles, . . .	London.
David James Murray, . . .	Ipswich.
Arthur Cyril Power, . . .	London.
William Frederick Earl Seymour, .	Adelaide.
Richard George Frank Warton, .	Cardiff.
Arthur Edward Williams, . .	London.

The President announced that, in accordance with the Rules of the Institution, the President, two Vice-Presidents, and five Members of Council, would retire at the ensuing Annual General Meeting; and the list of those retiring, all of whom offered themselves for re-election, was as follows:—

PRESIDENT.

William Anderson, D.C.L., F.R.S., . . Woolwich.

VICE-PRESIDENTS.

Sir Douglas Galton, K.C.B., D.C.L., F.R.S., London.
Sir James Ramsden, ... Barrow-in-Furness.

MEMBERS OF COUNCIL.

John Hopkinson, Jun., D.Sc., F.R.S., . London.
Samuel W. Johnson, ... Derby.
William Laird, ... Birkenhead.
John G. Mair-Rumley, ... London.
Edward P. Martin, ... Dowlais.

The following nominations had also been made by the Council for the election at the Annual General Meeting:—

Election as Member. MEMBERS OF COUNCIL.

1871. James Platt, ... Gloucester.
1872. Benjamin A. Dobson, . . . Bolton.
1876. Henry Shield, ... Liverpool.

The President reminded the Meeting that according to the Rules of the Institution any Member was now entitled to add to the list of candidates.

No other names were added.

The President announced that the foregoing names would accordingly constitute the nomination list for the election of officers at the Annual General Meeting.

The following Paper was then read and discussed:—

"Second Report of the Research Committee on the Value of the Steam-Jacket." Mr. HENRY DAVEY, *Chairman.*

Shortly before Ten o'clock the Meeting was adjourned to the following evening. The attendance was 97 Members and 55 Visitors.

The ADJOURNED MEETING was held at the Institution of Civil Engineers, London, on Thursday, 27th October 1892, at Half-past Seven o'clock p.m.; Dr. WILLIAM ANDERSON, F.R.S., President, in the chair.

The PRESIDENT gave notice, on behalf of the Council, that at the ensuing Annual General Meeting they would propose such amendments of the Articles of Association and the By-laws of the Institution as should be necessary for carrying into effect the following recommendations:—

"That a class of Associate Members be added, intermediate between the present classes of Members and Graduates. The Associate Members to be not under twenty-five years of age, and to have the same power of voting that is possessed by Members. This class to consist of those who, while engaged in such work as is connected with the practice or science of Engineering, in the opinion of the Council do not yet occupy positions of sufficient responsibility, or are otherwise not eligible, to become full Members.

"That a limit of age, say the completion of the twenty-sixth year, be set to continuance in the class of Graduates; and that before attaining the age of twenty-six years future Graduates be required to apply for election as Members, Associate Members, or Associates, if they desire to remain connected with the Institution.

"That candidates for admission as Graduates be required to furnish evidence of training in the principles as well as in the practice of Engineering.

"That among the qualifications for election into the class of Members be included the occupation during a sufficient period of a responsible position in connection with the practice or science of Engineering. Also that candidates for election as Members must be not under twenty-five years of age, instead of not under twenty-four as at present.

"That the entrance fees and annual subscriptions be modified in favour of Graduates and Associates.

"That the form of proposal, whereby at present the proposers recommend the class into which they wish the candidate to be admitted, be so altered as to omit this recommendation, and to leave entirely to the Council the determination of the class.

"That the number of votes necessary for election be four-fifths of those given, instead of three-fifths as at present.

"That certain provisions which applied solely to the time of the incorporation of the Institution in 1878, and which have now become obsolete, be omitted from the Articles of Association and the By-laws."

The following Paper was then read and discussed:—

"Experiments on the Arrangement of the Surface of a Screw-Propeller"; by Mr. William George Walker, of Bristol.

On the motion of the President a vote of thanks was unanimously passed to the Institution of Civil Engineers, for their kindness in granting the use of their rooms for the Meeting of this Institution.

The Meeting then terminated at Half-past Nine o'clock. The attendance was 77 Members and 41 Visitors.

SECOND REPORT OF THE RESEARCH COMMITTEE ON THE VALUE OF THE STEAM-JACKET.

Mr. HENRY DAVEY, *Chairman*.

The Research Committee appointed by the Council at the end of 1886, to investigate the Value of the Steam-Jacket, presented in October 1889 their First Report, containing tabulated results of previous experiments, obtained from published reports and other sources. Since that time other published results have been obtained, and some members of the Committee and others have carried out experiments. The additional information thus available is contained in this Second Report.

With regard to the work of the Committee themselves, besides collecting what information they could obtain from outside sources, they are now able to present in detail the records of five original sets of experiments, most of which have been specially made for contribution to this Report. In each case the experiments were made upon the same engine, both with and without steam in the jackets. The first set of experiments was made by Mr. John G. Mair-Rumley, on a compound jet-condensing beam pumping-engine at Hammersmith; the second, by the Chairman Mr. Henry Davey, and Mr. William B. Bryan, on a triple-expansion pumping-engine at Waltham Abbey; the third, by Lt.-Colonel English, Mr. Henry Davey, and Mr. Bryan Donkin, Jun., on a compound mill-engine at the Royal Arsenal, Woolwich; the fourth, by Professor W. Cawthorne Unwin, on the experimental engine at the City and

Guilds of London Central Institution, South Kensington; and the fifth, by Mr. Bryan Donkin, Jun., on an experimental vertical engine at Bermondsey.

In pages 470–479 of this Report there is given in Tables 52–56 an index summary of all the experiments which the Committee have thus far been able to record or to make. From these experiments it appears that the expenditure of a quantity of steam in an efficient jacket produces a saving of a greater quantity in the cylinder. The ratio between these two quantities is an important factor in this investigation. Unfortunately the jacket-water has not been recorded in many of the experiments of which the results have been collected; but in all the experiments made by members of the Committee it has been carefully measured and recorded. In the summary, column q gives the percentage saving in feed-water, that is the percentage less feed-water resulting from the use of the jackets; column p gives the actual saving in lbs. per indicated horse-power per hour; and column r gives the water condensed in the jackets in lbs. per indicated horse-power per hour. By an expenditure of r in the jackets there is $p + r$ less water passed through the cylinder, p being the credit balance of the account. The following examples are taken from Tables 54 and 56, pages 478 and 479.

a	q	$p + r$	r	Ratio $p + r$ to r
No.	Per cent.	Lbs.	Lbs.	
41	17·4	6·15	3·29	1·9 to 1
42	8·6	2·76	1·20	2·3 to 1
43	10·3	3·49	1·72	2·0 to 1
44	19·0	5·82	1·13	5·2 to 1

No. 44 shows that for every 1·13 lb. of steam expended in the jackets, there is 5·82 lbs. less feed-water passed through the cylinder, the net saving being thus 4·69 lbs.

It will be seen from the experiments that generally, the smaller the cylinder, the greater is the percentage of gain from the use of the jacket, arising doubtless from the fact that a small cylinder gives a larger jacket-surface for a given weight of steam passing through it than a larger cylinder does.

In some of the following experiments it was found possible to measure the consumption of coal as well as of feed-water; and as these figures have considerable practical value and interest they are in every case added to the results. Some of the experiments form therefore complete engine and boiler trials; but it must be remembered that the effect of the jackets is measured by the consumption of feed-water only and not of coal.

It has been thought desirable to add to this Report an appendix, page 453, containing some suggestions for the assistance of those who may undertake future experiments.

SUMMARIES OF SEVEN SETS OF EXPERIMENTS ON THE VALUE OF THE STEAM-JACKET.

SINGLE-CYLINDER CONDENSING ENGINES.

No. 35. *Record of two Experiments on same Engine with* HIGH-*pressure and with* LOW-*pressure Steam in the Jackets.*

Single-Cylinder Condensing Horizontal Engine. *Cylinder* 15 inches diameter, 30 inches stroke; the body and both ends of the cylinder were jacketed. Experiments made at Messrs. Rivolta's Electric Works, Milan, in February 1886, by Signor P. Guzzi. (College of Engineers and Architects in Milan, 1887.)

No experiment was made *without* steam in the jackets; but in the one case ordinary boiler-steam at a pressure of 48 lbs. per square inch above the atmosphere was circulated through the jackets; while in the other, steam was supplied to the jackets from a small Perkins boiler at a pressure of 187 lbs. per square inch above the atmosphere.

Pressure in Jackets, lbs. per square inch above atm.		**48 lbs.**	**187 lbs.**
Duration of Experiment	hours	7·2	6·3
Boiler Pressure, lbs. per square inch above atm.	lbs.	56·2	56·6
Revolutions per minute	revs.	79·0	77·4
Piston Speed, feet per minute	feet	395	387
Results. Indicated horse-power	I.H.P.	25·3	25·5
Feed-Water, lbs. per I.H.P. per hour, total	lbs.	23·85	19·85
" percentage less with higher-pressure steam in jackets	per cent.	..	**16·8 p.c.**

Single-Cylinder Condensing Engines (continued).

No. 36. *Record of two Experiments on same Engine*
WITH *and* WITHOUT *Steam in the Jackets.*

SINGLE-CYLINDER CONDENSING HORIZONTAL CORLISS ENGINE. *Cylinder* 21·81 inches diameter, 49·21 inches stroke; the body only was jacketed. Experiments made at Prague, in February 1888, by Professor Doerfel. (Zeitschrift des Vereines Deutscher Ingenieure, 1889.) Many other experiments are given in the same paper, showing the economy of the jacket to vary between 7 and 12½ per cent. for about the same cut-off.

Jackets, *With* or *Without* Steam		Without	With
Duration of Experiment	hours	6·4	7·3
Cyl. Initial Pressure, lbs. per sq. in. above atm.	lbs.	59·2	62·2
Number of Expansions		5	5
Revolutions per minute	revs.	56·0	55·7
Piston Speed, feet per minute	feet	551	548
Results. Indicated horse-power	I.H.P.	146	159
Jacket-Water, lbs. per I.H.P. per hour	lbs.	..	0·80
„ in percentage of feed-water	p.c.	..	4·0
Feed-Water, lbs. per I.H.P. per hour, total	lbs.	22·57	19·80
„ percentage less with steam in jackets		..	**12·3 p.c.**

Single-Cylinder Condensing Engines (continued).

No. 37. *Record of three Experiments on same Engine* WITH *and* WITHOUT *Steam in the Jackets.*

PAIR OF SINGLE-CYLINDER CONDENSING HORIZONTAL CORLISS ENGINES. *Cylinders* 25·06 inches diameter, 48 inches stroke; the body and both ends of each cylinder were jacketed. Experiments made at Messrs. J. and P. Coats' Thread Works, Paisley, in April 1889, by Mr. Michael Longridge. Steam was supplied to the jackets from the main steam-pipe, through three-quarter inch pipes. (Engine Boiler and Employers' Liability Insurance Co.; Chief Engineer's Report for 1889.)

Jackets in use		No Jackets	Ends only	All Jackets
Duration of Experiment	hours	4	7	7
Boiler Pressure, lbs. per sq. in. above atm.	lbs.	60	61	61
Initial ,, ,, ,,	lbs.	50	50	50
Number of Expansions		4·2	3·9	4·2
Revolutions per minute	revs.	65·0	65·0	65·1
Piston Speed, feet per minute	feet	520	520	521
Results. Indicated horse-power	I.H.P.	508	490	488
Jacket-Water, lbs. per I.H.P. per hour	lbs.	..	0·18	0·63
,, in p.c. of feed-water	p.c.	..	0·9	3·3
Feed-Water, lbs. per I.H.P. per hour	lbs.	19·77	19·52	19·27
,, p.c. less with steam in jackets		..	**1·3 p.c.**	**2·5 p.c.**

COMPOUND NON-CONDENSING ENGINE.

No. 38. *Record of two Experiments on same Engine*
WITH *and* WITHOUT *Steam in the Jackets.*

COMPOUND NON-CONDENSING ARMINGTON ENGINE. *Cylinders* 9·45 and 14·17 inches diameter, 10·04 inches stroke. Experiments made at Mulhouse, Alsace, in 1889, by M. Walther Meunier. (Société Industrielle de Mulhouse, 1890, vol. lx.) Steam was supplied by a locomotive boiler which was overworked. A brake was used to absorb the power. This is the same engine as for the set of Experiments No. 39, but here used without the condenser.

Jackets, *With* or *Without* Steam		**Without**	**With**
Boiler Pressure, lbs. per square inch above atm.	lbs.	117·3	117·3
Revolutions per minute	revs.	298	298
Piston Speed, feet per minute	feet	499	499
Results. Indicated horse-power	I.H.P.	83·6	84·4
Jacket-Water, lbs. per I.H.P. per hour	lbs.	..	1·18
„ in percentage of feed-water	p.c.	..	4·7
Feed-Water, lbs. per I.H.P. per hour, total	lbs.	26·29	25·25
„ percentage less with steam in jackets		..	**4·0 p.c.**

COMPOUND CONDENSING ENGINES.

No. 39. *Record of two Experiments on same Engine*
WITH *and* WITHOUT *Steam in the Jackets.*

COMPOUND CONDENSING ARMINGTON ENGINE. *Cylinders* 9·45 and 14·17 inches diameter, 10·04 inches stroke. Experiments made at Mulhouse, Alsace, in 1889, by M. Walther Meunier. (Société Industrielle de Mulhouse, 1890, vol. lx.) The engine was worked with a Körting condenser and without an air-pump. [*See also No.* 38.

Jackets, *With* or *Without* Steam		**Without**	**With**
Boiler Pressure, lbs. per square inch above atm.	lbs.	112	111
Revolutions per minute	revs.	302	302
Piston Speed, feet per minute	feet	505	505
Results. Indicated horse-power	I.H.P.	82·9	83·0
Jacket-Water, lbs. per. I.H.P. per hour	lbs.	..	0·85
„ in percentage of feed-water	p.c.	..	3·9
Feed-Water, lbs. per I.H.P. per hour, total	lbs.	22·85	21·57
„ percentage less with steam in jackets		..	**5·6 p.c.**

Compound Condensing Engines (continued).

No. 40. *Record of two Experiments on same Engine* WITH *and* WITHOUT *Steam in the Jackets.*

Pair of Compound Jet-Condensing Beam Pumping Engines, with four slide-valves to each cylinder. *Cylinders* 19 and 31 inches diameter, 44·6 and 66 inches stroke; all four bodies, and both ends of the two low-pressure cylinders, were jacketed. Experiments made at Copenhagen Water Works in February 1890 by Mr. F. Ollgaard. (Tekniske Forenings Tidsskrift, Copenhagen 1890.) The whole steam-supply to the cylinders passed through the jackets when in use.

Two other jacketed experiments also are recorded, in which the pressures in the two low-pressure jackets were reduced to about 14·25 lbs. per square inch, giving a mean consumption of 20·56 lbs. of feed-water per I.H.P. per hour, or a reduced economy of 13·8 per cent. due to this arrangement of jacketing.

Jackets, *With* or *Without* Steam		Without	With
Duration of Experiment	hours	8	8
Boiler Pressure, lbs. per square inch above atm.	lbs.	52·5	51·25
Number of Expansions		13·8	14·6
Revolutions per minute	revs.	19·1	24·0
Piston Speeds, feet per minute	feet	142 & 210	178 & 264
Results. Indicated horse-power	I.H.P.	65·8	81·2
Jacket-Water, lbs. per I.H.P. per hour	lbs.	..	—
,, in percentage of feed-water	p.c.	..	6·0
Feed-Water, lbs. per I.H.P. per hour, total	lbs.	23·84	19·41
,, percentage less with steam in jackets		..	**18·6 p.c.**

TRIPLE-EXPANSION CONDENSING ENGINE.

No. 41. *Record of two Experiments on same Engine* WITH *and* WITHOUT *Steam in the Jackets.*

TRIPLE-EXPANSION CONDENSING VERTICAL INVERTED ENGINE, on three uncoupled cranks, with cylinders provided with Meyer expansion gear. *Cylinders* 5, 8, and 12 inches diameter, 10, 10, and 15 inches stroke. All the bodies, ends, and valve-chests of the three cylinders were jacketed, as also were both receivers, and part of the steam-pipe, steam being admitted to the last three jackets throughout both of the experiments. Experiments made at Owens College, Manchester, in January and April 1889, by Professor Osborne Reynolds. (Institution of Civil Engineers, vol. xcix, 1889.) Steam was supplied to each of the fifteen jackets by a separate pipe, and all the jacket-water was drained through the first receiver into the boiler. In the high-pressure cylinder 64·7 per cent. of the internal surface is jacketed, in the intermediate cylinder 67·1 per cent., and in the low-pressure cylinder 75·2 per cent. Three independent brakes were used to absorb the power.

Jackets, *With* or *Without* Steam		**Without**	**With**
Duration of Experiment	hours	4·0	4·1
Boiler Pressure, lbs. per square inch above atm.	lbs.	188	192
Number of Expansions		15	26·5
Revolutions per minute	revs.	{322 & 320 & 276}	{230·5 & 298 & 299}
Piston Speeds, feet per minute	feet	{537 & 533 & 690}	{383 & 497 & 747}
Results. Indicated horse-power	I.H.P.	73·9	72·1
Coal, lbs. per I H.P. per hour	lbs.	1·62	1·33
„ percentage less with steam in jackets	p.c.	..	17·9
Jacket-Water, lbs. per I.H.P. per hour	lbs.	[2·14]*	3·29
„ in percentage of feed-water	p.c.	[13·0]*	24·3
Feed-Water, lbs. per I.H.P. per hour, total	lbs.	16·42	13·56
„ percentage less with steam in jackets		..	**17·4 p.c.**

* Jacket-Water from receiver and steam-pipe jackets only.

RECORDS OF FIVE SETS OF EXPERIMENTS ON THE VALUE OF THE STEAM-JACKET.

No. 42. EXPERIMENT ON A COMPOUND BEAM-ENGINE AT THE WEST MIDDLESEX WATER WORKS, HAMMERSMITH, BY MR. JOHN G. MAIR-RUMLEY.

Engine.—The engine on which the experiment was made, Fig. 1, Plate 69, is a compound jet-condensing beam pumping-engine made by Messrs. James Simpson and Co. of Pimlico. The diameters of the cylinders are 29 and 47·5 inches. Their strokes are 65·1 and 96 inches respectively. The diameters of the piston-rods are $4\frac{7}{8}$ and $5\frac{1}{8}$ inches respectively.

Only the body of each cylinder is jacketed, and steam is supplied to both jackets direct from the boiler at a pressure of 49 lbs. per square inch above the atmosphere.

The following shows the amounts of the jacketed and unjacketed portions of the inner surfaces of the cylinders :—

Cylinder.		Area of Inner Surface.		
		Jacketed.	Unjacketed.	Total.
High-pressure	body, square feet	41·18 sq. ft.	..	50·35 sq. ft.
	ends, square feet	..	9·17 sq. ft.	
	per cent. . .	81 8 p.c.	18·2 p.c.	100·0 p.c.
Low-pressure	body, square feet	99·48 sq ft.	..	124·01 sq. ft.
	ends, square feet	..	24·53 sq. ft.	
	per cent. . .	80·2 p c.	19·8 p.c.	100·0 p.c.

Details of Experiment.—The experiment consisted of two trials which were made at the West Middlesex Water Works, Hammersmith, on 21st, 22nd, and 23rd March 1888, by Mr. John G. Mair-Rumley. The first trial, during which steam was admitted into both jackets, commenced at 9.0 p.m. on 21st March, and was continued without stoppage until 5.0 a.m. on 22nd. The second trial, during which no steam was admitted into either jacket, commenced at 6.15 p.m. on 22nd March, and was continued without stoppage until 4.15 a.m. on 23rd. There were thus two separate trials, the first with steam in both jackets and lasting eight hours, and the second without steam in either jacket and lasting ten hours. All the other conditions were maintained as far as possible the same throughout both of the trials.

Feed-Water Measurement.—The feed-water was measured in two tanks; the smaller of the two was placed above the other, and was used as a measuring tank. Its capacity up to the overflow pipe was 395 lbs. of water at the feed temperature; and when this was allowed to run into the lower tank, to which the feed-pump of the engine was connected, the water-level in the lower tank was raised $8\frac{1}{4}$ inches as measured from a mark fixed near its upper edge. The number of times the upper tank was filled, and emptied into the lower tank, was checked by means of a counter which was read at the beginning and end of each trial.

Feed-Water Consumption, with *Steam in Jackets.*—At the commencement of the first trial, when steam was admitted into both jackets, the water-level in the lower tank was $14\frac{3}{4}$ inches below the datum mark. After the trial was started, a large amount of water was found in the indicator pipes (which were well lagged), and seemed to point to priming; it was therefore thought advisable to lower the water-level in the boilers, in order to obtain drier steam. No water was pumped into the boilers from 10.50 p.m. to 11.50 p.m. From the commencement of the trial at 9.0 p.m. until 10.50 p.m., or during 1 hour 50 minutes, the water-level in the boilers was kept constant, and 13½ tanks of water were fed into the boilers, being at the rate of 48·5 lbs. per minute. From 10.50 p.m. till 11.50 p.m. the feed was stopped, and the water-level in the boilers gradually fell. From 11.50 p.m. till 4.0 a.m., or during 4 hours 10 minutes, the water in the boilers was kept constant at this lower level, and 31 tanks of water were fed into the boilers, being at the rate of 49·0 lbs. per minute. From 4.0 a.m. to 5.0 a.m. the feed was increased; so that at 5.0 a.m., when the trial was stopped, the water in the boilers had reached the same level as it had at the commencement. During the last hour 12½ tanks of water were fed into the boilers, being at the rate of 82·3 lbs. per minute. At the end of the trial the water-level in the lower tank stood 3¼ inches higher than it did at the beginning, and this has been allowed for in the totals. The total feed-water used when steam was admitted into both jackets was thus 22,360 lbs. in eight hours, equivalent to 2,795 lbs. per hour, or 46·58 lbs. per minute.

Feed-Water Consumption, without *Steam in Jackets.*—In this trial steam was entirely shut off from both jackets, and the jacket drain-pipes were open to the air. The feed was continuous throughout the whole of the trial. The water-level in the boilers remained very nearly constant, and was precisely the same at beginning and end of the trial. During the trial 75 tanks of water were emptied into the lower tank; at the end the water-level in the lower tank stood 2⅞ inches higher than it did at the commencement, and this has been allowed for in the totals. The total feed-water used when no steam was admitted into either jacket was therefore 29,487 lbs. in ten hours, which is at the rate of 2,949 lbs. per hour, or 49·15 lbs. per minute.

Jacket-Water.—The drain-pipe from each jacket was made to deliver into a separate bucket. Each of these buckets held 31½ lbs. of water at the temperature at which it left the jacket; but as a certain amount was unavoidably lost, owing to splashing over the edge, 33 lbs. has been taken as the actual quantity of jacket-water represented by each bucketful. During the trial with steam in both jackets, 22 buckets or 726 lbs. of water were drained from the high-pressure jacket, and 27 buckets or 891 lbs. from the low-pressure jacket. These are equivalent to 90·75 lbs. per hour or 1·51 lbs. per minute from the high-pressure jacket, and 111·37 lbs. per hour or 1·86 lbs. per minute from the low-pressure; and to a total of 202·12 lbs. per hour or 3·37 lbs. per minute from both jackets.

Valve-Chest Water.—At 10.40 p.m., during the trial with steam in both jackets, the drain from the high-pressure valve-chest was opened, and the condensed water was collected and measured. From 10.40 p.m. to 12.10 a.m. 40 lbs. of water was condensed, from 12.10 to 1.30 a.m. 40 lbs., and from 1.30 to 5.0 a.m. 80 lbs. Thus 160 lbs. of water was condensed in 6 hours 20 minutes, equivalent to 25·26 lbs. per hour. During the whole of the trial without steam in either jacket, the condensed water from the valve-chest drain was collected and measured, and 41 lbs. was thus condensed in the 10 hours, equivalent to 4·1 lbs. per hour.

Pressures, &c.—The various pressure-gauges were read at regular intervals throughout both of the trials. The mean barometric pressure was 30·11 inches of mercury throughout the trial with steam in both jackets, and 29·52 inches of mercury throughout the trial without steam in either jacket. Indicator diagrams were taken from each end of each cylinder every half-hour throughout both of the trials. There were thus taken seventeen sets or 68 single indicator diagrams during the trial with steam in both jackets, and twenty-two sets or 88 single indicator diagrams during the trial without steam in either jacket. The set of indicator diagrams nearest to the mean for each of the two trials are shown superposed in Fig. 7, Plate 71. The following table gives the average mean effective pressure and indicated horse-power in each of the two cylinders throughout both of the trials:—

	Mean Effective Pressure. Lbs. per square inch.		Indicated Horse-Power.	
Date of Trial 1888	22 & 23 Mar.	21 & 22 Mar.	22 & 23 Mar.	21 & 22 Mar.
Jackets, *With* or *Without* steam	**Without**	**With**	**Without**	**With**
	lbs.	lbs.	I.H.P.	I.H.P.
High-pressure cylinder .	29·93	24·44	94·8	82·6
Low-pressure „ .	5·32	6·34	67·2	85·4
Total	12·80	12·45	162·0	168·0
	(reduced to low-p. cyl.)			

Temperatures.—The temperatures of the feed-water, injection-water, air-pump discharge, outer air, engine-room, and stoke-hold were all observed at regular intervals throughout both experiments. The injection-water temperature remained steady at 41° Fahr. throughout both of the trials.

The following table shows the mean temperatures obtained by calculation from these readings:—

Date of Trial	1888	22 & 23 Mar.	21 & 22 Mar.
Jackets *With* or *Without* Steam		**Without**	**With**
Temperature of Feed-Water	Fahr.	67·4°	75·2°
„ „ Injection-Water	Fahr.	41·0°	41·0°
„ „ Air-pump discharge Water	Fahr.	68·7°	69·1°
„ „ Outer Air	Fahr.	37°	44°
„ „ Engine-room	Fahr.	68°	69°
„ „ Stoke-hold	Fahr.	72°	75°

Boilers.—Steam was supplied from three Cornish boilers, each 28 feet long, and 6 feet diameter.

Results.—The results of the experiment are given in Table 42, pages 456–459; and the performance, and rates of feed-water and jacket-water consumption, are shown graphically in Plate 76. The plottings of the upper portions of this diagram speak for themselves. The bottom portion, representing the feed-water per indicated horse-power per hour, shows at a glance the excess of the unjacketed consumption over the jacketed consumption. In the latter the three dark bands, which are included in the total consumption, show at a glance the proportion of condensed steam collected from the high-pressure jacket, from the low-pressure jacket, and from the valve-chest. The position of these dark bands in the diagram is purely arbitrary for the sake of clearness; it is their thickness alone that is material.

No. 43. Experiment on a Triple-Expansion Engine at the Waltham Abbey Pumping Station of the East London Water Works, by Mr. Henry Davey and Mr. William B. Bryan.

Engine.—The engine on which the experiment was made, Fig. 3, Plate 70, is a triple-expansion surface-condensing engine of the ordinary inverted direct-acting marine type, made by Messrs. T. Richardson and Sons of Hartlepool; and was fully described and illustrated in "Engineering," 8th August 1890, pp. 158 and 162. The diameters of the cylinders are 18, 30·5, and 51 inches respectively. The stroke of each of the three cylinders is 36 inches. Each piston cross-head is connected to its crank by a single connecting-rod; and to a pump plunger by a pair of pump rods. There are thus three steam cylinders, three connecting-rods, three pairs of pump rods, and three pump plungers. Each of the three cylinders is provided with an ordinary slide-valve actuated by a separate eccentric on the crank-shaft. The slide-valve on the high-pressure cylinder is actuated direct by its eccentric, and is provided with a Meyer expansion-valve, by means of which the speed of the engine was regulated during the experiment. The slide-valves on the intermediate and low-pressure cylinders are actuated through variable-expansion links, the positions of which remained unchanged throughout the trial.

The crank-shaft is placed above the pumps, and the cranks rotate in the sequence—high, low, intermediate. The fly-wheel is placed towards one end of the crank-shaft, and beyond it there is fixed a crank actuating a pair of well-pumps by means of rocking quadrants. The bodies and both ends of all three cylinders are steam-jacketed, the cylinders forming liners in the body-jackets. Steam is supplied to these jackets through a pipe connected to the main steam-pipe on the boiler side of the stop-valve. The steam to the jackets enters at the top on one side and at the bottom on the opposite side; and in ordinary working the high-pressure jackets discharge directly into the boiler, while the intermediate and low-pressure jackets discharge through steam-traps into the hot-well. The jackets of the high-pressure cylinder receive steam at full boiler-pressure; and by means of reducing valves the pressures in the jackets of the intermediate and low-pressure cylinders are maintained a little higher than the pressures in their respective valve-chests. Each cylinder is therefore jacketed with steam a little above its own initial pressure.

The following are the areas of the inner surfaces of the cylinders, which are all wholly jacketed:—

Cylinder.	Area of Inner Surface. Wholly jacketed.		
	Body.	Ends.	Total.
	square feet	square feet	square feet
High-pressure	13·44	4·22	17·66
Intermediate	23·95	10·14	34·09
Low-pressure	40·05	28·37	68·42

Each of the three cylinders was provided with a Richards indicator, placed midway between the two ends and furnished with pipe connections and a three-way cock, so that a diagram from each end of the cylinder could be taken on the same paper. As a rule this is not a good arrangement; but in the present case the engine worked so slowly (only 23 revolutions per minute), and the indicator pipes were so large and so well lagged, that the diagrams may be taken as practically correct.

Details of Experiment.—The experiment consisted of two trials which were made at the Waltham Abbey pumping station of the East London Water Works on 6th and 7th December 1889, by Mr. Henry Davey and Mr. William B. Bryan. They began at 11.0 a.m. and were continued without stoppage until 7.0 p.m. on each of the two days. During the trial on 6th December steam was admitted to all the jackets; whilst on 7th December no steam was allowed to pass into any of the jackets. There were thus two eight-hour trials, the first with steam in all the jackets, and the second without steam in any of the jackets; all the other conditions remained as far as possible the same.

Feed-Water.—The feed-water was taken from the pumping main, and was measured in two feed-measuring tanks, each holding 120 gallons. Before the experiment commenced the tanks were carefully gauged by measuring water into them from a standard measure. The feed-water was pumped into the boiler through an economiser, the pump being supplied from one tank whilst the other was filling. All the feed-pipes were exposed to view.

In the first day's trial, when steam was admitted into the jackets, it was discovered shortly after the commencement that a little water was passing the safety-valve of the economiser. At 1.0 p.m., or exactly two hours after the commencement, measurements were begun to be made of this leakage, all the water that leaked past the valve being carefully collected and weighed. The leakage was found to be at the rate of 248 lbs. per hour, so that 496 lbs. has been taken as representing the leakage during the first two hours of the trial when the escaping water was not collected. The whole leakage from the commencement of the trial until the leak was stopped was thus 957 lbs., which has been deducted from the whole quantity that passed through the measuring tanks, in order to arrive at the total feed-water supplied to the engine. The mean temperature of the feed-water was 49·3° Fahr.

In the second day's trial, when no steam was admitted into the jackets, there was no leakage of any kind.

On its way to the engine the feed-water was made to pass through a meter, which was used to obviate the possibility of any error that might arise in counting the number of times the feed-tanks were filled and emptied. The reading of this meter was noted every twenty minutes.

Jacket-Water.—The discharge-pipe from the water-trap of each set of jackets was made to deliver into a separate bucket, and the weight of water in each bucket was determined from time to time by weighing on a sensitive weighing-machine. The amount of water obtained from each set of jackets was thus determined separately. It was possible to test from time to time whether or not the jackets were perfectly drained, as the water-traps passed a little steam if the drain-pipes were opened too much; and great care was taken to get all the jacket-water out without wasting any steam. The following shows the quantities of jacket-water condensed in the several jackets during the first day's trial:—

Jackets.	Condensed Water from Jackets.		
	During Trial.	Per Hour.	Per Minute.
	lbs.	lbs.	lbs.
High-pressure cylinder . .	895	111·9	1·87
Intermediate ,, . .	587	73·4	1·22
Low-pressure ,, . .	418	52·2	0·87
Total Jacket-Water . .	1,900	237·5	3·96

Measurement of Pressures, &c.—Every twenty minutes readings were taken of the pressure-gauges on each of the jackets, on each of the valve-chests, on the boiler, on the condenser, and on the pumping-main. Indicator diagrams were taken from each end of each cylinder every quarter of an hour. The reading of the counter was taken every twenty minutes. During each day's trial therefore, the gauges and the counter were each read twenty-five times, and thirty-three sets or 198 single indicator diagrams were taken. The set of diagrams nearest to the mean for each of the two trials are shown superposed in Plate 72. The following table gives the average mean effective pressure and indicated horse-power in each of the three cylinders throughout both of the trials :—

	Mean Effective Pressure. Lbs. per square inch.		Indicated Horse-Power.	
Date of Trial 1889 Jackets, *With* or *Without* steam	7 Dec. **Without**	6 Dec. **With**	7 Dec. **Without**	6 Dec. **With**
	lbs.	lbs.	I.H.P.	I.H P.
High-pressure cylinder .	56·24	40·69	57·7	41·6
Intermediate „ .	12·42	14·08	37·5	42·3
Low-pressure „ .	5·27	6·39	44·8	54·1
Total	16·46 (reduced to low-p. cyl.)	16·30	140·0	138·0

Boiler.—Steam was supplied by an externally-fired water-tube boiler made by the Babcock and Wilcox Co.

Coal.—Large and small Welsh coal, mixed in as nearly as possible equal proportions, was used throughout both of the trials. On 6th December when steam was admitted into all the jackets, the total coal used was 2,217 lbs., less 236 lbs. of clinker and ashes formed during the experiment, or 1,981 lbs. net. On 7th December, when no steam was admitted into any of the jackets, the total coal used was 2,580 lbs., less 235 lbs. of clinker and ashes, or 2,345 lbs. net. These are equivalent to 247·6 lbs. per hour or 4·13 lbs. per minute with steam in jackets, and 293·1 lbs. per hour or 4·88 lbs. per minute without.

Results.—The results of the experiment are given in Table 43, pages 456–459; and the performance, and rates of feed-water and jacket-water consumption, are shown graphically in Plate 77.

No. 44. Experiment on a Compound Horizontal Engine at the Royal Arsenal, Woolwich, by Lt.-Col. English, Mr. Henry Davey, and Mr. Bryan Donkin, Jun.

Engine.—This experiment was made on a compound horizontal surface-condensing engine, Fig. 4, Plate 70, driving the saw-mill in the Carriage Department at the Royal Arsenal, Woolwich. The engine was made by Messrs. Marshall, Sons and Co. of Gainsborough. The diameters of the cylinders are 18 and 32 inches. The stroke of each cylinder is 48 inches. The piston rods and tail rods are all 3½ inches diameter. The air-pump was originally driven from the tail rod of the low-pressure cylinder by means of a bell-crank lever; but previous to this experiment it had been disconnected, and was driven by a separate engine. At the same time the original jet-condenser had also been removed, and a surface-condenser substituted for it. Each of the two cylinders has four valves, two for steam admission and two for exhaust, and each pair of valves is driven by a single eccentric. The admission valves are of the double-beat type, provided with Proell's automatic expansion apparatus, for varying the cut-off automatically. The exhaust valves are cylindrical, and of the Corliss type. The two cylinders are placed side by side, with their centres 12 feet apart.

The intermediate receiver and the bodies of both cylinders are jacketed, the cylinders forming liners in the body jackets. The ends of the cylinders are not jacketed. The receiver jacket was not in use during the experiment. The steam-supply to the cylinders usually passes through the jackets on its way to the valve-chests; but in order to allow of the jacket-water being measured during the trial, the ordinary steam connections were removed and special steam-pipes substituted. By this means, during the trial with steam in jackets, steam was supplied to the jacket of the high-pressure cylinder by a separate pipe connected with the main steam-pipe on the boiler side of the stop-valve; and to the jacket of the low-pressure cylinder by a separate pipe from the receiver. The jacket-water from the two jackets and the drain-water from the receiver were discharged and collected separately by means of drain-pipes and water-traps. An air-cock was placed at the highest point of each jacket.

The following shows the areas of the jacketed and unjacketed portions of the inner surfaces of the cylinders, exclusive of passages:—

Cylinder.		Area of Inner Surface.		
		Jacketed.	Unjacketed.	Total.
High-pressure	body, square feet	18·84 sq. ft.	..	22·37 sq. ft.
	ends, square feet	..	3·53 sq. ft.	
	per cent.	84·2 p.c.	15·8 p.c.	100·0 p.c.
Low-pressure	body, square feet	33·51 sq. ft.	..	44·68 sq. ft.
	ends, square feet	..	11·17 sq. ft.	
	per cent. . . .	75·0 p.c.	25·0 p.c.	100·0 p.c.

Details of Experiment.—The experiment consisted of two trials, which were made at the Royal Arsenal, Woolwich, on 15th March and 17th May 1890, by Lt.-Colonel English, Mr. Henry Davey, and Mr. Bryan Donkin, Jun. In the first trial, made on 15th March and lasting for 5·97 hours, no steam was admitted to any of the jackets. In the second trial, made on 17th May and lasting for 5·57 hours, steam was admitted to the jacket of the high-pressure cylinder at a little under boiler pressure, and to the jacket of the low-pressure cylinder at a pressure of about 10 lbs. per square inch above atmosphere. The arrangement of steam-supply piping to cylinders, and all the other conditions, were maintained as far as possible the same throughout both of the trials. Steam was not admitted to the jacket of the receiver during either trial.

Owing to the special arrangements of steam piping which had to be made in order to permit of the jacket-water being measured, and the consequent increase of radiating surface, no doubt the efficiency of the engine was somewhat reduced. As however this applies equally to both of the trials, the percentage saving due to the two steam-jackets as determined by these two trials is probably little affected thereby. From another trial made on the same engine, but not here recorded, the increased saving in feed-water due to the use of the receiver jacket with boiler steam, in addition to the other two jackets supplied as above described, was found to be 1·7 per cent., or a total saving due to the three jackets of 19·0 + 1·7 = 20·7 per cent.

Feed-Water.—On its way to the boiler the feed-water was passed through tanks in which it was weighed. In the trial without steam in any of the jackets, the weight of water pumped into the boiler in 5·97 hours was found to be 16,268 lbs., which is at the rate of 2,725 lbs. per hour, or 45·4 lbs. per minute. This is equivalent to 7·59 lbs. of feed-water per lb. of coal. In the trial with steam in the two cylinder jackets, the weight of water pumped into the boiler was found to be 12,622 lbs. in 5·57 hours, which is at the rate of 2,267 lbs. per hour, or 37·8 lbs. per minute. This is equivalent to 7·51 lbs. of feed-water per lb. of coal.

Jacket-Water and Drain-Water.—The discharge from each jacket and from the receiver drain was made to deliver into a separate bucket, and the weight of water in each bucket was determined. During the trial with steam in the two cylinder jackets 235 lbs. of jacket-water was obtained from the jacket of the high-pressure cylinder, and 481 lbs. from the jacket of the low-pressure cylinder. This gives a total of 716 lbs. of jacket-water from both jackets in 5·57 hours, or 128·6 lbs. per hour, which is equivalent to 5·67 per cent. of the total feed-water.

The drain-water obtained from the receiver drain was 70 lbs. in 5·97 hours, or 11·7 lbs. per hour, during the trial without steam in any of the jackets; and 110 lbs. in 5·57 hours, or 19·8 lbs. per hour, during the trial with steam in the two cylinder jackets.

Measurement of Pressures, &c.—Every quarter of an hour readings were taken of the steam-pressure in the boiler and of the vacuum in the condenser. Each of these gauges was therefore read 25 times during the trial without steam in any of the jackets, and 23 times during the trial with steam in the two cylinder jackets. The pressure-gauges on the two steam-jackets were also read every quarter of an hour throughout the trial. Indicator diagrams were taken from each end of each cylinder every quarter of an hour by means of four Richards indicators. In this way 25 sets of diagrams, or 100 single diagrams in all, were obtained during the trial without steam in any of the jackets; and 23 sets of diagrams, or 92 diagrams in all, during the trial with steam in the two cylinder jackets. The engine-counter was also read every quarter of an hour throughout both of the trials. The set of indicator diagrams nearest to the mean for each of the two trials are shown superposed in Fig. 8, Plate 71. The following table gives the average mean effective pressure and indicated horse-power in each of the two cylinders throughout both of the trials:—

	Mean Effective Pressure. Lbs. per square inch.		Indicated Horse-Power.	
Date of Trial 1890 Jackets, *With* or *Without* steam	15 March **Without**	17 May **With**	15 March **Without**	17 May **With**
	lbs.	lbs.	I.H.P.	I.H.P.
High-pressure cylinder .	18·95	15·12	64·2	57·1
Low-pressure „ .	4·21	4·60	46·3	56·3
Total . . .	10·05 (reduced to low-p. cyl.)	9·26 (reduced to low-p. cyl.)	110·5	113·4

Temperatures.—During the trial without steam in any of the jackets, the temperature of the feed-water was observed 9 times, of the hot-well 23 times, and of the circulating and discharge water each 29 times. During the trial with steam in the two cylinder jackets, the temperature of the feed-water was observed 13 times, of the hot-well 20 times, and of the circulating and discharge water each 33 times. The following table shows the mean temperatures obtained by calculation from these readings :—

Date of Trial Jackets *With* or *Without* Steam	1890	15 March **Without**	17 May **With**
Temperature of Feed-Water	Fahr.	57·8°	82·7°
„ „ Hot-Well	Fahr.	67·8°	79·3°
„ „ Circulating Water	Fahr.	43·7°	58·8°
„ „ Discharge „	Fahr.	53·6°	85·3°

Boiler.—Steam was supplied by a Lancashire boiler made by Messrs. Clayton. It is 6 feet 6 inches diameter and 32 feet long, and the flues are 2 feet 6 inches diameter. The grate area is 35 square feet, and the total heating surface about 880 square feet.

Coal.—The coal used throughout both of the trials was Ravensworth Hartley, Durham, and had a calorific value of 14,780 thermal units per lb., which is equivalent to the evaporation of 15·3 lbs. of water from and at 212° Fahr. per lb. of coal. In the trial without steam in any of the jackets, the total coal used was 2,144 lbs. in 5·97 hours. This is equivalent to 359 lbs. per hour, or 6·0 lbs. per minute, and to 10·3 lbs. per square foot of grate area per hour. The quantity of clinker and ashes made during this trial was 188 lbs., equivalent to 8·8 per cent. of the total fuel. In the trial with steam in the two cylinder jackets, the total coal used was 1,680 lbs. in 5·57 hours. This is equivalent to 302 lbs. per hour, or 5·0 lbs. per minute, and to 8·6 lbs. per square foot of grate area per hour. The quantity of clinker and ashes made during this trial was 221 lbs., equivalent to 13·2 per cent. of the total fuel.

Results.—The results of the experiment are given in Table 44, pages 456–459; and the performance, and rates of feed-water and jacket-water consumption, are shown graphically in Plate 78.

Nos. 45 & 46. Experiments on the Experimental Engine at the City and Guilds of London Central Institution, South Kensington, by Professor W. Cawthorne Unwin.

Engine.—The engine on which the experiments were made, Fig. 2, Plate 69, is a two-cylinder horizontal surface-condensing experimental engine, capable of being worked either as a single-cylinder or as a compound engine. It was made by Messrs. Marshall, Sons and Co., of Gainsborough, and was fully described and illustrated in "Engineering," 16th November 1888, pp. 473–476. The diameters of the cylinders are 8·73 inches and 15·76 inches, and their common stroke is 22 inches. The high-pressure cylinder is provided with Hartnell expansion-gear, and the low-pressure with Meyer expansion-gear. The bodies of the cylinders are covered externally with felt and teak, and the ends are bright and not covered. The experimental arrangements involve greater complication of steam-pipes, and consequent loss by radiation, than is usual in ordinary engines. The engine is also at present worked at a steam pressure lower than that for which it was designed. The trials were made with a constant load on the brake.

Only the bodies and back ends of the cylinders are jacketed. The front ends and the receiver are not jacketed. Steam is supplied direct to the jackets by a separate pipe, and they are drained by gravitation into a reservoir under pressure, provided with a gauge-glass. The following shows approximately the areas of the jacketed and unjacketed portions of the inner surfaces of the cylinders:—

Cylinder.			Area of Inner Surface.		
			Jacketed.	Unjacketed.	Total.
High-pressure	body,	square feet	3·95 sq. ft.	0·90 sq. ft.	4·85 sq. ft.
	ends,	square feet	0·41 sq. ft.	0·41 sq. ft.	0·82 sq. ft.
	total,	square feet	4·36 sq. ft.	1·31 sq. ft.	5·67 sq. ft.
	per cent.	. .	76·9 p.c.	23·1 p.c.	100·0 p.c.
Low-pressure	body,	square feet	7·04 sq. ft.	2·15 sq. ft.	9·19 sq. ft.
	ends,	square feet	1·35 sq. ft.	1·35 sq. ft.	2·70 sq. ft.
	total,	square feet	8·39 sq. ft.	3·50 sq. ft.	11·89 sq. ft.
	per cent.	. .	70·6 p.c.	29·4 p.c.	100·0 p.c.

Two experiments, Nos. 45 and 46, were made on this engine, each consisting of two trials of about three hours' duration, with and without steam in the jackets. In Experiment No. 45 the engine was worked as a single-cylinder condensing engine, only the larger of the two cylinders being in use; in Experiment No. 46 both cylinders were used, and the engine was worked as a compound condensing engine. In the single-cylinder experiment, No. 45, there was a saving due to the steam-jackets of 17·0 per cent. in the feed-water, and in the compound experiment, No. 46, a saving of 7·3 per cent. The set of indicator diagrams nearest to the mean for each of the two trials in the single-cylinder experiment, No. 45, are shown superposed in Fig. 10, Plate 73; and those for the two trials in the compound experiment, No. 46, in Fig. 11. The conditions and results of both experiments are given in detail in Tables 45 and 46, pages 460–463; and the performances, and rates of consumption of feed-water and jacket-water, are shown graphically in Plate 79.

Nos. 47–51. Experiments on a Single-Cylinder Vertical Experimental Engine at Bermondsey, by Mr. Bryan Donkin, Jun.

These experiments were made at Bermondsey, on a single-cylinder vertical experimental engine, Figs. 5 and 6, Plate 70, under different working conditions and with various arrangements of jacketing. The cylinder is 6 inches diameter, and the stroke 8 inches. The slide-valve is provided with a Meyer variable-expansion valve. The body and both ends of the cylinder are jacketed, as is also the valve-chest cover; and steam is supplied to each of the four jackets direct from the boiler by a separate pipe. By a special arrangement, the water from the cylinder body-jacket was divided into two portions, and the weights of these are given separately. The first portion consists of the steam condensed on the inner vertical surface of the jacket, and is due to the heat passing through the walls into the cylinder, almost all of which is usefully employed in keeping the temperature of the cylinder nearly equal to that of the steam in the jacket. The other portion consists of the steam condensed on the outer vertical surface of the jacket, and is due to the heat uselessly radiated outwards owing to imperfect external covering. This loss should of course be reduced to a minimum; and in these experiments a layer of good non-conducting material was placed round the cylinder for this purpose.

Observations of the temperatures of the cylinder walls were made throughout the trials, following the method already published in Proceedings Inst. Civil Engineers, vol. c, 1890, page 347. These temperatures were taken by small thermometers inserted in $\frac{1}{8}$ inch holes, which were drilled into the metal walls and filled with mercury. It will be seen from Tables 47–51, pages 464–468, that the temperature of the cylinder walls was much higher with steam in the jackets than without.

Experiments were made not only with and without steam in the jackets, but also with all the jackets filled up with water, which was allowed to assume a normal temperature before commencing the trial; a trial was also made with a vacuum in the body jacket. When no steam is admitted into the jackets, the air in them gets gradually hotter, being heated by the cylinder walls, until it reaches a normal temperature. The same thing occurs when cold water is admitted into the jackets; it gradually becomes hotter, and then remains at a constant temperature.

Experiments were made at two different speeds—about 218 and 115 revolutions per minute. In order to shorten the duration of the trials, the engine was fitted with a surface condenser to measure the feed-water; and it was found that trials lasting for half an hour agreed satisfactorily with those of three hours' duration. The steam-pressure and cut-off were kept constant throughout the whole of the experiments, with the exception of No. 50 in which the cut-off was varied. A rope brake was used to absorb the power.

Five experiments were made on this engine, four condensing and one non-condensing. Experiment No. 47 consisted of four trials at about full speed, with steam in all the jackets, without steam in any of the jackets, with water in all the jackets, and with a vacuum in the body jacket. Experiment No. 48 consisted of two trials at about half speed, with and without steam in all the jackets. Experiment No. 49 consisted of two non-condensing trials at about full speed, with and without steam in all the jackets. Experiment No. 50 consisted of twelve trials, all at full speed, made in pairs, one unjacketed and one jacketed, with six different degrees of expansion, in order to determine how the changes in the ratio of expansion would affect the value of the steam-jackets. In experiment No. 51, steam was admitted to the body jacket and shut off again during the experiment, without stopping the engine, in order to ascertain the effect of this upon the speed of the engine, the load on the brake being kept constant. In all the experiments, the feed-water always included the whole of the jacket-water.

Experiment No. 47. Four trials with different Media in Jackets.—The details of this experiment are given in Table 47, page 464, and Figs. 41 to 44, Plate 80, and the set of indicator diagrams nearest to the mean for the first two trials are shown superposed in Fig. 14, Plate 75. These four trials were all made at a mean speed of about 218 revolutions per minute, condensing, and at a cut-off of 3–16ths.

The first trial was made without steam in any of the jackets, that is with hot air in all the jackets, and lasted for 34 minutes. The consumption of feed-water was at the rate of 41·23 lbs. per indicated horse-power per hour.

The second trial was made with steam in all the jackets, and lasted for 52 minutes. The consumption of feed-water, including jacket-water, was at the rate of 28·39 lbs. per indicated horse-power per hour, showing a saving in feed-water of 41·23 – 28·39 = 12·84 lbs. per indicated horse-power per hour, or a decrease in consumption of 31·1 per cent. due to the action of steam as compared with hot air in the four jackets. The mean temperature of the internal part of the cylinder walls was 28° higher with steam in the jackets than with hot air. The condensed water from the various jackets was as follows:—

Jacket.	Condensed water.
Body jacket, inner wall	0·69 lb. per I.H.P. per hour.
,, ,, outer wall	0·68 lb. ,,
Top-end jacket	0·15 lb.
Bottom-end jacket	0·41 lb.
Valve-chest jacket	0·18 lb.
Total Jacket-Water	2·11 lbs.

The total quantity of steam condensed in the jackets was therefore 2·11 lbs. per indicated horse-power per hour, which is equivalent to 7·43 per cent. of the total feed-water. Thus for every lb. of steam condensed in the jackets there was a net saving in feed-water consumption of 6·1 lbs. The following table shows the effect of the different jackets on the feed-water consumption, as determined by another experiment:—

Arrangement of Jacketing.	Feed-Water including Jacket-Water.		
	Consumption per I.H.P. per hour.	Saving due to Steam-Jackets.	
		per I.H.P. per hour.	percentage of 41·2.
	lbs.	lbs.	per cent.
Without steam in any of the jackets	41·2	..	..
With steam in two *end* jackets only	37·0	4·2	10·2
With steam in *body* jacket only .	30·1	11·1	26·9
With steam in *all four* jackets .	28·4	12·8	31·1

The third trial was made with hot water in all the jackets, and lasted for $22\frac{1}{2}$ minutes, the water in the jackets being allowed to assume a normal temperature before the observations were commenced. The consumption of feed-water was at the rate of 46·27 lbs. per indicated horse-power per hour, showing a difference in feed-water consumption of $46 \cdot 27 - 41 \cdot 23 = 5 \cdot 04$ lbs. per indicated horse-power per hour, or an increase of 12·2 per cent., due to the presence of hot water in the jackets, as compared with hot air.

In the fourth trial, which lasted for 26 minutes, a vacuum of 12·9 lbs. per square inch below the atmosphere was maintained in the body jacket by putting it in direct communication with the condenser, the other three jackets being open to the air. The consumption of feed-water was at the rate of 38·01 lbs. per indicated horse-power per hour, showing a saving in feed-water of $41 \cdot 23 - 38 \cdot 01 = 3 \cdot 22$ lbs. per indicated horse-power per hour, or a decrease of 7·8 per cent. due to maintaining a vacuum in the body jacket, as compared with hot air in all the jackets. This trial with a vacuum in the body jacket was made with the intention of lowering the temperature of the cylinder walls; but a little reflection showed that there was no medium to convey the heat from the hot metal walls to the cooler condenser. The walls were found to remain at a temperature of 264° Fahr. without parting with their heat, although exposed to the much lower condenser temperature of 130° Fahr. This has since been confirmed by further experiments, and shows a practical result which is not unimportant. The cylinder walls were rather hotter with a vacuum in the body jacket than with hot air in all the jackets, and external radiation was probably diminished.

A trial, not recorded in the tables, was also made on this engine, with the body jacket filled with mineral cylinder-oil kept at a temperature of 375° Fahr. by means of gas jets, and with steam in the other three jackets. The trial was made condensing, with a cut-off at 3-16ths. The consumption of feed-water, including the small quantity of jacket-water which was condensed during the trial, was at the rate of 28·86 lbs. per indicated horse-power per hour, which is slightly higher than in the second trial of experiment No. 47, the corresponding trial with steam in all the jackets.

Experiment No. 48. Two trials at half-speed with and without Steam in Jackets.—The details of this experiment are given in Table 48, page 465, and Figs. 45 to 48, Plate 80, and the set of indicator diagrams nearest to the mean in Fig. 12, Plate 74. The trials were made at a mean speed of about 116 revolutions per minute, condensing, and at a cut-off of 3-16ths. With the exception of the speed, which was a little more than half, all the conditions were the same as those for the first two trials of experiment No. 47.

The trial without steam in any of the jackets lasted for 36 minutes. The consumption of feed-water was at the rate of 46·25 lbs. per indicated horse-power per hour.

The trial with steam in all the jackets lasted for 28 minutes. The consumption of feed-water, including jacket-water, was at the rate of 30·43 lbs. per indicated horse-power per hour, showing a saving in feed-water of 46·25 − 30·43 = 15·82 lbs. per indicated horse-power per hour, or a decrease in the consumption of 34·2 per cent. due to the action of steam as compared with hot air in the jackets. The mean temperature of the internal part of the cylinder walls was 31° higher with steam in the jackets than without; and for these two trials the temperatures were a few degrees higher than in the first two trials of experiment No. 47, in which the speed was about double, all the other conditions being the same. The total condensed water from the jackets was as follows:—

Jacket.	Condensed water.
Body jacket, inner wall	0·68 lb. per I.H.P. per hour.
,, ,, outer wall	0·68 lb. ,,
Top-end jacket	0·32 lb.
Bottom-end jacket	0·50 lb.
Valve-chest jacket	0·25 lb.
Total Jacket-Water	2·43 lbs.

The total quantity of steam condensed in the jackets was therefore 2·43 lbs. per indicated horse-power per hour, which is equivalent to 7·99 per cent. of the feed-water. Thus for every lb. of steam condensed in the jackets there was a net saving in feed-water of 6·5 lbs.

Comparing these results with the first two trials of experiment No. 47, in which all the conditions were the same with the exception of speed, the effect of the steam-jackets at the reduced speed is seen to be increased by about 3 lbs. per indicated horse-power per hour, thus:—

Speed of Engine.	Saving due to Steam-Jackets.
218 revs. per minute	12·84 lbs. per I.H.P. per hour.
116 ,, ,, ...	15·82 lbs. ,,
Increase in saving at reduced speed .	2·98 lbs.

Experiment No. 49. *Two Non-condensing trials with and without Steam in Jackets.*—The details of this experiment are given in Table 49, page 465, and Figs. 49 to 52, Plate 80, and the set of indicator diagrams nearest to the mean in Fig. 13, Plate 74. The trials were made at about full speed. With the exception that in this case there was no vacuum in the condenser, all the conditions were the same as those for the first two trials of experiment No. 47.

The trial without steam in any of the jackets lasted for $38\frac{3}{4}$ minutes. The consumption of feed-water was at the rate of 62·64 lbs. per indicated horse-power per hour.

The trial with steam in all the jackets lasted for 44 minutes. The consumption of feed-water, including jacket-water, was at the rate of 35·69 lbs. per indicated horse-power per hour, showing a saving in feed-water of 62·64 − 35·69 = 26·95 lbs. per indicated horse-power per hour, or a decrease of 43·0 per cent. due to the action of steam as compared with air in the four jackets. The mean temperature of the internal part of the cylinder walls was 44° higher with steam in the jackets than without. The total condensed water from the jackets was as follows:—

Jacket.	Condensed water.
Body jacket, inner wall	0·75 lb. per I.H.P. per hour.
„ „ outer wall	0·81 lb. „
Top-end jacket	0·22 lb.
Bottom-end jacket	0·53 lb.
Valve-chest jacket	0·25 lb.
Total Jacket-Water	2·56 lbs.

The total quantity of steam condensed in the jackets was therefore 2·56 lbs. per indicated horse-power per hour, which is equivalent to 7·17 per cent. of the feed-water. Thus for every lb. of steam condensed in the jackets there was a net saving in feed-water consumption of 10·5 lbs.

Comparing these two non-condensing trials with the first two trials in experiment No. 47, which were condensing and had all the other conditions the same, the useful effect of the jackets was increased from 31·1 per cent. to 43·0 per cent. when the engine was worked without a vacuum in the condenser, thus:—

	Saving due to Steam-Jackets.	
Engine working	per I.H.P. per hour.	per cent.
Condensing	12·84 lbs.	31·1
Non-condensing	26·95 lbs.	43·0

Experiment No. 50. Twelve trials with and without steam in the jackets, at various ratios of expansion.—The details of these trials are given in Table 50, pages 466 and 467, and Plates 81 and 82. They were all made condensing, at a speed of about 218 revolutions per minute, and with a steam-pressure of about 50 lbs. per square inch above the atmosphere. The trials were made in pairs, jacketed and unjacketed, at six different degrees of cut-off—1–16th, 1–12th, 1–8th, 3–16ths, 5–16ths, and half-stroke—all the other conditions remaining as far as possible the same.

From Table 50 it will be seen that with the greatest number of expansions, 6·8 at 1–16th cut-off, the saving due to the jackets was 40·4 per cent., while with the smallest number of expansions, 1·8 at one-half cut-off, the saving was only 23·1 per cent.; or, speaking generally, the saving in steam consumption due to the use of steam-jackets increases with the number of expansions.

The great difference in the temperatures of the cylinder walls with and without steam in the jackets is clearly noticeable; this difference decreases considerably as the number of expansions decreases. The effect of steam-jacketing on the dryness-fraction of the steam in the cylinder, both at cut-off and at release, is distinctly marked.

At a speed of about 220 revolutions per minute, the cut-off giving the minimum steam consumption with steam in jackets was found to be at 1–8th of the stroke, equal to 4·8 expansions. The consumption of feed-water, including jacket-water, per indicated horse-power per hour in this case was 25·2 lbs., and the dryness-fraction of the steam in the cylinder at release was 96·6 per cent. At the same speed, without steam in the jackets, the best cut-off was found to be at 5–16ths of the stroke, equal to 2·6 expansions. Here the steam used was 38·7 lbs. per indicated horse-power per hour, and the dryness-fraction of the steam in the cylinder at release was 73·9 per cent.

In Plate 82 is shown graphically the variation of efficiency of the steam-jacket in lbs. of feed-water saved per indicated horse-power per hour at different ratios of expansion, for the same speed and steam pressure; also the variation of the cylinder-wall temperatures, and of the dryness-fraction of the steam at cut-off and release, at different ratios of expansion, both with and without steam in the jackets. It will be observed from the diagrams that the line of steam consumption tends to vary with that of the dryness-fraction at release; the steam consumption increasing with the wetness of the steam.

The drier the steam at release, the greater is the economy, the maximum economy being obtained with the steam practically dry at release.

Experiment No. 51. *To determine the effects of the Steam-Jacket on the speed of the engine, and temperature of the cylinder walls, &c.*—Throughout this experiment the condenser was kept in action, steam was cut off at 3–16ths, and the steam-pressure and load on the brake were kept constant. Only the body jacket was used, steam being shut off from the other three. The object of the experiment was to determine the effects on the speed of the engine and the temperature of the walls &c., when steam was admitted to the body jacket and shut off from it again without stopping the engine. Starting at 11.30 a.m. after normal working conditions had been reached, the engine was worked for eleven minutes without steam in the jacket, and allowed to attain a constant speed of 140 revolutions per minute and a constant temperature of the cylinder; after which, at 11.41 a.m., steam was admitted to the jacket without stopping the engine. Six minutes afterwards, at 11.47 a.m., when the engine had acquired an increased speed of 265 revolutions per minute and the temperatures had also again become constant, steam was shut off from the jacket, with the result that the speed and temperatures gradually decreased. Fifty minutes afterwards, at 12.37 p.m., the speed was found to have fallen to 190 revolutions per minute; and had the experiment been continued long enough, it would have decreased to 140 revolutions per minute, as at the commencement. Every two minutes throughout the experiment, observations were made of speed, jacket pressure, and the various principal temperatures; and these are plotted in Fig. 15, Plate 75. The principal results are also given in Table 51, page 468. It will be seen how quickly the temperatures in the cylinder increased after steam was admitted to the jacket; and on the contrary how they decreased, though less quickly, after the steam supply was shut off, the only heat then available being that shut into the jacket space and that remaining in the metal of the cylinder walls. The speed of the engine decreased considerably as this heat was gradually exhausted.

Superheating.—Subsequent experiments have been carried out using slightly superheated steam both in the jackets and in the cylinder. With only from 35° to 58° Fahr. of superheating, and using this steam in all four jackets as well as in the cylinder, there was an economy of about 26 per cent. due to the jackets, in comparison with the result obtained when using the same steam in the cylinder alone and shutting off the jackets. The engine was working condensing, at 1–16th cut-off and at a speed of 200 revolutions per minute. The increase in temperature of the cylinder walls when the jackets were in use was about 50° Fahr.

Conclusions.—With the same quality and temperature of steam, in these experiments, the hotter walls yielded the better results. With walls of the same temperature, the drier and hotter the steam the greater was the economy. The best result with all the jackets in use was obtained with 1–8th cut-off or 4·8 expansions; the consumption was 25·2 lbs. of steam per indicated horse-power per hour including jacket-water; and the dryness-fraction at release was 96·6 per cent., showing that in this case the exhaust steam was practically dry and neither superheated nor wet.

Appendix.

Suggestions for future experiments on the Steam-Jacket.

The date, place, and duration of each experiment should be recorded, and the names of the experimenters.

In the description of the engine it should be stated whether it is single-cylinder, compound, or triple-expansion, horizontal or vertical, condensing or non-condensing, and the diameters and strokes of the cylinders should be given, and the kind of valves mentioned.

The following particulars should also be furnished. Manner in which each cylinder and receiver is jacketed, giving dimensions for enabling the jacketed and unjacketed areas of the internal surfaces of each to be calculated. Particulars as to how the steam is supplied to the jackets, and how they are drained, giving the lengths and diameters of the supply pipes. Details of how the outer walls of the jackets are protected.

Full particulars of the volumes and surfaces of clearances and passages should be given for each cylinder.

A sufficient number of measuring tanks &c. should be provided, to allow the feed-water, and the discharges from the jackets, drain-pipes &c., to be measured separately.

Each body-jacket should be provided with a pressure-gauge; and also with two cocks, one at the highest point and the other at the lowest, to test for air, steam, or water during the experiments. Each of the other jackets should also be provided with an air-cock at its highest point, so that if air is present it may be allowed to escape from time to time.

A separate indicator should be used for each end of each cylinder, and the engine should be fitted with a speed-counter.

Before the experiments commence, the valves and pistons should be tested for leakage, by blocking the engines in different positions and opening the indicator cocks.

The body-jackets should also be tested for leakage at the usual steam-pressures, the pistons and cylinder covers being removed if possible, or by steam being admitted to the jackets and the indicator cocks opened before steam is admitted to the cylinders.

The quantity of steam condensed in the jackets when the engine is standing should be ascertained by measuring the discharge from each jacket in a given time, when supplied with steam under the normal conditions.

Throughout the whole of the experiments the quantities of feed-water, circulating or injection water, jacket-water, drain-water, leakage from valves or stuffing-boxes, and water from the steam separator should be measured. The discharge from each jacket should be taken separately, and the temperatures of the feed-water and of the circulating or injection water should be noted.

The readings of the speed-counter and the pressure-gauges should be all noted at regular intervals. Sets of indicator diagrams should be taken at regular intervals, each set consisting of a diagram from each end of each cylinder, taken as far as practicable simultaneously. The highest and lowest pressure in each cylinder should be ascertained from the diagrams and recorded, for allowing the range of temperature in each cylinder to be calculated. The mean effective pressures and ratio of expansion should be calculated from the diagrams and recorded, and the indicated horse-powers worked out. The percentage of moisture in the steam should be recorded if known. The kind of lubricant used for the cylinders should be stated, and the same quantity per hour should as far as possible be supplied during comparative trials with and without steam in the jackets.

TABLES 42–44 (*continued to page* 459).

Three Sets of Experiments, Nos. 42, 43, *and* 44,

at Hammersmith, Waltham Abbey, and Woolwich Arsenal, by Mr. Davey,

1	Description of Engine . . . .
2	Place of Experiments . . . .
3	Experiments conducted by . . . .
4	Jackets *With* or *Without* steam . . . .
5	Date of Experiment . . . .
6	Duration of Experiment . . . . hours
7	Cylinder diameter, high-pressure . . . . . . inches
8	„ „ intermediate . . . . . . inches
9	„ „ low-pressure . . . . . . inches
10	Stroke, length . . . . inches
11	Mean pressure in boiler above atmosphere . lbs. per square inch
12	„ „ „ high-pressure jacket above atm., lbs. per square inch
13	„ „ „ intermediate „ „ „ lbs. per square inch
14	„ „ „ low-pressure „ „ „ lbs. per square inch
15	Mean effective pressure, total reduced to low-pr. cyl. lbs. per sq. in.
16	Number of Expansions . . . .
17	Revolutions per minute, mean . . . . revs.
18	Piston speeds, feet per minute, mean . . . . . . feet
19	Indicated horse-power, mean total . . . . . I.H.P.

(*continued on next page*) TABLES 42–44.

WITH *and* WITHOUT *Steam in Jackets,*

Mr. Mair-Rumley, Mr. Bryan, Lt.-Col. English, and Mr. Donkin.

No. 42. Beam Compound. Hammersmith. Mr. J. G. Mair-Rumley.		No. 43. Triple Expansion. Waltham Abbey. Mr. H. Davey and Mr. W. B. Bryan.		No. 44. Horizontal Compound. Woolwich Arsenal. Lt.-Col. English, Mr. H. Davey, and Mr. B. Donkin, Jun.		1 2 3
Without	**With**	**Without**	**With**	**Without**	**With**	4
22 & 23 Mar. 1888	21 & 22 Mar. 1888	7 Dec. 1889	6 Dec. 1889	15 March 1890	17 May 1890	5
10·0	8·0	8·0	8·0	5·97	5·57	6
29·0	29·0	18·0	18·0	18·0	18·0	7
..	..	30·5	30·5	..	..	8
47·5	47·5	51·0	51·0	32·0	32·0	9
65·1 & 96·0	65·1 & 96·0	36·0	36·0	48·0	48·0	10
49·7	49·0	130·0	130·0	50·4	50·2	11
..	49·0	..	129·0	..	46·3	12
..	..	..	28·0	..	..	13
..	49·0	..	5·5	..	10·3	14
12·80	12·45	16·46	16·30	10·05	9·26	15
—	—	22	30	9·4	12·6	16
14·80	15·78	23·0	22·9	57·06	63·62	17
160·6 & 236·8	171·2 & 252·5	138·0	137·4	456·5	509·0	18
162·0	168·0	140·0	138·0	110·5	113·4	19

TABLES 42–44 (*continued from preceding page*).

Three Sets of Experiments, Nos. 42, 43, *and* 44,

at Hammersmith, Waltham Abbey, and Woolwich Arsenal, by Mr. Davey,

1	Description of Engine . . . .
2	Place of Experiments . . . .
3	Experiments conducted by . . . .
4	Jackets, *With* or *Without* Steam . . . .
20	Feed-Water, total used during trial . . . . . . lbs.
21	,, per hour, total . . . . lbs.
22	,, per indicated horse-power per hour, total . . lbs.
23	,, percentage less with steam in jackets . . per cent.
24	Jacket-Water, total during trial . . . . lbs.
25	,, per hour . . . . lbs.
26	,, from high-pressure jacket, per I.H.P. per hour . lbs.
27	,, ,, intermediate ,, ,, ,, ,, ,, . lbs.
28	,, ,, low-pressure ,, ,, ,, ,, ,, . lbs.
29	,, total from all jackets ,, ,, ,, ,, . lbs.
30	,, ,, in percentage of feed-water . . per cent.
31	Coal, total used during trial . . . . lbs.
32	,, per indicated horse-power per hour . . . . . lbs.

(concluded from page 456) TABLES 42–44.

WITH *and* WITHOUT *Steam in Jackets,*

Mr. Mair-Rumley, Mr. Bryan, Lt.-Col. English, and Mr. Donkin.

No. 42. Beam Compound. Hammersmith. Mr. J. G. Mair-Rumley.		No. 43. Triple Expansion. Waltham Abbey. Mr. H. Davey and Mr. W. B. Bryan.		No. 44. Horizontal Compound. Woolwich Arsenal. Lt.-Col. English, Mr. H. Davey, and Mr. B. Donkin, Jun.		1 2 3
Without	With	Without	With	Without	With	4
29,487	22,360	19,284	17,053	16,268	12,622	20
2,949	2,795	2,410	2,132	2,725	2,267	21
18·20	16·64	17·22	15·45	24·68	19·99	22
..	**8·6**	..	**10·3**	..	**19·0**	23
..	1,617	..	1,900	..	716	24
..	202·1	..	237·5	..	128·6	25
..	0·54	..	0·81	..	0·37	26
..	..	..	0·53	..	..	27
..	0·66	..	0·38	..	0·76	28
..	1·20	..	1·72	..	1·13	29
..	7·2	..	11·1	..	5·7	30
—	—	2,345	1,981	2,144	1,680	31
—	—	2·09	1·79	3·25	2·66	32

TABLES 45 & 46 (*continued to page 463*).—*Experiments on a Horizontal Surface-Condensing Engine* WITH *and* WITHOUT *Steam in Jackets, at South Kensington, by Professor Unwin.*

			No. 45. SINGLE-CYLINDER.		No. 46. COMPOUND.	
1	Engine working *Single-Cylinder* or *Compound*					
2	Jackets, *With* or *Without* Steam		Without	With	Without	With
3	Date of Trial	1890	6 Feb.	20 Feb.	12 March	27 March
4	Duration of Trial	hours	3·09	2·75	2·69	3·20
5	Number of Expansions (nominal)		4	4	7·23	9·29
6	Cylinder, diameter	inches	15·76	15·76	{8·73, 15·76}	{8·73, 15·76}
7	Stroke, length	inches	22	22	22	22
8	Piston, mean area	square inches	192·33	192·33	{57·43, 192·33}	{57·43, 192·33}
9	Volume swept out by piston per stroke	cubic feet	2·4485	2·4485	{0·7311, 2·4485}	{0·7311, 2·4485}
10	Mean boiler-pressure above atmosphere	lbs. per sq. inch	60·56	59·60	66·73	67·80
11	„ jacket „ „	lbs. per sq. inch	..	59·60	..	{67·80, 67·80}
12	„ effective „ in cylinders	lbs. per sq. inch	17·13	17·46	{31·34, 12·68}	{26·49, 14·32}
13	„ „ „ total reduced to low-p. cyl.	lbs. per sq. inch	17·13	17·46	22·04	22·23
14	„ back-pressure in condenser, absolute	lbs. per sq. inch	2·95	2·28	2·02	2·04

TABLES 45 & 46 (*continued on next page*).—*Experiments on a Horizontal Surface-Condensing Engine* WITH *and* WITHOUT *Steam in Jackets, at South Kensington, by Professor Unwin.*

1	Engine working *Single-Cylinder* or *Compound*		No. 45. SINGLE-CYLINDER.		No. 46. COMPOUND.	
2	Jackets, *With* or *Without* Steam		Without	With	Without	With
15	Revolutions, total during trial	revs.	20,851	16,786	15,125	18,453
16	„ per minute, mean	revs.	112·40	101·73	93·66	96·11
17	Piston speed, feet per minute, mean	feet	412·1	373·0	343·4	352·4
18	Coal burnt per indicated horse-power per hour	lbs.	3·53	2·94	2·30	2·13
19	Indicated horse-power	I.H.P.	41·14	37·96	{18·73 25·38}	{16·24 29·41}
20	„ „ mean total	I.H.P.	41·14	37·96	44·11	45·65
21	Brake „ „	B.H.P.	32·79	30·91	36·56	39·62
22	Horse-power expended in engine friction	H.P.	8·35	7·05	7·55	6·03
23	Percentage of total I.H.P. expended in engine friction	per cent.	20·3	18·6	17·1	13·2
24	Feed-Water, total used during trial	lbs.	4,088	2,786	2,500	2,851
25	„ per indicated horse-power per hour, total	lbs.	32·14	26·69	21·06	19·52
26	„ „ „ „ minute, total	lbs.	0·536	0·445	0·351	0·325
27	„ percentage less with steam in jackets	per cent.	..	**17·0** p.c.	..	**7·3** p.c.
28	Steam condensed, total during trial, excluding jacket-steam	lbs.	4,088	2,590	2,500	2,500
29	„ „ per indicated horse-power per hour	lbs.	32·14	24·81	21·06	17·12
30	„ „ „ „ „ minute	lbs.	0·536	0·414	0·351	0·285

TABLES 45 & 46 (*continued from preceding page*).—*Experiments on a Horizontal Surface-Condensing Engine* WITH *and* WITHOUT *Steam in Jackets, at South Kensington, by Professor Unwin.*

1	Engine working *Single-Cylinder* or *Compound*		No. 45. SINGLE-CYLINDER.		No. 46. COMPOUND.	
2	Jackets, *With* or *Without* Steam		Without	With	Without	With
31	Condensing Water, total during trial	lbs.	92,380	58,670	55,380	67,050
32	„ „ per indicated horse-power per hour	lbs.	726·3	562·0	466·5	459·0
33	Jacket-Water, total during trial	lbs.	..	196	..	351
34	„ per indicated horse-power per hour	lbs.	..	1·88	..	2·40
35	„ „ „ minute	lbs.	..	0·031	..	0·040
36	„ in percentage of feed-water	per cent.	..	7·04	..	12·31
37	Mean temperature of hot-well	Fahr.	137·8°	124·8°	122·5°	121·3°
38	„ „ feed-water	Fahr.	100·0°	100·0°	100·0°	100·0°
39	„ „ boiler steam	Fahr.	308·0°	307·0°	313·5°	314·3°
40	Heat taken up by feed-water per lb.	thermal units	1,107·9	1,107·6	1,109·6	1,109·8
41	„ „ „ condensing water per lb.	thermal units	43·5	44·1	42·5	37·8
42	„ given „ condensed steam „	thermal units	37·8	24·8	22·5	21·3
43	„ transferred from Jacket steam to Cylinder steam per indicated horse-power per minute	thermal units	..	27·9	..	35·8

TABLES 45 & 46 (*concluded from page 460*).—*Experiments on a Horizontal Surface-Condensing Engine* WITH *and* WITHOUT *Steam in Jackets, at South Kensington, by Professor Unwin.*

		No. 45. SINGLE-CYLINDER.		No. 46. COMPOUND.	
1	Engine working *Single-Cylinder* or *Compound*				
2	Jackets, *With* or *Without* Steam	Without	With	Without	With
	Heat Account, from Boiler Steam.	Thermal Units per I.H.P. per minute.			
44	Heat in steam passing through Cylinders	593·80	458·50	389·50	316·30
45	„ „ „ „ „ Jackets	..	34·30	..	44·40
46	„ total, in steam passing from Boiler	593·80	492·80	389·50	360·70
	Heat [illegible]t, from Discharged Water, etc.	Thermal Units per I.H.P. per minute.			
47	Heat turned into Work	42·75	42·75	42·75	42·75
48	„ lost given to Condensing W[illegible]er	526·40	413·20	330·20	289·20
49	„ „ in [illegible] [illegible]d [illegible]m above feed temperature . . .	20·26	10·27	7·90	6·07
50	„ „ „ [illegible]er „ „ „ . . .	..	6·42	..	8·57
51	„ „ by radiation	8·50	9·30	9·70	9·34
52	„ [illegible]ed for [illegible]ve or negative)	− 4·11	+ 10·86	− 1·05	+ 4·77
53	„ total, passing from Engine (= line 46)	593·80	492·80	389·50	360·70
54	Mechanical Efficiency of Engine (line 21 ÷ line 20) . per cent.	79·7	81·4	82·9	86·8
55	Thermal „ „ (line 47 ÷ line 46) . per cent.	7·2	8·7	11·0	11·9
56	Theoretical „ *perfect* Engine with same temperature limits . . . . . . per cent.	28·4	31·1	32·7	33·1
57	Relative Efficiency of Engine (line 55 ÷ line 56) . . per cent.	25·4	28·0	33·6	36·0

TABLE 47.—*Experiment on a Single-Cylinder Vertical Surface-Condensing Engine* WITH *and* WITHOUT *Steam, &c., in Jackets, at Bermondsey, by Mr. Bryan Donkin, Jun.*

1	Medium in jackets		Air.	Steam.	Water.	Vacuum.*
2	Duration of trial	hour	0·57	0·87	0·38	0·43
3	Valve-chest pressure, lbs. per square inch above atmosphere	lbs.	49·7	50·1	50·2	50·2
4	Steam-Jacket „ „ „ „ „	lbs.	0·0	51·8	53·6	-12·9
5	Number of expansions		3·77	3·75	3·75	3·77
6	Revolutions per minute, mean	revs.	220	216·3	218	213·5
7	Piston speed, feet per minute, mean	feet	293	288	291	285
8	Indicated horse-power, mean	I.H.P.	6·56	6·83	6·32	6·36
9	Mean temperature in body-jacket space	Fahr.	252·5°	298·5°	247·0°	248·0°
10	„ „ of cylinder wall, in six holes	Fahr.	260·5°	290·5°	249·5°	264·3°
11	„ „ „ „ „ 0·06 inches from piston	Fahr.	255·8°	284·0°	250·0°	266·0°
12	„ „ initial in cylinder	Fahr.	—	297·0°	—	—
13	„ „ terminal „	Fahr.	—	223·9°	—	—
14	„ „ during steam stroke in cylinder	Fahr.	—	260·4°	—	—
15	Percentage of feed-water present in cylinder as steam at cut-off	per ct.	42·6	61·2	—	—
16	„ „ „ „ „ at release	per cent.	63·7	89·2	—	—
17	Jacket-Water, lbs. per indicated horse-power per hour, total	lbs.	..	2·11	..	..
18	„ in percentage of feed-water	per cent.	..	7·43	..	..
19	Feed-Water, lbs. per indicated horse-power per hour, total	lbs.	41·23	28·39	46·27	38·01
20	„ percentage less with steam in jackets	per cent.	..	31·1 p.c.	..	..

* Body jacket in communication with condenser, air in other three jackets.

TABLES 48 & 49.—*Experiments on a Single-Cylinder Vertical Surface-Condensing Engine* WITH *and* WITHOUT *Steam in Jackets, at Bermondsey, by Mr. Bryan Donkin, Jun.*

	Engine working Condensing or Non-condensing		No. 48. CONDENSING.		No. 49. NON-CONDENSING.	
1	Jackets *With* or *Without* Steam		Without	With	Without	With
2	Duration of trial	hour	0·60	0·47	0·65	0·73
3	Valve-chest pressure, lbs. per square inch above atmosphere	lbs.	49·5	50·2	49·9	50·2
4	Steam-Jacket " " " " "	lbs.	0·0	51·2	0·0	53·3
5	Number of expansions		3·75	3·75	3·75	3·75
6	Revolutions per minute, mean	revs.	116·8	114·9	220·7	212·4
7	Piston speed, feet per minute, mean	feet	156	153	294	283
8	Indicated horse-power, mean	I.H.P.	3·92	3·72	3·80	4·19
9	Mean temperature in body-jacket space	Fahr.	250·0°	299·0°	246·0°	299·5°
10	" " of cylinder wall, in six holes	Fahr.	262·5°	295·0°	250·0°	297·0°
11	" " " " 0·06 inches from piston	Fahr.	260·0°	291·0°	252·0°	296·0°
12	" " initial in cylinder	Fahr.	292·2°	296·3°	289·2°	297·7°
13	" " terminal "	Fahr.	235·3°	228·2°	230·1°	221·1°
14	" " during steam stroke in cylinder	Fahr.	263·7°	262·2°	259·6°	259·4°
15	Percentage of feed-water present in cylinder as steam at cut-off	per cent.	33·0	57·6	42·1	72·1
16	" " " " " at release	per cent.	59·1	88·3	71·5	95·2
17	Jacket-Water, lbs. per indicated horse-power per hour, total	lbs.	..	2·43	..	2·56
18	" in percentage of feed-water	per cent.	..	7·99	..	7·17
19	Feed-Water, lbs. per indicated horse-power per hour, total	lbs.	46·25	30·43	62·64	35·69
20	" percentage less with steam in jackets	per cent.	..	**34·2** p.c.	..	**43·0** p.c.

TABLE 50 (*continued on next page*).

Experiment on a Single-Cylinder Vertical Surface-Condensing Engine, WITH *and* WITHOUT *Steam in Jackets,*

	Trial letter	*a*	*b*	*c*
1	N Jackets Not in use ; J all four Jackets in use	N	J	N
2	Duration of trial . . . . minutes	33·25	58·0	30·5
3	Valve-chest pressure, lbs. per square inch above atmosphere lbs.	49·0	49·9	49·5
4	Steam-Jacket „ lbs. per square inch above atmosphere lbs.	..	55·2	..
5	Cut-off in percentage of stroke . per cent.	6·25	6·25	8·33
6	Number of expansions	6·8	6·8	6·0
7	Revolutions per minute, mean . . revs.	219	222	222
8	Piston speed, feet per minute, mean . feet	292	296	296
9	Indicated horse-power, mean . . I.H.P.	4·06	4·36	4·48
10	Mean temperature in body-jacket space Fahr.	225°	301°	227°
11	„ „ of cylinder wall Fahr.	232°	298°	237°
12	„ „ „ „ „ 0·06 ins. from piston, Fahr.	236°	297°	240°
13	„ „ inside cylinder . Fahr.	217°	307°	224°
14	Percentage of feed-water present in cyl. as steam } at cut-off, per cent.	35·4	65·9	33·9
15	„ „ „ „ at release, per cent.	59·0	107·9	62·8
	Jacket-Water, lbs. per I.H.P. per hour.		lbs.	
16	„ from inner surface of body-jacket	..	0·57	..
17	„ outer surface of body-jacket	..	0·68	..
18	„ „ other three jackets . .	..	1·07	..
19	„ total from all four jackets . .	..	2·32	..
20	Jacket-Water, total, in percentage of feed-water per cent.	..	8·5	..
21	Feed-Water, total per indicated horse-power per hour lbs.	45·6	27·2	44·6
22	„ lbs. saved per lb. of jacket-water with steam in jackets lbs.	..	7·9	..
23	„ percentage less with steam in jackets per cent.	..	40·4	..

(continued from preceding page) TABLE 50.

with different Rates of Expansion,
at Bermondsey, by Mr. Bryan Donkin, Jun.

d	*e*	*f*	*g*	*h*	*i*	*k*	*l*	*m*	
J	N	J	N	J	N	J	N	J	1
97·75	28·0	92·25	34·25	52·25	31·0	41·75	24·25	22·0	2
49·3	49·8	49·3	49·8	50·2	50·7	48·5	49·1	51·1	3
51·3	..	53·4	..	51·9	..	50·0	..	52·4	4
8·33	12·5	12·5	18·75	18·75	31·25	31·25	50·0	50·0	5
6·0	**4·8**	**4·8**	**3·7**	**3·7**	**2·6**	**2·6**	**1·8**	**1·8**	6
219	214	217	220	216	214	219	224	205	7
292	285	289	293	288	285	292	299	273	8
5·18	5·46	5·86	6·56	6·83	7·77	8·27	9·61	9·81	9
295°	238°	297°	252°	299°	267°	303°	260°	296°	10
295°	246°	292°	260°	290°	270°	290°	272°	288°	11
294°	246°	287°	255°	284°	274°	281°	275°	284°	12
299°	232°	299°	244°	278°	250°	281°	255°	279°	13
58·2	38·1	66·5	42·6	61·2	50·5	70·8	58·5	77·4	14
92·2	63·4	96·6	63·7	89·2	73·9	85·1	68·5	87·9	15
lbs.		lbs.		lbs.		lbs.		lbs.	
0·51	..	0·53	..	0·69	..	0·64	..	0·49	16
0·64	..	0·51	..	0·61	..	0·53	..	0·45	17
1·42	..	1·25	..	0·81	..	0·79	..	0·54	18
2·57	..	2·29	..	2·11	..	1·96	..	1·48	19
9·6	..	9·1	..	7·4	..	6·8	..	4·8	20
26·7	41·0	25·2	41·2	28·4	38·7	28·8	39·9	30·7	21
7·0	..	6·9	..	6·1	..	5·1	..	6·2	22
40·1	..	**38·5**	..	**31·1**	..	**25·6**	..	**23·1**	23

TABLE 51.

Experiment on a Single-Cylinder Vertical Surface-Condensing Engine,

WITH *and* WITHOUT *Steam in Jackets,*

at Bermondsey, by Mr. Bryan Donkin, Jun.

		a.m.	a.m.	p.m.
1	Time of observations *	11.41	11.47	12.37
2	Body jacket, steam *Off* or *On* . . .	Off	On	Off
3	Steam-Jacket pressure, lbs. per square inch above atmosphere . . . . lbs.	0·0	55·0	7·5
4	Revolutions per minute, mean . . revs.	140	265	190
5	Piston speed, feet per minute, mean . feet	187	353	253
6	Indicated horse-power, mean . I.H.P.	4·00	7·51	5·06
7	Temperature in body jacket . Fahr.	240°	301°	246°
8	„ of jacket wall, mean Fahr.	233°	295°	240°
9	„ of cylinder wall „ . Fahr.	250°	284°	250°
10	„ in cylinder, maximum Fahr.	239°	253°	242°
11	„ „ „ minimum Fahr.	235°	252°	240°
12	„ „ „ (in steel cup)† maximum Fahr.	256°	286°	262°
13	„ „ „ (in steel cup)† minimum Fahr.	253°	286°	261°

* Trial began at 11.30 a.m.; the first recorded observations were taken eleven minutes after at 11.41 a.m., and steam was then admitted to jacket. Six minutes after, at 11.47 a.m., the second recorded observations were taken, and steam was again shut off from jacket. Fifty minutes after this, at 12.37 p.m., the third recorded observations were taken. Only the body jacket was in use. See page 452.

† Proceedings Inst. C.E., vol. c, 1890, pages 348 and 349, Figs. 1 and 2.

TABLES 52-56.

INDEX AND

SUMMARY OF EXPERIMENTS

ON THE

VALUE OF THE STEAM-JACKET.

TABLE 52 (*continued on next page*).

Index and Summary of Experiments on the Value of the Steam-Jacket.

Single-Cylinder Non-Condensing Engines.

a	*b*	c	*d*	*e*	*f*	*g*
Reference to Proceedings.		Authority.	Date of Experiment.	Description of Engine.	Diameter of Cylinder.	Parts of Cylinder Jacketed.
No. of Experiment.	Page.					
	Vol. 1889				inches	
1	705	Thomas and Laurens	1846	—	—	—
2	705	M'Kay	1859	—	2·0	—
*3	706	Farey and Donkin	1874	Beam	7·31	Body and bottom end.
	706		1874	,,	7·31	Do.
4	707	Cornut	1874	Corliss	—	—
	707		1874	,,	—	—
5	707	Do.	1874	Ingliss	—	—
6	708	Loring and Emery	1875	Marine	34·1	Body and both ends.
*7	709–10	Borodin	1881–3	Locomotive	16·54	Do.
	710		1881–3	,,	16·54	Do.
*8	713–14	Delafond	1883	Corliss	21·65	Body only.
	711–13		1883	,,	21·65	Do.
	Vol. 1892					
49	449 & 465	Donkin	1890	Vertical	6·0	Body, ends, and valve-chest.

* In No. 3 the lower line is the mean of two experiments with about the same number of expansions. In No. 7 the upper line is the mean of four, and the lower line of two. In No. 8 each line is the mean of six.

TABLE 52 (*continued from preceding page*).

Index and Summary of Experiments on the Value of the Steam-Jacket.

Single-Cylinder Non-Condensing Engines (*concluded*).

a	*h*	*i*	*j*	*k*	*l*	*m*	*n*	*p*	*q*	*r*
No. of Expt. in Procs.	Revolutions per minute.	Piston speed per minute.	No. of Expansions.	Indicated Horse-Power.	Steam Pressure per square inch above atmosphere		Total Feed-Water.			Jacket-Water per I.H.P. per hour.
					in Boiler.	in Jackets.	per I.H.P. per hour *without* Jackets.	Saving due to Jackets. per I.H.P. per hour.	Saving due to Jackets. Percentage of *n*.	
	revs.	feet		I.H.P.	lbs.	lbs.	lbs.	lbs.	per cent.	lbs.
1	—	—	—	120	—	—	—	—	8 to 10	—
2	203	271	—	—	115	—	—	—	23·0	—
*3	48·3	210	2·7	6·41	45	45	71·69	15·34	21·4	8·32
	48·4	211	1·0	7·54	45	45	72·89	9·28	13·0	7·45
4	—	—	18 to 21	78·9	—	—	23·87	4·66	19·5	—
	—	—	7 to 8	136	—	—	21·25	2·93	13·8	—
5	—	—	10	—	—	—	32·00	6·12	19·1	—
6	51·3	256	3·5 & 4·1	201	68·5	—	29·66	3·05	10·3	0·85
*7	99·4	391	3·4	50·8	56·8	—	33·28	4·09	12·2	0·79
	93·4	368	5·1	51·6	77·8	—	32·16	5·17	16·0	0·88
*8	60·8	439	1·0 to 3·9	158	59·3	—	30·04	1·83	6·8	0·25
	61·6	445	4·4 to 7·1	171	99·4	—	27·28	4·60	16·8	0·54
49	212·4	283	3·75	4·19	50·2†	53·3	62·64	26·95	43·0	2·56

* In No. 3 the lower line is the mean of two experiments with about the same number of expansions. In No. 7 the upper line is the mean of four, and the lower line of two. In No. 8 each line is the mean of six. † In valve-chest.

TABLE 53 (*continued to page 475*).

Index and Summary of Experiments on the Value of the Steam-Jacket.

Single-Cylinder Condensing Engines.

a	b	c	d	e	f	g
Reference to Proceedings.						
No. of Experiment.	Page.	Authority.	Date of Experiment.	Description of Engine.	Diameter of Cylinder.	Parts of Cylinder Jacketed.
	Vol. 1889				inches	
9	715	Th mas and Laurens.	1842	—	—	—
10	715	Combes . . .	1843	—	15·35	—
11	716	Isherwood . .	1860	Vertical . .	5·25	Body, ends, and valve-chest.
12	716	Do. . . .	1860	Beam . .	90	Body only.
13	717	Donkin . . .	1870	Cornish Bull .	70	Do.
14	717	Farey and Donkin .	1870	Beam . .	7·31	Body and bottom end.
15	718	Hirn and Hallauer .	1873	Corliss . .	20·06	—
*16	718	Rennie . . .	1873	Beam . .	26	—
17	719	Emery . . .	1874	Marine . .	25	—
*18	719–20	Loring and Emery .	1875	" . .	34·1	Body and both ends.
19	721	Meunier and Keller .	1878	Corliss . .	24	Do.
	721		1878	" . .	24	Do.

* No. 16 is the mean of two, and No. 18 of four experiments.

TABLE 53 (*continued on next page*).

Index and Summary of Experiments on the Value of the Steam-Jacket.

Single-Cylinder Condensing Engines (continued).

a	*h*	*i*	*j*	*k*	*l*	*m*	*n*	*p*	*q*	*r*
No. of Expt. in Procs.	Revolutions per minute.	Piston speed per minute.	No. of Expansions.	Indicated Horse-Power.	Steam Pressure per square inch above atmosphere		Total Feed-Water.			Jacket-Water per I.H.P. per hour.
							per I H.P. per hour *without* Jackets.	Saving due to Jackets.		
					in Boiler.	in Jackets.		per I.H.P. per hour.	Percentage of *n*.	
	revs.	feet		I H.P.	lbs.	lbs.	lbs.	lbs.	per cent.	lbs.
9	—	—	6	16	53	—	—	—	30 to 32	—
10	—	—	20	6 to 7	50 to 58	50 to 58	50·9	21·2	41·7	2·97
11	61·4	102	5·3	1·13	38·2	38·2	79·38	24·75	31·2	12·76
12	8·09	158	1·7	361	12	—	36·37	0·98	2·7	0·60
13	6	111	4	161	36	—	—	—	[12·9]†	—
14	39·8	173	1·8	9·33	45	45	55·54	10·78	19·4	7·72
15	55	383	10	83·1	80	—	23·6	5·6	23·7	0·68
*16	37	185	3·4	406	16·4	—	—	—	[31·8]‡	—
17	53·8	215	5·1	116	79·5	—	26·25	3·10	11·8	—
*18	41·3 to 68·7	206 to 343	1·5 to 7·3	95·3 to 348	13·1 to 71·6	—	29·06	3·38	11·0	0·49 to 2·1
19	50	400	5·7	158	67·2	—	24·76	5·31	21·5	0·67
	50·4	403	12·4	100·3	68·1	—	27·20	8·23	30·3	0·91

* No. 16 is the mean of two, and No. 18 of four experiments.

† Percentage less thermal units rejected per I.H.P. per minute. ‡ Percentage less *Coal* per I.H.P. per hour.

TABLE 53 (*continued from preceding page*).

Index and Summary of Experiments on the Value of the Steam-Jacket.

Single-Cylinder Condensing Engines (continued).

a Reference to Proceedings. No. of Experiment.	*b* Reference to Proceedings. Page.	*c* Authority.	*d* Date of Experiment.	*e* Description of Engine.	*f* Diameter of Cylinder.	*g* Parts of Cylinder Jacketed.
	Vol. 1889				inches	
20	722	Fletcher	1877	Horizontal	30	Body and both ends.
21	722	Farcot	1879	Bède Corliss	39·37	Body only.
22	723	Mair-Rumley	1882	Beam Pumping	32	Body and bottom end.
*23	727–28	Delafond	1883	Corliss	21·65	Body only.
	724–28		1883	„	21·65	Do.
	723–27		1883	„	21·65	Do.
	Vol. 1892					
35	421	Guzzi	1886	Horizontal	15	Body and both ends.
36	422	Doerfel	1888	„ Corliss	21·81	Body only.
37	423	Longridge	1889	„ „	25·06	Body and both ends.
45	442, & 460–3	Unwin	1890	Horizontal	15·76	Body and back end.
47	445 & 464	Donkin	1890–91	Vertical	6	Body, both ends, and valve-chest.
48	448 & 465	Do.	1890	„	6	

* In No. 23 the first line is the mean of four experiments, the second of ten, and the third of seven, grouped according to their number of expansions.

TABLE 53 (*concluded from page 472*).

Index and Summary of Experiments on the Value of the Steam-Jacket.

Single-Cylinder Condensing Engines (*concluded*).

a	*h*	*i*	*j*	*k*	*l*	*m*	*n*	*p*	*q*	*r*
					Steam Pressure per square inch above atmosphere		Total Feed-Water.			
								Saving due to Jackets.		
No. of Expt. in Procs.	Revolutions per minute.	Piston speed per minute.	No. of Expansions.	Indicated Horse-Power.	in Boiler.	in Jackets.	per I.H.P. per hour *without* Jackets.	per I.H.P. per hour.	Percentage of *n*.	Jacket-Water per I.H.P. per hour.
	revs.	feet		I.H.P.	lbs.	lbs.	lbs.	lbs.	per cent.	lbs.
20	53·2	372	4·7	192	55 to 60	—	28·51	4·41	15·5	0·81
21	27·7	327	15	166	61·8	61·8	22·38	3·83	17·1	—
22	20·3	223	4·3	123	42	—	26·46	4·40	16·6	1·09
*23	60·1	434	2·3 to 4	172	49·8	—	20·09	0·29	1·4	0·26
	59·3	428	4·4 to 8·2	154·2	70·4	—	19·72	1·87	9·2	0·32
	59·9	432	10 to 12	121·7	88·8	—	21·46	3·90	17·9	0·50
35	77·4	387	—	25·5	56·6	187	[23·85]†	[4·0]‡	[16·8]‡	—
36	55·7	548	5	159	62·2	—	22·57	2·77	12·3	0·8
37	65·1	521	4·2	488	61	50	19·77	0·50	2·5	0·63
45	101·7	373	4	38·0	59·6	59·6	32·14	5·45	17·0	1·88
47	216·3	288	3·75	6·83	50·1§	51·8	41·23	12·84	31·1	2·11
48	114·9	153	3·75	3·72	50·2§	51·2	46·25	15·82	34·2	2·43

* In No. 23 the first line is the mean of four experiments, the second of ten, and the third of seven, grouped according to their number of expansions.

† Feed-Water with low-pressure steam in jackets. ‡ Saving due to high-pressure steam in jackets. § In valve-chest.

TABLE 54 (*continued to page 479*).

Index and Summary of Experiments on the Value of the Steam-Jacket.

Compound Condensing Engines.

a Reference to Proceedings. No. of Experiment.	*b* Page.	*c* Authority.	*d* Date of Experiment.	*e* Description of Engine.	*f* Diameter of Cylinders.	*g* Parts of Cylinders Jacketed.
	Vol. 1889				inches	
24	730	Hirn	1854–5	Beam	16·37 & 29·41	Bodies and bottom ends.
25	730	Farey and Donkin	1859	"	7·31 & 14·0	Do.
26	731	Do.	1868	"	13·19 & 24·0	Do.
	731		1868	"	13·19 & 24·0	Do.
27	731	Do.	1870	"	7·31 & 14·0	Do.
28	732	[illegible]	1873	"	20·19 & 31·25	Low-p. cyl. and receiver, the [illegible] always.
29	732	Emery	1874	Marine	16 & 25	—
*30	733–4	Farey, Donkin, and Salter	1875–81	Tandem	6 & 10	High-pressure body only.
	733–4		1875–81	"	6 & 10	Low-pressure body only.
	733–5		1875–81	"	6 & 10	Both bodies only.
31	736	Mair	1881	Woolf Beam	22 & 34	All bodies and ends.
32	736	Do.	1881	" "	24·5 & 38	Both bodies only.
33	737	Kennedy and Donkin	1887	Tri-cyl. Vertical	19 & 25 & 25	Three bodies only.
*34	738–45	[illegible]	—	Tandem	6 & 10	Both bodies only.
	738–45		—	"	6 & 10	Do.

* In No. 30 the first line is the mean of three experiments, the second of seven, and the third of eight, grouped according to the arrangement of the jacketing.

In No. 34 the upper line is the mean of five experiments, and the lower of six, grouped according to their number of expansions.

TABLE 54 (*continued on next page*).
Index and Summary of Experiments on the Value of the Steam-Jacket.
Compound Condensing Engines (*continued*).

a	*h*	*i*	*j*	*k*	*l*	*m*	*n*	*p*	*q*	*r*
No. of Expt. in Procs.	Revolutions per minute.	Piston speed per minute.	No. of Expansions.	Indicated Horse-Power.	Steam Pressure per square inch above atmosphere		Total Feed-Water.			Jacket-Water per I.H.P. per hour.
							per I.H.P. per hour *without* Jackets.	Saving due to Jackets.		
					in Boiler.	in Jackets.		per I.H.P. per hour.	Percentage of *n*.	
	revs.	feet		I.H.P.	lbs.	lbs.	lbs.	lbs.	per cent.	lbs.
24	23·5	194 & 261	4·3	[102·6]†	55·8	—	—	—	[23·5]‡	[0·92]‡
25	36·3	158 & 218	5·5	19·8	41	41	44·48	12·74	28·6	2·84
26	32·5	213 & 292	11	46·2	40·9	40·9	32·74	10·23	31·2	3·86
	32·6	214 & 294	12	37·2	41	4	32·74	4·47	13·7	2·81
27	43·7	190 & 262	10·3	16·5	45	45	39·49	15·24	38·6	5·67
28	25	215 & 295	13	119	60	—	—	—	4 to 5	—
29	53·2	213	7	99·2	80·2	—	23·04	2·71	11·7	—
*30	98	196	7·0 to 10·8	7·86	42	—	—	—	[25·5]§	3·78
	98	196	5·4 to 15·8	9·89	42	—	—	—	[29·0]§	4·26
	98	196	4·6 to 13·1	9·46	42	—	—	—	[33·9]§	5·55
31	19·6	141 & 216	15·8	75·2	47·3	—	26·62	9·28	34·9	1·72
32	34	232 & 374	9·6	268	71·8	—	19·24	1·68	8·7	1·35
33	55	220	7	168	80	—	23·93	3·68	15·4	1·32
*34	98·4	197	3·5 to 6·4	10·96	58·7	—	36·96	7·67	20·1	2·16
	97·5	195	7·6 to 11·1	11·06	73·9	—	36·56	8·97	23·9	1·79

* In No. 30 the first line is the mean of three experiments, the second of seven, and the third of eight, grouped according to the arrangement of the jacketing. In No. 34 the upper line is the mean of five experiments, and the lower of six, grouped according to their number of expansions. † *Brake* H.P. ‡ Per *Brake* H.P. § Percentage less thermal units rejected per I.H.P. per minute.

TABLE 54 (*continued from preceding page*).—*Index and Summary of Experiments on the Value of the Steam-Jacket. Compound Condensing Engines (continued).*

a	b	c	d	e	f	g
Reference to Proceedings.		Authority.	Date of Experiment.	Description of Engine.	Diameter of Cylinders.	Parts of Cylinders Jacketed.
No. of Experiment.	Page.					
	Vol. 1892				inches	
39	424	Meunier . . .	1889	Armington .	9·45 & 14·17	—
40	425	Ollgaard . . .	1890	Beam Pumping .	19 & 31	Both bodies and Low-p. ends.
42	427, & 456-9	Mair-Rumley. . .	1888	„ „ .	29 & 47·5	Bodies only.
44	437, & 456-9	English, Davey, & Donkin	1890	Horizontal .	18 & 32	Do.
46	442, & 460-3	Unwin	1890	„ . .	8·73 & 15·76	Bodies and back ends.

TABLE 55 (*continued on next page*).—*Compound Non-Condensing Engine.*

a	b	c	d	e	f	g
	Vol. 1892				inches	
38	424	Meunier . . .	1889	Armington .	9·45 & 14·17	—

TABLE 56 (*continued on next page*).—*Triple-Expansion Condensing Engines.*

a	b	c	d	e	f	g
	Vol. 1892				inches	
41	426	Reynolds . . .	1889	Vertical inverted	5 & 8 & 12	All bodies, ends, and receivers.
43	432, & 456-9	Davey and Bryan . .	1889	„ „	18 & 30·5 & 51	All bodies and ends.

TABLE 54 (*concluded from page 476*).—*Index and Summary of Experiments on the Value of the Steam-Jacket. Compound Condensing Engines* (*concluded*).

a	*h*	*i*	*j*	*k*	*l*	*m*	*n*	*p*	*q*	*r*
No. of Expt. in Procs.	Revolutions per minute.	Piston speed per minute.	No. of Expansions.	Indicated Horse-Power.	Steam Pressure per square inch above atmosphere.		Total Feed-Water.			Jacket-Water per I.H.P. per hour.
					in Boiler.	in Jackets.	per I.H.P. per hour *without* Jackets.	Saving due to Jackets. per I.H.P. qr hour.	Saving due to Jackets. Percentage of *n*.	
	revs.	feet		I.H.P.	lbs.	lbs.	lbs.	lbs.	per cent.	lbs.
39	302	505	—	83·0	111	—	22·85	1·28	5·6	0·85
40	24·0	178 & 264	14·6	81·2	51·25	39	23·84	4·43	18·6	—
42	15·8	171 & 252	—	168	49	49	18·20	1·56	8·6	1·20
44	63·6	509	12·6	113·4	50·2	46·3 & 10·3	24·68	4·69	19·0	1·13
46	96·1	352·4	9·3	45·65	67·8	67·8	21·06	1·54	7·3	2·40

TABLE 55 (*continued from preceding page*).—*Compound Non-Condensing Engine* (*concluded*).

a	*h*	*i*	*j*	*k*	*l*	*m*	*n*	*p*	*q*	*r*
	revs.	feet		I.H.P.	lbs.	lbs.	lbs.	lbs.	per cent.	lbs.
38	298	499	—	84·4	117·3	—	26·29	1·04	4·0	1·18

TABLE 56 (*continued from preceding page*).—*Triple-Expansion Condensing Engines* (*concluded*).

a	*h*	*i*	*j*	*k*	*l*	*m*	*n*	*p*	*q*	*r*
	revs.	feet		I.H.P.	lbs.	lbs.	lbs.	lbs.	per cent.	lbs.
41	230 & 298 & 299	383 & 497 & 747	26·5	72·1	192	192	16·42	2·86	17·4	3·29
43	22·9	137·4	30	138·0	130	129 & 28 & 5·5	17·22	1·77	10·3	1·72

Discussion.

The President considered the report now read was a substantial addition to the valuable records which the Research Committees had presented; and he was sure the members would join in thanking the Committee, especially the Chairman, Mr. Henry Davey, for the pains and trouble they had devoted to the work. The other members of the Committee were—Professor Archibald Barr, D.Sc., Mr. William B. Bryan, Mr. Bryan Donkin, Jun., Lt.-Colonel English, Mr. Jeremiah Head, Professor Alexander B. W. Kennedy, F.R.S., Mr. Michael Longridge, Mr. Norman Macbeth, Mr. John G. Mair-Rumley, Mr. William Parker, and Mr. James W. Restler.

Mr. Henry Davey said there had been a discussion in the Committee as to whether this report should contain any conclusions drawn from the experiments, or any theories founded upon them; but it had been considered that at the present stage of the enquiry it would be better to submit the bare facts as they had been ascertained, leaving it to this meeting to advance theories and criticisms, which would probably lead to further and more valuable facts being ascertained. It appeared to himself that but little economy was derived from the steam-jacket, beyond that which resulted from effecting what had been pointed out with reference to Mr. Donkin's experiments (page 452), namely obtaining the steam practically dry at release. All the experiments tended to show that this was the great secret of the economy of the jacket. There was also probably one other small element of economy in the jacket, which had not been touched upon in the report, and was seldom alluded to in discussions on the subject: namely the effect of the jacketing in keeping the cylinder in a truer condition than it would be without a jacket. Taking the extreme case of a Cornish engine working without a jacket, the upper end of the cylinder was at a much higher temperature than the lower, which was in consequence naturally a trifle smaller than the upper end; but when the jacket was applied, it tended to correct this inequality, and to make the cylinder more uniform in size throughout. Similarly in a double-acting engine the cylinder would be somewhat smaller in the middle of its length, and larger at the ends. Though this was probably an extremely small matter as far as the economy was concerned, yet it was a practical point. The Committee were still pursuing their researches, and had already materials from several further experiments, while other trials would also be made as soon as engines suitable for the purpose were available for proceeding with them. The results would be embodied in a third report.

Mr. Bryan Donkin, Jun., considered that it would be desirable to apply complete steam-jackets to the cylinders of locomotive engines; and suggested that the Committee should make some experiments on the application of steam-jackets not only to locomotive engines but also to modern marine engines.

Some of the conclusions which he had arrived at from his own experiments recorded in this report might perhaps be of interest, and were as follows. The quality of the steam—whether a mixture of steam and water, or superheated—had a great influence on the results. When the quality of the steam remained the same, the temperature of the internal passages and of the cylinder walls touched by the steam and exposed to the exhaust had also a considerable effect. To produce the maximum economy in a given engine, the question seemed to be whether the steam should be allowed to escape as a mixture of steam and water or as steam into the condenser or atmosphere. The less water in the steam, the better was the effect, either with or without jackets; and if the entering steam was sufficiently superheated to impart to the walls a temperature high enough to prevent condensation, a more economical result was arrived at, similar to or greater than that obtained by efficient jacketing. The thermometer placed inside the engine cylinder in these experiments proved that the temperature was much hotter with steam in the jackets than without; the hotter temperature tended to diminish the amount of fog or moisture in the cylinder. In some of the jacketed experiments the steam at release was even superheated. If the cylinder walls were heated so that they were hotter than the incoming steam, there was comparatively little

(Mr. Bryan Donkin, Jun.)

condensation during admission, and consequently little water to be re-evaporated during expansion. On the other hand, with the colder unjacketed walls there was greater condensation during admission, and considerable re-evaporation at the expense of the heat in the walls. At release there was much less water present with jackets than without. It appeared therefore that, in order to obtain the greatest economy, the steam escaping into the condenser or into the air should neither be superheated nor contain water. When escaping it ought to contain the minimum amount of water, so as to abstract the smallest quantity of heat from the cylinder walls. Professor Dwelshauvers-Dery of Liége had pointed out the great importance of the dryness-fraction of the steam at release. The pressure or the temperature in the jackets should therefore be so arranged, for a given set of conditions, that the mixture leaving the last cylinder should be dry and not superheated. In compound or triple-expansion engines, the latter proviso was not so important, in regard to the cylinders preceding the low-pressure, as there was no disadvantage in sending superheated steam from one cylinder into another. With the same quality and temperature of steam in these experiments, the hotter walls yielded the better results. With walls of the same temperature, the less the water in the mixture, the greater was the economy. As stated in page 451, the best result at 220 revolutions per minute and with all the jackets in use was obtained with 1-8th cut-off or 4·8 expansions; the consumption was then 25·2 lbs. of steam per indicated horse-power per hour, including jacket-water; and the dryness-fraction at release was 96·6 per cent., showing that in this case the exhaust was not superheated and contained very little water. One effect of the steam-jacket was to diminish considerably the action of the internal surface of the cylinder walls, so that there was much less alternate heating and cooling of the metal at every revolution; or in other words the heat penetration per steam stroke was much reduced by raising the temperature of the metal. The less the weight of metal heated up, the less was the condensation and the greater the economy. For maximum economy it was not so much a question of steam in the jackets, as of some method of increasing the temperature of the internal parts of the cylinder, covers, and passages. For the purpose of this investigation therefore, instead of a study of the value of the steam-jacket, a more correct designation would be a research for the most economical temperature of cylinder walls. The whole question seemed to turn on the temperature of the metal. Steam in a jacket at a proper pressure and temperature was only a convenient and ready way of raising the metal temperature nearly to or above that of the steam used inside the cylinder. There were of course many other ways of bringing about the desired end.

The President said he was informed that some years ago, as a matter of scientific research, Mr. Donkin had tried the effect of a lavish use of oil on the inside of the cylinder walls, with a view to prevent or reduce the condensation of the steam, and to avoid the need of a jacket.

Mr. Donkin said that was so, but a large quantity of oil was required to obtain as good a result as with steam-jackets.

Lt.-Colonel English, referring to the statement (page 420) that generally, the smaller the cylinder, the greater was the percentage of gain from the use of the steam-jacket, said this was certainly correct so far as it went; but the actual dimensions were only one of the elements in the initial condensation which occurred in an unjacketed cylinder. It would perhaps be a little more correct to say that, the greater the initial condensation in a cylinder, the greater would be the percentage of gain from the use of the jacket.

Mr. D. B. Morison said he had lately made experiments in connection with steam-jackets, the object of which was to determine the relative transmission of heat from steam through cast-iron cylinder-liners of varying thickness to water boiling at atmospheric pressure. The apparatus he had used for this purpose was shown in elevation and section in Fig. 60, Plate 83. The outer cylinder was of cast-iron, covered with non-conducting material on the sides, but

(Mr. D. B. Morison.)
with the bottom uncovered. Within this cylinder was placed another cylinder or liner, also of cast-iron, jointed at the top and bottom. The space between the two was arranged as a steam-jacket, which received steam from a boiler through the valve V; and the water condensed in the jacket was drained into the water trap W. Throughout the trials the water level in the trap was kept constant at half glass, so that there was no possible accumulation of water in the jacket. Water was placed inside the inner cylinder up to the level of the top of the jacket; and this level was carefully maintained throughout the trial by feed-water admitted from the small tank T overhead, into which it had been accurately weighed. Each experiment lasted two hours. The steam pressure in the jacket could be adjusted with great accuracy, and was maintained throughout each trial within half a pound per square inch of the intended pressure. The trials were made with pressures of from 10 to 80 lbs. per square inch above the atmosphere, in successive steps of 10 lbs.

The actual internal diameter of the inner cylinder or liner in the first set of trials was 9·19 inches, and the mean external diameter 11 inches; so that the mean thickness of metal was 0·9 inch, or practically an inch. The interior was bored in a lathe, the surface being similar to that of an ordinary cylinder bored in the ordinary manner. The external surface was the rough casting as it came from the mould. For the second and third sets of experiments the liner was further bored out internally to 10·16 and 10·62 inches diameter, thus reducing the thickness of metal to 0·42 and 0·19 inch respectively.

The results of the three series of experiments were given in the accompanying Table 57 (page 485), and were plotted in the form of diagrams in Figs. 61 to 63, Plate 84, in which the abscissæ were temperatures above 212° Fahr.; and the ordinates marking the successive increments of 10 lbs. pressure, at which the experiments were made, were drawn at the temperatures corresponding with the several pressures. The vertical scale in Fig. 61 represented pounds of water evaporated per hour from and at 212°. The black dots represented the several results, and were clearly seen to

TABLE 57.—*Transmission of Heat from Steam-Jacket through Cast-Iron Cylinder-Liners of varying thickness.*

External Surface of Liner 11 inches diameter, 3·84 square feet.

Internal Surface of Liner:—

9·19 inches diameter, 3·007 square feet.
10·16 " " 3·324 " "
10·62 " " 3·477 " "

Pounds of Water Evaporated per hour from and at 212° Fahr. Figs. 61 to 63, Plate 84.	Thickness of metal.	Steam Pressure in Jacket, pounds per square inch above atmosphere; and corresponding Temperature Fahr.							
		10 lbs. 240·1°	20 lbs. 259·3°	30 lbs. 274·4°	40 lbs. 287·1°	50 lbs. 298·0°	60 lbs. 307·5°	70 lbs. 316·1°	80 lbs. 324·1°
	Inch.	Lbs.	Lbs.	Lbs.	Lbs.	Lbs.	Lbs.	Lbs.	Lbs.
Actual Evaporation.	0·90	20½	37¾	51	62¼	72	80¼	88	95
	0·42	35	63	84¾	102¼	119	133	145½	157
	0·19	47½	87½	119	145¼	168	188	206	222
Per degree of difference in temperature between Steam and Water.	0·90	0·73	0·80	0·82	0·83	0·84	0·84	0·85	0·85
	0·42	1·24	1·33	1·36	1·36	1·38	1·39	1·40	1·40
	0·19	1·69	1·85	1·90	1·93	1·95	1·97	1·98	1·98
Per degree of difference, and per square foot of internal surface of liner.	0·90	0·243	0·265	0·272	0·276	0·279	0·280	0·282	0·282
	0·42	0·373	0·400	0·409	0·409	0·415	0·418	0·422	0·422
	0·19	0·486	0·532	0·547	0·555	0·562	0·567	0·569	0·569

(Mr. D. B. Morison.)
range themselves fairly along three distinct straight lines. The position of the vanishing point of the three lines, at a temperature somewhat above 212°, would be accounted for by radiation from the funnel which surmounted the cylinder, and which was not lagged; and the slight divergences of the dots from the straight lines might be due to variations in temperature of the atmosphere, the experiments having occupied several days, and having been conducted in an open shed, through which there was a slight breeze blowing. The reason why the three highest dots were all of them above the uppermost line might be that the ebullition was so fierce under the action of the high-pressure steam through the thinnest liner that a slight priming must have taken place. In Fig. 62 was plotted the evaporation per hour and per degree of difference in temperature between the steam and the water; and in Fig. 63 the evaporation per hour per degree of difference, and per square foot of internal surface of the liner.

The figures and the diagram, Fig. 63, practically proved that, taking as unity the evaporation per degree of difference in temperature between the steam and the water and per square foot of internal surface of the liner, corresponding with the greatest thickness of metal, the evaporation corresponding with half the thickness was about half as much again, and with a quarter of the thickness was about double; the actual ratios varied only from 1·53 to 1·48 with the medium thickness, and from 2·00 to 2·02 with the thinnest liner. The irregularities of the dots were seen to be so small that the experiments as far as they went might apparently be entirely relied on, and the mean results might be considered accurate. The experiments showed so conclusively the greater transmission of heat as the thickness of the cylinder liner was reduced, that there was no doubt greater efficiency would be obtained with thin cast-steel liners ribbed in the jacket space than with the thick cast-iron liners at present used.

A reason why steam-jackets were not popular in marine engines was the difficulty experienced in keeping them in working order, free from accumulation of water, so as to be always in a condition of maximum efficiency. Two of the best plans for jackets that were supplied direct with boiler steam were to drain them either direct back into the boiler by gravitation, or else direct into a feed-water heater or into the hot well. An arrangement was shown in Fig. 64, Plate 85, in which the high-pressure and intermediate jackets of a triple-expansion engine were drained through an evaporator into a simple form of feed-water heater placed on the suction side of a feed-pump. Steam from the boiler was supplied at B into the top of the high-pressure jacket, at the bottom of which a gauge-glass G was put upon the drain pipe, and a small drain-cock below the glass was opened just enough to drain off all the water that was condensed, and to pass a little steam as well, so as to prevent any possibility of accumulation of water in the glass. The steam was taken from the bottom of the high-pressure jacket through the reducing valve R to the top of the intermediate jacket; and the drain pipe from the bottom of the intermediate jacket was joined with that from the high-pressure, and led into the heating coils of the evaporator E. The evaporator was adjusted to supply the auxiliary feed necessary, and was consequently kept working continuously; both the high-pressure and intermediate jackets were thus drained continuously and automatically. The efficiency of the jackets was also materially increased by this arrangement, as all the steam required by the intermediate jacket circulated first through the high-pressure jacket, and all the steam required by the evaporator circulated first through both jackets. Another advantage was the maintained efficiency of the jackets, owing to the non-accumulation of air in either of them. The steam evaporated outside the coils, together with the water condensed inside them, was then led into the small feed-water heater H, and was sufficient to heat the feed up to about 150°, or just about as hot as the feed-pumps could take it; between the pumps and the boiler the feed could be further heated, if so desired. As an alternative arrangement, the steam from the intermediate jacket could of course be discharged direct into the feed-heater, or direct into the hot well, without passing through the evaporator, whenever it was desired to examine or clean the evaporator. There was no economy in discharging the steam generated in the evaporator into the condenser,

(Mr. D. B. Morison.)
and not much in passing it through the valve-chest of the low-pressure cylinder. Consequently the advantages of the arrangement shown in Fig. 64, Plate 85, were the automatic draining of the high-pressure and intermediate jackets, the heating of the feed-water to about 150°, and the production of fresh auxiliary feed-water at the least possible cost; the only heat lost was that which was radiated, and that which was contained in the brine discharged from the evaporator.

In a steam-jacketed cylinder the heat from the steam in the jacket passed both inwards into the cylinder and outwards through the outer wall of the jacket. In the ordinary design of triple-expansion engine for moderate power the high-pressure cylinder only was fitted with a liner; and this was but seldom used as a steam-jacket, for the reason already mentioned that trouble had been experienced with jackets in the past, and owners preferred to have an engine at a lower cost without jackets. If however an engine could be arranged with an efficient jacket at a small cost, the result would obviously be advantageous. The design of jacket from which the greatest duty could be obtained was one that had steam at both sides, being itself surrounded by a receiver such as had been commonly adopted in the early days of two-cylinder compound engines, in which the exhaust steam from the high-pressure cylinder flowed past and around the outside of this cylinder on its way to the steam-chest of the low-pressure cylinder. The objection to that arrangement was that the whole of the exhaust steam from the high-pressure cylinder took the shortest course it could find through the receiver to the low-pressure steam-chest: with the result that only a small portion of it came into actual contact with the jacket, and a large portion of the jacket heating-surface was practically ineffective. In Figs. 65 to 68, Plates 86 and 87, were shown two arrangements of a circulating receiver surrounding the high-pressure jacketed cylinder of a triple-expansion engine; assuming the high-pressure cylinder to be forward of the intermediate, Figs. 65 and 66 showed the high-pressure valve-chest placed aft of its own cylinder, or between the high-pressure and intermediate cylinders; while Figs. 67 and 68 showed the high-pressure valve-chest arranged forward of its own cylinder. The circulating receiver here shown was designed on the principle that, the more complete the circulation over a heating surface, the greater was the quantity of heat transmitted. By means of suitable partitions cast within the receiver, the steam was compelled to follow a zigzag course, passing uniformly over the entire heating surface of the jacket on its way to the next cylinder, and to become thoroughly mixed up: so that not only did it abstract a larger amount of heat from the jacket by reason of its flowing over a larger extent of heating surface, but also the whole of the steam being continuously intermingled was thoroughly dried before entering the next cylinder. The arrangement shown in Figs. 67 and 68 was rather more complete and elaborate than that in Figs. 65 and 66, guiding the flow of the steam more thoroughly over the whole of the outer heating surface of the jacket.

By this arrangement it was not intended to re-evaporate the actual water formed during expansion in the high-pressure cylinder, inasmuch as this water was drained off by a water trap connected with the intermediate steam-chest; but the object was to re-evaporate the mist or finely divided moisture that was suspended in the steam exhausting from the high-pressure cylinder, and by completely drying and superheating this steam to bring it into a condition of maximum efficiency for the intermediate and low-pressure cylinders. In an ordinary triple-expansion engine, when the steam was not superheated either in the high-pressure cylinder or in the high-pressure jacket, it was not possible for the high-pressure exhaust to be free from water in large quantities; and the effect of the jacket in regard to the interior of the high-pressure cylinder was to a great extent limited to drying the cylinder walls, and thereby diminishing initial condensation. It was during expansion however that water was formed; and the greater the work done in a high-pressure cylinder, the greater was the amount of water produced. In an unjacketed triple-expansion engine indicating 1,000 horse-power and having a high-pressure cylinder about 23 inches diameter and 42 inches stroke, the actual water discharged from the high-pressure exhaust was from 1½ to 2 gallons per minute; and under such conditions it was evident that a large

(Mr. D. B. Morison.)
quantity of moisture in a finely divided state must also be mixed with the steam. The actual water should undoubtedly be drained off as soon as possible, and should not be allowed to flow through the ports and passages of the intermediate and low-pressure cylinders. The moisture in suspension in the steam should be re-evaporated, and pure steam only should be admitted into the intermediate cylinder. The pressure of the high-pressure exhaust in a triple-expansion engine with 160 lbs. boiler-pressure was about 55 lbs. per square inch above the atmosphere, corresponding with a temperature of about 302° Fahr.; whilst the pressure in the jacket was 160 lbs., corresponding with 371°. With the arrangement of circulating receiver the exhaust steam with its suspended particles of water not only passed over a surface heated to 371°, but in its passage the direction of its flow was varied, and it was continually mixed up, whereby the whole of the water particles were re-evaporated, with the result that the intermediate cylinder received pure steam only. Though he was not able to state the actual amount of saving, he could say that the engines fitted with the circulating receiver which he had described were much more economical than those without it.

Professor W. Cawthorne Unwin considered the Committee were to be congratulated on having exhibited in such a clear way the large number of results they had collected, and on having so thoroughly tabulated them and illustrated them by diagrams. He was sorry however that they had taken the view of their duties which their Chairman had mentioned (page 480); if they had realized their immunity, individually and collectively, from any consequences which might ensue, they might have expressed some opinions on the action of steam-jackets. Had they done so, there would perhaps have been a fuller opportunity for discussing the many difficult points which arose. It was striking that, amidst a great variety of conditions and sizes of engines, they seemed to have found no case in which the steam-jacket had done harm. The gain due to it was sometimes more and sometimes less; but he had not noticed in the report any case where condensation of steam in the jacket had not been balanced by a saving of steam in the cylinder at least equal in amount.

It had been urged (page 480) that the object to be aimed at in the use of a steam-jacket was that the steam in the cylinder should be dry at release: dry indeed all through the stroke if it could be kept so, but at any rate dry at release. This should not however be put forward as a principle; it was merely a partial deduction from the general conclusion that condensation due to the action of the cylinder walls was prejudicial. The phrase "wet steam" was commonly used; and properly understood, it was a convenient phrase, conveying no erroneous impression. But it was apt to lead to a serious blunder, inasmuch as there was really no such thing as wet steam. There was such a definite thing as superheated steam, and also as saturated steam. But wet steam was merely a mixture of steam and water; and to say that steam was to be dry at release was merely to say that, if it could be prevented, there must not be water in a cylinder at release, because, if there were, it was pretty sure to be most of it on the cylinder wall; and if there were water on the cylinder wall at release, it meant that there would be evaporation and cooling of the wall during exhaust. In any engine evaporation from the cylinder wall during exhaust was a most serious evil, cooling the cylinder wall which had to be heated again by condensation during admission. Thus the reasons were obvious enough for having the steam dry at release, and if possible throughout the stroke.

The two experiments of his own which were included in the report, pages 442–3, were not selected experiments, but were simply the two which happened to have been taken nearest to the time of the Committee applying to him for information on this subject. An experimental engine however was attended with a good many disadvantages, which did not apply to ordinary engines. All sorts of arrangements were added to it for measuring, which did not add to its efficiency. Nevertheless it appeared that a tolerably good result had been obtained, when the size and speed of the engine were considered. No doubt the smaller the engine, the greater was the advantage of the jacket. Such a result seemed remarkable for this

(Professor W. Cawthorne Unwin.)

reason amongst others, that, the smaller the engine, the larger in proportion was the waste involved in the additional appliances, such as steam-traps &c., for measuring the jacket condensation. The report showed however pretty clearly that, the smaller the engine, the greater was the efficiency of the jacket. No doubt this was a consequence of the larger cylinder-surface in small engines per pound of steam used. It was curious that in practice steam-jackets should be put on big engines, where they seemed to do but little good, and should be left off smaller engines, where they were so much more advantageous.

Mr. John G. Mair-Rumley, Member of Council, mentioned that the experiment which he had made on the waterworks pumping engine at Hammersmith (page 427) had been carried out during the ordinary working of the engine. With regard to the apparent priming at the beginning of the trial with steam in the jackets (page 428), he believed this was due to the fact that in the ordinary working of the engine it was the practice as a precaution against carelessness to keep the water at a somewhat higher level in the boilers than he had generally found desirable for making a trial. On finding that the steam came off rather wet, he had obtained permission to lower the water level about 2½ inches by stopping the feed for an hour, as described in the report. The same explanation applied to the much larger quantity of water drained off from the high-pressure valve-chest (page 430) in the trial made with steam in the jackets than in that made without.

Mr. Charles Cochrane, Past-President, pointed out that in page 424 there was recorded an experiment, No. 38—not one of those made by the Committee—in which the jacket water amounted to 1·18 lb. per indicated horse-power per hour, whilst the economy in feed-water was only 1·04 lb. Although however in this example there had been more water consumed in the jacket than was saved in the feed, yet even here the jacket had done no harm, because the jacket steam was supplied from the feed-water, and therefore the jacket water was itself included in the feed-water, the total amount of which was 1·04 lb. less per indicated horse-power per hour with steam in the jacket than without.

Sir Frederick Bramwell, Bart., Past-President, mentioned a practical test, wholly unpremeditated and unconcerted, of the value of steam-jackets, which had occurred in his own experience. Some thirty years ago he had designed a tandem compound engine for some saw-mills in Bankside, with steam-jacketed cylinders high enough up for the jacket water to be returned by gravitation to the boiler. After the engine had worked ten years, it happened one day that it could not be made to go. Having been sent for, he found on examination that everything appeared to be right; and he had no notion what was the matter, until by good luck he put his hand upon the return pipe from the jackets and felt that it was quite cold. On being taken off, it was found that after all those years the return pipe had become choked with a deposit of lime. It was cleaned, and put on again, and then the engine worked as well as ever.

Another thing that he had done did not appear to have been done in the experiments now reported. In a steam-jacketed engine he had made an experiment as to the amount of water delivered from the jacket per hour when the engine was not at work, and then when it was at work. The larger amount in the latter case he presumed was due to the work that was being done in the cylinder, absorbing heat and rendering the cylinder colder; while the smaller amount when the engine was standing represented the heat that was lost through imperfect cleading of the jacket, and was therefore the preventable amount. In the Committee's report he did not see that any experiment of that kind had been made; and he thought it was desirable to separate as far as possible the jacket water due to imperfect cleading from the jacket water which must necessarily be produced, however perfectly the jackets were cleaded.

Another matter on which he did not notice any statement in the report was the point at which it became unsafe to have a steam pressure in the jacket above the pressure prevailing in the cylinder: that is, unsafe with regard to the cutting of the cylinder surface from its being too dry. He remembered an invention being brought to

(Sir Frederick Bramwell, Bart.)
England a long time ago by Mr. Wethered of Baltimore, called "combined" or "mixed" steam, which consisted in injecting into the steam-pipe from the boiler to the engine a certain proportion of steam which had passed through a superheater (Proceedings 1860, page 35). The plan had been tried on one of the old paddle-steamers belonging to the government, and much economy had been produced. But the endeavour to develop it still further by superheating the whole of the steam had been attended with the result that the cylinders were cut to pieces on account of their being too dry and hot. In those days combined steam had been thought to be a good thing, but it had now been wholly given up.

The circulating receiver which had been described by Mr. Morison (page 488) reminded him forcibly of the principle of what was known as the "hot pot," introduced many years ago by Mr. Cowper in connection with compound engines, for heating up the exhaust steam from the high-pressure cylinder prior to its admission into the low-pressure. That contrivance consisted of an intermediate vessel, steam-jacketed in a peculiar manner, so that the steam on its entrance into the vessel from the high-pressure cylinder did not come into contact with the jacket, but did so most thoroughly on its way out of the vessel to the low-pressure cylinder. This plan had been applied in 1870 to a pair of compound engines fitted by Messrs. Rennie in H.M.S. "Briton," with the result that the consumption of coal had been brought down to as low as 1·3 lb. per indicated horse-power per hour (Proceedings 1872, page 153).

Mr. John Phillips said he had himself been an engineer in one of the steamers fitted on Wethered's plan mentioned by Sir Frederick Bramwell (page 494), and remembered that great advantage had been found in that mode of using superheated steam. There were two sets of stop-valves, one for admitting the ordinary steam from the boiler, and the other for the superheated steam. Pending some question about the right to use the latter, strict orders had been given not to open the superheated-steam valves; but it so happened that during the voyage they were found to have worked open of themselves, with the result of an increase in economy. The stokers certainly were much pleased with this result, and altogether the plan proved highly satisfactory. So far as that experience went, he could speak with some confidence of superheating. Those were the days of jet condensers, when it had been the custom to note by a mark on the injection cock how much injection water was required; and he remembered, as soon as engines had the superheater fitted, how great was the surprise at finding that considerably less injection water was required, although at the same time the engine went faster. From the conclusions arrived at in connection with the present report, he presumed the reason had been that with the superheater there was less actual dampness in the steam, and consequently less injection water was required to bring down the temperature of the exhaust steam. In the value of superheating he had a strong belief; the question was how it could best be done. In the engines of the Royal Mail Steamship Company, in which superheated steam had been largely used, great care had been taken, by placing thermometers at the valve-jackets and also at the superheaters, to avoid risk of getting the steam too hot. The steam pressure used was about 15 lbs. per square inch above atmosphere; and at that pressure the steam was never allowed to be heated over 330° Fahr. If it got hotter, the hemp or cotton packings were liable to burn; but there had never been any trouble in the cylinders. The slide-valves had been troublesome at first; but by having them made of first-class good hard metal all difficulty had been got over. Although it was called superheated steam, and was really so on leaving the superheater, yet by the time it got to the valve-jacket it was often not any hotter than the temperature due to the pressure of steam employed.

Of steam-jacketing the pistons he had not seen any mention made in the report. Some years ago he remembered being told by the late Mr. J. L. K. Jamieson of some vessels that had been wanting a little in speed, and their speed had been increased to the required amount by simply putting boiler steam inside the pistons. It appeared to him that, if the cylinder were jacketed both at top and at bottom and also round the sides, the piston and piston-rod might properly be

(Mr. John Phillips.)

jacketed also. The piston-rod passing in and out of the cylinder must have a large effect in condensing the steam in the cylinder.

When he was at Messrs. Rennie's works he remembered observing the effect resulting from the use of a steam-jacket on the shop engine, which at the time of observation and without steam in the jacket was making 32 revolutions per minute, and indicating 41¼ horse-power. Five minutes after the steam had been admitted into the jacket, the engine was making 35 revolutions per minute, and indicating 49½ horse-power. The engine was that referred to in the report in Table 53, page 472, No. 16. On several occasions he had measured the quantity of water drained from the jacket, and it came out near to some of the examples recorded in the report, namely about 2¼ to 2½ lbs. per indicated horse-power per hour. That was with a low pressure of steam; but on the whole the value of the jacket in regard to economy was remarkable, the saving of coal amounting to over 30 per cent.

Another matter which he wished to mention in respect to jacketing was the jacketing of the steam-pipes from the boilers. Much was often heard about boilers priming; and not long ago in the trial of the "Tartar" (Proceedings 1890, page 233), where there was a large boiler but little steam used, there had been thought to be priming; but before admitting this to be the case, he should have liked to look into some of the details, such as the length of the steam-pipes, and how they and the stop-valves were covered. It appeared to him that a large amount of condensation might take place in the transit of steam from the boiler to the high-pressure valve-chest; but that was a totally different thing from priming. Water was carried into the cylinder by the steam; but it had left the boiler not as water but as steam, and had been condensed afterwards on its way to the cylinder. If therefore jacketing was of value, as he thought the report showed that it was, why not jacket the steam-pipes as well, through which the steam was led from the boilers into the cylinders?

Mr. Donkin said that the jacketing of pistons had been already done, and the advantage proved, and published in the Committee's first report (Proceedings 1889, page 721).

Mr. William Schönheyder, referring to the statement in page 420 that the smaller cylinders seemed to give the greater advantage on account of their larger jacket-surface in proportion to the weight of steam passing through them, thought this representation should be qualified because the element of time was omitted; and that the proportion should rather be considered of the jacket surface to the weight of steam passing through the cylinder per hour, and to the mean difference of temperature between the inside of the cylinder and the jacket. Whether a smaller cylinder gave better or worse results than a larger depended entirely upon the amount of surface in the jacket for drying up a certain quantity of steam per hour; and this again depended upon the difference of temperature between the inside and outside of the cylinder wall.

Experiment No. 35 recorded in page 421 was particularly interesting on account of the remarkably large economy obtained by having so much higher a temperature in the jacket than in the cylinder. That was a condition so seldom met with in practice as to render the experiment especially deserving of attention. An attempt in the same direction had been made more than thirty years ago, in 1860 or 1861, by the late Mr. David Thomson, who had had strong notions as to the advantage of a higher temperature in the jacket than in the cylinder. But at that time no doubt too little was known as to the quantity of steam condensed in the jacket; hence he had provided only a small boiler for the steam supply to the jacket, the heating surface being only about one-tenth of that in the main boilers. Although there was a high pressure of steam in the supplementary boiler, yet the moment the steam-valve was opened not only did the pressure disappear, but nearly the whole of the water in the small boiler disappeared also.

In experiment No. 37 (page 423) the diameter of the steam-pipes which supplied the jackets was given as only $\frac{3}{4}$ inch. For the size stated of the engine, such small pipes were of course inadequate to supply sufficient steam to the jackets. It would be interesting to know whether the poor economy obtained in that instance by the use of jackets was due to the want of sufficient pressure of steam in the jackets, and therefore to the rather low temperature in the

(Mr. William Schönheyder.)
jackets. If the pressure in the jackets had been taken at the time, it would be well for it to be recorded in the report. In experiment No. 43 at the East London Water Works it was mentioned (page 433) that the pressures in the several jackets were kept slightly higher than the initial pressures in the respective cylinders; and he thought it was probably due mainly to this circumstance that the economy in feed-water was so small as only 10·3 per cent. Marine engines he believed were generally worked in that manner, notwithstanding it was well known that no considerable economy could be gained in that way.

In experiment No. 44 at the Royal Arsenal it was stated in page 438 that an air-cock was placed at the highest point of each jacket. This appeared to him to be an error, because it must have been known from the time of James Watt that when air and steam were mixed together the air always occupied the lowest place. Were it not so, it would be impossible to obtain a good vacuum in a condensing engine; for the air-pump bucket, after sucking up the water which had accumulated from the previous stroke, next drew in from the bottom of the condenser all the air, or at least as much as it could get according to the size of the air-pump &c. If the air accumulated on the top of the steam, there ought to be a separate pump for drawing off the water at the bottom and a separate pump for drawing off the air at the top of the condenser; but he did not suppose that any such plan had ever been recommended. As the result partly of experiments by Regnault he believed, the weight of vapour had been found to be only about five-eighths of that of air at the same temperature and pressure. Hence if there was no strong circulation of steam in each jacket, and if air was not drawn off from the lower part, it seemed possible that the jacket might become almost full of air, with no steam passing into it, and therefore no steam being condensed. For this reason it appeared to him that perhaps some of the experiments made by the Committee might be a little incorrect, although the error was on the safe side, because there was no doubt they had really got the economy stated; but still more economy might have been obtained if the air had been drawn away from the lower part of the jacket, instead of from the highest point.

In regard to the amount of steam condensed in the jacket when an engine was at work and when it was not at work (page 493), he noticed in page 446 that in one of the ingenious experiments made by Mr. Donkin the steam condensed against the inner wall of the body jacket had been separated from that condensed against the outer wall, with the remarkable result that the amount condensed was practically the same, namely 0·69 lb. per indicated horse-power per hour on the inner wall and 0·68 lb. on the outer: showing that as much steam was condensed outwards as inwards.

Professor T. Hudson Beare said he had recently carried out thirteen trials on the same experimental engine at University College with which the experiments had been made by Professor Kennedy that were recorded in the first report of the steam-jacket Research Committee (Proceedings 1889, pages 738–745). As the results of his own trials had already been published in full ("Engineering," 25 December 1891), he did not purpose saying more about them now than was necessary for drawing attention to one or two points. The engine had been undergoing an overhaul shortly before the trials, and he had taken advantage of the opportunity for making a set of trials with the engine entirely unlagged: first without steam in the jackets, and then with. The result was an economy of feed-water in the jacketed trials of 12·3 per cent. over the unjacketed; and the steam passing through the jackets was roughly about 12 per cent. of the total feed. After the engine had been carefully lagged with felt and wood, trials were made without steam in the jackets, and the result was an economy of 4·5 per cent. over the unlagged engine, showing that 4·5 per cent. had been lost through external radiation when the engine was not lagged. The jacketed trials gave a saving in feed-water of from 15·9 to 21·1 per cent. over the unlagged unjacketed trials; but over the trials with no steam in the jackets and with the engine lagged the economy was from 13·8 to 19·9 per cent. The jackets therefore showed

(Professor T. Hudson Beare.)
distinctly better results when used on a lagged cylinder than when in use on an unlagged cylinder. These results pointed he thought to one consideration which had not been referred to in the report: namely that all the trials when there was no steam in the jackets had necessarily been made with a double wall, that is, the wall of the liner and the wall of the casing, or the inner and outer walls of the jacket; they had thus been air-jacket trials. In that case therefore there must of course have been considerably less radiation from the steam inside the cylinder than there would be in a cylinder with only a single wall. It would be a good plan, he thought, if a trial could be carried out with two cylinders of about the same size, and with the speed and other conditions the same, one of the cylinders having no liner and the other having a liner, but with the steam-jacket not in use on the latter. In that way he thought it would be found out how much of the benefit of the steam-jacket when in use was due to its merely preventing radiation direct into the external air from the cylinder steam.

As to the amount of steam condensed in the jackets when the engine was standing, in comparison with the amount condensed when it was running (page 493), in his own experiments after the running trial had been completed the engine had stood an hour with steam in the jackets, and the amount of steam condensed in the jackets had been carefully recorded. As already published, the results in all cases showed a large quantity of steam condensed in the jackets while the engine was standing. It would be a great advantage, he thought, if trials could be carried out with cylinders much larger than some of those recorded in the report; for he believed it must necessarily be the case that a smaller cylinder would derive more benefit from jacketing than a larger, for the reason already mentioned (page 497) that the weight of steam passing through the smaller cylinder per hour bore a smaller proportion to the surface warmed by the jackets. This he thought accounted for the much greater gain shown with the small experimental engine used by Mr. Donkin, in contrast with the results of the experiments carried out by other members of the Committee on much larger engines. Speed was also a most important factor; and he hoped that some future experiments would be made with the same engine running at different speeds. With his own experimental engine at University College he intended to try this if possible during the coming winter.

The President asked Mr. Aspinall whether he had had any experience of steam-jacketed locomotive cylinders.

Mr. John A. F. Aspinall, Member of Council, replied that he had not; but he should be happy to make some steam-jacket experiments on locomotives, as he considered it highly desirable for the Institution to have records of such trials, for showing whether the advantages resulting from the use of steam-jackets in the other classes of engines dealt with in the Committee's report, and in the reports of the marine-engine trials, could be realized also in locomotive practice.

Mr. Davey said the Committee were very much obliged to Mr. Aspinall for his kind offer, of which they would gladly avail themselves.

Professor Alexander B. W. Kennedy, Vice-President, felt particularly interested in what had been said by Mr. Morison (page 488) as to utilizing the outside surface of the steam-jacket for heating the exhaust steam in the receiver, and especially as to securing a thorough circulation of the receiver steam over the heating surface of the jacket. Inside the jacket itself he had found a difficulty in getting such an efficient circulation as he considered there ought to be. Notwithstanding that there appeared to be every facility for the steam to circulate, he had still failed in certain cases to get the jackets to do anything better in working than just to save as much steam in the cylinders as was used in the jackets themselves. It was true Mr. Morison's plan was to circulate a larger quantity of steam in the receiver outside the jacket; but to get the steam to behave itself exactly as it ought inside a narrow jacket seemed sometimes to cause trouble, and certainly in one or two cases where he had tried to do this it had been a failure.

The PRESIDENT asked whether in those cases there had been an air-cock at the bottom of the jacket.

Professor KENNEDY replied that there had not, nor had there been one at the top. He should have to adopt Mr. Schönheyder's idea (page 498), and see what could be done by placing an air-cock at the bottom of the jacket.

The measurement of the feed-water as carried out in these trials he thought had been accurate; but some of the remarks made in the report showed the great difficulty experienced in carrying out feed-water measurements when the water had to be measured into the boiler, as it had had to be in experiments Nos. 42 and 43 (pages 428 and 434), and as it had to be in most practical trials. In this matter, as in so many others, the late Mr. Willans he believed had been right in urging that it was much more accurate, whenever it could possibly be done, to measure the water as it came out of the engine: that is, to measure it by a surface-condenser of some kind. The latter plan got rid of the difficulty which had been already mentioned (page 496), that if the water was measured into the boiler the engine was unavoidably debited with the water condensed in steam-pipes, and with priming water, as well as with all leakages in boiler or pipes, all drain water, and also water blown off through safety-valves as steam, if any. No doubt by the use of separators, and by weighing back carefully all the water taken from them as well as all the water taken from drains, some of these difficulties might be partly obviated; but they always left very sensible possibilities of inaccuracy. If the water was measured by condensing the steam after it had left the engine, and if the steam-pipes and steam-chest were kept properly and thoroughly drained, these sources of error were to a large extent eliminated; because in this case it would not matter if even a certain amount of steam was sent to waste through the drains: it would not be debited to the engine. Attention had been drawn by Mr. Phillips to a very important and somewhat neglected matter, by pointing out that the water often found in steam cylinders might not be priming water at all, but might be merely water condensed in the pipes or in the cylinder itself, long after the steam had left the boiler as steam.

The use of superheated steam, in regard to which Sir Frederick Bramwell had recalled the fear formerly entertained of damage to the cylinders by scoring, now appeared under certain circumstances to be perfectly practicable, according to his own experience. In testing recently the Serpollet boiler he had found that, even when the steam was superheated to such an extent that in the exhaust pipe there was still as much as 20° or 30° Fahr. of superheat, there was absolutely no trouble in the cylinder. He had seen a cylinder opened which had been running for months with steam superheated to that extent, and it was in just as good a condition as any other cylinder. The result was singular, and he did not profess to explain it; but it would appear from this experience that superheated steam must have been too hastily condemned formerly, or must have been found to be a failure through some causes which were not fully understood.

Mr. JOHN I. THORNYCROFT, referring to the difficulty mentioned by Professor Kennedy (page 501) of getting the jacket steam to work as efficiently as it should do, had noticed that in French marine engines advantage was sometimes taken of the opportunity, where steam was wanted also for auxiliary engines, to pass it through the jackets of the main engines; in that way a pretty good current of steam was obtained through the jackets, which probably took out with it any air, besides ensuring that no water accumulated there. With regard to what had been said by Mr. Morison (page 488) about using the jacket to heat or dry the receiver steam, theoretically it appeared to him difficult to imagine that it could be useful to employ high-pressure steam in re-evaporating moisture and making steam at a lower pressure. There was no doubt that an advantage was obtained; but it could not consist simply in producing the steam at a lower pressure than that at which it had originally been made in the boiler. With regard to the heating of the cylinder walls to so high a temperature with high-pressure steam as to cause cutting, he had never had any trouble with the walls of jacketed cylinders, even with a much higher pressure of steam in the jacket than in the cylinder; but then the amount of work the engine was doing had been small. The experiments described in the report seemed to him remarkable

(Mr. John I. Thornycroft.)
in showing the saving by the use of the steam-jacket to be so large in small engines, and then to taper off as the engines increased in size. Some of the engines did not run fast; and in large engines working fast he should expect that the economy would taper off to an exceedingly small percentage. But there was a great deal to be said in favour of steam-jackets for marine engines, where they had to run comparatively slowly, like those of the navy, where the normal working was not at full power but at a small fraction of it. When such engines were running slowly, there was probably a great saving in jacketing; but these experiments appeared to him to indicate that with full power in a large engine there would not be any great saving. He would refrain however, as the Committee had done, from putting forward any theory at the present stage of the enquiry, because he thought it would be better to leave theories in abeyance until the Committee had given the results of their further experiments.

Mr. Jeremiah Head, Past-President, thought one of the most important statements contained in the report was that given in page 451, where it appeared that at the speed there mentioned and with steam in the jackets the cut-off giving the minimum steam consumption was found to be at 1-8th of the stroke, equal to 4·8 expansions; while at the same speed but without steam in the jackets the best cut-off was found to be at 5-16ths of the stroke, equal to 2·6 expansions. This seemed to him to be exceedingly important to bear in mind, because there were so many instances in practical work where engines were required to be either linked up, as in the case of locomotives, or worked with variable expansion, as in the case of rolling-mill engines and others. Therefore the fact that the use of the jacket seemed to put the engine under so much more favourable conditions for working at a high rate of expansion would naturally inspire much more confidence in resorting to linking up in the one case, or to using variable expansion in the other.

The President asked how it was that cutting off at 1-8th of the stroke gave only 4·8 expansions.

Mr. Donkin replied that it was a small engine, and therefore the clearance had a larger effect in reducing the actual amount of expansion.

Professor Unwin hoped the time would come for trying superheated steam again. The common objection as to its burning up the lubricant in the cylinder had never seemed to him to have much force. In ordinary practice such a case would hardly be likely to occur as that mentioned by Professor Kennedy (page 503), where the steam was so highly superheated that even at the exhaust it was found to be still superheated. With any ordinary extent of superheating, the cooling action of the cylinder wall would cause the steam to become saturated the moment it entered the cylinder. All that the superheat could do would be to warm the cylinder wall a little; and he thought it would not be possible in practice to have superheated steam in the cylinder during the stroke, even though the initial steam had undergone a considerable amount of superheating before admission.

Mr. Davey said that, as regarded the amount of condensation in the steam-jacket that was due to the loss by radiation from its outer surface, in the recorded trials in which he had himself taken part it had not been practicable to put the engine under such a condition as would enable this loss to be ascertained. The East London Water Works engine had had to be tried in regular work; and it would not have been convenient to stop the work in order to ascertain the jacket condensation while standing. The subject however would be borne in mind for a future opportunity of carrying out such experiments, wherever it might present itself. It was only in Mr. Donkin's small experimental engine that trials had been made (pages 446, 449, and 450), from which information could be recorded upon this point.

With regard to the risk of cutting the cylinder surfaces from overheating them by putting too hot steam into the jackets, at a higher pressure than the initial pressure used in the cylinder, he thought there was no difficulty to be anticipated in that direction. As

(Mr. Henry Davey.)

had been pointed out by Professor Unwin (page 505), there was always initial cooling, even in cylinders with steam-jackets. In other engines than those dealt with in the report he had tried high-pressure steam both in the high and the low-pressure jackets, hotter than the initial steam of the high-pressure cylinder, without experiencing any difficulty from cutting the cylinders. When experiments on the use of superheated steam had been carried out years ago, the engines had not the excellent metallic packing they now had; and the difficulty he believed had been more with the packing itself than either with the piston or with the cylinder. Moreover the cylinders and pistons were probably made of much better metal now than they had been then. If the use of superheated steam were now to be reverted to, he thought the difficulties would not again be met with that had been experienced years ago when it was tried and discarded. There was a great deal to be done he believed in the direction of superheated steam; and if it could be used sufficiently superheated without cutting the cylinders, it would be possible he considered to get the same economy from superheated steam that was now got from the steam-jacket, or probably even a superior economy.

The circulation of the steam in the jackets had been pretty good in the experiments recorded in the report. Engines generally were not so constructed as to secure perfect circulation; and this he considered was the reason for so many failures in the application of steam-jackets, which were unable to perform their intended functions for want of proper circulation of the steam, and consequently failed to give an economical result. It was not easy to arrange for the circulation in the jacket in the position in which the cylinders were frequently placed in relation to the steam supply from the boilers; but in the experiments now reported particular attention had been paid to see that there was a proper circulation in the jackets, as far as possible under the circumstances. Although in page 438 mention was made only of the extra precaution that had been taken in placing an air-cock at the highest point of each jacket, there was no special necessity for one at the bottom, because the drain pipe, for water was at the bottom, and there was therefore always an exit for the air from that place. Test cocks both at the top and at the bottom were useful, and they were both recommended (page 454) among the suggestions which had been appended to the report.

Mr. W. C. Carter, being unable to attend the meeting, wrote that it appeared to him that steam-jackets must necessarily be considerably more efficient on the ends of a cylinder than on the body, for the following reasons. The heating effect of the jacket is required as much as possible for the entering steam, and as little as possible for the exhaust. The cover jacket is clearly more effective upon the former, and less so upon the latter, owing to their comparative volumes and also to the difference in the duration of exposure. The converse is the case with the body jacket. These facts are shown in the experiments recorded in the report. Thus in experiment No. 37 (page 423), out of the whole saving of 2·5 per cent. of feed-water due to all the jackets, as much as 1·3 per cent. or more than half was due to the end jackets, in spite of the fact that the surface of one end was presumably not more than one-eighth of that of the body. For equal areas therefore the efficiency of the end jacket in this instance would appear to be rather more than eight times that of the body jacket. This conclusion is also confirmed by the results given in page 447, where the saving of feed-water due to all the jackets is 12·8 lbs. per indicated horse-power per hour, of which 4·2 lbs. are due to the end jackets. Assuming the surface of one end to be here about one-fifth of that of the body, these results would give a ratio of say 1·64 to 1 for the comparative efficiencies. Although this latter ratio, the key to which may be found in the first paragraph on page 420, is not so striking as the former ratio of 8 to 1, it is sufficiently so to make it a matter of regret that the trials of engines possessing end jackets do not all comprise comparative results with and without the end jackets. The table in page 446, giving the amount of water condensed in the different jackets, might easily be revised to show the

(Mr. W. C. Carter.)
comparative condensation in the body and end jackets per unit of area. It is doubtful whether results obtained from engines requiring from 35 to more than 60 lbs. of feed-water per indicated horse-power per hour, as in experiment No. 49 (page 450), are of much practical utility.

Too much stress can hardly be laid upon the necessity for the most careful attention to steam-jackets in working, and upon the importance of proper draining, which should be automatic if possible. Doubtless the shyness with which steam-jackets are regarded by many marine engineers arises from experience of instances similar to one that came under the writer's notice, where an old chief engineer of an under-power coasting steamer was using the steam-jacket to condense water for washing purposes. Even in steamers so successful as the "Majestic" and "Teutonic" of the White Star line, the writer believes the jackets are simply air spaces; and so far as he is aware there are no reliable results extant showing a clear gain from steam-jackets in use at sea.

In the unrecorded trial with the oil-jacket (page 448), it is difficult to understand why, assuming steam at about 65 lbs. absolute pressure per square inch, the great excess of temperature in the body jacket, if maintained, should not have shown an increase of efficiency due to superheating and diminished liquefaction. The disappointing result seems to indicate defective circulation, preventing a sufficiently rapid restoration of the heat abstracted from the oil by the cylinder walls. The matter is one of great interest to the writer, who has recently been compelled temporarily to abandon experiments in this direction with a liquid of high boiling point, owing to the circulating appliances all proving inadequate. So far as they went however his experiments showed an efficiency increasing with the temperature; and when circumstances permit, the writer hopes to obtain data of interest, which he will be happy to place at the service of this Institution.

Mr. George R. Dunell wrote that at one time he had been engaged on a fairly complete set of trials as to the value of superheating, and had had a small steam-yacht built, largely for this purpose. The general result had been that he found great additional efficiency was obtained by superheating, and that this could be carried to high temperatures without scoring cylinders or valve-faces: an experience which he had afterwards confirmed with larger vessels. In a tug for which he had designed the machinery, and which had a coil superheater, the steam had on one occasion become so hot, through shortness of water, that the felt covering had been burnt off the steam pipe; in spite of this, neither the cylinders nor the valves were injured. The investigation of the value of superheating had subsequently been pursued in a more complete manner by Mr. Isherwood; the results were contained in an elaborate report * made to the United States government, of which he had much pleasure in presenting copies to the Library of the Institution.

Mr. Alfred Saxon wrote that in his opinion a reasonable explanation of the better results gained in steam-jacketing smaller cylinders over those obtained with larger lay in the fact that the radiation of the heat from the jacket would penetrate further in proportion into the interior space of a smaller cylinder; and it would be found that in proportion to the size of the cylinder a greater quantity of the steam admitted would be dry at release from a smaller cylinder than could possibly be the case from a larger. The larger the cylinder, the less effective towards its centre would be the penetration of the heat from the jacket.

Mr. John G. Hudson wrote that the full efficiency of jackets was frequently not realised, whilst mechanical difficulties were experienced, through want of sufficient attention being given to the by no means simple problem of their complete and continuous voiding of water and air. The removal of the air presented the greatest difficulty; and being in his opinion best effected by a continuous circulation through the jackets, he attached great

* Report of a Board of United States Naval Engineers on the Herreshoff system of Motive Machinery as applied to the steam-yacht Leila; 3 June 1881.

(Mr. John G. Hudson.)

importance to this point, and had given special attention to it. In a set of horizontal triple-expansion engines,* constructed by Messrs. Hick, Hargreaves and Co., he had adopted the following arrangement, with a view to meeting the above requirements. There are four cylinders, two being low-pressure, all of which, together with the two intermediate receivers, had to be jacketed with steam at the full boiler-pressure of 160 lbs. To drain each of these six jackets separately by any of the usual methods would have led to a good deal of complication, and would have called for much attention in working. They are therefore connected in series, the bottom of the first to the upper part of the next, and so on throughout the six. The first jacket, on the high-pressure cylinder, takes steam from the boiler side of the main stop-valve; and the collective drains from the last low-pressure cylinder are discharged into a drain-receiver, fitted with pressure and water-gauges and standing on the engine-room floor in full sight of the attendant. The drain-receiver is voided of water by a hand blow-through at starting, and by a small supplementary feed-pump worked off the engine when running; and of air by a connection to the first intermediate receiver. This connection is fitted with a diaphragm pierced with a small orifice, giving a constant leakage from the drain-receiver into the first intermediate receiver, sufficient to remove any air which might otherwise accumulate in the drain-receiver, without at the same time passing steam enough to interfere appreciably with the proper mode of working of the engine. The air and steam thus added to the working steam discharged from the high-pressure cylinder into the first receiver pass of course through the remaining cylinders, and the air is finally discharged by the air-pump. This arrangement has worked satisfactorily, requiring little attention, and proving convenient for warming up the cylinders and receivers before starting. There is no noise from "water hammer" either at starting or subsequently, such as it was feared might possibly be the case; and the pressure in the drain-receiver is within 3 lbs. of the boiler pressure. The efficiency of the jackets may be

* Described in "Industries," 12 Feb. 1892, page 152.

judged from the results of a two days' trial, during which the feed, as measured by meter and including the jacket water, averaged only 11·23 lbs. per hour per indicated horse-power. This result may be compared with that obtained from a set of unjacketed triple-expansion Corliss engines * by the same makers, working at about the same pressure, which were shown by tank measurement during a careful trial to use 12·79 lbs. The present result is 12·2 per cent. better; and this gain in economy is not greater, judging by the Committee's report, than might be expected to result from the use of jackets working under such favourable conditions as to temperature and circulation.

In the above instance the engines are of comparatively small size, namely 325 indicated horse-power; and the system of jacketing employed would hardly be suitable to larger engines, if only on the ground that it would not generally be prudent to subject ordinary cast-iron jackets of large diameter to the full boiler-pressure. For larger engines, the writer has arrived at the conclusion that the balance of advantages lies with the plan of supplying each cylinder with steam through its own jacket, as is so frequently done on the Continent and in America; and he has adopted it for several sets of large engines, including 3,000 I.H.P. triple-expansion, and 1,600 † and 1,200 I.H.P. compound engines, as well as others smaller. It is true that the plan is open to the obvious objection that some of the water condensed in the jackets may be carried into the cylinders; but this may be minimised, if not wholly prevented, by a suitable arrangement of the inlets and outlets of the jackets. Evidence that the objection is not a serious one is furnished by the exceptionally economical results obtained during trials made on the Continent of engines jacketed on this plan; and the writer has been informed on good authority that one of the leading French engine-builders claims for this method a gain of 6 per cent. in economy, as compared with the more usual method of jacketing, when applied to single-cylinder

---

* Described in "The Engineer," 31 July 1891, page 88; and 26 Aug. 1892, page 171.

† Described in "Engineering," 25 Nov. 1892, page 662.

(Mr. John G. Hudson.)
engines. Further evidence that the plan cannot be very uneconomical is supplied by the Committee's example No. 40; and it is to be regretted that in example No. 44 advantage was not taken of the engine's having been arranged for working in this way, to test it under its ordinary working conditions, as well as under the modified conditions under which it was actually tested; the comparison would have been most instructive. Whatever may be the economical value of the plan, it has great practical advantages. By preventing the possibility of admitting steam into the cylinder without first or simultaneously heating the jacket, or into the upper part of the jacket whilst the lower is filled with lukewarm water, it removes the serious risk of injury from unequal expansion to which large cylinders are liable when jacketed in the ordinary way. It ensures a vigorous circulation through the jackets, and the consequent removal of any air; and automatically graduates the pressure in each successive jacket, without the use and risk of the reducing valves ordinarily employed. Incidentally it adds usefully to the receiver capacity, and simplifies the pipe connections; and in horizontal four-valve Corliss engines it allows the unsightly overhead steam-pipes to be dispensed with.

Referring to the statement (page 503) that in French marine engines the steam supply to auxiliary engines is sometimes taken from the jackets of the main engines, the same arrangement has been used by the writer for a pair of large land engines provided with independent air-pump engines; and he considers it advantageous. Unfortunately the opportunity rarely occurs in land practice, the use of auxiliary engines being quite exceptional.

The progressive experiments made by Mr. Morison (page 484), in evaporating water by steam through walls of varying thickness, appear to the writer, like the progressive speed-trials of steamers, to be of special value, and to supply information which could not be obtained from many times the number of isolated examples. By showing the additional steam temperature needed to transmit the same quantity of heat through a wall of increased thickness, these experiments furnish data for dividing the total difference of temperature into the two separate differences needed to overcome respectively the surface resistances and the internal resistance to the passage of heat. When analysed with this view, the experiments plotted in the diagrams, Plate 84, are found by the writer to agree closely with the two following inferences drawn from them:—firstly, as regards the combined resistances of the two surfaces of the cast-iron wall, that one degree Fahr. of difference sufficed to pass from 651 to 737 units of heat per square foot per hour, the quantity passed increasing slowly with the rate of evaporation; and secondly, as regards the internal resistance of the cast-iron, that one degree of difference sufficed to pass from 409 to 392 units of heat per square foot per hour through one inch of metal, the quantity passed falling off slightly as the rate of evaporation increased, while the difference required varied directly as the thickness. The total difference of temperature needed to produce a given rate of evaporation would of course be the sum of the two separate differences thus found. These conclusions indicate a somewhat lower rate of internal conduction of heat through cast-iron than is usually assumed.

EXPERIMENTS ON THE ARRANGEMENT OF THE SURFACE OF A SCREW-PROPELLER.

By Mr. WILLIAM GEORGE WALKER, of Bristol.

The Experiments about to be described were projected and carried out by the author during 1891 in making some trials with a Screw-Propeller invented by the late Mr. B. Dickinson, of Messrs. Navin and Co., London; and to his assistance he is especially indebted for enabling them to be undertaken. The primary object of the experiments was to ascertain the efficiency of screw-propellers having the same diameter, pitch, and area, but different numbers of blades, and with the blades in various positions on the boss: in other words, of screw-propellers differing from one another only in the arrangement of their virtual surface.

Screws.—Seven screws were tried, which are lettered A to G, all 38½ inches diameter with bosses 7¾ inches diameter. They may be divided into two series: the first comprising the five screws A B C D E; and the second the two remaining screws F G. In the five screws A to E the pitch was 64¼ inches, and the aggregate developed area of the blades was 395 square inches; A had two blades, and B C D E had each four blades. In the two screws F G the pitch was 71¾ inches, and the developed area of the blades was 381 square inches; F had three blades, and G six. The arrangement of the blades is shown in Figs. 1 to 7, Plate 88, looking forwards. The blades are of forged steel, and were pressed by cast-iron moulds into their required pitch-shape. The bosses are of cast-iron, and the blades are keyed on. If the blades, which are inclined slightly aft, are projected on a plane parallel to the axis, their leading and following edges are parallel, as shown in the side elevations, Figs. 8 and 9.

The first series A to E, Figs. 1 to 5, Plate 88, consists essentially of two double-bladed propellers, each having its two blades opposite to each other. The stern shafting was made to receive both these propellers at the same time, one being placed immediately in front of the other and touching it, Fig. 8. The forward screw *f* was permanently keyed for the time being on the shaft; while the after screw *a* had seven keyways in the boss, and was tried in seven different positions in reference to the forward screw; five of these positions form the screws A to E, and the two other positions are not considered in this paper.

The second series F and G, Figs. 6 and 7, Plate 88, consists of two three-bladed propellers, which were fixed on the shaft in the same manner as those in the first series. It will be noticed that the blades of the second series are narrower or shorter than those of the first; but the thickness at the roots and tips, and also the coefficient of surface-friction, are practically the same in both series.

Screw A, Fig. 1, Plate 88, was formed by bringing the leading edge of the after portion *a* into contact with the following edge of the forward portion *f*, thus forming a two-bladed propeller of the ordinary kind; the edges were a good fit, and two correct helicoidal blades were thereby formed. In screw B, Fig. 2, the after portion was placed immediately behind the forward. Screw C, Fig. 3, was formed from screw A, Fig. 1, by turning the after blades forwards in the direction of rotation, into the position shown in Fig. 3, so that the after blades rotate in advance of the forward; the roots of the leading edges of the forward blades and of the following edges of the after blades are here in the same straight lines, parallel to the axis. In screw D, Fig. 4, the four blades are at right-angles to one another; but this is not a propeller of the ordinary kind, because the alternate blades revolve in different planes. Screw E, Fig. 5, was obtained by turning the after portion backwards, in the contrary direction to the rotation, into a position just behind the forward portion, so that the after portion rotates immediately in the wake of the forward portion; this is not unlike an ordinary two-bladed propeller, but having a gap in the middle of the width of each blade. Screw F, Fig. 6, was arranged in a similar manner to screw A, thus forming a three-bladed propeller of the ordinary kind. In screw G, Fig. 7, the six blades are placed at equal angles with one another, and the alternate blades revolve in different planes.

Vessel.—The yacht "Ethel" on which the experiments were made is owned by Mr. G. A. Newall of Bristol, to whom also the author's thanks are due for much valuable assistance and advice. It was built and engined by his firm in 1886. It is 55 feet long and 9 feet beam; the mean draught during the experiments was 3 ft. 3 ins., corresponding with a displacement of 18½ tons. There are two saloons on deck, one fore and one aft. The engines are vertical compound surface-condensing of the ordinary marine pattern. The diameters of the high and low-pressure cylinders are 7 and 14 inches respectively, and the stroke is 9 inches; the high-pressure cylinder is placed forward; the piston-rods are 1⅝ inch diameter. The boiler is placed forward of the engines, and is 5 ft. diameter and 6 ft. 6 ins. length; the working pressure is 100 lbs. per square inch. Before commencing the experiments, the engines though in good condition were thoroughly overhauled. The indicator pipes were fitted carefully and with as few bends as possible. The fire-grate area of the boiler was enlarged, as were also the areas of the uptakes and funnel, in order to enable the boiler to maintain a constant pressure of steam at all speeds.

Mode of Experiment.—The trials were made on the Avon on the regatta course at Saltford, about midway between Bath and Bristol. This was an exceptionally suitable place, as there is a perfectly straight run of nearly a mile, the sides of the course are parallel, and the width and depth of the water ample, and it is also well shaded. The width is about 100 feet, the depth about 14 feet, and the sectional area of the river about 1,200 square feet, while the greatest immersed section of the yacht is 20·4 square feet for the displacement of 18½ tons. During the time the runs were made, the course was entirely free from traffic; the water was smooth, and the speed and depth of the current varied but little. The experiments were made on a base line of 3,400 feet, or 0·64 mile. Two upright poles were erected at each end of the base line, one twenty yards behind the other in a line at right-angles to the stream. Each screw was tried at six successive speeds, namely 4, 5, 6, 7, 7½, and 8 miles per hour, as nearly as possible. For each speed the yacht was run over the course six times, three in each direction, and the mean was taken. The supply of steam was regulated by the throttle-valve, which was locked for each experiment, the reversing gear being fully open. Two indicators were employed, one on each cylinder; and each was connected with both top and bottom. Three sets of diagrams were taken during each run, or eighteen for each experiment, making altogether seventy-two single diagrams. The same set of springs, excepting a broken one, were employed throughout the experiments. The counter was placed in a convenient position on deck, and motion was given to it by a rod from the engine cross-head; it could be stopped or started without interfering with the engines. The working pressure in the boiler was constant at 100 lbs. per square inch throughout the experiments. It was found at first that in the single instance of full speed the boiler pressure was inclined to vary somewhat. This objection was overcome by the aid of forced draught, derived from a small fan, which was driven off the screw shafting. The condenser vacuum was constant at 27 inches of mercury. In making the runs, while the vessel was passing the two ranging poles at either extremity of the base line, a chronometer and the counter were started or stopped, as the case might be, at the moment when the two poles were in a straight line with a point on deck. The time was taken to a third of a second. In order to ensure having the same weight on board during the runs, the coal was weighed as it was used, and its weight was replaced at the end of each experiment. The same steersman and engineer were employed throughout the experiments; and the readings were taken by a staff of four observers.

The total indicated horse-power and the revolutions per minute were obtained exactly for the speeds of 4, 5, 6, 7, 7½, and 8 miles per hour, by plotting the experimental results and drawing fair curves through them, as shown for screw D in Figs. 31 to 35, Plates 94 and 95, and then scaling off the quantities from the curves corresponding with these speeds. The equivalents of these speeds in knots are 3·47, 4·34, 5·20, 6·07, 6·51, and 6·94 knots. The results of the seven screws, A to G, together with their analysis, are given in Tables 1 to 6, pages 528 to 533.

The efficiency of the screws has been examined by the method employed by Mr. Isherwood, the late Chief Engineer to the navy of the United States. It is not unlike the method proposed by the late Dr. Froude (Inst. Naval Architects 1876, page 167), whereby he calculated the indicated thrust from the piston pressures, and gave a number of elements which must be subtracted from the indicated thrust in order to ascertain the ship's true resistance. As the screws in each series of the present experiments have exactly the same pitch, diameter, and physical surface, they have therefore the same coefficient of friction; and although some of the quantities taken in the analysis are perhaps only approximately correct, such as the coefficient of friction, the experiments are nevertheless thoroughly suitable for comparison with one another, and are free from the objections which generally apply to propellers having slightly different dimensions, although meant to be the same.

For convenience the mean pressures of the compound engine have been referred to the high-pressure cylinder. The effective mean pressure is therefore the actual mean pressure P in the high-pressure cylinder, together with the mean pressure p in the low-pressure multiplied by the ratio 4·083 of their capacities; that is, the effective mean pressure is $P + p \times 4{\cdot}083$.

The indicated horse-power developed by the engine has been resolved into the following five constituent parts:—

1. The power to overcome the friction of the unloaded engine, which includes the duty of the air-pump and the feed-pump.
2. The power to overcome the friction due to the working load.
3. The power to overcome the friction of the screw blades.
4. The power expended in the slip of the screw.
5. The power necessary for the propulsion of the vessel.

The net horse-power applied to the shaft is obtained by deducting from the total indicated horse-power the power to overcome the friction of the engine. The steam pressure necessary to overcome the friction of the engine was equal to 9 lbs. per square inch, and was constant at all speeds. Following Mr. Isherwood (Inst. Civil Engineers, vol. cii, page 136), the friction due to the working load was taken at 7½ per cent. of the net power; and the blade friction at 0·45 lb. per square foot of blade surface when moving in its helical path at a velocity of 10 feet per second, and for other speeds in the ratio of the squares of those speeds to the square of 10. If the first three of the above quantities, Nos. 1 to 3, are subtracted from the total indicated horse-power, there remains a quantity, the sum of Nos. 4 and 5, absorbed in the action and reaction of the screw, which is the actual power applied to the screw less the surface resistance of the blades. From this remainder was subtracted the slip, No. 4; and the final remainder was then the power required to propel the vessel, or the useful work, No. 5. This final power divided by the net power is a measure of the efficiency of the screw. The respective steam-pressures may be calculated from the powers.

The thrust was obtained from the following expression:—thrust in lbs. × speed of vessel in feet per minute = 33,000 × horse-power expended in propulsion of vessel.

Examination of Results.—It will be seen that screws C and G were the most efficient in the first and second series respectively. Screws B, C, and D, were exactly equal for all speeds up to six miles per hour; screw A was also equal to them for speeds up to five miles per hour. Screw C, the most efficient, ranged from equality at these speeds up to 2·02, 2·59, 1·45, and 4·24 per cent. greater efficiency at the maximum speed than the screws A, B, D, and E respectively. Screw G was more efficient than screw F, at all speeds except the minimum, when they were equal; the difference in efficiency was 1·53 per cent. at the maximum speed of eight miles per hour. It is not unreasonable to suppose that, if a position of blades had been tried in the second series similar to that of screw C in the first, a greater efficiency would have been obtained. It will be seen that at the lower speeds the results of the screws in each series were not affected either by the number or by the position of the blades. The blades of screws B and E evidently affected the water injuriously for each other at the higher speeds.

The screws attained an efficiency of about 70 per cent. at the minimum speed, with a slip of about 22 per cent.; and afterwards fell off, with further increase of slip. The screws are in themselves bad, inasmuch as their maximum efficiency should have been at the working speed of about 7½ miles per hour, but the screw space was insufficient to allow for more blade area; this however will not in the slightest affect the object of the experiments.

In order to try the effect of friction of the blades, screw D was tested with a much rougher surface, presenting if anything greater roughness than that of very coarse sand. The revolutions are plotted in Fig. 31, Plate 94. The loss of power due to the increased surface-resistance was two-fold; for in addition to the extra pressure necessary to rotate the screw at the same number of revolutions, the slip was greater, showing that the transverse component or rotary motion of the water was increased, thereby reducing the sternward or thrust component.

From the accompanying figures it is seen that the sum of the mean pressures employed in the propulsion and slip of the screws is practically equal for the screws in each series, for the same speeds; in other words the turning moments necessary to maintain the rotation

Sum of the Pressures of Slip and Propulsion,
in lbs. per square inch referred to high-pressure cylinder.

Miles per hour.	4	5	6	7	7½	8
Screw.	Lbs.	Lbs.	Lbs.	Lbs.	Lbs.	Lbs.
A	13·36	18·52	27·43	44·76	55·81	66·59
B	13·36	18·52	27·44	44·76	55·82	66·60
C	13·36	18·52	27·44	44·76	55·82	66·60
D	13·36	18·52	27·44	44·73	55·82	66·50
E	—	—	—	—	—	66·49
F	14·99	20·69	30·70	50·03	62·48	74·52
G	14·99	20·71	30·71	50·03	62·48	74·52

of the screws, when the vessel was going at the same speed, are equal. Therefore it may be clearly laid down that with screws of equal pitch and equal area of blades, when propelling a vessel at a certain speed, the pressure on the pistons will be equal, whatever may be the state of the water in which the screws work, and the number of revolutions, and therefore the slip; for the revolutions will regulate themselves to the existing circumstances of the liquid fulcrum against which the blades act. That such is the case is seen from the fact that, while the pressures have remained constant, the revolutions have varied to a considerable extent.

On comparing the above pressures in the first and second series, they are found to vary in the direct ratio of the pitch, namely as 1·00 to 1·12, thus showing that for equal speed the turning moment varies in the direct proportion of the pitch. This law is fully demonstrated in Mr. Blechynden's paper, on the reaction and efficiency of the screw-propeller (North-East Coast Inst. of Engineers and Shipbuilders, 1887, vol. iii, page 179); in which he also says the effect of the surface is the same, irrespective of the number of blades into which it is divided. On this point he does not appear to have made any experiments, but accepts the results of Mr. Isherwood, which however were not sufficiently extended to warrant the assumption.

Discussion of Results.—The results show that, the kind and aggregate extent of surface remaining the same, it is advantageous to increase the number of blades. Hence the blades will become proportionately narrower, and this is undoubtedly the cause of increased efficiency; that is, narrower blades are proportionately more effective than wider ones. For when the two-bladed and three-bladed propellers were divided respectively into four blades at right angles to one another (screw D) and into six blades (screw G), the slip was reduced for the same speed, with exactly the same steam pressure; consequently the efficiency is greater. This conclusion appears to be somewhat contrary to prevailing opinions; but the author is not acquainted with any trials of propellers having six or more blades. The above remark refers only to smooth water; in a rough sea the advantage of many blades appears obvious. In his Experiments in Aerodynamics, published by the Smithsonian Institution, Washington, 1891, Mr. S. P. Langley has shown (page 106) that with horizontal planes moving horizontally in air the time in falling is greater for those planes whose extension from front to back is small compared with their length measured perpendicularly to the line of advance: in other words, narrower planes have less slip than wider ones.

It does not seem difficult to find an explanation for the foregoing result. The normal pressure on a blade moving obliquely through water does not occur at the centre of the width, but at a position about one-fourth or one-third of the width from the leading edge. If a sheet of paper or metal is moved through air or water with its surface inclined to the direction of motion, the leading edge will suffer considerable deflection compared with the following, and will tend to assume a position at right-angles to the line of motion. The leading face of a propeller blade is thus more effective than the following. In a paper read before the Institution of Naval Architects in 1890, Mr. Howden has noticed this important fact, for he says (page 237), "it is evident that the efficiency of the blade decreases towards the after edge. The first six inches on the entering side are more effective than the last six inches on the after side." It is thus chiefly by the leading side of the blade that the inertia of the water is overcome, and its acceleration produced, probably in the form of an impact, thereby giving the water a greater velocity than that of the screw itself. This would be a cause for apparent negative slip, the water flowing through the propeller faster than is accounted for by the product of the pitch and revolution. There must then be a region of reduced pressure on the working face of the blade. That there is a vacuum in the wake of a blade, caused by the slip and by the body of the blade displacing the water, is proved by the performance of screw E; for with exactly the same steam pressure the revolutions increased to a marked extent, showing that the after section was exerting little or no thrust.

It may be well to mention the case of a propeller 10 feet diameter, which was brought under the author's notice after some five or six months' continual voyaging to the Mediterranean. In Fig. 20, Plate 91, is shown an expanded view of the working face, of which the portion L on the leading edge was clean scoured; the part H was covered with marine growth some inches long; while the part K on the after edge had a short growth; the greater part of the back of the blade was covered with grass. This seems to show that only a small part of the area of the blade was effective; and affords an argument in favour of narrow blades.

In accounting for the superiority of screw C, it will be noticed that the blades of each portion, when the vessel is in motion, necessarily cut into water undisturbed with respect to the screw, which was probably not the case with the other screws, whose blades injuriously disturbed the water for one another. Moreover the spent water from the forward blades of screw C, being discharged with a velocity at the back of the after blades, destroys the partial vacuum behind the after blades, thus augmenting the thrust by preventing the water from tending to run round to the back of the after blades.

Practical Trials.—Previous to carrying out the above experiments, some trials were made with propellers on the principle of screw C, which is the form invented by Mr. Dickinson, by whose permission the results are here given, being of interest for comparison with those subsequently obtained from the experiments. After several successful trials on steam launches, two vessels belonging to Messrs. Weatherley Mead and Hussey of London, the "Herongate" and the "Belle of Dunkerque," were kindly offered by the owners for further trial of the screw. These vessels are sister ships in every respect, built in 1887 by Messrs. Short Brothers, and engined on the triple-expansion plan by Mr. John Dickinson of Sunderland, each of 525 tons register. The ordinary propeller, shown in Figs. 10 and 11, Plate 89, was of cast-iron, 10 ft. 9 ins. diameter, 15 feet pitch, with four blades and 32 square feet area; the Dickinson screw, shown in Figs. 12 to 14, was of cast-steel, 10 ft. 6 ins. diameter and 15 feet pitch, made as a double three-bladed screw, with six blades and 30 square feet area.

In Table 7 (page 534) is given a record of the first two deeply laden voyages of these two vessels to Spanish ports in the Mediterranean, the ordinary propellers being exactly the same on both vessels. The following are the mean results —

Comparison of Ordinary four-bladed and Dickinson six-bladed Propeller. Average results of Table 7.

Steamer	Belle of Dunkerque.	Herongate.	
Propeller	Ordinary.	Ordinary.	Dickinson.
Speed, knots per hour	8·738	8·393	8·743
Revolutions of screw per minute	83·5	82·0	73·5
Boiler pressure, lbs. per square inch	152·5	149·0	152·5
Vacuum, inches of mercury	25·0	24·0	25·0
Draught of steamer, feet and inches	12 8½	12 8¾	12 10
Coal burnt in 24 hours, tons	7½	6¾	6½
Knots run per ton of coal	28·0	29·8	32·4
Slip of screw, per cent.	29·2	30·8	19·6
Allowance for foul bottom, knots per hour	0·33		0·50

When running with the Dickinson propeller the condition of the "Herongate" was very foul. On the first voyage, from London to Cullera, she started with a three months' growth on a coating of tar. On her return she made a voyage to the Baltic; after which she started on her voyage from Cardiff to Tarragona with a growth of five months and a half. Returning thence, she was found to be extremely foul, being covered from keel to water line with grass three to four inches long; she was therefore cleaned and painted, and thus commenced with a clean bottom her voyage from London to Denia with the ordinary propeller. On her return, having been aground, she was again docked and painted with antifouling composition, before starting on her next voyage from London to Valencia. The "Belle of Dunkerque" had a growth of one month and two months and a half, on commencing her first and second voyage respectively.

In the navigation of the "Herongate" it was found that the Dickinson propeller afforded greater command of the vessel, through improved steering and greater rapidity in evolution, and power of stopping and starting more promptly; and there was a reduction in vibration. It was also reported that, when running high on a ballast trim, with the propeller not thoroughly submerged, the efficiency was even more pronounced, inasmuch as the screw possessed a firmer grip of the water, and held the vessel steady to the wind.

The "Herongate" was also run from Upper Blyth buoy in Sea Reach (the second reach below Gravesend) to the North Sand Head lightship (the North Goodwin lightship), first with the ordinary and afterwards with the six-bladed propeller. These trials were conducted by Mr. George R. Dunell, and the following results were obtained:—

Comparative Trials of the "Herongate" with Ordinary four-bladed and Dickinson six-bladed Propeller.

Propeller	Ordinary.	Dickinson.
Duration of trial, hours and minutes	5 40	4 27
Distance by log, knots	50	42
Speed, knots per hour	8·82	9·43
Revolutions of screw per minute	75·3	73·3
Boiler pressure, lbs. per square inch	129·2	150·6
Vacuum, inches of mercury	25·6	25·6
Mean total Indicated Horse-power	328·6	389·3
Displacement, tons	1,107	1,112
Slip of screw, per cent.	20·8	13·0
Coefficient of performance *	223	232

$$\text{* Coefficient of performance} = \frac{\sqrt[3]{\text{Displacement in tons}^2} \times (\text{Speed in knots})^3}{\text{Mean total I.H.P.}}$$

It must be remembered in comparing these two performances that the vessel had a foul bottom when tried with the Dickinson propeller, whilst when running with the ordinary propeller she was just out of dock after cleaning.

Trial on Tug Boat.—On the completion of the experiments at Saltford, the author carried out a trial on the tug boat "Frank

Stanley," which was kindly placed at his disposal by Captain Roberts of Bristol. The object of this trial was more especially to ascertain how far a screw with narrow blades was suitable for general towing. The ordinary four-bladed propeller of the tug is shown in Figs. 15 and 16, Plate 90; the Dickinson screw employed is shown in Figs. 17 to 19, having six blades in two portions. The tug is 70 feet length, and 14 feet beam, with a draught of 5 feet forward and 7½ feet aft. Both screws were of cast-iron and 6 feet diameter; the ordinary four-bladed had a pitch of 9 feet and an expanded blade-area of 13·58 square feet; the pitch of the Dickinson screw was 9 ft. 6 ins., and its blade area 12·12 square feet. The following were the results of the trial:—

Comparative Trials of the Tug "Frank Stanley" with Ordinary four-bladed and Dickinson six-bladed Propeller.

Propeller	Ordinary.	Dickinson.
Speed, knots per hour	10·8	11·1
Revolutions of screw per minute	136·7	128·0
Boiler pressure, lbs. per square inch	58·5	58·0
Vacuum, inches of mercury	15·0	15·0
Slip of screw, per cent.	11·1	7·5

After three months' continuous work in all seas and weathers Captain Roberts reported as follows respecting the towing qualities of the six-bladed propeller:—"While the towing power of the tug is greater, by an amount I should think approaching six or seven horse-power, the action of the propeller on the water is also smoother, and it appears to have a firmer grip of the water, which is especially noticeable when getting under weigh, the required speed being obtained almost immediately on starting the engine."

These practical trials on the "Herongate" and the "Frank Stanley" form a strong corroboration of the somewhat more scientific trials at Saltford, demonstrating the superiority of narrow blades, which is obtained by increasing their number and arranging them on the principle of the screw C, Fig. 3, Plate 88. The slip is thereby reduced, with even less blade-area, as is seen from an examination of the results. The practical objections are overcome by making the propeller in two portions, which are more easily manufactured and handled, and can be fitted on without the often tedious necessity of withdrawing the stern shafting; all that is required in case of accident is to replace the injured portion. This principle applies advantageously to vessels of light draught; for by increasing the number of portions the diameter of the screw may be proportionately diminished, and the required blade-area obtained, with a uniform pitch. That the steering is much improved is obvious, in consequence of the greater distribution of water discharged against the rudder; whereby also the vibration is reduced to a considerable extent.

In conclusion, it will be understood that the author does not wish to advocate a particular kind of screw-propeller, or even to assert that any one form is the best, as there are so many considerations affecting the question; but he desires rather to draw attention to several important points bearing upon the subject.

TABLE 1.—*Screw* A, *two blades, pitch* 64¼ *inches. Fig.* 1, *Plate* 88.

Speed, miles per hour	4	5	6	7	7·5	8
Revolutions of screw per minute	83·75	109·11	137·91	172·00	190·30	209·88
Slip of screw, per cent.	21·4	24·7	28·4	33·06	35·1	37·3
Indicated Horse-Powers	I.H.P.	I.H.P.	I.H.P.	I.H.P.	I.H.P.	I.H.P.
Propulsion	1·496	2·591	4·610	8·771	11·733	14·915
Slip of screw	0·407	0·849	1·828	4·332	6·346	8·872
Friction of screw blades	0·069	0·154	0·312	0·605	0·816	1·100
Friction due to load	0·159	0·291	0·517	1·111	1·532	2·018
Net Power	2·131	3·885	7·297	14·819	20·427	26·905
Friction of engine	1·282	1·671	2·113	2·634	2·916	3·215
Total Power	3·413	5·556	9·410	17·453	23·343	30·120
Mean Pressures per square inch*	Lbs.	Lbs.	Lbs.	Lbs.	Lbs.	Lbs.
Propulsion	10·50	13·95	19·64	29·96	36·22	41·75
Slip of screw	2·86	4·57	7·79	14·80	19·59	24·84
Friction of screw blades	0·48	0·83	1·33	2·07	2·52	3·08
Friction due to load	1·11	1·57	2·33	3·79	4·73	5·65
Friction of engine	9·00	9·00	9·00	9·00	9·00	9·00
Total Mean Pressure	23·95	29·92	40·09	59·62	72·06	84·32
Thrust, lbs.	140·2	194·3	288·1	469·9	586·6	699·1
Coefficient of Performance †	85·91	103·07	105·15	90·03	82·79	77·87
Efficiency, per cent.	70·53	66·69	63·17	59·19	57·44	55·43

* Referred to high-pressure cylinder.

† Coefficient of Performance = $\sqrt[3]{\text{Displacement in tons}^2} \times (\text{Speed in knots})^3 \div$ Total I.H.P.

TABLE 2.—*Screw* B, *four blades, pitch* $64\frac{1}{4}$ *inches. Fig.* 2, *Plate* 88.

Speed, miles per hour	4	5	6	7	7·5	8
Revolutions of screw per minute	83·75	109·11	136·98	172·10	192·33	211·99
Slip of screw, per cent.	21·4	24·7	27·9	33·07	35·8	37·9
Indicated Horse-Powers	I.H.P.	I.H.P.	I.H.P.	I.H.P.	I.H.P.	I.H.P.
Propulsion	1·496	2·591	4·613	8·775	11·731	14·923
Slip of screw	0·407	0·849	1·785	4·335	6·541	9·107
Friction of screw blades	0·069	0·154	0·307	0·605	0·846	1·134
Friction due to load.	0·159	0·291	0·543	1·112	1·550	2·040
Net Power	2·131	3·885	7·248	14·827	20·668	27·204
Friction of engine	1·282	1·671	2·098	2·636	2·946	3·247
Total Power	3·413	5·556	9·346	17·463	23·614	30·451
Mean Pressures per square inch*	Lbs.	Lbs.	Lbs.	Lbs.	Lbs.	Lbs.
Propulsion	160	13·95	19·79	29·97	35·84	41·36
Slip of screw	2·86	4·57	7·65	14·79	19·98	25·24
Friction of screw blades	0·48	0·83	1·32	2·07	2·58	3·14
Friction due to load.	1·11	1·57	2·33	3·79	4·74	5·66
Friction of engine	9·00	9·00	900	9·00	9·00	9·00
Total Mean Pressure	23·95	29·92	40·09	59·62	72·14	84·40
Thrust, lbs.	140·2	194·3	288·3	469·9	586·5	699·5
Coefficient of Performance †	85·91	103·07	105·87	89·98	81·85	77·03
Efficiency, per cent.	70·53	66·69	63·64	59·18	56·75	54·86

* Referred to high-pressure cylinder.

† Coefficient of Performance = $\sqrt[3]{\text{Displacement in tons}^2} \times (\text{Speed in knots})^3 \div$ Total I.H.P.

TABLE 3.—*Screw C, four blades, pitch 64¼ inches. Fig. 3, Plate 88.*

Speed, miles per hour	4	5	6	7	7·5	8
Revolutions of screw per minute	83·75	109·11	136·98	167·76	186·20	203·00
Slip of screw, per cent.	21·4	24·7	27·9	31·3	33·7	35·2
Indicated Horse-Powers	I.H.P.	I.H.P.	I.H.P.	I.H.P.	I.H.P.	I.H.P.
Propulsion	1·496	2·591	4·613	8·779	11·728	14·911
Slip of screw	0·407	0·849	1·785	4·000	5·961	8·099
Friction of screw blades	0·069	0·154	0·307	0·562	0·768	0·995
Friction due to load	0·159	0·291	0·543	1·081	1·496	1·946
Net power	2·131	3·885	7·248	14·422	19·953	25·951
Friction of engine	1·282	1·671	2·098	2·570	2·852	3·109
Total Power	3·413	5·556	9·346	16·992	22·805	29·060
Mean Pressures per square inch*	Lbs.	Lbs.	Lbs.	Lbs.	Lbs.	Lbs.
Propulsion	10·50	13·95	19·79	30·75	37·01	43·16
Slip of screw	2·86	4·57	7·65	14·01	18·81	23·44
Friction of screw blades	0·48	0·83	1·32	1·96	2·42	2·88
Friction due to load	1·11	1·57	2·33	3·79	4·72	5·63
Friction of engine	9·00	9·00	9·00	9·00	9·00	9·00
Total Mean Pressure	23·95	29·92	40·09	59·51	71·96	84·11
Thrust, lbs.	140·2	194·3	288·3	470·3	586·4	698·9
Coefficient of Performance †	85·91	103·07	105·87	92·47	84·94	80·71
Efficiency, per cent.	70·53	66·69	63·64	60·87	58·77	57·45

* Referred to high-pressure cylinder.

† Coefficient of Performance = $\sqrt[3]{\text{Displacement in tons}^2} \times (\text{Speed in knots})^3 \div \text{Total I.H.P.}$

TABLE 4.—*Screws D and E, four blades, pitch* $64\frac{1}{4}$ *inches. Figs. 4 and 5, Plate 88; and 31 to 35, Plates 94 and 95.*

Screw	D	D	D	D	D	D	E
Speed, miles per hour	4	5	6	7	7·5	8	8
Revolutions of screw per minute	83·75	109·11	136·98	169·30	188·31	208·00	218·00
Slip of screw, per cent.	21·4	24·7	27·9	31·9	34·4	36·7	39·6
Indicated Horse-Powers	I.H.P.	I.H.P.	I.H.P.	I.H.P.	I.H.P.	I.H.P.	I.H.P.
Propulsion	1·496	2·591	4·613	8·776	11·736	14·902	14·897
Slip of screw	0·407	0·849	1·785	4·110	6·154	8·639	9·770
Friction of screw blades	0·069	0·154	0·307	0·577	0·795	1·070	1·233
Friction due to load	0·159	0·291	0·543	1·091	1·515	1·995	2·100
Net Power	2·131	3·885	7·248	14·554	20·200	26·606	28·000
Friction of engine	1·282	1·671	2·098	2·593	2·884	3·186	3·338
Total Power	3·413	5·556	9·346	17·147	23·084	29·792	31·338
Mean Pressures per square inch*	Lbs.	Lbs.	Lbs.	Lbs.	Lbs.	Lbs.	Lbs.
Propulsion	10·50	13·95	19·79	30·46	36·62	42·10	40·16
Slip of screw	2·86	4·57	7·65	14·27	19·20	24·40	26·33
Friction of screw blades	0·48	0·83	1·32	2·00	2·48	3·02	3·32
Friction due to load	1·11	1·57	2·33	3·79	4·72	5·64	5·66
Friction of engine	9·00	9·00	9·00	9·00	9·00	9·00	9·00
Total Mean Pressure	23·95	29·92	40·09	59·52	72·02	84·16	84·47
Thrust, lbs.	140·2	194·3	288·3	470·1	586·8	698·5	698·3
Coefficient of Performance †	85·91	103·07	105·87	91·63	83·72	78·73	74·80
Efficiency, per cent.	70·53	66·69	63·64	60·33	58·09	56·00	53·21

* Referred to high-pressure cylinder.

† Coefficient of Performance = $\sqrt[3]{\text{Displacement in tons}^2} \times (\text{Speed in knots})^3 \div \text{Total I.H.P.}$

TABLE 5.—*Screw* F, *three blades, pitch* $71\frac{3}{4}$ *inches. Fig.* 6, *Plate* 88.

Speed, miles per hour	4	5	6	7	7·5	8
Revolutions of screw per minute	76·4	100·45	127·31	154·90	172·48	192·21
Slip of screw, per cent.	23·0	26·7	30·6	33·5	36·0	38·7
Indicated Horse-Powers	I.H.P.	I.H.P.	I.H.P.	I.H.P.	I.H.P.	I.H.P.
Propulsion	1·500	2·594	4·616	8·771	11·739	14·918
Slip of screw	0·448	0·945	2·036	4·418	6·602	9·457
Friction of screw blades	0·056	0·127	0·259	0·469	0·646	0·894
Friction due to load	0·163	0·297	0·560	1·107	1·539	2·049
Net Power	2·167	3·963	7·471	14·765	20·526	27·318
Friction of engine	1·170	1·538	1·950	2·372	2·642	2·945
Total Power	3·337	5·501	9·421	17·137	23·168	30·263
Mean Pressures per square inch*	Lbs.	Lbs.	Lbs.	Lbs.	Lbs.	Lbs.
Propulsion	11·54	15·17	21·30	33·27	39·99	45·61
Slip of screw	3·45	5·52	9·40	16·76	22·49	28·91
Friction of screw blades	0·43	0·75	1·20	1·78	2·20	2·73
Friction due to load	1·25	1·74	2·58	4·19	5·24	6·26
Friction of engine	9·00	9·00	9·00	9·00	9·00	9·00
Total Mean Pressure	25·67	32·18	43·48	65·00	78·92	92·51
Thrust, lbs.	140·6	194·5	288·5	469·9	586·9	699·3
Coefficient of Performance †	87·86	104·10	105·06	91·69	83·42	77·52
Efficiency, per cent.	69·22	65·45	61·78	59·40	57·19	54·61

* Referred to high-pressure cylinder.

† Coefficient of Performance = $\sqrt[3]{\text{Displacement in tons}^2} \times (\text{Speed in knots})^3 \div \text{Total I.H.P.}$

TABLE 6.—*Screw G, six blades, pitch* $71\frac{3}{4}$ *inches.* *Fig.* 7, *Plate* 88.

Speed, miles per hour	4	5	6	7	7·5	8
Revolutions of screw per minute	76·4	100·00	124·90	151·00	168·08	187·34
Slip of screw, per cent.	23·0	26·4	29·3	31·8	34·3	37·2
Indicated Horse-Powers	I.H.P.	I.H.P.	I.H.P.	I.H.P.	I.H.P.	I.H.P.
Propulsion	1·500	2·594	4·616	8·771	11·740	14·922
Slip of screw	0·448	0·930	1·909	4·089	6·133	8·838
Friction of screw blades	0·056	0·126	0·231	0·434	0·597	0·828
Friction due to load	0·163	0·295	0·547	1·077	1·497	1·993
Net Power	2·167	3·945	7·303	14·371	19·967	26·581
Friction of engine	1·170	1·532	1·913	2·313	2·574	2·869
Total Power	3·337	5·477	9·216	16·684	22·541	29·450
*Mean Pressures * per square inch*	Lbs.	Lbs.	Lbs.	Lbs.	Lbs.	Lbs.
Propulsion	11·54	15·25	21·71	34·12	41·04	46·80
Slip of screw	3·45	5·46	9·00	15·91	21·44	27·72
Friction of screw blades	0·43	0·74	1·08	1·70	2·07	2·60
Friction due to load	1·25	1·73	2·57	4·19	5·26	6·25
Friction of engine	9·00	9·00	9·00	9·00	9·00	9·00
Total Mean Pressure	25·67	32·18	43·36	64·92	78·81	92·37
Thrust, lbs.	140·6	194·5	288·4	470·0	586·9	699·4
Coefficient of Performance †	87·86	104·55	107·37	94·18	85·74	79·51
Efficiency, per cent.	69·22	65·75	63·20	61·03	58·80	56·14

* Referred to high-pressure cylinder.

† Coefficient of Performance = $\sqrt[3]{\text{Displacement in tons}^2} \times (\text{Speed in knots})^3 \div \text{Total I.H.P.}$

TABLE 7.

Voyages of Two Steamers to Spanish ports in the Mediterranean, with Ordinary four-bladed Propeller, Figs. 10 *and* 11, *and Dickinson six-bladed Propeller, Figs.* 12 *to* 14, *Plate* 89.

Steamer	Belle of Dunkerque.		Herongate.			
Propeller	Ordinary Propeller.		Ordinary Propeller.		Dickinson Screw.	
Voyage	Cardiff to Alicante.	Cardiff to Almeria.	London to Denia.	London to Valencia.	London to Cullera.	Cardiff to Tarragona.
Distance run, knots	1458	1288	1611	1[illegible]8	1632	1660
Time, hours	165	149	184	184	181	196
Speed, knots per hour	8·836	8·640	8·755	8·02	9·016	8·470
Revolutions of screw per minute	83	84	85	79	74	73
Boiler pressure, lbs. per square inch	152	153	155	143	150	155
Vacuum, inches of mercury	25	25	24	24	25	25
Draught of steamer, feet and inches	12 7	12 10	12 8½	12 9	12 8½	12 11½
Coal burnt in 24 hours, tons	7½	7½	7	6½	7	6
Knots run per ton of coal	28·27	27·76	30·02	29·66	30·91	33·88
Slip of screw, per cent.	28·0	30·5	30·4	31·3	17·7	21·6

Discussion.

Mr. Walker showed models of the ordinary and the Dickinson propellers used on the "Herongate" (page 523); and a photograph of the yacht "Ethel" (page 516), on which his own experiments had been made.

The paper made no pretence to dealing exhaustively with the subject of the arrangement of screw-propeller surface; but was intended simply to place before the Institution for examination the observations made in the experiments recorded, which he should have been glad to carry much further, had he been able to afford time or money for the purpose. It was also intended to be suggestive on one or two important points. Although he had every confidence in the accuracy of these experiments as far as they went, nothing would give him greater pleasure than to see them extended on a much larger scale, in the light of which his views might be confirmed or corrected. Such an undertaking however he thought was beyond the means of a single person, and was rather the work of a research committee, similar to that whose report had been read and discussed yesterday evening. In the observations he had brought forwards in the paper, even though he wished the experiments could have been more conclusive, there was sufficient matter he thought for considerable discussion; and he hoped it might be useful in eliciting the valuable opinions of those conversant with so difficult and important a problem as that of the screw-propeller had all along been.

Mr. R. Edmund Froude wished to bear testimony to what appeared to him the unusual degree of scientific accuracy shown in the results of the trials of the yacht "Ethel" as compared with steam trials in general on ships. Nearly every precaution seemed to have been taken to ensure accuracy. In one particular respect the trials seemed to have had a distinct advantage over ordinary steam-trials: namely in the course on which they had been carried out, and the freedom which that course must have had from those variations in speed of current which formed so fruitful a source of error in

(Mr. R. Edmund Froude.)
ordinary steam-trials on ships in tidal waters. The measurement of speed in steam trials was far the most important element; all the other measurements, however accurately they might be obtained, in themselves were meaningless unless they could be referred to the speed which was associated with them. At the same time, although he had a high opinion of the accuracy and value of the results presented in this interesting paper, he was inclined to think that the conclusions which had been deduced from them by the author were to some extent both too precise and too general; and he would endeavour to give his reasons for so thinking.

The results were analysed in an elaborate and careful manner in the tables appended to the paper, and the analysis appeared to him to have been carried out on an unexceptionable principle; and it was not all experimenters who took the trouble to make that kind of analysis. At the same time he thought it had here been carried to rather a finer point than the data could justify. Taking for example one of the important elements which must come into account, namely the elimination of the screw-blade friction, this was well known to be always a most difficult matter; in fact it was more true to say that nobody yet knew how to do it correctly. The utmost that could be hoped was to get a kind of approximation, which might be called a conventional approximation, serviceable for certain purposes, but, it must not be forgotten, little more than a convention; it was only an approximation after all, and rather a crude one. This had not indeed been forgotten by the author of the paper, who in fact had expressly noticed it in page 518; but still in the treatment of the results it seemed to some extent to have been ignored. There were some points connected with the manner in which even that conventional approximation to the blade friction should be obtained, in regard to which it did not seem clear whether they had been quite legitimately treated. But without going further into those intricate matters, the coefficient of friction for the blades, which was admitted by the author (page 518) to be an uncertain quantity, had in his opinion been taken here decidedly too low, seeing that for the purposes of analysis the whole of the edgeways resistance of the blades should be included. In this opinion he was confirmed by noticing how small a proportion the horse-power calculated for the blade friction bore to the total horse-power. In a good screw, correctly designed for the work it had to do, and working with the slip ratio which gave the maximum of efficiency, the power consumed in the edgeways resistance of the blades should be just equal to the power wasted in the slip. This had been clearly proved in a paper read by his father before the Institution of Naval Architects in 1878 (page 51). In the tables appended to the present paper it would be seen that the amount calculated for the blade friction varied from about one-sixth to one-eighth only of the power wasted in the slip. Naturally, in screws such as those dealt with in the paper, which were working with a slip ratio much greater than that of the maximum efficiency, the blade friction was much less than equal to the power wasted in slip; at the same time he thought it could not be so much less as was here represented, particularly since the power represented as wasted in slip was estimated from what was called the "apparent slip," the true slip being probably from 30 to 40 per cent. greater. Therefore the power here estimated for the blade friction, he should say, was not much more at full speed than one-twelfth of the power actually wasted in the slip. No doubt where a screw was working at such a high slip-ratio as in these experiments, the element of edgeways resistance was rather small relatively to that of the loss by slip. This was a point to which he would recur (page 542) in considering the ultimate conclusion to be drawn from the results; and he mentioned it here merely in order to acknowledge that, inasmuch as the element of blade friction was comparatively so unimportant an item, it would be hypercritical to take exception to any moderate error in calculating the amount of this friction, were it not that—on examining some of the results to which special attention was drawn in the paper, and tracing these back to their origin in the tables in order to verify them with the degree of precision which the figures given in the paper appeared to claim,—it would be found that a great deal did turn on comparatively small differences in the calculated element of blade friction, which he considered could not be accepted as accurately ascertained.

(Mr. R. Edmund Froude.)

Next came the elimination of engine friction. Passing over the element of working friction (in the tables called "friction due to load"), which he could not recognise as much more than a guess, there remained the element of the initial friction or dead-load friction, which was here designated "friction of engine." This he understood from the author was ascertained by experiment, though it was not expressly so stated in the paper. It was put down in the tables at the round figure of 9 lbs. per square inch referred to the high-pressure cylinder, and was taken as constant throughout the whole series of experiments. In the main perhaps it was fairly constant; but he should certainly expect that it must have been subject to capricious variations above and below the average taken, which would be quite sufficient in amount to make considerable havoc with the decimals in the rather delicate remainder values of the calculations. Bearing in mind this constant round figure of 9 lbs. running through the appended tables from one end to the other, and then turning back to the table in page 520 showing figures all carried to two places of decimals, and observing the harmony they presented, which was vitiated by discrepancies of only about 1 in 2,000, it certainly seemed to him, however it might be explained, that the results here given were rather too good to be true.

If these trials had been made under his own superintendence, and he had been entrusted with the duty of reporting upon them, the conclusion he should have been contented to draw, believing that no more precise conclusion could legitimately be drawn, would have been the following. The figures given in the second line of the appended tables for the revolutions of the screw per minute, taken in connection with the figures of speed given in the first line, might be accounted as unusually accurate: quite exceptionally so, considering the immunity from error in speed measurement which he had already referred to (page 536); and the differences in these revolutions per minute under the different conditions tried had a certain importance taken alone. Lower down came the line of figures representing the mean steam-pressure per square inch in the cylinders, which might be taken as accurate, but not more accurate than were the measurements of the ordinates in the indicator diagrams whereby these mean pressures were arrived at. Between these mean pressures and the turning moment on the screw shaft, which was of course the result that was desired to be known, there lay the big element of engine friction, which naturally enlarged considerably the limits of error that must be considered to pertain to the measurement of mean pressure. On comparing these figures of steam pressure, it appeared that, within the limits of error in measurement, the steam pressure for given pitch of screw and given speed of vessel was the same in all the experiments. That was the result which it would be expected to find, on the principle of virtual velocities, if there were no such thing as edgeways resistance of screw blades, or if it could be assumed that the edgeways resistance of the screw blades was constant within the limits of error: which it must be remembered were wide limits, where the whole edgeways resistance bore only so small a proportion to the total resistance overcome by the steam pressure. The conclusion to be drawn seemed to him therefore to be that, within the limits of error, the edgeways resistance of the screw blades was constant, as it would be if the change in arrangement of blade area, with the concomitant change in revolutions per minute due to change in slip, was such as did not, within the limits of error, alter the edgeways resistance.

This was a conclusion which had a certain important bearing on the ultimate conclusion that would be drawn: and was a point which he would presently consider; but that aspect of the question he thought would be best introduced by referring to a matter to which some attention was drawn in pages 521 and 522 of the paper: namely the principle that a blade moving obliquely through fluid—water or air—experienced a greater resisting pressure by being made long transversely to the direction of motion, and narrow in the direction of motion. Of this principle a little too much he thought had been heard lately in connection with the recent publication of Mr. Langley's experiments in aerodynamics, since, true as it was, it was not really novel: certainly he himself well remembered how, when he commenced to study these subjects, his attention had early been called by his father to this very principle. The explanation of the principle was simple, but seemed to him to be slightly mystified in

(Mr. R. Edmund Froude.)
the paper, where it appeared to be suggested that, since in a blade moving obliquely through water the leading half of the blade did the lion's share of the work, it was only necessary to take away the following half in order to get the benefit of the high pressure on the leading part as mean for the whole. That was certainly not the right way of looking at the matter. The principle arose entirely from what Mr. Thornycroft had rather aptly termed the tendency of the water to "short-circuit" over the end of the blades. If a narrow thin long blade, like a lath or a harlequin's wand, were moving obliquely through water in a direction transverse to its length, and attention were confined to the middle portion of its length, leaving the ends out of account, it would be seen that the oblique blade created a complicated disturbance in the water, of which the dynamic upshot might be expressed correctly enough for the present purpose by saying that the deflection which was due to the obliquity of the blade was imparted to a layer of water of a certain thickness, that thickness being proportional to the width of the blade. Consequently when the blade was made wider or narrower, the layer of deflected water was made proportionally thicker or thinner, whereby the total pressure on the blade was proportionally increased or diminished: so that, whether the blade were made wider or narrower, the coefficient of pressure, that is the pressure per square inch, would not be altered thereby. Now leaving the middle portion of the blade, and considering the end portions, it would be found that towards the ends the disturbance in the water changed its character, in virtue simply of the escape of some of the water over the ends, whereby the pressure was diminished. Hence there was found on approaching the ends a decline in pressure, at first gentle, and more and more rapid as the end was neared. The radius of field, so to speak, of this decline in pressure was proportional to the width of the blade; and consequently if, starting with a blade of certain proportions, its width were increased, leaving its length unaltered, or in fact the proportion of width to length were anyhow increased, the proportion of the whole blade area which was subject to the diminution of the pressure would also be increased. Conversely, starting with a blade wide in proportion to its length, if the width was diminished or the

length increased, the coefficient of pressure was increased, at first considerably, but less and less as the influence of the ends declined in importance, until the increase in the coefficient of pressure became infinitesimal.

The application of this principle to a screw-propeller led to the conclusion that, if there were no concern with questions of structural strength, there would be no theoretical limit to the extent to which the process of multiplying and narrowing the blades might advantageously be carried. Beyond a certain point the gain would be insignificant; but the effect would be always in the direction of gain. But of course the considerations of structural strength did bring in a limit at an early stage, because in order to be strong enough for the work the blades had to be of a certain thickness, and the subdividing of the blade area did not enable the thickness to be reduced: in fact the narrower blades would if anything require to be actually thicker; and consequently by multiplying the number of the blades and narrowing them their bluntness was increased. The point at which it was desirable to stop in the process of multiplying and narrowing the blades seemed therefore to be the point at which the gain by the diminished slip—which the experiments, as pointed out by the author (page 521), had shown to arise from multiplying and narrowing the blades—was swallowed up by the increase in the edgeways resistance of the blades, due to their increased bluntness. This element of resistance-increase due to multiplying and narrowing was, to say the least, but slenderly represented in the conditions of the experiments recorded in the paper, where two blades were converted into four, or three into six, by simply cutting each blade in two. Hence there was little or no increase of edgeways resistance, and the effect of so limited a procedure was to exaggerate the case considerably in favour of multiplying the number of blades. There was also another circumstance which conduced to the same effect. As he had already noticed (page 537), the particular screw tried in these experiments was one working at such a high slip-ratio that the element of edgeways resistance of the blades was unusually unimportant compared with the element of slip. It was in precisely

(Mr. R. Edmund Froude.)
such a condition as this that the gain by diminished slip would be found to be the greatest, and the loss by increased edgeways resistance to be the smallest. Hence in this way also it happened that these particular experiments exaggerated the case in favour of a multiplicity of blades; and this being so, it was not at all surprising to find that in these experiments there should have been some gain due to converting two blades into four, or even three blades into six. The gain, as was seen, was not a large amount after all. Comparing the two blades with four, and again the three blades with six, the saving in horse-power was only about 3 per cent., which was no more than in ordinary measured-mile trials might possibly be covered merely by the limits of error.

The ultimate conclusion therefore to be drawn from these experiments, which was a highly valuable conclusion, appeared to him to be this: that it was shown by these unusually accurate and scientific experiments that no substantial advantage was to be gained by multiplying the number of blades beyond what was customary.

Mr. J. Macfarlane Gray observed that, from the comparative trials of the "Herongate" with an ordinary four-bladed and a Dickinson six-bladed propeller, the conclusion which seemed to be drawn in the paper (page 525) was that the six-bladed propeller was the better of the two. Looking at the recorded results however in the light of the usual rule of thumb applicable to similar cases, there did not appear to him to be any such difference. The rule generally taken was that, if two screw-propellers were acting equally well, the horse-power was as the cube of the speed. In the present example (page 525) the speed was increased from 8·82 to 9·43, or about one-fifteenth more. The cube of $1\frac{1}{15}$ was about $1\frac{1}{5}$; and on looking at the horse-power, it was seen to be increased from 328·6 to 389·3, which was again just one-fifth more. Therefore the efficiency of the two propellers, as it appeared to him, was just about the same.

Mr. John I. Thornycroft, concurring in the conclusion drawn by Mr. Froude, would add that to his mind these experiments, done so carefully and accurately, clearly showed that trials on a larger scale would be a mistake. The experiments showed that in a particular vessel fitted with a particular propeller a small advantage had been gained by multiplying the blades; and Mr. Froude had shown clearly in what way that advantage had been obtained. Some stress had been laid in the paper on the relative position of the blades. The propeller C, Fig. 3, Plate 88, which gave the greatest efficiency, had rather an unsymmetrical look; but he thought it was more symmetrical than it would at first appear. For the two portions of the propeller, one behind the other, were here working in different planes; and though at first sight they seemed to be working together, they were really working with a thin screw of water between them. This was not so in the propeller A, Fig. 1, in which the after portion could not properly be said to be behind the forward, because the blades were there in such a position that they represented only one continuous blade in working. If the after portion were turned through a quarter of a revolution, forming the symmetrical four-bladed propeller D, Fig. 4, each portion was then acting equally and independently of the other; and prior to the experiments he should have been inclined to think this would probably be the best position for the blades; but it was a difficult question to venture an opinion upon, because the subject was so complicated. As to experiments with screw-propellers on a large scale, in his own opinion there was not time to make them with any approach to completeness: with a large vessel a week's experiments he thought would not suffice to find out whether one screw was really better than another. In the yacht experiments described in the paper, the high slip-ratio to which Mr. Froude had called attention (page 537) showed that in this particular instance the screw was too small for the vessel; and hence it was that by multiplying the number of blades the advantage had been gained of lessening the slip. If any experiments were to be made, he agreed that in the first place they should be made on a small scale; and they might afterwards be verified by trials on a larger scale when the smaller experiments justified the expenditure. They were greatly indebted to the author for spending so much time and money on this investigation, and for his careful analysis of the results; but he thought he was not drawing a correct conclusion in

(Mr. John I. Thornycroft.)
inferring that the particular screw dealt with in the paper was necessarily better than a screw of more ordinary form. With regard to the strength of the screw, as had been pointed out by Mr. Froude (page 541), the blades manifestly became weaker as their number was increased. For his own part, if it were practicable to make a screw of only one blade, he should prefer it as being stronger than a screw with more blades. The least number of blades that could be used was two, but in order to diminish vibration it was often advantageous to use three; in fact the use of three blades appeared to have great advantages in most cases.

Mr. Sydney W. Barnaby had had the pleasure of knowing the late Mr. Dickinson and of seeing one of his propellers, and, at the request of Mr. Dunell, had witnessed a trial of a boat fitted with that propeller and with a small propeller of the ordinary kind, constructed to compare with it. The boat however he thought Mr. Dunell would agree was unsuitable for getting any fair comparison at all: she was a short bluff little launch, which went easily enough up to about six miles an hour, but after that stood up on end, so to speak, and could not be got to travel any faster. Hence any comparison between the two propellers was rendered futile just about at the point where it might otherwise possibly have begun to prove interesting. The trials described in the paper he agreed with Mr. Froude were remarkable as having been so carefully made. He was not quite sure whether the author had been well advised in following Mr. Isherwood's method so closely; as to the accuracy with which he had worked out his figures, he fortunately had not followed Mr. Isherwood's custom of giving the horse-power to five or six places of decimals. If the trials described in the paper were accepted as having been fairly and accurately made —apart from log-book reports as to the performance of a vessel on a long voyage, which were of little value—then some reason must be found why one form of the Dickinson propeller had shown a small advantage at certain speeds. At slow speeds it would be seen that there was absolutely no difference between the two-bladed propeller A, Fig. 1, Plate 88, and the propeller shifted round and made into a four-bladed propeller C, Fig. 3, or D, Fig. 4. The reason he thought was simply that the propeller was much too small. What practically decided the question of making a propeller with two or three or four blades was the diameter for which there was room in the ship. Supposing it was found that in a particular instance a propeller of 6 feet diameter was required to give the maximum efficiency, then if there was room for that diameter the screw would be made with only two blades or three; but if there was not room for a propeller of more than 3 feet diameter, then it must have four blades, or perhaps five or even six, although the latter numbers were unusual. Comparing the four-bladed Dickinson propeller C or D with the same propeller converted into the two-bladed form A, both having uniform pitch, the difference was that the two-bladed propeller had blades twice as broad. There was no doubt in his mind that a broad-bladed propeller should not have a uniform pitch. In order that the broader blades should fit the contracting stream of water passing through the propeller, he was strongly of opinion that they must be so curved as to have a slightly increasing pitch towards the following edge. If a propeller having two broad blades made with increasing pitch were tried against the Dickinson four-bladed propeller C, Fig. 3, he thought it would compare more favourably with the latter; but he should still expect the four-bladed screw to have an advantage on the yacht used in the author's experiments, because of the restricted diameter. As far as he understood the theory of the Dickinson propeller, it seemed to depend upon the principle which had been dealt with by Mr. Froude (page 540): namely that the pressure upon the leading half of the blade was greater than upon the following half; and so it seemed to be thought that by cutting the blade in half the greater pressure would be got upon both of the two half-blades: as in the story of the man who was offered a stove that would save half the coal, and so bought two to save it all.

Negative slip was referred to in page 522 as accounted for by the water probably getting a greater velocity than that of the screw itself. Here he thought the author was right, though he did not agree with him that the higher velocity was due to impact between the blade

(Mr. Sydney W. Barnaby.)
and the water. Twenty-six years ago, in a paper read before the Institution of Naval Architects (Transactions 1866, page 121), Sir Edward Reed had said that his friend and assistant, Mr. Crossland, had suggested the idea that negative slip must be accounted for by the fact that the water ran away from the screw: that is, that it had a greater backward velocity relatively to the vessel than was measured by the product of the revolutions of the screw multiplied by the pitch; and this fact had then been put down to elasticity of the water. That explanation however he thought was wrong; for it had now been demonstrated that water did act in this way, not through its elasticity but through other causes which had been explained by Mr. Froude: water did flow backwards from the screw faster than the speed of the screw itself measured in the ordinary way; and this would account for negative slip, as he had himself shown in his paper read to the Institution of Civil Engineers in 1890 (vol. cii, page 74). Thus the interesting facts brought forward in the paper might be interpreted by everyone in his own way, if not perhaps quite in the same way in which they had been interpreted by the author.

Mr. Henry Shield said that, without agreeing with all that was claimed for the propeller on which the interesting experiments described in the paper had been made, his own experience of propellers during the last two or three years had led him rather to sympathise with the author's thinly veiled suspicion that his views would not be generally received. It seemed to himself that one great difficulty in connection with screw-propellers had always been the tendency to generalize from trials of a particular specimen. He therefore wished to avoid generalization and theories as far as possible; and would simply state what had come under his own observation with regard to the passenger ferry-boats crossing the Mersey between Liverpool and Birkenhead, and the tugs which towed on the river. For these vessels had been adopted the Myers propeller, shown in Figs. 23 to 30, Plates 92 and 93, the form of which he was not prepared to defend theoretically, and in which he candidly confessed he did not quite believe when he first tried it. By means of this propeller it had been found that the very results stated in the paper had been attained: namely, vibration had been avoided in a remarkable manner; an increase of speed had been gained; and above all, such a control had been obtained over vessels which were dangerous and somewhat difficult to manage in the crowded estuary of the Mersey that they were now controlled with perfect ease and regularity. The captains of these boats were willing to give evidence that they could stop any of them at full speed in its own length; and although he did not put this forward as his own evidence, there was no doubt that a greatly improved command over the boats had been attained, which had astonished no one more than himself. These results could be easily verified by any one crossing the Mersey on the two large twin-screw ferry-boats called the "Mersey" and the "Wirral," which now carried the passenger traffic between Liverpool and Birkenhead. The two boats differed only in this respect, that in the "Wirral" the ordinary propellers had been replaced by the Myers propellers shown in Figs. 27 to 30, Plate 93. These might be described as six-bladed propellers with the tips of the blades united in three pairs in the way shown in the drawings, and also in the model which he had brought to illustrate the construction, because it was so difficult to explain its principles. The immediate consequence of the change of propellers, so far as that large ferry-boat was concerned, had been an entire cessation of vibration. The result was perfectly uncontrovertible: the vibration had previously been very great, and now on the "Wirral" there was none. Prior to the change on the "Wirral" the manager of the ferry-boats had rigged up an instrument to measure this vibration, and it was found to be something serious; but when the ordinary propellers were done away with and the new propellers adopted, the vibration absolutely ceased, and the vessel was now under complete control when stopping and starting. There was claimed for the new propellers a certain additional speed; but he did not attach much importance to this, because both boats had previously been doing the service in the allotted time in fine weather. It was in the facility of starting from the landing-stage in rough weather that the Myers propellers gave the advantage to the "Wirral."

TABLE 8.—*Trials of Steam Tug "Liberator," on measured mile at Waterloo, near Liverpool, with Ordinary and Myers Propeller.*

Tug:—length 90 feet, breadth 18 feet, draft about 7¾ feet.

Engine:—cylinders 16 and 32 inches diameter, 20 inches stroke. Propeller geared two revolutions to one of engine.

Ordinary Propeller, Figs. 21 *and* 22, *Plate* 91.

Diameter 6½ feet, pitch 7½ feet, three blades right-handed, expanded surface 13 square feet, projected 10½ square feet.

1890.	Trial.	Revolutions of Engine per minute.	Speed, knots per hour.		Slip of Propeller.	Run of Tug.		Tide. With or Against tug.
			Tug.	Propeller.		Out or In.	Duration.	
	No.	Revs.	Knots.	Knots.	Per cent.		Minutes.	
September 24.	1	71½	9·23	10·60	12·8	Out	6½	With
	2	68½	7·50	10·14	26·1	In	8	Against
	3	66	8·88	9·76	9·0	Out	6¾	With
	4	68	7·50	10·06	25·5	In	8	Against
	5	65	8·57	9·62	11·0	Out	7	With
	6	67	7·50	9·90	24·3	In	8	Against
	7	67	8·57	9·90	13·5	Out	7	With
	8	66	8·00	9 76	18·0	In	7½	Against
Mean		67⅛	8·22					

1890.	Trial.	Revolutions of Engine per minute.	Mean Effective Pressure. Lbs. per square inch.		Indicated Horse-power.		
			High-pressure cylinder.	Low-pressure cylinder.	High-pressure cylinder.	Low-pressure cylinder.	Total.
			Lbs.	Lbs.	I.H.P.	I.H.P.	I.H.P.
Sep. 24	4	68	60·4	6·9	81·8	38·0	119·8

TABLE 8 (*continued*). *Myers Propeller, Figs.* 23 *to* 26, *Plate* 92.

Diameter 6 feet 4 inches, pitch 7 feet, three double blades right-handed, expanded surface 18 square feet, projected $14\frac{1}{2}$ square feet.

1891.	Trial.	Revolutions of Engine per minute.	Speed, knots per hour. Tug.	Speed, knots per hour. Propeller.	Slip of Propeller.	Run of Tug. Out or In.	Run of Tug. Duration.	Tide. With or Against tug.
July	No.	Revs.	Knots.	Knots.	Per cent.		Minutes.	
9	9	70	8·88	9·66	8·0	Out	$6\frac{3}{4}$	None
9	10	69	8·57	9·54	10·3	In	7	None
Mean		$69\frac{1}{2}$	8·72					
July 25.	11	70	7·50	9·66	22·4	Out	8	Against
	12	69	9·60	9·54	− 0·5	In	$6\frac{1}{4}$	With
	13	$69\frac{1}{2}$	6·92	9·60	28·0	Out	$8\frac{1}{4}$	Against
	14	$69\frac{1}{2}$	10·00	9·60	− 4·0	In	6	With
	15	$68\frac{1}{2}$	6·66	9·47	30·4	Out	9	Against
	16	$67\frac{1}{2}$	10·00	9·33	− 7·0	In	6	With
Mean		69	8·45					

1891.	Trial.	Revolutions of Engine per minute.	Mean Effective Pressure. Lbs. per square inch. High-pressure cylinder.	Mean Effective Pressure. Lbs. per square inch. Low-pressure cylinder.	Indicated Horse-power. High-pressure cylinder.	Indicated Horse-power. Low-pressure cylinder.	Indicated Horse-power. Total.
July			Lbs.	Lbs.	I.H.P.	I.H.P.	I.H.P.
9	10	69	64·4	7·74	90·3	43·3	133·6
25	13	$69\frac{1}{2}$	59·2	6·36	83·5	35·9	119·4
25	14	$69\frac{1}{2}$	59·2	5·87	83·5	33·2	116·7

(Mr. Henry Shield.)

With regard to towing, he had brought these new propellers forwards because they seemed to him to substantiate the representation in the paper, that by increasing the number of the blades an advantage of some value could really be gained. By the substitution of the Myers propeller on the tug "Arrow," which was working near Liverpool, its towing power had been increased from about eight barges to eleven; and in the case of the "Liberator," which was a tug belonging to the Weston Point Steam Tug Co., the captain gave it as his opinion that he could tow 25 per cent. more barges, and in fact he had done it very nearly, under practically the same conditions as before as to steam pressure and so on. In Figs. 21 and 22, Plate 91, was shown the ordinary propeller previously on the "Liberator," and in Plate 92 the Myers propeller substituted for it; and in Table 8 (pages 548–9) were given the particulars of the speed trials of the tug on the measured mile at Waterloo near Liverpool, before and after the change. With the new propeller he had himself nothing to do, except that he had taken it up somewhat against advice; and he had done so because, although he did not know much about it and was sure nobody else did, he did not see how it could possibly be known whether there was or was not anything in it except by trial. It had now been tried on a number of boats, and he believed in every instance an increase of speed had been found, together with a great advance in towing power, and above all a command of the vessel which could not be over-estimated in a crowded estuary.

When it had been determined to apply the new propeller to the twin-screw ferry-boat "Wirral," he had made patterns for the twin Myers propellers in cast-iron, in order that from these patterns the propellers themselves might be cast in manganese bronze. The ordinary propellers in cast-iron taken off the "Wirral" were each 7 ft. 9 ins. diameter and 14 ft. 2½ ins. pitch, with three blades having an expanded surface of 18 square feet. The new propellers, also in cast-iron, shown in Plate 93, were only about half the weight, 6 ft. diameter and 12 ft. pitch, having an expanded surface of 20 square feet or 14½ square feet projected. They were tried during six or seven weeks, in order to settle whether they had the right pitch or surface. During that period they worked without the slightest hitch, brought down the vibration to nothing, and greatly improved the handiness of the boat. As it was found that some improvements could be made in the pitch and the surface, new Myers propellers in cast-iron of slightly altered dimensions were made and put on the "Wirral," and were now working there. At the same time the twin pattern Myers propellers taken off the "Wirral" were fitted to two tugs of the Original Mersey Steam Tug Co., called the "Knight of the Cross" and the "Knight of St. John," with the result that the owners certified an increase of a knot in speed, as well as a considerable increase in towing power. With one of these tugs an extraordinary experience had occurred, showing the great strength that the uniting of the tips of the blades gave to the propeller. It happened on one occasion that the propeller struck against a rock and made about half-a-dozen beats upon it; and instead of every blade being stripped as had been expected, it was found that no harm had been done, except that the surface of the blades was a little jagged and uneven.

The experience with these Myers propellers he thought to some considerable extent justified the author of the paper in the conclusion he had drawn, that under certain conditions the increased number of blades did give advantages, more especially in towing and in places where great command of the boat was wanted. The new propellers on the Mersey boats had not yet been put on any of the ocean steamers, for he had not yet found any superintendent engineer with sufficient confidence in them to make a trial on such a steamer; but he saw no reason why they should not be employed with equal success on a large scale.

Mr. George R. Dunell said the comparative trials of the "Herongate," to which reference was made in the paper (page 525), were only part of a series of experiments which he had made for Mr. Dickinson. When he had first been asked to take the matter up, he had told Mr. Dickinson that from the data already collected he thought this propeller was not better than any ordinary propeller; nevertheless Mr. Dickinson wished him to go on with the trials and find out

(Mr. George R. Dunell.)
whether he was right or not. In the "Herongate" trials he thought the opinion he had expressed beforehand had been borne out; and Mr. Macfarlane Gray had already pointed out (page 542) that the efficiency of the Dickinson propeller was about the same as that of the ordinary screw. If there had been any great advantage either way, it would have proved only that one particular propeller on one particular ship was better than another propeller on the same ship; and that was not a very valuable result to obtain. The experiments on the "Herongate" were not at all of a scientific nature, and were not intended to be so; and he should himself certainly not have thought them worth including in the paper.

Some other trials of this propeller had been made, which perhaps were a little more interesting than those on the "Herongate," simply from the fact that they had been made with a torpedo boat which Mr. Yarrow, from his love of research, had kindly placed at Mr. Dickinson's disposal for the purpose, along with the services of the crew, and the stores and coal. Mr. Yarrow had himself previously tried to persuade Mr. Dickinson that success was not at all likely to attend his exertions and outlay; nevertheless a special Dickinson propeller was made, having its blade-area, diameter, and pitch corresponding with the design of screw adopted by Mr. Yarrow in his ordinary torpedo-boat practice; and by Mr. Yarrow's permission some particulars could now be furnished of the trial. The draught of water was 2 feet 5 inches, and the displacement of the boat 14 tons. The diameter of each propeller was 3 feet $6\frac{1}{2}$ inches, and the pitch 4 feet; the developed area of the blades was 351 square inches. The Yarrow propeller had three blades, and the Dickinson four blades arranged like screw C in Fig. 3, Plate 88. The boiler pressure was 160 lbs. per square inch. The revolutions of the Yarrow propeller were 540 per minute, and of the Dickinson 516·6 or $4\frac{3}{4}$ per cent. less. The result was that the Yarrow propeller gave a speed of 17·112 knots, and the Dickinson a speed of 15·800 knots or $7\frac{2}{3}$ per cent. less. There was thus a distinct advantage for the Yarrow propeller. The apparent slip with the Yarrow screw was 19·7 per cent., and with the Dickinson 22·5 per cent. Other trials were also made, but the above might be taken as fair examples.

In the "Herongate" experiments quoted in page 525 of the paper it should be remarked that the term "slip of screw" was used in its conventional sense of "apparent slip," which did not allow for the motion of the water produced by that of the vessel. Moreover the coefficient of performance there given was not the one originally devised for the comparison of propellers; but was more for the performance of the ship itself. It had the advantage however in the present instance of being a coefficient that Mr. Dickinson had understood, and it was the one which he had liked to use; and as the other conditions were equal for the two propellers, and there was small difference between the speeds, it doubtless did give a fair comparison.

In the torpedo-boat trial the difference in the revolutions of the Yarrow screw and the Dickinson screw bore upon what had been said in the paper (page 522) and also by Mr. Froude (page 540), as to the pressure and consequent friction of the blades. In that trial there were two propellers of the same diameter, pitch, and area, and with equal depth of submersion; and the steam pressure was the same. The Yarrow propeller with three blades made 23·4 revolutions per minute more than the Dickinson four-bladed; and this was due to the different arrangement of the blades, because the other conditions were alike. Mr. Dickinson's theory, that the leading or forward third of the blade was the efficient part, appeared to be adopted in the paper, and was probably true generally. But it did not follow that any propeller could be made in which the whole of the blade surface should be as efficient as that forward portion. It had to be borne in mind that the frictional resistance of blade surface in passing through water was also greater over the leading or forward part.

After thus criticising an invention upon which he had been asked to experiment, he wished to add that, wholly apart from its merits, the late Mr. Dickinson deserved the thanks of all interested in steam navigation; out of pure love of the subject he had freely devoted his considerable means to its investigation, without hope of return.

Mr. Henry M. Rounthwaite suggested a general consideration, which in a launch, or in a small yacht such as that on which the author's trials had been made, might account for a rather better efficiency of a six-bladed propeller. The observed tendency of a right-handed propeller to swing the ship round on her centre in one direction, and of a left-handed propeller to do the same in the opposite direction, was accounted for by the fact that in the lower half of their revolution the screw blades got a better hold from acting in deeper water and therefore against a greater resistance than in the upper half. The same explanation he thought would account for the vibration set up by a two-bladed propeller: every time the revolving pair of blades passed through the vertical position there was a statical couple tending to bend the shaft and also to cause a small launch to pitch endways. Both the vibration and the pitching would tend to reduce the efficiency of the propeller, because the resistance of the ship would be greater. The extreme limits of this principle would be, on the one hand, to use a single segmental blade, which would give the most violent pitching and the greatest amount of vibration; and on the other hand, to have a continuous helical surface forming a complete revolution of the screw, which would be the most free from either action. It was towards the latter limit that a six-bladed propeller made somewhat of an approach; and he thought therefore this consideration would tend in some measure to explain the better results obtained by the six-bladed propeller, and its more uniform action, owing to the fact that the pressures on the blades were practically always more nearly in equilibrium. It was only by such experiments as those described in the paper, made as a check upon theories, that progress could be realised. Generally the results obtained in the experiments seemed to himself to be those that might have been expected; and the considerations on which they depended were clearly indicated by the author in his discussion of results in pages 521–3 of the paper.

Professor Alexander B. W. Kennedy, Vice-President, noticed that in the tables appended to the paper, while the friction of the engine running empty had been assumed as constant, that is, as represented by a constant steam pressure which had no doubt been taken as the mean obtained from a number of experiments, the engine friction due to the load had been arrived at in accordance with Mr. Isherwood's idea that it kept on increasing with the load. It might sound at first as if it were wrong to controvert this view; but he had himself made many experiments with different kinds of engines, in which the load outside the engine could be measured accurately, because it was an electrical load; and he had found practically without exception the results represented by the diagram, Fig. 37, Plate 96. Taking equal horizontal and vertical scales for horse-power, and setting off as a diagonal straight line the net or useful horse-power, which might also be called the brake horse-power or the useful work done, it was found, on plotting the indicated horse-power at each of the points of observation, that the line representing the indicated horse-power ran generally parallel with the line of net or brake horse-power. As the load increased, the power necessary to drive the engine with its load did not vary beyond the limits of the experimental errors in the indicator diagrams; in the great majority of cases the two lines were sensibly parallel. As all engineers who had experimented must have found, it was not so easy to get the ordinate AB, representing the power necessary to drive the engine empty, as it was to get the vertical difference CD between the two lines at any other point, because to run the engine empty at a perfectly steady speed was most difficult. For any period such as five minutes the speed would be steady, that is, the mean speed during one such period would be the same as during another; but it would not be quite steady during the few seconds when the indicator diagrams were being taken; it would be either increasing or decreasing. It was so difficult to prevent the speed from either increasing or decreasing momentarily when the engine was running empty, that there was much more variation in the determinations of the power represented by the ordinate AB than in the differences between the brake horse-power and the corresponding indicated horse-power at other points. So long as there was any load on the engine these differences could be determined with such a degree of accuracy as produced in the diagram the sensible

(Professor Alex. B. W. Kennedy.)
parallelism of the line BD representing the indicated horse-power with the straight line AC representing the brake horse-power. In the paper therefore he thought that the sum of the two quantities put down for friction due to load and for friction of empty engine, which sum in Table 1 varied from 10·11 to 14·65 lbs. per square inch on the high-pressure piston, might in fact have been taken as much more nearly constant; at any rate that had been his own experience with engines of much the same kind as those dealt with by the author, although of somewhat larger dimensions. He fully shared the feeling expressed by previous speakers as to the usefulness of experiments so carefully conducted and so completely recorded as these.

Mr. John Phillips considered that in the propeller shown in Fig. 20, Plate 91, it would have been much better if the inner portion or base of the blade, which had been found by the author to be covered with marine growth some inches long (page 523), had been covered up by a larger boss, such as had always been advocated by Mr. Griffiths. Furthermore a screw blade of the shape shown in Fig. 20 had the disadvantage of being so weak that he thought practically it would soon come to grief. The trussed propeller described by Mr. Shield (page 551) seemed to him a remarkably strong construction, as had been proved by its having stood knocking against a rock with impunity.

Mr. Walker said his primary object had been to make comparative tests; nevertheless he thought the results were not far from correct as absolute and not merely comparative results. The total indicated horse-power might be divided into two parts, internal and external: the internal being that due to the friction of the unloaded engine; and the external that due to the load, called in the paper the net power. The efficiency of the screw was the ratio of the useful work it performed to the work expended upon it. In this ratio should clearly not be included the friction of the engine, which was apart from the present problem.

With regard to the five constituent parts into which he had divided the total indicated horse-power, various doubts had been expressed as to the accuracy of some of the quantities. Two only of these he thought were open to doubt. One was the power due to the surface friction of the blades, which however could be calculated within small limits; and considering that the total power spent in the friction of the blades was small, any error in the calculation would form such a minute percentage of the total indicated horse-power that it might indeed be left out of account in these experiments, because any error would be the same for each screw. The other quantity—namely the friction due to the working load—was of course open at first sight to doubt. While an important point had been raised by Professor Kennedy (page 555) in considering this quantity constant, he thought that the conditions of experiment and of analysis were somewhat different. The matter could have been tested by placing a thrust dynamometer on the screw shaft, and so obtaining the useful horse-power expended, which was the product of the thrust multiplied by the speed of the vessel; and the useful horse-power thus obtained could have been compared with the useful horse-power calculated in the analysis. If these two quantities were equal, it could have been affirmed that the various quantities in the analysis were correct. These two methods had been employed by Mr. Isherwood, who had obtained exactly the same useful or propulsion horse-power by both, having in his calculation taken the friction due to the working load at 7½ per cent. of the net power. As Mr. Isherwood's experiments had been made on a yacht 56 feet long, only one foot longer than the yacht on which his own experiments described in the paper were made, and the engines also indicated about the same power, he had felt justified in taking these quantities as trustworthy.

The power to overcome the friction of the unloaded engine had been ascertained by disconnecting the screw shafting, and driving the engine at various speeds. The steam pressure was practically constant, and equal to about 8 lbs. per square inch; which he had then raised to the round figure of 9 lbs. so as to allow for the friction

(Mr. Walker.)

of the screw shafting. No doubt there must be a slight error, but it was exactly the same in each case.

The power due to the surface friction of the blades was considered by Mr. Froude (page 536) to have been taken too low in the paper. It certainly had been taken low, when compared with that for propellers of ordinary construction; but the experimental propellers differed greatly from ordinary propellers of the same diameter. The developed area of the blades was about 20 per cent. less than that of the yacht's ordinary screw: which was due to the screw space being insufficient for a propeller of equal area when made in sections; and this was the cause of the high slip-ratio, and the consequent small waste of power due to the surface friction of the blades. He agreed with Mr. Froude (page 537) that in a good screw when working at its maximum efficiency the power spent in blade friction should equal that wasted in slip; but the experimental screws were not tried at a speed or a slip-ratio which gave their maximum efficiency, and therefore from this point alone the power spent in blade friction should be less than that wasted in slip, as had been noticed by Mr. Froude. Also in the experimental propellers the element of edgeways resistance was practically nil: meaning thereby the edgeways resistance due to the bluntness or thickness of the blades, and apart from the edgeways resistance caused by the skin or surface friction of the blades; this division of the total edgeways resistance was simply a conventional one, to which he would again refer (page 561) in considering the ultimate conclusion of Mr. Froude. It had been clearly proved that in these propellers this edgeways resistance was not increased by multiplying the number of blades, as was shown by the fact that the sum of the steam pressures due to slip and propulsion was practically equal (page 520); in other words the turning moment was not increased. This had been recognized by Mr. Froude (page 539) when he concluded that the edgeways resistance was a constant quantity which conclusion confirmed the absence of edgeways resistance due to bluntness. Taking into account therefore the above facts, the blade friction he thought had not been assumed too low, and was what might be expected under the circumstances. The marked equality of the sum of the steam pressures due to slip and propulsion was of course owing to the comparative nature of the experiments, in which any slight error would be the same in every case.

The speeds, revolutions, and steam pressures were considered by Mr. Froude to be accurate (page 538); and these quantities were employed in the construction of the formula representing the coefficient of performance, which could be used in the comparison of the screws, instead of the efficiency ratio, whereby the comparison would be free from any doubts attending the latter.

In connection with the oblique action of a screw blade, and the consequent pressure upon it, he had referred in the paper to Mr. Langley's experiments because they seemed to him to bear upon this question; and as they had been published only last year, he was not aware of their having been much referred to already in connection with the subject of the present paper. With regard to the increased efficiency of the narrower blades, there was another point which might be remembered. Referring for illustration to screw A, Fig. 1, Plate 88, the width at the tips was about 16 inches; in one complete revolution each blade passed over a certain number of different sections of water, which number might be ascertained by dividing its helical path by its width; and its resistance to rotation was due to the inertia of those sections of water. In the four-bladed propeller D however, Fig. 4, each blade would pass over double the number of sections of water; and again, agreeably with Mr. Thornycroft's opinion (page 543), each portion was acting independently of the other, thus forming four helical paths instead of two.

The advantage of multiple blades had been considered by Mr. Froude from one point of view only: namely the slight gain due to the narrowing process (page 541). But there were several other points which must be borne in mind, and which had not been fully taken advantage of in the experiments. The remarks of Mr. Rounthwaite (page 554), with which he was inclined to concur, had reference to the fact that the current in which the screw worked was not a uniform one; it varied in different parts of the screw's revolution, which caused vibration in the screw, and affected the manœuvring powers of the vessel. These effects would be more

(Mr. Walker.)

pronounced in an ordinary sea voyage, where the vessel was subjected to pitching, which did not occur in the smooth-water trials at Saltford; but even here the vibration had been reduced in the case of multiple blades equally spaced round the boss, and the steering had been improved. In this connection it was desirable to refer to Mr. Froude's experimental paper read before the Institution of Naval Architects in 1883 (vol. xxiv, page 235), the gist of which had again been stated in the discussion upon Mr. Barnaby's paper (Inst. C. E. 1890, vol. cii, page 95), namely that the varying current in which the screw worked had no effect upon its efficiency. This conclusion had been arrived at by Mr. Froude by first trying the model screw in undisturbed water, and then behind a model of the ship; but it appeared to himself that those smooth-water experiments, valuable as they were, had not included the points at issue, inasmuch as the model screw was not attached to a shaft connected with the model ship, and could not therefore except through the water give its vibratory motion to the model ship. Also the effect on the manœuvring power of the vessel was not taken into account, the model being pulled by dynamometric apparatus in a straight line forwards.

The ultimate conclusion drawn by Mr. Froude (page 542) from the experiments described in the present paper—namely that no substantial advantage could be gained by multiplying the number of blades beyond what was customary—seemed to be arrived at as a consequence of the increase in edgeways resistance, which resistance was governed by considerations of the structural strength of the blades. Considerations of structural strength he agreed did at an early stage limit the multiplying process; but he thought there were several points connected with edgeways resistance which might tend somewhat to modify Mr. Froude's conclusion, or at any rate had an important bearing on the subject. The augmentation of edgeways resistance due to the multiplying process was considered by Mr. Froude (page 541) to be caused by increase in bluntness or thickness of the blades. By this he understood to be meant headway resistance to the blades, consequent on their leading edges encountering the impulses of the water; and thus in multiplying the number of blades the number and thickness of the leading edges were also multiplied. Under certain conditions he believed the existence of edgeways resistance due to bluntness—that is, headway resistance—could be disputed on the stream-line principle, which was that the only resistance to a submerged ship-shape body was surface friction. The opinion had been expressed by Mr. Isherwood (Inst. C. E. 1890, vol. cii, page 137) that "no gain could be accomplished by making the blades thinner, because, the blades being wholly submerged and tapering almost to a line from the centre to the edges, no power was expended in their displacement of water; the only resistance they opposed to movement was that of their wetted surface." If a screw, of uniform pitch relative to the working face, were moving through water without slip relatively to that face, the only resistance to that face would be surface friction, while the round part of the back would encounter the impulses of the water. If however the blade were moving obliquely, or slip were taking place, the back would not encounter the same number of impulses; and on a certain slip being reached they would disappear, but with the probable formation of an eddy on the following portion of the back of the blade, as had probably been the case with the experimental propellers. Strictly speaking, edgeways resistance was due to a negative pressure, consequent upon the pressure on the stem and stern of a moving ship not being equal. Under certain conditions, such as a high slip, he thought the thickness of the blades could be increased, especially towards the following edge at the back of the blade, without increasing the adverse impulses due to the edgeways motion; and might in certain cases prove beneficial, by causing the transverse section of the blades to approach nearer to that of a ship-shape section when moving in their helical path. In order to test the effect of the thickness of the blades he had recently carried out some trials with model screws of 14 inches diameter, rotating in a fixed position, and in air instead of in water. The models were thus tried under conditions widely different from ordinary screw-propellers, and practical conclusions could not therefore be drawn from them. The screws were of two kinds: in one the transverse

(Mr. Walker.)
section of the blades was similar to that of an ordinary propeller, only much thicker in proportion; in the other kind the thickness was reduced to that of a thin plate, the other dimensions being identical. Screws of two, three, and six blades were tried at progressive revolutions ranging from 800 to 1,800 per minute; and it was found that the screws with the thicker blades were more efficient than those with the thinner blades. Thus these experiments, so far as they went and under their peculiar conditions, gave a contrary result to what might be expected to occur from the remarks of Mr. Froude (page 541) on the effect of the bluntness or thickness of the blades, inasmuch as the efficiency of these model screws in air was augmented by increasing the thickness of the blades. In a published account of some similar experiments made before his own he had met with practically the same results.

While perhaps it could hardly be expected that engineers and shipbuilders would conduct trials on the scale of those described in the paper, he thought trials of a similar kind would prove beneficial if made on even a few vessels of different types and modern construction; and he agreed with Mr. Thornycroft (page 543) that it would be better if trials were first made on a small scale. From model screws he thought many valuable data had already been obtained, which had been important in the elementary aspect of the subject. But there were still many points connected with the engines and hull that required development, such as friction and the motion of the wake, of which nothing definite could as yet be stated; and without this information much of the value of the model experiments was lost.

In Mr. Barnaby's view (page 545) he agreed, namely that a propeller with broad blades should not have a uniform pitch; or in other words, broad blades of uniform pitch were not efficient. The recommendation of an increasing pitch for the after portion of the blade seemed to recognize the existence of a region of reduced pressure, as stated in the paper (page 522). But while some advantage might be obtained by using an increasing pitch, the transverse section of a blade of this kind would be far from ship-shape; and the turning moment he thought would be increased, owing to the back pressure which might be expected to occur at the back of the blade in the portion having the increased pitch.

With the propeller described by Mr. Shield he could readily understand that a reduction of vibration would take place, owing probably to the joining of the tips of the blades, whereby their stiffness would naturally be increased.

One important point, which he had hoped would have been referred to in the discussion, was the statement in the paper (page 520) that the maximum efficiency of the screws experimented upon should have occurred at the working speed of the yacht, namely 7½ miles an hour. It appeared to him an important consideration that the maximum efficiency of any screw should occur at the working speed of the vessel. It might be said that, taking any screw, a vessel could be tried with it through a range of speed sufficient for obtaining a maximum point in the curve of efficiency, and then the speed corresponding with the maximum efficiency of the screw would be the proper speed for the vessel. Model screws had a maximum efficiency of about 70 per cent.; but larger screws on vessels had a somewhat higher efficiency. The general shape of the efficiency curve was almost the same for all screws; from zero at zero thrust it rose to a maximum at a certain speed, and afterwards fell off with further increase in speed. In the experiments described in the paper of course only a portion of the curve was obtained, as shown in Fig. 35, Plate 95; and in Fig. 36, Plate 96, was given an idea of its shape continued backwards to zero, though this portion must not be taken as correct but only as an illustration. The object therefore was so to design a screw that the maximum efficiency should occur at the intended working speed of the vessel. Whether a screw was a bad or a good one, the efficiency curve plotted from a series of trials on the vessel for which it was intended would enable the screw to be so modified as to have its maximum efficiency at the desired working speed of that particular vessel. At the most it would be requisite to make only two propellers: the first as the experimental one, which might even possibly happen to turn out to be itself the best; or if not, then a second, modified in accordance with the results obtained from the trials of the first. In many cases a new set of blades only

(Mr. Walker.)

might be needed. This important subject had been dealt with in a masterly paper by Mr. Froude before the Institution of Naval Architects in 1886 (page 263); and also by Mr. Thornycroft and Mr. Barnaby (Inst. C. E. 1890, vol. cii, page 76). Mr. Froude had made numerous experiments and constructed many efficiency curves, and shown how it was possible from those experiments to design a screw having its maximum efficiency at any given speed. The experiments had been made with small model screws working in undisturbed water; and therefore certain allowance must be made for the following current and also for the friction of the engine. They were perfectly correct for designing a screw to drive what Mr. Froude had called a "phantom" ship, and also to be driven by what Mr. Barnaby had called a "phantom" engine: that is, a vessel which had no following current, and an engine which had no friction. For applying those results to practice, certain modifications had to be made, depending on the kind of engines and form of vessel; and an element of doubt was thus met with, which he thought would be much reduced by making the experiments on the vessel itself. This of course would hardly be necessary in the case of vessels that were of a similar kind to those from which results had already been obtained.

In reference to the trial mentioned by Mr. Dunell (page 552) with Mr. Yarrow's torpedo boat, it was not his wish to advocate any special kind of screw propeller; and he had drawn attention to the "Herongate" trials because he thought six-bladed propellers on ocean-going vessels were rare, and he had every confidence in the figures he had given. The trial on Mr. Yarrow's torpedo boat had not been referred to in the paper, partly because he had not had anything to do with it himself, and partly because so many trials of a similar nature had been made at different times. Interesting as the torpedo-boat trial was, he was afraid that scientific conclusions could not be drawn from it, because two different propellers were used, and as their physical surface was not the same, their respective coefficients of blade friction were probably not the same. It must be remembered that only one form of the Dickinson screw was tried, and at only one speed. From the ordinary three-bladed propeller used by Mr. Yarrow the Dickinson screw had been designed by dividing the three blades into four, and arranging them on the principle of screw C, Fig. 3, Plate 88, the whole being carried out under Mr. Dickinson's own supervision. By that procedure the speed of the boat was found to be reduced to the extent of more than one knot. There was a great difference between that result and those obtained by Mr. Isherwood, who had found that, if the area of the ordinary blades were subdivided into a suitable number of narrower blades, the efficiency of the propeller would not be affected; whereas Mr. Dickinson, on dividing the three-bladed area into four blades, had found that the efficiency fell off to so remarkable an extent. Then as to the pitch of 4 feet for the screw of 3 feet 6½ inches diameter, from his own experience a Dickinson screw of that diameter required a larger pitch. Another point with regard to the trial on the Yarrow torpedo boat was that the bay room was not large enough to allow of the after half of the Dickinson propeller being placed sufficiently far astern of the forward half; the blades consequently over-lapped one another to a certain extent, instead of leaving adequate clear way between them to prevent their injuriously disturbing the water for one another. It appeared to him that Mr. Dickinson had not gone far enough into the details of the construction, notwithstanding that the propeller had been designed on his own principle. While therefore he did not question the accuracy of the torpedo-boat trials, he demurred to their being taken as conclusive. From this trial he believed Mr. Dickinson had come to the conclusion that the portions near the roots of the propeller blades were detrimental to their efficiency, especially in multiple blades, owing to the water not being able to flow through; but that with single, not twin, screws this was not of great importance on cargo or passenger boats, as those portions then worked in a partial vacuum (see Fig. 20, Plate 91, and page 523). In Mr. Yarrow's boat the whole of the propeller was exposed to the full current, and consequently those parts increased the resistance of the boat by reason of the adverse current, the narrow spacing of the blades preventing the water from flowing through; some modification would therefore be required in vessels of that class.

(Mr. Walker.)

The coefficient of performance adopted in the tables had been referred to by Mr. Dunell (page 553) as not the coefficient originally devised for comparing propellers, but as one that had suited Mr. Dickinson. So far as his own experience went, he believed this coefficient had been used in the government trials for the last twenty years, and was the general coefficient still employed in all important trials, and taught in all modern text-books on the subject.

In conclusion it would be noticed that the paper had other points of interest in addition to the arrangement of surface: such as the analysis of the indicated horse-power into its various constituent elements, with a view more especially to arriving at the useful horse-power, that is, the horse-power expended upon the propulsion of the vessel. There was no other method that he knew of for arriving at this power, except by the aid of the thrust dynamometer. Another important object of the paper was to draw attention to the efficiency curves, by means of which, after having put a vessel through a series of trials, a screw of maximum efficiency for the working speed of the vessel could be designed.

The PRESIDENT remembered being once told by a high authority on the subject that anything crooked would do for a screw. Whatever amount of truth there might be in so sweeping an assertion, he was sure the members would all unite in thanking Mr. Walker for the trouble he had taken in making these experiments, and in bringing this paper before the Institution. With the highly interesting discussion to which it had given rise, it had added materially to the facts from which the true theory of the screw-propeller would some day be evolved.

MEMOIRS.

Henry John Marten was born on 3rd February 1827 at Plaistow, Essex. His grandfather, Robert Humphrey Marten, was much interested as a director in various London waterworks and in the Thames Tunnel; and his grandson inherited his liking for engineering. He was articled to Mr. Thomas Wicksteed, who did much for introducing the Cornish engine for waterworks pumping in London. During his apprenticeship he was resident engineer at the Hull Water Works, and afterwards at the Wolverhampton Water Works. At the latter he remained as engineer for the company after the expiration of his articles; and there changed the mode of supply from the intermittent to the constant system, of which he was an early and strong advocate. As the supply at Wolverhampton was given by the working of specially designed engines, which automatically adjusted their speed to the supply needed, experiments and tests were able to be made, which have often been quoted, as to the demand under the constant system for every hour of the twenty-four; these set at rest the fear that the mains laid out for the intermittent system would not be able on the constant system to supply during the hours of greatest demand. In 1856 (Proceedings, page 7) he gave a full description of the Wolverhampton pumping engines to this Institution, of which he became a Member in 1853. Waterworks at many of the South Staffordshire and neighbouring towns engaged his attention for some years, until he had to take charge as partner of the Parkfield Furnaces, near Wolverhampton; and much of the information that he gathered for his own guidance was given to the Institution in 1859 in a paper on hot-blast ovens (Proceedings, pages 62 and 97). His experience in pumping from mines enabled him to assist at the application for rather novel legislation, which was embodied in the South Staffordshire Mines Drainage Act; and he subsequently became one of the three arbitrators to the commission appointed for carrying out the provisions therein enacted. He had a large practice as a parliamentary witness on such subjects as waterworks, drainage, and river conservancy, in regard to which his experience as engineer to the River Severn and to the Staffordshire and Worcestershire Canal made him familiar with the matters involved in such enquiries; and his last public appearance was before the Royal Commission upon Metropolitan water supply, in support of the case put forward by the Thames Conservancy for storage reservoirs in the upper Thames basin. He also constructed sewage works in various parts of the country, and had some important work in hand for the improvement of canals and waterways, notably the improvement works now being carried out on the River Severn; many such works on the Continent he had visited with a view to carrying out similar works for facilitating the water traffic of the Midland district;. and he was preparing a paper on the subject for this Institution, which his failing health prevented him from completing. His death took place at his residence at Codsall, near Wolverhampton, on 3rd November 1892, in the sixty-sixth year of his age. For some years previously his eldest son, Mr. E. D. Marten, had been his partner.

James Salkeld Robinson was born at Rochdale on 1st February 1849, being the eldest son of the late John Robinson (Proceedings 1878, page 13), and grandson of the founder of the firm of Messrs. Thomas Robinson and Son, makers of wood-working machinery. At an early age he entered his father's business, and by a long course of practical training gained a thorough knowledge of the construction and working of the special class of machinery there made, and was associated with the introduction of the many improvements in wood-working machines which have been brought out by his firm. On the conversion of the business in 1880 into a private company, he became chairman, having with him on the board of direction his three younger brothers. In 1882 he was placed on the commission of the peace for the county of Lancaster. His death took place at Rochdale on 14th July 1892, at the age of forty-three. He became a Member of this Institution in 1876.

George Ryder was born on 10th June 1839 at Turner Bridge, Tong, near Bolton. For several years he was associated with Mr. William Ryder, of Bolton, in making tools, fluted rollers, spindles, flyers, &c. The Ryder forging machine now so extensively used was for the most part his work, and many of the machines for making the fluted rollers used in the machinery for spinning cotton were perfected by him. In 1865 he founded the firm of Thomas Ryder and Son, Turner Bridge Iron Works, for the same business, which he successfully carried on until his decease, employing a large number of hands. For many years he was managing director of the Bolton Union Spinning Co.; and for six years he served on the Town Council of Bolton. His death took place at his residence at Tong on 5th December 1892, in the fifty-fourth year of his age. He became a Member of this Institution in 1883.

Christopher James Schofield was born at Manchester on 17th March 1832. In early life he devoted himself to the invention of a machine for cutting fustian, an operation which is still done by hand. The experiments with the machine not being so successful as he wished, he commenced business as a chemical manufacturer, and constructed the largest vitriol chambers in use at that time. In connection with this trade he invented an annular revolving furnace for rendering the production of soda ash a continuous process, the revolving trough or bed being charged on one side of the stationary brickwork casing, and discharged by scrapers fixed at the opposite side. Recently he introduced an apparatus for the concentration of sulphuric and other acids, by which the fracture of the glass retorts employed was to be prevented. His connection with engineering consisted principally in the interest he took in the management of large works. He was a director of the Ashbury Railway Carriage and Iron Co., of Messrs. Charles Cammell and Co., of the Chatterley Co., and of Messrs. Andrew Knowles and Sons, and was largely concerned in other steel and iron works and collieries. He was owner of the Bedworth collieries, near Nuneaton, the machinery of which he almost entirely renewed with appliances of the most modern kind. He was a justice of the peace for the county of

Lancaster. His death took place at Whalley Range, near Manchester, on 8th January 1892, in the sixtieth year of his age. He became an Associate of this Institution in 1875.

Dr. ERNST WERNER VON SIEMENS was born at Lenthe, near Hanover, on 13th December 1816, being the eldest brother of the late Sir William Siemens (Proceedings 1884, page 69). In 1834 on the completion of his education at the gymnasium of Lübeck, he entered the Prussian artillery as a volunteer, and upon receiving his officer's commission in 1839 devoted himself to the study of chemistry and physics. His first experience was somewhat unfortunate, for an explosion, caused by a preparation of phosphorus and chlorate of potash, burst the drum of his right ear; and having previously met with a similar accident to the other ear, he was for a time stone deaf. In 1840 his studies were again interrupted by a sentence to five years' imprisonment for acting as second in a duel. Being allowed to continue his experiments in prison, he successfully plated a silver spoon with gold; and being pardoned after a month's detention, he completed his experiments so far as to introduce in 1841 a process for electro-gilding and silvering. In 1842 in conjunction with his brother William he devised a differential governor (Proceedings 1853, page 75). In 1844 he was appointed superintendent of the artillery workshops in Berlin, where he turned his attention to telegraphy; and in 1845 he invented the dial and printing telegraph instruments. While still a military officer he founded in 1847, in conjunction with Mr. Halske, a manufactory in Berlin for the production of apparatus and materials of all kinds required for telegraphic purposes. In 1846 he adopted gutta-percha, which was just then becoming known, for the insulation of wires laid underground; and constructed a screw-press machine for covering wires with gutta-percha made plastic by heat, and without a seam. In 1848 in conjunction with his brother-in-law, Professor C. Himly, he laid the first electric submarine mines to protect the town of Kiel, and thereby saved it from being bombarded by the Danish fleet. In the same year he was deputed by the Prussian government to lay the first great underground telegraph line from Berlin to Frankfort-on-the-Main; and in 1849 another from Berlin to Cologne, Aix-la-Chapelle, and Verviers. Leaving the army and government service, he then devoted himself to scientific pursuits, and to the management of the telegraph factory, of which the operations extended until branches had to be established in London, St. Petersburg, Vienna, and Paris. In 1856 he brought out a system of duplex telegraphy, by which two messages could be sent simultaneously along a single wire. About the same time he published an elaborate treatise on submarine telegraphy, showing how it was affected by the retardation of the current, and in what way testing for faults was to be conducted. The mercury standard of electric resistance, which was brought out by him, though neglected for many years, is now about to be adopted, with a change of dimension, as the legal standard in this country. He also invented the pneumatic system for the despatch of messages. The crowning practical discovery of his life was embodied in the dynamo, of which he published an account in 1866. This was followed in 1867 by a proposal for electric railways, which however had to lie dormant for many years before being realised. His varied qualities as a man of science, inventor, engineer, and practical man of business, were recognized by the Emperor Frederick, who raised him to the rank of nobility in 1888. He was also a knight of the highest scientific Prussian Order in the country, and in 1860 received an honorary doctor's degree from the University of Berlin. His death took place in Berlin from inflammation of the lungs on 6th December 1892, in the seventy-sixth year of his age. He became a Member of this Institution in 1888; and was also a member of the leading scientific societies in his own country and abroad.

INDEX.

1892.

LIVERPOOL MOTIVE POWER. *Plate I.*

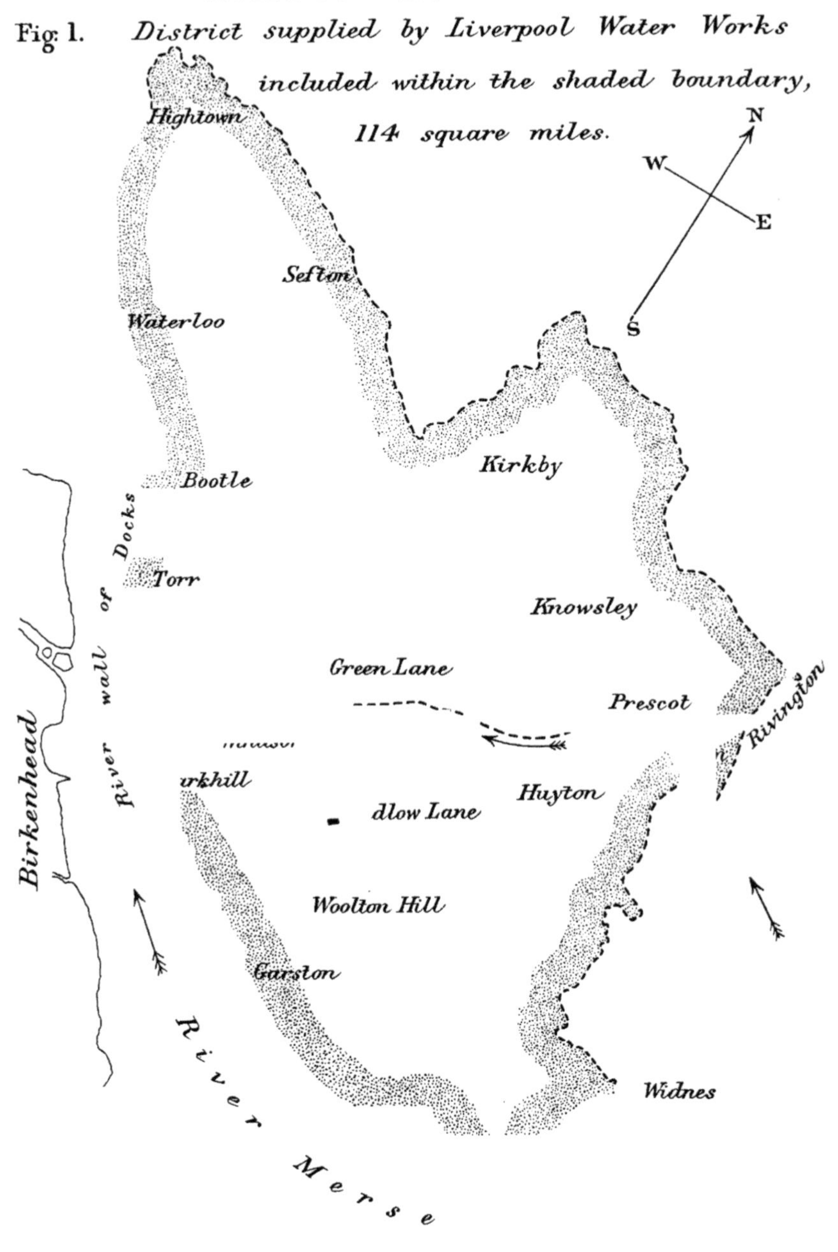

Fig. 1. *District supplied by Liverpool Water Works included within the shaded boundary, 114 square miles.*

LIVERPOOL MOTIVE POWER. *Plate 2.*

Fig: 2. *Pumping Stations and Town Reservoirs of the Liverpool Water Works.*

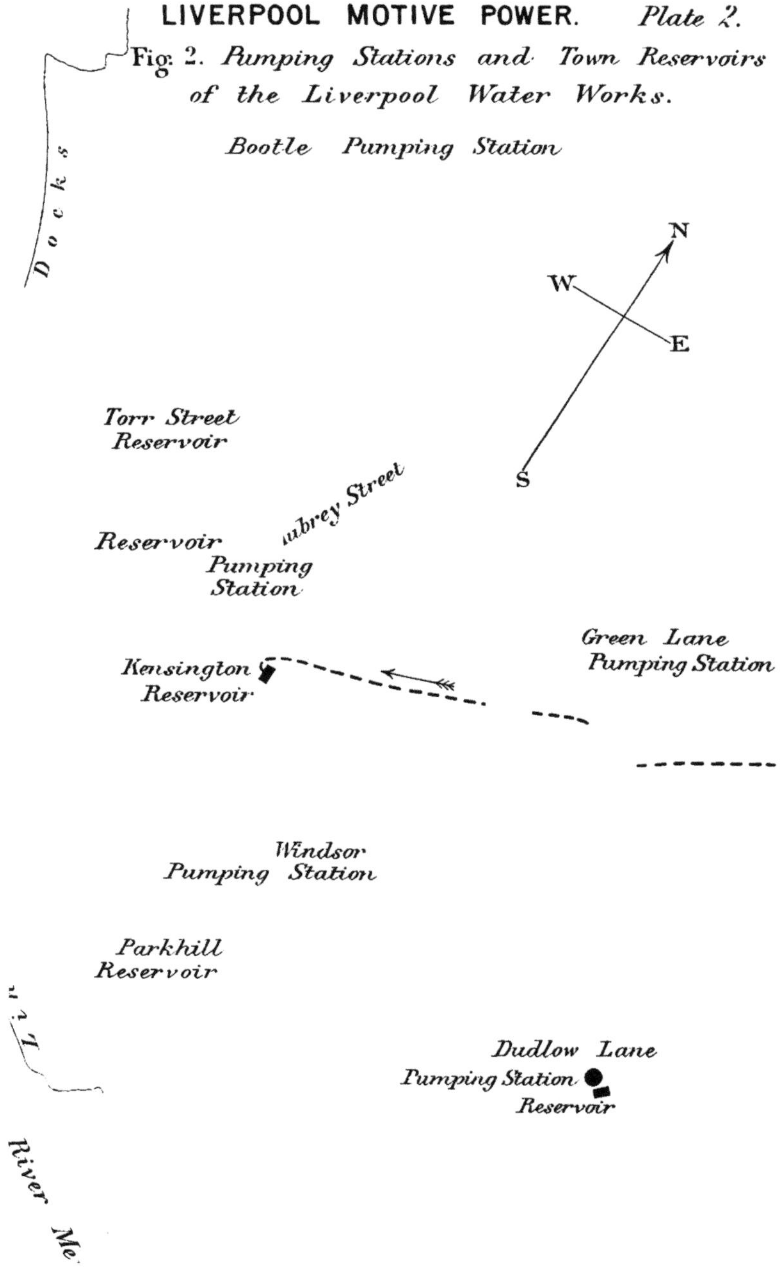

Bootle Pumping Station

Docks

N
W
E
S

Torr Street Reservoir

ubrey Street

Reservoir

Pumping Station

Green Lane Pumping Station

Kensington Reservoir

Windsor Pumping Station

Parkhill Reservoir

Dudlow Lane Pumping Station

Reservoir

River Me

Fig. 3. *Comparative Cost of High-pressure and Low-pressure Water for equivalent Power. See Table 3.*

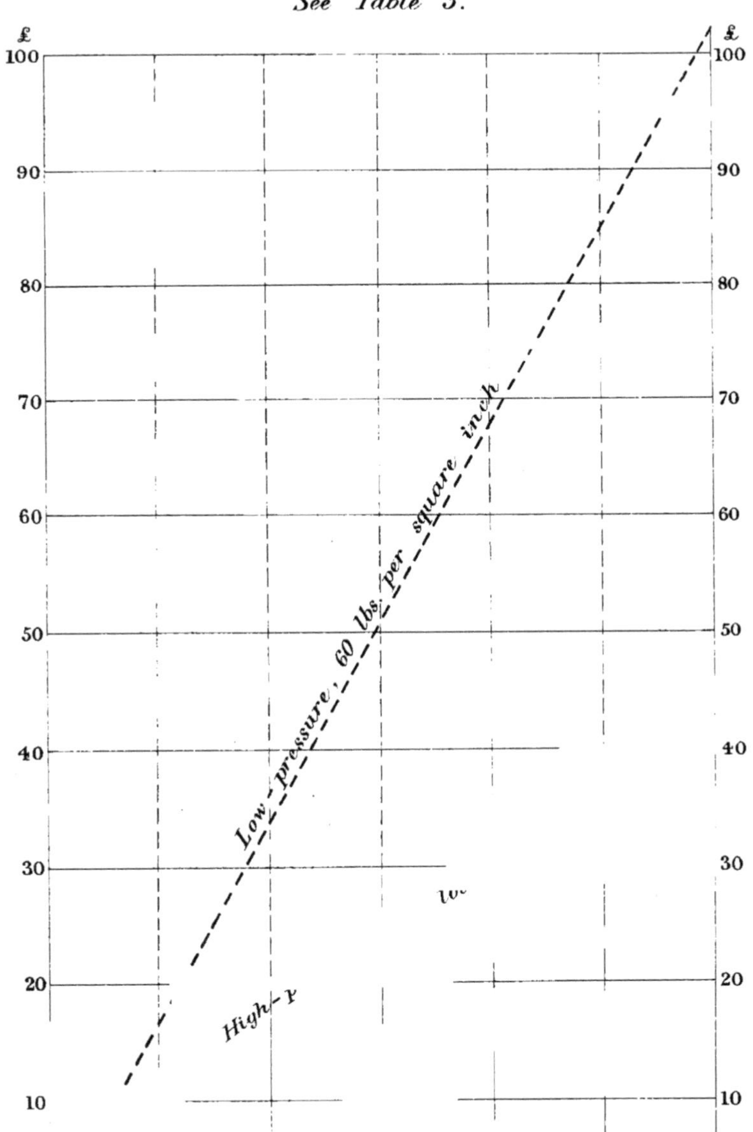

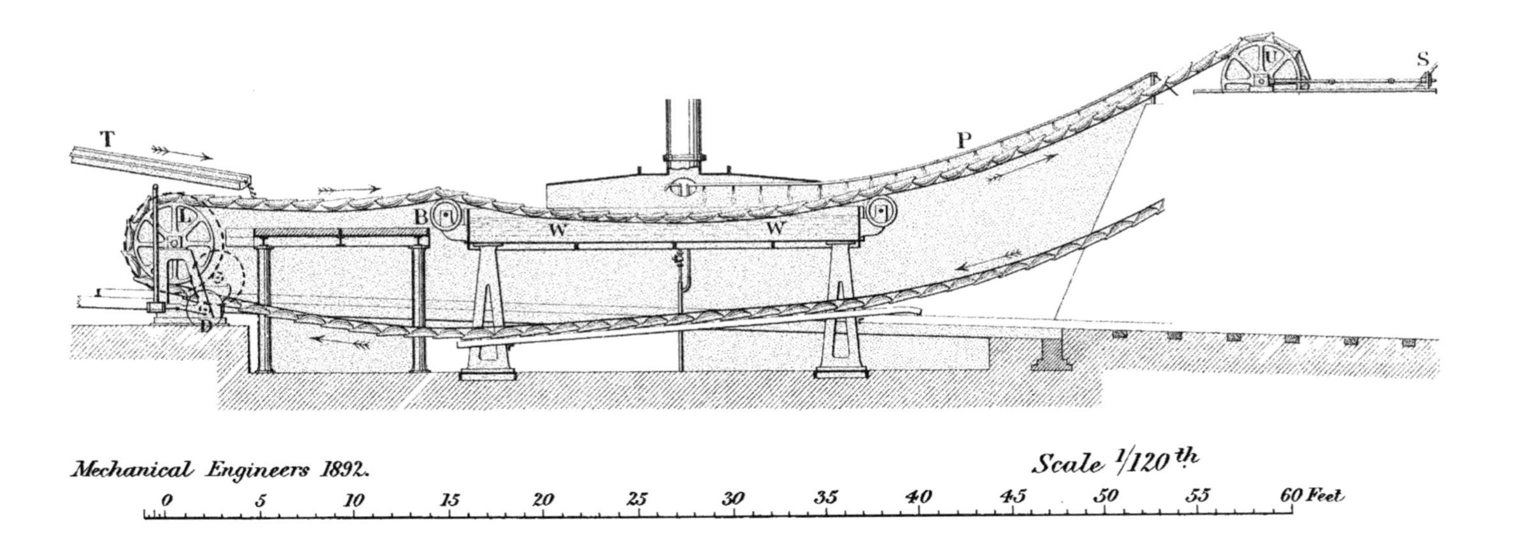

Mechanical Engineers 1892.

Pl

Fig: 2.

Endless - Chain Slag Machine.

ngitudinal Section of Water -

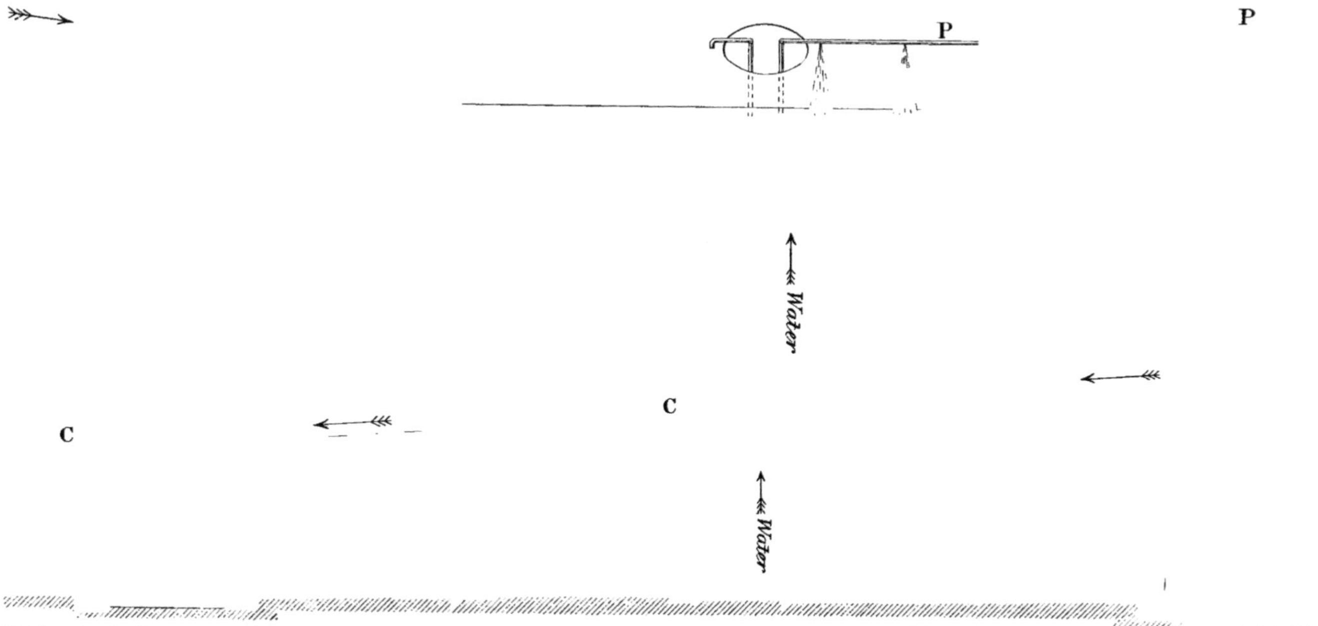

P

P

Water

C

C

Water

Mechanical Engineers 1892.

Scale 1/40th.

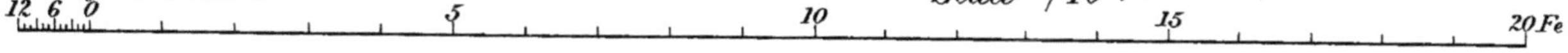

Inches 12 6 0 5 10 15 20 Fe

DISPOSAL OF SLAG. *Plate 6.*

Endless-Chain Slag Machine.

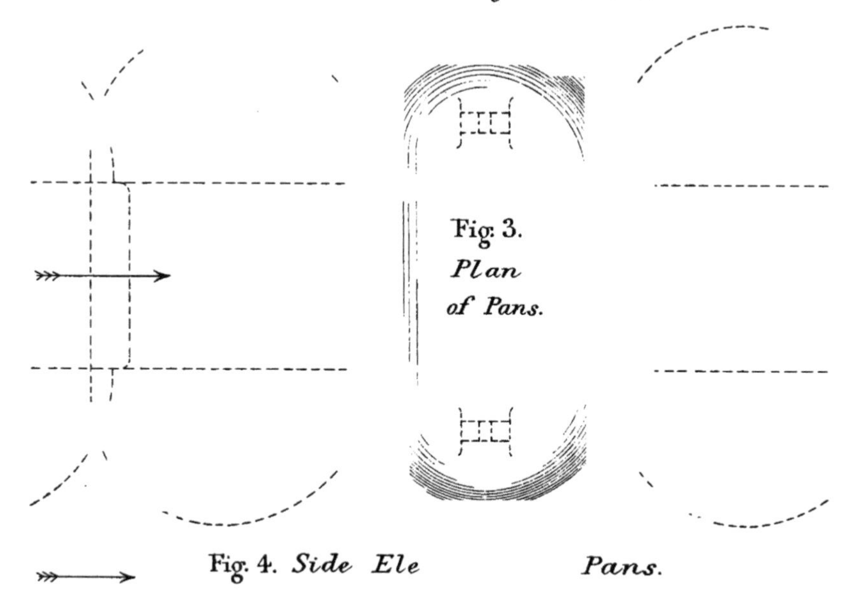

Fig. 4. *Side Ele* *Pans.*

Fig: 5.
Inverted Plan
of Pans.

DISPOSAL OF SLAG. *Plate 7.*

Endless-Chain Slag Machine.

Fig. 6. *Transverse Section of Pan.*

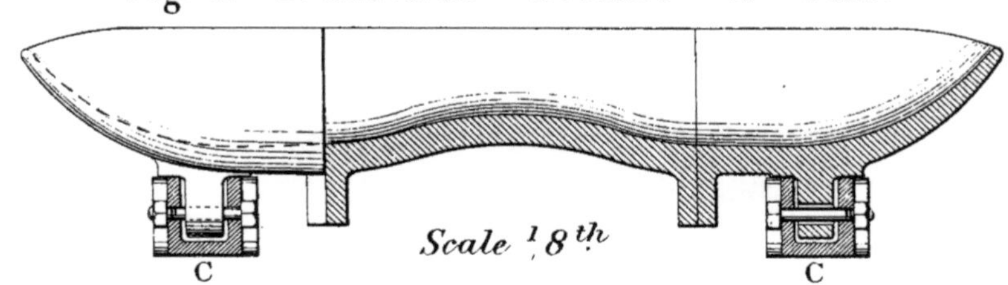

Fig. 7. *Longitudinal Section of Pan.*

Fig. 8. Fig. 9.

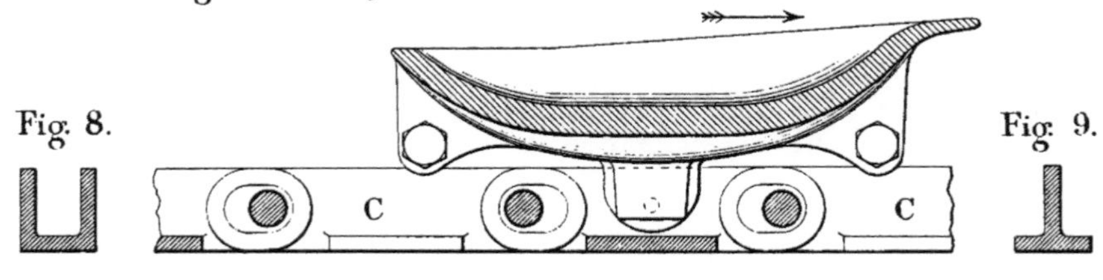

Hog-back Casting. Scale 1/16th

Fig. 10. *Transverse Sectional Elevation.*

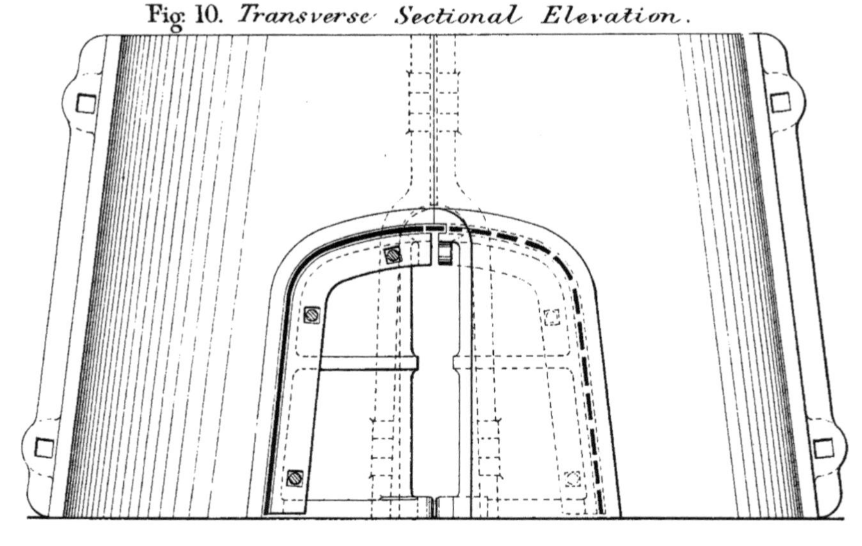

Scale 1/16th

Ins. 12 6 0 1 2 3 4 Feet

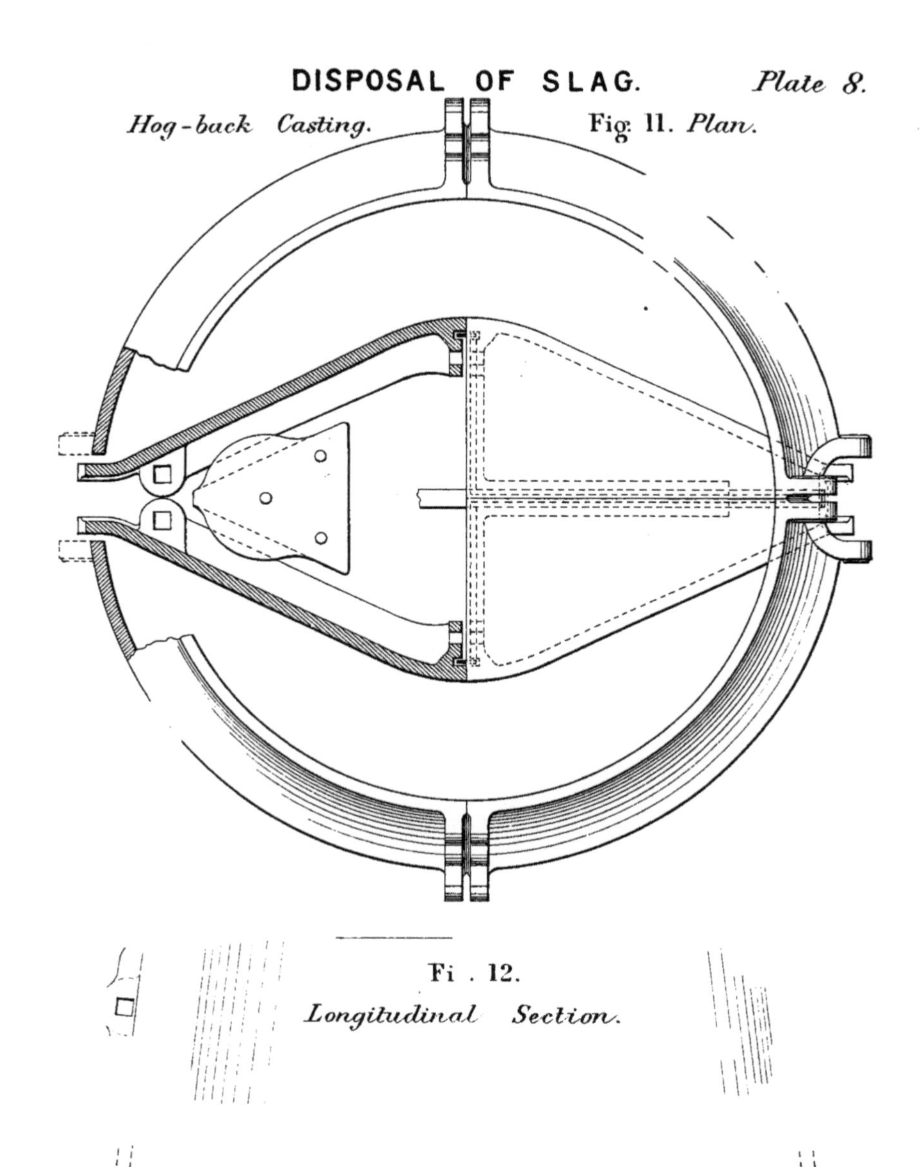
DISPOSAL OF SLAG.
Plate 8.
Hog-back Casting.
Fig. 11. Plan.
Fi . 12.
Longitudinal Section.

Institution of Mechanical Engineers.

Record of Progress from commencement.

2,200 2,200

2,000 2,000

1,500

1,000

500

Institution of Mechanical Engineers.

Record of Progress from commencement.

£	£
22,000	22,000
	20,000
15,000	15,000
	10,000
	5,000

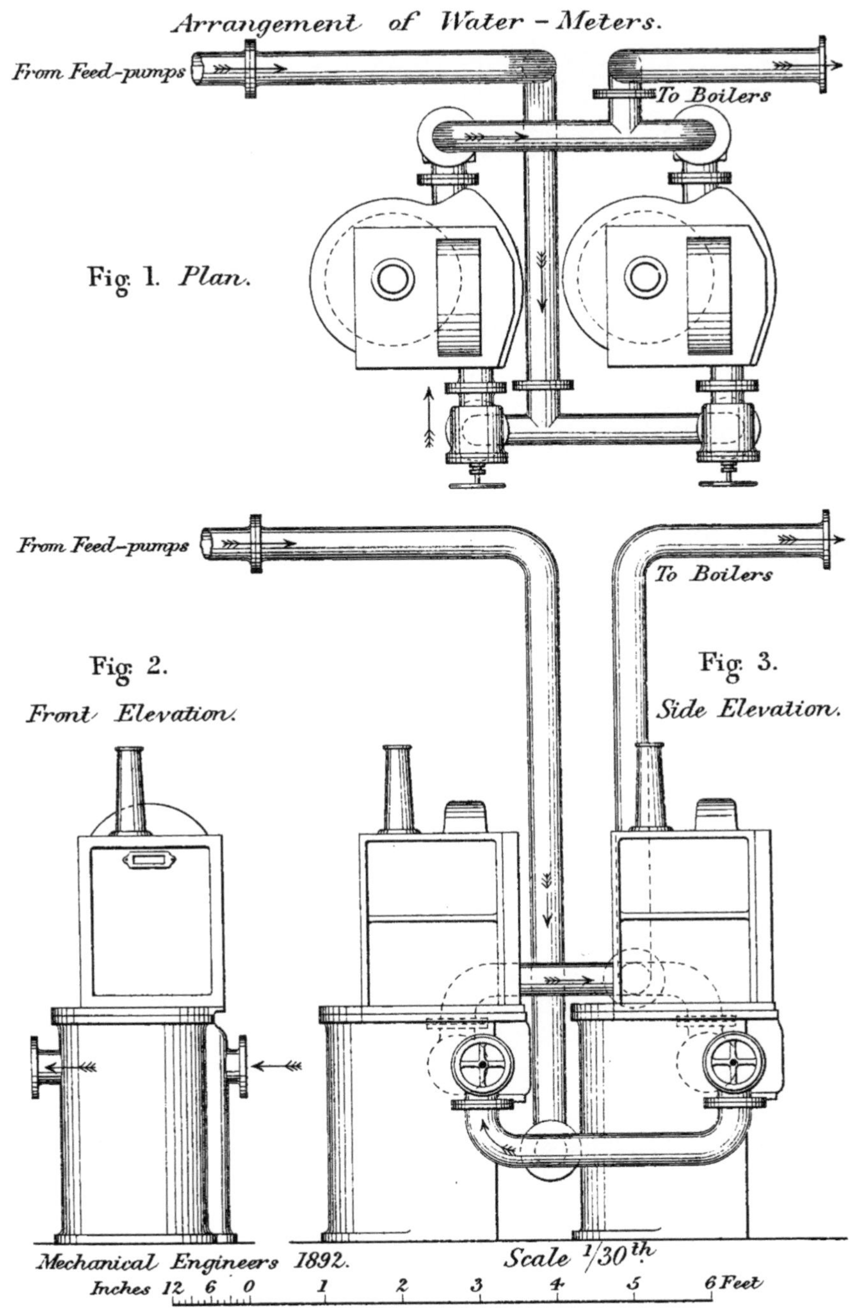
Trial of the "Ville de Douvres."
Arrangement of Water-Meters.
From Feed-pumps
To Boilers
Fig. 1. Plan.
From Feed-pumps
To Boilers
Fig. 2.
Front Elevation.
Fig. 3.
Side Elevation.
Mechanical Engineers 1892.
Scale 1/30th.
Inches 12 6 0 1 2 3 4 5 6 Feet

Trial of the "Ville de Douvres."

Fig. 4. *Total Fuel, Water, and Revolutions; and Temperature of Feed-Water.*

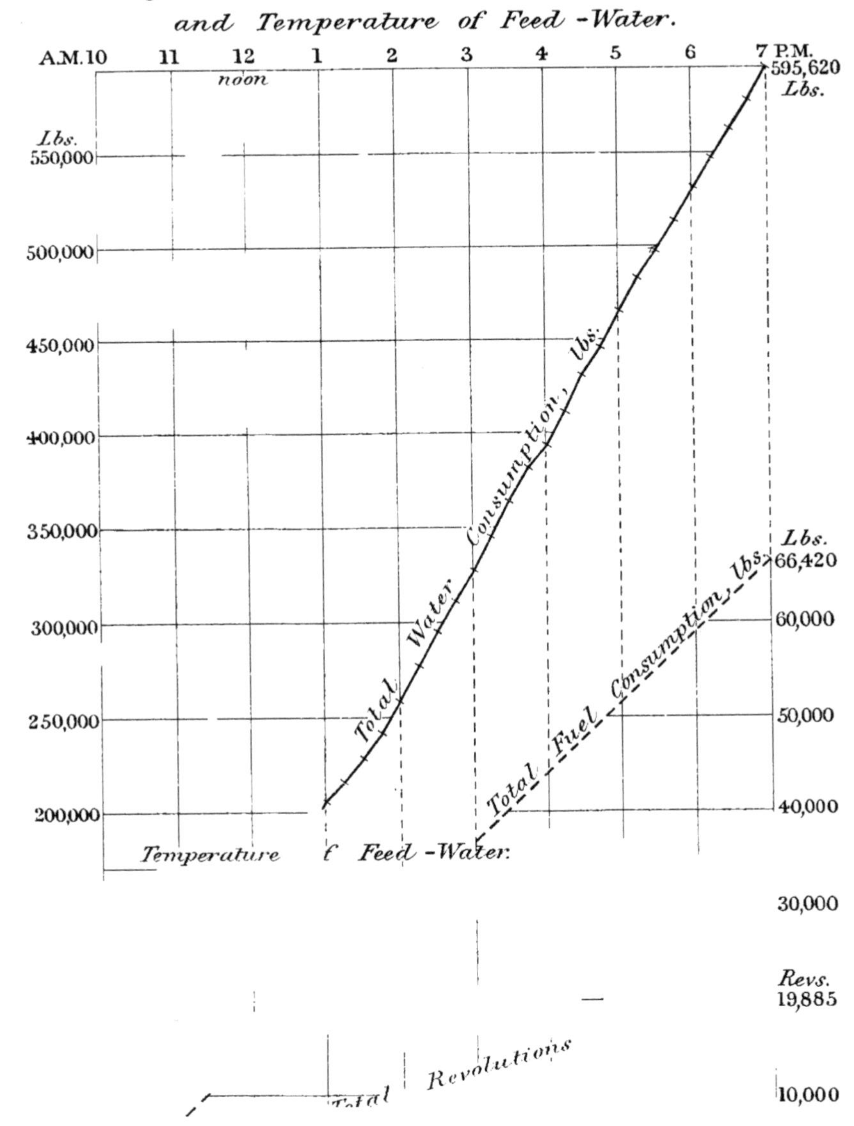

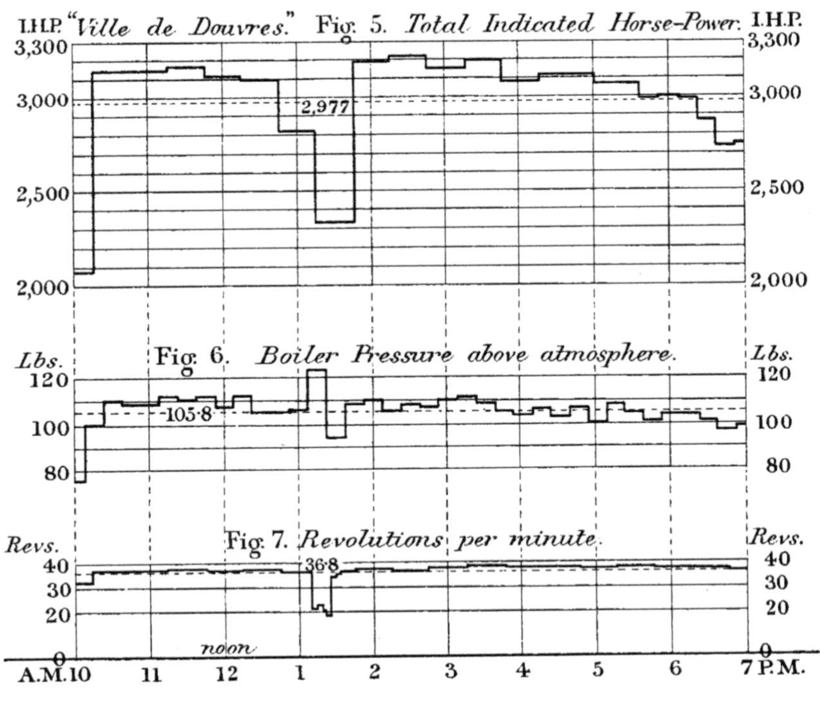
I.H.P. "Ville de Douvres." Fig. 5. Total Indicated Horse-Power. I.H.P.
3,300
3,000
2,977
2,500
2,000
Lbs. Fig. 6. Boiler Pressure above atmosphere. Lbs.
120
100
105·8
80
Revs. Fig. 7. Revolutions per minute. Revs.
40
30
36·8
20
0
noon
A.M.10 11 12 1 2 3 4 5 6 7 P.M.

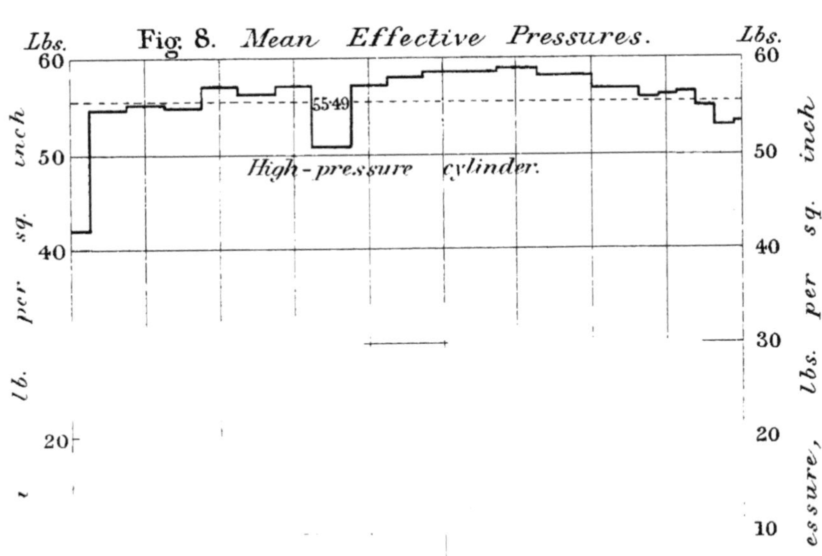
Lbs. Fig. 8. Mean Effective Pressures. Lbs.
60
55·49
50
High-pressure cylinder.
40
30
20
10
lb. per sq. inch
Pressure, lbs. per sq. inch

"*Ville de Douvres*" *Indicator Diagrams, Set 16.*

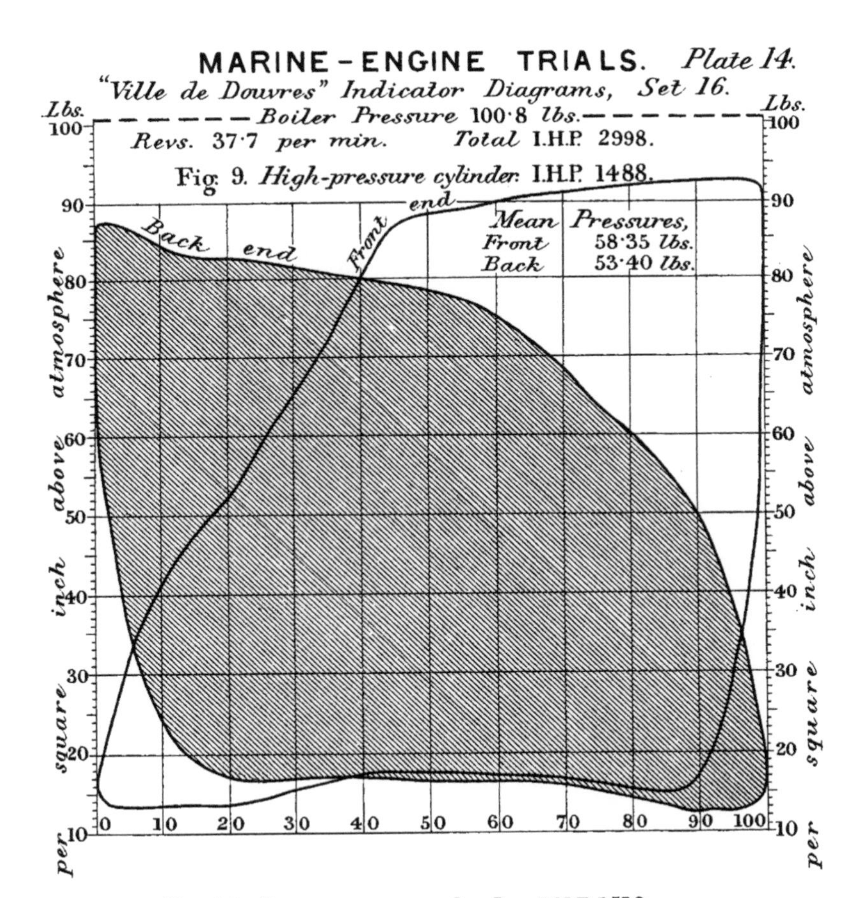

Fig. 10. *Low-pressure cylinder.* I.H.P. 1510.

Mean Pressures, Front 14·65 *lbs.*

Back 1

end

10

5

Pressure, lbs. per

Trial of the "Ville de Douvres."

Fig. 11. *Mean Indicator Diagrams.*

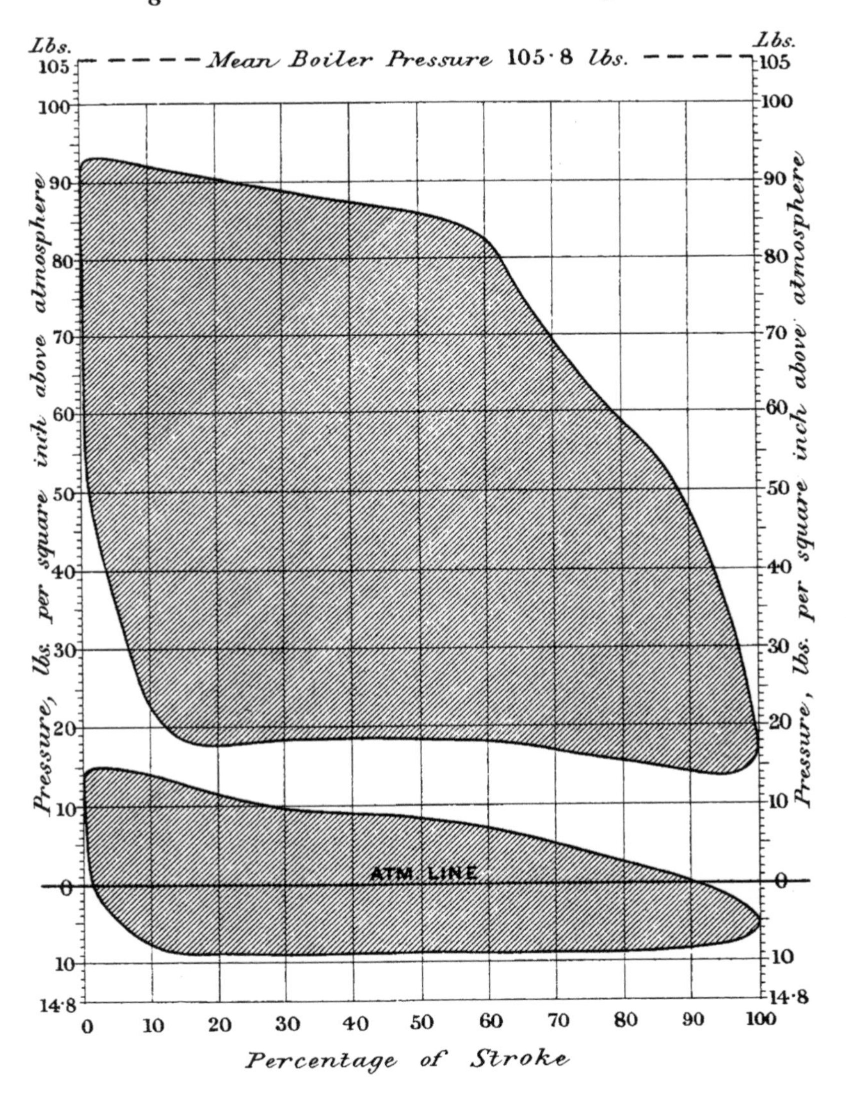

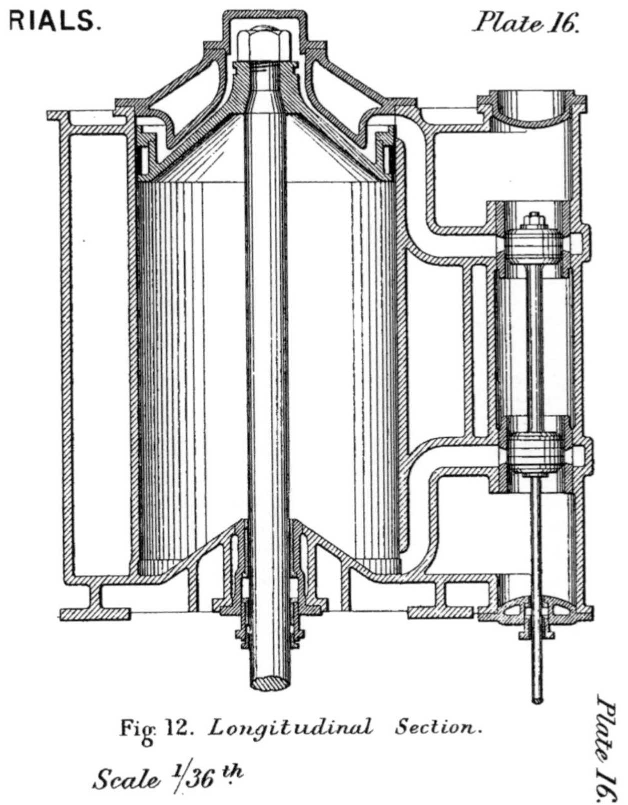

Fig. 12. *Longitudinal Section.*

Scale 1/36th.

"Ville de Douvres."

·s 1892.

Plate 16.

Cylinders of the "Ville de Douvres." Fig. 13. *Sectional Plan.*

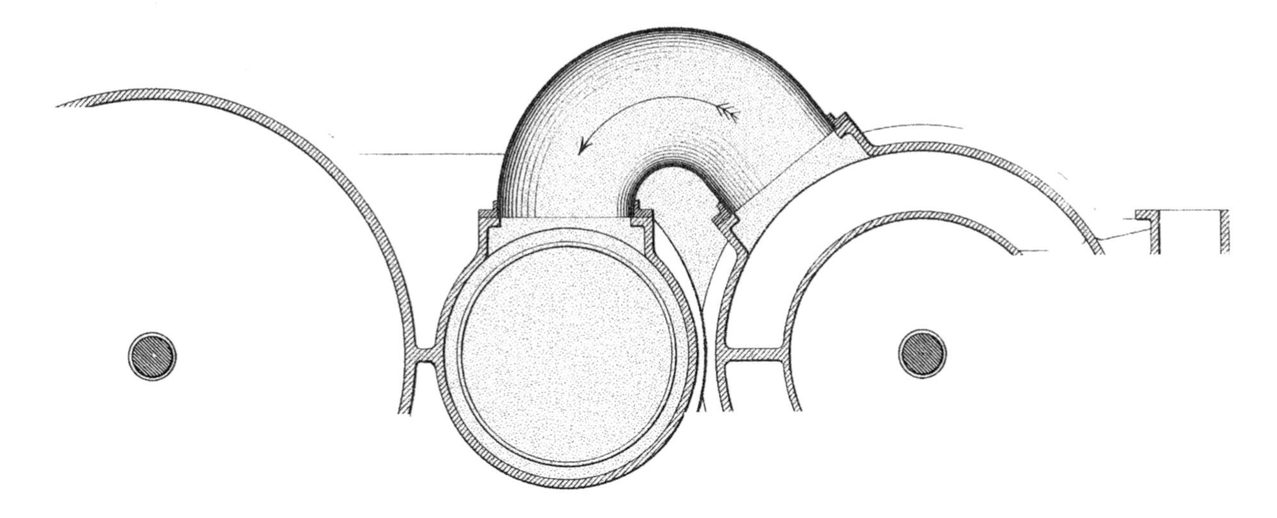

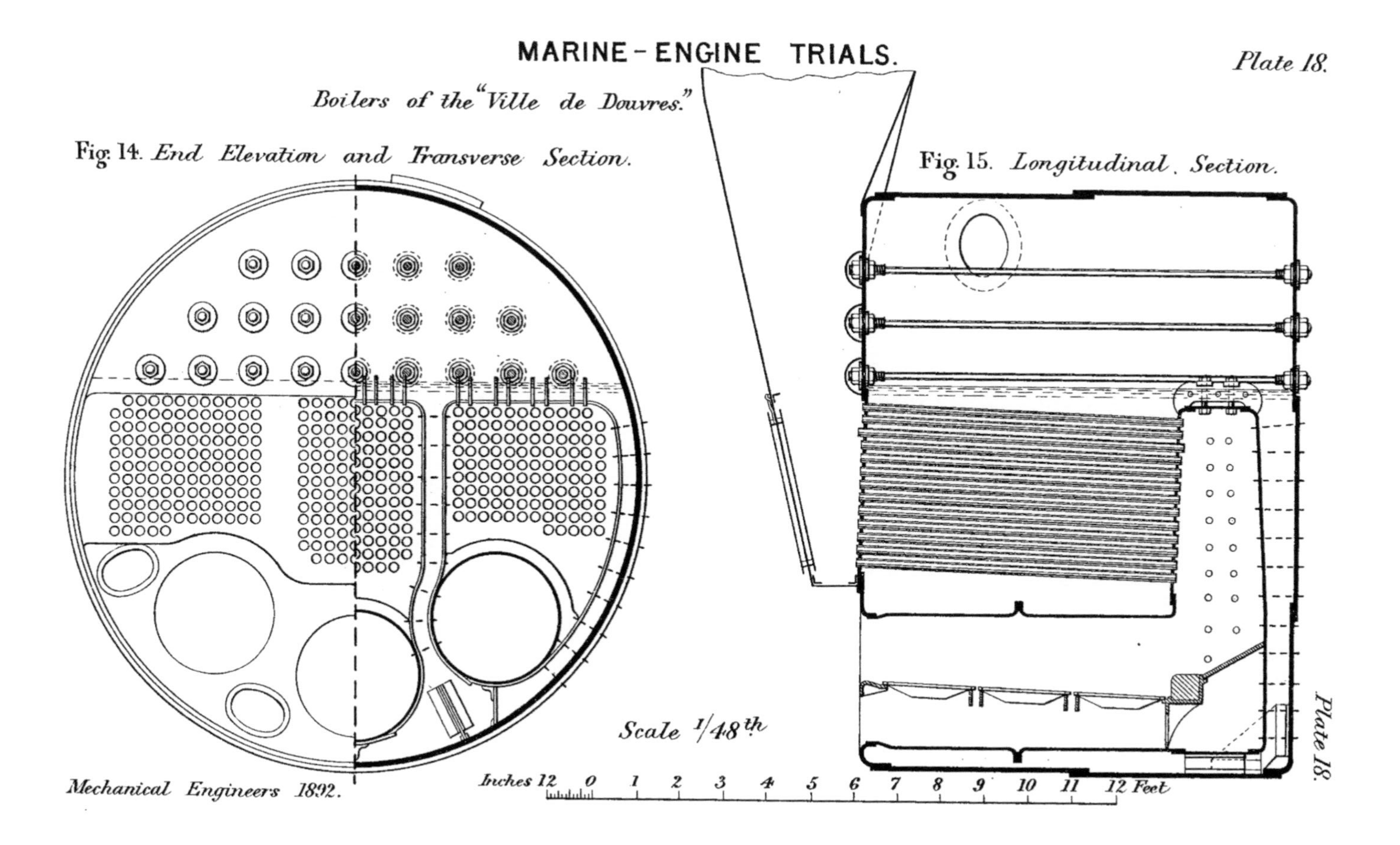
MARINE - ENGINE TRIALS.
Boilers of the "Ville de Douvres."
Fig. 14. End Elevation and Transverse Section.
Fig. 15. Longitudinal. Section.
Scale 1/48th
Inches 12 0 1 2 3 4 5 6 7 8 9 10 11 12 Feet
Plate 18.

MARINE-ENGINE TRIALS.

"Ville de Douvres".

Fig. 16. *General Arrangement of Engines and Boilers.*

Scale 1/120th

0 5 10 15 20 25 30 35 40 45 50 55 60 65 Feet

MARINE-ENGINE TRIALS.

"*Ville de Douvres.*"

Fig. 17. *Condenser and Air-pump.*

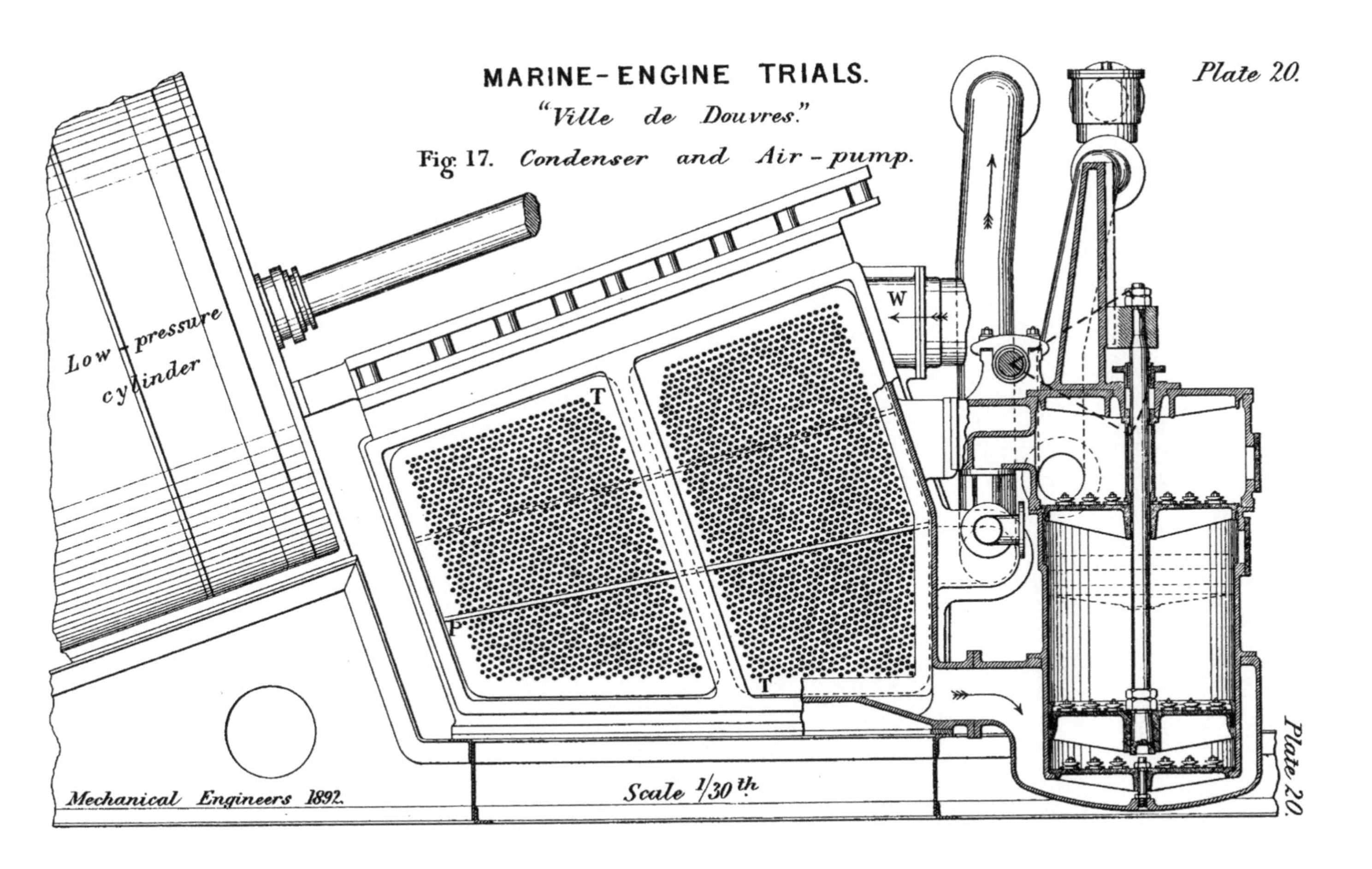

"*Ville* *Douvres*." Fig. 18. *Transverse Section of* *'enser.*

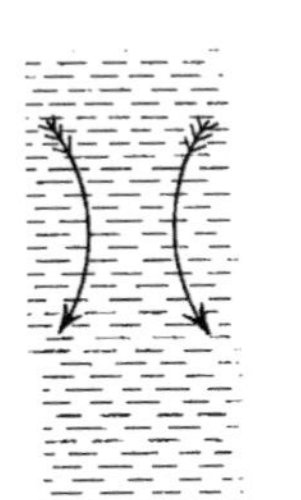

Mechanical Engineers 1892.

Scale 1/30th.

Inches 12 6 0 1 2 3 4 5 6 7 8 9 10 11 12 13 14 15 *Feet*

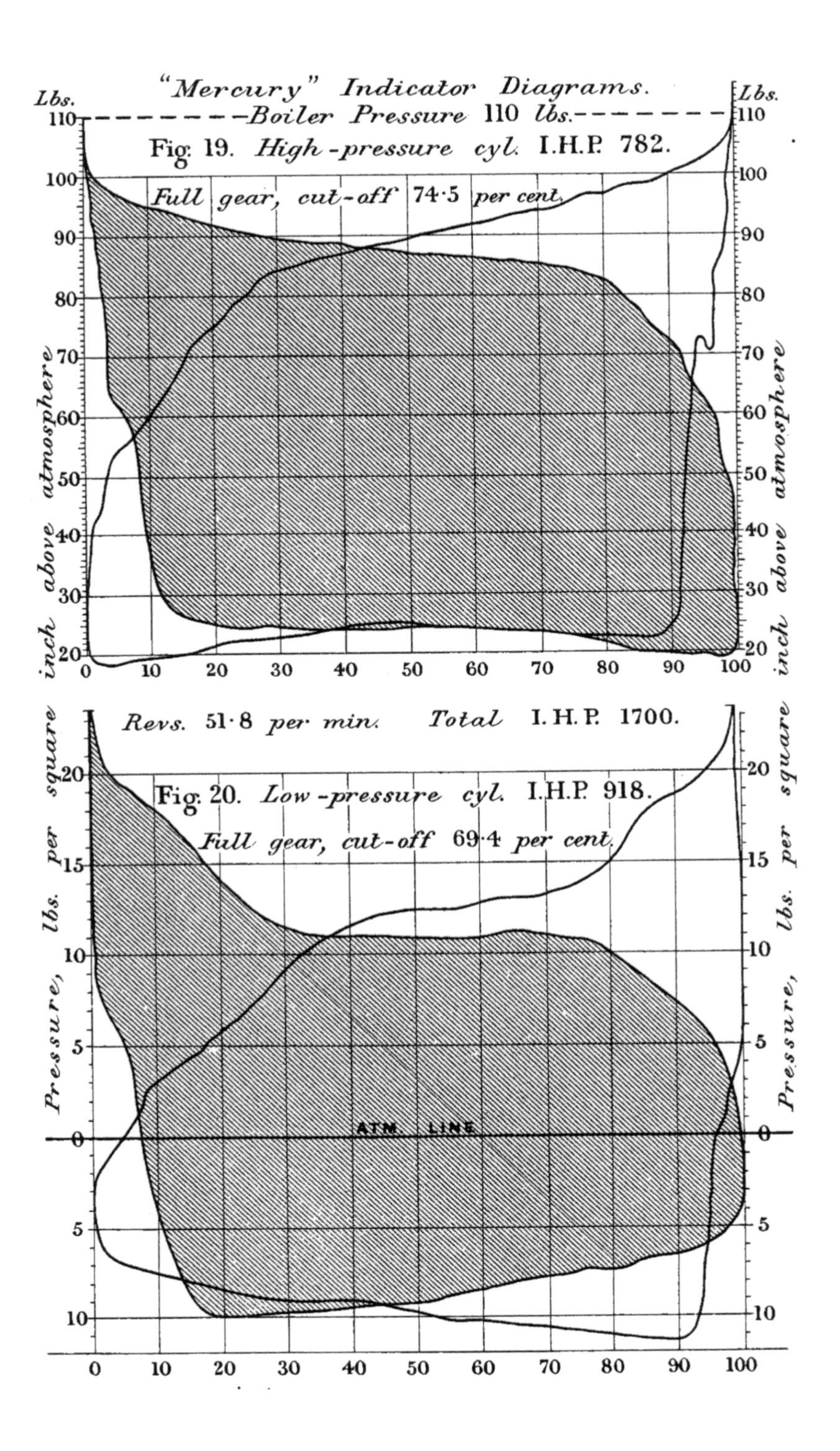

"Mercury" Indicator Diagrams.
Lbs.
Lbs.
Boiler Pressure 110 lbs.
Fig. 19. High-pressure cyl. I.H.P. 782.
Full gear, cut-off 74·5 per cent.
Pressure, lbs. per square inch above atmosphere
Revs. 51·8 per min. Total I. H. P. 1700.
Fig. 20. Low-pressure cyl. I.H.P. 918.
Full gear, cut-off 69·4 per cent.
ATM. LINE

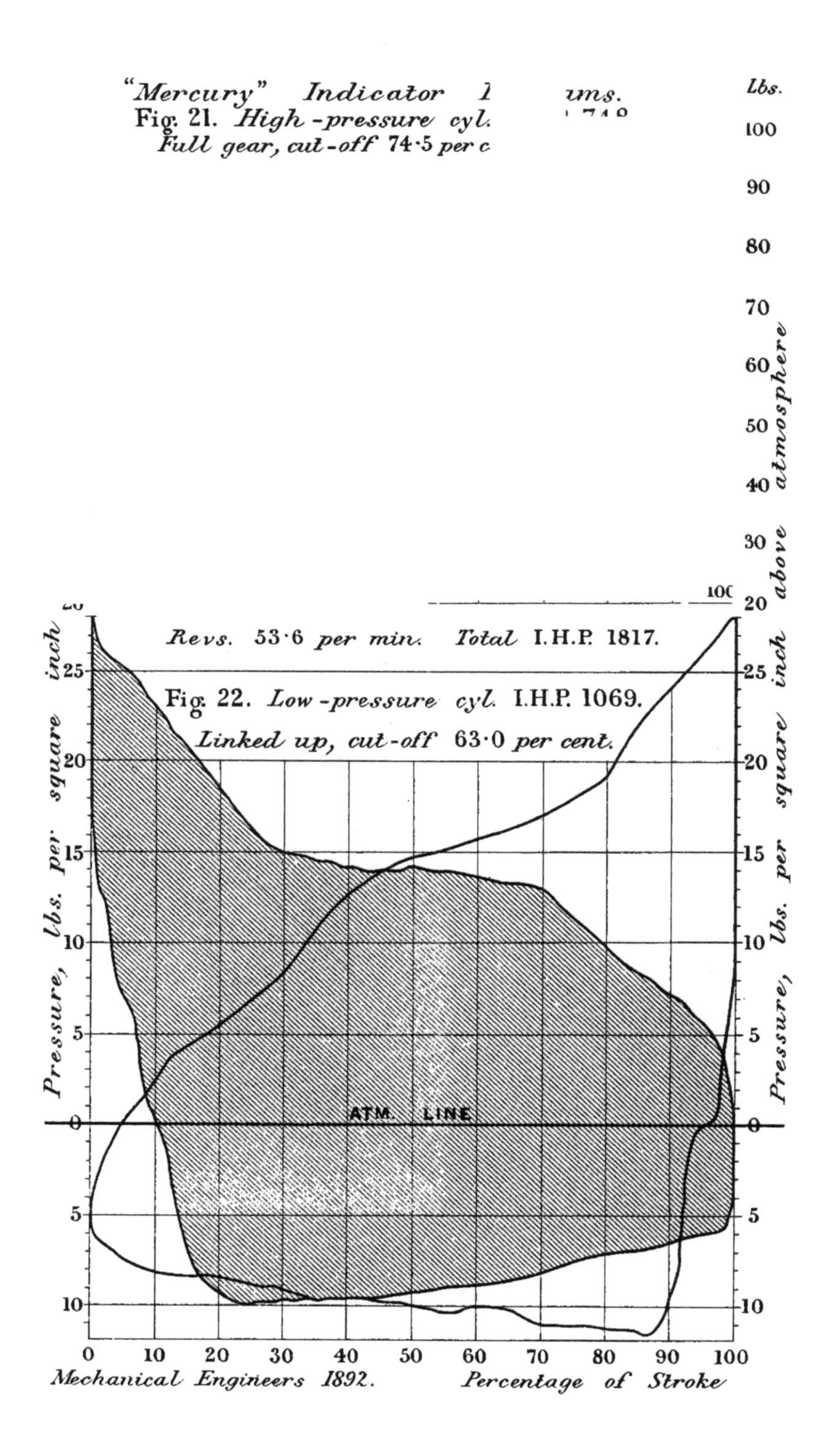
"Mercury" Indicator
Fig. 21. High-pressure cyl.
Full gear, cut-off 74·5 per c
Lbs.
100
90
80
70
60
50
40
30
20
atmosphere
above
Revs. 53·6 per min. Total I.H.P. 1817.
Fig. 22. Low-pressure cyl. I.H.P. 1069.
Linked up, cut-off 63·0 per cent.
ATM. LINE
Pressure, lbs. per square inch
25
20
15
10
5
0
5
10
0 10 20 30 40 50 60 70 80 90 100
Mechanical Engineers 1892.
Percentage of Stroke

MARINE - ENGINE TRIALS.

Fig. 23. *"Mercury"*

Expansion of Indicator Diagrams,
Plates 22 *and* 23.

V VV *volume of Steam*
converted into lbs. weight.

Dotted lines, low-pressure gear linked up.

Full lines, full gear.

Fig. 24.

Coal Consumption.

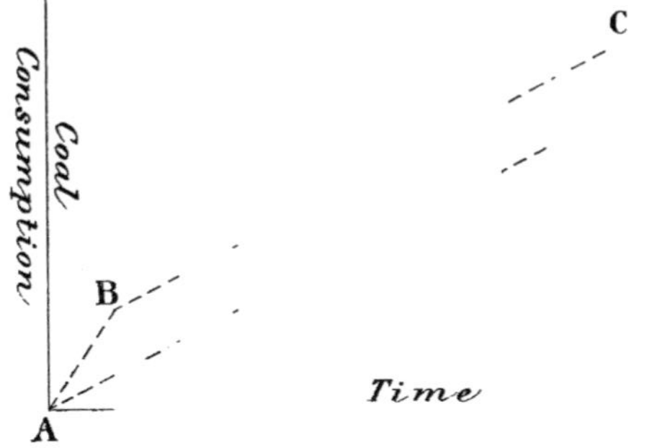

Fig. 1. *Observed and Calculated percentage of steam supply condensed at cut-off.*

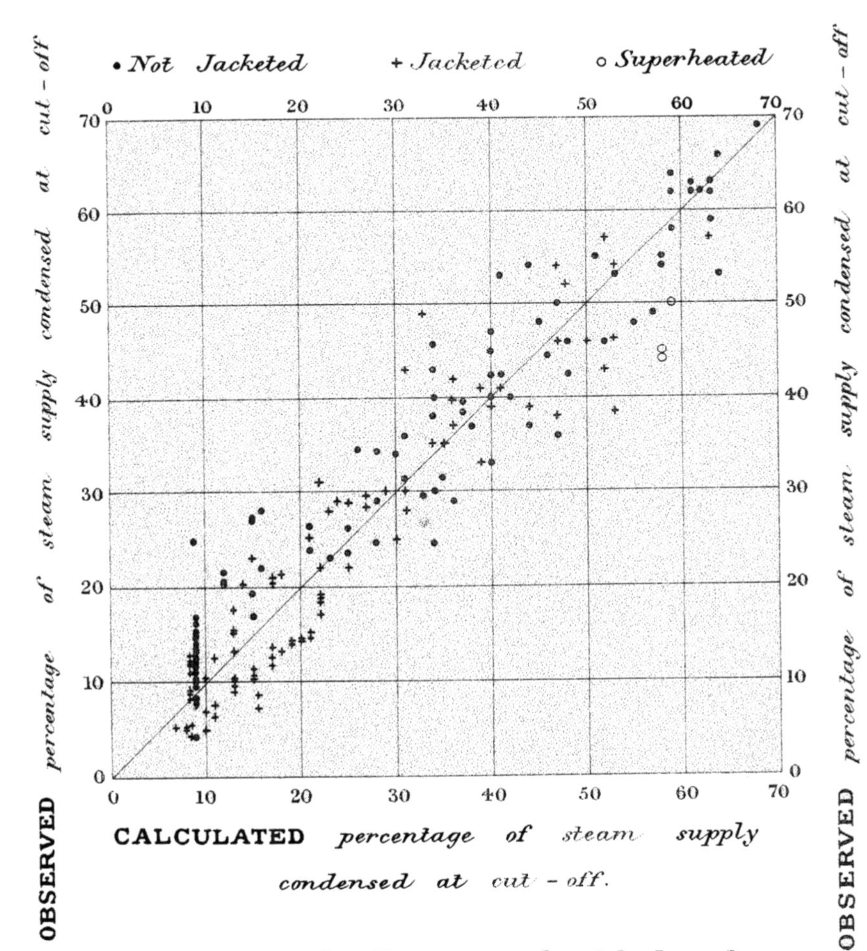

Fig. 2. *Proposed Experimental Cylinder Cover.*

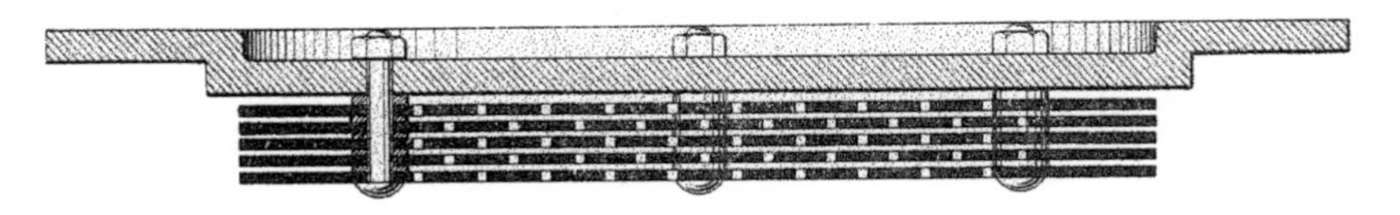

Fig. 1.

"Centurion."

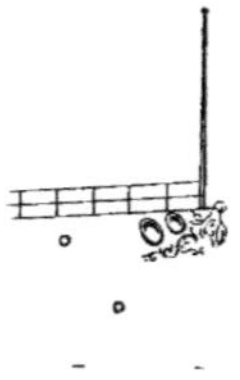

Scale 1/600th.

Feet 10 0 50 100 150 200 250 300 Feet

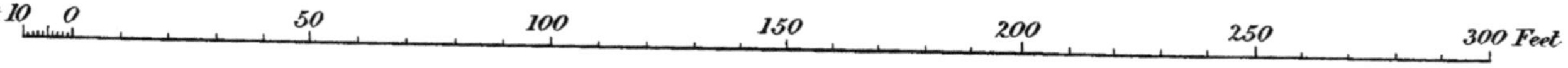

SHIPBUILDING IN PORTSMOUTH DOCKYARD.

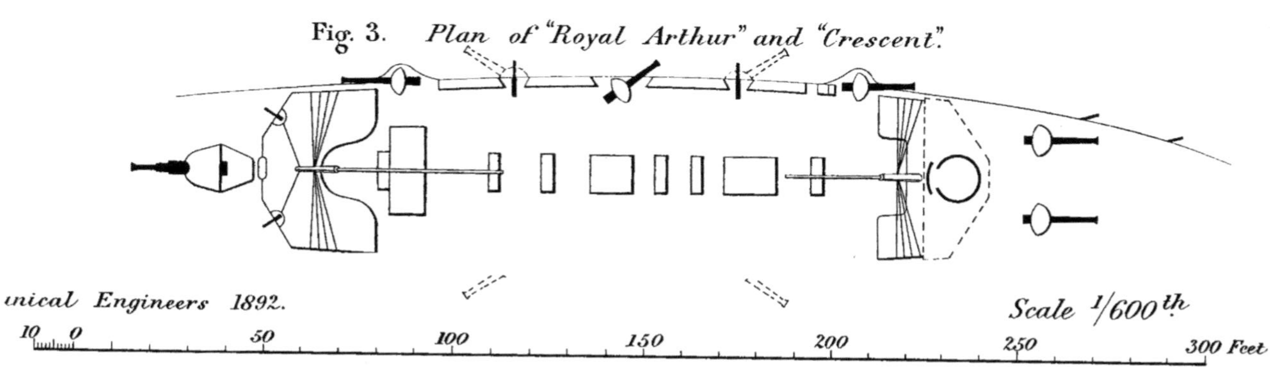

Fig. 2. *Plan of "Centurion".*

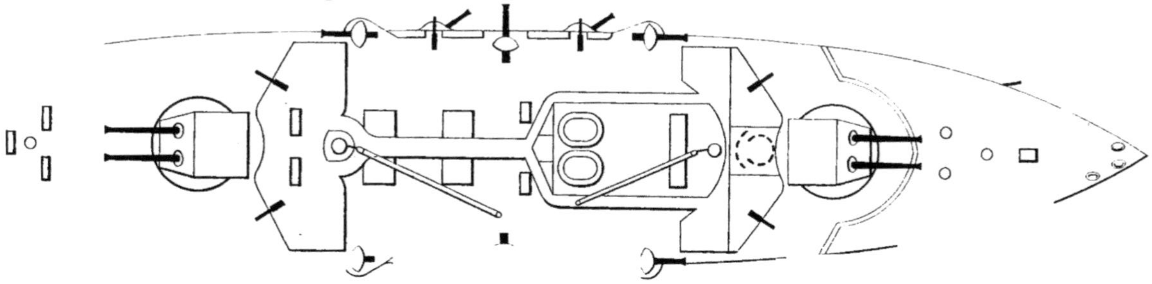

Fig. 3. *Plan of "Royal Arthur" and "Crescent".*

Plate 27.

ILDING IN PORTSMOUTH DOCKYARD.

Fig. 4.

"Royal Arthur" and "Crescent."

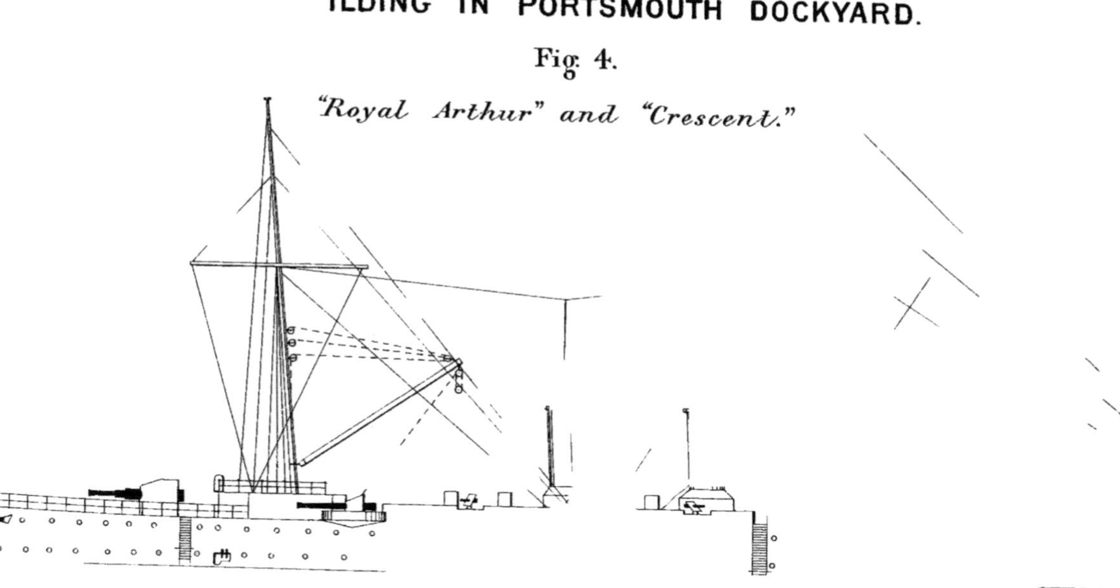

Plate 28.

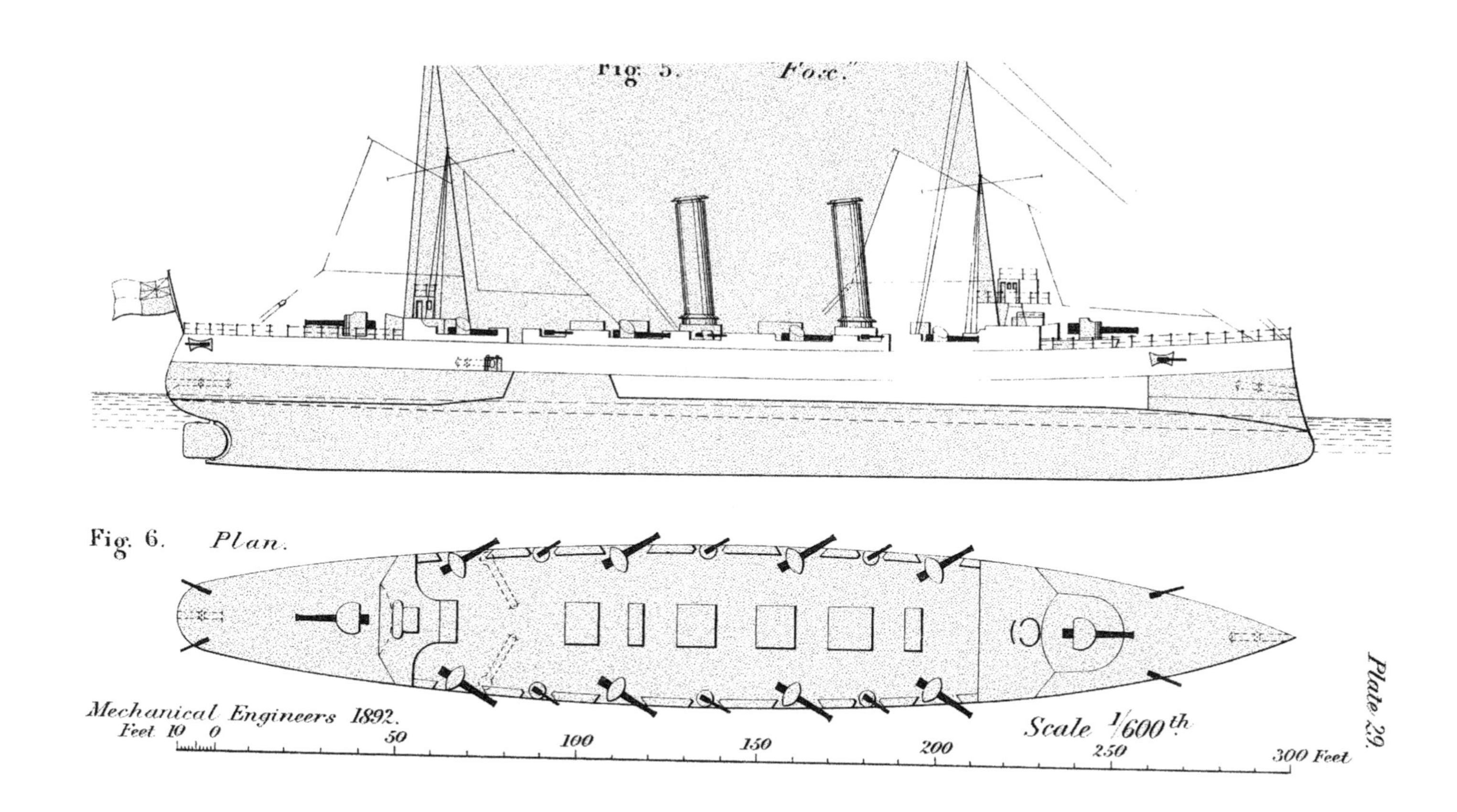
Plate 29.
Fig. 5. "Fox."
Fig. 6. Plan.
Mechanical Engineers 1892.
Scale 1/600th
Feet 10 0
50
100
150
200
250
300 Feet

Midship Sections.

Fig. 7. "Centurion."

Fig. 8. "Royal Arthur" and "Crescent."

Fig. 9. "Fox."

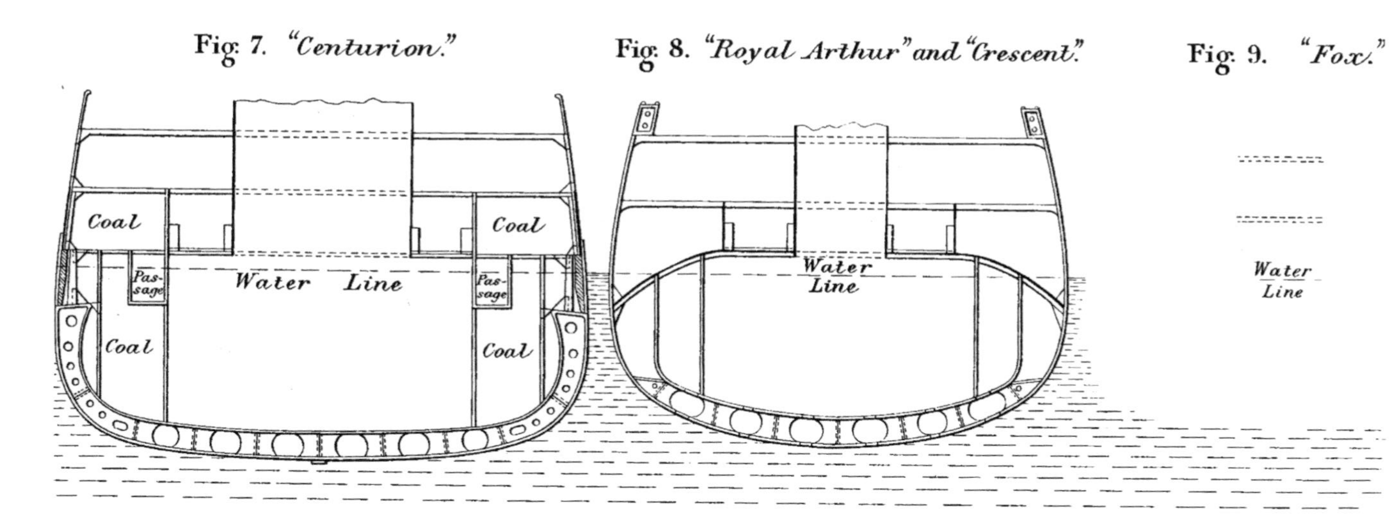

Mechanical Engineers 1892.

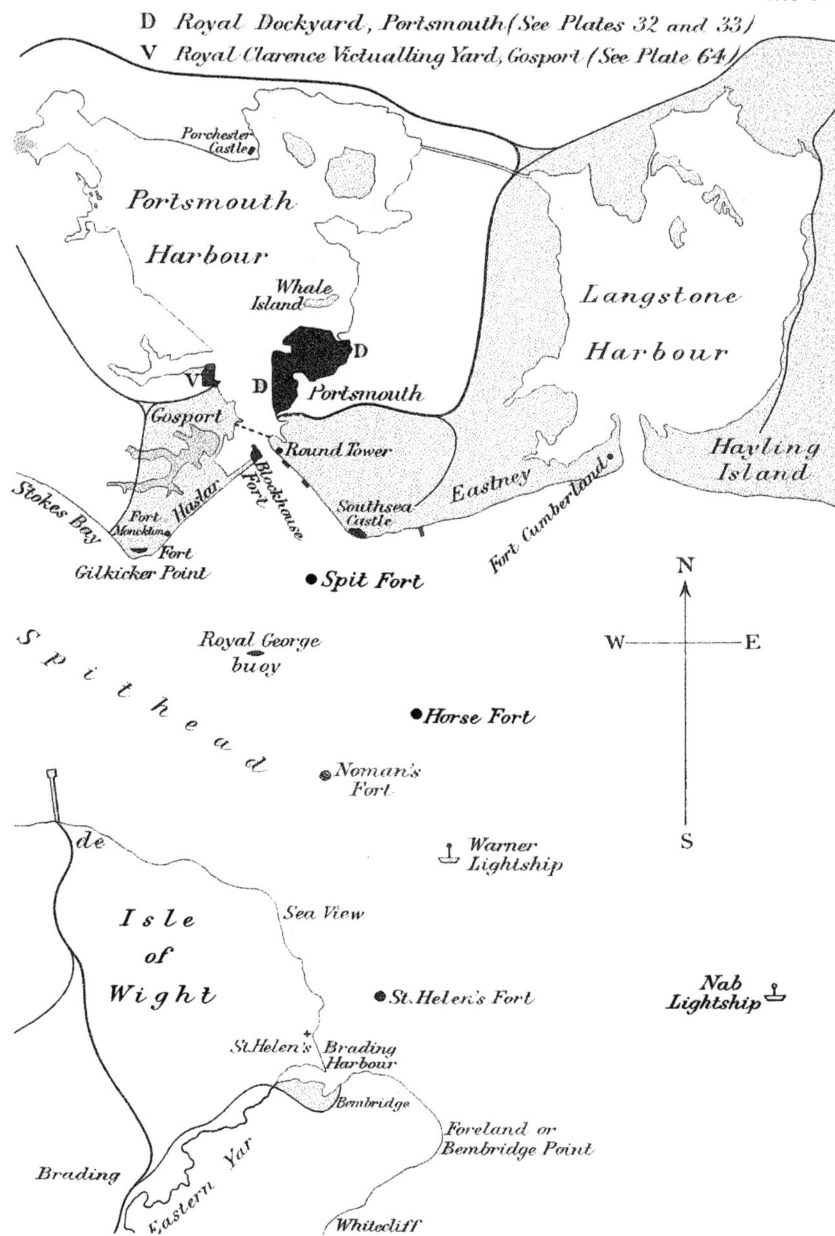
PORTSMOUTH.
Plate 31.
D Royal Dockyard, Portsmouth (See Plates 32 and 33)
V Royal Clarence Victualling Yard, Gosport (See Plate 64)
Porchester Castle
Portsmouth Harbour
Whale Island
D
V
D
Portsmouth
Gosport
Round Tower
Blockhouse Fort
Haslar
Stokes Bay
Fort Monckton
Fort
Gilkicker Point
Southsea Castle
Eastney
Fort Cumberland
Langstone Harbour
Hayling Island
Spit Fort
N
W
E
S
Spithead
Royal George buoy
Horse Fort
Noman's Fort
Warner Lightship
de
Isle of Wight
Sea View
St. Helen's Fort
Nab Lightship
St. Helen's
Brading Harbour
Bembridge
Foreland or Bembridge Point
Brading
Eastern Yar
Whitecliff

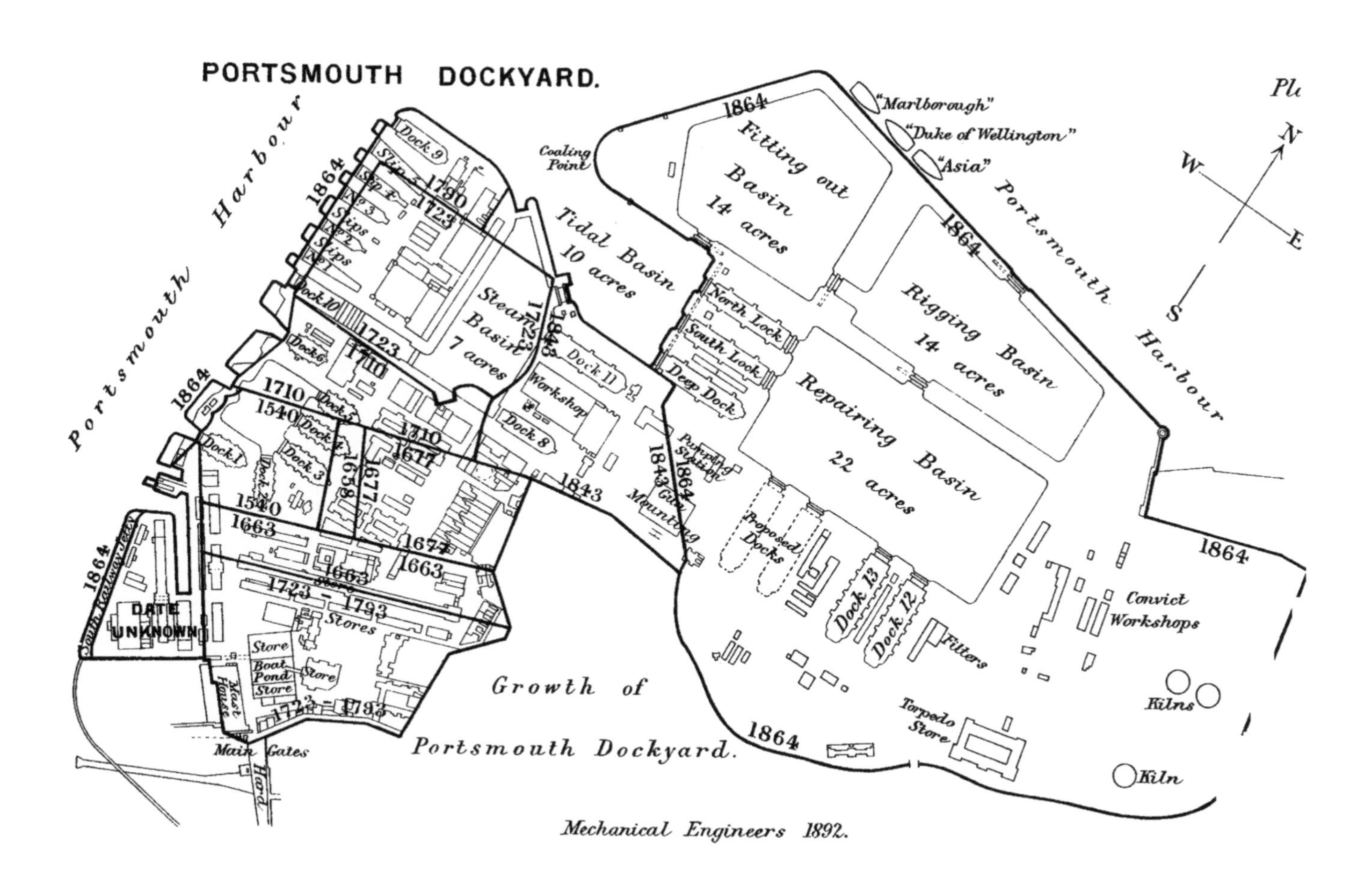
PORTSMOUTH DOCKYARD.
Portsmouth Harbour
"Marlborough"
"Duke of Wellington"
"Asia"
N
E
S
W
Coaling Point
Fitting out Basin 14 acres
Rigging Basin 14 acres
Repairing Basin 22 acres
Tidal Basin 10 acres
Steam Basin 7 acres
North Lock
South Lock
Deep Dock
Pumping Station
Gun Mounting
Proposed Docks
Dock 13
Dock 12
Filters
Convict Workshops
Kilns
Kiln
Torpedo Store
Dock 11
Workshop
Dock 8
Dock 9
Slip 5
Slip 4
No. 3 Slips
No. 2 Slips
No. 1
Dock 10
Dock 6
Dock 5
Dock 4
Dock 3
Dock 2
Dock 1
Stores
Store
Boat Pond
Mast House
Main Gates
Hard
South Railway Jetty
DATE UNKNOWN
1864
1843
1790
1723
1710
1677
1663
1658
1540
1793
Growth of Portsmouth Dockyard.
Mechanical Engineers 1892.

PORTSMOUTH DOCKYARD.

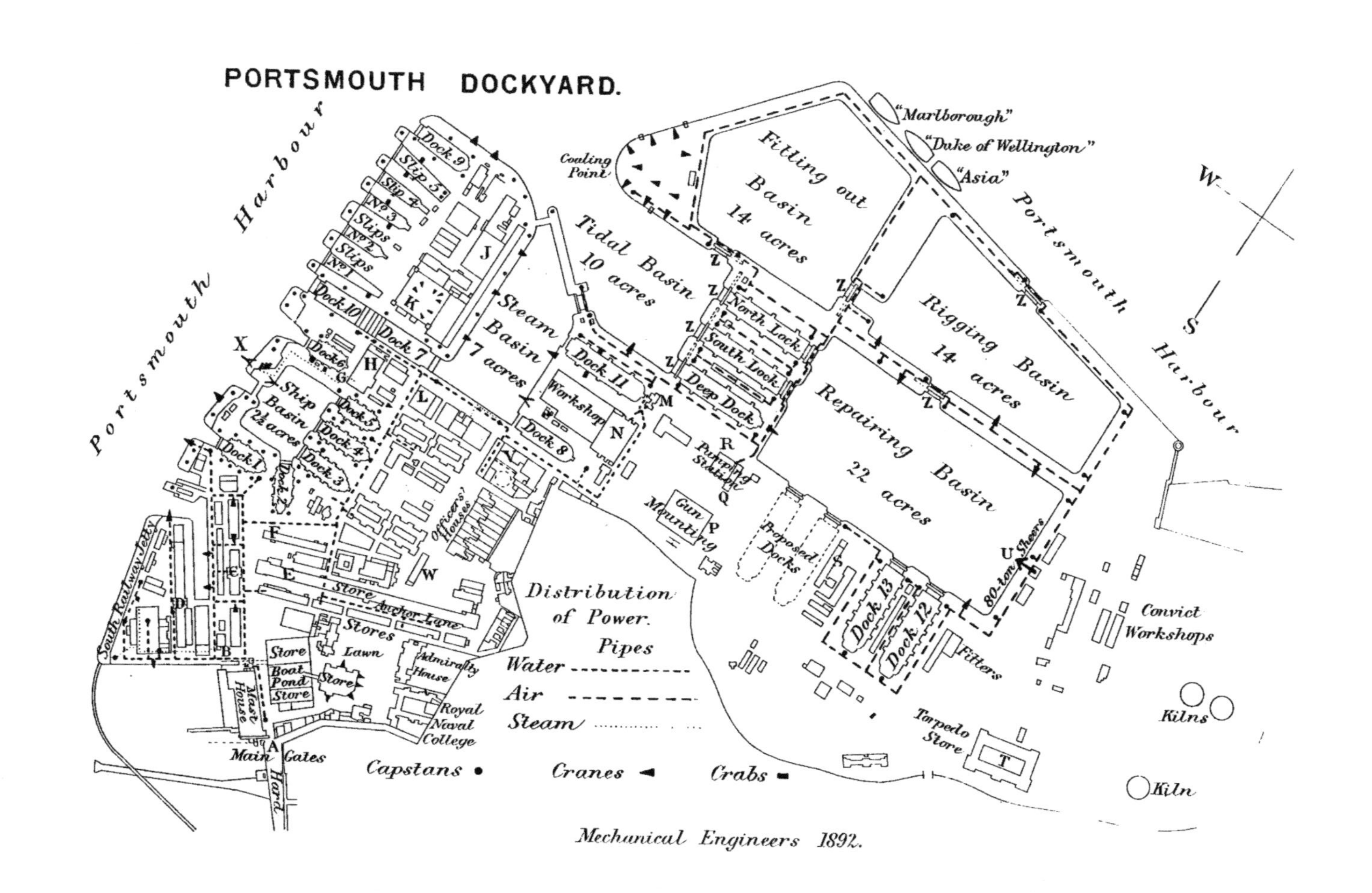

DOCKYARD

Caisson. Fig. 1. *Plan.*

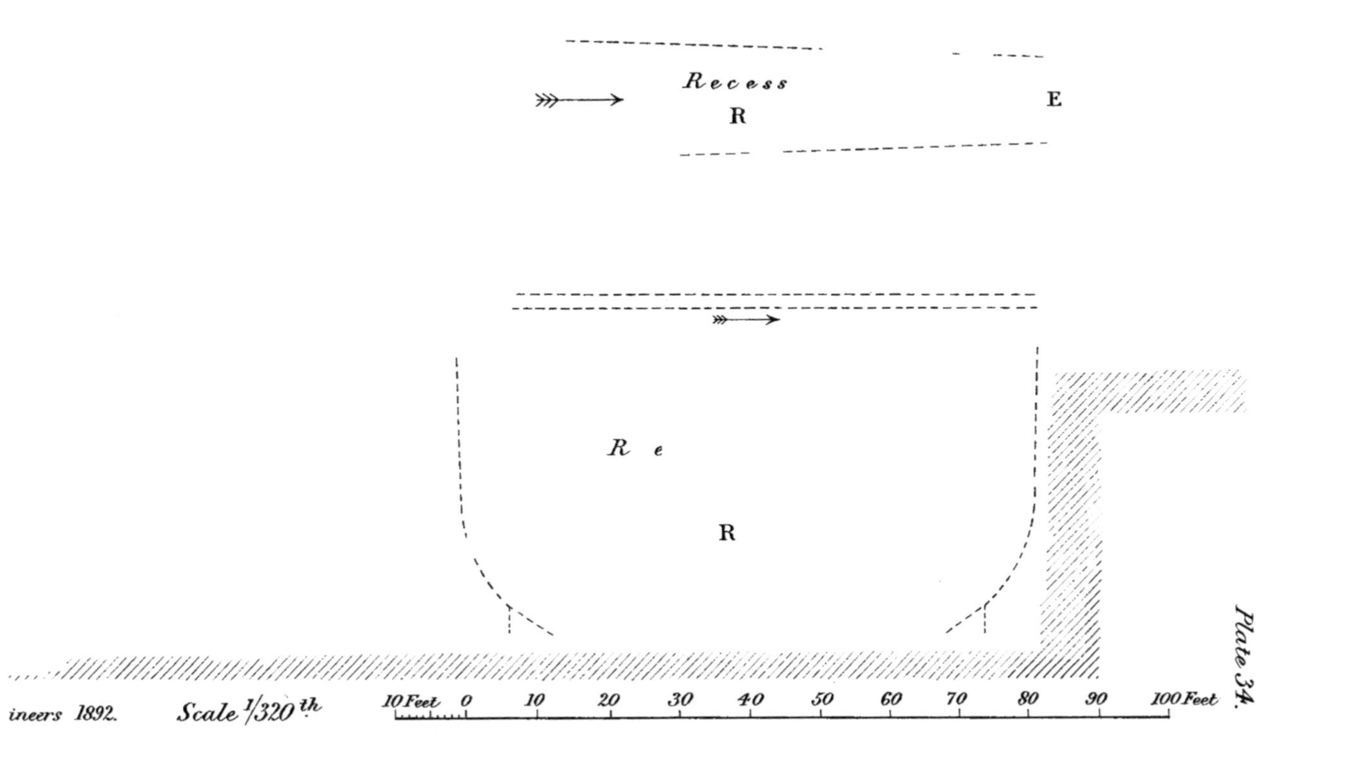

ineers 1892. *Scale 1/320th.*

DOCKYARD LIFTING A

Fig. 3. *Plan of upper Platform*

Fig. 4. *Longitudinal Section of upper half of Sliding Caisson.*

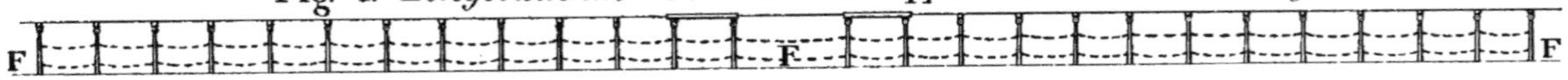

F

F

F

Scale. 1/16

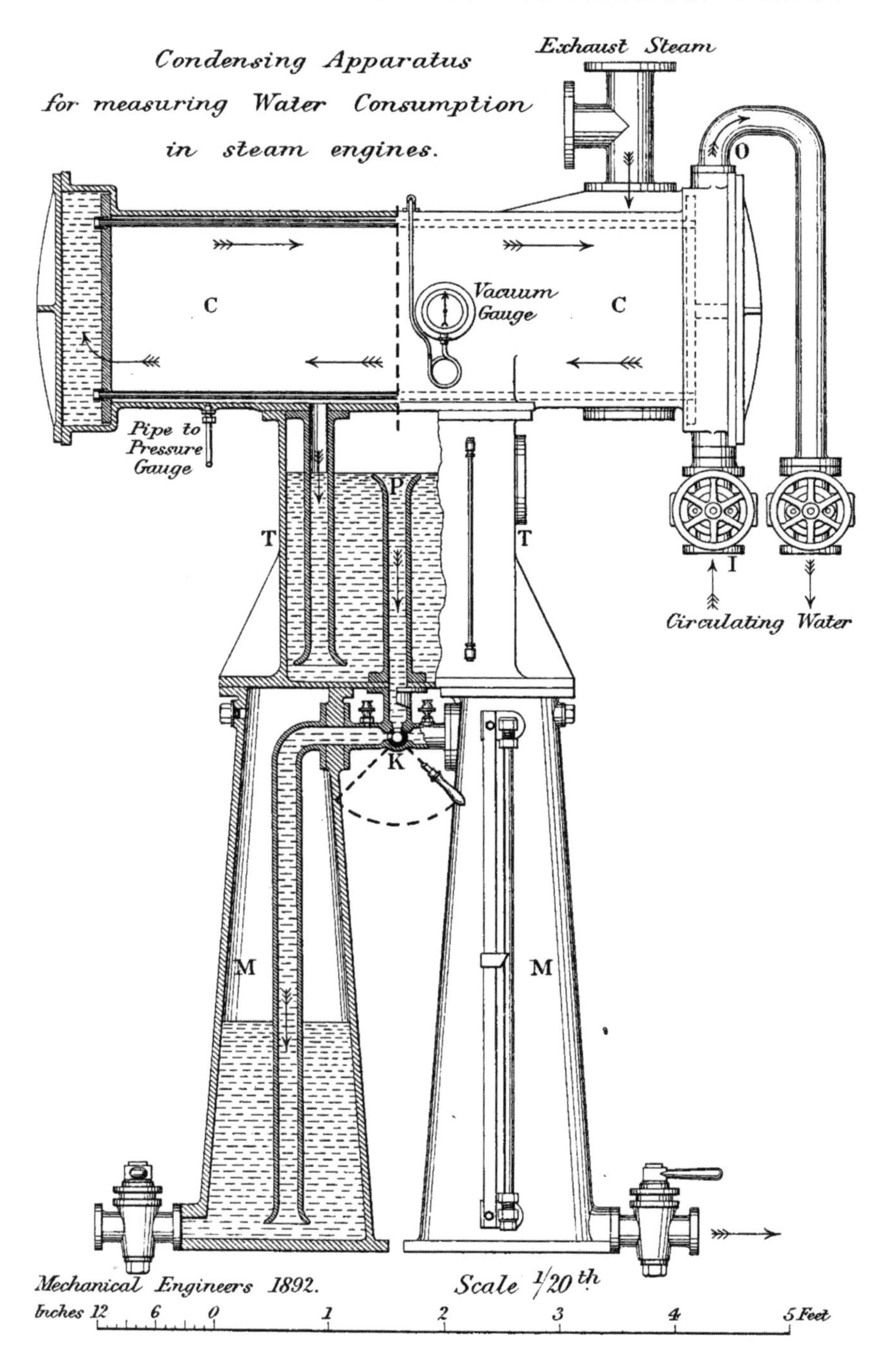
Condensing Apparatus
for measuring Water Consumption
in steam engines.
Exhaust Steam
O
C
Vacuum Gauge
C
Pipe to Pressure Gauge
P
T
T
I
Circulating Water
K
M
M
Mechanical Engineers 1892.
Scale 1/20th
Inches 12 6 0 1 2 3 4 5 Feet

AMPTON PIER.

Fig: 1. *Plan.*

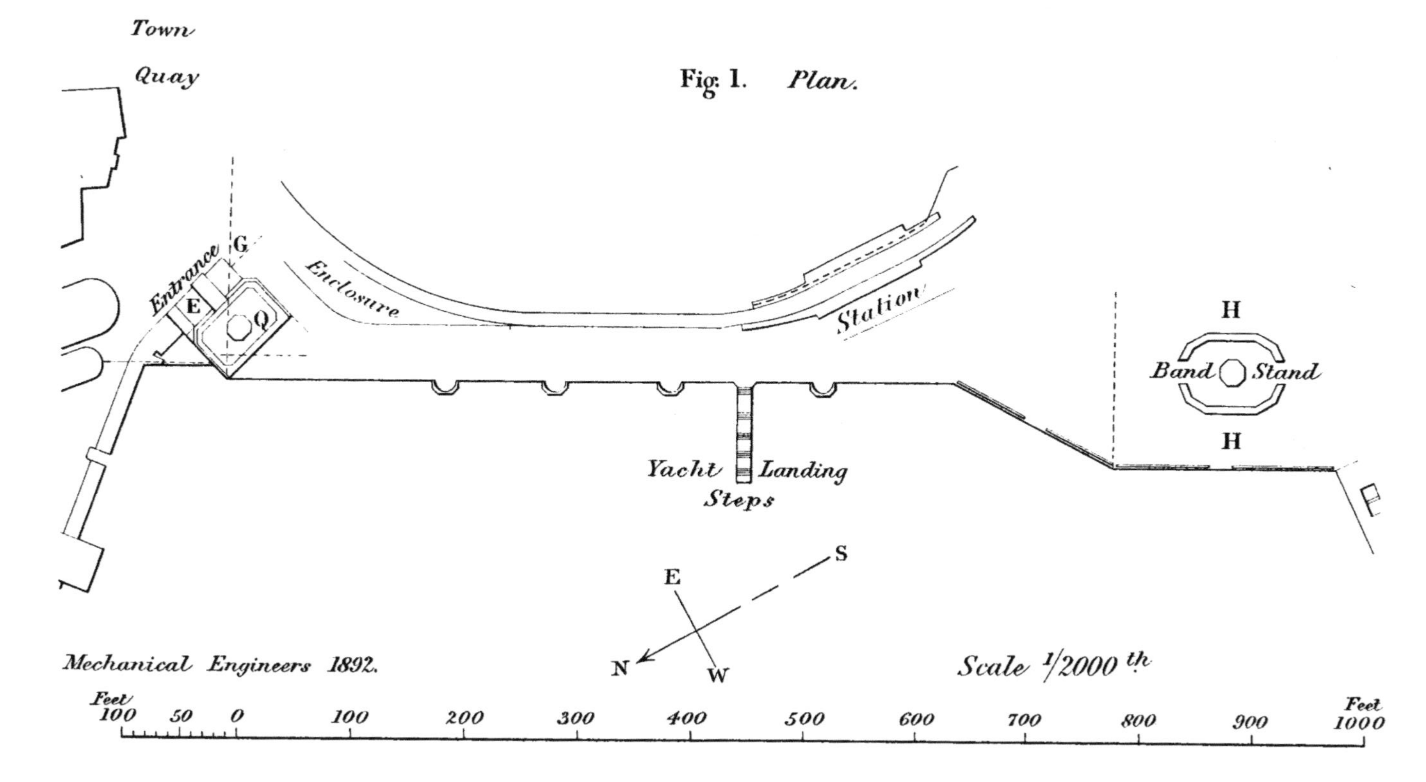

Mechanical Engineers 1892.

SOUTHAMPTON PIER.

Fig. 2. *Plan of either Arm.*

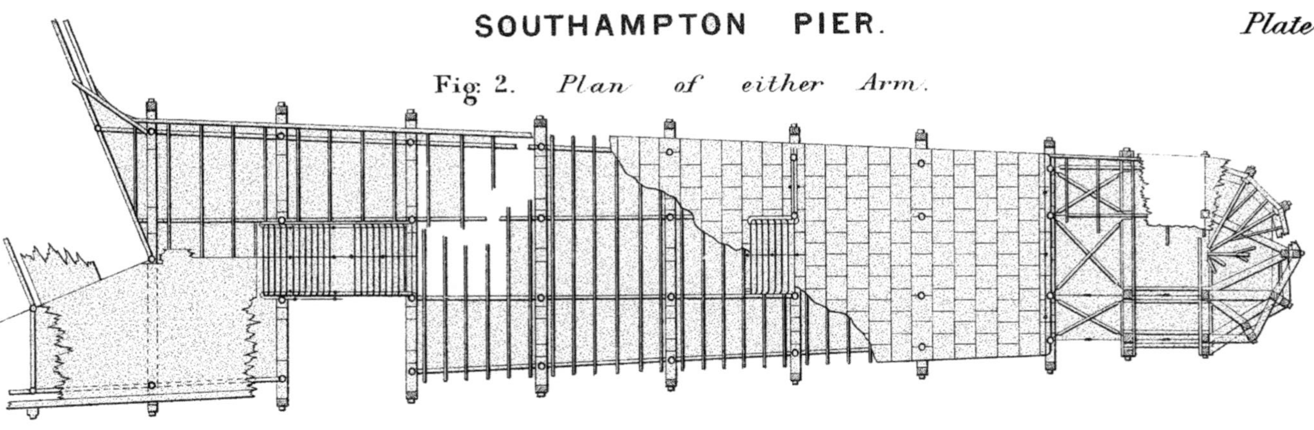

Fig. 3. *Longitudinal Section of either Arm.*

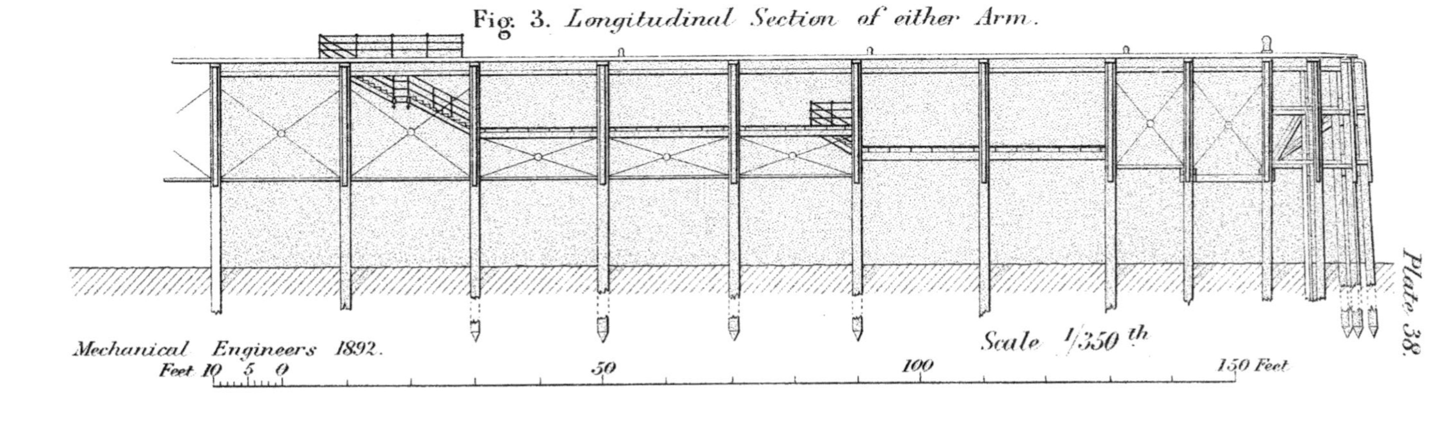

Mechanical Engineers 1892.

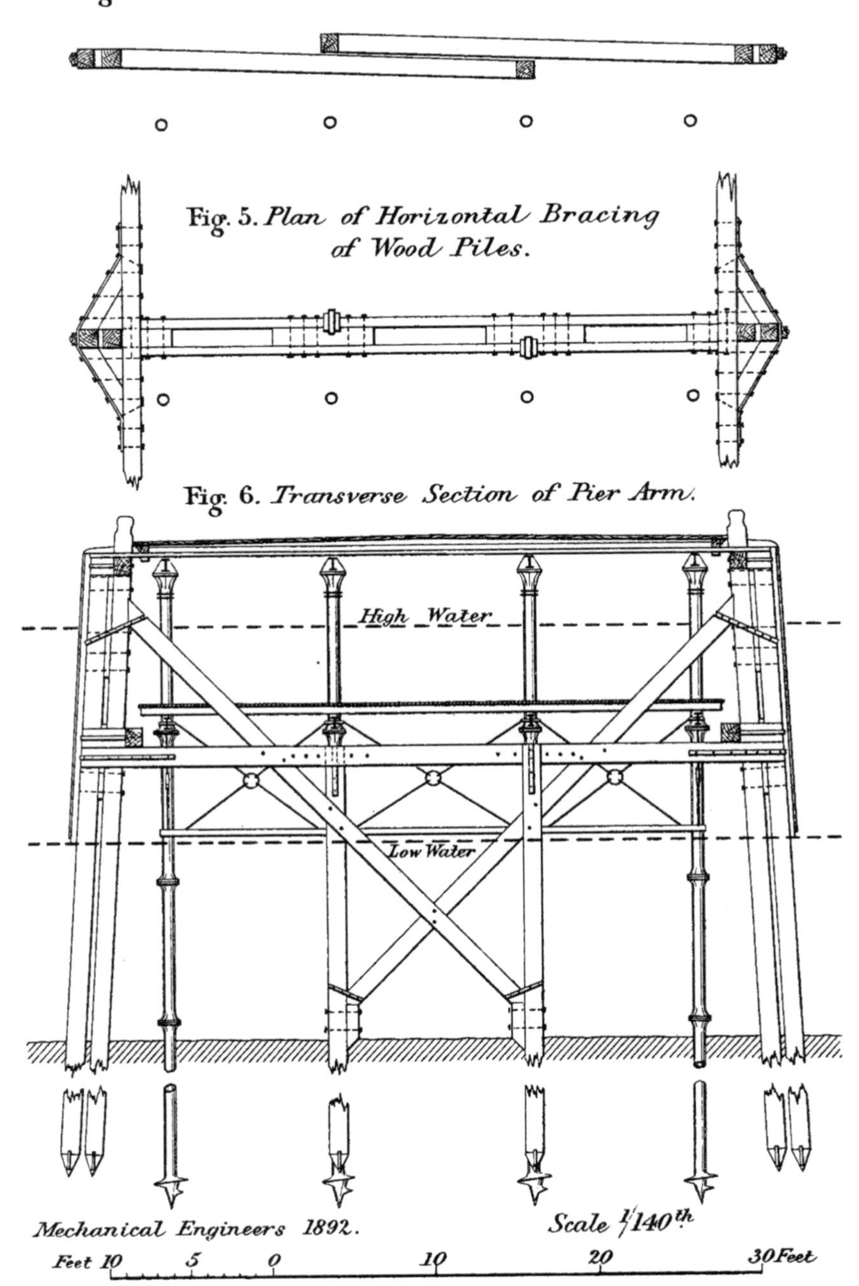

Mechanical Engineers 1892. *Scale 1/140th.*

SOUTHAMPTON PIER.

Plate 40.

Fig. 7. *Transverse Section of Pier Head.*

Fig. 8.

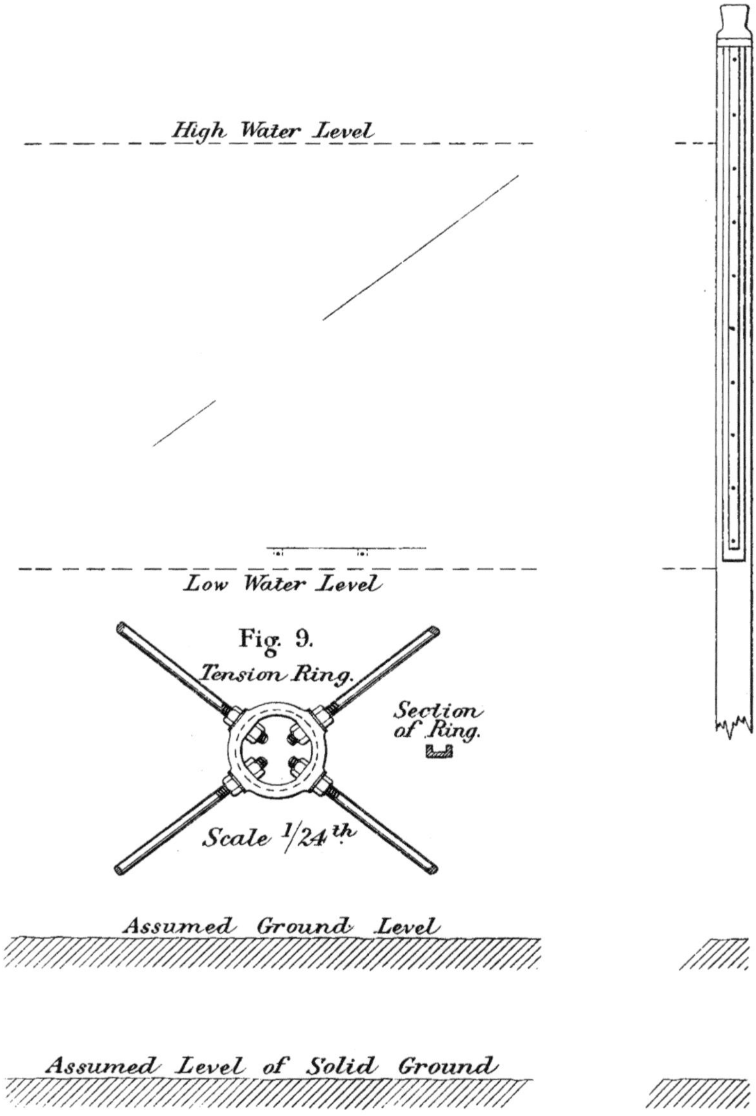

High Water Level

Low Water Level

Fig. 9.

Tension Ring.

Section of Ring.

Scale 1/24th

Assumed Ground Level

Assumed Level of Solid Ground

SOUTHAMPTON PIER. *Plate 41.*

Girders, Cast-Iron Piles, and Caps.

Fig. 10.

Fig. 11.

Fig. 13.

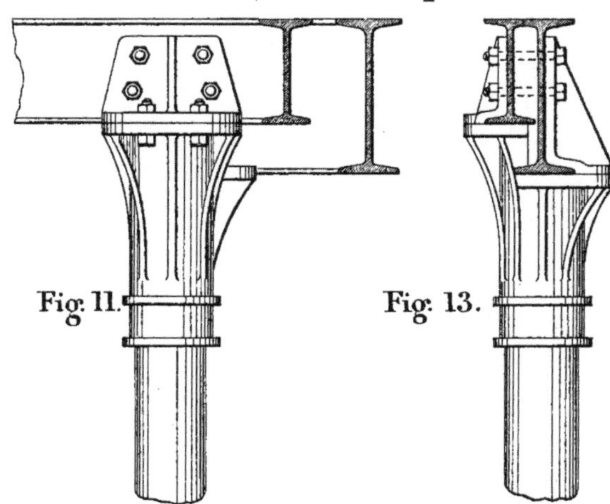

Fig. 17.

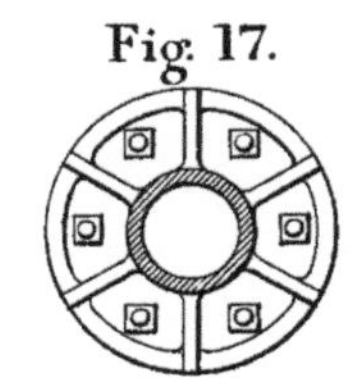

Fig. 12.

Plan of Fig. 11.

Fig. 18.

Fig. 14.

Fig. 15.

Fig. 16.

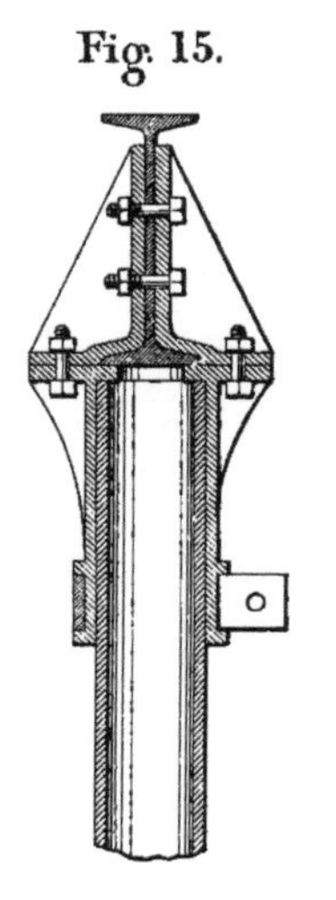

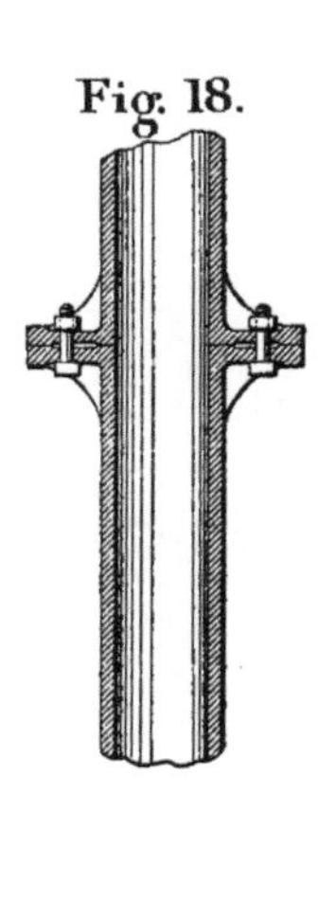

SOUTHAMPTON PIER. *Plate 42.*

Bracings and Wood Piles.

Fig. 22.

Fig. 23.

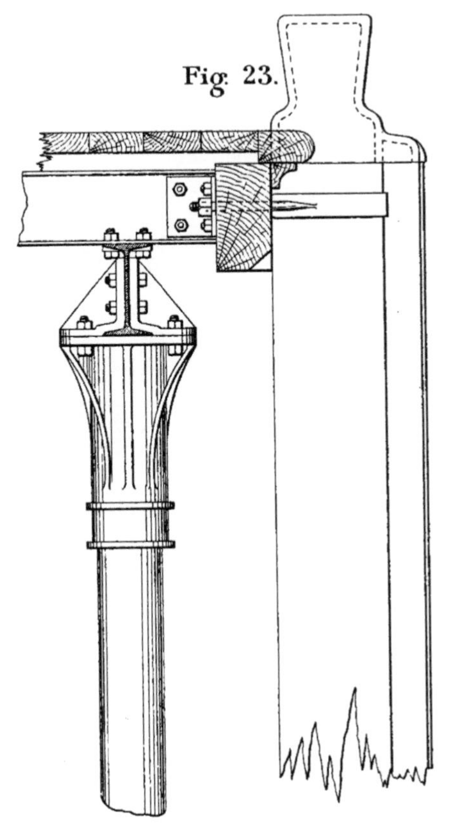

orizontal Bracings.

Fig. 19. *Plan.*

Fig. 20. *vation.*

Fig. 21. *Plan of Diagonal Stiffening Ties of Horizontal Bracings.*

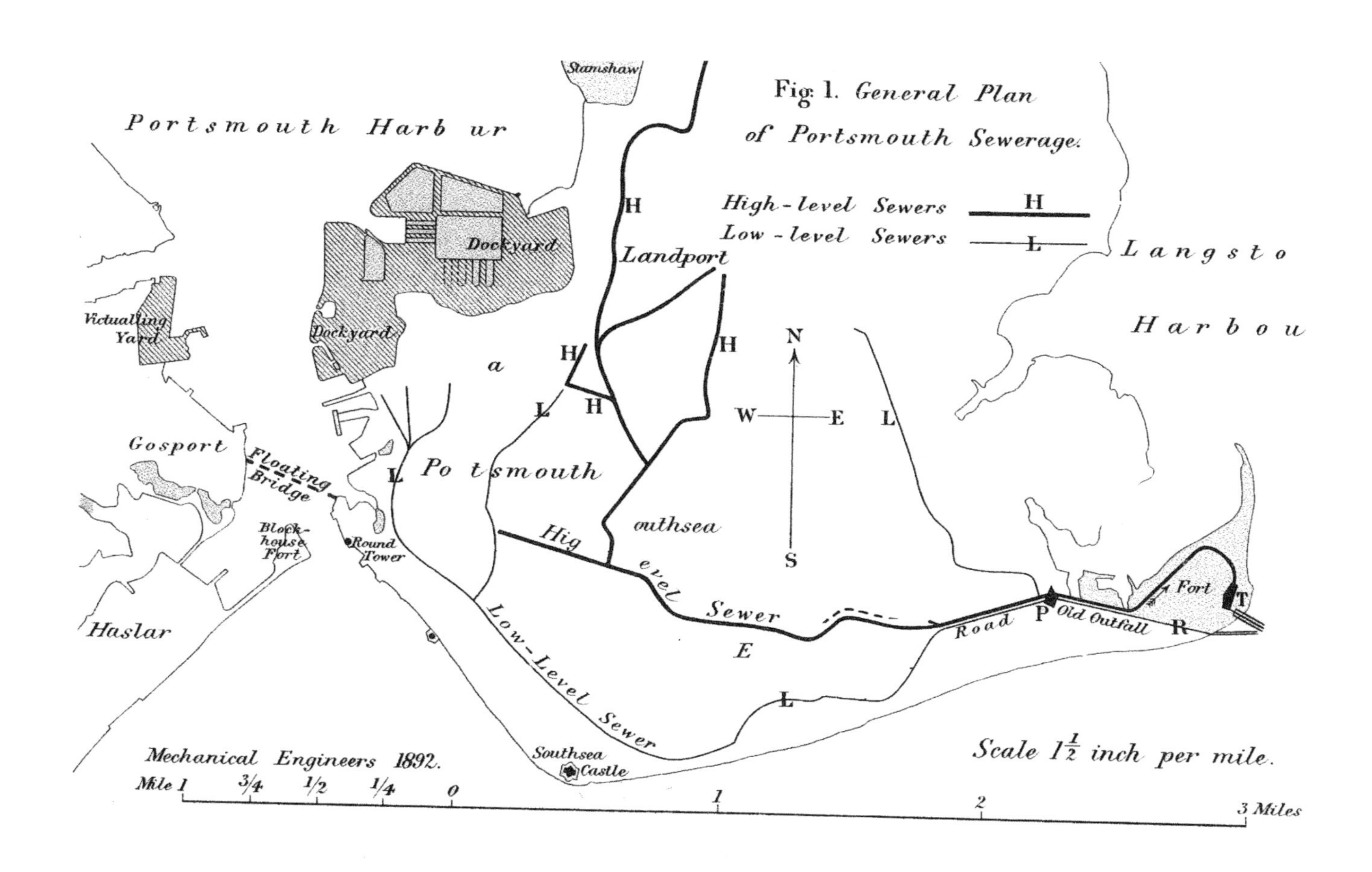

Fig. 1. General Plan of Portsmouth Sewerage.

SEWAGE OUTFALL WORKS. *Plate 44.*

Fig. 2. *Ebb Currents out of Langstone Harbour.*

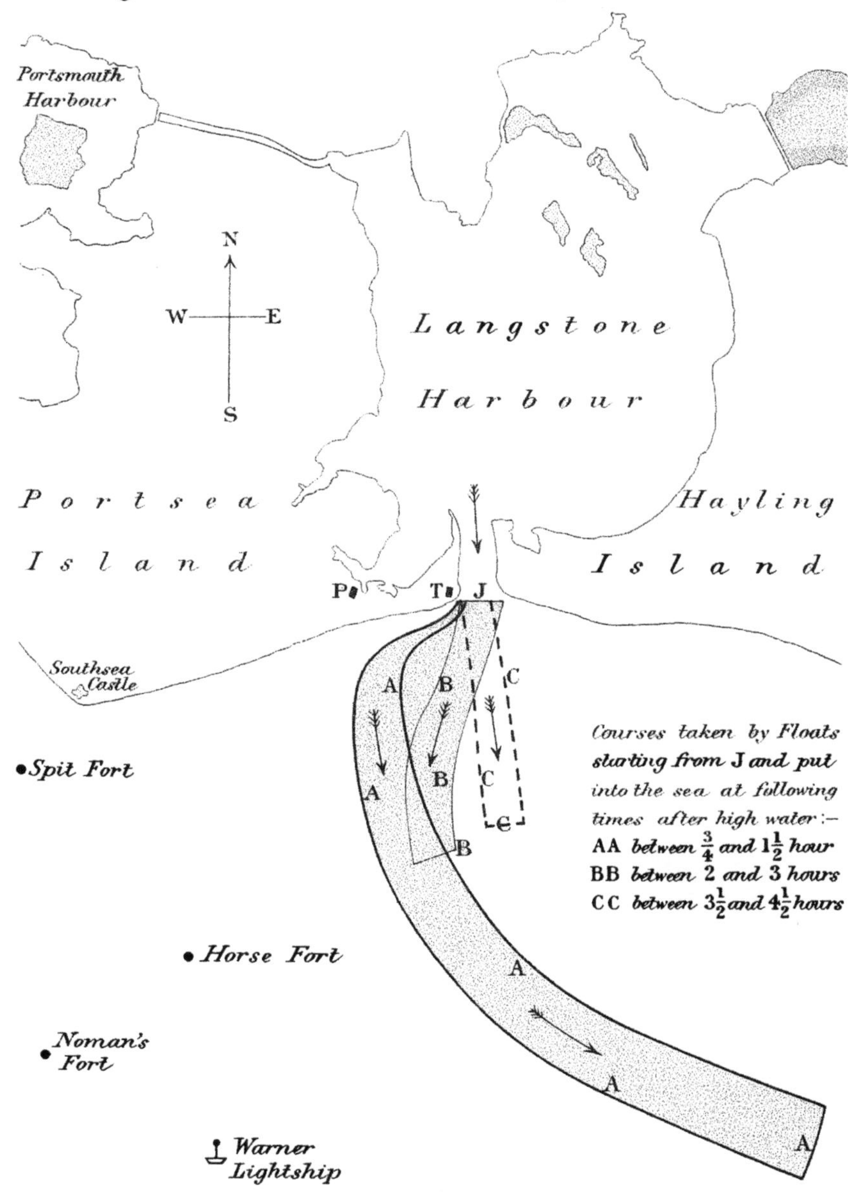

Pla

Storage Tank. Fig. 3. Plan.

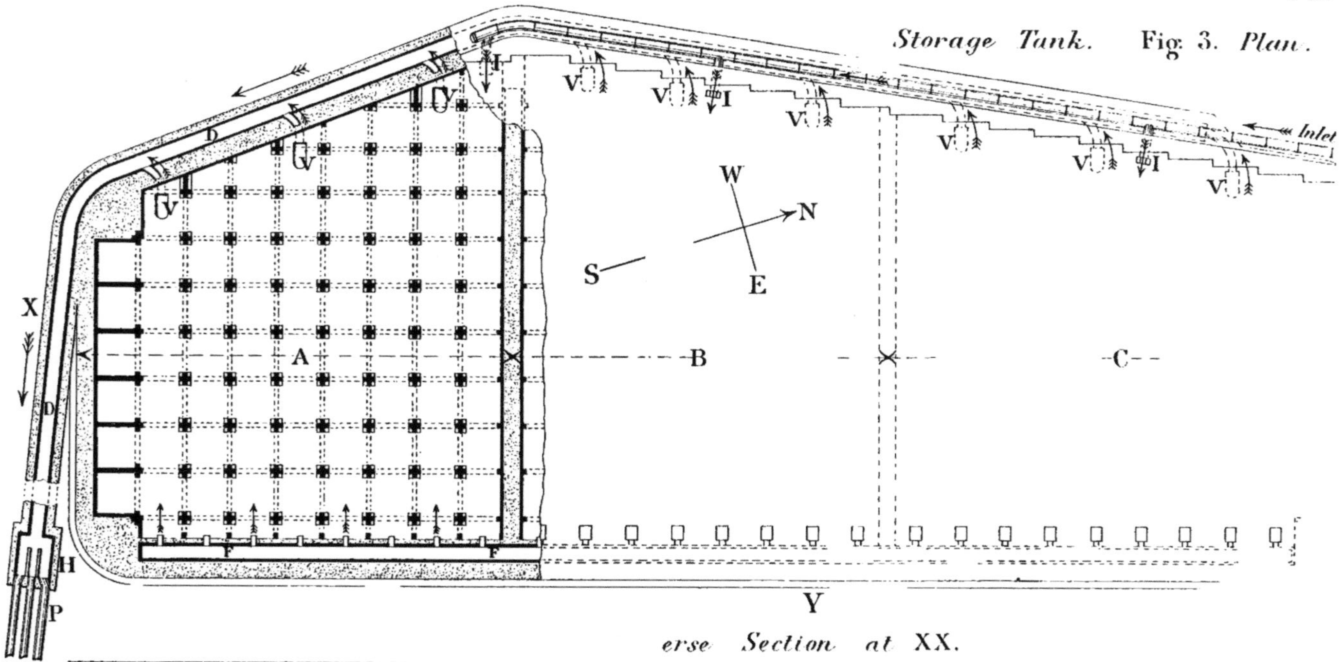

Inlet

V V I V V V I V

W N S E

X A B C

D F H P Y

erse Section at XX.

A B O

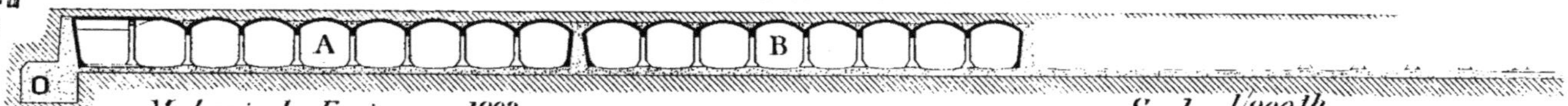

Mechanical Engineers 1892.

Scale 1/800th

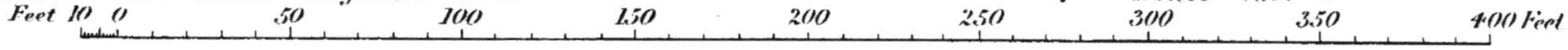

Feet 10 0 50 100 150 200 250 300 350 400 Feet

SEWAGE OUTFALL WORKS.

Storage Tank. Fig. 5. Longitudinal Section. Scale 1/150th

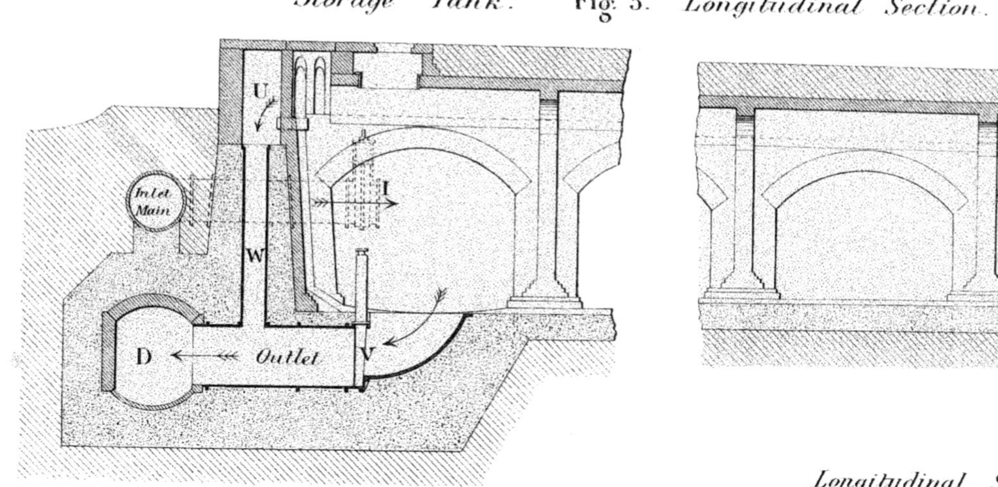

Fig. 6.

Longitudinal Section at YY, Fig. 3, Plate 4

Scale 1/800th

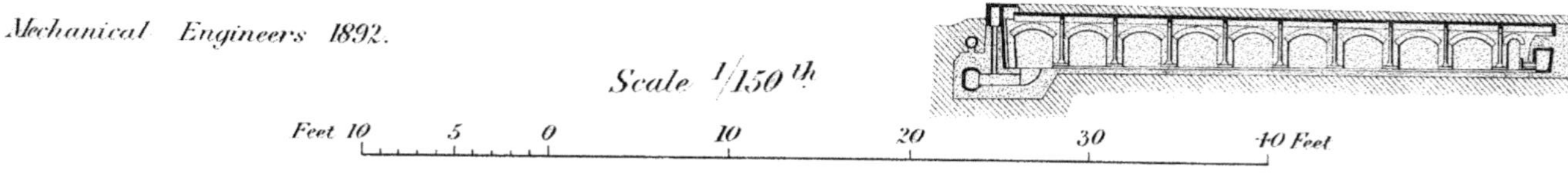

Scale 1/150th

Feet 10 5 0 10 20 30 40 Feet

Storage Tank.

Fig. 7. *Test Load of* $11\frac{3}{4}$ *tons on Arch of two half-brick rings.*

Fig. 8. *Autographic Diagram of Filling and Discharge of Storage Tank.*

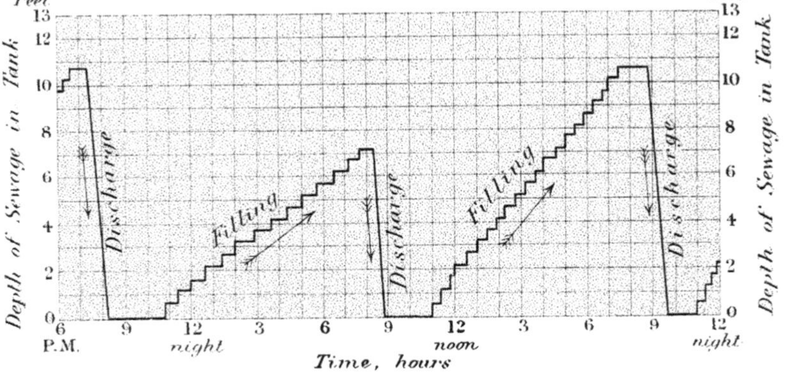

Feet
13
12
10
8
6
4
2
0

Depth of Sewage in Tank

Discharge
Filling
Discharge
Filling
Discharge

6 P.M. 9 12 *night* 3 6 9 12 *noon* 3 6 9 12 *night*

Time, hours

SEWAGE OUTFALL WORKS. *Plate 4*

Fig. 9. *Turbine for opening Outlet Valve.*

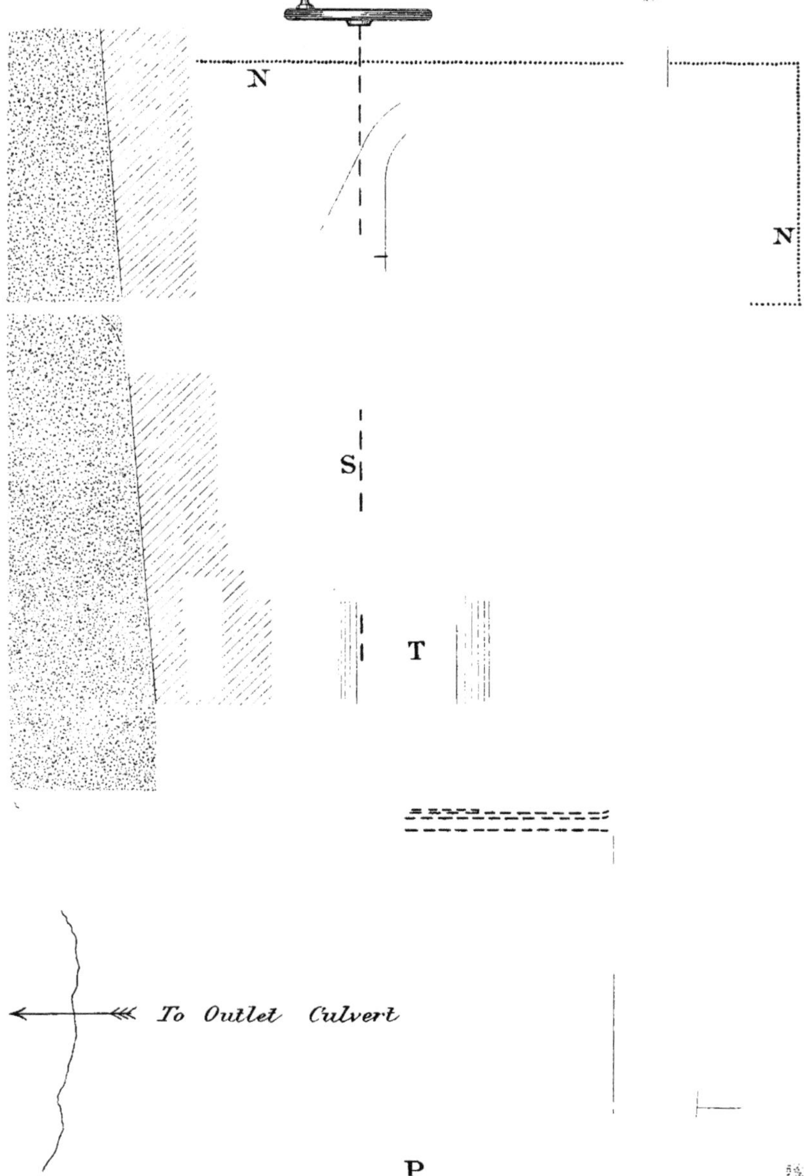

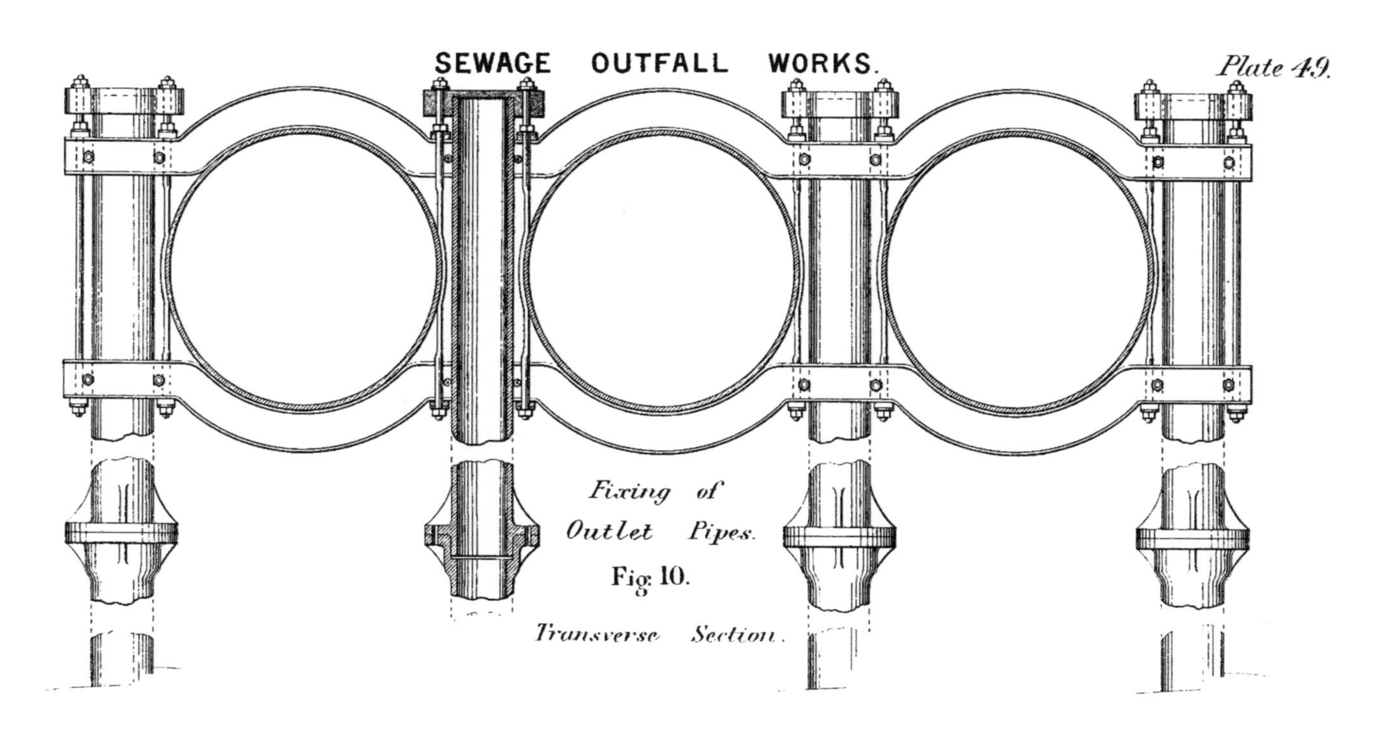
SEWAGE OUTFALL WORKS.
Fixing of Outlet Pipes.
Fig. 10.
Transverse Section.

Plate

Fig. 11. *Plan of Engine-House premises.*

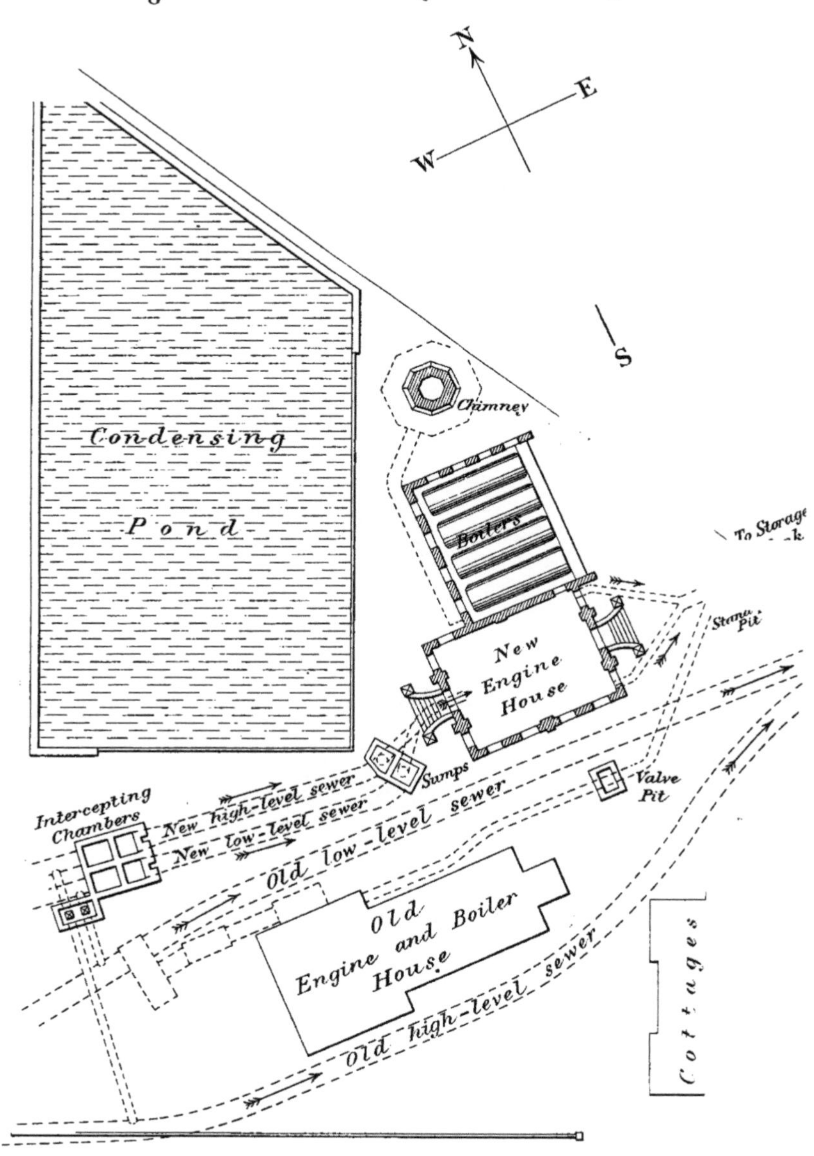

 Scale 1/600th.

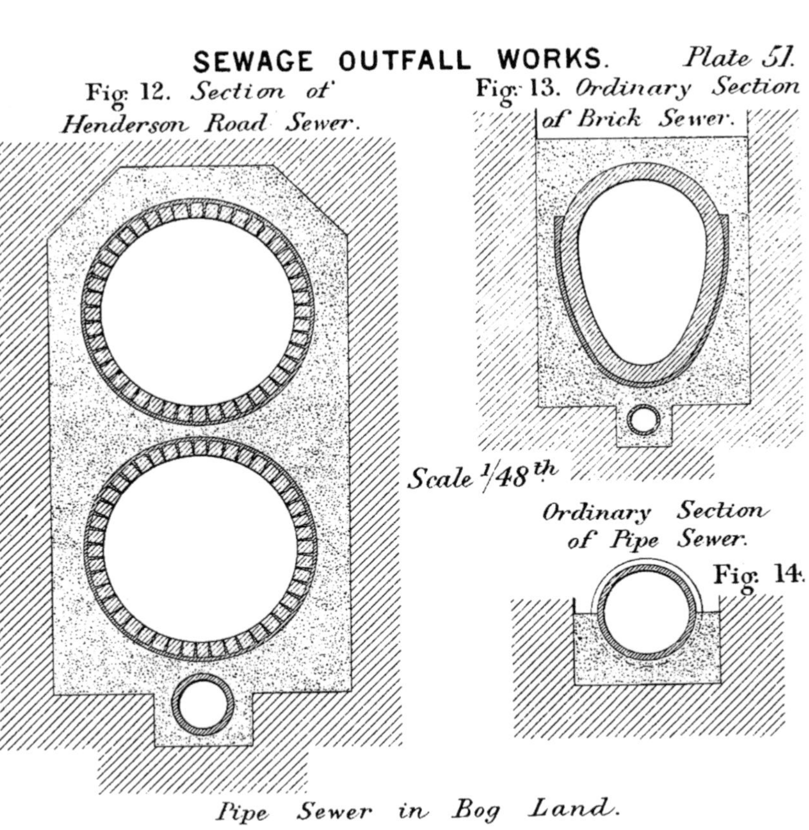
SEWAGE OUTFALL WORKS.
Plate 51.
Fig. 12. Section of Henderson Road Sewer.
Fig. 13. Ordinary Section of Brick Sewer.
Scale 1/48th.
Ordinary Section of Pipe Sewer.
Fig. 14.

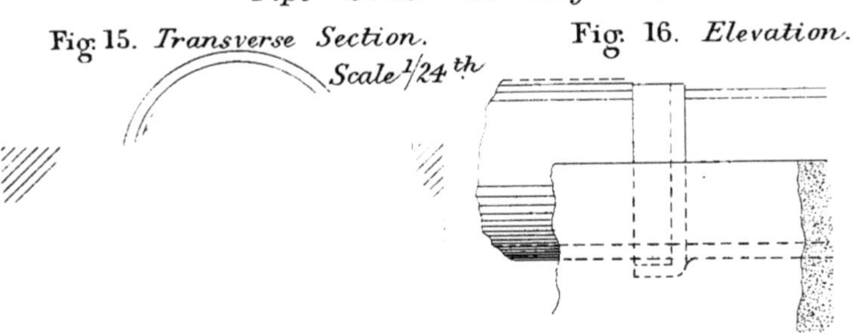
Pipe Sewer in Bog Land.
Fig. 15. Transverse Section.
Scale 1/24th
Fig. 16. Elevation.

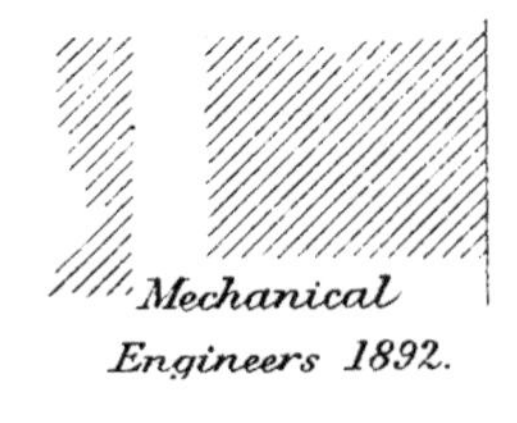
Mechanical Engineers 1892.

Scale 1/24t.

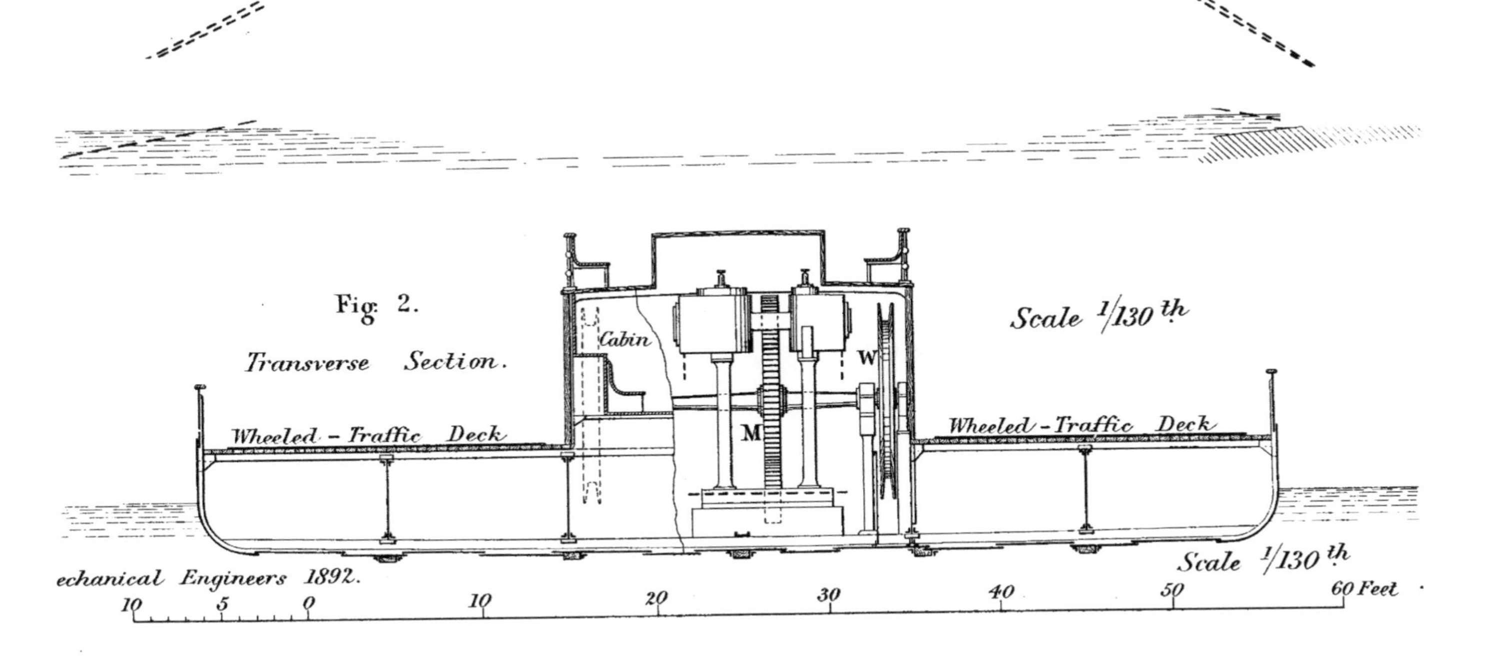
Scale 1/260th.
Fig. 2.
Transverse Section.
Scale 1/130th.
Cabin
W
M
Wheeled - Traffic Deck
Wheeled - Traffic Deck
Scale 1/130th.
echanical Engineers 1892.
10 5 0 10 20 30 40 50 60 Feet

FLOATING BRIDGE.

New Floating Bridge between Portsmouth and Gosport. Fig. 3. *Plan.*

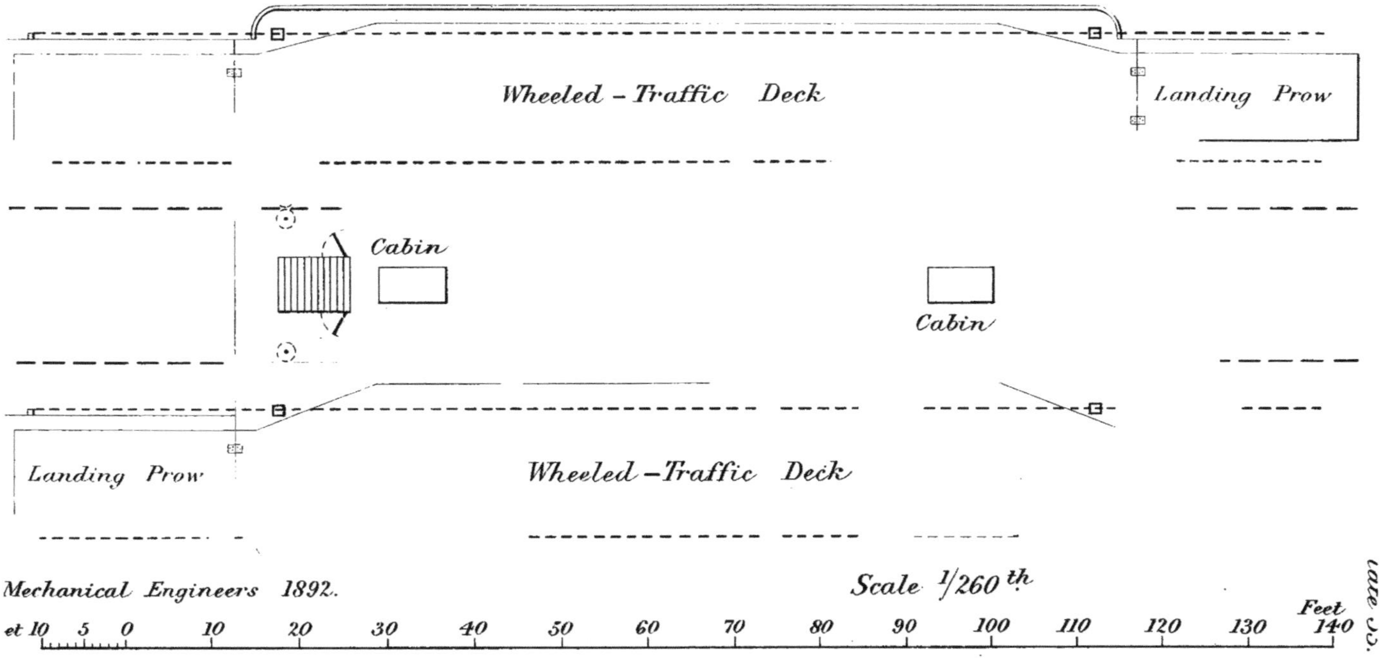

Mechanical Engineers 1892.

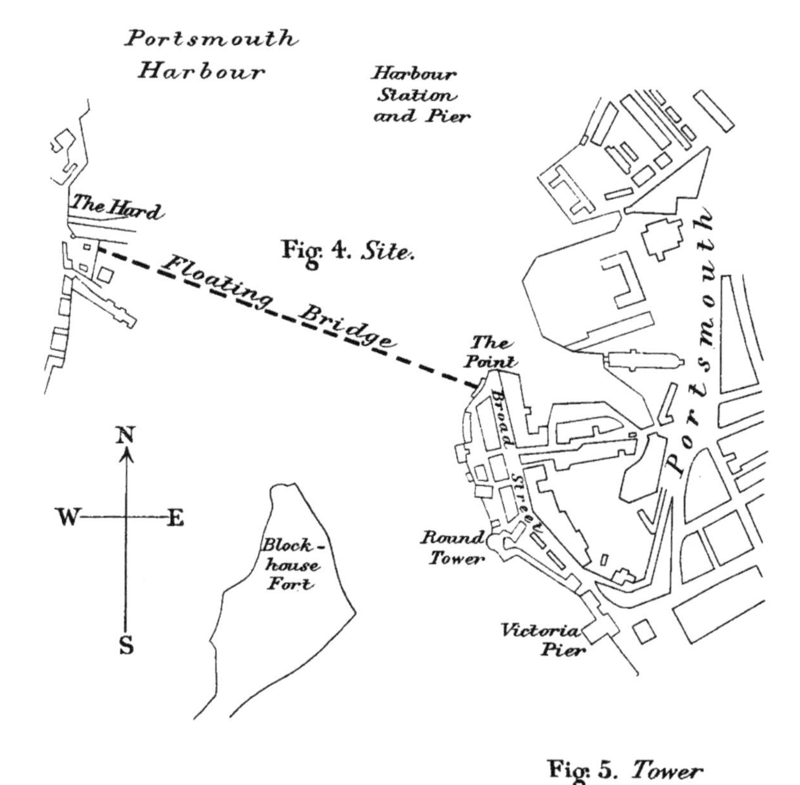

Fig. 4. *Site.*

Fig. 5. *Tower Opening Bridge.*

Fixed Bridge for Foot Traffic

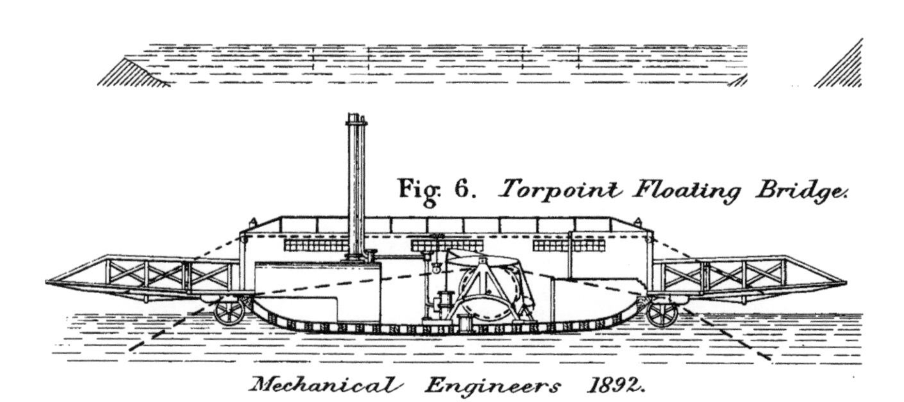

Fig. 6. *Torpoint Floating Bridge.*

Mechanical Engineers 1892.

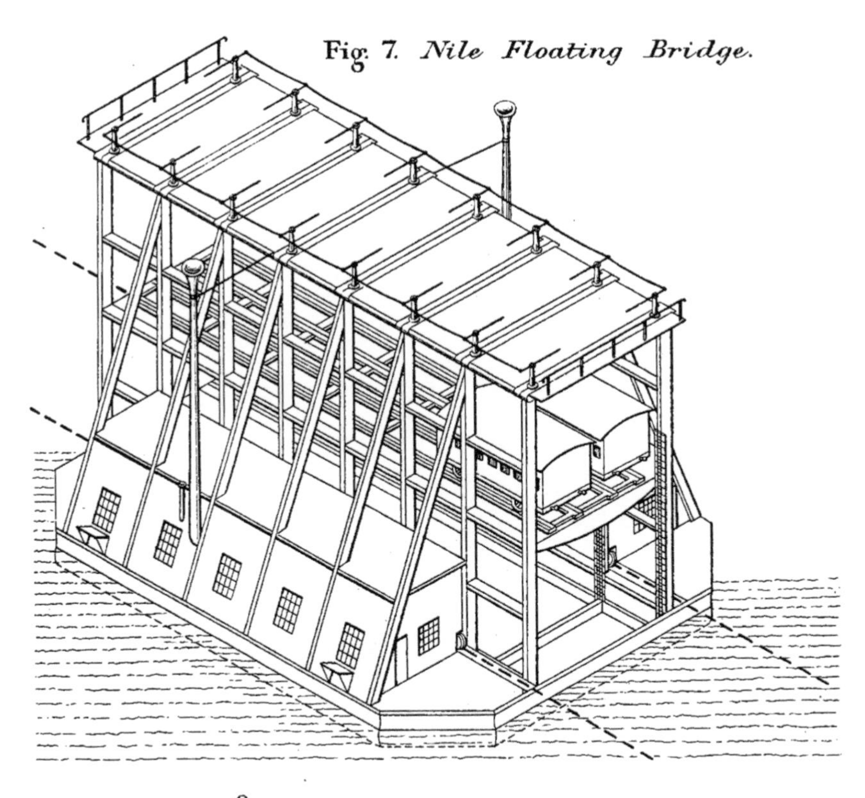

Fig. 7. *Nile Floating Bridge.*

Fig. 8. *St Malo Traversing Bridge at low water.*

High Water
at spring tides

Plate 56.

SEWAGE AND REFUSE WORKS.

General Plan of southern portion of Southampton.

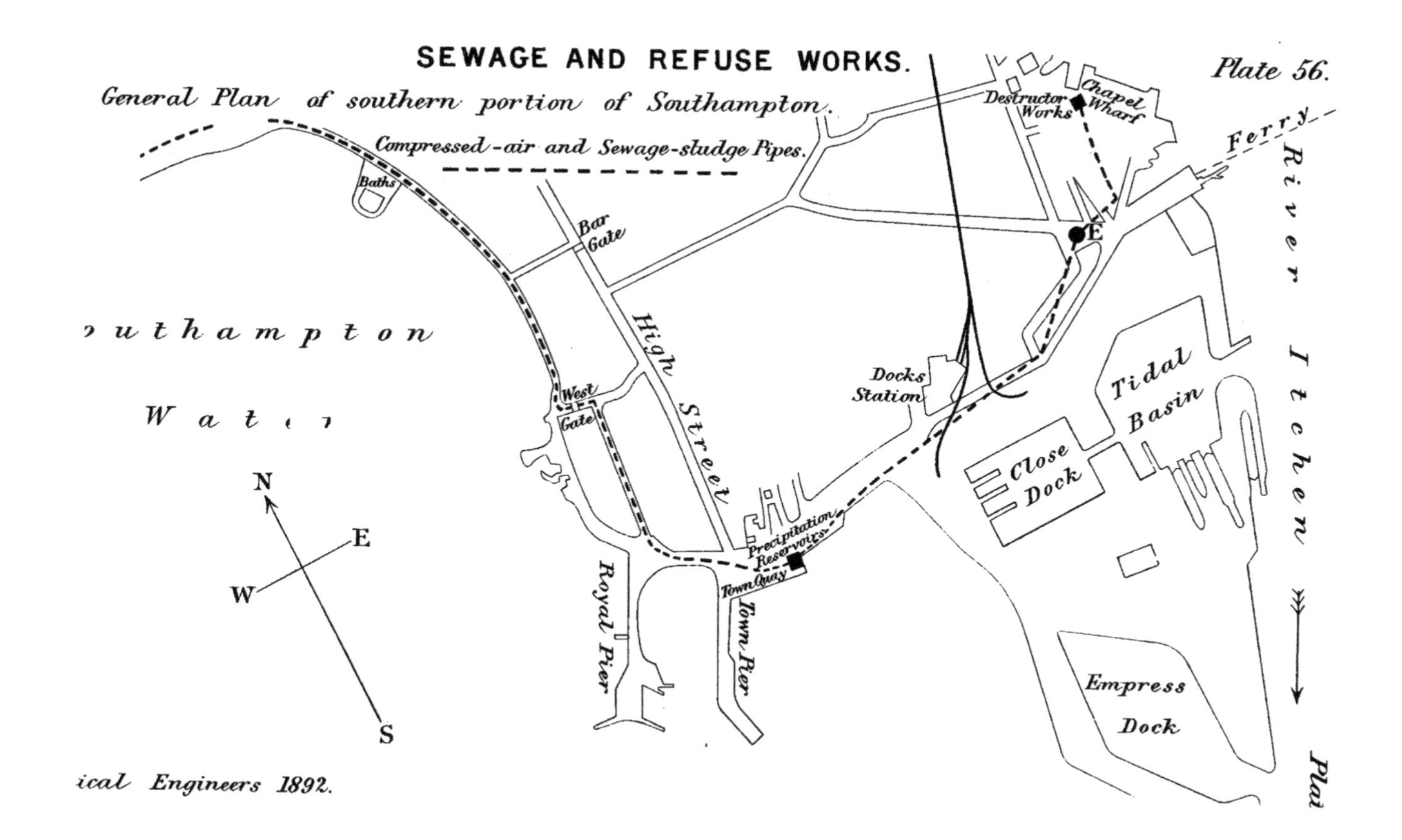

ical Engineers 1892.

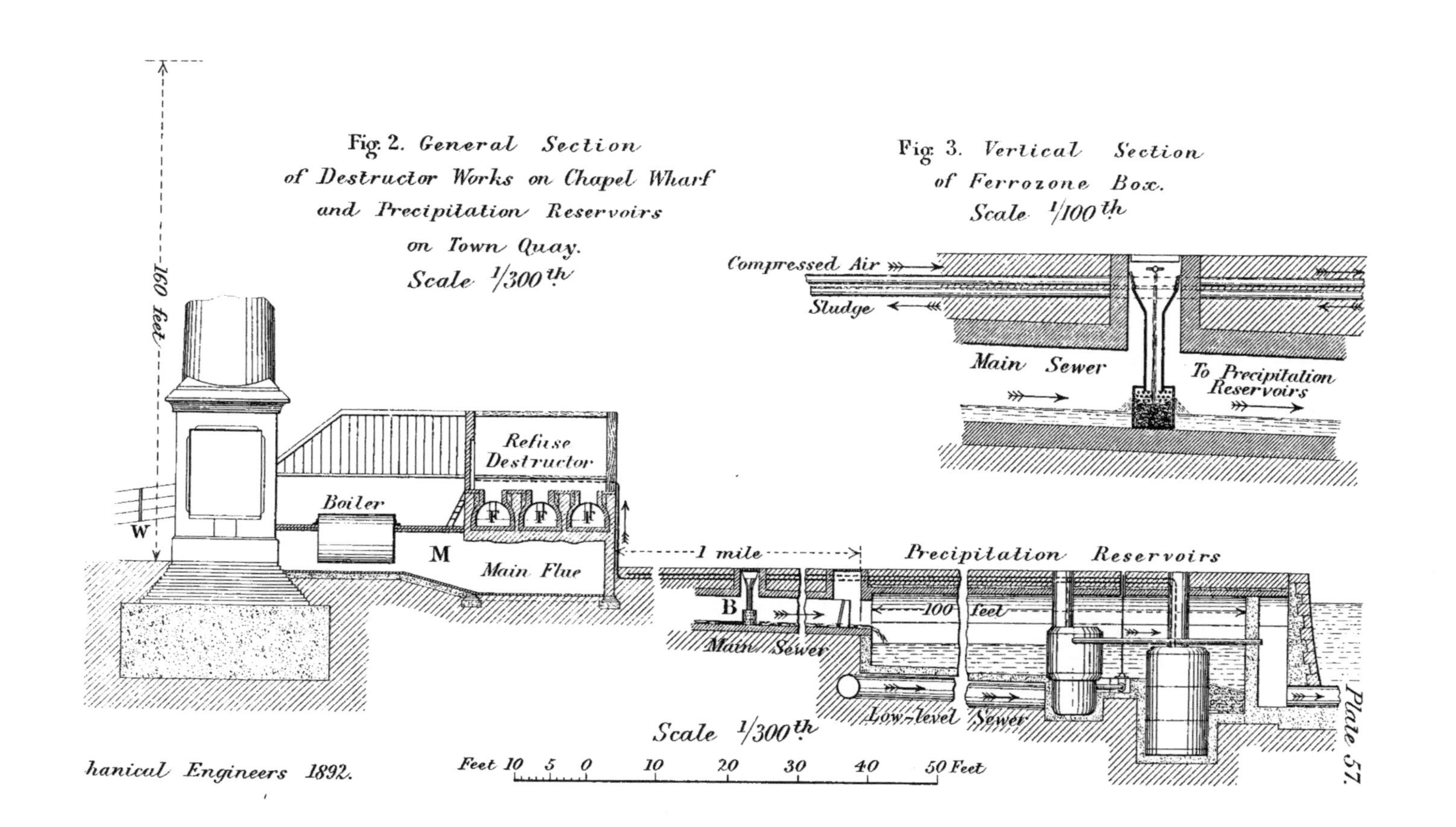

Fig. 2. General Section of Destructor Works on Chapel Wharf and Precipitation Reservoirs on Town Quay. Scale 1/300th

Fig. 3. Vertical Section of Ferrozone Box. Scale 1/100th

Fig. 4. *Plan of Destructor Works at Chapel Wharf. Scale 1/200th.*

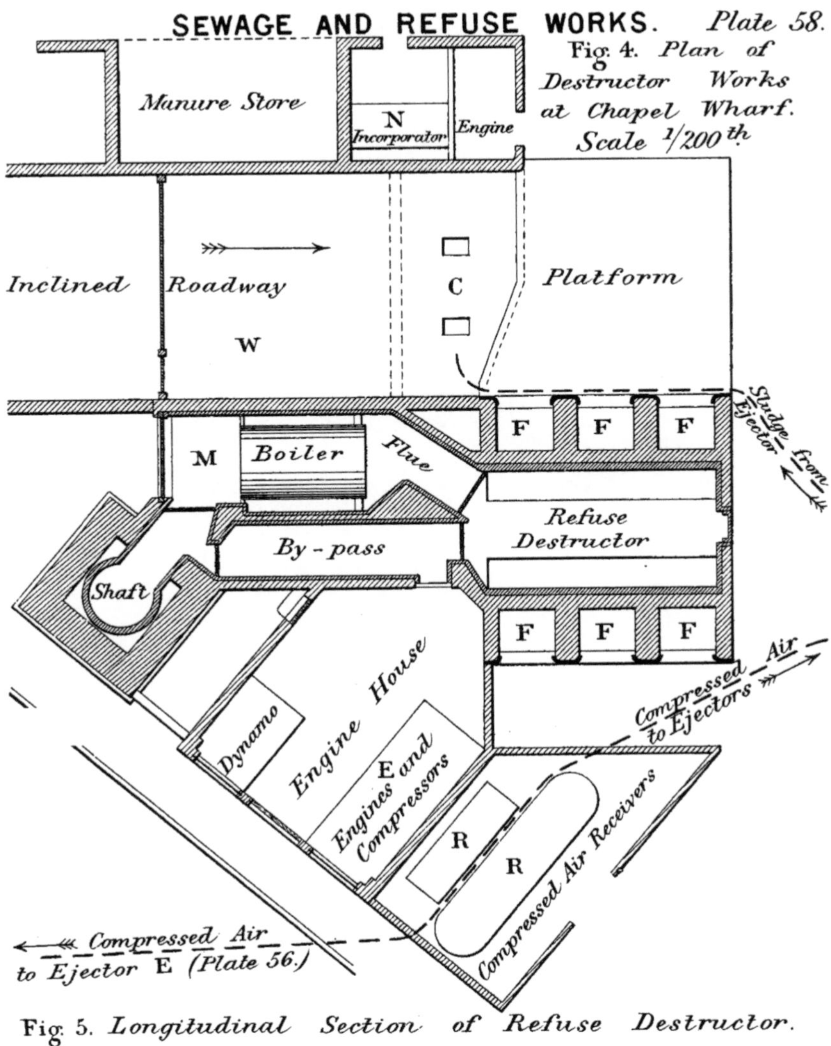

Fig. 5. *Longitudinal Section of Refuse Destructor. Scale 1/80th*

To

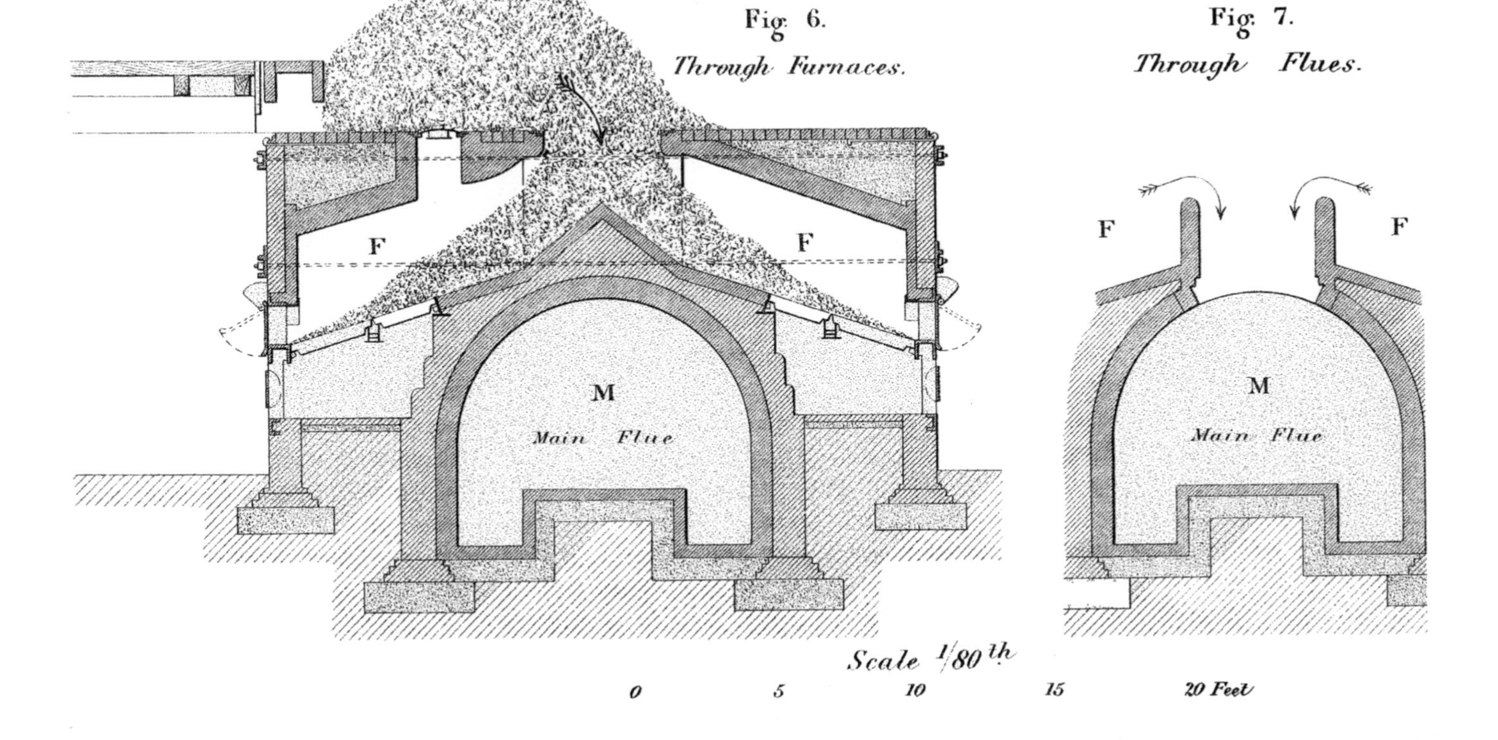
SEWAGE AND REFUSE WORKS.
Plate 5.
Transverse Sections of Refuse Destructor.
Fig. 6.
Through Furnaces.
Fig. 7.
Through Flues.
F
F
F
F
M
Main Flue
M
Main Flue
Scale 1/80th
0
5
10
15
20 Feet

Precipitation Reservoirs.

Fig. 8. *Sectional Plan.*

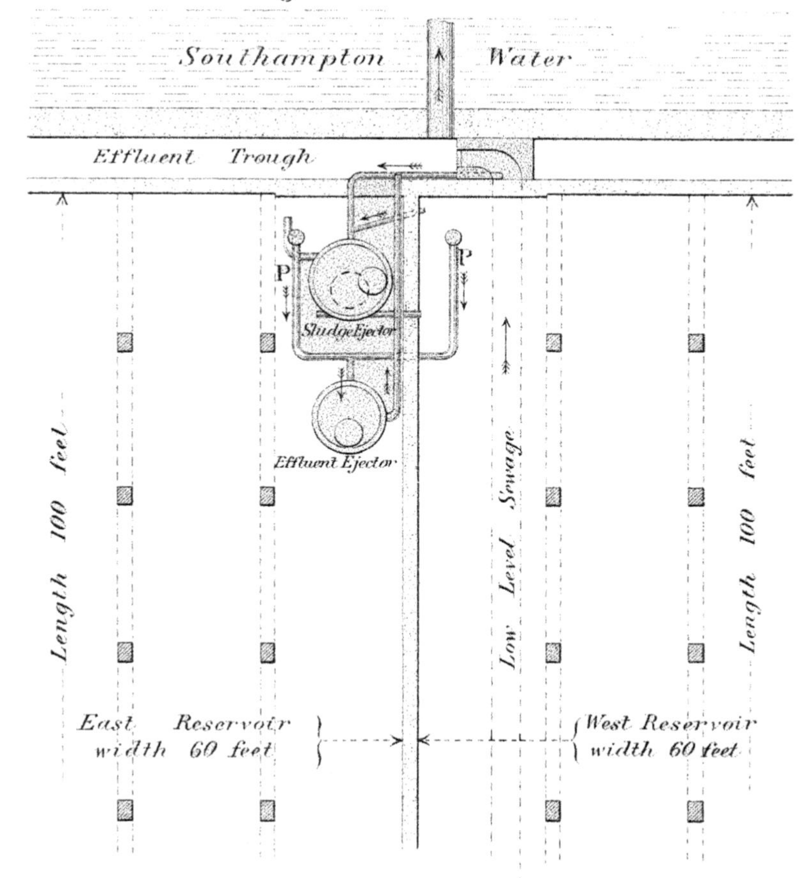

Fig. 9. *Transverse Section.*

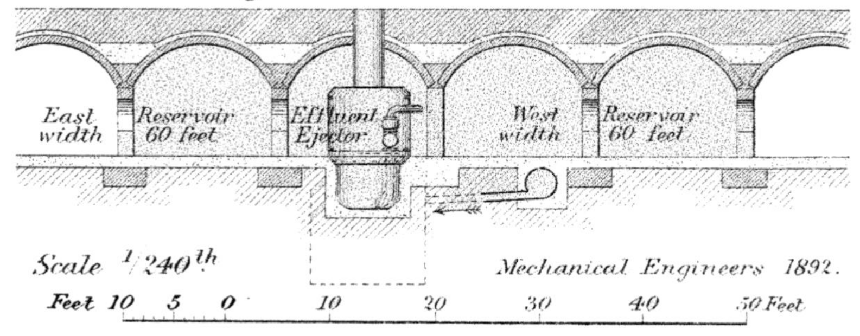

Scale 1/240th

Mechanical Engineers 1892.

Feet 10 5 0 10 20 30 40 50 Feet

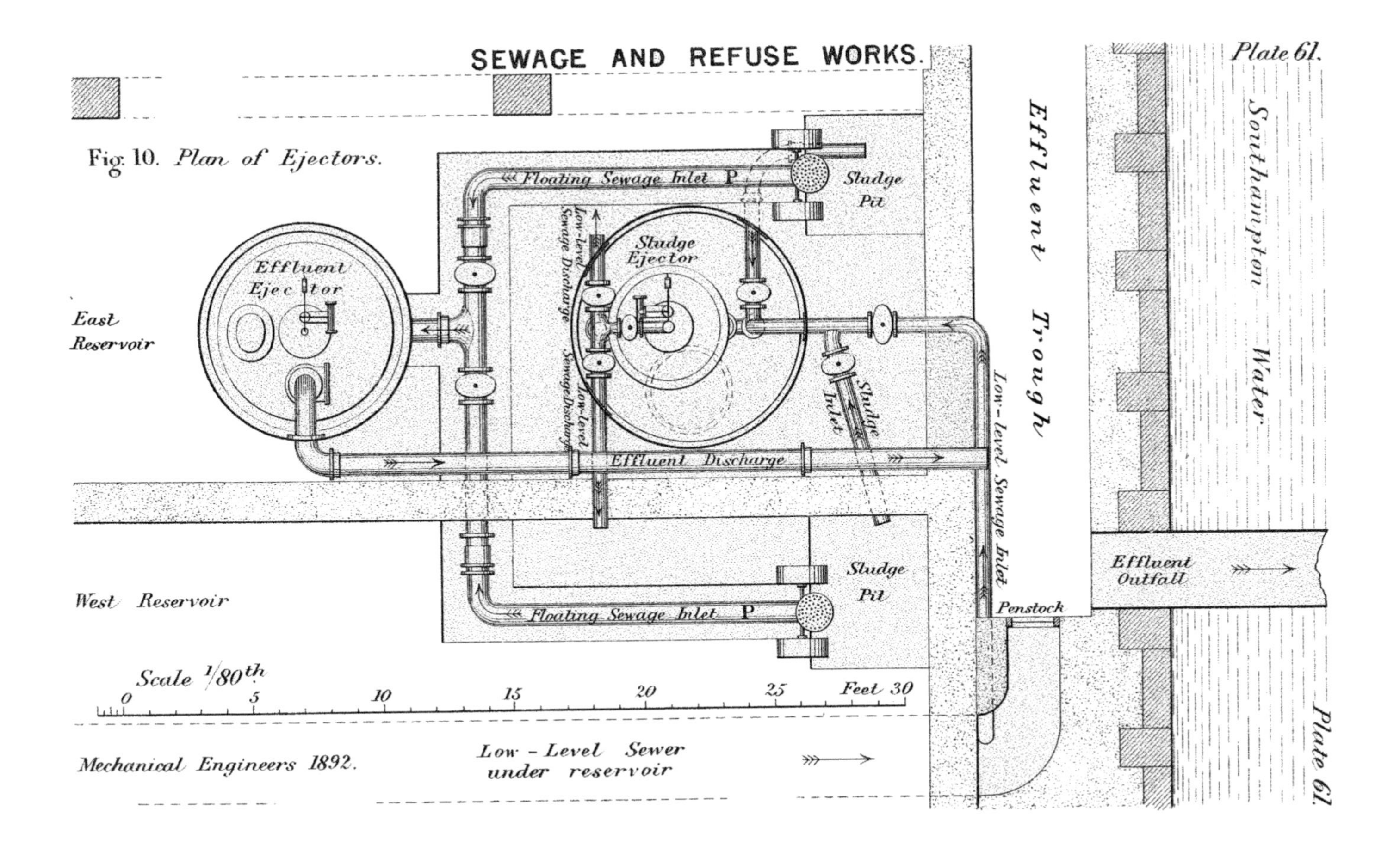

SEWAGE AND REFUSE WORKS.
Plate 61.
Fig. 10. Plan of Ejectors.
Floating Sewage Inlet
P
Sludge Pit
Low-level Sewage Discharge
Sludge Ejector
Effluent Ejector
East Reservoir
Low-level Sewage Discharge
Sludge Inlet
Effluent Trough
Southampton Water
Low-level Sewage Inlet
Effluent Discharge
Effluent Outfall
Penstock
Sludge Pit
West Reservoir
Floating Sewage Inlet
P
Scale 1/80th
0
5
10
15
20
25
Feet 30
Mechanical Engineers 1892.
Low-Level Sewer under reservoir
Plate 61.

SEWAGE AND REFUSE WORKS.

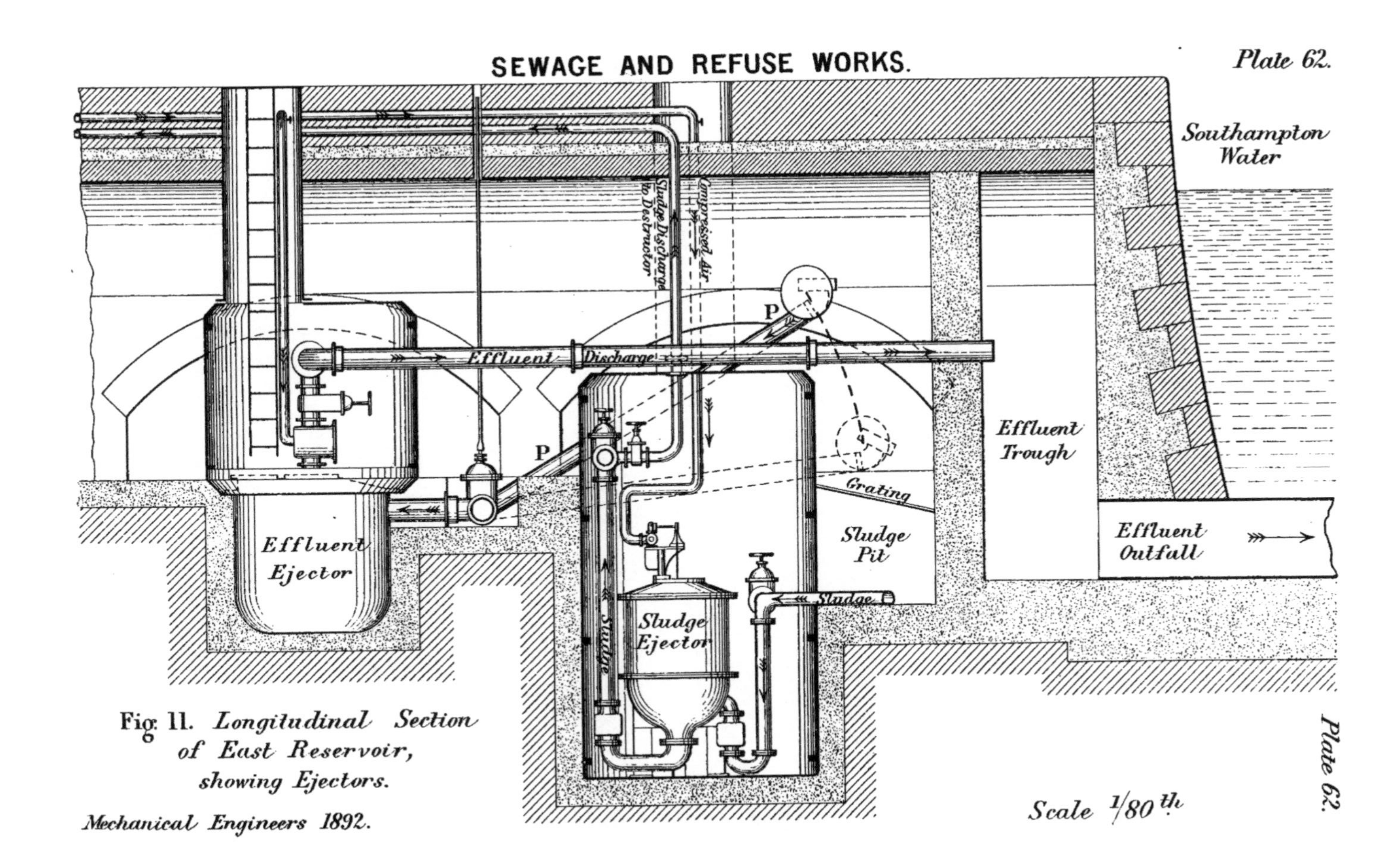

Fig. 11. *Longitudinal Section of East Reservoir, showing Ejectors.*

Scale 1/80th

Mechanical Engineers 1892.

Plate 62.

Fig. 12. *Vertical Section of Pneumatic Ejector.*

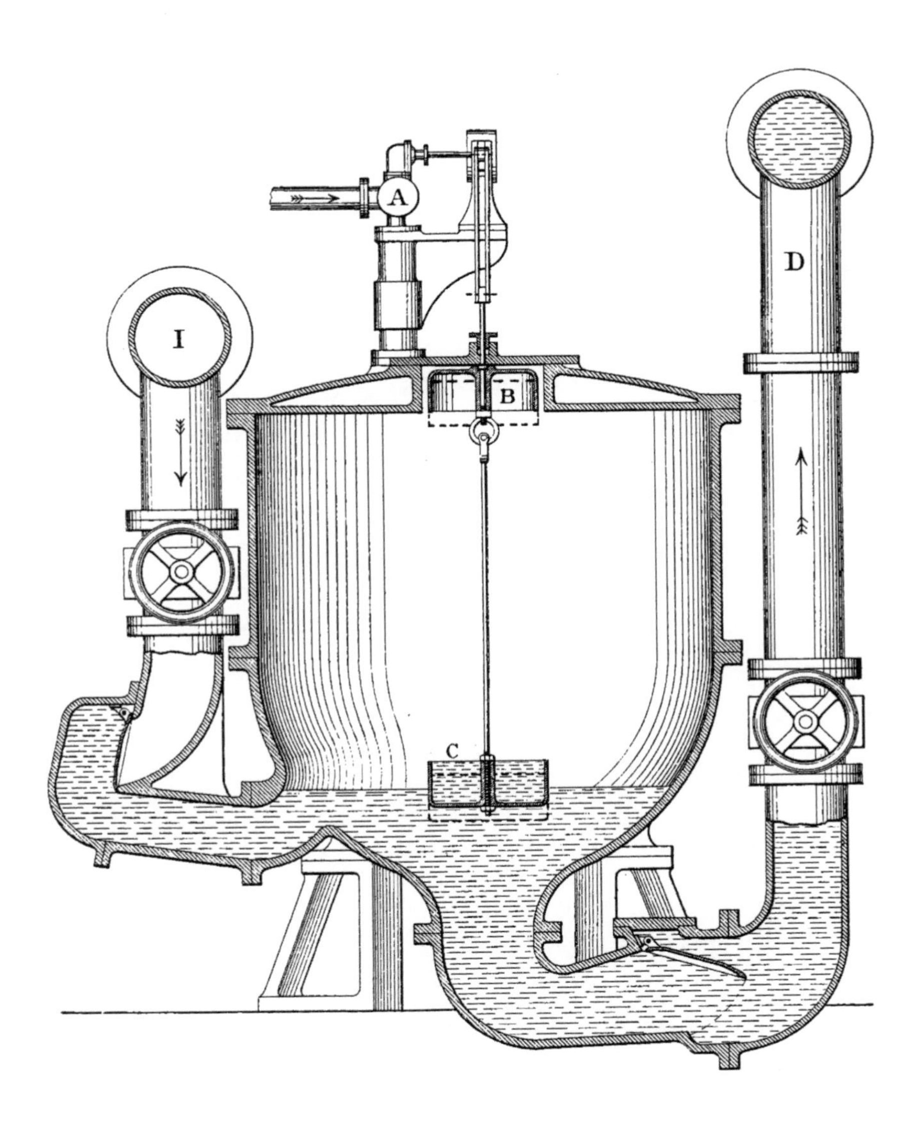

ROYAL CLARENCE VICTUALLING YARD. *Plate 6*

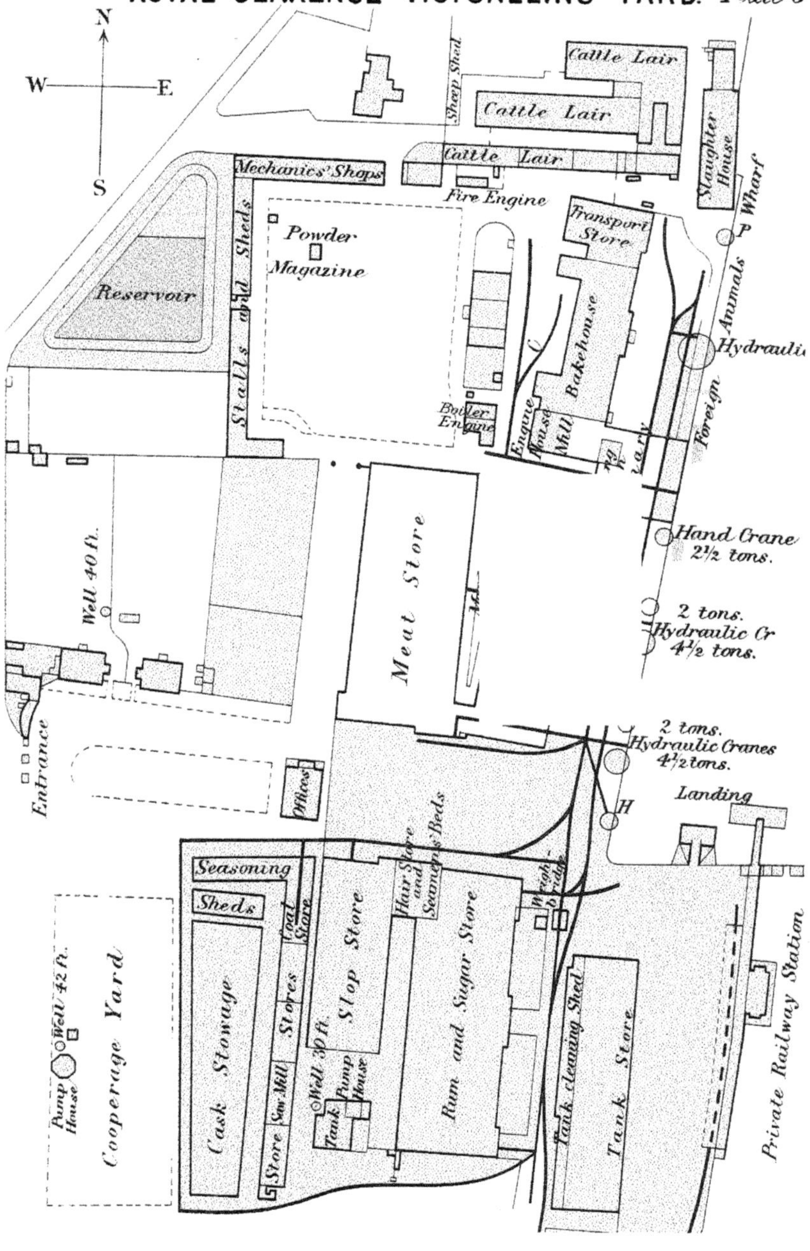

N
W
E
S
Sheep Shed
Cattle Lair
Cattle Lair
Cattle Lair
Slaughter House
Mechanics' Shops
Fire Engine
Transport Store
Powder Magazine
Reservoir
Stalls and Sheds
Bakehouse
Boiler Engine
Engine House
Mill
Animals Wharf
P
Hydraulic
Foreign
Hand Crane 2½ tons.
2 tons.
Hydraulic Cr 4½ tons.
Well 40 ft.
Meat Store
2 tons.
Hydraulic Cranes 4½ tons.
Landing
H
Entrance
Offices
Seasoning
Sheds
Coal Store
Hair Store and Seamen's Beds
Slop Store
Rum and Sugar Store
Weigh bridge
Tank cleaning Shed
Tank Store
Private Railway Station
Pump House
Well 42 ft.
Cooperage Yard
Cask Stowage
Stores
Saw Mill
Store
Well 30 ft.
Tank
Pump House

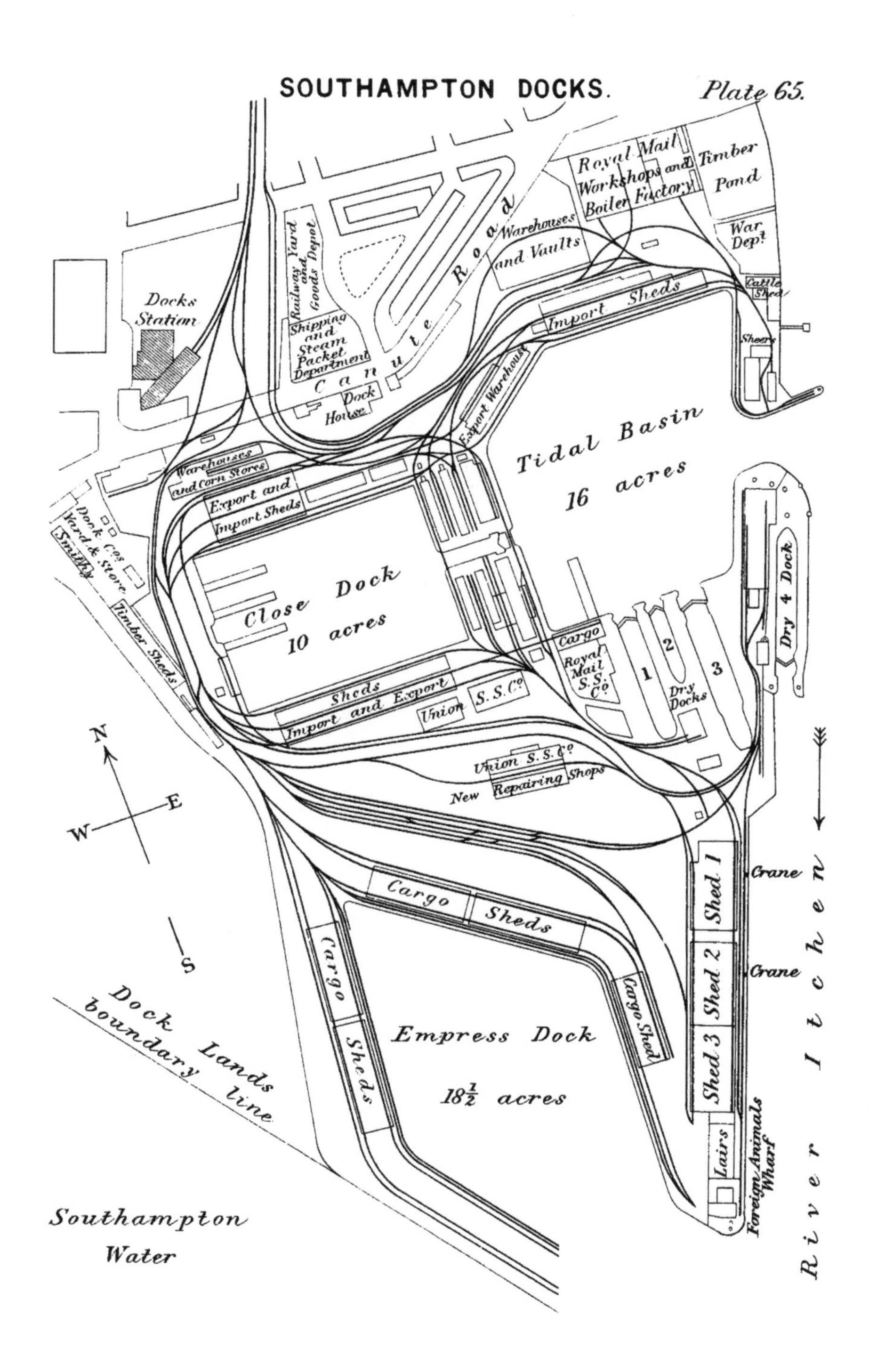
SOUTHAMPTON DOCKS.
Plate 65.
Royal Mail Workshops and Boiler Factory
Timber Pond
War Dept.
Warehouses and Vaults
Canute Road
Railway Yard and Goods Depot
Shipping and Steam Packet Department
Dock House
Docks Station
Cattle Shed
Sheers
Import Sheds
Export Warehouse
Tidal Basin
16 acres
Warehouses and Corn Stores
Export and Import Sheds
Dock Co's Yard & Store
Smithy
Timber Sheds
Close Dock
10 acres
Sheds
Import and Export
Union
S. S. Co.
Cargo
Royal Mail S. S. Co.
1
2
3
Dry Docks
Dry 4 Dock
Union S. S. Co.
New Repairing Shops
N
E
W
S
Cargo Sheds
Cargo Sheds
Cargo Shed
Shed 1
Shed 2
Shed 3
Crane
Crane
Empress Dock
18½ acres
Lairs
Foreign Animals Wharf
River Itchen
Dock Lands boundary line
Southampton Water

STEAMSHIP REPAIRING SHOPS. *Plate 66.*

Union Steamship Company's

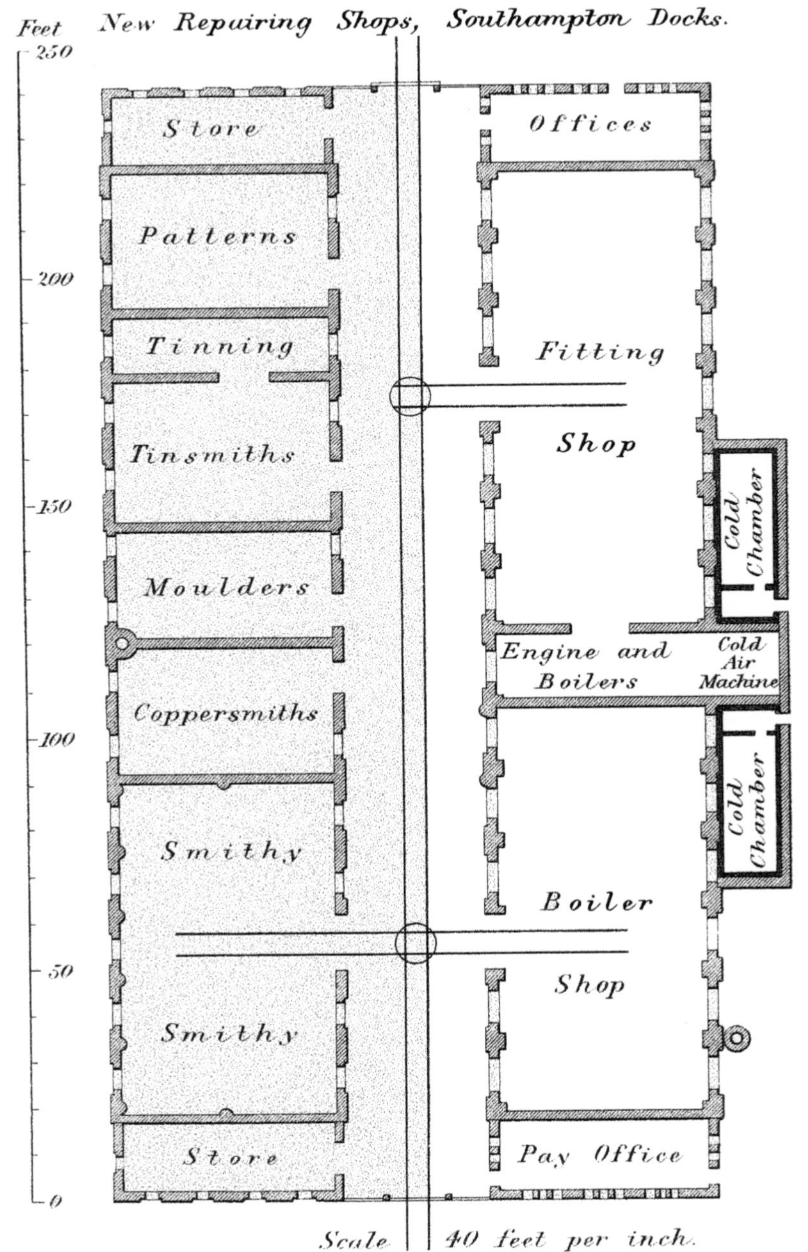

Mechanical Engineers 1892.

EASTLEIGH WORKS.

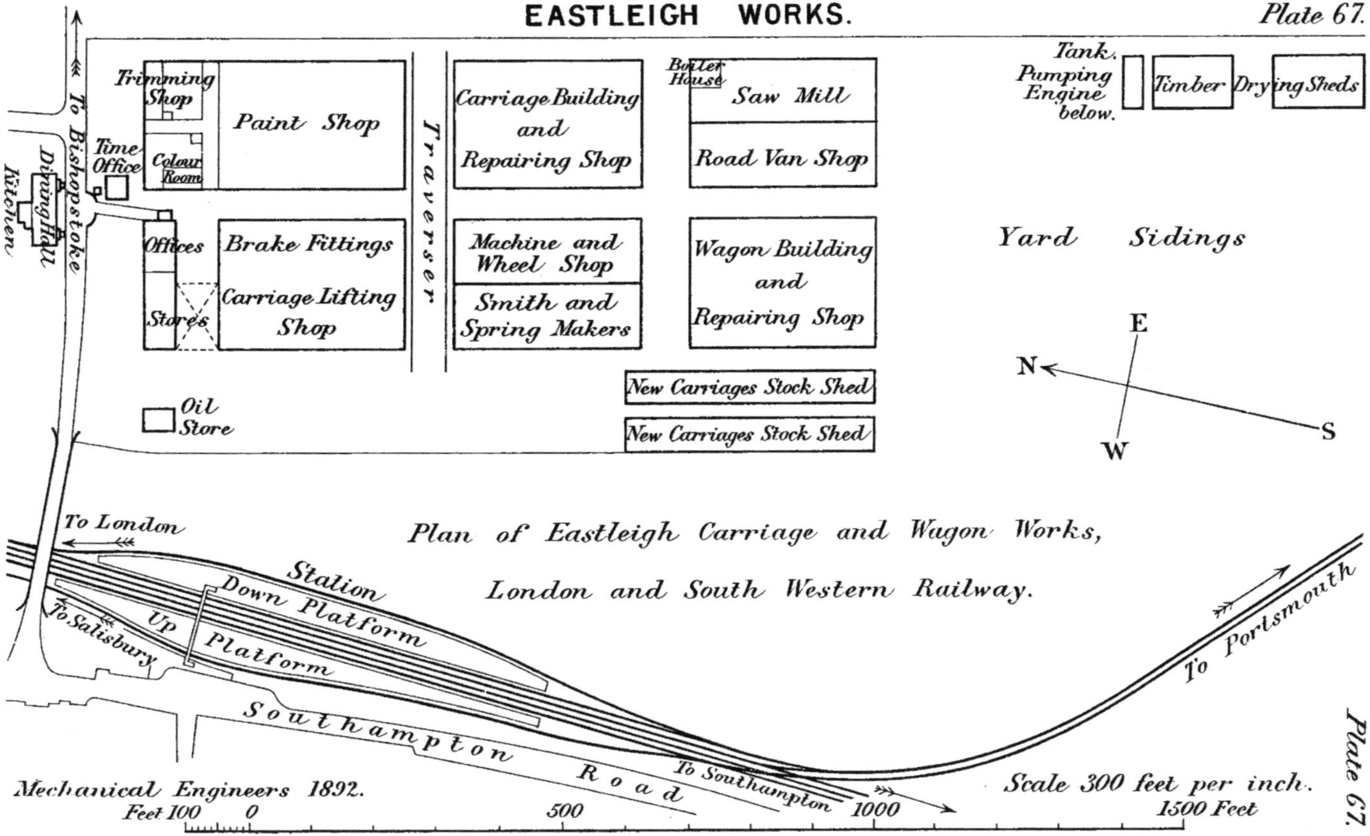

Trimming Shop
Colour Room
Paint Shop
Time Office
Offices
Stores
Brake Fittings
Carriage Lifting Shop
Traverser
Carriage Building and Repairing Shop
Machine and Wheel Shop
Smith and Spring Makers
Boiler House
Saw Mill
Road Van Shop
Wagon Building and Repairing Shop
New Carriages Stock Shed
New Carriages Stock Shed
Oil Store
Tank. Pumping Engine below.
Timber Drying Sheds
Yard Sidings
N
E
S
W
To Bishopstoke
Dining Hall
Kitchen
To London
Station
Down Platform
Up Platform
To Salisbury
Southampton Road
To Southampton
To Portsmouth

Plan of Eastleigh Carriage and Wagon Works,
London and South Western Railway.

Scale 300 feet per inch.
Feet 100 0 500 1000 1500 Feet

Plate 67.

BRIGHTON LOCOMOTIVE WORKS.

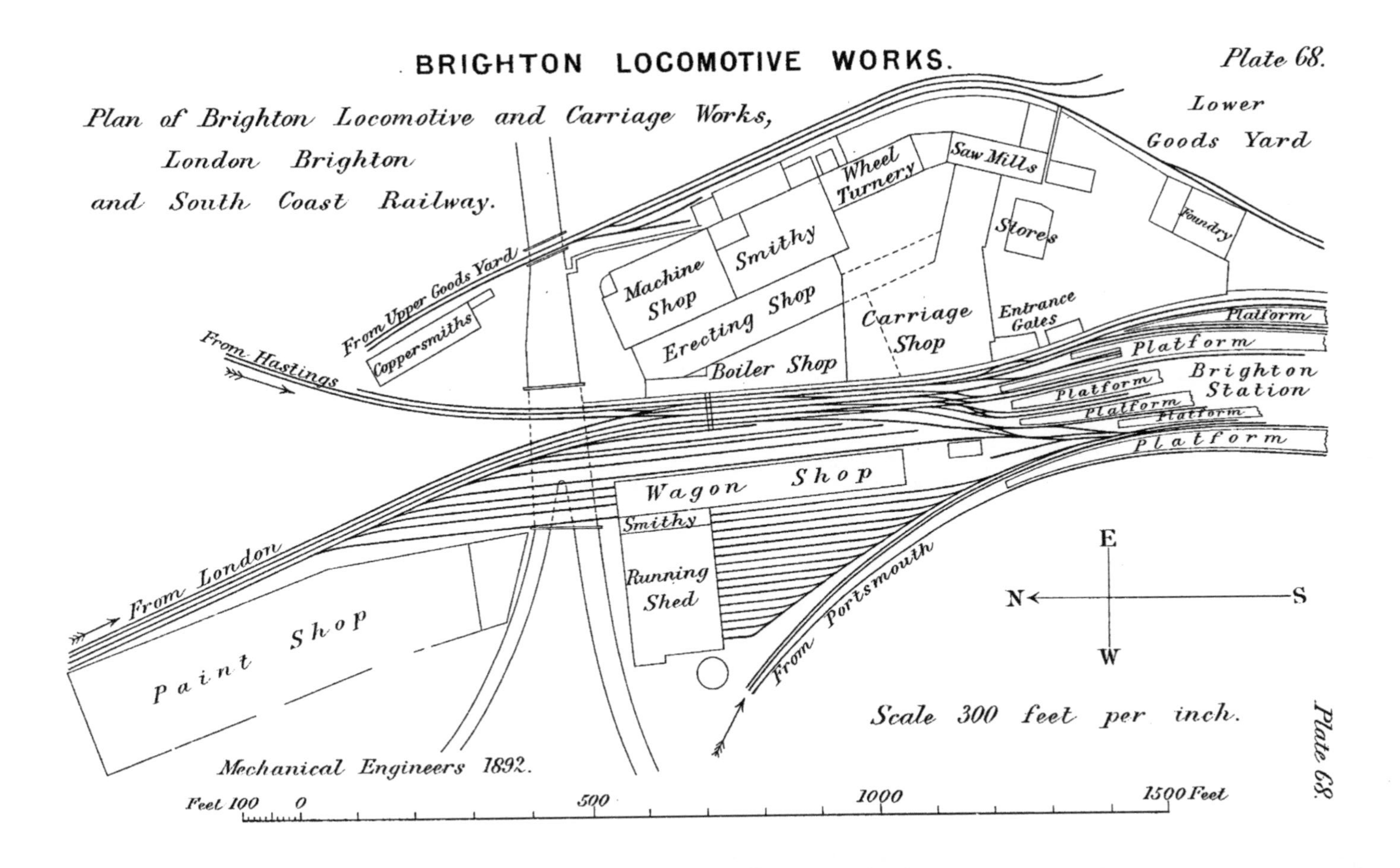
Plan of Brighton Locomotive and Carriage Works,
London Brighton
and South Coast Railway.
Lower
Goods Yard
Foundry
Saw Mills
Stores
Wheel Turnery
Smithy
Machine Shop
Erecting Shop
Boiler Shop
Carriage Shop
Entrance Gates
Platform
Platform
Platform
Platform
Platform
Platform
Brighton Station
From Upper Goods Yard
Coppersmiths
From Hastings
Wagon Shop
Smithy
Running Shed
From London
Paint Shop
From Portsmouth
E
N
S
W
Scale 300 feet per inch.
Mechanical Engineers 1892.
Feet 100
0
500
1000
1500 Feet

Plate 68.

VALUE OF STEAM-JACKET.

Fig: 1. *Beam Compound Pumping Engine. Hammersmith. Experiment 42.*

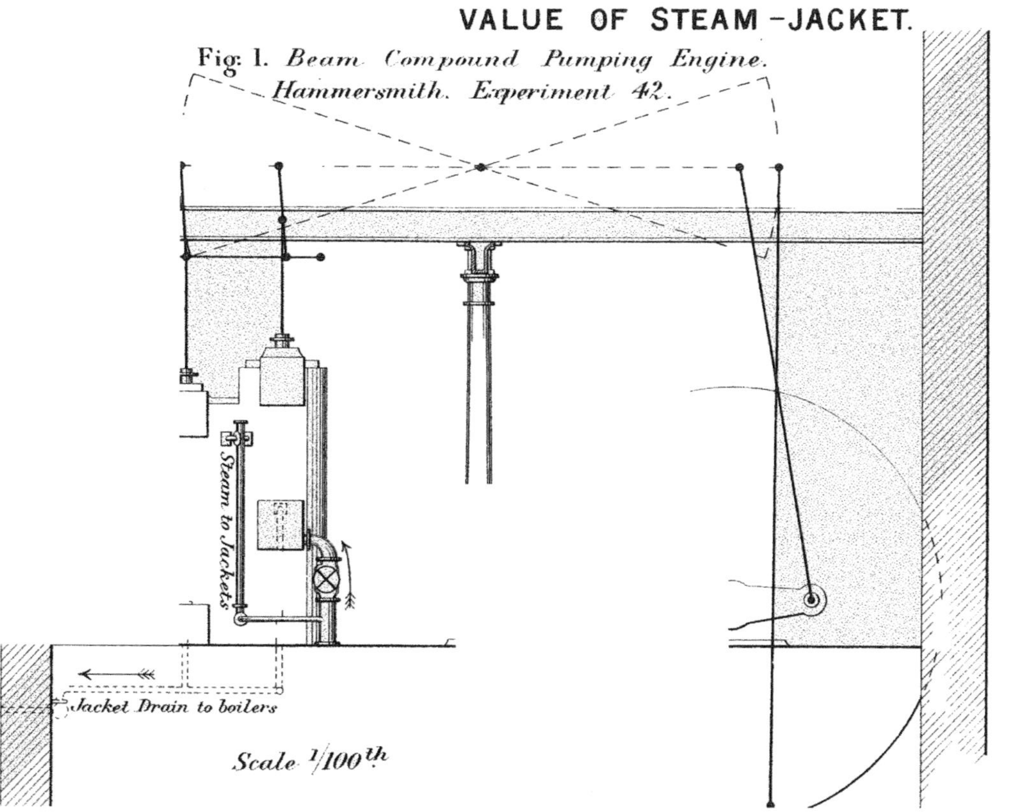

Scale 1/100th

Fig: 2. *Two-Cylinder Hor Experimental E South Kensin Experiments 45*

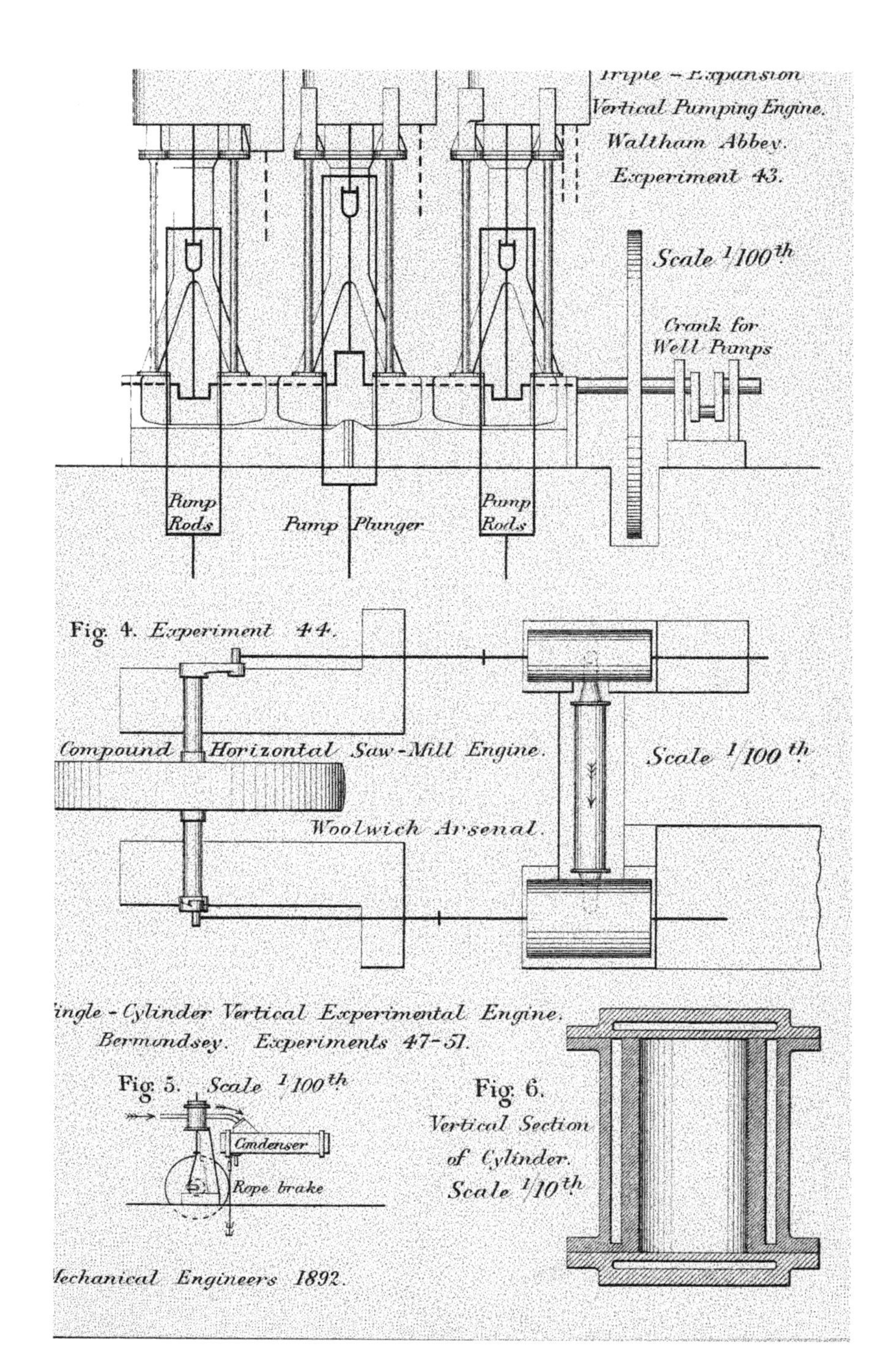
Triple - Expansion
Vertical Pumping Engine.
Waltham Abbey.
Experiment 43.
Scale 1/100th
Crank for
Well Pumps
Pump
Rods
Pump Plunger
Pump
Rods
Fig. 4. Experiment 44.
Compound Horizontal Saw-Mill Engine.
Scale 1/100th
Woolwich Arsenal.
ingle - Cylinder Vertical Experimental Engine.
Bermondsey. Experiments 47-51.
Fig. 5. Scale 1/100th
Condenser
Rope brake
Fig. 6.
Vertical Section
of Cylinder.
Scale 1/10th
lechanical Engineers 1892.

Beam Compound Pumping Engine. Hammersmith.

Fig. 7. *Indicator Diagrams nearest mean. Experiment 42.*

Jackets with or without steam		WITHOUT	WITH	
Mean Effective Pressures	*High-p. cyl.*	29·48	24·00	*lbs. per sq. inch*
	Low-p. cyl.	5·45	6·40	
	Revs. per min.	14·80	15·78	
	Total I.H.P.	162·30	167·40	

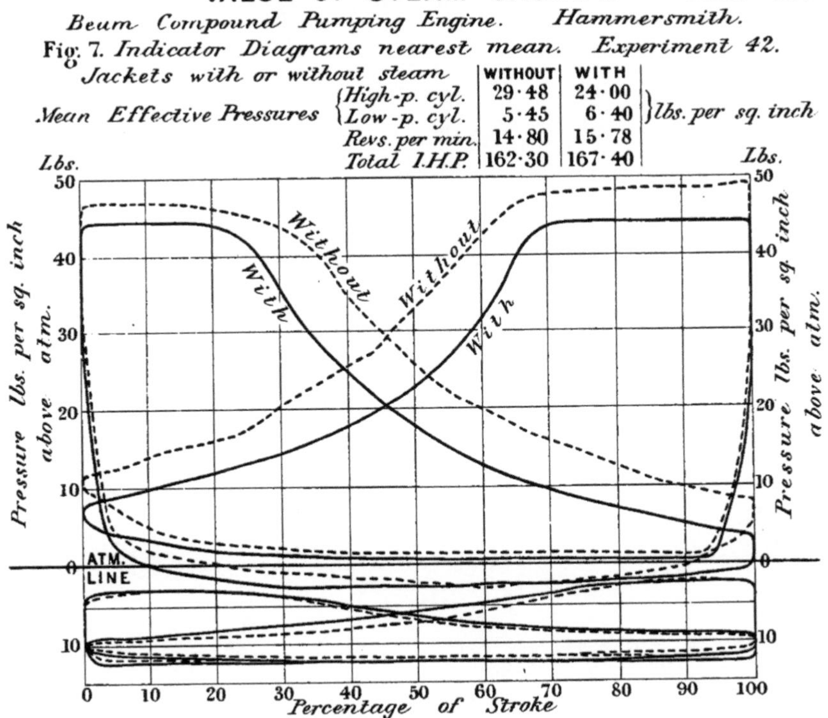

Compound Horizontal Saw-Mill Engine. Woolwich Arsenal.

Fig. 8. *Indicator Diagrams nearest mean. Experiment 44.*

Jackets with or without steam		WITHOUT	WITH	
Mean Effective Pressures	*High-p. cyl.*	18·83	15·12	*lbs. per sq. inch*
	Low-p. cyl.	4·20	4·60	
	Revs. per min.	57·06	63·62	
	Total I.H.P.	110·00	113·40	

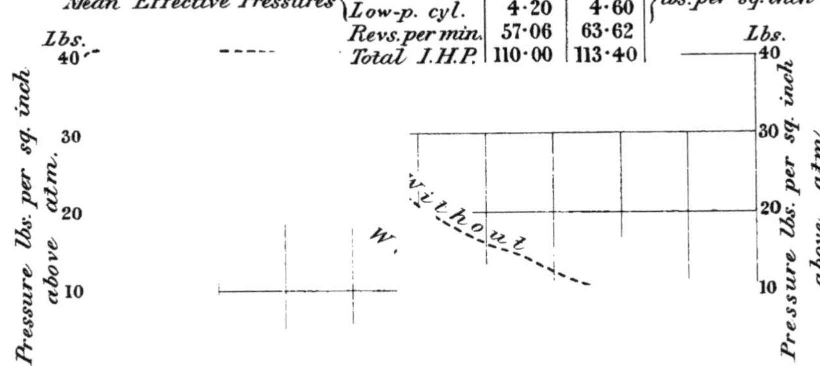

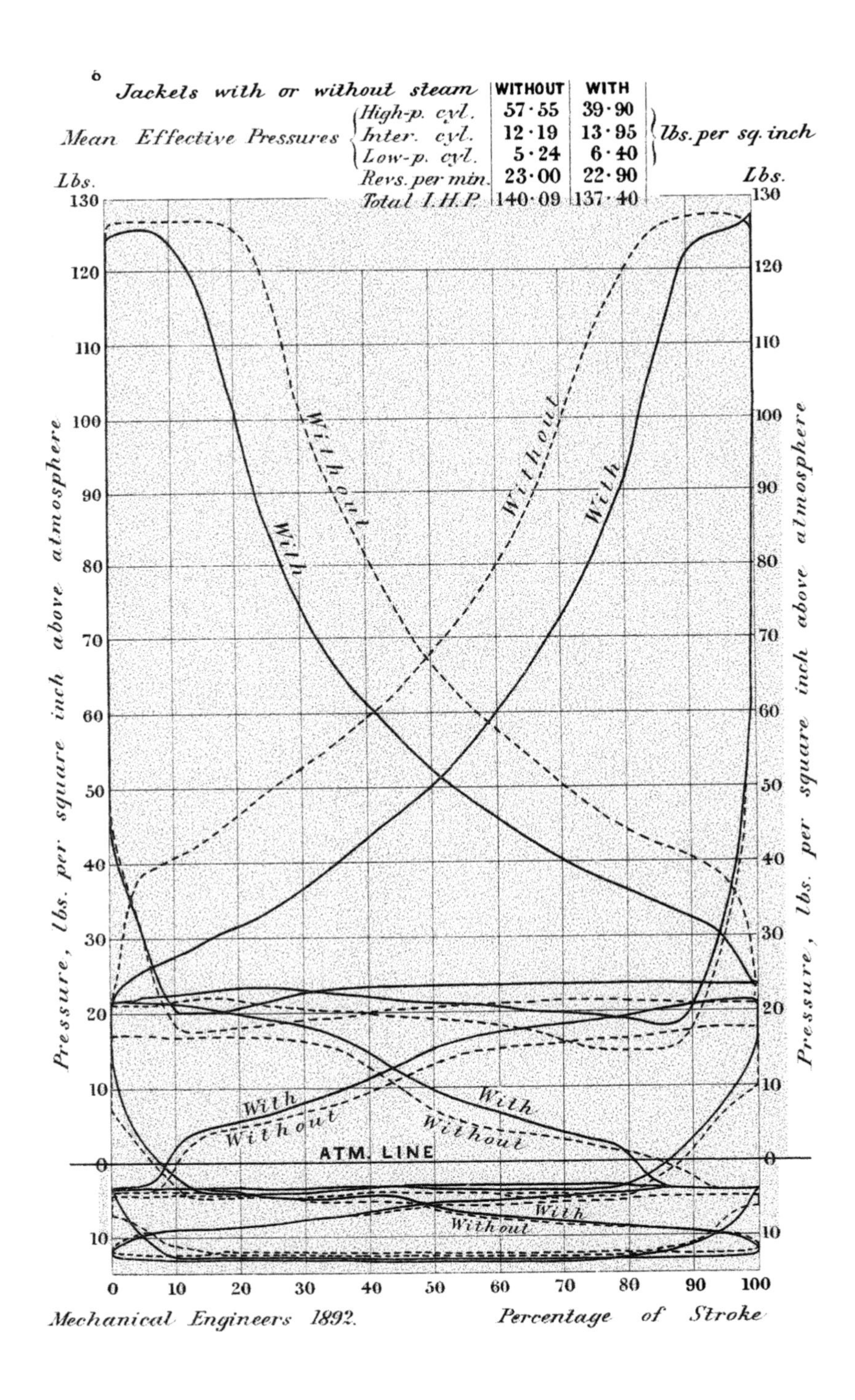
6
Jackets with or without steam
WITHOUT
WITH
Mean Effective Pressures
High-p. cyl. 57·55 39·90
Inter. cyl. 12·19 13·95
Low-p. cyl. 5·24 6·40
lbs. per sq. inch
Revs. per min. 23·00 22·90
Total I.H.P. 140·09 137·40
Lbs.
Lbs.
130
120
110
100
90
80
70
60
50
40
30
20
10
0
10
Pressure, lbs. per square inch above atmosphere
Pressure, lbs. per square inch above atmosphere
Without
With
Without
With
With
Without
With
Without
ATM. LINE
With
Without
0
10
20
30
40
50
60
70
80
90
100
Mechanical Engineers 1892.
Percentage of Stroke

VALUE OF STEAM-JACKET. *Plate 73.*

Two-Cylinder Horizontal Experimental Engine. South Kensington.

Fig. 10. *Indicator Diagrams nearest mean. Experiment 45. Single Cylinder.*

Jackets with or without steam	WITHOUT	WITH	
Mean Effective Pressures	17·10	17·49	*lbs. per sq. inch*
Revs. per min.	112·40	101·73	
I. H. P.	41·07	38·02	

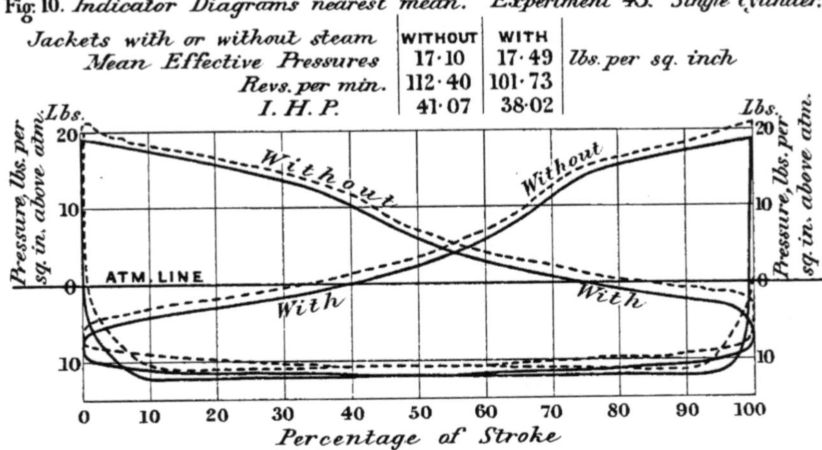

Fig. 11. *Indicator Diagrams nearest mean. Experiment 46. Compound.*

Jackets with or without steam		WITHOUT	WITH	
Mean Effective Pressures	*High-p. cyl.*	30·60	26·40	*lbs. per sq. inch*
	Low-p. cyl.	12·85	14·34	*lbs. per sq. inch*
	Revs. per min.	93·66	96·11	
	Total I.H.P.	44·01	45·64	

Lbs. 65 60 50 40 30 20 10

Pressure, lbs. per sq. inch above atm.

Lbs. 65 60 50 40 30 20 10

Pressure, lbs. per sq. inch above atm.

VALUE OF STEAM-JACKET. *Plate 74.*

Single-cylinder Vertical Experimental Engine. Bermondsey.

Fig. 12. *Indicator Diagrams nearest mean. Experiment 48. Condensing.*

Jackets with or without steam	WITHOUT	WITH	
Mean Effective Pressures	29·98	29·01	*lbs. per sq. inch*
Revs. per min.	116·80	114·90	
I.H.P.	3·92	3·72	

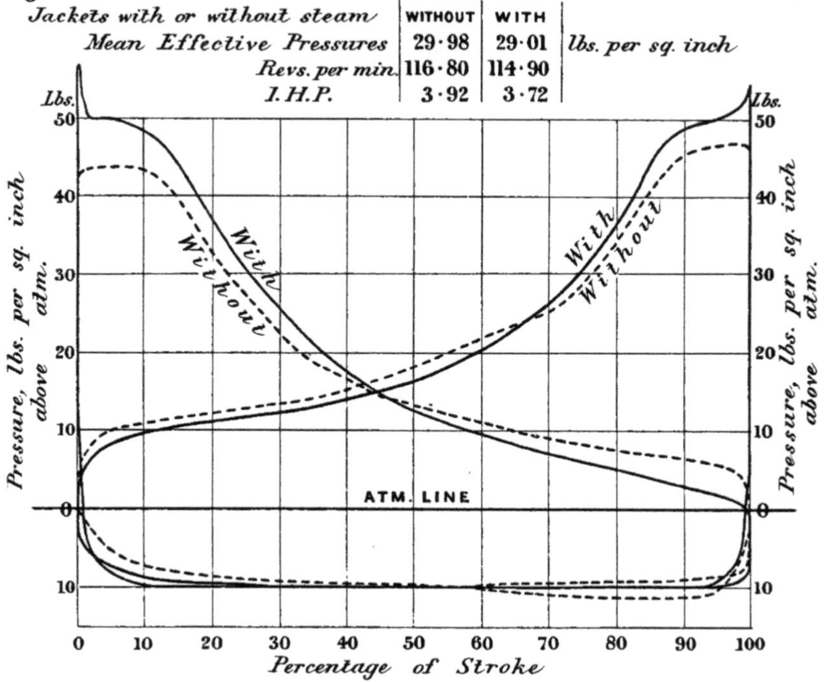

Fig. 13. *Indicator Diagrams nearest mean. Experiment 49. Non-Condensing.*

Jackets with or without steam	WITHOUT	WITH	
Mean Effective Pressures	15·42	17·66	*lbs. per sq. inch*
Revs. per min.	220·70	212·40	
I.H.P.	3·80	4·19	

Lbs.
50
40
30
20
10

essure, lbs. per sq. inch above atm.

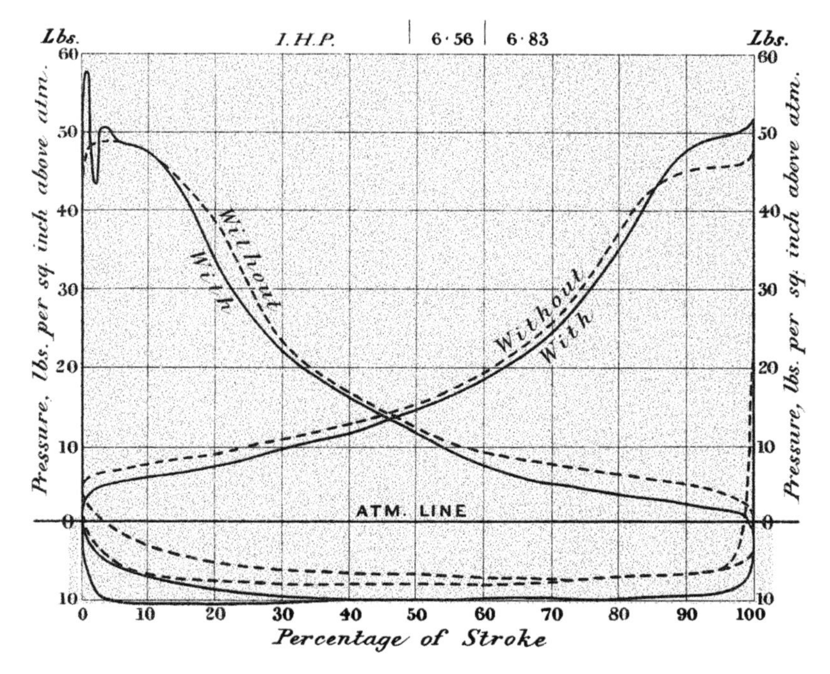

Lbs. Fig. 15. *Experiment 51. Condensing. Body Jacket only used.* Lbs.

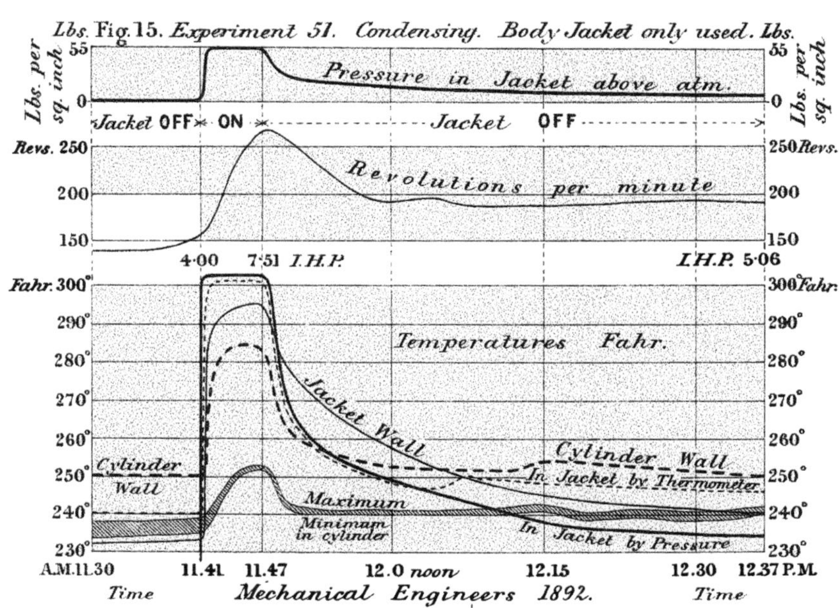

Mechanical Engineers 1892.

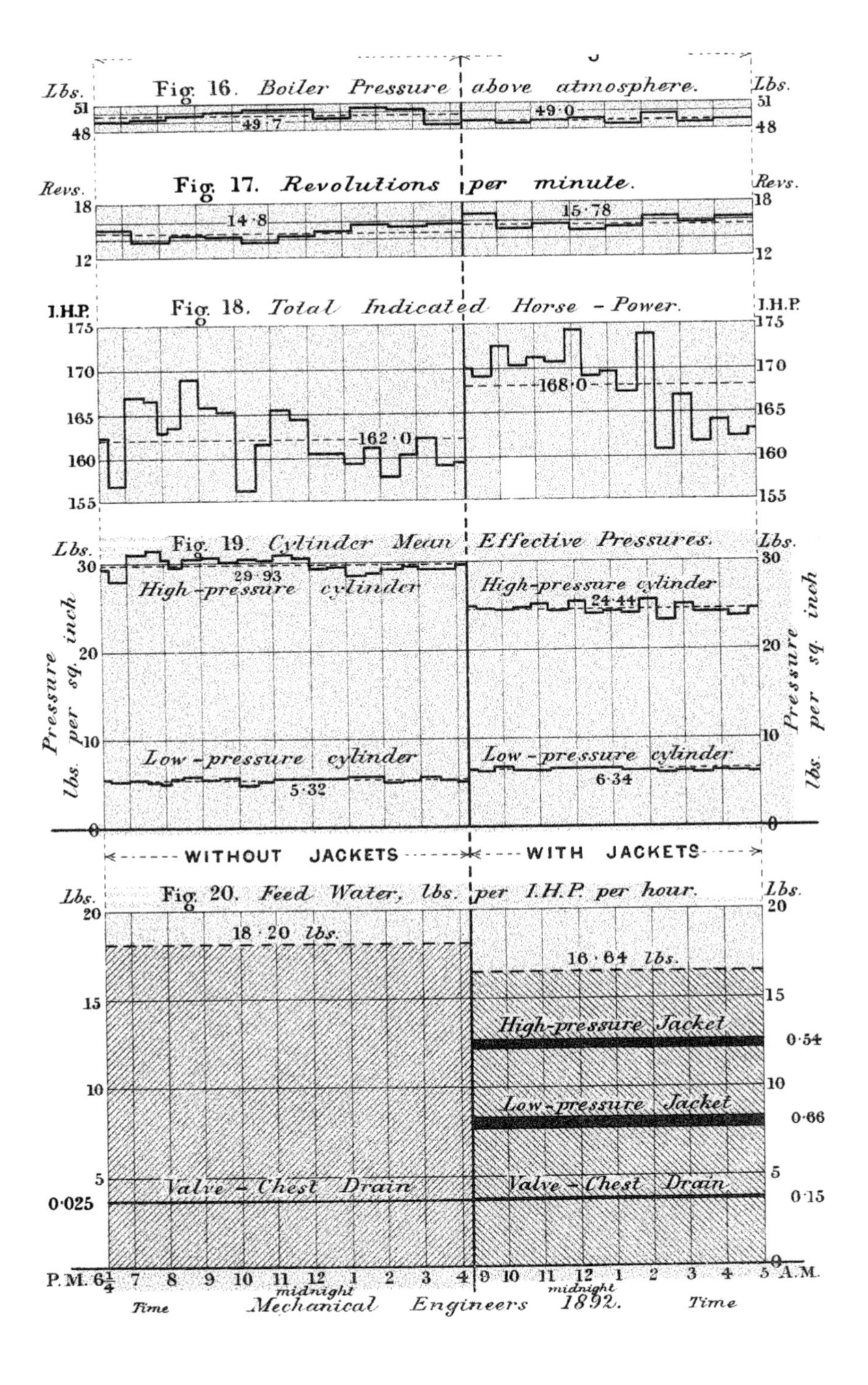
Fig. 16. Boiler Pressure above atmosphere.
Lbs.
51
48
49·7
49·0
Fig. 17. Revolutions per minute.
Revs.
18
12
14·8
15·78
Fig. 18. Total Indicated Horse-Power.
I.H.P.
175
170
165
160
155
162·0
168·0
Fig. 19. Cylinder Mean Effective Pressures.
Lbs.
30
20
10
0
Pressure lbs. per sq. inch
High-pressure cylinder
29·93
24·44
Low-pressure cylinder
5·32
6·34
WITHOUT JACKETS
WITH JACKETS
Fig. 20. Feed Water, lbs. per I.H.P. per hour.
Lbs.
20
15
10
5
0
18·20 lbs.
16·64 lbs.
High-pressure Jacket
0·54
Low-pressure Jacket
0·66
Valve-Chest Drain
0·025
0·15
P.M. 6¼ 7 8 9 10 11 12 1 2 3 4 9 10 11 12 1 2 3 4 5 A.M.
midnight
Time
Mechanical Engineers 1892.

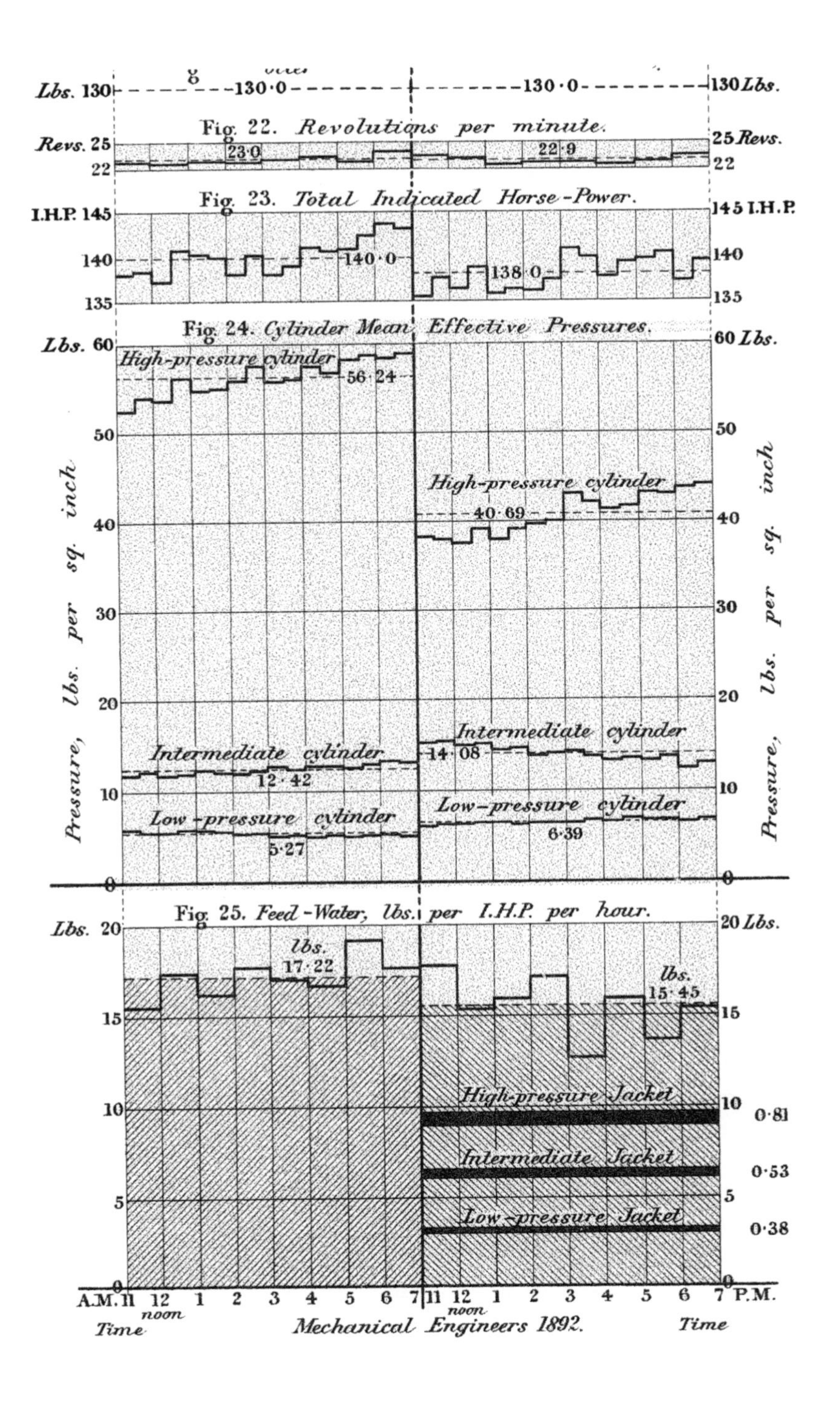
Lbs. 130
130·0
130·0
130 Lbs.
Fig. 22. Revolutions per minute.
Revs. 25
22
23·0
22·9
25 Revs.
22
Fig. 23. Total Indicated Horse-Power.
I.H.P. 145
140
135
140·0
138·0
145 I.H.P.
140
135
Fig. 24. Cylinder Mean Effective Pressures.
Lbs. 60
60 Lbs.
High-pressure cylinder
56·24
High-pressure cylinder
40·69
Intermediate cylinder
12·42
Intermediate cylinder
14·08
Low-pressure cylinder
5·27
Low-pressure cylinder
6·39
Pressure, lbs. per sq. inch
50
40
30
20
10
0
Fig. 25. Feed-Water, lbs. per I.H.P. per hour.
Lbs. 20
20 Lbs.
lbs.
17·22
lbs.
15·45
15
10
5
0
High-pressure Jacket
0·81
Intermediate Jacket
0·53
Low-pressure Jacket
0·38
A.M. 11 12 noon 1 2 3 4 5 6 7
11 12 noon 1 2 3 4 5 6 7 P.M.
Time
Time
Mechanical Engineers 1892.

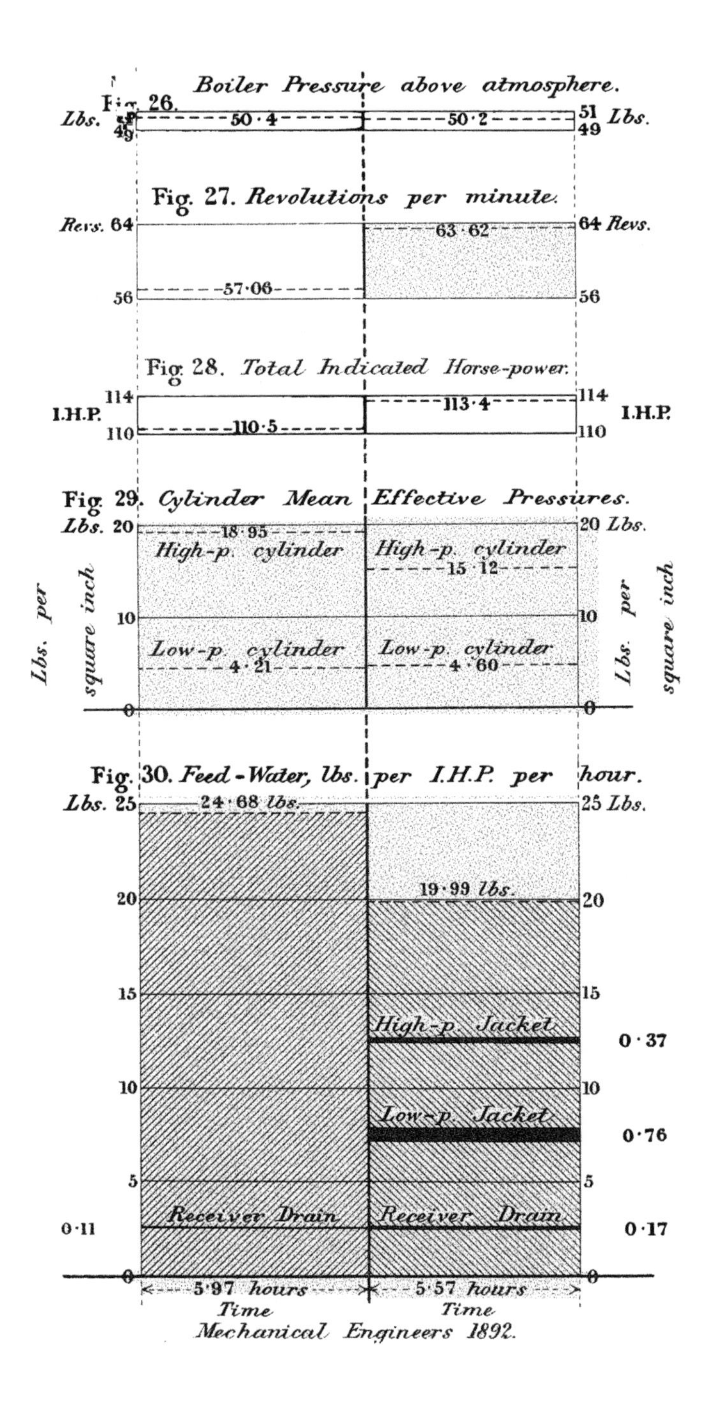

Fig. 26. Boiler Pressure above atmosphere.
Lbs.
50·4
50·2
51
49
Lbs.
Fig. 27. Revolutions per minute.
Revs. 64
63·62
64 Revs.
57·06
56
56
Fig. 28. Total Indicated Horse-power.
114
I.H.P.
110
110·5
113·4
114
I.H.P.
110
Fig. 29. Cylinder Mean Effective Pressures.
Lbs. 20
18·95
High-p. cylinder
High-p. cylinder
15·12
20 Lbs.
10
10
Low-p. cylinder
Low-p. cylinder
4·21
4·60
Lbs. per square inch
Lbs. per square inch
0
0
Fig. 30. Feed-Water, lbs. per I.H.P. per hour.
Lbs. 25
24·68 lbs.
25 Lbs.
19·99 lbs.
20
20
15
15
High-p. Jacket
0·37
10
10
Low-p. Jacket
0·76
5
5
Receiver Drain
Receiver Drain
0·11
0·17
0
0
5·97 hours
5·57 hours
Time
Time
Mechanical Engineers 1892.

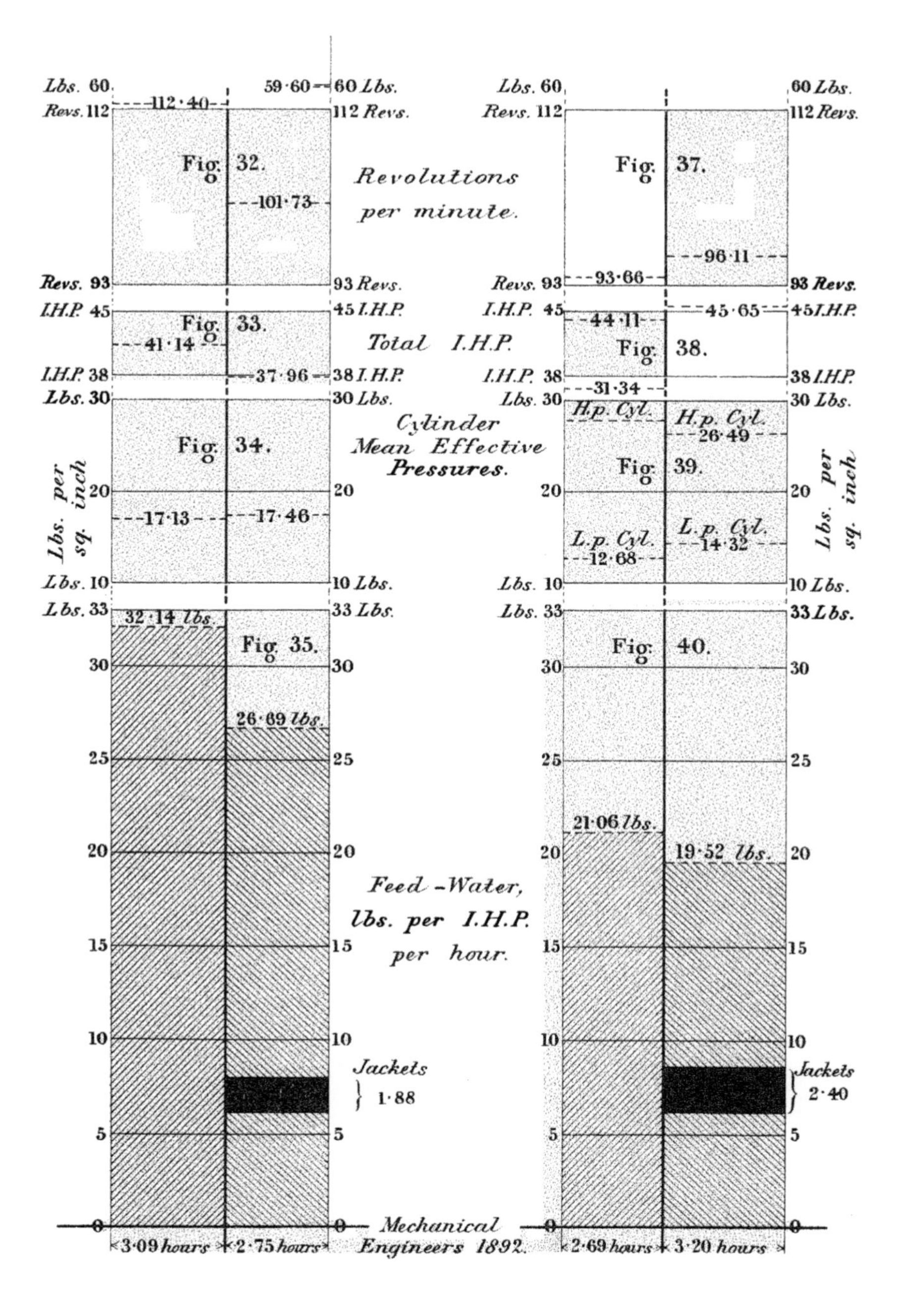

Fig. 32.
Revolutions per minute.
112·40
101·73
59·60
Fig. 33.
Total I.H.P.
41·14
37·96
Fig. 34.
Cylinder Mean Effective Pressures.
Lbs. per sq. inch
17·13
17·46
Fig. 35.
32·14 lbs.
26·69 lbs.
Feed-Water, lbs. per I.H.P. per hour.
Jackets 1·88
3·09 hours
2·75 hours
Mechanical Engineers 1892.
Fig. 37.
93·66
96·11
Fig. 38.
44·11
45·65
31·34
Fig. 39.
H.p. Cyl.
26·49
L.p. Cyl.
12·68
14·32
Fig. 40.
21·06 lbs.
19·52 lbs.
Jackets 2·40
2·69 hours
3·20 hours

VALUE OF STEAM-JACKET. *Plate 80.*

Single-cylinder Vertical Experimental Engine. Bermondsey.

Experiment 47. *Condensing.* — *Experiment 48.* *Condensing.* — *Experiment 49.* *Non-Condensing.*

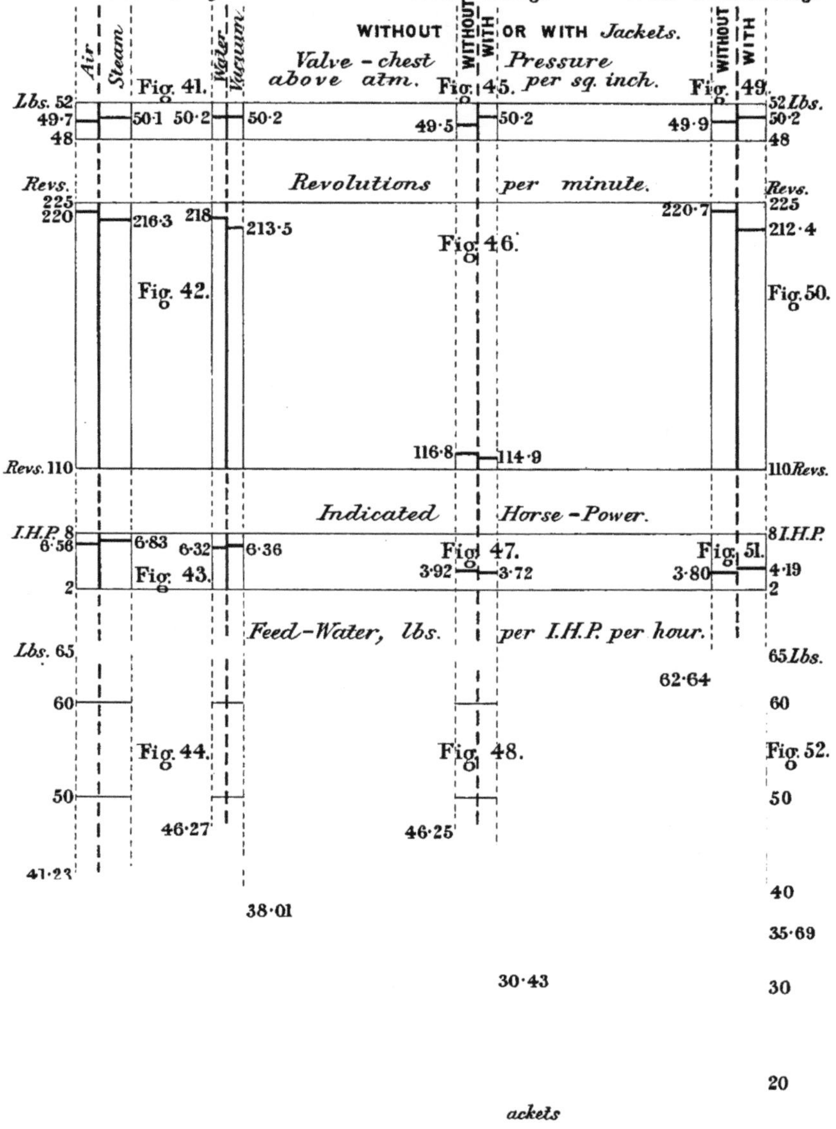

Air · Steam · Water · Vacuum — WITHOUT · WITH — WITHOUT · WITH

WITHOUT OR WITH *Jackets.*

Valve-chest Pressure above atm. per sq. inch.

Fig. 41. Fig. 45. Fig. 49.

Lbs. 52 — 49·7 — 48 — 50·1 — 50·2 — 50·2 — 49·5 — 50·2 — 49·9 — 50·2 — 52 *Lbs.* — 48

Revolutions per minute.

Revs. 225 — 220 — 216·3 — 218 — 213·5 — 220·7 — 212·4 — 116·8 — 114·9 — *Revs.* 110 — 110 *Revs.*

Fig. 42. Fig. 46. Fig. 50.

Indicated Horse-Power.

I.H.P. 8 — 6·56 — 6·83 — 6·32 — 6·36 — 3·92 — 3·72 — 3·80 — 4·19 — 2 — 8 *I.H.P.*

Fig. 43. Fig. 47. Fig. 51.

Feed-Water, lbs. per I.H.P. per hour.

Lbs. 65 — 65 *Lbs.* — 62·64 — 60 — 50 — 46·27 — 46·25 — 41·23 — 40 — 38·01 — 35·69 — 30·43 — 30 — 20

Fig. 44. Fig. 48. Fig. 52.

ackets

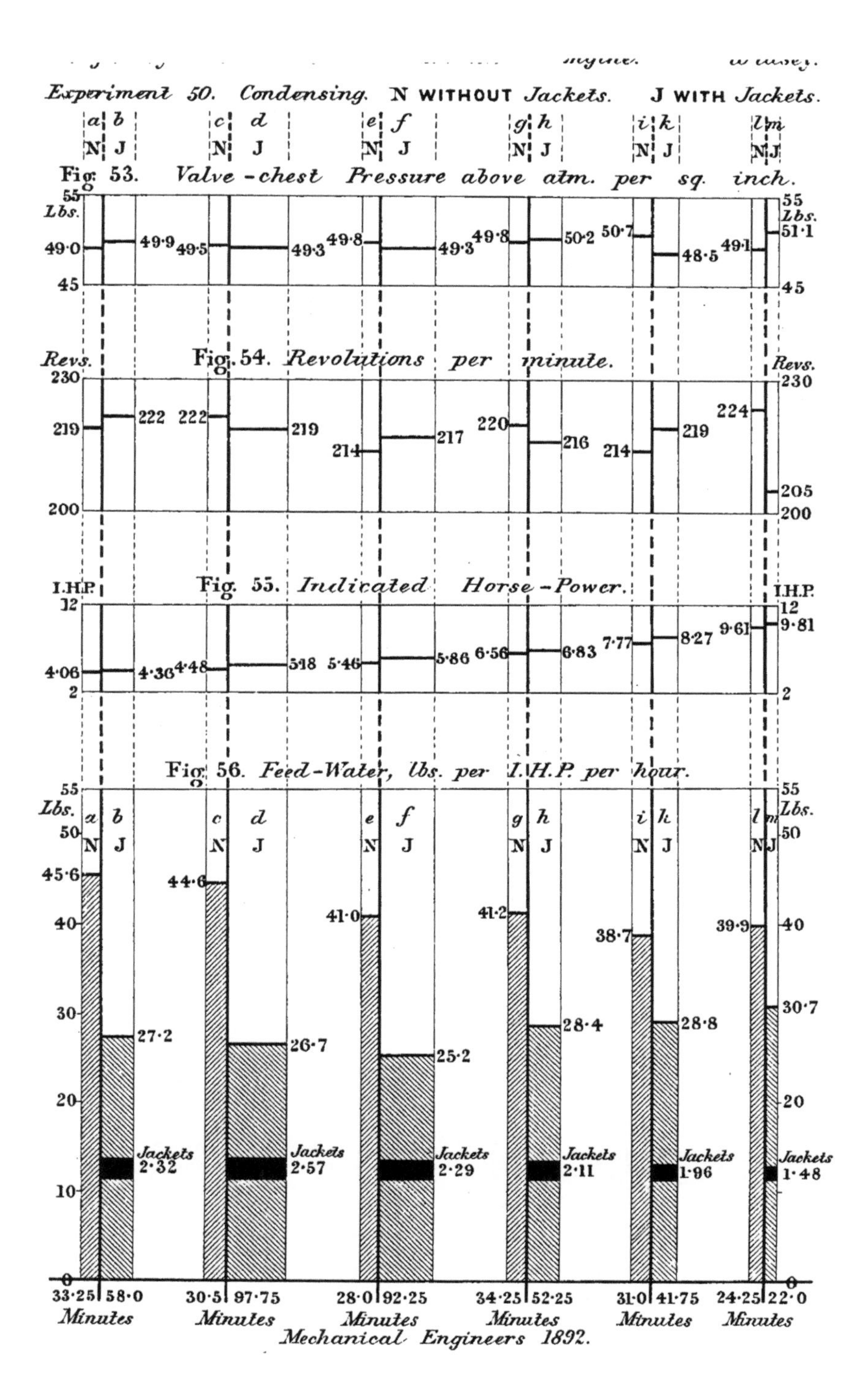

Mechanical Engineers 1892.

VALUE OF STEAM-JACKET. *Plate 82.*

Single-cylinder Vertical Experimental Engine. Bermondsey.
Experiment 50. Condensing.

Fig. 57. *Mean Temperature of Cylinder Wall.*

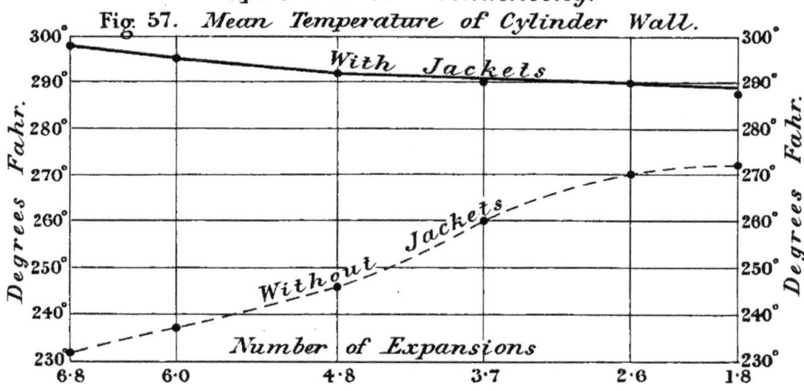

Fig. 58. *Percentage of Feed-water in cylinder as steam at cut-off and at release.*

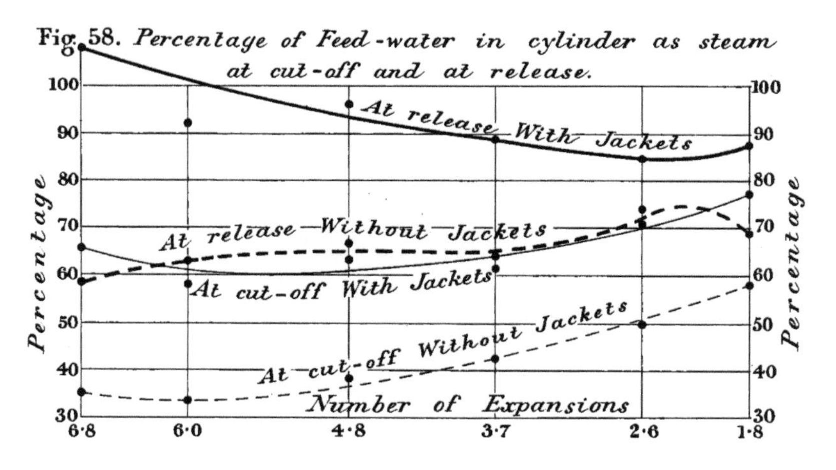

Lbs. Fig. 59. *Feed-Water, lbs. per I.H.P. per hour.* *Lbs.*

45 45

40 40

VALUE OF STEAM-JACKET. *Plate 83.*

Fig. 60. *Experimental Apparatus for determining Transmission of Heat from Steam-Jacket through Cast-Iron Cylinder-Liners of varying thickness.*

VALUE OF STEAM-JACKET. *Plate 84.*

Transmission of Heat from Steam-Jacket through Cast-iron Cylinder-Liners of varying thickness.

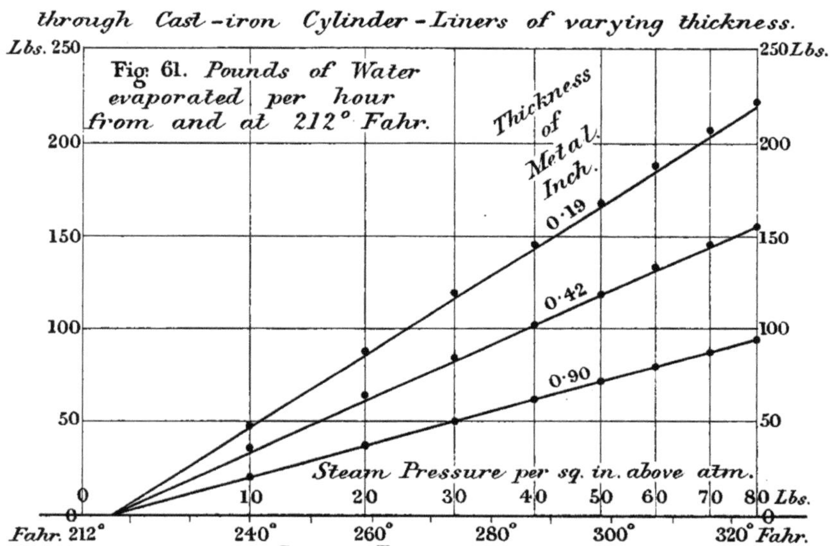

Fig: 61. *Pounds of Water evaporated per hour from and at 212° Fahr.*

Fig: 62. *Evaporation per hour and per degree of difference in Temperature between Steam and Water.*

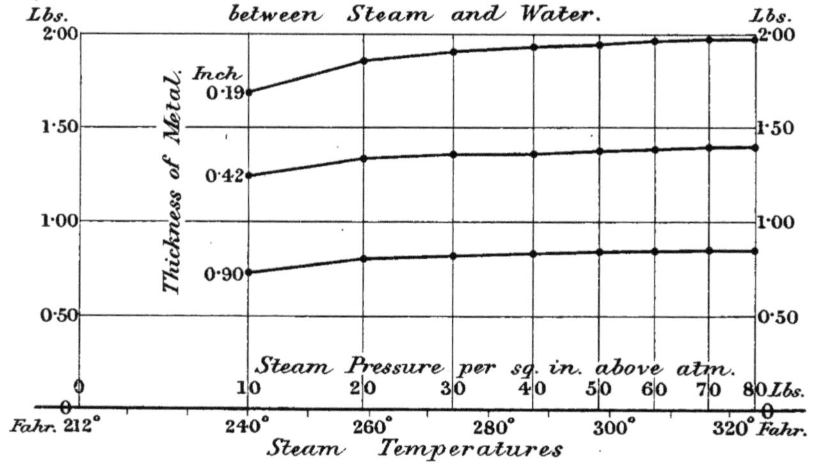

Lbs. 2·00 · 1·50 · 1·00 · 0·50 · 0

Thickness of Metal. Inch: 0·19 · 0·42 · 0·90

Steam Pressure per sq. in. above atm. 10 20 30 40 50 60 70 80 Lbs.

Fahr. 212° 240° 260° 280° 300° 320° Fahr.

Steam Temperatures

Fig: 63. *Evaporation per hour per degree of difference, and per square foot of internal surface of liner.*

0·75 Lb.

Fig: 64. *Draining of Steam-Jackets through Evaporator into Feed-Water Heater.*

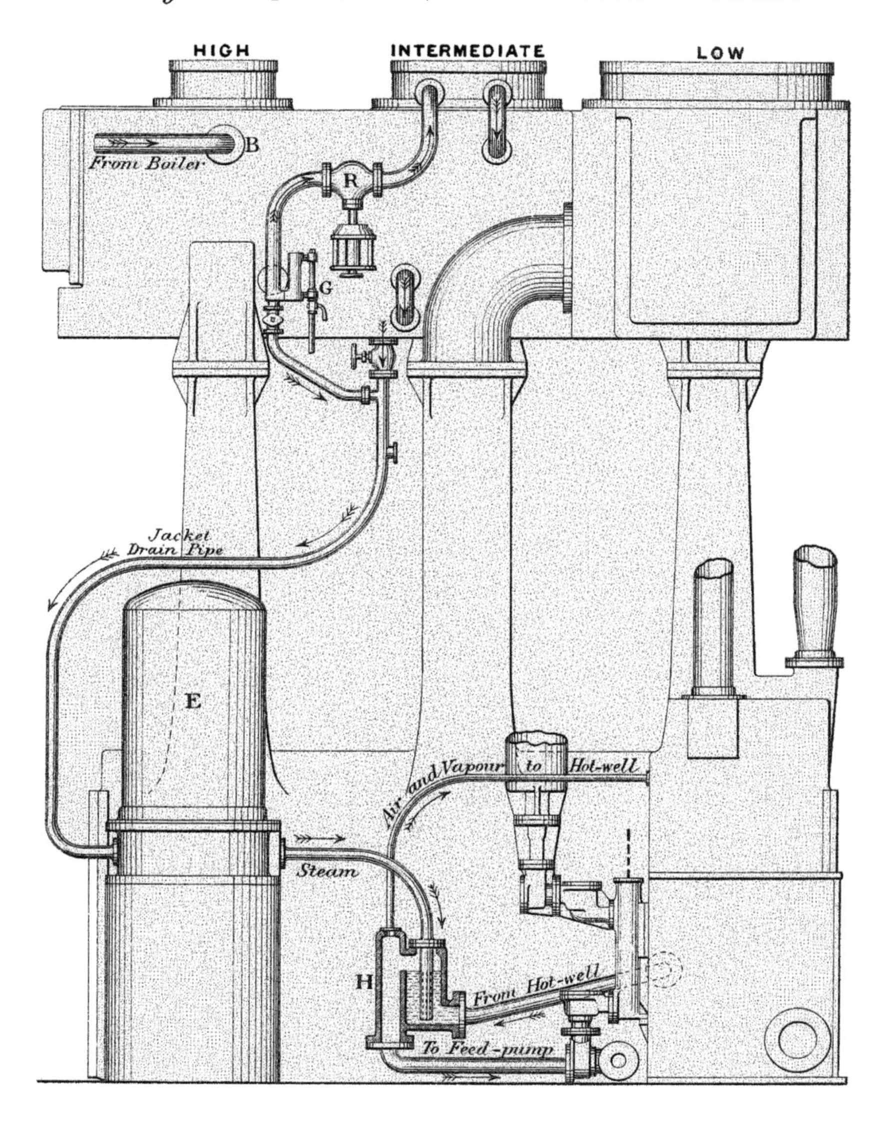

Mechanical Engineers 1892.

High-pressure Jacketed Cylinder with Circulating Receiver.

Fig: 65. *Sectional Plan.*

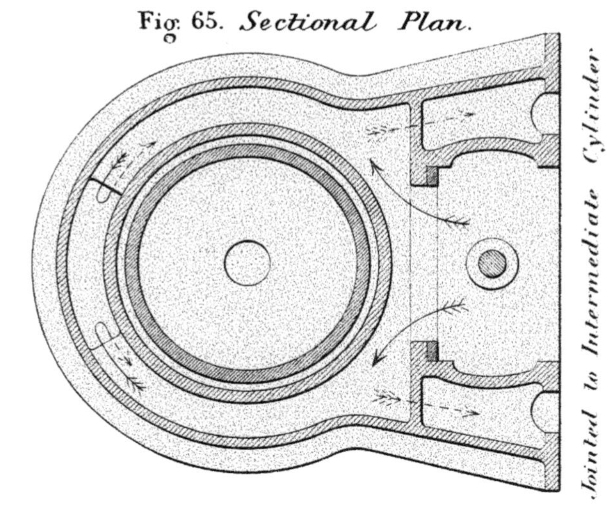

Fig: 66. *Vertical Section.*

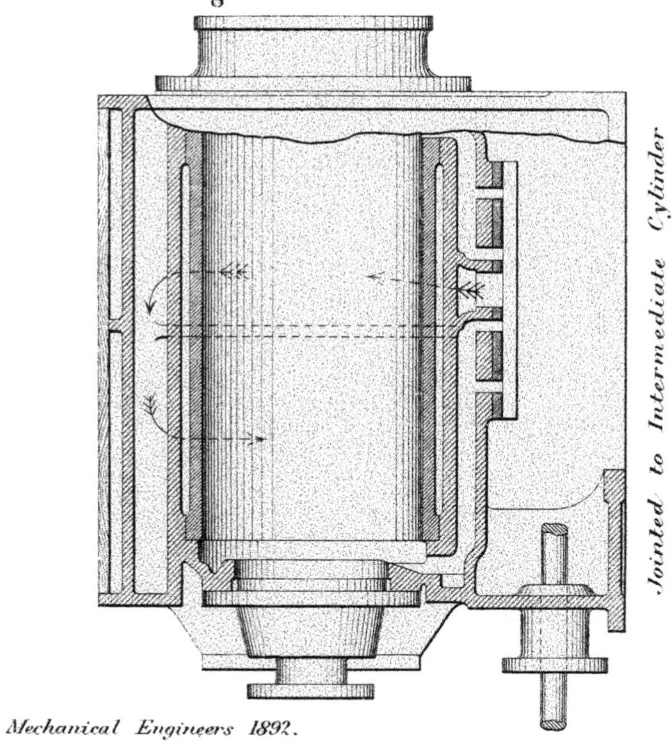

High-pressure Jacketed Cylinder with Circulating Receiver.

Fig. 67. *Sectional Plan.*

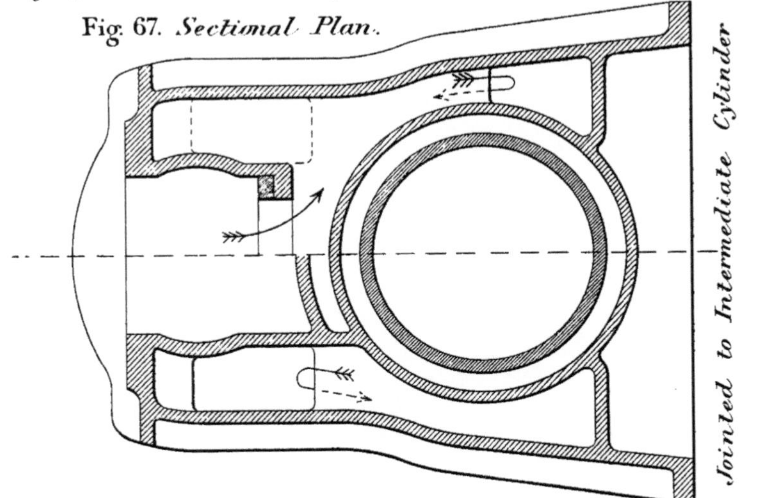

Fig. 68. *Vertical Section.*

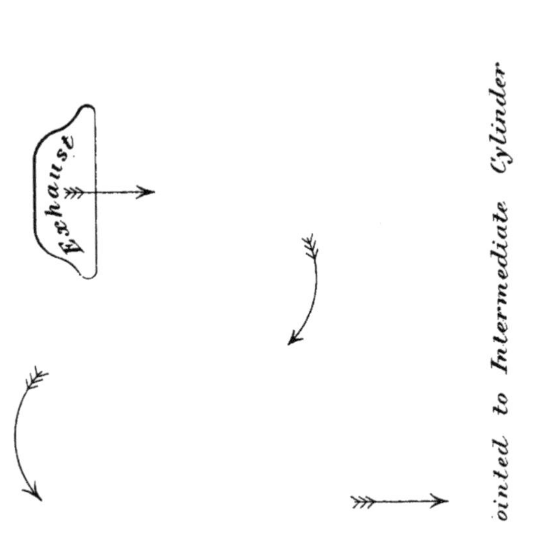

SCREW - PROPELLER SURFACE.

Yacht "Ethel." Arrangement of blades, looking forwards.

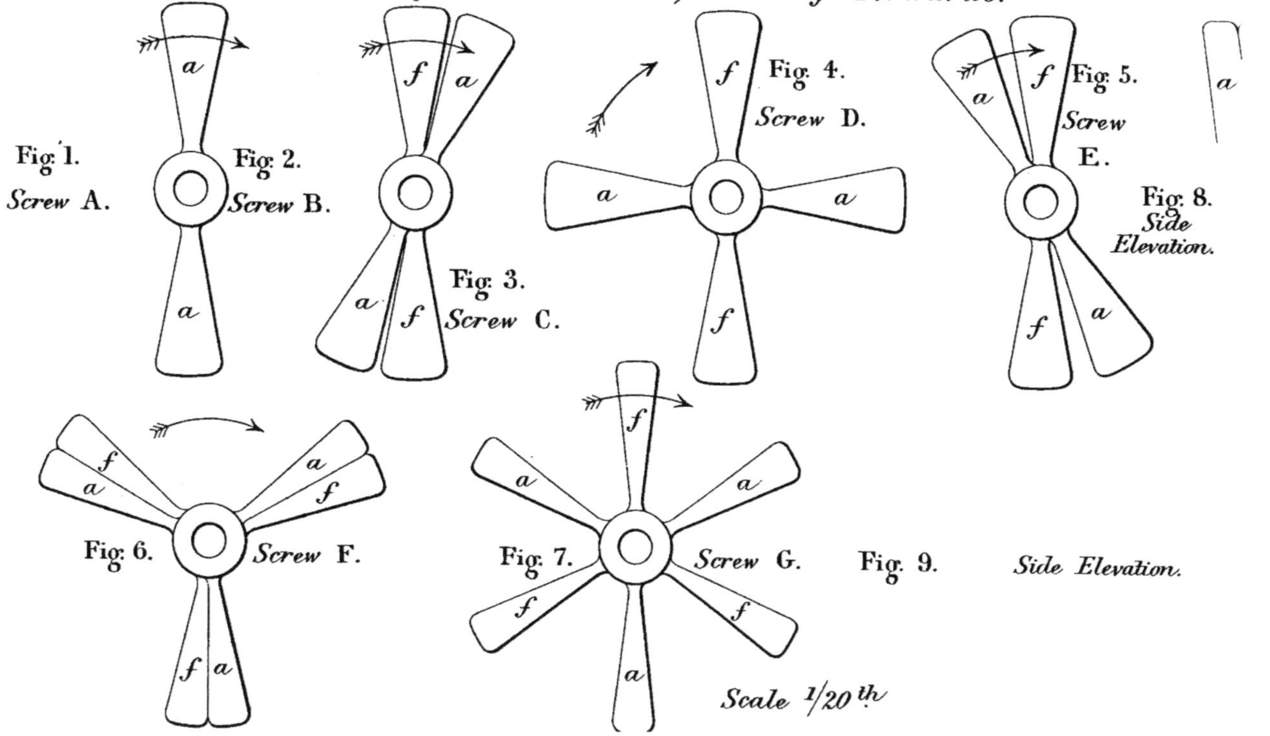

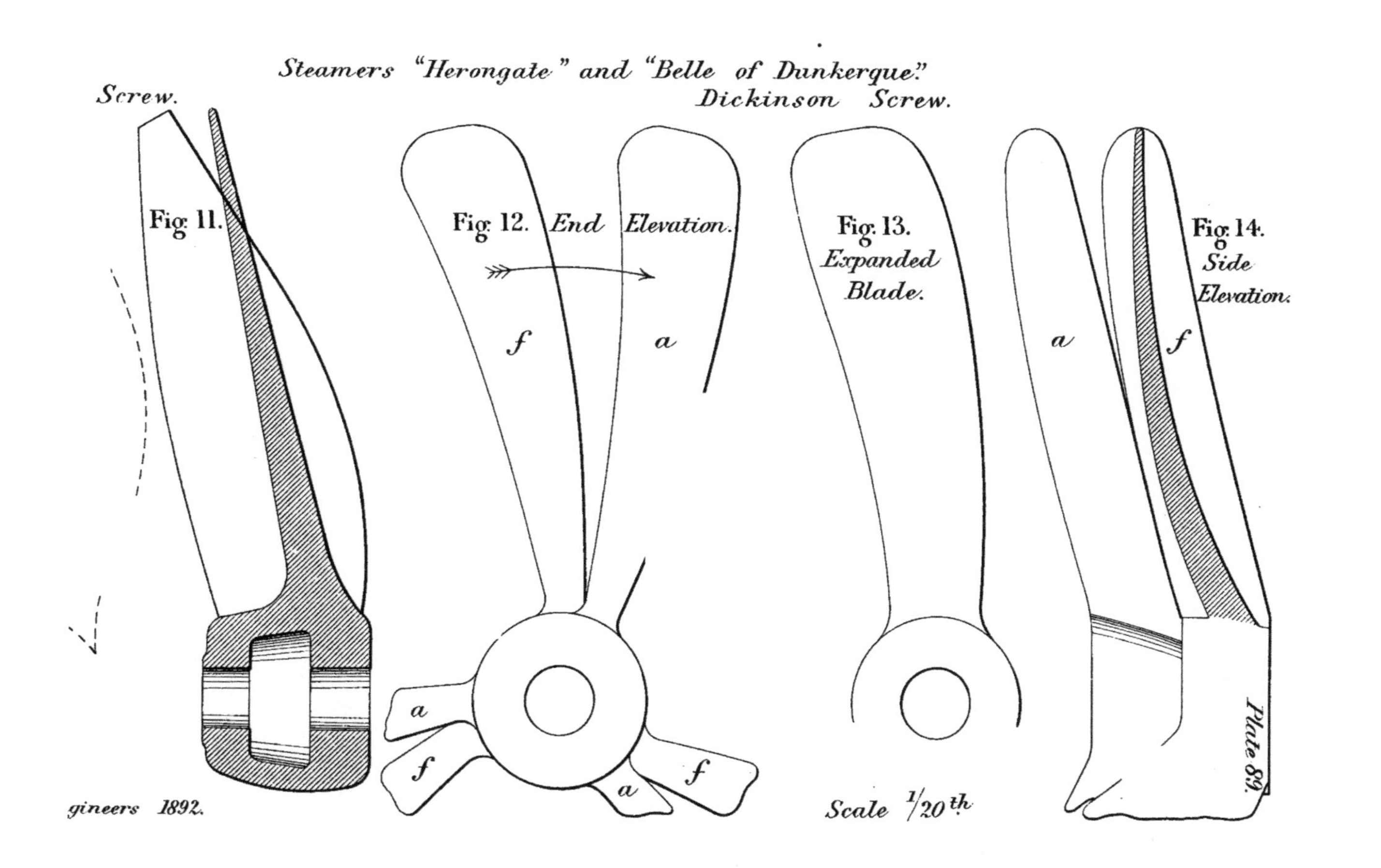
Steamers "Herongate" and "Belle of Dunkerque."
Dickinson Screw.
Screw.
Fig. 11.
Fig. 12. End Elevation.
f
a
a
f
a
f
Fig. 13.
Expanded
Blade.
Fig. 14.
Side
Elevation.
a
f
Plate 89.
gineers 1892.
Scale 1/20th.

Tug "Frank Stanley."

inary Screw.

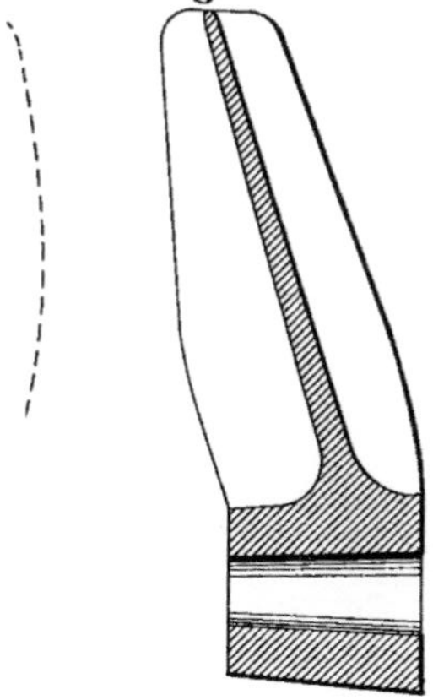

Dickinson Screw.

Fig. 16.

Fig. 17. End Elevation.

f

a

a

f

f

a

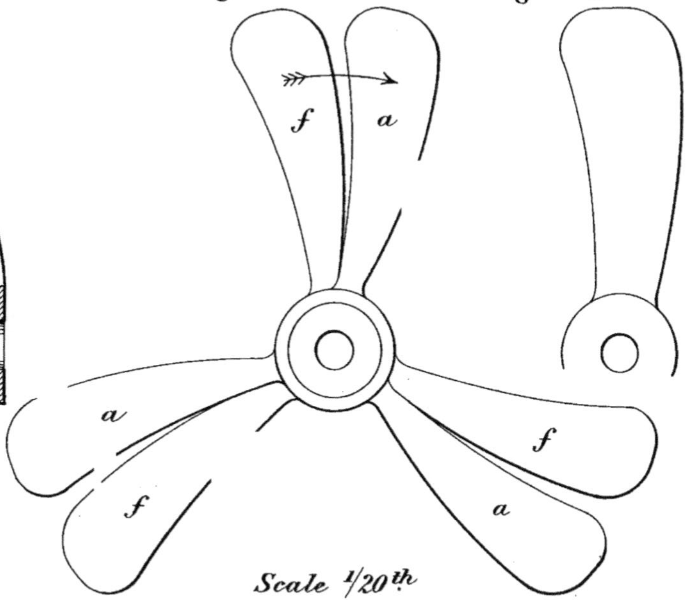

Fig. 18. Expanded Blade.

Fig. 19.

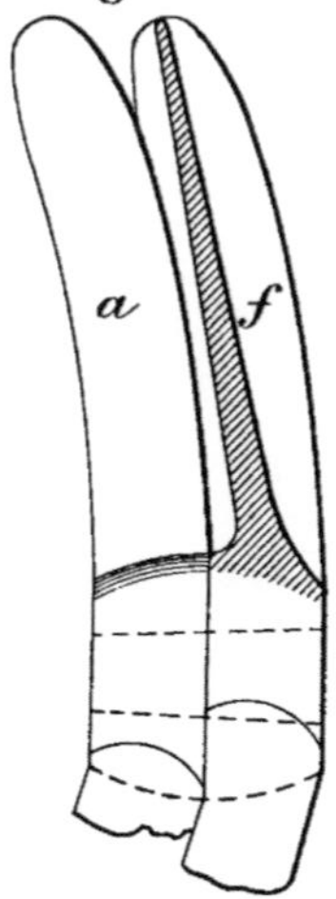

a

f

Side Elevation.

Scale 1/20th.

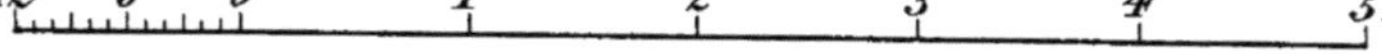

Inches 12 6 0 1 2 3 4 5 Feet

SCREW-PROPELLER SURFACE.

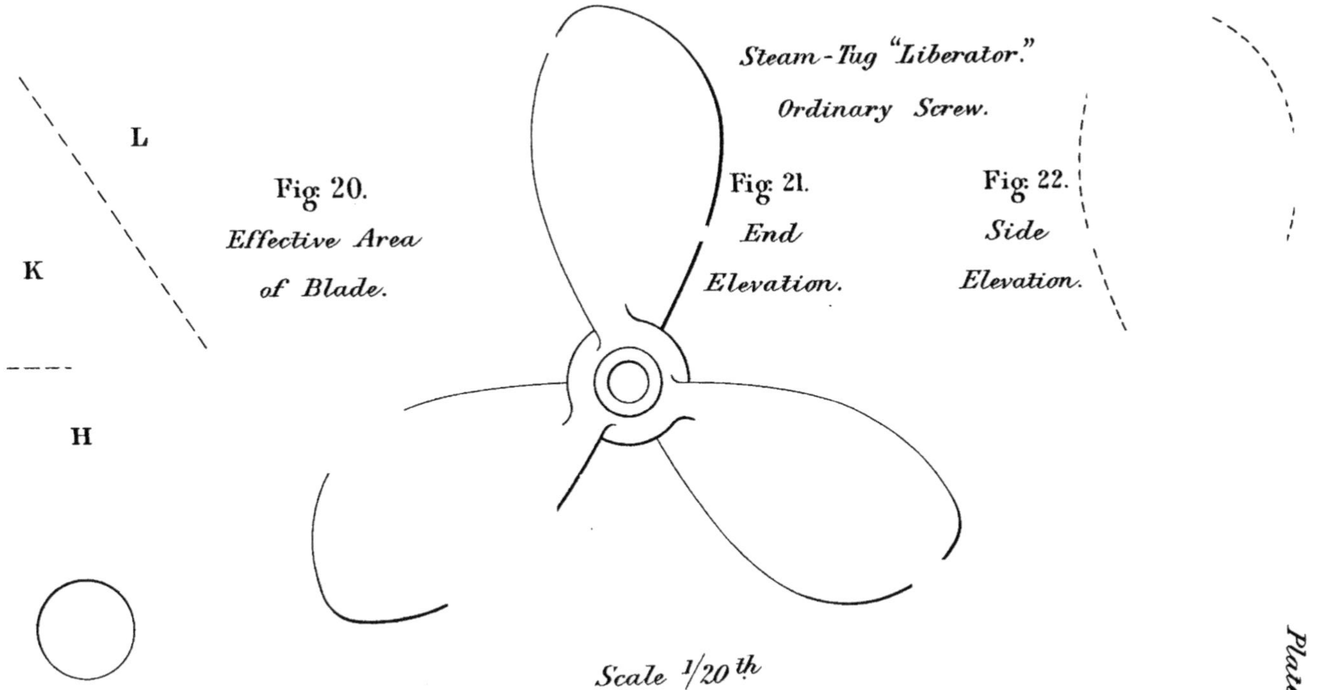

Steam-Tug "Liberator."

Ordinary Screw.

Fig: 20. *Effective Area of Blade.*

Fig: 21. *End Elevation.*

Fig: 22. *Side Elevation.*

L

K

H

Scale 1/20th

SCREW-PROPELLER SURFACE.

Steam-Tug "Liberator." Myers Propeller.

Fig. 23.

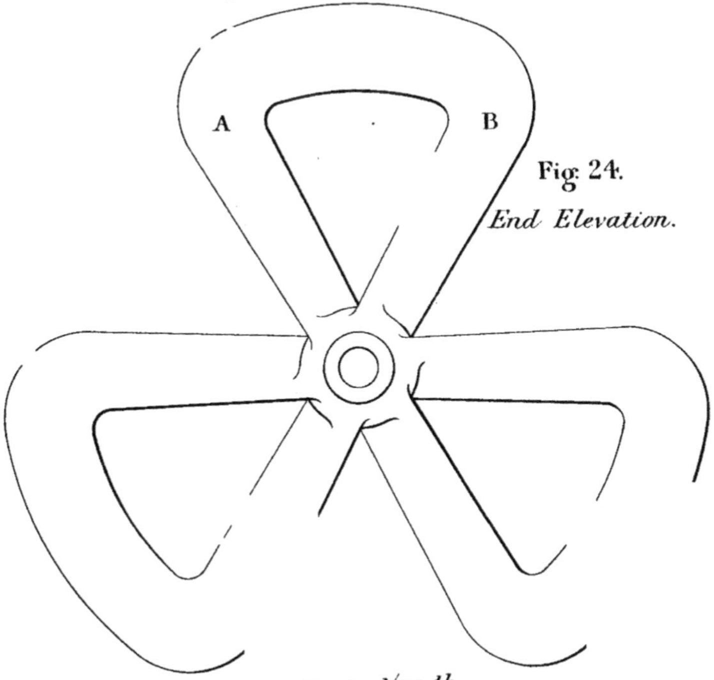

Fig. 24. *End Elevation.*

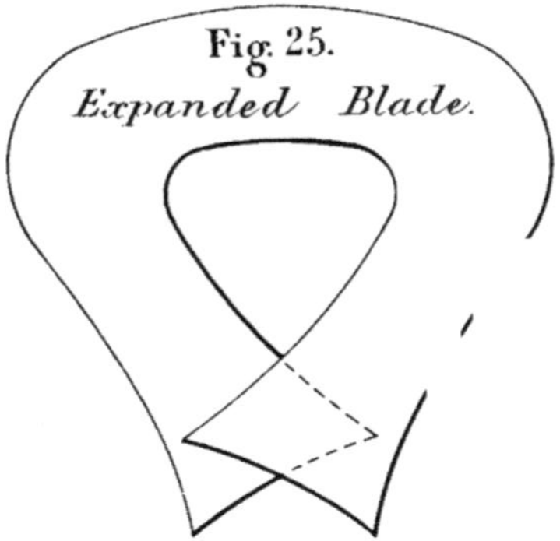

Fig. 25. *Expanded Blade.*

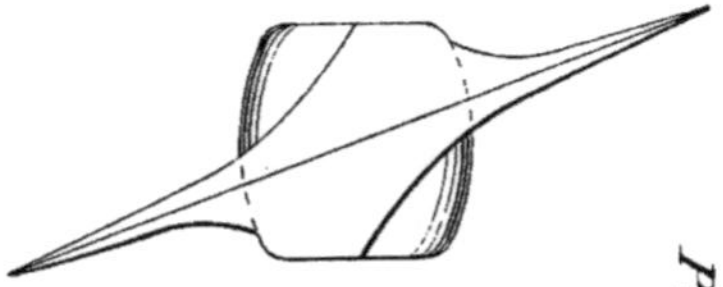

Fig. 26. *Plan.*

Scale 1/20 th

SCREW - PROPELLER SURFACE.

Twin - Screw Ferry - boat "Wirral." Myers Propeller.

Fig. 27.

Side Elevation.

Fig. 28.

End Elevation.

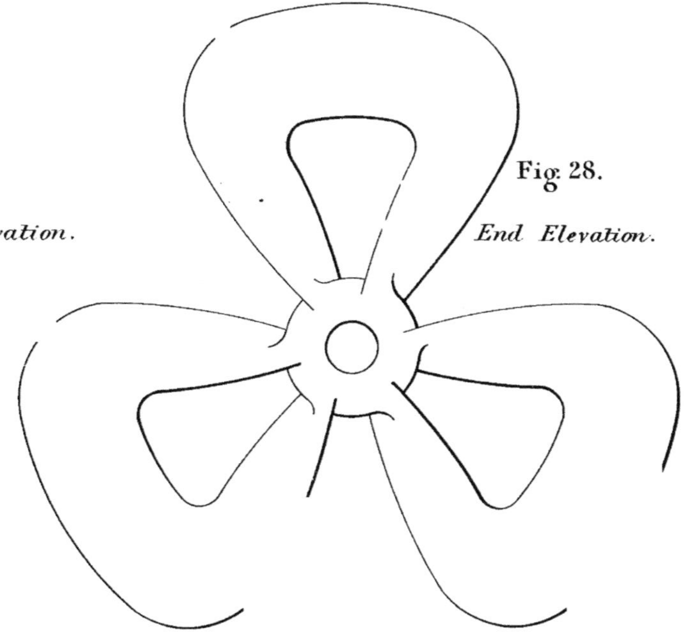

Fig. 29.

Section of Blade.

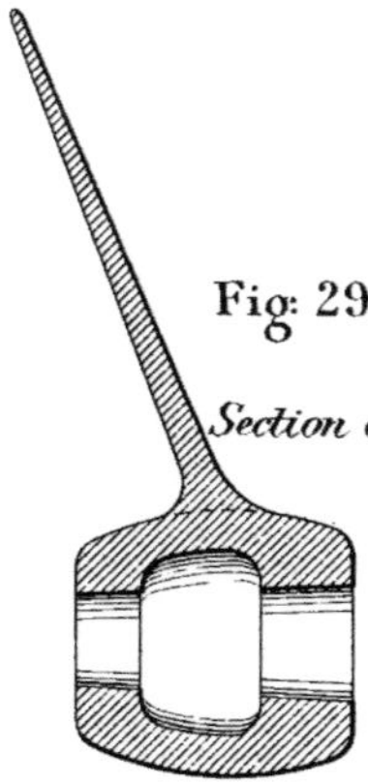

Fig. 30. *Plan.*

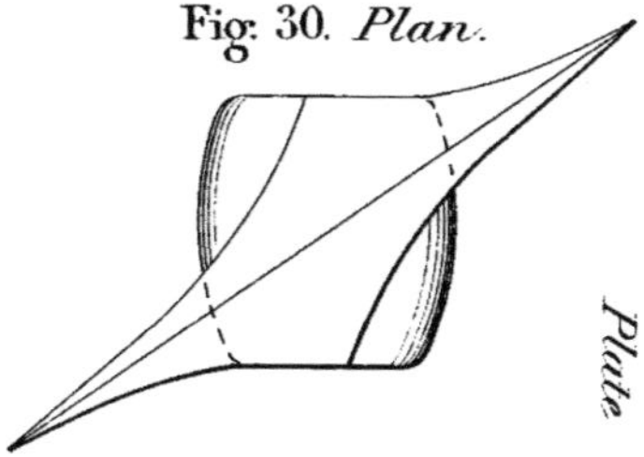

Scale 1/20 th.

Plate

Screw D, four blades, pitch 64¼ inches. Table 4.

Fig. 31. *Revolutions of Screw per minute.*

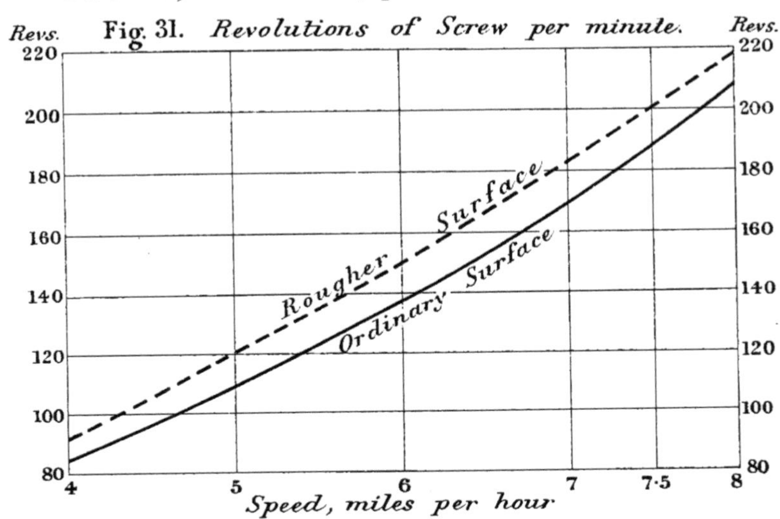

Fig. 32. *Total Indicated Horse-Power.*

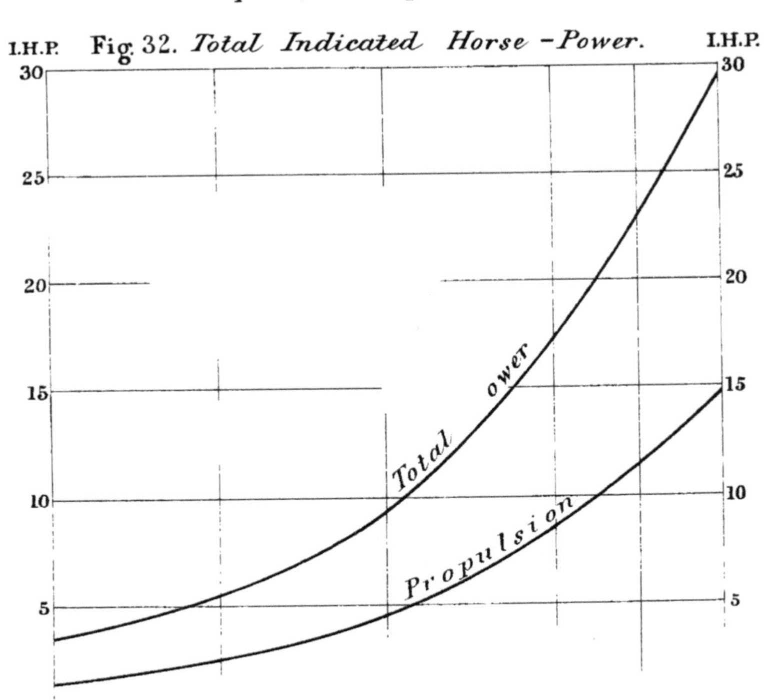

Screw D, four blades, pitch 64¼ inches. Table 4.

Fig. 33. *Thrust, lbs.*

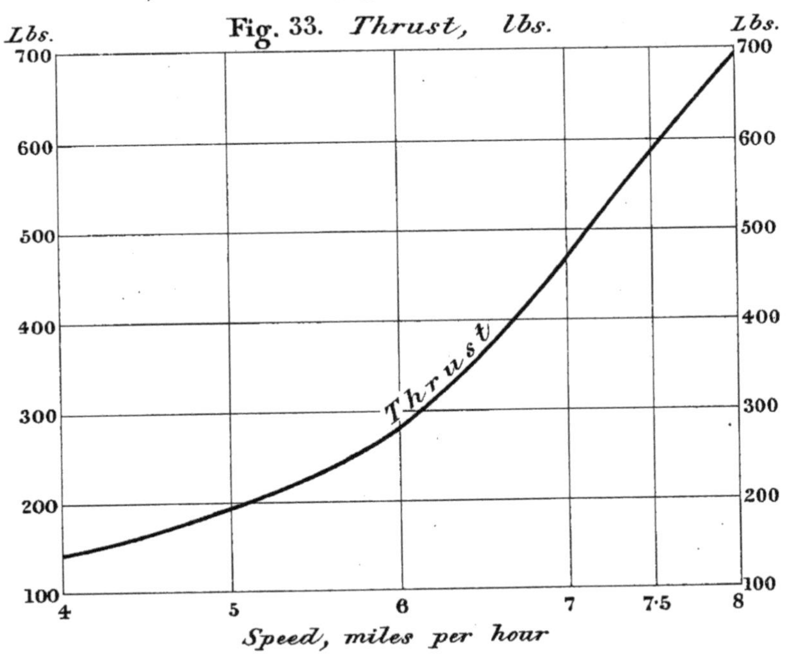

Fig. 34. *Coefficient of Performance* $= \dfrac{\sqrt[3]{\text{Displ. tons}^2} \times \text{knots}^3}{\text{Total I.H.P.}}$

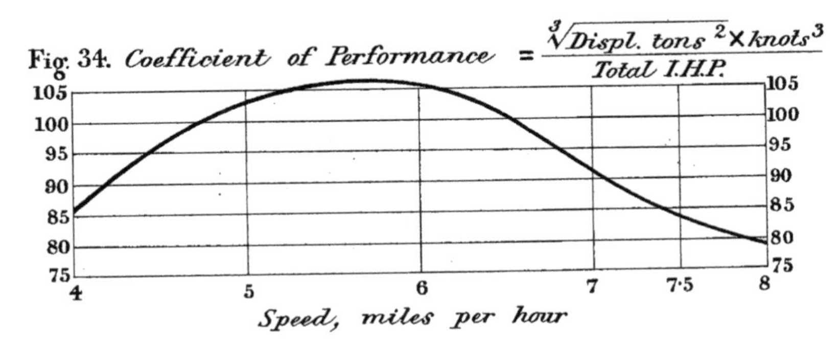

Fig. 35. *Efficiency, per cent.*

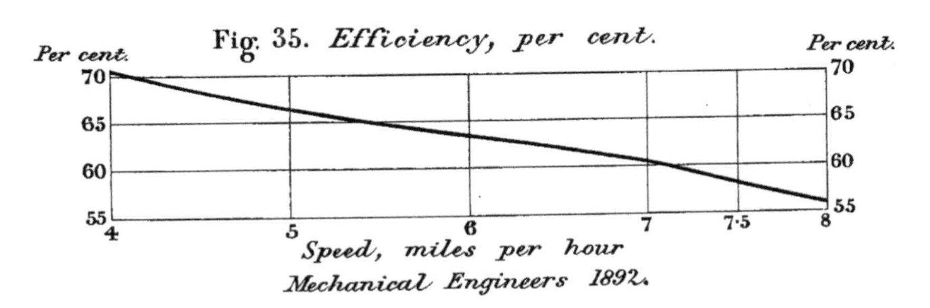

Mechanical Engineers 1892.

Lightning Source UK Ltd.
Milton Keynes UK
UKHW010948211118
332724UK00008B/177/P